# Ocean Dynamics

NIPA® GENX ELECTRONIC RESOURCES & SOLUTIONS P. LTD.
New Delhi-110 034

## About the Author

**Prof. (Dr.) A.S.N. Murty** was born in a hamlet in Vizianagaram district of Andhra Pradesh in 1952. Prof. Murty was a meritorious student throughout his School and further studies. Prof. Murty educated through merit scholarship right from the secondary to graduation and stood class first and in first two ranks in college and university studies. As Murty came from economically weaker section and meritorious, he has been awarded free boarding by Simhachalam Devasthanam Choultry in Vizianagaram when he studied in M.R. College during graduate studies. Later Prof. Murty obtained his M.Sc (Tech) in Meteorology and Oceanography from the Andhra University, Visakhapatanam, AP, in 1976 and Ph.D from the School of Marine Sciences of Cochin University of Science and Technology (CUSAT) Koch, Kerala under the guidance of Prof. G.S. Sharma, an eminent oceanographer and Director of School of Marine Science in CUSAT. Prof. Murty started his professional career first as U.G.C Research Scholar in CUSAT during 1976-1980 and later as Lecturer in Oceanography way back in 1980 in Marine Science Department of Berhampur University and subsequently rose to higher levels of Reader and professor in 1985 and 1999 respectively. Prof. Murty guided 10 Ph.D degrees and over 15 M.Phil degrees and published over 30 research papers in national and international journals. Prof. Murty earlier published a text book in Physical Oceanography in 2010 by APH Publishing Corporation, New Delhi. Prof. Murty has been deputed to Ethiopia to work as professor under United Nations Development Program while in service at Berhampur University during 2003-05. Later after retirement in 2012 again he joined the Department of Meteorology and Hydrology of Arbaminch University, Ethiopia and started many new courses like M.Sc climate change, Meteorology, Agricultural meteorology and served for the full development of these programs. Prof. Murty proved himself to be a good teacher and an active academician and worked as Head, chairman P.G. Council, Coordinator OASTC (Ocean Atmospheric Science and Technology Cell) of Department of Ocean Development (Government of India) and In-charge Vice-chancellor at Berhampur University. Prof. Murty worked as guest professor and selection committee member to select teachers in several universities in India after retirement and serving the students and Research scholars in various universities by guidance and supplying the lecture notes through online publication in Academia. https://berhampurunivrsitybhanjabihar.academia.edu/AyapillaMurty

# Ocean Dynamics
## Theory and Exercises with Solutions

**A.S.N. Murty**
Professor, Head, Chairman P.G. Council
Coordinator, OASTC (Ocean Atmospheric Science and Technology Cell)
Department of Ocean Development (Government of India)
and
In-charge Vice-chancellor
Berhampur University.

**NIPA® GENX ELECTRONIC RESOURCES & SOLUTIONS P. LTD.**
New Delhi-110 034

**NIPA® GENX ELECTRONIC RESOURCES & SOLUTIONS P. LTD.**

101,103, Vikas Surya Plaza, CU Block
L.S.C. Market, Pitam Pura, New Delhi-110 034
Ph : +91 11 27341616, 27341717, 27341718
E-mail: newindiapublishingagency@gmail.com
Website: www.nipabooks.com

***For customer assistance, please contact***
Phone: + 91-11-27 34 17 17
Fax: + 91-11-27 34 16 16
E-Mail: feedbacks@nipabooks.com



ISBN: 978-93-95319-72-0



Composed and Designed by NIPA®.

# Preface

This book on Ocean Dynamics: Theory and Exercises with Solutions, though is a compilation from many sources, it mainly forms part of my personal teaching material at Berhampur University, University of Hyderabad and Arbaminch University. This book will be highly useful for graduate and post graduate students of Oceanography, physical oceanography, meteorology, atmospheric sciences, Aeronautical, Agricultural and space meteorlogy and many other related fields in civil and ocean engineering. Special interest in this book is the provision of many exercises and their solutions under each chapter for better understanding and applications.

This book covers almost all the important topics in physical oceanography like currents with and without friction, upwelling and downwelling, water masses and TS-analysis, waves and tides, Estuaries, internal waves, seiches and storm surges and Rossby and Kelvin waves and related topics. Particular interest and attraction in this book is inclusion of recent developments on North Indian Ocean circulation under chapter 1.

In addition to my personal knowledge and experience in teaching and research, I have taken the help of many recent text books and journals. I express my deep sense of gratitude and indebtedness to various journals and text books I used while writing this book which was properly cited and given reference wherever needed and used. I also consulted many of my teachers at Andhra University and Cochin University of Science and Technology, scientists at national organizations and my colleagues at various universities and Prof. Mark Jury at Arbaminch University. I, therefore, gratefully acknowledge my indebtedness to all my teachers, colleagues, friends at Andhra University, and more importantly my own research scholars and students of Berhampur University who are now in eminent positions in various national and international organizations who helped me either directly or indirectly while writing this book.

My special regards and source of inspiration is my Ph.D guide, Prof. G.S. Sharma, Director, School of Marine Sciences, Cochin University of Science and Technology, though he is no more now. My regards also are to Prof. V.P,

Subrahmanyam, Prof. A.R. Subramaniam, Prof. I. Subbaramayya, Prof. B.S.R. Reddy, Prof. K.P.R. Vitthalamurty of Andhra Unoiversity, who always remain in my mind as a special source of inspiration as my teachers. My special thanks are to my class mate and friend Dr. G.H.S.L.V. Prasada Rao who motivated me to publish this book with the present NIPA publishers.

I am greatly indebted to Andhra University for getting the initial studies and training at M.Sc (Tech) and Ph.D at Cochin University of Science and Technology. My special gratefulness is to Berhampur University for appointing me and providing me a base to do service and acquire and improve my knowledge in this field which also gave me an opportunity to grow to the highest positions of Professor, Chairman P.G. Council, Chairman Board of studies, Coordinator O.S.T.C and in-charge Vice-chancellor.

Finally my special thanks are to my wife Santhi and sons A.S. Bhargav & A.S. Gouthamwithout whose help, encouragement and inspiration at every level of writing this book would not have been successfully completed.

**A.S.N. Murthy**
Author

## Contents

# 1

# Introduction

## Water characteristics and current structure in the North Indian Ocean

The two limbs of Arabian Sea and Bay of Bengal form parts of North Indian Ocean (NIO) on the left and right respectively. Though the Arabian Sea and the Bay of Bengal are located in the same latitudinal belt and receive the same amount of solar radiation, oceanographically these two basins exhibit remarkable differences. The reason for it is Bay of Bengal is shallow and less saline due to heavy river water discharge and Arabian Sea is deep and cold and more saline. It is cold because more upwelling in the coastal parts of west coast of India and Somalia and more saline as it is directly connected to Red sea and Persian Gulf. So Bay of Bengal offers more low pressure systems.

### 1.1. Variation of SST in NIO

Mean monthly variation of Sea Surface Temperature (SST) over Arabian Sea (AS: $08^0$ N – $25^0$ N, $40^0$ E – $75^0$ E) and Bay of Bengal (BB: $08^0$ N - $25^0$ N, $80^0$ E – $95^0$ E) were presented in Fig.1.1. While SST is bimodal with peaks ($\geq 29^0$ C) in May and October, minimum is in January with values $26^0$ C and about $27^0$ C in Arabian Sea and Bay of Bengal respectively. Note that Bay of Bengal remains warmer than the Arabian Sea throughout the year as Arabian Sea has strong upwelling and is deeper than Bay of Bengal.

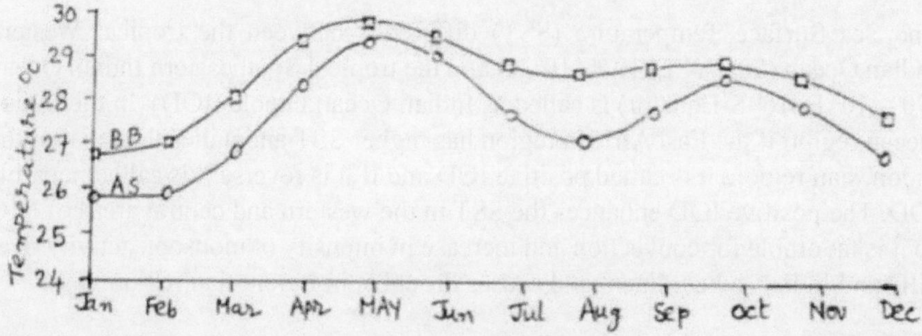


**Fig.1.1:** Monthly variation of Sea Surface Temperature (SST) over Arabian Sea (AS) and Bay of Bengal (BB)

The mean seasonal spatial patterns of SST for winter (January-February) and monsoon (June-September) are presented in Figs.1.2 A and B respectively. During winter season (Fig. 1.2A), the SSTs in the Arabian Sea have south-north gradient with region of cold waters (<24$^0$ C) lying in the northern Arabian Sea. The thermal ridge runs from southwest to northeast along 5$^0$ S to 70$^0$ E. During monsoon season (Fig. 1.2B), the Arabian Sea starts cooling and has coldest regions (SST < 25.5$^0$ C) along Somalia and Arabian coasts. SST gradient over the Arabian Sea changes from south-north to west– east. Climatological thermal high, which is south of the equator in northern winter, shifts northwards during monsoon season and lies at 65$^0$ E - 75$^0$E. A cold region (<27.5$^0$ C) along the Kerala coast is also observed, indicating a region of upwelling during monsoon season. In contrast, SSTs all over the Bay of Bengal remain above 28.5$^0$ C and in some pockets even temperatures are above 29$^0$ C. However, the temperature gradient remains south to north. The availability of vast areas with temperatures greater than 28.5$^0$ C could be one of the reasons responsible for the formation of number of low pressure systems in the Bay of Bengal than the Arabian Sea.

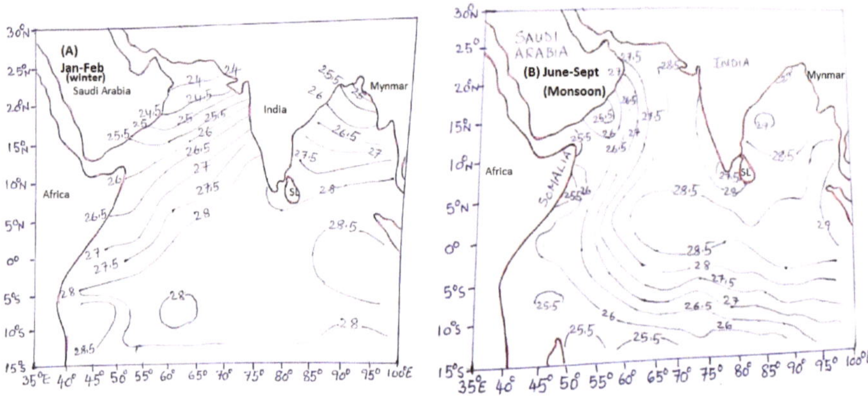


**(A)** **(B)**

**Fig.1.2:** The mean seasonal spatial patterns of SST for **(A)**: winter (January-February) and **(B):** monsoon (June-September) in NIO.

The Sea Surface Temperature (SST) difference between the tropical Western Indian Ocean (50$^0$-70$^0$ E, 10 $^0$S-10$^0$ N) and the tropical Southeastern Indian Ocean (90$^0$-110$^0$ E, 10$^0$ S-Equator) is called as Indian Ocean Dipole (IOD). In the Indian Ocean region if the East African region has higher SST anomalies than that of the Indonesian region, it is called positive IOD and if it is reverse it is called negative IOD. The positive IOD enhances the SST in the western and central areas of NIO so it is favorable for convection and increase in intensity of monsoon activity over NIO and Indian subcontinent and causes droughts in Indonesian region.

## 1.2. Variation of SSS in NIO

The mean sea surface salinity (SSS) variation is shown in Table 1.1 and Fig.1.3. While the Arabian Sea is highly saline ( > 36 psu), the Bay of Bengal is low saline (< 32 psu) (Table 1.1). The reason for this contrast is the Arabian Sea is linked to Red Sea and Persian Gulf. Here evaporation greatly exceeds precipitation and so results in higher salinities. As many mighty rivers like Ganges, Brahmaputra, Irrawaddy, Godavari, and others drain into the Bay of Bengal, it is less saline. Bay of Bengal is partially linked with the Pacific Ocean and receives per year of fresh water as runoff from these major rivers (Subramanian,1993). Table 1.1 gives the ranges of variation of salinity in the Arabian Sea and Bay of Bengal during North East (NE) and Southwest (SW) monsoons respectively (Oldemar, 2022).

**Table 1.1:** Salinity variation in NIO

| S.No | Sea | NE Monsoon | | | SW Monsoon | | |
|---|---|---|---|---|---|---|---|
| | | Jan | Feb | Mar | Jul | Aug | Sep |
| 1. | Arabian Sea (North to south) ($25^0$ N-$0^0$) | 36.5-35.25psu | 36.25-35.0 psu | 36.25-34.75 | 36.5-35.25psu | >36-35psu | 36.25-35.5psu |
| 2. | Bay of Bengal (North to south) ($25^0$ N-$0^0$) | 31.0-33.75psu | 31.0-34.5psu | 32.25-34.0psu | 32.0-34.0psu | 31.0-33.5psu | 25.0-34.0psu |

The spatial variation of SSS (Fig.1.3) also shows that in the Bay of Bengal (BoB) during June to September, the salinity is much lower (33 psu) in the northern part in comparison to that in the Arabian Sea (>35 psu). This low salinity is because of the fresh water runoff from rivers like Ganga and Brahmaputra. While the salinity, in general, decreases from south to north in Bay of Bengal, it increases from north to south and westwards in Arabian Sea with a range of over 2-3 psu in both. Starting from the Head of the Bay with the lowest salinity of <31psu, it increases southward up to15$^0$ S latitude up to 34.5psu with a terminus at 65$^0$ E in Bay of Bengal.The incursion of Pacific waters into Indian Ocean can be seen from the layering of isohalines between 34 and 34.5 psu between 10 to15$^0$ S and extending up to 65$^0$ E. Similarly incursion of Red Sea and Persian Gulf waters into Arabian Sea is seen from 36 and 36.5 psu.

Thus the Sea surface salinities vary between 36.5psu to 35.5psu with a peak of the order of 38 psu in Arabian Sea, it varies between 29.5psu to 33psu in the Head of the Bay (Gangetic region) and 31psu to 33.5psu in the central Bay of Bengal.

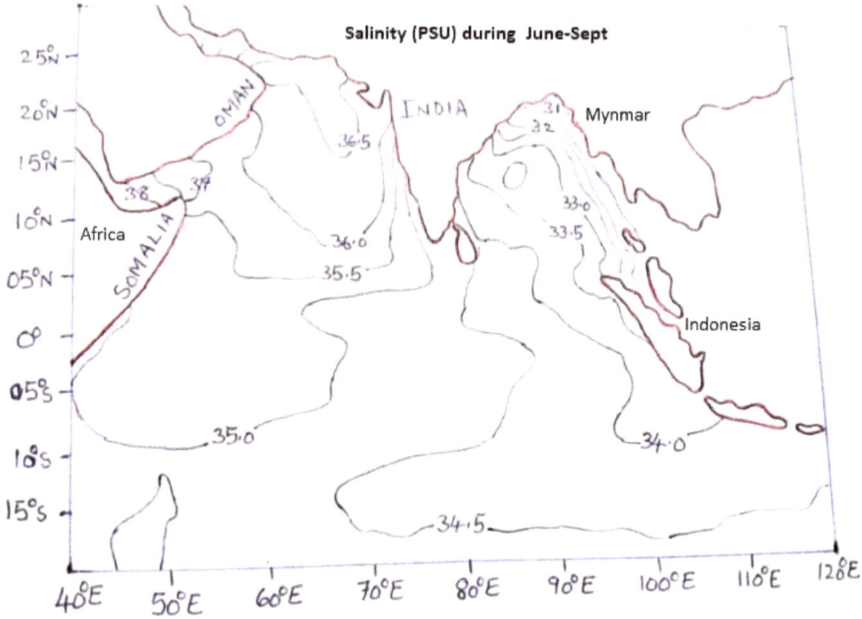


**Fig.1.3:** Salinity distribution in North Indian Ocean during monsoon season

The reversal of winds and currents during the monsoons also causes the exchange of waters and salt between Bay of Bengal and Arabian Sea. For example, the East India Costal Current (EICC) carrying low salinity water from the Bay of Bengal to the westward flowing North Equatorial Current (NEC) through North East Monsoon Current (NEMC) into the Arabian Sea during the winter monsoon season (January–April) (Figure 1.4). Similarly the exchange of mass takes place during summer monsoon (June-Sept) between Arabian Sea and Bay of Bengal through Southwest Monsoon Current (SWMC) and from west to east of NIO through Somali current (SC) and Equatorial Counter Current (ECC). The West India Coastal Current (WICC) feeds into Sri Lankan dome.

The seasonal reversal of monsoon winds and the differences in precipitation (P), evaporation (E) and river runoff (R) makes the Bay of Bengal a net precipitation basin and the Arabian Sea a net evaporative basin (Sengupta et al., 2006). In order to understand the balance of salt and mass, there is a need to study the dynamics of ocean salt distribution as detailed below.

$$\frac{\partial S}{\partial t} = -U\frac{\partial S}{\partial x} - V\frac{\partial S}{\partial y} + h^{-1}\frac{dh}{dt}H\left(\frac{dh}{dt}\right)\left(S - S_{z=h}\right) + h^{-1}\left(E - P\right)S \qquad (1)$$

where S is salinity, U the zonal component of velocity, V the meridional component of velocity, 'h' the depth of the upper mixed layer, $S_{z=h}$ the salinity just below the mixed layer depth, H the Heaviside step function, E the evaporation, and P the precipitation.

In order to understand the relative roles of geostrophic and Ekman components in the Indian Ocean on the magnitudes and effects in the transport of Sea surface salinity (SSS), the horizontal advection is split into Ekman and geostrophic current components.

The geostrophic component of the horizontal salt transport is estimated from

$$U_{gs} = -U_g \frac{\partial S}{\partial x} \text{ and } V_{gs} = -V_g \frac{\partial S}{\partial y} \qquad (2)$$

where S is salinity, $U_g$ and $V_g$ are surface geostrophic currents computed respectively from

$$U_g = -\frac{g}{f}\frac{\partial \zeta}{\partial y} \text{ and } V_g = \frac{g}{f}\frac{\partial \zeta}{\partial x} \qquad (3)$$

where g is acceleration due to gravity (9.8 $ms^{-2}$), f is the Coriolis parameter (2Ω sinϕ).

Ekman wind drift component of the horizontal salt transport is estimated from

$$U_{es} = -U_e \frac{\partial S}{\partial x} \text{ and } V_{es} = -V_e \frac{\partial S}{\partial y} \qquad (4)$$

Where $U_e$ and $V_e$ are the surface Ekman drift computed respectively from

$$U_e = \frac{\tau_y}{\rho_0 f h} \text{ and } V_e = \frac{-\tau_x}{\rho_0 f h} \qquad (5)$$

where 'h' is 5 m depth, selected as the depth of stability maximum in the salinity stratified layer, $\rho_0$ is the water density (1023 kg $m^{-3}$). Finally, surface salinity transport is computed from

$$F_S = v\,H\,\Delta S \qquad (6)$$

Where Fs is the salt transport in psu $m^2s^{-1}$, v is the magnitude of current velocity in $ms^{-1}$, H is the first model layer depth (~5 m) and ΔS is the salinity anomaly in psu. ΔS is determined from $\Delta S = S - \overline{S}$, where $\overline{S}$ is the annual mean Sea Surface Salinity.

The Ekman drift component is found to be more important than the geostrophic component in the exchange of salt between the Arabian Sea and Bay of Bengal.

## 1.3. Variation of current structure in NIO

Generally the circulation in major oceans consists of gyres, equatorial current system, and western and eastern boundary currents. A clockwise gyre in Northern Hemisphere and an anticlockwise gyre in the Southern Hemisphere is common. Equatorial current system comprises of westward flowing North and South Equatorial Currents and an eastward flowing Equatorial Counter current and in between them an eastward flowing Equatorial Under Current (EUC) at about 50 to 100 m depth. While the poleward flowing western Boundary currents are stronger, equatorward flowing eastern boundary currents are weaker in the Northern Hemisphere. The same is true in the Southern Hemisphere also but with a reverse circulation. However, in the case of Indian Ocean, this kind of circulation is partly true. While the south Indian Ocean behaves like that of the major oceans, the North Indian Ocean (NIO) circulation is controlled by the monsoons. During North East (NE) monsoon (January–April), the North Indian Ocean circulation is same as that of the major oceans of Atlantic and Pacific but it is quite different during Southwest (SW) monsoon (June-September) (Fig.1.4 A, B). The differences in current structure in NIO during SW monsoon are (i) the westward flowing North Equatorial Current (NEC) vanishes and in its place Southwest monsoon current (SWMC) occurs. (ii) The clockwise gyre extends to full extent of NIO. (iii) The equatorial counter current (ECC) and Equatorial Under Current joins together and flows as eastward SWMC (Fig.1.4 B). The differences of circulation in NIO between NE and SW are given in the table1.2.

The NEC during the NE monsoon flows with a speed of about 0.3 m $s^{-1}$ from the Malacca straits to southern Sri Lanka through the Andaman Sea. The easterly moving intense Indian Equatorial Jet (also called Wyrtki Jet, WJ in Fig.1.4 A) flowing between 2° N and 2° S during the transition period before the onset of the NE monsoon along the equator (Wyrtki 1973) with velocities of 0.7 m $s^{-1}$ or more results in deepening the thermocline along the Sumatra coast and uplifting the thermocline along the African coast. It reaches its peak in November with velocities of 1.0–1.3 m $s^{-1}$ and disappears in early January. Later when the southwest monsoon is fully established, during July to September, the NIO is dominated by the eastern flowing SWMC right from Southern Arabian Sea to the Andaman Sea through the Bay of Bengal (Fig.1.4B).

**Table 1.2:** The differences of currents in NIO between NE and SW monsoon seasons

| S.No | Abbrevia-tion | Name of the current | NE monsoon | SW monsoon | Direction | Remarks |
|---|---|---|---|---|---|---|
| 1 | SEC | South Equatorial current | Exists | Exists | Westward | |
| 2. | ECC | Equatorial Counter Current | Exists | Exists | Eastward | |
| 3. | NEC | North Equatorial Current | Exists | Does not exist | | Eastward flowing SWMC exists in the place of NEC |

| | | | | | | |
|---|---|---|---|---|---|---|
| 4. | EACC | East Africa Coastal Current | Exists in S.H | Exists in S.H | Equatorward | Flows along the east coast of Africa from to equator |
| 5. | NEMC | Northeast Monsoon current | Exists: | NEMC does not exist: | NEMC flows westward. SWMC flows eastward | in its place eastward flowing SWMC exists |
| 6. | SC | Somali current | Exists. flows southward (equatorward) | Exists. Flows northward | | |
| 7. | WICC | West India Coastal Current | Exists. Flows northward | Exists. Flows southward | | |
| 8. | EICC | East India Coastal Current | Exists. Flows southward | Exists. Flows northward | | |
| 9. | LH | Laccadive High | Exists | Does not Exist | | In its place LL exists |
| 10. | LL | Laccadive Low | Does not exist | Exists | | In its place LH exists |
| 11 | GW | Great whirl | Does not exist | Exists | | GW exists off Somalia |
| 12 | WJ | Wyrtki Jet | Exists | Exists | Eastward flowing | |

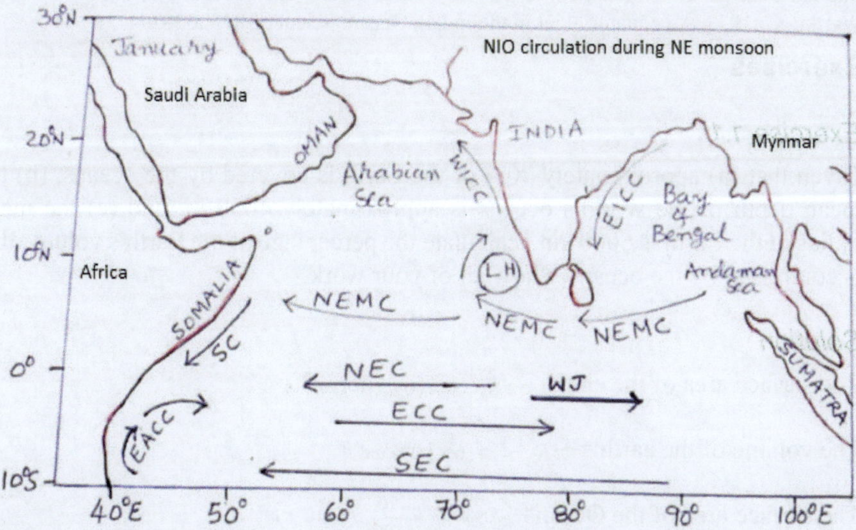


**Fig.1.4 (A):** North Indian Ocean (NIO) circulation during North East monsoon season (January)

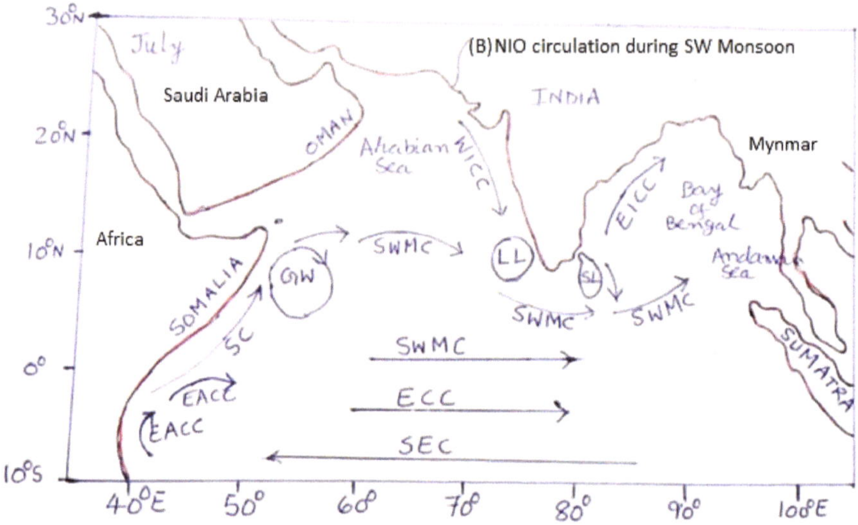


**Fig.1.4 (B):** North Indian Ocean (NIO) circulation during Southwest monsoon season (July)

## References

Oldemar de Oliveira Carvalho (2022): Salinity in the Surface of the Indian Ocean, Marine Science, 10(1): 1-8. DOI: 10.5923/j.ms.20221001.01

Sengupta, D., G. N. Bharath Raj, and S. S. C. Shenoi (2006), Surface freshwater from Bay of Bengal runoff and Indonesian Throughflow in the tropical Indian Ocean, Geophys. Res. Lett., 33, L22609, doi:10.1029/ 2006GL027573.

Subramanian, V., 1993. Sediment load of the Indian Rivers, Curr. Sci., 64, 928-930.

Wyrtki, K., 1973. An equatorial jet in the Indian Ocean. Science 181, 262–264.

## Exercises

### *Exercise 1.1*

Given that (a) approximately 70% of the Earth is covered by the oceans, (b) the mean depth of the world's oceans is approximately 4000 m; and (c) the mean radius of the Earth is 6000 km., calculate the percentage of the Earth's volume that is comprised by the oceans. Show all of your work.

### *Solution*

The surface area of the earth = $\pi R_E^{\ 2} = \pi(6\times10^6)^2 m^2$ ;

The volume of the earth = $\pi R_E^{\ 3} = \pi(6\times10^6)^3 m^3$

The surface area of the Ocean = $70\%\pi R_E^{\ 2} = \frac{70}{100}\pi(6\times10^6)^2 m^2$

Volume of Oceans = $\frac{70}{100}\pi(6\times10^6)^2 m^2 \times 4\times10^3 m$

Percentage of oceans in the earth's volume

$$= \frac{70}{100}\frac{100}{\pi(6\times10^6)^3 m^3}\pi(6\times10^6)^2 m^2 \times 4\times10^3 m = 4.7\times10^{-3}$$

### *Exercise 1.2*

Give an estimate of the mean wind-stress at the ocean-atmosphere interface if the wind speed over the region of the Antarctic Circumpolar Current is about 10 $ms^{-1}$. (Take $C_D = 1.2 \times 10^{-3}$, $\rho_a = 1.2$ kg/$m^3$,

### *Solution*

**wind stress:** $\tau = \rho_a C_d U^2 = 1.2$ $kgm^{-3} \times 1.2 \times 10^{-3} \times 10^2$ $m^2$ $s^{-2} = 1.44 \times 10^{-1} = 0.14$ $kgm^{-1}$ $s^{-2} = 0.14$ $Nm^{-2} = 0.14$ Pascals

### *Exercise 1.3*

If a unidirectional wind, of 10 $ms^{-1}$ , blows over an initially motionless sea water column of 10 m thickness, then calculate the flow velocity in this layer after 3 hours. (Take $\rho = 1027$ $kgm^{-3}$)

### *Solution*

The wind stress will transfer momentum to the water column of depth h according to the balance $\rho h\frac{du}{dt} = \tau$, where 'u' is the (uniform) velocity of the layer. From this equation we can estimate $u(t) \approx u(0) + \frac{\tau}{\rho h} t$ ➔ $u(t) \approx 0 + \frac{0.14 kgm^{(-1)} s^{(-2)}}{1027 kgm^{(-3)} \times 10m} \times 3 \times 3600s = 0.15ms^{-1}$

So if, t= 3 hours that $u(t) = 0.15ms^{-1}$

### *Exercise 1.4*

Prove that exactly half the area of the earth lies equatorward of latitude.

### *Solution*

$$\text{Area} = \int_0^2 \int_{30^0 S}^{30^0 N} R_E^2 cos\varphi d\varphi d\lambda = 2\pi R_E^2 \int_{30^0 S}^{30?^0 N} cos\varphi d\varphi$$

$$= 2\pi R_E^2 (sin\varphi) ||_{-30^0}^{+30^0} = 2\pi R_E^2 \{0.5 - (-0.5)\} = 2\pi R_E^2$$

As the area of the sphere is $4\pi R_E^2$, between $0^0$ and $30^0$ N or $30^0$ S, it is half of the area of the sphere ($2\pi R_E^2$).

### *Exercise 1.5*

How many days would it take for a ship traveling in the oceans eastward along $40^0$ S at a mean speed of 20 knots to circumnavigate the globe? Assume there is no frictional resistance due to wind and ocean currents (Take radius of the earth as $6.37 \times 10^6$ m).

### *Solution*

The length of the latitude circle at$40^0$ s is

$l = 2\pi R_E \cos 40^0 = 2\pi \times 6.37 \times 10^6$ m $\times \cos 40^0 = 30.66 \times 10^6$ m $= 30.7 \times 10^3$ km

The time (t) taken by the ship to travel around $40^0$S latitude circle is the distance divided by the velocity.

Velocity 20 kts = $10ms^{-1}$ since $1ms^{-1} \approx 2$knots

$$t = \frac{30.66 \times 10^6 m}{10 ms^{-1}} = 30.7 \times 10^5 s = \frac{30.7 \times 10^5}{3600} = 852.8 \text{ hours} \approx 35.5 \text{ days}$$

### *Exercise 1.6*

How far below the ocean surface of the water does a diver experience a pressure of 2 atmospheres (i.e., a doubling of the ambient atmospheric pressure due to the weight of the overlying sea water.(Take $\rho_w = 1027kgm^{-3}$)

### *Solution*

At the depth where the pressure is 2 atmospheres, half the pressure is due to the weight of the overlying air and the other half (1 atm) is due to the weight of the overlying water, which is equal to the density times the depth times g. The density of sea water given is $\rho_w = 1027kgm^{-3}$. Pressure of the sea water, P = $\rho gz$ = 1 atmosphere = 1000mb = $10^6$ dynes $cm^{-2} = 10^5$ pascals

| 1 Pascal (Pa) | $1\ Nm^{-2}$ | $10\ dyn\ cm^{-2}$ |
|---|---|---|
| 1 mb | $10^3\ dyn\ cm^{-2}$ | $10^2\ Nm^{-2}$ |

$$\therefore z = \frac{P}{\rho g} = \frac{10^5 Nm^{-2}}{1027 kgm^{-3} \times 9.8 ms^{-2}} \approx 9.94m$$

### *Exercise 1.7*

If the Ocean area is consisted of an incompressible fluid whose density was everywhere equal to that observed at sea level ($1027kgm^{-3}$) how deep would it have to be to account for the observed mean surface pressure of $\sim 10^5$ pascals.

### *Solution*

If 'h' is the depth of the of the ocean area, p = $\rho gh = 10^5$ pascals ➔ $h = \frac{p}{\rho g}$

| 1 newton (N) | $1\ kgms^{-2}$ |
|---|---|
| 1 Pascal (Pa) | $1\ Nm^{-2}$ |

$10^5$ *pascals* $= 10^5\ Nm^{-2} = 10^5\ kgms^{-2}\ m^{-2} = 10^5\ kgm^{-1}\ s^{-2}$

$$h = \frac{p}{\rho g} = \frac{10^5 kgm^{-1}s^{-2}}{1027 kgm^{-3} \times 9.8 ms^{-2}} \approx 9.94m$$

### *Exercise 1.8*

The equatorward flow in the trade wind region is of the order of 1 $ms^{-1}$ averaged around the circumference of the earth at15$^0$ N and15$^0$ S and it extends through a layer extending from sea-level up to around the 150 dB pressure surface in the oceans. Estimate the equatorward mass flux into the equatorial zone due to this circulation.

### *Solution*

The equatorward mass flux across the15$^0$ N, in units of kg $s^{-1}$, is given by

$$\oint_{15^0 N} \int_0^{150dB} \rho_w v dz dx$$

where $\rho_w$ is the density of the ocean water, v is the meridional (northward) velocity component, the line integral denotes an integration around the15$^0$ N latitude circle and the vertical integral is from sea-level up to the depth of the 150 decibar surface. To evaluate the integral, we make use of the relations

$1 = \oint_{15^0 N} dx = 2\pi R_E \cos 15^0 = 2\,\pi \times 6.37 \times 10^6$ m $\times \cos 15^0$ (where $l$ is the length of the latitude circle) and $m = \int_0^{150dB} \rho_w \, dz = \dfrac{dp}{g} = \dfrac{(p150 - po)}{g} = \dfrac{(150-0)dB}{9.8}$, (where $m$ is the mass per unit area in the layer)

The mass flux, per unit length along the latitude circle is mv. As the flux enters the equatorial region from 15$^0$ S and15$^0$ N, we have to multiply the length of the latitude circle by 2 (from (15$^0$ S to 0$^0$ and from 0$^0$ to15$^0$ N). We obtain the total mass flux

$$F_m = 2 \times 1 \times m \times v = 2 \times 2\,\pi \times 6.37 \times 10^6 \text{ m} \times \cos 15^0 \times \frac{(150)dB}{9.8ms^{-2}} \times 1ms^{-1}$$

(Note 1 dB = 100mb = $10^5$ dyn c $m^{-2}$ = $10^4$ N $m^{-2}$ = $10^4$ $kgms^{-2}$ $m^{-2}$)

| 1 Pascal (Pa) | 1 $Nm^{-2}$ | 10 dyn $cm^{-2}$ |
|---|---|---|
| 1 mb | $10^3$ dyn $cm^{-2}$ | $10^2$ $Nm^{-2}$ |
| 1 newton (N) | 1 $kgms^{-2}$ | |
| 1 Pascal (Pa) | 1 $Nm^{-2}$ | |

$F_m = 2 \times 2\pi \times 6.37 \times 10^6 \text{ m} \times \cos 15^0 \times \dfrac{(150) \times 10^4 kgms^{-2} m^{-2}}{9.8ms^{-2}} \times 1ms^{-1} = 1183.5 \times 10^{10}$ $kgs^{-1}$ = 1.18 × 10 13 $kgs^{-1}$

### *Exercise 1.9*

During September, October and November the mean surface pressure over the northern hemisphere increases at a rate of ~1 hPa per 30-day month. Calculate the mass averaged northward velocity across the equator that is required to account for this pressure rise.

### *Solution*

The required mass transport Fm is equal to the rate of pressure change, divided by g, times the area of the hemisphere ($2\pi R_E^2$), or

$$F_m = \frac{1}{g}\frac{\delta p}{\delta t} \times 2\pi R_E^2 = v_m \oint \int_0^\infty \rho dx\, dz$$

Where δt is 1 month which has 30 days = 30 × 24 × 60 × 60 = 2.592 ×10⁶. Noting that $\oint dx = 2\pi R_E$ and $\int_0^\infty \rho dz = \frac{p_0}{g}$, we can write

$$\frac{1}{g}\frac{\delta p}{\delta t} \times 2\pi R_E^2 = 2\pi R_E \times v_m \times \frac{p_o}{g}$$

Solving we obtain,

$$v_m = \frac{\delta p \times R_E}{p_o \delta t} = \frac{100 \times 6.37 \times 10^6}{10^3 \times 2.592 \times 10^6} = 2.46 mms^{-1}$$

### *Exercise 1.10*

The eastward flowing equatorial jet (Wyrtki Jet) in the region between $2^0$ N to $2^0$ S is of the order of 1.3 averaged around the circumference in a layer extending from sea-surface to around 50m in the Indian Ocean. Estimate the mass flux of the Wyrtki Jet in the region in kg $s^{-1}$ as well as Sverdrups.

### *Solution*

The eastward mass flux across the $2^0$ N, in units of kg $s^{-1}$, is given by

$$-\oint_{2^0 N} \int_0^{50m} \rho_w u dz dx$$

where $\rho_w$ is the density of ocean water(1027kg$m^{-3}$), u is the eastward velocity component, the line integral denotes an integration around the $2^0$ N latitude circle and the vertical integral (dz) is from sea-level up to the depth of the 50 m surface. To evaluate the integral, we make use of the relations

$l = \oint_{2^0 N} dx = 2\pi R_E cos 2^0 = (2\pi \times 6.37 \times 10^6 m \times cos 2^0)$ (where $l$ is the length of the latitude circle) and

$m = \int_0^{50m} \rho_w dz = 1027$ kg$m^{-3}$ × 50m, (where $m$ is the mass per unit area in the layer)

As the flux flowing the region from $2^0$ S and $2^0$ N, we have to multiply by the length of the latitude circle by 2 (from ($2^0$ S to $0^0$ and from $0^0$ to $2^0$ N). So the total mass flux, $F_m$ is

$F_m = 2 \times l\,(m) \times m\,(kgm^{-2}) \times v(ms^{-1}) =$

$= 2 \times 2\pi \times 6.37 \times 10^6\ m \times cos 2^0 \times 1027\ kgm^{-3} \times 50m \times 1.3ms^{-1}$

$= 5340334 \times 10^6\ kg\ s^{-1} = 53 \times 10^{11}\ kg\ s^{-1}$

We know one Sverdrup = $10^6$ m$^3$ s$^{-1}$ and $\rho_w$ = 1027kgm$^{-3}$

So the mass flux in sverdrups =

$$\frac{53\times10^{11}\,kgs^{-1}}{\rho_w}=\frac{53\times10^{11}\,kgs^{-1}}{1027kgm^{-3}} = 0.052 \times 10^{11}\ m^3\ s^{-1} = 52 \times 10^8\ m^3\ s^{-1}$$

## Exercise 1.11

(a)If the monsoon winds are of the order of 10 ms$^{-1}$ averaged around the circumference of the Arabian Sea region at $0^0$ N and$10^0$ N, and extends through a layer extending from sea-level up to around the 850 hPa pressure surface in the oceans, estimate the eastward mass flux in the North Indian Ocean due to this circulation. (b)What do you call this eastward mass flux? (c) Compare this mass with that computed in the exercise 1.10.

## Solution

(a)The eastward mass flux across the$10^0$ N, in units of kg s$^{-1}$ , is given by

$$\oint_{10}\int_{1000}^{850hPa} vdzdx$$

where $\rho_a$ is the density of air, v is the meridional (northward) velocity component(monsoon winds are meridional in North Indian Ocean), the line integral denotes an integration around the $10^0$ N latitude circle and the vertical integral is from sea-level up to the depth of the 850 hecta pascal surface. To evaluate the integral, we make use of the relations

$l=\oint_{10^0 N} dx = 2\pi R_E \cos 10^0 = 2\pi \times 6.37 \times 10^6\ m \times \cos 10^0$ (where $l$ is the length of the latitude circle) and

$m=-\int_{1000}^{850dB}\rho a dz=-\frac{dp}{g}=-\frac{(p_{850}-p_o)}{g}=-\frac{(850-1000)hPa}{9.8}=-\frac{150hPa}{9.8}$, (where $m$ is the mass per unit area in the layer)

The mass flux entering the region from $0^0$ and$10^0$ N per unit length along the latitude circle is mv. So the total mass flux, $F_m$ is

$$F_m = l \times m \times v = 2\pi \times 6.37 \times 10^6\ m \times \cos 10^0 \times \frac{(150)hPa}{9.8ms^{-2}} \times 10ms^{-1}$$

(Note 1 hPa= 100Pa = $10^2$ N m$^{-2}$ = $10^2$ kgms$^{-2}$ m$^{-2}$)

| 1 Pascal (Pa) | 1 Nm$^{-2}$ | 10 dyn cm$^{-2}$ |
|---|---|---|
| 1 mb | $10^3$dyn cm$^{-2}$ | $10^2$ Nm$^{-2}$ |
| 1 newton (N) | 1 kgms$^{-2}$ | |
| 1 Pascal (Pa) | 1 Nm$^{-2}$ | |

$F_m = 2\pi \times 6.37 \times 10^6$ m $\times \cos 10^0 \times \dfrac{(150)\times 10^2 kgms^{(-2)} m^{(-2)}}{9.8ms^{(-2)}} \times 10$ms$^{-1} \approx 603 \times 10^9$ kgs$^{-1} \approx 6 \times 10^{11}$ kgs$^{-1}$

(b) This eastward mass flux is called Southwest monsoon current (See figure 1.4b)

(c) Comparison of this mass with that computed in the exercise 1.10:

Mass in the exercise 1.10: $53 \times 10^{11}$ kg s$^{-1}$, mass in exercise 1.11= $6 \times 10^{11}$ kg s$^{-1}$

$$\frac{53\times 10^{11} kgs^{-1}}{6\times 10^{11} kgs^{-1}} = 8.8 \approx 9$$

This means Wyrtki jet is 9 times more than the Southwest monsoon current.

### *Exercise 1.12*

(a) The equatorward flow in the trade wind region is of the order of 1 ms$^{-1}$ averaged around the circumference of the earth at$15^0$ N and$15^0$ S, and it extends through a layer extending from sea-level up to around the 850 hPa pressure surface. Estimate the equatorward mass flux into the equatorial zone due to this circulation.

(b) In the oceanographic literature, the unit of mass transport is called as the Sverdrup (Sv) (which is equal to one million cubic meters per second, $10^6$ m$^3$ s$^{-1}$). The transport of ocean water by the Gulf Stream is estimated to be of the order of 150 Sv. Compare the transport by the Gulf Stream with the transport by the trade winds estimated in (a) (Take the density of sea water as$10^3$Kgm$^{-3}$).

### *Solution*

The equatorward mass flux across the$15^0$ N, in units of kg s$^{-1}$, is given by

$$\oint_{15^0 N} \int_{1000}^{850hPa} \rho_a v dz dx$$

where $\rho_a$ is the density of air, v is the meridional (northward) velocity component, the line integral denotes an integration around the$15^0$ N latitude circle and the vertical integral is from sea-level up to the depth of the 850 hecta Pascal surface. To evaluate the integral, we make use of the relations

$l = \oint_{15^0 N}$ dx $= 2\pi R_E \cos 15^0 = 2\pi \times 6.37 \times 10^6$ m $\times \cos 15^0$ (where $l$ is the length of the latitude circle) and $m = -\int_{1000}^{850dB} \rho_a dz = -\dfrac{dp}{g} = -\dfrac{(p_{850} - p_{1000})}{g} = \dfrac{(150)hPa}{9.8}$, (where $m$ is the mass per unit area in the layer)

The mass flux, per unit length along the latitude circle is mv. As the flux enters the equatorial region from$15^0$ S and$15^0$ N, we have to multiply by the length of the latitude circle by 2 (from ($15^0$ S to $0^0$ and from $0^0$ to$15^0$ N) we obtain the total mass flux

$$F_m = 2 \times l \times m \times v = 2 \times 2\pi \times 6.37 \times 10^6\ m \times \cos 15^0 \times \frac{(150)\text{hPa}}{9.8\text{ms}^{-2}} \times 1ms^{-1}$$

(Note 1 hPa = 100 Pa = $10^2$ N $m^{-2}$ = $10^2$ $kgms^{-2}$ $m^{-2}$)

| 1 Pascal (Pa) | 1 $Nm^{-2}$ | 10 dyn $cm^{-2}$ |
|---|---|---|
| 1 mb | $10^3$ dyn $cm^{-2}$ | $10^2$ $Nm^{-2}$ |
| 1 newton (N) | 1 $kgms^{-2}$ | |
| 1 Pascal (Pa) | 1 $Nm^{-2}$ | |

$$F_m = 2 \times 2\,\pi \times 6.37 \times 10^6 \text{ m} \times \cos 15^0 \times \frac{(150)\times 10^2\,kgms^{-2}m^{-2}}{9.8ms^{-2}} \times 1\text{ms}^{-1}$$

$= 1183.5 \times 10^8$ $kgs^{-1}$ $= 1.18 \times 10^{11}$ $kgs^{-1} \approx 1.2 \times 10^{11}$ $kgs^{-1}$

(b) Given Gulf Stream transport = 150 Sv =

$= 150 \times 10^6$ $m^3$ $s^{-1} \times 10^3$ $Kgm^{-3} = 150 \times 10^9$ $Kgs^{-1} = 1.5 \times 10^{11}$ $Kgs^{-1}$

Comparing this result with (a) is : $1.2 \times 10^{11}$ $kgs^{-1}$

This means (a) and (b) are comparably equal in order of magnitude

## Exercise 1.13

A heavy tropical storm downpours 20 cm of rainfall in a region of the ocean in which the salinity is 35.00 g $Kg^{-1}$ and the mixed layer depth is 50 m. Assuming that the water is well mixed, by how much does the salinity decrease?

## Solution

The volume of water in a column extending from the surface of the ocean to the bottom of the mixed layer is increased by a factor: $\frac{Rainfall}{mixed\ layer\ thickness} = \frac{20cm}{50m} = \frac{0.2m}{50m} = 4\times 10^{-3}$ = 1.004 kg of rain water added and (ignoring the small difference between the densities of salt water and fresh water) the mass of the water in the column increases by a corresponding amount. The mass of salt dissolved in the water remains unchanged. Hence, the salinity drops to:

$$\frac{35.00g\ of\ salt}{1.004kg\ of\ rain\,water} = 34.86g\ Kg^{-1}$$

## Exercise 1.14

If the mass of Greenland ice sheet is $5 \times 10^3$ $kgm^{-2}$, estimate how much the sea level would rise if the entire Greenland ice sheet were to melt.(Take area of the Greenland as 0.35 times of the earth's surface area.)

### *Solution*

From Appendix 1:

Area of earth is $5.10 \times 10^{14}$ m$^2$, Ocean surface area is $3.61 \times 10^{14}$ m$^2$. Let density of water is $10^3$ kgm$^{-3}$

The mass of the Greenland ice sheet is equal to its mass per unit area averaged over the surface of the Earth times the area of the Earth :

$5 \times 10^3$ kgm$^{-2}$ $\times 5.10 \times 10^{14}$ m$^2 = 2.55 \times 10^{18}$ kg

If the ice cap were to melt, this mass would be distributed uniformly over the ocean-covered area of the Earth's surface. Let *x* is the sea level rise, then we can write:

(Area of oceans) × (density of water) x = mass of ice sheet ➔

$3.61 \times 10^{14}$ m$^2 \times (10^3$ kgm$^{-3}$) x = $2.55 \times 10^{18}$ kg

$$\therefore \text{x} = \frac{2.55\times10^{18}\,kg}{3.61\times10^{14}\,m^2\times(10^3\,kgm^{-3})} = 0.71\times10\text{m} \approx 7 \text{ m}$$

Thus if the entire Greenland ice melts, the sea level would rise by 7 meters.

### *Exercise 1.15:*

By how much global sea level would rise if both the Greenland and the Antarctic ice sheets were to entirely melt? (Take the mass of Greenland ice sheet as $5 \times 10^3$ kgm$^{-2}$, mass of Antarctic ice sheet as $53 \times 10^3$ kgm$^{-2}$

### *Solution*

From Appendix 1:

Area of earth is $5.10 \times 10^{14}$ m$^2$, Ocean surface area is $3.61 \times 10^{14}$ m$^2$, Let density of water is$10^3$ kgm$^{-3}$

The mass of the Greenland and Antarctic ice sheet is equal to its mass per unit area averaged over the surface of the Earth times the area of the Earth :

($5 \times 10^3$ kgm$^{-2}$ + $53 \times 10^3$ kgm$^{-2}$) $\times 5.10 \times 10^{14}$ m$^2$ = $29.5 \times 10^{18}$ kg

If the ice cap were to melt, this mass would be distributed uniformly over the ocean-covered area of the Earth's surface. Let *x* is the sea level rise, then we can write:

(Area of oceans)×(density of water) x = mass of ice sheet ➔

$3.61 \times 10^{14}$ m$^2 \times (10^3$ kgm$^{-3}$) x = $29.5 \times 10^{18}$ kg

$$\therefore \text{x} = \frac{29.5\times10^{18}\,kg}{3.61\times10^{14}\,m^2\times(10^3\,kgm^{-3})} \approx 8.17\times10\text{m} \approx 81.7 \text{ m}$$

# 2

# Equation of Motion (Forces)

## 2. The Coordinate system

The coordinate system used is X component is towards east (Zonal component), Y component is towards north of the globe (meridional component) and Z component is down below the oceans in the vertical as shown below in Fig.2.1:

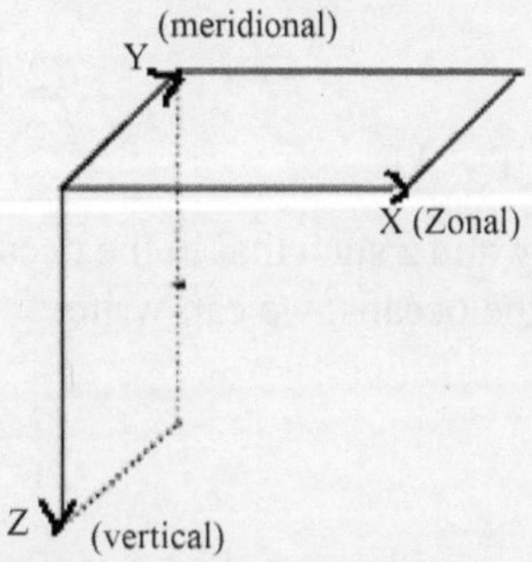


**Fig. 2.1:** Coordinate system in oceans

## 2.1. Equation of motion in the Oceans

To understand the movement of a ocean particle, We have to consider different forces. The forces that are working are pressure gradient force, coriolis force, acceleration due to gravity and friction. The summation of all these forces is balanced by total force. As we consider unit mass, the force is equal to acceleration. The total equation is called equation of motion.

If Newton's 2nd Law is applied to water we can write :

$$\sum F = ma$$

But oceanographers use acceleration instead of force to avoid mass i.e., force per unit volume and per unit mass.

$$\therefore a = \frac{\sum F}{m} = \frac{1}{\rho}\frac{\sum F}{m}$$

Acceleration per unit mass can be written as:

$$a = \frac{du}{dt} = \frac{1}{\rho}\sum F$$

To understand the movement of a ocean particle, we have to consider different forces. The different forces act on a parcel of water in the oceans (ΣF) are pressure gradient, Coriolis,Acceleration due to gravity and friction. So we can write:

$$\frac{du}{dt} = \sum F = F_P + F_C + F_g + F_f$$

The forces that are working are pressure gradient force ($F_p$), coriolis force ($F_c$), acceleration due to gravity ($F_g$) and friction ($F_f$). The summation of all these forces is balanced by total force. As we consider a unit mass, the force is equal to acceleration. The total equation of motion is:

Total acceleration = pressure gradient + coriolis + acceleration due to gravity + friction

$$\frac{dv}{dt} = -\alpha\nabla P + 2\Omega K \times V + g + F \tag{2.1}$$

Where $V = iu + jv + kw$, $P = ip_x + jp_y + kp_z$, $F = iF_x + jF_y + kF_z$

Resolving equation (2.1) into three components of x,y and z such that in the ocean x is east, y is meridional and z is vertically down in the oceans, we can write:

$$\frac{du}{dt} = -\alpha\frac{\partial p}{\partial x} + 2\Omega v\sin\varphi + 2\Omega w\cos\varphi + F_x \tag{2.2}$$

$$\frac{dv}{dt} = -\alpha\frac{\partial p}{\partial y} - 2\Omega u\sin\varphi + F_y \tag{2.3}$$

$$\frac{dw}{dt} = -\alpha\frac{\partial p}{\partial z} + 2\Omega u\cos\varphi + g + F_z \tag{2.4}$$

### 2.1.1. Scaling the equation of motion

In order to avoid the complicated and non- significant terms in the equations of 2.2 to 2.4 scaling is useful. Scaling means the weight or amount of each parameter will be calculated from each force and the unimportant terms will be neglected basing on the weight. These weights are obtained from general measurements on an average in the oceans which are given in the table (2.1) as below:

**Table 2.1:** Dimensional values for scaling the equation of motion

| S.No | Parameter | Symbol | Size | Unit |
|---|---|---|---|---|
| 1. | Horizontal Length | L | $10^6$ | Meters (m) |
| 2. | Vertical Length | H | $10^3$ | m |
| 3. | Horizontal speed (for u, v) | U | $10^{-1}$ | $ms^{-1}$ |
| 4. | Vertical speed | W | $10^{-4}$ | $ms^{-1}$ |
| 5 | Time | T | $10^6$ | $ms^{-1}$ |
| 6 | Coriolis parameter at $45^0$ latitude | f | $10^{-4}$ | $s^{-1}$ |
| 7 | Acceleration due to gravity | g | $10^1$ | $ms^{-2}$ |
| 8 | Specific volume | α | $10^{-3}$ | $m^3\ kg^{-1}$ |
| 9 | Pressure at $10^3$ meters | P | $10^7$ | Pascals |
| 10 | Horizontal eddy coefficients of viscosity | $A_x$,$A_y$ | $10^5$ | $m^2\ s^{-1}$ |
| 11 | Vertical eddy coefficient of viscosity | $A_z$ | $10^{-1}$ | $m^2s^{-1}$ |
| 12 | Kinematic molecular viscosity | $\mathbf{v = k} = \dfrac{\mu}{\rho}$ | $10^{-6}$ | $m^2s^{-1}$ |
| 13 | $2\Omega\cos\varphi$ | f | $10^{-4}$ | $s^{-1}$ |

## 2.1.2. Scaling the vertical component of equation of motion

Expanding the equation (2.1) and substituting the values from the above table 2.1, we can write as:

We know in differentiation: Total derivative = Local term {first bracket} + convective term {second bracket}

$$\frac{\mathrm{d}}{\mathrm{dt}}=\left\{\frac{\partial}{\partial t}\right\}+\left\{\ddot{u}\frac{\partial}{\partial x}+\ \frac{\partial}{\partial y}+\ \frac{\partial}{\partial z}\right\} \tag{2.5}$$

Expanding equation (2.4) using the equation (2.5) ,left hand side $\left(\dfrac{dw}{dt}\right)$ of equation (2.4) will become:

$$\frac{dw}{dt}=\frac{\partial w}{\partial t}+u\frac{\partial w}{\partial x}+v\frac{\partial w}{\partial y}+\mathrm{w}\frac{\partial w}{\partial z} \tag{2.6}$$

Substituing equation (2.6) in equation (2.4) we can write

$$\frac{\partial w}{\partial t}=u\frac{\partial w}{\partial x}+v\frac{\partial w}{\partial y}+\mathrm{w}\frac{\partial w}{\partial z}=-\alpha\frac{\partial p}{\partial z}+2\Omega u\cos\varphi+g+F_z \tag{2.7}$$

The frictional term can be written as:, $\mathrm{F_z} = A_z\dfrac{\partial^2 w}{\partial z^2}$ (2.8)

Substituting equation (2.8), in equation (2.7):

$$\frac{\partial w}{\partial t}+u\frac{\partial w}{\partial x}+v\frac{\partial w}{\partial y}+w\frac{\partial w}{\partial z}=-\alpha\frac{\partial p}{\partial z}+2\Omega u\cos\varphi+g+A_z\frac{\partial^2 w}{\partial z^2} \tag{2.9}$$

Equation (2.9) can be scaled by substituting the values from the table 2.1, we can write serially as:

$$\frac{W}{T}+\frac{UW}{L}+\frac{VW}{L}+\frac{W^2}{H}=\frac{-aP}{H}+fU+g+A_z\frac{W^2}{H^2}$$

Substituting the corresponding weights

$$\frac{10^{-4}}{10^6}+\frac{10^{-1}.10^{-4}}{10^6}+\frac{10^{-1}.10^{-4}}{10^6}+\frac{(10^{-4})^2}{(10^3)^2}$$

$$=-\frac{10^{-3}.10^7}{10^3}+10^{-4.}10^{-1}+10^1+\frac{10^{-1}.10^{-4}}{(10^3)^2}$$

$10^{-10}+10^{-11}+10^{-11}+10^{-11}=-10^{+1}+10^{-5}+10^{+1}+10^{-11}$

Comparing all these terms $10^{-5}$ and $10^{-11}$ are negligibly small on the right hand side while on the left side $10^{-10}$ is higher than $10^{-11}$. So retaining higher order terms we can get

$$\frac{dw}{dt}=-\alpha\frac{\partial p}{\partial z}+g \tag{2.10}$$

Similarly we can scale the X and Y component equations and neglect the small terms and write as

$$\frac{du}{dt}=-\alpha\frac{\partial p}{\partial x}+fv \tag{2.11}$$

$$\frac{dv}{dt}=-\alpha\frac{\partial p}{\partial y}-fu \tag{2.12}$$

## 2.2. Hydrostatic equation

If friction is neglected in equation (2.4), the vertical component of equation of motion (2.10) under non-accelerated motion, turns out to be hydrostatic equation.

$$\frac{dw}{dt}=-\alpha\frac{\partial p}{\partial z}+g+0=0$$

$$\therefore\frac{\partial p}{\partial z}=\rho g \tag{2.13}$$

This is called the hydrostatic equation. This can be written as

$$\alpha dp=gdz\ or\ \varphi=\int d\varphi=\int\alpha dp=\int gdz \tag{2.14}$$

$\phi$ is called geopotential and the equigeopotential lines are also called level surfaces and on these surfaces potential energy (gz) is conserved.

## 2.3. Barotropic and Baroclinic modes

If the isobaric surfaces and level surfaces are parallel, it is called barotropic mode and if they are inclined at each other it is called baroclinic mode where the motion is maximum.

*Barotropic flow* occurs when levels of constant pressure in the ocean are always parallel to the surfaces of constant density. *Baroclinic flow* occurs when levels of constant pressure are inclined to surfaces of constant density. In general, the variation of flow in the vertical can be decomposed into a barotropic component which is independent of depth, and a baroclinic component which varies with depth.

An isobaric surface in a fluid is one on which the hydrostatic pressure is constant, while an isopycnal (or isosteric) surface is one on which the density of the fluid is constant. When the density of water is a function of pressure alone, the isobaric and isopycnal surfaces are parallel to each other. This condition is called barotropic field of mass. On the other hand if the density is a function of pressure as well as temperature and varies horizontally with them, the isobaric and isopycnal surfaces are no more parallel but inclined to each other and then the mass field is said to be baroclinic. The water is stationary in the barotropic field of mass and it is not in baroclinic. In the ocean upper 1000 m water is baroclinic and deep water is barotropic.

It should be noted that relative to geopotential (level) surfaces, the isobaric and isopycnal surfaces have opposite slopes in the baroclinic case while they are parallel in case of barotropic as shown in the Fig. 2.2. Thus the ocean can be considered as a two layer structure with upper baroclinic and deep barotropic layer.

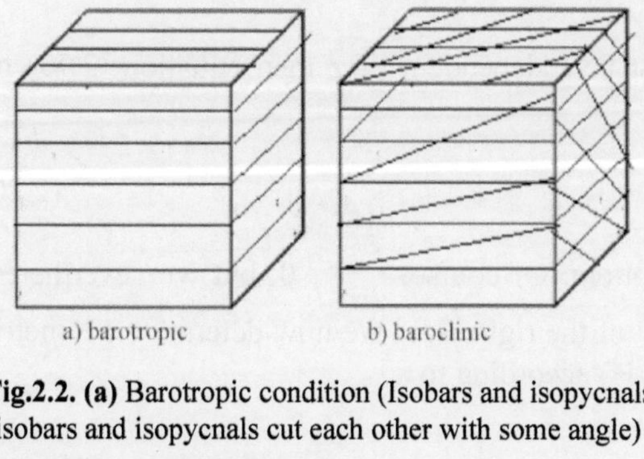


**Fig.2.2. (a)** Barotropic condition (Isobars and isopycnals are parallel. **(b)** Baroclinic condition (isobars and isopycnals cut each other with some angle)

### Barotropic equations

The term, barotropic, generally applies to a condition where the fluid density is a function of pressure only. Here we will be a bit more restrictive and require that

$\rho = \rho_0$ = constant. With reference to Fig. 2.3 we now integrate the hydrostatic equation

$$\frac{\partial p}{\partial z} = -\rho_0 g \tag{2.15}$$

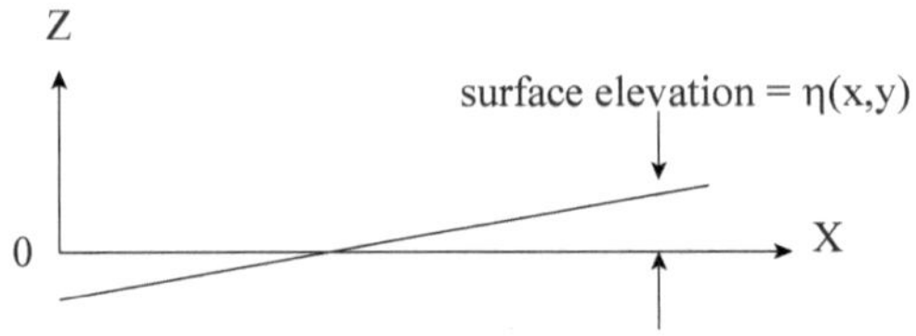


**Fig. 2.3:** The surface, Z = 0, represents the geoid; that is, a surface everywhere perpendicular to the gravity vector. The sea surface elevation, η, is measured relative to the geoid.

Using the boundary condition, $p$ = atmospheric pressure = $p_a$ at $z = \eta$. The result is

$$p = p_a \rho_o g[\eta(x,y)-z] \tag{2.16}$$

where $\eta = \eta(x,y)$ is the sea surface elevation. Neglecting variations in $p_a$ the important result emerges that $\frac{\partial p}{\partial x} = \rho_o g\frac{\partial \eta}{\partial x}$ *and* $\frac{\partial p}{\partial y} = \rho_o g\frac{\partial \eta}{\partial y}$ so that using geostrophic equation

$$-fv = -\frac{1}{\rho_o}\frac{\partial p}{\partial x} \text{ and } fu = -\frac{1}{\rho_o}\frac{\partial p}{\partial y} \tag{2.17}$$

We can write as $-fv = -g\frac{\partial \eta}{\partial x}$ *and* $fu = -g\frac{\partial \eta}{\partial y}$ (2.18)

According to the Sverdrup's relation we know

$$u\frac{\partial f}{\partial x} + v\frac{\partial f}{\partial y} = f\frac{\partial w}{\partial z} \tag{2.19}$$

Since $\eta = \eta$ (x,y) and $\frac{\partial w}{\partial z}$ must be independent of z then equation (2.19) on integrating

$$f\frac{\text{w(x,y,n)-w(x,y,-H)}}{H} = u\frac{\partial f}{\partial x} + v\frac{\partial f}{\partial y} \tag{2.20}$$

Where z = -H is at the ocean bottom. Of course, $\frac{\partial f}{\partial x}$ = 0, but we nevertheless retain the term and the symmetry of the right side. We now determine geometric relations for w(x,y, η) and w(x,y,-H) according to

$$w(\eta) = u\frac{\partial \eta}{\partial x} + v\frac{\partial \eta}{\partial y}$$

$$w(-H) = -u\frac{\partial H}{\partial x} - v\frac{\partial H}{\partial y} \tag{2.21}$$

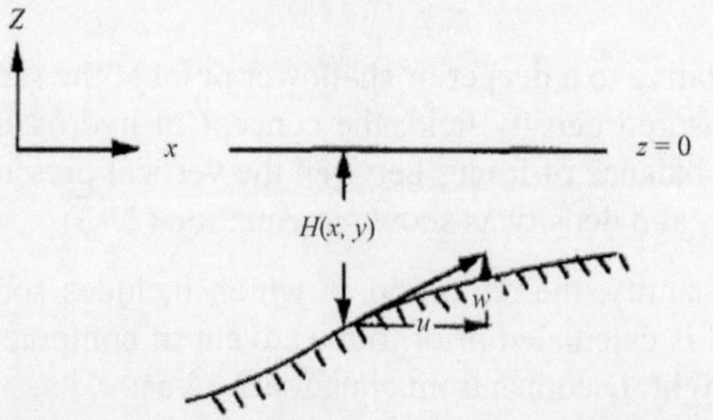


**Fig.2.4.** Illustrating the relation $w/U = -\frac{\partial H}{\partial x}$ at the bottom of the ocean for the case where $\frac{\partial H}{\partial y} = 0$. Variations in sea surface slope may be neglected relative to the much larger variations in depth

The relation {equation(2.21)÷U} = $-\frac{\partial H}{\partial x}$, for the case where $\frac{\partial H}{\partial y} = 0$ is illustrated in Fig.2.4. It may be shown that the sea surface slope is much smaller than typical bottom slopes so that the contribution of w(x,y,η) to equation (2.20) may be neglected in comparison with w(x,y, – H) Thus,

$$\frac{f}{H}\left(u\frac{\partial H}{\partial x} + v\frac{\partial H}{\partial y}\right) = u\frac{\partial f}{\partial x} + v\frac{\partial f}{\partial y}$$

Or rearranging, we can write as; $u\frac{\partial}{\partial x}\left(\frac{f}{H}\right) + v\frac{\partial}{\partial y}\left(\frac{f}{H}\right) = 0$ (2.22)

Equation (2.22) may also be written vectorially as V.∇h $\left(\frac{f}{H}\right)$ = 0 where ∇h = i $\frac{\partial}{\partial x} + j\frac{\partial}{\partial y}$ is the horizontal gradient operator. Thus the velocity vector is perpendicular to the gradient of $\frac{f}{H}$ and therefore parallel to contours of constant $\frac{f}{h}$ as illustrated in the Fig.2.5

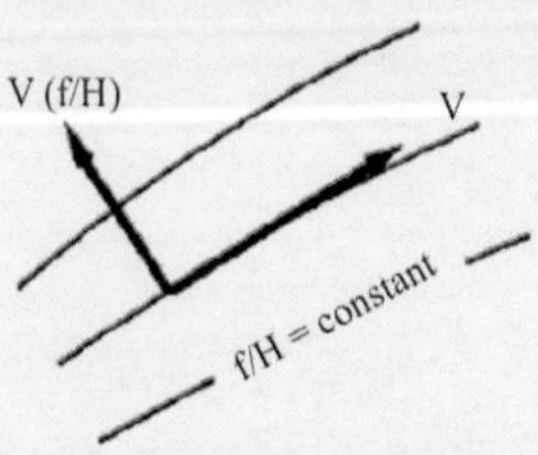


**Fig. 2.5.** Barotropic and geostrophic flow over a basin where both *f* and *H* are functions of the horizontal space coordinates.

## 2.4. Dynamic height

To calculate pressure at a point relative to a deeper or shallower point at the same horizontal location, from the measured density field, the concept of hydrostatic balance can be used, which is the balance of forces between the vertical pressure gradient and the product of gravity and density as shown in equation (2.13).

Dynamic height is a historical quantity, the definition of which includes some anachronisms owing to the way it is calculated prior to the advent of computers. Thus the definition of dynamic height, D comes from equation 2.14 as:

$$10\Delta D = g\Delta Z$$

$$10(D_2 - D_1) = g\int_1^2 dZ$$

Thus D would be exactly the height z if g = 10 m/sec$^2$ and the "10" had units of m/sec$^2$.

If $\Delta z$ = 1 meter, $\Delta D = (g\,\Delta z)/10 = 1$ m$^2$/sec$^2$ = 1 dynamic meter.

To calculate D from density, equation (2.14) can be used.

$$D = -\frac{1}{10}\int_{p_1}^{p_2} \alpha dp$$

Then you can see how the density comes into work- out the pressure change from one depth to another.

## 2.5. Centrifugal force

The two forces, *centrifugal* and *Coriolis* are to be considered on a rotating earth (Fig.2.6). We must acknowledge these because we want to measure motion relative to the rotating earth - this is an "accelerated reference frame" just like the floor of an elevator.

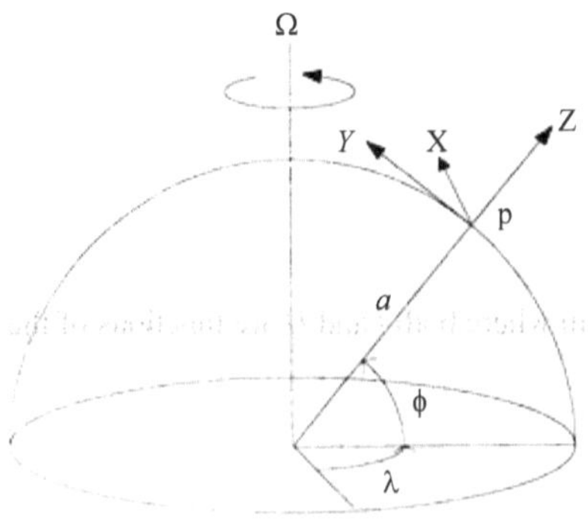


**Fig. 2.6.** Generation of angular velocity & centrifugal force at point P on the surface of the earth along with the components on a rotating earth

When you sit on a merry-go-round, you feel thrown radially outward. This is due to centrifugal force, which is equal to:

Centrifugal acceleration = $(\Omega)^2$ a,

where 'a' is the distance from the center (radius) and omega($\Omega$) is the angular velocity such that $\Omega = 2\pi/T$ where T is the time taken to complete one full rotation, Ø is the latitude and λ is longitude in Fig.2.6.

Applying this to the earth taking the radius (a) = 6000 km,it comes as omega($\Omega$) = $2\pi$ /86400 sec, $(\Omega)^2$a = 3.17 cm/sec$^2$. This would be the centrifugal force at the equator, where it is maximum. At the poles, it is of course 0, since the radius to the axis of rotation is zero. (So the actual force is $(\Omega)^2$a cos $\phi$. Compare this centrifugal force with acceleration due to gravity of 980 cm/sec$^2$, which pulls downward toward the earth's center. Therefore, on the earth, you weigh 0.3% less at the equator than at the pole.

If the solid earth were a true sphere, the motionless ocean would be 20 km deeper at the equator than at the pole. We don't see anything like this. Therefore, we conclude: the solid earth itself is spheroidal, with the equatorial radius about 20 km larger than the polar radius.

*Thus the horizontal gravity + horizontal centrifugal force = 0. So we can neglect centrifugal force because the earth itself is spheroidal.*

## 2.6. Coriolis force

This apparent (or pseudo) force exists because we all live on an accelerating reference frame (a rotating planet).Since our coordinate system is rotating with the planet, it is indistinguishable from a 'real' force, and we will treat it as a fictitious force.

If a missile is fired from the equator, of the earth towards North Pole, it is deflected to the right hand side as shown by 'p' in Fig 2.7. Similarly if a missile is fired from the equator towards the south pole it deflects towards its left hand side as shown by 'q' in Fig.2.7. Thus any particle moving in Northern Hemisphere is deflected towards its right hand side and in Southern Hemisphere it deflects towards left hand side.

The action of the Coriolis force is in the following way:

- Acts only to the moving objects
- Action is proportional to the speed
- It makes objects go to the right in the Northern Hemisphere
- It makes objects go to the left in the Southern Hemisphere
- It is the biggest force at the poles, and goes to zero on the equator, if the motion is parallel to the earth's surface (not vertical).

Since the earth rotates through $2\pi$ radians in one Siderial day (23.93447 hours), the angular speed of the earth is

$$\Omega = \frac{2\pi}{23.9447hr} \times \frac{hr}{3600s} = 7.292 \times 10^{-5}\, raans\, s^{-1}$$

## 2.6.1. The Action of Coriolis force

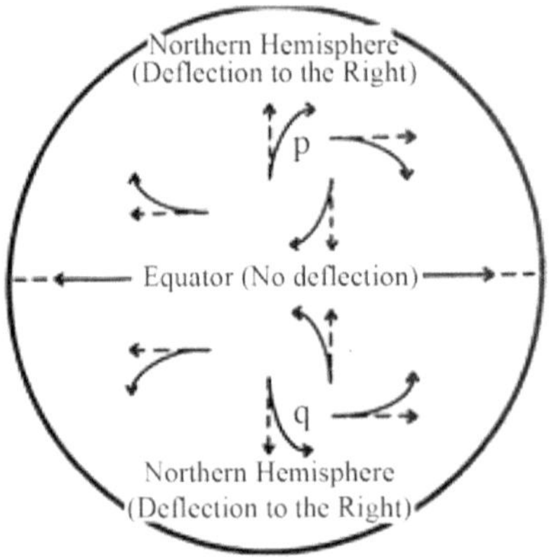


**Fig. 2.7.** Action of Coriolis force. Dashed arrow initial moving object, curved arrow Coriolis action

## 2.6.2. The Derivation of Coriolis acceleration

Newton's law depends entirely upon the nature of coordinate system relative to the surface on which it is acting. So on the surface of the earth, this coordinate system also rotates with the earth about its axis of rotation. As a result the departures of acceleration under such a rotating coordinate system have to be worked out. The coordinates on the rotating system of earth taken are positive 'x' along the latitude towards east, positive 'y' along the meridian towards the North Pole and 'z' as the depth into the oceans as shown in Fig.2.1.

In the same way let us consider a coordinate system with the coordinates x', y' and z' and as shown in Fig.2.8. Let the coordinate system is rotated by an angle $\Omega$t and the new coordinates are x,y and z respectively. You can see the details in the right side of Fig.2.8. From the rotational coordinate system we can write

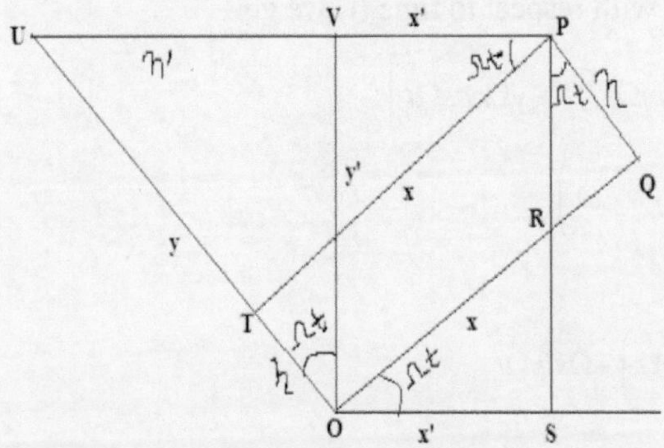


**Fig. 2.8:** Rotating coordinate system

From triangle ORS, OR = $\frac{X'}{\cos \Omega t}$,

From triangle PQR, QR = OQ-OR = $X - \frac{X'}{\cos \Omega t}$; PR = $\frac{\eta}{\cos \Omega t}$

From triangle PQR, $\sin \Omega t = \frac{QR}{PR} = \frac{X - \frac{X'}{\cos \Omega t}}{\frac{\eta}{\cos \Omega t}} \Rightarrow X' = X \cos \Omega t - \eta \sin \Omega t$

Similarly from triangle PTU, PT = X, From triangle OUV, TU =

$\frac{\eta'}{\cos \Omega t} - \eta \left( \because OT = \eta, OU = \frac{\eta'}{\cos \Omega t} \right)$

From triangle PTU, $\tan \Omega t = \frac{TU}{PT} = \frac{\frac{\eta'}{\cos \Omega t} - \eta}{X} = \frac{\sin \Omega t}{\cos \Omega t}$

$\therefore \eta' = X \sin \Omega t + \eta \cos \Omega t$

It may be noted that the letter 'Y' may be read in place of η then, we get the following equations.

$$X' = X \cos \Omega t - Y \sin \Omega t$$

$$Y' = X \sin \Omega t + Y \cos \Omega t \quad (2.23)$$

$$Z' = Z$$

Similarly in place of X, Y, and Z if forces $F_x$, $F_y$, and $F_z$ are considered we can write

$$F_x' = F_x \cos \Omega t - F_y \sin \Omega t$$

$$F_y' = F_x \sin \Omega t + F_y \cos \Omega t \quad (2.24)$$

$$F_z' = F_z$$

If we differentiate equation (2.23) with respect to time (t) we get

$$\frac{\partial x'}{\partial t} = cos\Omega t\frac{\partial x}{\partial t} + x.\Omega(-\sin\Omega t) - \left| sin\Omega t\frac{\partial y}{\partial t} + y\Omega cos\Omega t\right|$$

$$\frac{\partial x'}{\partial t} = cos\Omega t\left[\frac{\partial x}{\partial t} - y\Omega\right] - sin\Omega t\left(\frac{\partial y}{\partial t} + x.\Omega\right)$$

Differentiating once again:

$$\frac{\partial^2 x'}{\partial t^2} = cos\Omega t\left[\frac{\partial^2 x}{\partial t^2} - \Omega\frac{\partial y}{\partial t}\right] + \left(\frac{\partial x}{\partial t} - y\Omega\right) - \Omega sin\Omega t$$

$$-\left[sin\Omega t\left(\Omega\frac{\partial x}{\partial t} + \frac{\partial^2 y}{\partial t^2}\right) + \left(x\Omega + \frac{\partial y}{\partial t}\right)\Omega cos\Omega t\right]$$

$$\frac{\partial^2 x'}{\partial t^2} = cos\Omega t\left[\frac{\partial^2 x}{\partial t^2} - 2\Omega\frac{\partial y}{\partial t} - \Omega^2 x\right] - sin\Omega t\left[\frac{\partial^2 y}{\partial t^2} + 2\Omega\frac{\partial x}{\partial t} - \Omega^2 y\right] \quad (2.25)$$

Put $\frac{\partial x'}{\partial t} = \dot{x}', \frac{\partial^2 x'}{\partial t^2} = \ddot{x}', \frac{\partial y'}{\partial t} = \dot{y}', \frac{\partial^2 y'}{\partial t^2} = \ddot{y}'$

Then we can write equation (2.25):

$$(\ddot{x}' = cos\Omega t[\ddot{x} - 2\Omega\dot{y} - \Omega^2 x] - sin\Omega t[\ddot{y} + 2\Omega\dot{x} - \Omega^2 y] \quad (2.26)$$

According to Newton's law, acceleration is double derivative,

$$a = F/m = \frac{\partial^2 x}{\partial t^2} \quad (2.27)$$

So combining equations 2.24 and 2.27 we write or writing equation (2.24) in terms of force: $F_{x'} = F_x \cos\Omega t - F_y \sin\Omega t$

So acceleration is divided by mass: $\frac{F_{x'}}{m} = \frac{F_x}{m}cos\Omega t - \frac{F_y}{m}sin\Omega t$ (2.28)

Since $\ddot{x}' = \frac{F_{x'}}{m}$

Equations 2.26 and 2.28 can be equated. After equating and comparing the coefficients of CosΩt and SinΩt we get as below:

$$\frac{F_x}{m} = \ddot{x} - 2\Omega\dot{y} - \Omega^2 x$$

$$\frac{F_y}{m} = \ddot{y} + 2\Omega\dot{x} - \Omega^2 y \quad (2.29)$$

Similarly can be written $\ddot{Z} - \frac{F_z}{m} = 0$

Equations (2.29) can be written as

$$\ddot{x} - \frac{F_x}{m} = \Omega^2 x + 2\,\Omega\,\dot{y}$$

$$\ddot{y} - \frac{F_y}{m} = \Omega^2 y - 2\,\Omega\dot{x} \tag{2.30}$$

Equations (2.30) are Newton's second law transformed to a rotating coordinate system. The first of the terms on the right hand side ($X\Omega^2$, $Y\Omega^2$) denote the centrifugal accelerations pointing radially outward. The second terms on the right hand side (2 $\dot{Y}X, 2\dot{X}\Omega$) represent accelerations owing to the combined effects of rotation of the coordinate axes and motion of the particle, relative to the rotating coordinate system. These are called Coriolis accelerations. The action of this Coriolis acceleration is:

1. It acts only perpendicular to the plane of the moving particle.
2. It acts right hand side in the N.H and left hand side in the S.H of the moving particle.
3. It comes into being only for a moving particle but not for a fixed particle.
4. It does not change the magnitude of the velocity but changes only the direction of the moving particle as it acts perpendicular to the plane of moving particle.

Thus the centrifugal and Coriolis accelerations are extra terms obtained due to non-inertial (non-Newtonian) frame of reference.

Hence Coriolis accelerations are: $\ddot{x} = 2\Omega\dot{y}$ and $\ddot{y} = -\,2\Omega\dot{x}$ (2.31)

### 2.6.3. The Spherical coordinate system

To obtain an accurate description of geophysical processes, equations should be written in spherical coordinates. Thus, as illustrated in Figure 2.9, the position of a point is determined by specifying the radial distance from the earth's center, *r*, latitude, $\phi$, and longitude, $\lambda$. Mathematically, the values of $\phi$ range from $-\frac{\pi}{2}$ (south pole) to $+\frac{\pi}{2}$ (north pole) whereas the values of $\lambda$ range from $-\pi <$ to $+\pi$. However one conventionally refers to a mid-latitude point in the northern hemi sphere, for example, as $\phi = 45°$N whereas, in the southern hemisphere, $\phi = 45°$S. Also one refers to a point in the eastern hemisphere, for example, as = $90^0$E whereas, in the western hemisphere, $\lambda = 90^0$ W. The meridian, $\lambda = 0°$, passes through Greenwhich, England and the meridian, $\lambda = 180^0$E = $180^0$W, is the International dateline. The earth's rotational vector $\Omega$, is aligned with the axis through the poles ($\phi = 90^0$N, $90^0$S); its approximate magnitude is $\Omega$, = $2\pi$ day$^{-1}$ = $0.7272 \times 10^{-4}$ s$^{-1}$. A more exact value which accounts for the earth's yearly rotation around the Sun is $0.7292 \times 10^{-4}$ s$^{-1}$. Discussion of physical processes is simplified if one uses rectangular coordinates. Results will be qualitatively correct but quantitatively more or less approximate depending on the scale of the problem. Later, all concepts and results can be recast in spherical coordinates to restore accuracy.

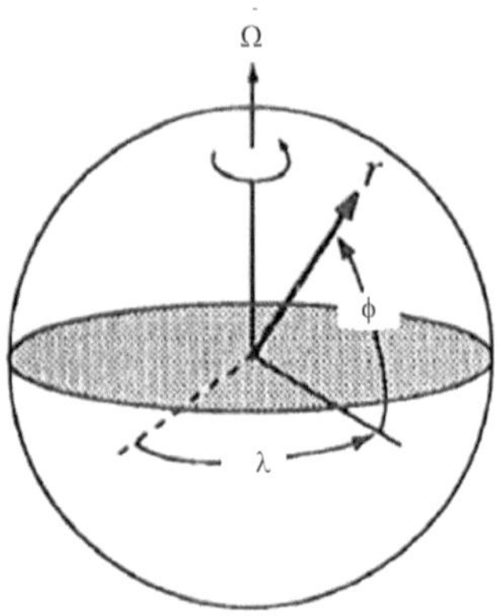


**Fig. 2.9.** Spherical coordinate system

We shall now apply these results on a rotating spherical earth, in which x is tangent eastward, y meridionally northward and z radially upward into the atmosphere as shown in Fig.2.10. Let 'P' be a point on the earth's surface which is rotating with the angular velocity Ω, at a distance of 'acosφ' from the axis of rotation where 'a' is the radius of the earth and 'φ' is the latitude. From Fig.2.10, $\Omega_x = \Omega \text{ Cos } 90 = 0$, $\Omega_y = \Omega \text{ Sin } (90\text{-}\varphi) = \Omega \text{ Cos } \varphi$, $\Omega_z = \Omega \text{ Sin}\varphi$.

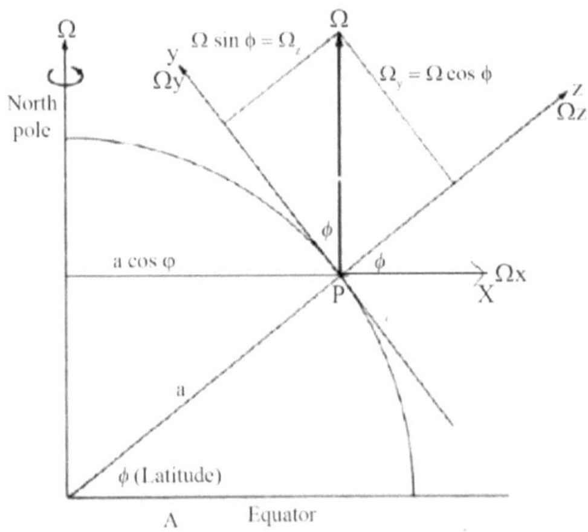


**Fig. 2.10.** Components on a rotating earth

To find the general expression for Coriolis acceleration at any latitude, consider the Fig.2.11 below:

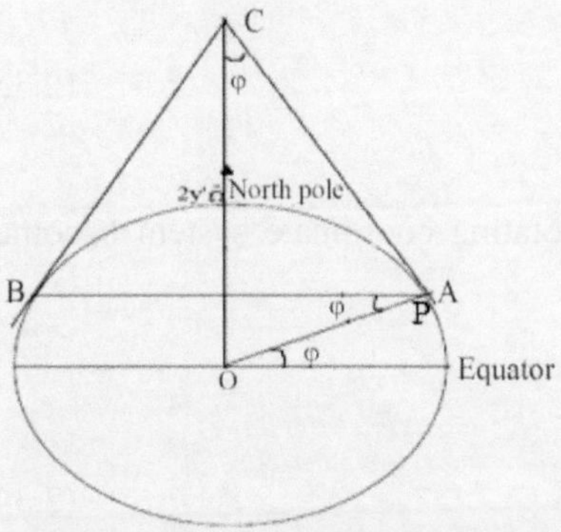


**Fig. 2.11.** A cone subtending on a rotating earth above the North pole

In Fig.2.11, let at point 'P' on the latitudinal circle AB will subtend a cone ACB with the angle2ϕ. We have calculated the magnitude of the deviating force at the pole OC = 2ẏΩ in equation (2.23). Now we have to find this deviating force at any point 'A' which is equal to OA.

From the triangle OAC, OA = OC sin ϕ. So OA = 2Ωẏ sin ϕ, where $\dot{y} = \dfrac{\partial y}{\partial t} = v$

So OA = 2Ωv sinϕ

If the velocity of the particle is V= iu +jv+kw then the Coriolis acceleration is curl of the vector. The direction of the vector Ω is such that when the fingers of the right hand twist (curl) about the axis of rotation in the same sense as the motion, the thumb points in the direction of Ω.

In Fig. 2.11, let 'a' is the position vector of the moving particle 'P' as measured from the origin of coordinates (centre of the earth). So the vector (Ω×a) has the magnitude 'Ω a sinφ', which is just the linear speed (V) of the particle. Therefore V = Ω×a. Similarly if the magnitude of velocity (V) is considered as the position vector, then the acceleration is Ω×V. Thus we got the Coriolis accelerations in equation (2.31) as $-2\dot{X}\Omega and - 2\dot{Y}\Omega$, where $\dot{X} and \dot{Y}$ are the velocities $(\dfrac{dx}{dt}, \dfrac{dy}{dt})$. This means the Coriolis accelerations on the surface of the earth can be written as C = -2Ω×V.

$$C = -2\Omega\times V = -2\Omega \begin{vmatrix} i & j & k \\ 0 & \cos\varphi & \sin\varphi \\ u & v & w \end{vmatrix} = -2\Omega \begin{vmatrix} i & j & k \\ \Omega_x & \Omega_y & \Omega_z \\ u & v & w \end{vmatrix} \quad (2.32)$$

where u,v, w are the x,y,z components of the velocity. Expanding the determinant and separating the different components we get,

$C_x$ = -2Ω w cosφ + 2Ωv sinφ

$C_y$ = -2Ωu sinφ (2.33)

$C_z$ = 2Ω u cos φ

The Newton's second law of motion on a rotating coordinate system becomes from equation 2.30 as:

$$\ddot{X} - \frac{F_x}{m} = 2\Omega v \,\mathrm{Sin}\,\varphi - 2\Omega w \,\mathrm{Cos}\,\varphi$$

$$\ddot{Y} - \frac{F_y}{m} = -2\Omega u \,\mathrm{Sin}\,\varphi \tag{2.34}$$

$$\ddot{Z} - \frac{F_z}{m} = 2\Omega u \,\mathrm{Cos}\,\varphi$$

Finally 2Ωw cosϕ and 2Ωu cosϕ are neglected as they (centrifugal accelerations) are relatively very small compared to the rest and so the Coriolis accelerations are as below:

$$\ddot{X} - \frac{F_x}{m} = 2\Omega v \,\mathrm{Sin}\,\varphi$$

$$\ddot{Y} - \frac{F_y}{m} = -2\Omega u \,\mathrm{Sin}\,\varphi$$

$$\ddot{Z} - \frac{F_z}{m} = 0$$

## The characteristic time scale

The horizontal component f = 2Ω sin ϕ of the Coriolis acceleration is the most important term affecting the ocean circulation. Its characteristic time scale is the inertial timescale

$$\tau_f = \frac{1}{f}$$

Near the equator this time scale increases rapidly and $\tau_f$ reaches a minimum at the poles. To determine its interpretation in terms of vorticity, consider a fluid parcel moving on the sphere in the northern hemisphere at a latitude $\phi_0$; the Coriolis acceleration deflects its path to the right which gives a contribution (Fig.2.11)to the vertical component of the vorticity with a magnitude $f_0$ = 2Ωsin $\phi_0$. The inertial time scale is about $10^4$ s at 45°N.

If the horizontal scale of the motion is so small that we can take a constant $f_0$ = 2Ωsinϕ, we say we use the f-plane approximation. Introduction of a local coordinate y = ($\phi$-$\phi_0$)$r_0$ and then a Taylor series near $\phi_0$ gives

$f = f_0 + \beta_0 y + \mathcal{O}(\phi - \phi_0)^2 \ \beta_o = \frac{2\Omega}{r_o} \cos\phi_0$

This linear meridional variation of f is part of the β-plane approximation. The characteristic time scale associated with the meridional variation of the Coriolis acceleration is

$\tau_\beta = f_1 = f_0 + \beta_0 L$

An interpretation in terms of vorticity is again possible (Fig. 2.11a,c). A water parcel that moves northward in the northern hemisphere is deflected more strongly at a more northerly position. This gives a contribution to the vertical component of the vorticity with a magnitude of $\beta_0$ L. At 45°N, a value of $\tau_\beta$ is $10^4$ s for a flow with a length scale L = 5000 km.

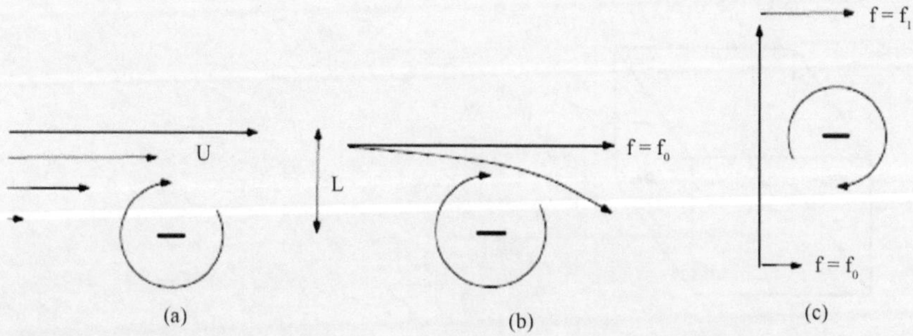


**Fig. 2.11.** Illustration of **(a)** the advective time scale $\tau_a$, **(b)** the inertial time scale $\tau_f$ and **(c)** the time scale $\tau_\beta$ where $f_1 = f_0 + \beta_0 L$

## The advective time scale

Let L and U be characteristic horizontal length and velocity scales of a particular ocean flow. For example, L could be the width of an ocean basin (L = O ($10^6$) m) and U a maximum depth-averaged horizontal velocity in the basin {U = O ($10^{-2}$ $ms^{-1}$)}. The advective time scale $\tau_a$ associated with the flow is then given by

$$\tau_a = \frac{L}{U}$$

This indicates the time scale on which the ocean flow with velocity U changes on the length scale L. Typical values for the basin scale flow above would be = $10^8$ s ≈ 3 years. The advective time scale also has an interpretation in terms of vorticity, which is related to velocity gradients. Consider a flow with a gradient in the horizontal velocity U over a length scale L. This gradient provides a contribution to the vertical component of the vorticity vector of magnitude U/L = 1/ $\tau_a$ (Fig. 2.11,a).

The average depth of the ocean, say D, is much smaller than the horizontal scale of a typical ocean flow. This introduces a small parameter $\delta = D/L$, the so-called aspect ratio; for basin scale flows with $L = 10^6$ m and $D = 10^3$ m the value of $\delta \approx 10^{-3}$.

## 2.7. The Pressure gradient force

According to Fig.2.12, pressure gradient force between faces 1 and 2 is: $F = -(P_2 - P_1)dy.dz = -\Delta P dy.dz$. The force per unit volume in the X direction is $\frac{\Delta P}{\Delta X}$. Because of this force, water moves from one area to another. Thus a Pressure *Gradient* is a measure of how the pressure is changing as one moves in the horizontal direction.

It is simply: $-\frac{\Delta P}{\Delta X}$

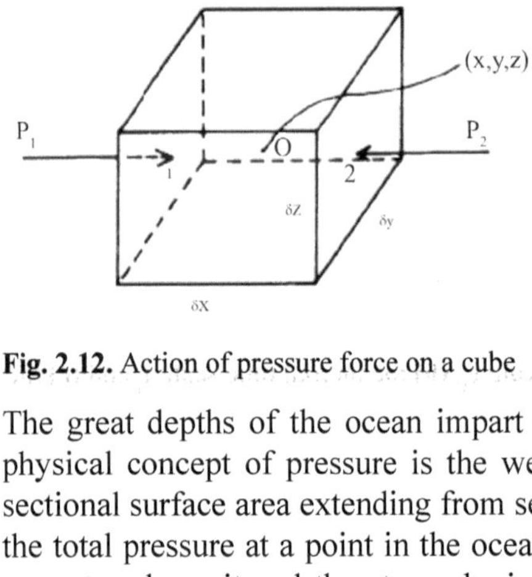


**Fig. 2.12.** Action of pressure force on a cube

The great depths of the ocean impart considerable pressure on the water. The physical concept of pressure is the weight of water in a column of unit cross sectional surface area extending from sea surface to the required depth. Although the total pressure at a point in the ocean would be due both to the weight of the seawater above it and the atmospheric pressure, unless otherwise specified, the pressure at a point in the ocean is taken to be just that due to the seawater, not the seawater plus atmosphere. The units are $N/m^2$ = 1 Pascal = $10^{-5}$ bar = 10 dyne $cm^{-2}$. Pressure is usually measured with a strain gauge transducer.

Quite often we measure pressure but report it as depth assuming that 1 m depth equals to about 1 decibar or $10^4$ Pa. The total hydrostatic pressure at any location is

$$P_z = \int_0^z -\rho_z g dz$$

For example, a given water density of 1020 kg $m^{-3}$, a depth increment of 1 m and $g = 9.8$ m $s^{-2}$ yields a pressure of $1.02 \times 10^4$ Pa.

## 2.8. The Derivation of Pressure gradient force

Consider a rectangular volume having the sides $\delta_x$, $\delta_y$ and $\delta_z$ fixed in a coordinate system relative to the solid earth as shown in Fig.2.13. Then the force in the x-direction due to the atmospheric pressure is, $p\delta_y\, \delta_z$, where 'p' is the atmospheric pressure and $\delta_y\, \delta_x$ is the area of the face it is acting.

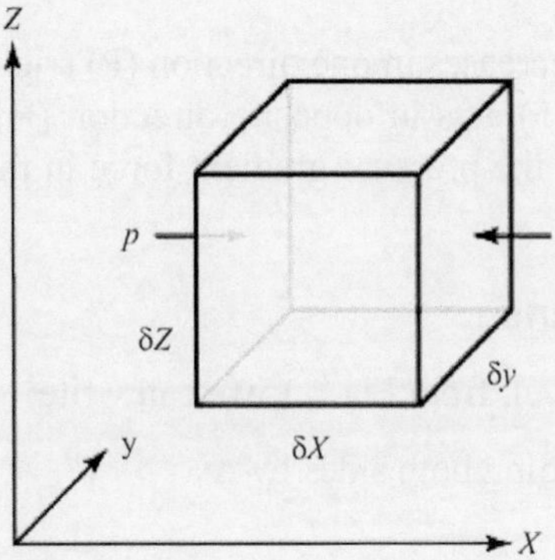


**Fig. 2.13.** Pressure force acting on a rectangular volume element

Similarly the pressure force acting on the opposite face would be - $\left(p+\frac{\partial P}{\partial x}\delta x\right)\delta y\delta z$

The net force acting in the x direction (i) would be

$(ip\delta y\delta z)+\left\{-i(p+\frac{\partial p}{\partial x}\delta x)\delta y\delta z\right\}=-i\frac{\partial p}{\partial x}\delta x\delta y\delta z$, pressure force per unit volume= $-i\partial p/\partial x$ because the volume of the cube is $\delta x\delta y\delta z$.

Similarly considering in other directions the total pressure force would be

$$-\left\{i\frac{\partial p}{\partial x}+j\frac{\partial p}{\partial y}+k\frac{\partial p}{\partial z}\right\}$$

The force per unit mass = $-\frac{1}{\rho}\left\{i\frac{\partial p}{\partial x}+j\frac{\partial p}{\partial y}+k\frac{\partial p}{\partial z}\right\}$ = $-\alpha\nabla p$ (2.35)

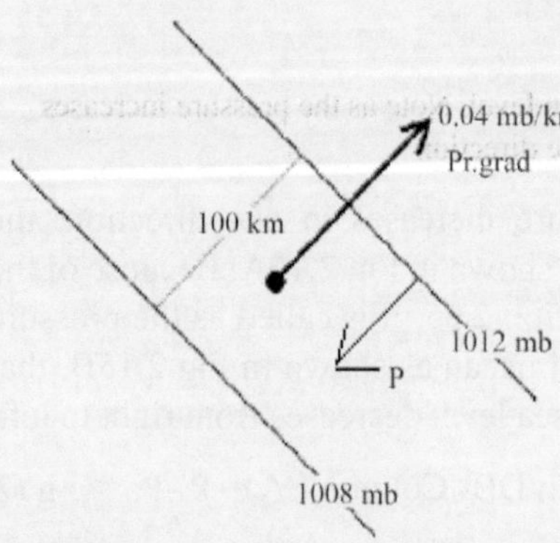


**Fig. 2.14.** While the actual pressure variation, P is from right to left, the pressure gradient force ($-\alpha\nabla p$) acts in the opposite direction (left to right)due to negative sign in the equation.

The magnitude of pressure gradient as per Fig. 2.14 is as below:

$$\nabla p \cong \frac{\Delta p}{\Delta n} = -\left[\frac{1012mb - 1008mb}{100km}\right] = -0.04mb/km$$

The minus sign indicates that as the pressure decreases in one direction (P) (right to left), the pressure force (∇p in equation 2.35) acts in opposite direction (left to right) as shown in Fig.2.14. This is called as the pressure gradient force in the atmosphere.

## 2.9. The Pressure gradient in the Oceans

With regard to the pressure variation and sea level, from Fig.2.15we can write:

$\Delta P = P_1 - P_2 = \rho\, g\, z - \rho\, g\, (Z + \Delta Z) = -\rho\, g\Delta z$.. Dividing both sides by $\Delta x$

$$\frac{\Delta P}{\Delta X} = -\rho g \frac{\Delta z}{\Delta x} = -\rho g \tan\theta$$

Thus as the pressure increases in one direction, the pressure force acts in opposite direction in oceans rather the pressure gradient acts down the gradient as shown in Fig.2.15.

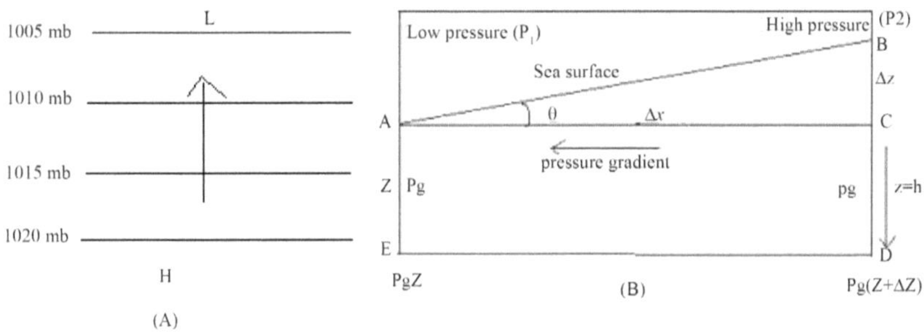


**Fig. 2.15.** Relation between pressure variation and sea level. Note as the pressure increases from top to bottom, pressure force acts in the opposite direction

The minus sign indicates that as the pressure increases in one direction, the pressure force acts in the opposite direction as shown in Fig.2.15A. Because of the presence of gradient (del,∇) operator in equation 2.35, it is called as the pressure gradient force. The same is true in the case of ocean as shown in Fig.2.15B, that is, as pressure increases from left to right, the sea level decreases from right to left.

From Fig.2.15 (b) we can write: $\Delta P$ = (Column DB- Column EA)= $P_2 - P_1 = \rho\, g\, (Z + \Delta Z) - \rho\, g\, Z = \rho\, g\Delta Z$

$$-\frac{\Delta P}{\Delta X} = -\rho g \frac{\Delta Z}{\Delta x} = -\rho g \tan\theta$$

## The Rossby Number

A convenient measure of the magnitude of the acceleration compared to the Coriolis force may be obtained by forming the ratio of the characteristic scales for the acceleration and the Coriolis force terms: $\frac{u_o{}^2}{\Delta x}\times\frac{1}{fu_0}=\frac{u_0}{f_0 L}=R_0\ (\because \Delta x = L)$. This ratio is a non-dimensional number called the *Rossby number* after the Swedish meteorologist C. G. Rossby (1898–1957) and is designated by $R_o=\frac{u_o}{Lf_o}$. Where $u_0$ is the characteristic velocity and $f_o$ is the characteristic Coriolis parameter. The smallness of the Rossby number is a measure of the validity of the geostrophic approximation.

It is obtained from the balance of forces between pressure gradient, Coriolis and centrifugal forces. In N.H cyclone the balance is

Pressure gradient = Coriolis+ centrifugal

$$\therefore -a\frac{\partial p}{\partial n}=fv+\frac{v^2}{R}$$

If we write under characteristic scales and taking the mod and divide throughout by fv

$$\therefore \left|-a\frac{\partial p}{\partial n}-fv\right|=\frac{v^2}{R}\ \Rightarrow \text{dividing by fv} \Rightarrow \frac{a}{fv}\frac{\partial p}{\partial n}+1=\frac{v}{fR}=R_o$$

$\frac{v}{fR}$ is called Rossby number ($R_o$) and is represented in characteristic scales as

$$R_o=\frac{U}{fL}$$

i) If $R_0 \ll 1$, then the centripetal acceleration is extremely small compared to the Coriolis acceleration, so the balance of forces is in between pressure gradient and Coriolis. Then it is called geostrophic motion.

ii) If $R_0 \approx 1$, then all the three forces are important, then it is called gradient motion.

iii) If $R_0 \gg 1$, then Criolis term is negligible compared to the centripetal acceleration and so Coriolis acceleration is negligible. This balance between pressure gradient and centripetal accelerations is called cyclostrophic motion.

Thus the *Rossby* number is a non-dimensional number defined in characteristic scales as

$$R_o=\frac{u_o}{Lf_o} \tag{2.36}$$

where, once again, $u_0$ is a scale characteristlc of ocean velocities; $L$ is a length scale, characteristic of horizontal variation of properties in the ocean *and f* is the scale of the *Coriolis* parameter. For large scale, oceanic circulation, we have $L$ = 4000 km. A typical velocity scale might be $u_0 \sim 10 cms^{-1}$ and $f_0 \sim 10^{-4} s^{-1}$. Thus, $R_0 \sim 10^{-4}$, a rather small number. If one reviews the order of magnitude of terms in equations of motions, one sees that the ratio of the advective terms to the *Coriolis* terms is

$$R_o = \frac{u_o^2 / L}{u_o f_o}$$

and therefore the advective terms can be neglected for small $R_0$.

These detailed equations of geostrophic, gradient and cyclostrophic motions are discussed in the next chapter.

## References

Neumann G., and W.J. Pierson. 1966. Principles of Physical Oceanography. New Jersey:Prentice-Hall.

Open University 1989 Ocean Circulation. Oxford: Pergamon Press,Oxford,.

Pond S. & G.L, Pickard 1983 Introductory Dynamical Oceanography, Pergamon Press,

Tolmazin D. 1985. Elements of Dynamical Oceanography. Boston: Allen and Unwin.

## Exercises

### *Exercise 2.1*

Find the value of Ω

### *Solution*

Since the earth rotates through 2 radians in one Siderial day (23.93447 hours), the angular speed of the earth is

$$\Omega = \frac{2\pi}{23.9447 hr} \times \frac{hr}{3600 s} = 7.292 \times ?10^{-5} radianss^{-1}$$

### *Exercise 2.2*

Find the value of Coriolis parameter, $f = 2\,\Omega\,\sin\phi$ at $45^0$ latitude

### Solution

$f = 2\,\Omega \sin\phi = 2 \times 7.292 \times 10^{-5} \sin\phi \approx 14.6 \times 10^{-5} \sin\phi = 14.6 \times 10^{-5} \times 1.414$

$= 10.32 \times 10^{-5} \approx 10^{-4}$

### Exercise 2.3

An approaching hurricane causes a uniform, 4-m sea level rise along a north-south oriented coastline relative to a point located 100-km directly offshore at the edge of the continental shelf is the magnitude of the pressure change and pressure gradient force caused by the hurricane.

### Solution

We know hydrostatic equation: $\frac{\Delta P}{\Delta Z} = \rho g$. So the elevation of sea level ($Z = \zeta$) due to storm can be written as : $\Delta \varsigma = \frac{\Delta P}{\rho g}$. So the pressure change = $\phi$

$$\Delta P = \rho g \Delta \varsigma = 4m \times 10^3\, kgm^{-3} \times 10ms^{-2} = 4\times 10^4\, kgm^{-1}s^{-2} = 4\times 10^4\, newtons$$

Pressure gradient acceleration (force per unit mass):

$$= \frac{1}{\rho}\frac{\Delta P}{\Delta x} = g\frac{\Delta Z}{\Delta x} = 10ms^{-2} \times \frac{4m}{100\times 10^3\, m} = 4\times 10^4\, ms^{-2}$$

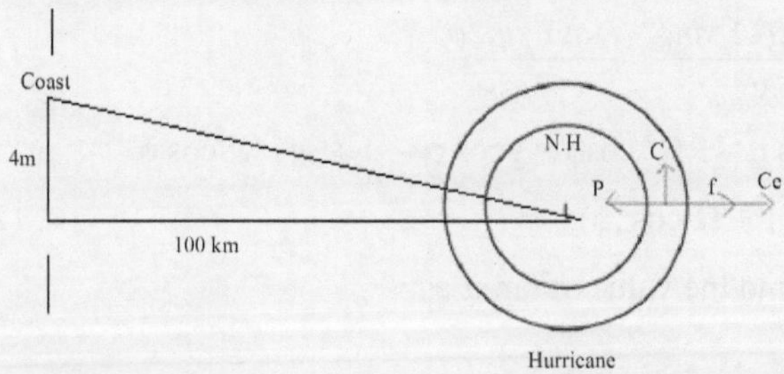


### Exercise 2.4

Find the angle between true gravity and apparent gravity at any latitude, $\phi$

### Solution

True gravity (g') is directed to the center of the earth where as the apparent gravity (g) acts along the plumb line at that latitude.

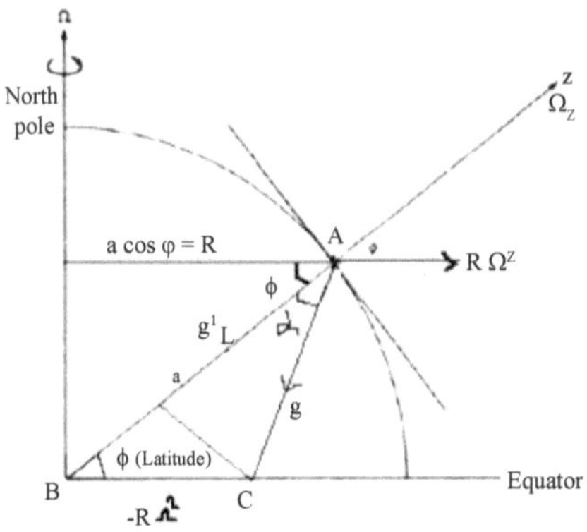


Let 'a' is radius of the earth, α is the angle between g and g'ϕ, is the latitude and R = a cosϕ is the distance of the point from axis of rotation.

Then from triangle ABC,

$$\frac{AC}{sin\phi}=\frac{BC}{\sin a}=\frac{AB}{sin\{180-(\phi+\alpha)\}}\Rightarrow\frac{g}{sin\phi}=\frac{-R\Omega^2}{sin\alpha}=\frac{g'}{sin(\phi+\alpha)}$$

$$\therefore\frac{sin\phi}{g}=\frac{sin\alpha}{-R\Omega^2}$$

$$\text{or } sin\alpha=\frac{-R\Omega^2 sin\phi}{g}=\frac{-acos\phi\Omega^2 sin\phi}{g}=\frac{-a\Omega^2 sin2\phi}{2g} \quad (1)$$

From triangle ABC, r = gcos α – RΩ² cosϕ = gcos α – *a* cosϕ. Ω² cos ϕ

$$\text{gcos }\alpha = a + a\,\Omega^2\cos^2\phi = a(1+\Omega^2\cos^2\phi) \quad (2)$$

Divide Equation (1) by (2) find the value of tan α as:

$$tan\alpha=\frac{-\Omega^2 sin2\phi}{2(1+\Omega^2 cos^2\phi)}$$

Putting the value of g= 980cm $s^{-2}$ at $\phi = 45^0$

$$\tan\alpha=\frac{-\Omega^2}{2+\Omega^2}$$

Putting the value of $\Omega = 7.292 \times 10^{-5}$ radians $s^{-1}$;

$\tan\alpha = 2.66 \times 10^{-9}$ or $\alpha = 1.52 \times 10^{-7}$

## *Exercise 2.5*

What is the centrifugal force of the earth exerted on a typical 50 kg person at latitude 43°N? What is the value of the Coriolis parameter 'f' in units of $sec^{--1}$? If a person runs the 30 m dash in 10 sec, what is the Coriolis force and centrifugal force felt by the person? What is the respective ratios of these two "forces" to the person's weight, mg? What inference do you get from this?

### Solution

To determine the *north-south* Coriolis effect on the ocean, consider the motion of a fluid parcel moving eastward with a speed 'u' along a line of latitude, $\phi$.

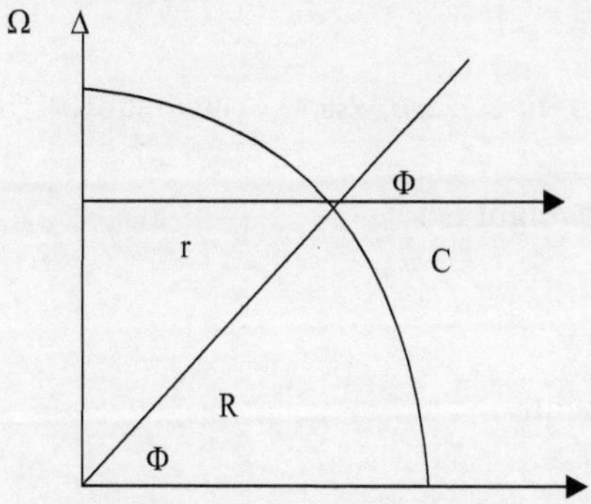


The tangential speed (V) of the Earth is

$V = \Omega\, r = \Omega\, R \cos \phi$

Where $\Omega$ is the Earth's rotation rate (earth's angular velocity); R is the earth's radius; and r is radius of gyration.

The magnitude of the radial centrifugal force (C) of the earth on the mass 'm' is:

$$C = \frac{mV^2}{R} = \frac{m(\Omega R\cos\varphi)^2}{R} = mR(\Omega\cos\varphi)^2$$

We know $\Omega = 7.3 \times 10^{-5}$ red/sec; R = radius of earth = 6368 km = 6368 × $10^3$ m. Given m = 50 kg and $\varphi = 43^0$N. So Radial Centrifugal force is :

i) $$C = \frac{mV^2}{R} = \frac{m(\Omega R\cos\varphi)^2}{R} = mR(\Omega\cos\varphi)^2$$

$$= 50kg \times 6368 \times 10^3\, m(7.3\times10^{-5}\, s^{-1} \cos 43)^2 = 0.91 kgms^{-2}$$

ii) Corisolis parameter = $f = 2\Omega\sin\varphi = 2\times\times 7.3\times10^{-5}\, s^{-1}\sin 43 = 9.95\times10^{-5}\, s^{-1}$

iii) So centrifugal force on the moving body with velocity 'u' is $C_e = \dfrac{m(V+u)^2}{r}$

But as per the figure, the horizontal component of the centrifugal force, $C_h$ is

$$C_h = m\frac{(V+u)^2}{R\cos\varphi}\sin\varphi = \frac{mV^2\sin\varphi}{R\cos\varphi} + \frac{m2Vu\sin\varphi}{R\cos\varphi} + \frac{mu^2\sin\varphi}{R\cos\varphi}$$

The first term on the right is centrifugal force at rest and is balanced by earth's ellipticity. The second term on the right is the coriolis force. The third term on the right is the added centrifugal force due to parcel motion.

The velocity of the runner given is = u = 30m/10s = 3m/s.

iv) So centrifugal force

$$= \frac{mu^2 \sin\varphi}{R\cos\varphi} = \frac{mu^2 \tan\varphi}{R} = \frac{50kg \times 9(ms^{-1})^2 \tan 43}{6368 \times 10^3 m} = 0.07 \times 10^{-3} kgms^{-2}$$

v) Coriolis force =

$$\frac{m2Vu\sin\varphi}{R\cos\varphi} = \frac{m2\Omega Rco\varphi u\sin\varphi}{R\cos\varphi} = m2\Omega u\sin\varphi = 50kg \times 2 \times 7.3 \times 10^{-5} s^{-1} \times 3ms^{-1} \times \sin 43 = 1493.6 \times 10^{-5} kgms^{-2}$$

Ratio of Centrifugal force (iv) with person's weight is :

$$\frac{0.07 \times 10^{-3} kgms^{-2}}{50kg} = 14 \times 10^{-7} ms^{-2}$$

Ratio of Coriolis force (V) with person's weight:

$$\frac{1493.6 \times 10^{-5} kgms^{-2}}{50kg} = 29.87 \times 10^{-5} ms^{-2}$$

The inference is action of centrifugal force is much smaller than that of the Coriolis force on the mass.

### *Exercise 2.6*

If a soccer player kicks a ball a horizontal distance of 100 m in the positive x direction at $30^0$ latitude in 4 sceonds, by how much distance is it deflected horizontally due to earth's rotation. Find also the angle of deflection.

### *Solution*

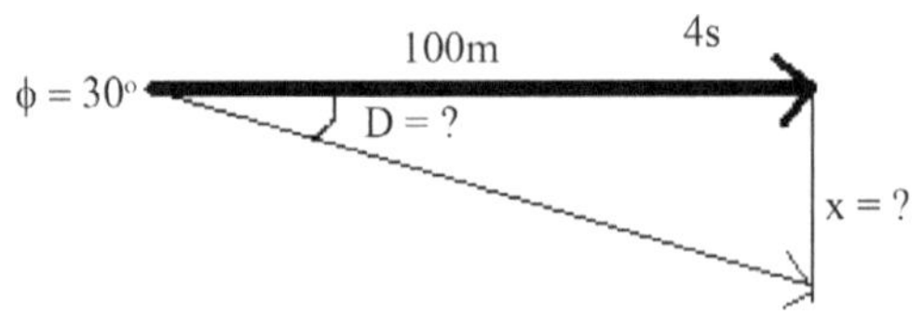


We know the Coriolis acceleration from equation of motion as: $\frac{du}{dt} = fv_0$

$\int du = fv_0 \int dt$ ; on integrating w.r.t 't', we get: $u = \frac{dx}{dt} = fv_0 t \Rightarrow$ on further integration

$dx = fv_0 t dt \Rightarrow x = fv_0 \frac{t^2}{2}$; Given $V_0$ = 100m per 4 seconds or $V_0$ = 100/4 = 25 m/s, t = 4s, $\phi = 30^0$

Hence $x = fv_0 \frac{t^2}{2} = 2 \times 7.3 \times 10^{-5} \times 25 \times \frac{4^2}{2} \times \sin 30 = 1460 \times 10^{-5} m = 1.46cm$

Angle of deflection : $\tan\theta = \frac{x}{100} = \frac{1460\times10^{-5}m}{100m} = 2\times10^{-4}$ $\therefore \theta = tan^{-1}(2\times10^{-4}) = 0.01$

### *Exercise 2.7*

Calculate the deflection due to the Coriolis acceleration of a baseball travelling a distance of 60 ft at a speed of 50 mph. (Take f = $10^{-4}$ $S^{-1}$. Assume that the velocity perturbations due to the Coriolis accelerations are small. Find the Rossby number for this case and mention its significance.

### *Solution*

Look at the Figure and in the similar way of exercise (2.6):

$$\frac{du}{dt} - fv \Rightarrow \int du = \int fv\, dt;\ u = fv_0\, t \Rightarrow \frac{dx}{dt} = fv_0 \Rightarrow \int dx = \int fv_0\, dt$$

$$\therefore x = fv_0\frac{t^2}{2} = fy\frac{t}{2} (\because v_0 t = y) \qquad (2.7.1)$$

Given : $V_0 = 50$ mph $= \frac{50\times220\times3\times8}{60\times60}$ ft/sec; f = $10^{-4}$; y = 60 ft

From equation (2.7.1): $x = fy\frac{t}{2} = 10^{-4}\times60\,ft\times\frac{t}{2}$ (2.7.2)

Again using equation (2.7.1): $\therefore x = fv_0\frac{t^2}{2} = 10^{-4}\times\frac{50\times220\times3\times8}{60\times60}\times\frac{t^2}{2}$ (2.7.3)

Equating equations (2.7.2) and (2.7.3):

$$x = 10^{-4}\times60\,ft\times\frac{t}{2} = 10^{-4}\times\frac{50\times220\times3\times8}{60\times60}\times\frac{t^2}{2}$$

$$\Rightarrow t = \frac{60\times60\times60}{50\times220\times3\times8} = \frac{216000}{264000} = 0.818\sec onds \qquad (2.7.4)$$

Substituting the value of 't' in eqn (2.7.4) in eqn (2.7.2)

$$x = fy\frac{t}{2} = 10^{-4}\times60\,ft\times\frac{0.818}{2}$$

$$= 24.54\times10^{-4}\,ft = 24.54\times10^{-4}\times12 inch = 294.48\times10^{-4} in = 0.0294 in = 0.03 inch$$

Rossby number (equation 2.36) in the given problem =

$$R_0 = \frac{U}{fL} = \frac{50mph}{10^{-4}s^{-1}\times60\,ft} = \frac{50\times0.447041ms^{-1}}{10^{-4}s^{-1}\times60\times\frac{1}{3.281}m} = \frac{73.34}{60}\times10^4 = 1.22\times10^4 = 12,200$$

As the Rossby number is very very higher (12,200) than 1, Coriolis term is negligible and the balancing forces are pressure gradient and centrifugal accelerations only in this case which is under cyclostrophic condition.

### *Exercise 2.8*

Calculate Coriolis force (f), relative vorticity ($\zeta$) and Rossby Number ($R_0$) for a flow having a width of 200 km directed northward with a maximum velocity of $2ms^{-1}$ at $27^0$N latitude

### *Solution*

i) $f = 2\Omega \sin\varphi = 2\times 7.3\times 10^{-5}\times \sin 27 = 6.63\times 10^{-5} s^{-1}$

ii) Relative vorticity $\zeta = \frac{\partial v}{\partial x} - \frac{\partial u}{\partial y}$ but the flow is in the meridional direction.

So u will vanish. Hence $\zeta = \frac{\partial v}{\partial x} = \frac{2ms^{-1}}{200\times 10^3 m} = 1\times 10^{-5} s^{-1}$

iii) Rossby Number ($R_0$) : In N.H. cyclone the balance of forces are pressure gradient = Coriolis + centrifugal

$$\left|-\alpha\frac{\partial p}{\partial n}\right| = |fv| + \left|\frac{v^2}{R}\right| \rightarrow \left|-\alpha\frac{\partial p}{\partial n} - fv\right| = \left|\frac{v^2}{R}\right|$$

If we divide by fv :

$$\frac{\alpha}{fv}\frac{\partial p}{\partial n} + 1 = \frac{v}{fR} = R_0$$

$$R_0 = \frac{v}{fR} = \frac{2ms^{-1}}{6.63\times 10^{-5} s^{-1}\times 200\times 10^3 m} = 0.15$$

### *Exercise 2.9*

If a ballistic missile first travelled 1000 km due eastward at 43$^0$N latitude, when it is fired at a speed of 500 m/s, by how much angle and displacement is it deflected due to the Coriolis force?

### *Solution*

Given: $\theta$ = 43$^o$N, u of missile = 500m/s, distance travelled =1000km = $10^6$m; We have to find: lateral deflection (y).

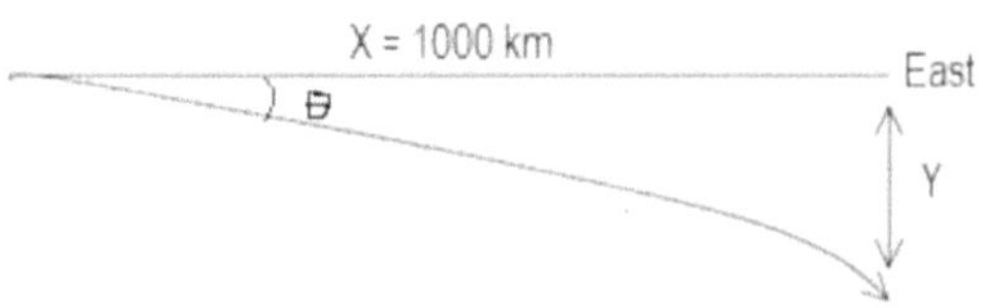


$$\text{We know } \frac{dv}{dt} = -fu = -2\times 7.3\times 10^{-5}\times \sin 43 s^{-1}\times 500 ms^{-1} = -49.78\times 10^{-3} = -0.0497 ms^{-2}$$

$$\int dv = -0.0497\int dt \Rightarrow v = -0.0497t = \frac{dy}{dt} \Rightarrow \int dy = -0.0497\int t dt \Rightarrow y = -0.0497\frac{t^2}{2}$$

What is 't' value: speed given, for 500 m = 1second, for 1000km ($10^6$m) = $10^6$ /500= 2000 seconds

So $y = -0.0497\dfrac{t^2}{2} = -0.0497 \times \dfrac{(2\times10^3)^2}{2} = -994\times10^2 m = -99.4km$

The minus sign indicates the deflection is in the negative 'y' direction.

Angle of deflection $\tan\theta \approx \dfrac{y}{x}$ $\dfrac{99.4}{1000}$ 0.0994 , θ = 5.676°

### *Exercise 2.10*

An aircraft flying eastward at u = 1000 nautical miles/hour at latitude 60°N relative to a rotating earth. Neglecting second order effects (e.g. $u^2$), compute the deflection of local apparent vertical as observed by a person on the airplane. How much does the inclusion of the second order term change this result?

### *Solution*

We know : $\dfrac{du}{dt} = fv = 2\times7.3\times10^{-5}\times\sin 60 s^{-1}\times500ms^{-1} = 63\times10^{-3}ms^{-2}$

This means in one second the rate of change of velocity is $0.87 \times 10^{-3}$ $ms^{-1}$

$$du = 0.87\times10^{-3}\int dt \Rightarrow u = 0.87\times10^{-3}t = \frac{dx}{dt} \Rightarrow \int dx = 0.87\times10^{-3}\int tdt \Rightarrow x = 0.87\times10^{-3}\frac{t^2}{2}$$

As the plane velocity is given for one hour, In one hour, the displacement is:

$$x = 0.87\times10^{-3}\frac{t^2}{2} = 0.87\times10^{-3}\frac{3600^2}{2} = 5638m$$

If the non linear term is included:

$$\frac{du}{dt} = \frac{\partial u}{\partial t} + u\frac{\partial u}{\partial x} \Rightarrow u\frac{\partial u}{\partial x} = \frac{du}{dt} - \frac{\partial u}{\partial t}$$

$$\partial u = \frac{5638m}{3600s^{-1}} = 1.57ms^{-1}$$

$$u\frac{\partial u}{\partial x} = 500ms^{-1}\times\frac{1.57ms^{-1}}{5638m} = 0.138ms^{-2}$$

If second order term is included the change in acceleration is 0.138 m $s^{-2}$

### *Exercise 2.11*

Consider the movement of a solid sphere (mass = m) on a frictionless, rotating plane, with a constant Coriolis parameter = f. If the sphere is given an initial northward velocity $V_0$ at an *initial time* (**t = 0**), then at time = t :

a) what is the zonal (east-west) velocity?

b) how far east (or west) has the sphere moved?

c) how far north has the sphere moved?

d) Assuming an initial northward velocity of $V_0$= 200 cm/sec, t = 5 days, m = 1gm, and f = $10^{--4}$sec$^{-1}$, what are the *numerical values* for the above questions ? Do your answers make sense?

### *Solution*

We know from the previous problems:m $\frac{du}{dt} = fv; \Rightarrow \int du = \frac{1}{m} \int fv\, dt;$

a) $u = \frac{1}{m} fv_0 t$ (1)

$$\Rightarrow \frac{dx}{dt} = \frac{1}{m} fv_o t \Rightarrow \int dx = \int \frac{1}{m} fv_o t dt$$

b) $\therefore x = \frac{1}{m} fv_0 \frac{t^2}{2}$ (2)

c) $x = \frac{1}{m} fy \frac{t}{2} (\because v_0 t = y)$

$\Rightarrow y = \frac{2mx}{ft}$ (3)

d ) i): from equation (1): $u = \frac{1}{m} fv_0 t = \frac{1\times10^{-4}s^{-1}\times200cms^{-1}\times5\times24\times60\times60}{1gm}$ 86..4 m/s per gram

ii) from equation (2): the distance sphere moved:

$\therefore x = \frac{1}{m} fv_0 \frac{t^2}{2} = \frac{1\times10^{-4}S^{-1}\times200cms^{-1}\times(5\times24\times60\times60S)^2}{1gm\times2} = 18662.4$ km per gram

iv) From equation (3) how far north the sphere moved;

$$y = \frac{2mx}{ft} = \frac{2\times1gm\times18662.4kmgm^{-1}}{10^{-4}s^{-1}\times5\times24\times60\times60s} = 864km$$

### *Exercise 2.12*

A base ball is thrown due south 40 m at 30° N in 1 second. A) How much is the displacement B) How much is the angle of deflection from the direction of throw? C) Estimate the Rossby number.

### *Solution*

This exercise is like Exercise 2.6. so refer for details.

$$\therefore x = fv_0 \frac{t^2}{2} = 2\times7.3\times10^{-5}\times\sin 30\times40\times\frac{1}{2} = 146\times10^{-5}$$

$$\tan\theta = \frac{146\times10^{-5}}{40} = 3.65\times10^{-5}$$

$$R_0 = \frac{v}{fR} = \frac{40ms^{-1}}{2\times7.3\times10^{-5}\times\sin 30s^{-1}\times40m} = 13,698$$

### Exercise 2.13

P.T. usha, 60 kg runner runs 100 m in 10 seconds at London (51.5074° N, 0.1278° W) Olympics. If she runs due east, estimate her apparent reduction in weight due to the vertical action of Coriolis force.

### Solution

$\frac{dw}{dt} = 2\Omega u\cos\varphi - g \; or \; (2\Omega u\cos\varphi)m = mg$ ; The weight reduction due to vertical component of Coriolis force (mg) =

$(2\Omega u\cos\varphi)m = 2\times7.3\times10^{-5}s^{-1}\times10ms^{-1}\times\cos(51.5)\times60kg = 545\times10^{-4}kgms^{-2}$

Players actual weight = 60kg × 9.8ms$^{-2}$ = 588 kg m s$^{-2}$

So reduction in weight = 588 – 0.0545 = 587.945kg m s$^{-2}$

So her apparent weight = 587.945/9.8 = 59.99 kg

Her loss of weight = 60-59.99 = 0.01 kg = 10grams

### Exercise 2.14

What is the centrifugal force exerted on a typical 50 kg person at latitude 43$^0$ N? What is the value of the Coriolis parameter f in units of sec$^{-1}$? If the person runs a 30 m dash in 10 sec, what is the Coriolis force felt by the person? What is the respective ratios of these two "forces" to the person's weight, mg? Which is more and by how much?

### Solution

we know the equation of motion: $\ddot{X} = X\Omega^2 + 2\dot{Y}X + F_x/m$

Centrifugal force = $X\Omega^2 = 30m\times(7.3\times10^{-5}s^{-1})^2 = 1598\times10^{-10}ms^{-2}$ per unit mass

Centrifugal force on 50 kg =

$1598\times10^{-10}ms^{-2}\times50kg = 799\times10^{-8}kgms^{-2} = 0.799\times10^{-5}newtons$

Coriolis parameter = $2\times(7.3\times10^{-5}s^{-1})\times\sin 43 = 9.95\times10^{-5}s^{-1}$

Coriolisfoce = $2\times(7.3\times10^{-5}s^{-1})\times\sin 43\times3ms^{-1} = 29.87\times10^{-5}ms^{-2}$

Ratio of Centrifugal force = $1598\times10^{-10}ms^{-2}\div50kg = 31.96\times10^{-10}ms^{-2}kg^{-1}$

Ratio of Coriolis force:

$2\times(7.3\times10^{-5}s^{-1})\times\sin 43\times3ms^{-1}\div50kg = 29.87\times10^{-5}ms^{-2}\div50kg = 0.5974\times10^{-5}ms^{-2}kg^{-1}$

This means the contribution of coriolis force is 1869 times greater than centrifugal force on the person's weight.

### *Exercise 2.15*

If the velocity field is defined as:

u (x,y,z,t) = $5t^2 + 3x + 2y$

v (x,y,z,t) = 0 and w(x,y,z,t) = 0

find the following:

(a) Compute an expression for the total derivative of the above velocity field. Show all work.

(b) Compute the total derivative of the velocity field at t = 2, x = 3, and y = 3. Show all work.

(c) What is the ratio of the "local" acceleration to the "advective" accelerations?

### *Solution*

a) We know $\frac{du}{dt} = \frac{\partial u}{\partial t} + u\frac{\partial u}{\partial x} + v\frac{\partial u}{\partial y} + w\frac{\partial u}{\partial z}$ (1)

In this $\frac{du}{dt}$ = total derivative; $\frac{\partial u}{\partial t}$ = Local derivative and $\left(u\frac{\partial u}{\partial x} + v\frac{\partial u}{\partial y} + w\frac{\partial u}{\partial z}\right)$ = advective or convective derivative.

As, u (x,y,z,t) = $5t^2 + 3x + 2y \rightarrow \frac{\partial u}{\partial t} = 10t;\ u\frac{\partial u}{\partial x} = 3(5t^2 + 3x + 2y);\ v\frac{\partial u}{\partial y} + \frac{\partial u}{\partial z} = 0$ (2)

From equations (1) and (2)

$$\frac{du}{dt} = 10t + 3(5t^2 + 3x + 2y) \rightarrow 15t^2 + 10t + 9x + 6y$$

b) at t = 2, x = 3, and y = 3. $\frac{du}{dt} = 15t^2 + 10t + 9x + 6y = (15\times2^2) + (10\times2) + (9\times3) + (6\times3) = 125$

c) Ratio of the "local" acceleration to the "advective" accelerations

$$= \frac{\frac{\partial u}{\partial t}}{u\frac{\partial u}{\partial x}} = \frac{10t}{3(5t^2+3x+2y)} = \frac{20}{105} = \frac{4}{21} = 0.1905$$

# 3

# Inertial Currents

Under horizontal frictionless motion, we have mainly two types of motion. One is accelerated (inertial) motion and the other is non accelerated (geostrophic) motion.

## 3.1. Inertial Motion

Inertial motion is typically excited when the surface currents are impulsively excited by wind forcing. If the wind were to remain steady—then the currents would eventually reach equilibrium and the time-dependent term would go to zero. But if the winds blew for a few hours and then turned off then the current would be still in motion for some time even if there would be no forcing due to turning off winds. If we are far from boundaries we can neglect pressure gradients and friction (bottom is also far away).

This is a simple case of accelerated currents without friction. The main assumption is that the isobaric surfaces are horizontal and so pressure is constant in the horizontal. Hence the pressure gradient vanishes.

$$\therefore \ \frac{du}{dt} = fv \qquad (3.1.1)$$

$$\frac{dv}{dt} = -fu \qquad (3.1.2)$$

Multiply (3.1.1) by u and (3.1.2) by v and add then we get

$$u\frac{du}{dt} + v\frac{dv}{dt} = 0 \text{ i.e} \frac{1}{2}\frac{d}{dt}\left(u^2 + v^2\right) = 0 \ or \ \frac{1}{2}\frac{d}{dt}\left(V_H{}^2\right) = 0$$

where $V_H = iu + jv$

This means $V_H$ is constant implies that water particle moves with constant speed.

$$\frac{1}{2}\left(\frac{\partial u^2}{\partial t} + \frac{\partial v^2}{\partial t}\right) = \frac{\partial (KE)}{\partial t} = 0, KE = \text{constant} \qquad (3.1.3)$$

This Equation (3.1.3) states that the kinetic energy (KE= $\frac{1}{2}mv^2$, *for unit mass*) remains constant in inertial motion. Even though the flow is accelerating (du/dt and dv/dt are not zero). What happens is that current speed remains constant but changes directions and a particle embedded in this would simply go around in a circle as shown in fig. 3.1.1

Then according to Newton, acceleration will result from change of direction under constant velocity. This can be easily seen by multiplying the equation (3.1.1) by v and (3.1.2) by u and subtract one from the other

$$v\frac{du}{dt}-u\frac{dv}{dt}=f\left(v^2+u^2\right)=fV_H{}^2$$

$$v^2\frac{d}{dt}\left(\frac{u}{v}\right)=fV_H^2 \tag{3.1.4}$$

If $\alpha$ is the angle between the x-axis and the direction of current then the velocities are $u = V_H\cos\alpha$ and $v = V_H\sin\alpha$ as shown in Fig.3.1.1.

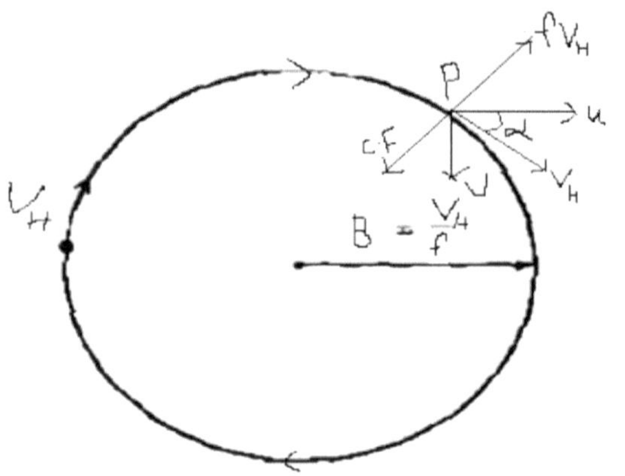


**Fig. 3.1.1.** Inertial circle and the associated forces.

$\therefore \frac{u}{v}=\cot\alpha$, substituting this in equation (3.1.4) and simplifying we get

$$\frac{d}{dt}\cot\alpha=\frac{f}{\sin^2\alpha} \text{ on differentiating we get } \frac{d\alpha}{dt}=-f \tag{3.1.5}$$

$\therefore \alpha = -ft$

$\therefore u = V_H\cos\alpha = V_H\cos(ft)$

$\therefore v = V_H\sin\alpha = -V_H\sin(ft)$

These are the velocities of particles in the northern hemisphere oceans traveling clock wise. They are called gyres.

Equation (3.1.5) states that the moving water particle changes its direction at a constant rate of latitude ($\phi$). These are called inertial currents and the circle is called inertial circle.

If the radius of inertial circle is say B, then the balance of forces of point 'P' are centrifugal force $\left(\frac{V_H^2}{B}\right)$ and Coriolis force (f.$V_H$)

$$\therefore \left(\frac{V_H^2}{B}\right) = f \cdot V_H, \quad \therefore B = \frac{V_H}{f} \tag{3.1.5a}$$

this is the radius of inertial circle.

The time period of the inertial circle (T) is the circumference of the inertial circle divided by its velocity

$$\therefore T = \frac{2\pi B}{V_H} = \frac{2\pi}{f} = \frac{\pi}{\Omega \sin\varphi} \tag{3.1.6}$$

Since ($f = 2\,\Omega\, sin\, \phi$)

Equation 3.1.1 can be written as $\frac{\partial u}{\partial t} = fv$ (3.1.1)

and Equation 3.1.2 can be written as $\frac{\partial v}{\partial t} = -fu$ (3.1.2)

Taking the time derivatives of equation 3.1.1 and 3.1.2 we get:

$$\frac{\partial^2 u}{\partial t^2} = f\frac{\partial v}{\partial t}, \tag{3.1.1a}$$

$$\frac{\partial^2 v}{\partial t^2} = -f\frac{\partial u}{\partial t} \tag{3.1.2b}$$

Substituting equation (3.1.2) in equation (3.1.1a) we get:

$$\frac{\partial^2 u}{\partial t^2} = -f^2 u \tag{3.1.7}$$

Similarly Substituting equation (3.1.1) in equation (3.1.2b) we get:

$$\frac{\partial^2 v}{\partial t^2} = -f^2 v \tag{3.1.8}$$

Equations 3.1.7 and 3.1.8 are wave equations, and have solutions that can use either the sine or cosine functions. Let's use the sine function, then

$$u = A\ \text{Sin}\ (\Omega t) \tag{3.1.9}$$

where 'A' is the amplitude of the wave and Ω is the wave's frequency

Double differentiating equation (3.1.9) with respect to 't' we get

$$\frac{\partial^2 u}{\partial t^2} = -A\Omega^2\ \text{Sin}\ \Omega t = -u\Omega^2 \tag{3.1.10}$$

On equating eqn (3.1.10) with eqn (3.1.7) and simplifying we get:

$$\Omega^2 = f^2 \text{ or } \Omega = f \tag{3.1.11}$$

This means that the frequency of the wave is the Coriolis frequency f.

From equation 3.1.6, we can write, f = 2π/T where T is the local inertial period or the time it takes a particle to complete a single loop.

So we can write eqn (3.1.9) using eqn (3.1.11) as:

$u = ASin\ (ft)$

By inspection of equation of motion (equations 2.1 and 2.2) we can see that if u varies with the Sin (*ft*) then v (which is the derivative of u) must vary with *-Cos(ft)*.

Thus $u = A\ Sin\ (ft)$ and $v = -ACos(ft)$

This can be written in complex notation, where u = u* iv, this yields

$u = Ae^{ift}$

Where $e^{ift} = Cos(ft) + i\ Sin\ (ft)$

which traces out a circle over period T=2π/f as per Fig.3.1.1.

Note that in the southern Hemisphere f is negative, so the rotation is in the opposite sense (counter-clockwise).

In reality a storm that excites inertial oscillations has a finite length scale. This leads to spatial variability in the strength of inertial oscillations. This variability leads to horizontal flow divergences (and convergences) which then excites internal waves at or near the inertial frequency. The propagation of these waves away from the generation site will have a downward component and this can increase vertical shears in the main oceanic thermocline and reduce the Richardson number to a point where mixing may occur. Downward propagating near-inertial waves is an important mechanism that transmits the surface wind forcing to the main oceanic thermocline. Furthermore, these inertial waves interact with the mean flow fields that can focus the energy in regions of negative vorticity.

The period of inertial circle at different latitudes is

At poles = 11.97 hours

At 30° = 24 hours

At 45° = 16.93 hours

At 0° = infinity

Thus it decreases with increase of latitude ϕ.

## Exercises

### Exercise 3.1

What are the periods (hours) and radii (meters) of the inertial circles of particles with respective speeds and latitudes of:

a) 100 cm s $^{-1}$at 10° N latitude?

b) 1 cm s $^{-1}$at 45° N latitude?

### Solution

a) 100 cm s $^{-1}$at 10° N latitude: We know:

$$R = \frac{V}{f} = \frac{100cms^{-1}}{2\times7.3\times10^{-5}\times\sin10s^{-1}} = 39.44\times10^5cm = 39440m$$

We know : $T = \frac{2\pi R}{V} = \frac{2\pi}{f} = \frac{2\times\pi}{2\times7.3\times10^{-5}\times\sin10s^{-1}} = 24736 \text{ second} = \frac{24736^5}{3600} = 7\text{hours}$

b) Similarly for 1 cm s $^{-1}$at 45° N latitude:

$$R = \frac{V}{f} = \frac{1cms^{-1}}{2\times7.3\times10^{-5}\times\sin45s^{-1}} = 9686cm = 96.86m = 97m$$

$$T = \frac{2\pi R}{V} = \frac{2\pi}{f} = \frac{2\times\pi}{2\times7.3\times10^{-5}\times\sin45s^{-1}} = 0.61\times10^5S = 16.96\,hours = 17\,hours$$

### Exercise 3.2

The period of inertial motion was observed to be 12 hours and an array of current meters measured the velocity field to be approximately horizontally homogeneous. At 12.00 noon, the current was 10 cm s $^{-1}$ eastward. a) At what latitude were the observations taken? b) What is the velocity field at 3pm, 6pm, 9pm and midnight?

### Solution

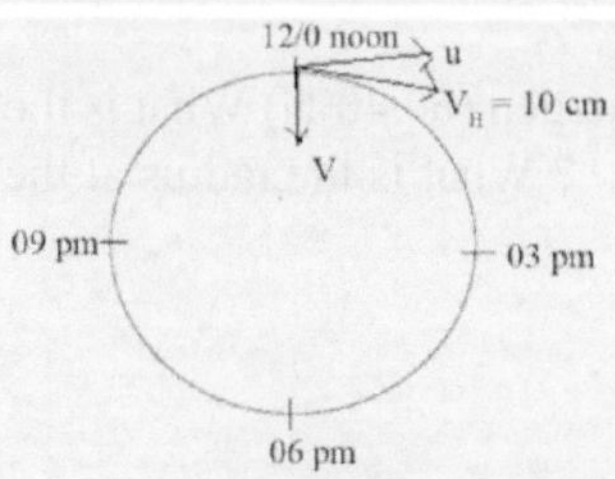


a) Given the time period of inertial circle is $T = \frac{2\pi}{f} = 12hours \Rightarrow f = \frac{\pi}{6}$

$$\Rightarrow f = \frac{2\pi}{12} \Rightarrow 2\times\frac{2\pi}{24}\times\sin\varphi = \frac{2\pi}{12} \Rightarrow \sin\varphi = 1 \Rightarrow \varphi = 90^0$$

b) We know inertial velocities as : $u = V_H cos(ft)$ and $\upsilon = -V_H sin(ft)$

i) At 12 noon, $u_{12} = V_H cos(0) = V_H = 10cms$ and $v = -V_H sin(ft) = - V_H sin(0) = 0$

ii) At 3 pm, $u_3 = V_H \cos(\frac{\pi}{6} \times 3) = 10cm\, s^{-1} \times \cos(\frac{\pi}{2}) = 9.996$ and

$$v_3 = -V_H \sin(ft) = -10\sin(\frac{\pi}{2}) = -0.274$$

iii) At 6 pm, $u_6 = V_H \cos(\frac{\pi}{6} \times 6) = 10cms^{-1} \times \cos(\pi) = 9.9849$ and

$$v_6 = -V_H \sin(ft) = -10\sin(\frac{\pi}{6} \times 6) = -0.548$$

iv) At 9 pm, $u_9 = V_H \cos(\frac{\pi}{6} \times 9) = 10cms^{-1} \times \cos(\frac{3\pi}{2}) = 9.9661$ and

$$v_9 = -V_H \sin(ft) = -10\sin(\frac{\pi}{6} \times 9) = -0.8215$$

v) At 12 mid night, $u_{12} = V_H \cos(\frac{\pi}{6} 12) = 10cms^{-1} \times \cos(2\pi) = 9.9399$ and

$$v_{12} = -V_H \sin(ft) = -10\sin(\frac{\pi}{6} \times 12) = -1.0944$$

### *Exercise 3.3*

What is the volume transport in $m^3$ /sec and in sverdrups of a current with velocity 10 cm/sec over a width of 200 km and a depth of 5000m?

### *Solution*

First convert all in meters. Volume transport is the product of velocity and area,= $10 \times 10^{-2} \times 200 \times 10^3 \times 5000 = 10^7 m^3 S^{-1} = 10$ sverdrups (since $10^6 m^3 S^{-1} = 1$ Sv).

### *Exercise 3.4*

What is the period of inertial rotation at the latitude Bergen (60.40°N) What is the radius of the inertial circle if the flow speed is $0.1 ms^{-1}$ ? What is the radius at the North Pole or at the Equator?

### *Solution*

From equation (3.1.6):

a) $T = \frac{2\pi}{f} = \frac{2\pi}{2 \times 7.3 \times 10^{-5} sin(60.40^0) s^{-1}} = 49512.8 \approx 49513s \approx \frac{49513s}{3600} = 13.75 hours$

b) Radius of inertial circle from equation (3.1.5a):

$$B=\frac{V_H}{f}=\frac{0.1ms^{-1}}{2\times7.3\times10^{-5}sin(60.40^0)s^{-1}}\approx788m$$

c) At North Pole: $B=\frac{V_H}{f}=\frac{0.1ms^{-1}}{2\times7.3\times10^{-5}sin(90^0)s^{-1}}\approx685m$

d) At the equator: $B=\frac{V_H}{f}=\frac{0.1ms^{-1}}{2\times7.3\times10^{-5}sin(0^0)s^{-1}}\approx\frac{0.1}{0}=\infty$

## Exercise 3.5

Figure 3.5.1 shows some famous current meter measurements taken in the Baltic Sea (at 57.8⁰ N) by Gustaffson and Kullenberg during August 17 – 24, 1933. a) What is the theoretical period of the inertial oscillations at this location? b)What is the typical speed of the inertial current? c)What do you infer by examining the figure with regard to the change in the circles? d)What is the reason for the change in size of the circles with time

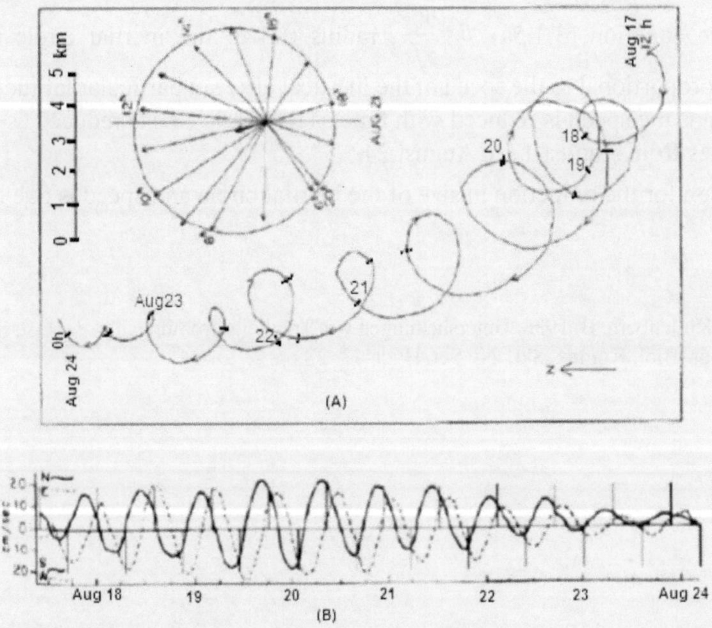


**Fig. 3.5.1.(A):** Current measurements in the Baltic in 1933. The plot show trajectories (progressive vector diagram) based on current meter measurements at a fixed point. The time interval between each tick mark is 12 hours and the numbers denote the dates from August 17 to 24.The scale is the length (Gustaffson and Kullenberg, 1936).**(B)**: A 7-day time series of the eastward (E; dashed) and northward (N; solid) measured currents in August (18-24),1933 indicating strong circular inertial motion. (Neumann & Pierson)

### *Solution*

From equation (3.1.6):

a) $T = \frac{2\pi}{f} = \frac{2\pi}{2\times 7.3\times 10^{-5} sin(57.8^0)s^{-1}} \approx 50876s \approx\approx \frac{50786s}{3600} = 14.10 hours$

b) To calculate typical speed of this current: Let the largest length between tick marks is 5 km and the smallest length between tick marks is 1 km as per the scale given in the figure such that each has elapsed 12 hours, so

$$= \frac{1}{2}\left[\frac{1km}{12hrs} + \frac{5km}{12hrs}\right] \approx \frac{1}{24}[6km] = \frac{1}{4} kmhr^{-1} = 0.25 kmhr^{-1} = \frac{0.25\times 1000m}{3600} = 0.069 ms^{-1} \approx 7cms^{-1}$$

Another way is from Fig 3.5.1B: average between highest and lowest if we take : $\frac{20+5}{2} = 12.5cms^{-1}$ for north, $\frac{9+2}{2} = \frac{11}{2}$ = 5.5cms$^{-1}$ for East

$\therefore$ Net = 12.5 – 5.5 = 7cms$^{-1}$

c) From the equation (3.1.5a) $B = \frac{V_H}{f}$, radius (B) of the inertial circle is directly proportional to the speed of the inertical circle at particular latitude. This means the speed is reduced with time as the circle size is reduced as it progresses from August 17 to August 24.

d) The reason for the reduction in size of the inertial circle and speed is due to friction.

## References

Gustafsson, T. & Kullenberg, B. 1936: Untersuchungen von Trägheitsströmungen in der Ostsee. — Sv. Hydr.-Biol. Komnz. Skr., Ny Ser. Hydr. 13. 28 pp.

# 4

# Geostrophic & Meandering Currents

## 4.1. Geostrophic Motion

It is a non-accelerated frictionless horizontal motion where pressure gradient balances coriolis acceleration as shown in Fig.4.1 in both the hemispheres.

From equations (2.11) and (2.12) for non-accelerated frictionless horizontal motion, we can write

$\alpha \frac{\partial p}{\partial x} = fv$ and $\alpha \frac{\partial p}{\partial y} = -fu$ squaring these two and adding we get

$$\alpha \sqrt{\left(\frac{\partial p}{\partial x}\right)^2 + \left(\frac{\partial p}{\partial y}\right)^2} = f\sqrt{u^2 + v^2}$$

$$\alpha \frac{\partial p}{\partial n} = fV_H \ \ or \ \ V_H = \frac{\alpha}{f}\frac{\partial p}{\partial n} \tag{4.1}$$

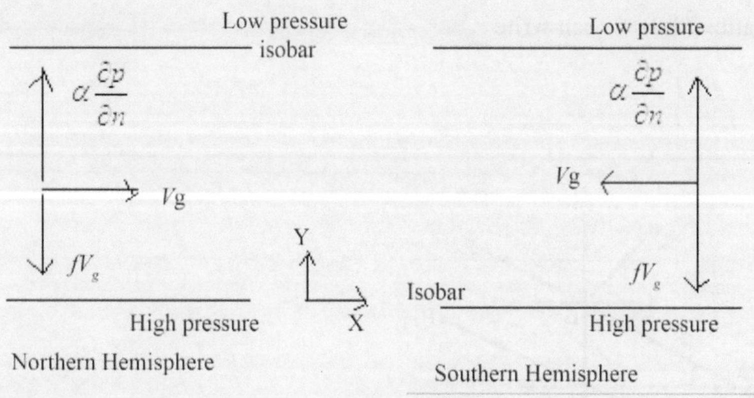


**Fig. 4.1.** Balance of Coriolis and pressure gradient forces

### 4.1.1. Helland-Hansen's computation of slope and currents

The theoretical geostrophic equation (4.1) is not practicable to measure the slope of the sea surface. So the distribution of density is used as the density variation in the oceans is related to the slope of the sea surface.

### 4.1.2. Slope of the isobaric surface

Under barotropic conditions the slope of the isobaric surface is constant ($\theta_1 = \theta_2 = \theta$) with the level surface or geopotential surface whereas under baroclinic conditions this slope is different at different depths ($\theta_1 \neq \theta_2$) as shown in Fig. 4.2.

The geostrophic currents can be calculated from Fig.4.2

If the balance of forces at 'B' is taken

$$\alpha \frac{\partial p}{\partial n} \cos\theta = g$$

$$\alpha \frac{\partial p}{\partial n} \sin\theta = fu$$

$$\therefore u = \frac{g}{f} \tan\theta \tag{4.2}$$

If the subsurface layer $Z_1$ to $Z_o$ is considered, we can write the slope. Let $\theta_1$ is the inclination of the isobaric surface $P_1$ at $Z_1$ depth and let $Z_o$ is the reference level which coincides with the geopotential surface. So this will be considered as the level of no motion as shown in Fig.4.2. Let the height of this isobaric surface from the reference level is $h_A$ and $h_B$ and the densities be $\rho_A$ and $\rho_B$ respectively at stations A and B. Let 'L' is the horizontal distance between the two stations A and B.

$$\therefore \tan\theta_1 = \frac{h_B - h_A}{L}$$

Using equation (4.2) we can write

$$u_1 = \frac{g}{f}\left[\frac{h_B - h_A}{L}\right] \tag{4.3}$$

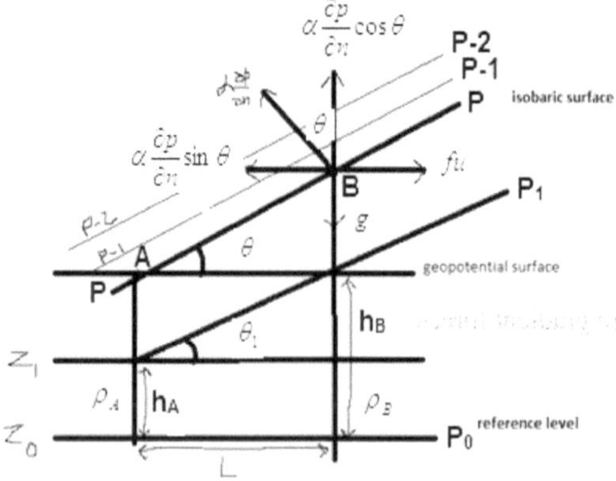


**Fig.4.2.** Helland-Hansen's computation of slope currents

The difference in hydrostatic pressure between $P_1$ and $P_o$ at station A and B is same because at both the same stations same isobaric surfaces.

$\therefore$ *pressure* difference at 'A' = pressure difference at 'B'

$\therefore (P_{1A} - P_o) = (P_{1B} - P_o)$

since $P_{1A} = P_{1B}$ and $P = \rho gh$ ,

$\rho_A g h_A = \rho_B g h_B$

$$\therefore\ h_A = h_B . \frac{\rho_B}{\rho_A} \qquad (4.4)$$

Substituting (4.4) in (4.3) we get

$$u_1 = \frac{g}{fL} h_B \left[1 - \frac{\rho_B}{\rho_A}\right] \qquad (4.5)$$

This is called Helland-Hansen's equation. This shows that the geostrophic current depends on the density distribution. The direction of current is taken as denser water (low topography) to the left in the Northern hemisphere and to the right hand side in the southern hemisphere.

### 4.1.3. Practical method of computation of relative currents

To compute currents at successive depths in a series of manner a little modified formula of Helland- Hansen (1930) is used. It works out in the same lines but uses dynamic depths which can be conveniently computed from the temperature salinity data that is regularly collected in the oceans.

As per the Fig. 4.3 let the two stations A and B are at a distance of L on the sea surface AB. Let the isobaric surfaces $P_1$ and $P_2$ are inclined at angles $\theta_1$ and $\theta_2$ with respect to equi-geopotential surfaces $\phi_1$ and $\phi_2$ respectively and let $v_1$ and $v_2$ are the currents at these surfaces. Let the depths $Z_1$, $Z_2$, $Z_3$ and $Z_4$ of the respective points as indicated in the Fig. 4.3.

We know from equation (4.2)

$fv_1 = g\ tan\theta_1$ and $fv_2 = g\ tan\theta_2$

$$\therefore f(v_1 - v_2) = g\,(tan\theta_1 - tan\theta_2) = g\left[\frac{B_1C_1}{A_1C_1} - \frac{B_2C_2}{A_2C_2}\right]$$

We know $B_1C_1 = B_1C_2 - C_1C_2$ and $B_2C_2 = B_1C_2 - B_1B_2$

$\therefore B_1C_1 - B_2C_2 = B_1B_2 - C_1C_2$ we also know $A_1C_1 = A_2C_2 = L$

$$\therefore f(v_1 - v_2) = \frac{g}{L}[B_1B_2 - A_1A_2](\because C_1C_2 = A_1A_2)$$

We can write $B_1B_2 = BB_2 - BB_1 = Z_3 - Z_1$

And $A_1A_2 = AA_2 - AA_1 = Z_4 - Z_2$

$$\therefore f(v_1 - v_2) = \frac{g}{L}\left[(Z_3 - Z_1) - (Z_4 - Z_2)\right]$$

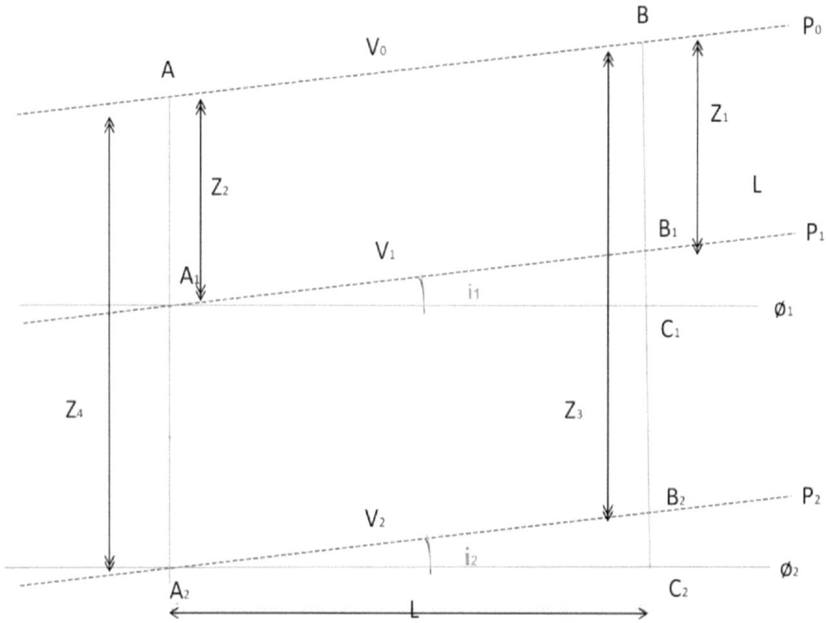


**Fig. 4.3.** Calculation of relative currents

Now from hydrostatic equation and geopotential we know

$$\int_{Z_1}^{Z_3} g dz = \int_{P_1}^{P_2} \alpha dp = \int_{P_1}^{P_2} (\alpha_{35,0,P} + \delta_A) dp$$

Similarly for station B we can write $\int_{Z_2}^{Z_4} g dz = \int_{P_1}^{P_2} \alpha dp = \int_{P_1}^{P_2} (\alpha_{35,0,P} + \delta_B) dp$

Substituting all the above we get

$$\frac{g}{L}\left[(Z_3 - Z_1) - (Z_4 - Z_2)\right] = \frac{1}{L}\left[\int_{P_1}^{P_2} \delta_A dp - \int_{P_1}^{P_2} \delta_B dp\right]$$

$$\therefore f(v_1 - v_2) = \frac{1}{L}\left[(\Delta\varphi_A) - (\Delta\varphi_B)\right]$$

Where ($\Delta\phi_A$ and $\Delta\phi_B$) are called geopotential anomalies at A and B

$$\therefore (v_1 - v_2) = \frac{10}{fL}[(\Delta\varphi_A) - (\Delta\varphi_B)] \quad (4.6)$$

The conversion factor 10 is used to convert the geometric meter (dz) to dynamic meter (D) since d $\phi$ = gdz = 10dz as geopotential is measured in dynamic meters. So these depths are called dynamic depths.

If L is in meters, $\delta$ in $cm^3g^{-1}$ and P in decibars the current will be in $ms^{-1}$.

## References

Helland-Hansen, B. 1930. Physical oceanography and meteorology. Michael Sam North Atlantic Deep-Sea Exped, 1910, Rept. Sci. Results, v. 1, 115+ 102 pp.

## Exercises

### *Exercise 4.1*

a) Calculate the alongshore (parallel to shore) and offshore (perpendicular to shore) components of the current measured as 20 cm $s^{-1}$westward along Puri coast (east coast of India) on 23 March, taking into account of the fact that the Puri coastline is oriented along a 65 degrees true North compass heading. Explain with figure. (b) If a fishing vessel is sunk 10 km away from Puri coast, how far from the coast the rescue time will search for bodies after 24 hours assuming the currents have the same intensity and direction throughout on this day due to strong sea breeze.

### *Solution*

a) Currents:

Along shore current (parallel to coast) = -20 cos $25^0$= -18.13 $cms^{-1}$

Off shore current (perpendicular) = -20 cos $65^0$ = – 8.45 $cms^{-1}$

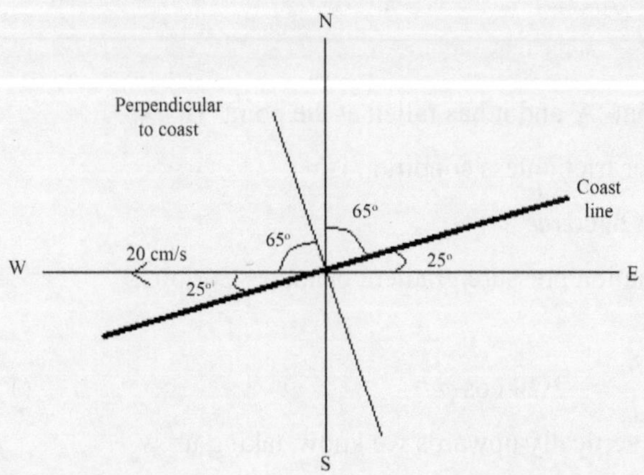

b) Rescue team's work along and offshore the distance:

After 24 hours, the distance bodies carried by the currents from the point of sinking are:

i) Along shore: $18.13 cms^{-1} \times 24 \times 60 \times 60 \times 60 = 1566432 cm \approx 15.7 km$.

From the coast the distance carried = point of sunk + distance carried by the current

= 20 km + 15.7 km = 35.7 km

ii) Similarly perpendicular to coast offshore =

$18.45 cms^{-1} \times 24 \times 60 \times 60 = 730080 cm \approx 7.3$ km.

So from the coast = 20 km + 7.3 km = 27.3 km

### *Exercise 4.2*

(a)A particle is thrown vertically upwards with a velocity $w_0$ at latitude $\phi$. Show that when it returns to the ground it will be at a $\dfrac{4\Omega cos\phi w_0^3}{3g^2}$ distance of to the west of its starting point. Assume there is no frictional resistance in the atmosphere. (b) Calculate the displacement from its initial position for a projectile fired upwards on the equator with a velocity of $500 ms^{-1}$.

### *Solution*

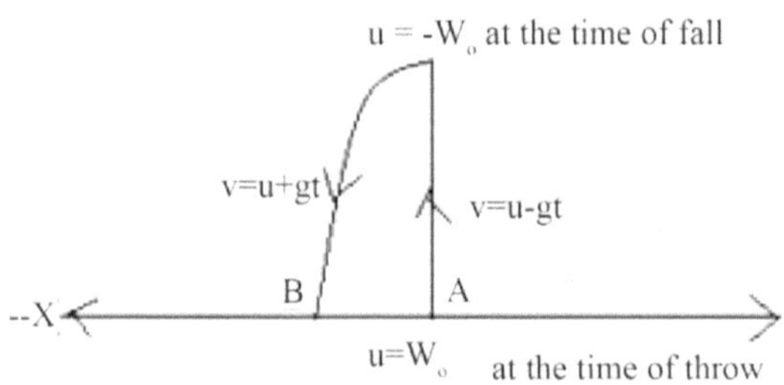


Let the particle is thrown from 'A' and it has fallen at the point 'B'

The equation of motion under frictionless condition is

$$\frac{du}{dt} = -\alpha\frac{\partial p}{\partial x} + 2\Omega v\sin\varphi - 2\Omega w\cos\varphi$$

Under geostrophic approximation pressure gradient balances Coriolis,

$$\alpha\frac{\partial p}{\partial x} = fv$$

so the terms remained are: $\dfrac{du}{dt} = -2\Omega w\cos\varphi$ (1)

When the particle is thrown vertically upwards we know, taking u= $w_0$

$v = u - gt \Rightarrow w = w_0 - gt$ (2)

$$\frac{dz}{dt} = w = -gt \Rightarrow \int_{Z=h}^{Z=0} dz = -g\int_0^t t\,dt \Rightarrow (0-h) = -g\frac{t^2}{2} \Rightarrow or\, t = \left[\frac{2h}{g}\right]^{\frac{1}{2}}$$

Substituting the value of 't' in equation (7)

$$x = \Omega g\frac{t^3}{3} = \frac{\Omega g}{3}\left[\frac{2h}{g}\right]^{\frac{1}{2}} \tag{9}$$

If 'h'=500 m, $x = \frac{\Omega g}{3}\left[\frac{2h}{g}\right]^{\frac{3}{2}} = \frac{7.3\times10^{-5}\times9.8}{3}\left[\left(\frac{2\times500}{9.8}\right)^{\frac{1}{2}}\right]^{3} = 0.25m$

### *Exercise 4.4*

What is the centrifugal force exerted on a typical 50 kg person at latitude $43^0$ N? What is the value of the Coriolis parameter f in units of $sec^{-1}$? If the person runs a 30 m dash in 10 sec, what is the Coriolis force and centrifugal force felt by the person? What is the respective ratios of these two "forces" to the person's weight, mg? Which is more and by how much?

### *Solution*

We know the equation of motion from equation 2.29 as: $\ddot{X} = X\Omega^2 + 2\dot{Y}X + F_x/m$

Centrifugal force = $X\Omega^2$ = 30m × (7.3 × $10^{-5}$ $s^{-1}$)$^2$ = 1598 × $10^{-10}$ $ms^{-2}$ per unit mass

Centrifugal force on 50 kg = 1598 × $10^{-10}$ $ms^{-2}$ × 50kg = 799 × $10^{-8}$ $kgms^{-2}$ = 0.799 × $10^{-5}$ newtons

Coriolis parameter = 2 × (7.3 × $10^{-5}$ $s^{-1}$) × sin 43 = 9.95 × $10^{-5}$ $s^{-1}$

Coriolis foce = 2 × (7.3 × $10^{-5}$ $s^{-1}$) × sin 43 × 3$ms^{-1}$ = 29.87 × $10^{-5}$ $ms^{-2}$

Ratio of Centrifugal force =1598 × $10^{-10}$ $ms^{-2}$ ÷ 50kg = 31.96 ×$10^{-10}$ $ms^{-2}$ $kg^{-1}$

Ratio of Coriolis force:

2 × (7.3 × $10^{-5}$ $s^{-1}$) × sin 43 × 3$ms^{-1}$ ÷ 50kg = 29.87 × $10^{-5}$ $ms^{-2}$ ÷ 50kg = 0.5974 ×$10^{-5}$ $ms^{-2}$ $kg^{-1}$

This means the contribution of coriolis force is 1869 times greater than centrifugal force on the person's weight.

(See the difference in working with exercise 2.5)

### Exercise 4.5

Calculate the geostrophic velocity if the isobars are parallel with a pressure difference of 4 mb with a spacing of 200 km at latitude $40^0$ (Given the ocean water density as 1025 kg/m$^3$ and $2\Omega = 14.6 \times 10^{-5}$ radians/sec.

### Solution

Given : $\Delta P = 4mb = 400\ N/m^2 = 400 kgm^{-1}\ s^{-2}$; $L = 200\ km = 200 \times 10^3 m$; $\phi = 40^0$; $\rho_w = 1025\ kg/m^3$

We know the equation:

$$V_g = \frac{1}{\rho f}\frac{\partial p}{\partial n} = \frac{400kgm^{-1}s^{-2}}{1025kgm^{-3} \times 2 \times 7.3 \times 10^{-5} \times \sin 40 s^{-1} \times 200 \times 10^3 m} = 0.02ms^{-1} = 2cms^{-1}$$

### Exercise 4.6

If the Gulf Stream flows near the surface northward at 200 cm/sec in North Atlantic Ocean along east coast of North America as shown in Figure below, (a) What pressure gradient is required geostrophically to balance this flow at latitude $\phi = 30^0$ N? (recall that dimensions of pressure gradient are dynes/cm$^2$-cm).(b) Given a 50 km-wide Gulf Stream of uniform flow, what is the pressure difference across the stream in dynes/cm$^2$? And in decibars? This corresponds to how many meters (or centimeters) of excess water height? (C) Which side of the stream is higher pressure, east or west?

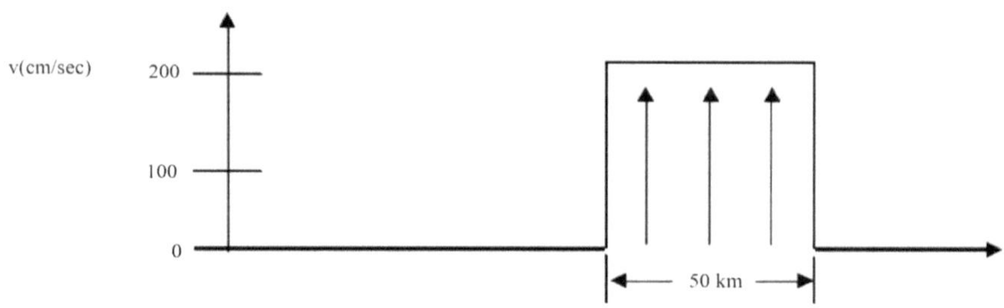


### Solution

We know the balance of forces for geostrophic flow as:

a) $$\frac{\partial p}{\partial n} = f\rho V_g = (2 \times 7.3 \times 10^{-5} \times \sin 30 s^{-1}) \times (1gmcm^{-3})200cms^{-1} = 14.6gcm^{-2}s^{-2} = 14.6dynes$$

($\because$ *1 dyne = 1 gmcms*$^{-2}$)

b) If $\Delta$ N = 50 km, Pressure difference =

$$\Delta P = f\rho V_g \Delta N = 14.6 \times 10^{-2} kgm^{-2}s^{-2} \times 50 \times 10^3 m = 73 \times 10^2 kgm^{-1}s^{-2} = 73 \times 10^2 Nm^{-2}$$

$$= 73 \times 10^2 \times 10^5 = 73 \times 10^7 dynes$$

1 mb = $10^3$ dynes/cm$^2$; so $73\times10^7\times10^{-3}=73\times10^4 mb=73\times10^2 db=7300 meters$

c) As pressure gradient balances coriolis and coriolis is right side of the flow, pressure gradient is higher on the left side (west).

### Exercise 4.7

i) Calculate the geostrophic velocity and direction between two stations A & B at a depth of 1000 db from the sea surface with reference to 2000 db level surface having densities = 26.5 and = 26.2 respectively at both the stations at 30$^0$N latitude. The two stations are separated by 100 km in North Indian Ocean such that B is east of A. ii) What is the height difference between these two stations? iii) If these two stations are in South Indian Ocean what is the current direction? (Take f = 2 × 7.292 × $10^{-5}$ rad/sec).

### Solution

i) From Fig. 2.16 and Equation 2.50: $u=\frac{g}{fL}h_B\left[1-\frac{\rho_B}{\rho_A}\right]$

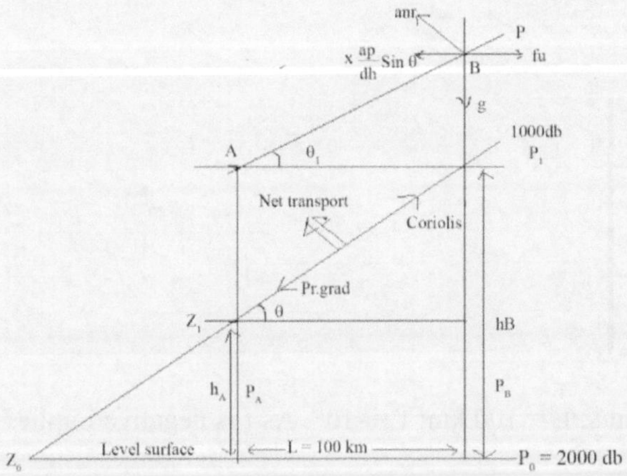


Here $h_B = Z_0 - Z_1$ = 2000-1000 = 1000 db = $10^3$ m (since 1 db = 1m);

f = 27.3$10^{-5}$ Sin 30 =7.3$10^{-5}$

L = 100 × $10^3$ m

$$u=\frac{9.8}{7.3\times10^{-5}}\times\frac{1000}{100\times10^3}\left[1-\frac{1.0262}{1.0265}\right]=\frac{2.94}{7.49}=0.4m/s$$

Direction of this current is into the paper as shown in Figure as low topography is to the left hand side.

ii) Height difference between A & B:

Again from using equation (4.3): $u=\frac{g}{f}\left[\frac{h_B-h_A}{L}\right]$

$$(h_B - h_A) = \frac{ufL}{g}$$

$$(h_B - h_A) = \frac{0.4 \times 2 \times 7.29 \times 10^{-5} \times 0.5 \times 100 \times 10^3}{9.8} = 0.3m$$

$h_B = 0.3\ h_A$; This means Station B is 0.3 m higher than station A.

iii) If these stations are in South Indian Ocean, Coriolis force action is towards left. This means current should be out of the paper.

## *Exercise 4.8*

Given that the Brazil Current (located off the east coast of South America) has surface velocities on the order of 65 cm/s (about half that of the Gulf Stream!) and an average width of ~100 km. Estimate the magnitude and direction of the sea surface slope across the Brazil Current (Take f = -10^-5^).

***Solution:***

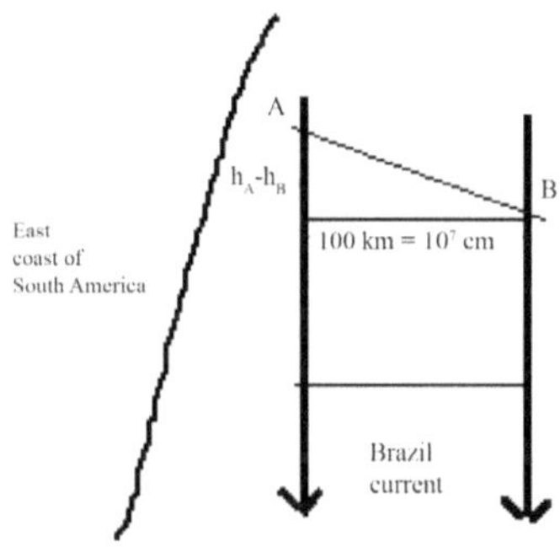


We know: Given : u = 65 cm/s; L = 100 km; f = $-10^{-5}$. As f is negative implies it is in S.H

$$(h_B - h_A) = \frac{ufL}{g}$$

$$(h_B - h_A) = \frac{65cms^{-1} \times (-10^{-5}s^{-1}) \times 100 \times 10^5 cm}{980cms^{-2}} = -6.63cm$$

As it is negative, $h_A = 6.63 + h_B$. This means $h_A$ is higher than $h_B$ by 6.63 cm. This means west side(station A) is higher than east side (Station B).

## *Exercise 4.9*

What is the dynamic height difference over a width of 100 km associated with a current speed of 100 cm/sec at 45$^0$ latitude?

### Solution

We know the equation for relative currents as: $v_1 - v_2 = \frac{10}{fL}[(\Delta\varphi_A) - (\Delta\varphi_B)]$

$$[(\Delta\varphi_A) - (\Delta\varphi_B)] = \frac{fl}{10}(v_1 - v_2) = \frac{2\times 7.3\times 10^{-5}\times \sin 45 s^{-1}\times 100\times 10^3 m}{10}\times\frac{100}{100}ms^{-1} = 1.032m = 103.2cm$$

### Exercise 4.10

The diagram below shows a two 2-layer water columns with different potential densities ($\sigma_\theta$). Calculate the dynamic height of the surface relative to 2000 decibars (in dynamic meters) for each water column.

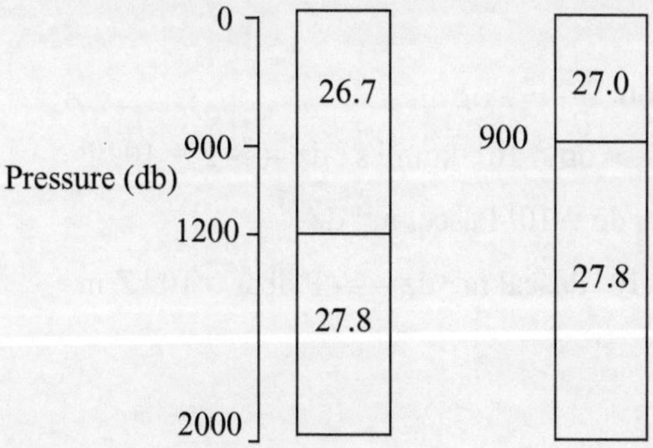


### Solution

We know 1db = $10^4$Pascals = $10^2$kg m$^{-1}$ s$^{-2}$; we also know 1 dynamic meter = 10 m$^2$ s$^2$

We further know: $\rho_\theta = (\sigma_\theta + 1000)$ kg m$^{-3}$

i) For water column between 0 to 1200 db: dD =

$$\int_0^{1200} \alpha_{1200} dp = \frac{1}{1026.7 kgm^{-3}}\times 1200\times 10^2 kgm^{-1}s^{-2} = 116.87 m^2 s^{-2} = \frac{116.8}{10} = 11.7 dm$$

ii) For water column between 1200 to 2000 db: dD =

$$\int_{1200}^{2000} \alpha_{2000} dp = \frac{1}{1027.8 kgm^{-3}}\times 800\times 10^2 kgm^{-1}s^{-2} = 77.8 m^2 s^{-2} = \frac{77.8}{10} = 7.78 dm$$

iii) For water column between 0 to 900 db: dD =

$$\int_0^{900} \alpha_{900} dp = \frac{1}{1027 kgm^{-3}}\times 900\times 10^2 kgm^{-1}s^{-2} = 87.63 m^2 s^{-2} = \frac{87.63}{10} = 8.7 dm$$

iv) For water column between 900 to 2000db: dD =

$$\int_{900}^{2000} \alpha_{2000} dp = \frac{1}{1027.8 kgm^{-3}}\times 1100\times 10^2 kgm^{-1}s^{-2} = 107.02 m^2 s^{-2} = \frac{107}{10} = 10.7 dm$$

## *Exercise 4.11*

i) Show using $Z = 10^{-4}$ P hydrostatic equation and taking $\rho = 1035$ kgm$^{-3}$ and g = 10m s$^{-2}$. ii) Calculate dynamic depth anomaly and relative currents at each level for the following stations which are separated by 100 km. (take $f = 10^{-4}$) relative to 2000db and 1500 db surfaces. Comment if station B is right of station A,

| Depth/pressure | P db | 0 | 50 | 100 | 200 | 500 | 1000 | 1500 | 2000 |
|---|---|---|---|---|---|---|---|---|---|
| Station A | $\delta_A \times 10^5$ cm$^3$ gm$^{-1}$ | 350 | 300 | 250 | 150 | 100 | 75 | 50 | 40 |
| Station B | $\delta_B \times 10^5$ cm$^3$ gm$^{-1}$ | 650 | 600 | 550 | 400 | 200 | 105 | 60 | 50 |

## *Solution*

i)We know from hydrostatic equation: $\frac{\partial p}{\partial z} = -\rho g$

$\longrightarrow$ dp = 1035 kgm$^{-3}$ $\times$ 10ms$^{-2}$ dz $\longrightarrow$ dp $\approx 10^4$ kgm$^{-2}$ s$^{-2}$ dz $\longrightarrow$ Z = $10^{-4}$ P

$\longrightarrow$ dp $\approx 10^4$ ( kgm$^{-1}$ s$^{-2}$) m$^{-1}$ dz $\longrightarrow$ dp = $10^4$ Pascal m$^{-1}$ dz

Since 1 decibar = $10^4$ Pascals; dp = $10^4$ Pascal m$^{-1}$ dz $\longrightarrow$ P dbar = $10^{-4}$ Z m

ii)

| P db | $\delta_A \times 10^5$ $cm^3$ $gm^{-1}$ | Mean | $\delta_A \times dp$ $cm^3\ gm^{-1}$ | $10^{-4}\ \delta_A$ | $\sum\delta_A\ dp = \Delta\phi_A$ | P dB | $\delta_B \times 10^5\ cm^3\ gm^{-1}$ | mean | $\delta_B \times dp$ | $10^{-4}\ \delta_B$ | $\sum\delta_B\ dp = \Delta\phi_B\ \Delta\phi_B - \Delta\phi_A$ | $\Delta\phi_B$-$\Delta\phi_A$ | $(V_2-V_1)$ | Abs currentw.r.t 2000 db $ms^{-1}$ | Abs currentw.r.t 1500 db $ms^{-1}$ |
|---|---|---|---|---|---|---|---|---|---|---|---|---|---|---|---|
| 0 | 350 | 325 | 16250 | 1.625 | 0 | 0 | 650 | 625 | 31250 | 3.125 | 0 | 0 | 0 | 1.57 | 1.525 |
| 50 | 300 | | | | 1.625 (0+1.625) | 50 | 600 | | | | 3.125 | 1.5 | 0.15 | 1.42 | 1.375 |
| | | 275 | 13750 | 1.375 | | | | 575 | 28750 | 2.875 | | | | | |
| 100 | 250 | | | | 3.000 (1.625 +1.375) | 100 | 550 | | | | 6.000 | 3.0 | 0.3 | 1.27 | 1.225 |
| | | 200 | 20,000 | 2.000 | | | | 475 | 47500 | 4.750 | | | | | |
| 200 | 150 | | | | 5.000(3+2) | 200 | 400 | | | | 10.750 | 5.75 | 0.575 | 0.995 | 0.950 |
| | | 125 | 37,500 | 3.750 | | | | 300 | 90000 | 9.000 | | | | | |
| 500 | 100 | | | | 8.750 (5+ 3.750) | 500 | 200 | | | | 19.750 | 11.0 | 1.1 | 0.47 | 0.425 |
| | | 87.5 | 43,750 | 4.375 | | | | 152.5 | 76250 | 7.625 | | | | | |
| 1000 | 75 | | | | 13.125 (8.75+4.375 | 1000 | 105 | | | | 27.375 | 14.25 | 1.425 | 0.145 | 0.1 |
| | | 62.5 | 31,250 | 3.125 | | | | 82.5 | 41250 | 4.125 | | | | | |
| 1500 | 50 | | | | 16.250 (13.125 + 3.1250 | 1500 | 60 | | | | 31.5 | 15.25 | 1.525 | 0.045 | 0 |
| | | 45 | 22,500 | 2.250 | | | | 55 | 27500 | 2.750 | | | | | |
| 2000 | 40 | | | | 18.5 (16.25 + 2.25) | 2000 | 50 | | | | 34.2 | 15.7 | 1.57 | 0 | -0.045 |

$$(V_2 - V_1) = \frac{(\Delta\phi_B - \Delta\phi_A)}{fL} = \frac{1.5}{10^{-4} \times 10^5 m} = 0.15ms^{-1} \text{ at 50 meters}$$

To calculate the absolute currents from the reference surface of 2000 db, then the value of $(V_2 - V_1)$ is to be brought to the sea surface and all the$(V_2 - V_1)$ values are to be subtracted from the value of 2000m (i.e., 1.57 $ms^{-1}$)

If station B is right of station A, Station B is higher height than station A

## Exercise 4.12

The mean geostrophic current through Palk Strait is 0.2 $ms^{-1}$ towards east. (i) What is the slope of the sea surface? (ii) Is the mean sea level higher on the Indian side or the Sri Lankan side (Given the latitude of Palk Strait as $5^0$ N and the width is 50 km).

## Solution

i) We know the geostrophic equation (2.47): $\therefore u = \frac{g}{f} \tan\theta$

$$\text{Or } \therefore \tan\theta = \frac{fu}{g} = \frac{2 \times 7.3 \times 10^{-5} \times \sin 5 \times 0.2}{9.8} = 0.025 \times 10^{-5}$$

So the slope of the sea surface is $0.025 \times 10^{-5}$

ii) We know from Exercise (4.7) and equation (4.3),

$$(h_B - h_A) = \frac{ufL}{g} \Rightarrow \frac{(h_B - h_A)}{L} = \frac{uf}{g} = \tan\theta$$

$(h_B - h_A) = L \tan\theta$; So $(h_B - h_A) = 50 \times 10^3 \times 0.025 \times 10^{-5} = 0.0125$ m = 1.25 cm

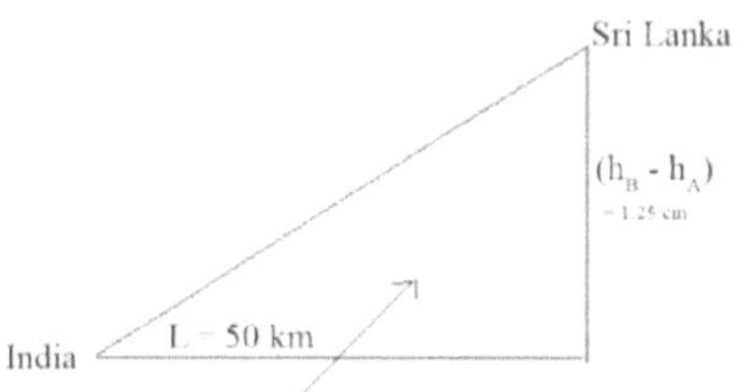


As low topography to the left, India should be on the left side of the current as shown by the arrow in the figure and the sea level is lower by 1.25 cm to the Sri Lankan waters.

## Exercise 4.13

What is the volume transport in $m^3$ /sec and in Sverdrups of a current with velocity 10 cm/sec over a width of 200 km and a depth of 5000m?

## Solution

First convert all in meters. Volume transport is the product of velocity and area,= $10 \times 10^{-2} \times 200 \times 10^3 \times 5000 = 10^7 m^3 S^{-1} = 10Sv$

### *Exercise 4.14*

What is the dynamic height difference over a width of 100 km associated with a current speed of 100 cm/sec at 45°latitude?

### *Solution*

(use eqn 4.6 and calculate $[\Delta\phi_A - \Delta\phi_B]$ =20.64 dynamic meters)

### *Exercise 4.15*

A typical change in the sea surface height across the Gulf Stream is approximately 1 m. Given that the Gulf Stream is approximately 100 km in width, what is a typical sea surface slope (in degrees please!) across the Gulf Stream. Draw a diagram as part of your answer and show all work.

### *Solution*

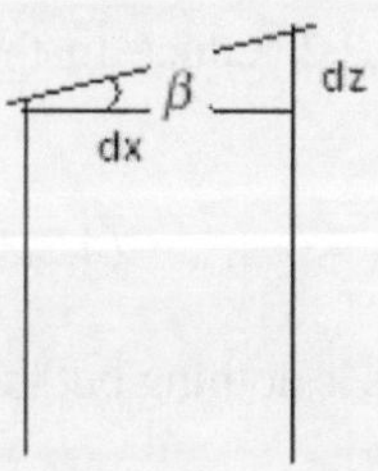


$$tan\beta = \frac{dz}{dx} = \frac{1m}{100km} = \frac{1}{10^5} = 0.00001$$

$\beta = \tan^{-1}(0.00001) = 5.72$ $\beta = 0^0\ 0'\ 2.06''$

## 4.2 Gradient or Meandering Currents

### 4.2.1 Balance of forces in a cyclone and anticyclone:

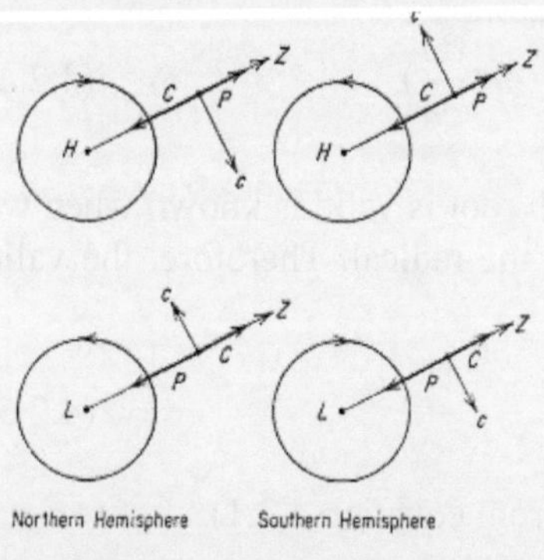


**Fig. 4.2.1.** Balance of forces in gradient motion for a cyclone and anticyclone. P denotes the pressure gradient, C denotes Coriolis and Z the Centrifugal force.

Fig.4.2.1 shows the balance of forces in a cyclone and anticyclone in N.H and S.H respectively. Which clearly shows that pressure gradient is more than the Coriolis for cyclonic motion (Low) and Coriolis is more than the pressure gradient for anticyclonic motion (High).

Balance of forces under anticyclone is:

$$f.C = -\alpha\frac{\partial p}{\partial n} + \frac{C^2}{r} \tag{4.2.1}$$

Balance of forces under cyclone is:

$$f.C = -\alpha\frac{\partial p}{\partial n} - \frac{C^2}{r} \tag{4.2.2}$$

From the observation of equations 4.2.1 and 4.2.2, it is clear that in the case of cyclonic flow, the centrifugal force augments the Coriolis force and for anti-cyclonic flow it augments the pressure gradient force.

In the absence of a pressure gradient, equations 4.2.1 and 4.2.2 reduce to the balance between Coriolis and centrifugal forces as

$$f.C = \pm\frac{C^2}{r} \tag{4.2.3}$$

Where + sign is for anticyclone and minus sign for cyclone. This is nothing but the inertial motion and 'r' is the inertial radius.

Considering the equation for cyclone (negative case), equation 4.2.2 can be written as:

$$-r\alpha\frac{\partial p}{\partial n} = rfC + C^2$$

$$C^2 + rfc + r\alpha\frac{\partial p}{\partial n} = 0$$

This is a quadratic equation in C and its roots are

$$\mathrm{C} = \frac{-rf \pm \sqrt{r^2 f^2 - 4r\alpha\frac{\partial p}{\partial n}}}{2} = \frac{-rf}{2} \pm \frac{1}{2}\sqrt{r(rf^2 - 4\alpha\frac{\partial p}{\partial n})} \tag{4.2.4}$$

Out of the two roots of the equation 4.2.4, which root is valid is known when we put $\alpha\partial p/\partial n=0$, C=0 taking the positive sign for the radical. Therefore, the valid equation for the cyclone is:

$$\mathrm{C} = \frac{-rf}{2} + \sqrt{\frac{f^2r^2}{4} - r\alpha\frac{\partial p}{\partial n}} \tag{4.2.5}$$

And the corresponding flow for anticyclone is (from equation 4.2.1).

$$C^2 - rfc + r\alpha\frac{\partial p}{\partial n} = 0 \tag{4.2.6}$$

solving this quadratic equation, we get two more roots again, taking the condition when $\alpha\partial p/\partial n = 0$, C=0 when the radical is taken as negative. So the valid root for anticyclone is:

$$C = \frac{rf}{2} - \sqrt{\frac{f^2r^2}{4} + r\alpha\frac{\partial p}{\partial n}} \tag{4.2.7}$$

Thus out of the four roots of equations (4.2.4) and (4.2.6), only two are valid such that the radical for the cyclonic flow is positive and negative for anti-cyclonic flow. From the equation 4.2.5, it is obvious that if $r\alpha\frac{\partial p}{\partial n} \geq \frac{f^2r^2}{4}$ the roots will be imaginary. Therefore the pressure gradient in an anticyclone theoretically cannot exceed the Coriolis force. Thus in order to maintain a real value of 'C' for a current around a high pressure, it is necessary that

$$\left|\frac{\partial p}{\partial n}\right| \leq \left|\frac{r}{\alpha}\Omega^2 \sin^2\Phi\right|$$

That is, in a high the magnitude of the pressure gradient may not exceed a certain value determined largely by the latitude φ and distance from the centre (r). Which means as r becomes small; the pressure gradient also becomes small. This is the reason for the small pressure gradient and light currents over a high pressure (anticyclone). In a cyclone (low) there is no such restriction and the pressure gradient and currents can be very large near the centre.

### 4.2.2. Comparison of Geostrophic and Gradient Motions

We know the equation of geostrophic current from equation (4.1) as

$$-\alpha\frac{\partial p}{\partial n} = f.v_g$$

$$or V_g = -\frac{1}{f\rho}\frac{\partial p}{\partial n}$$

We know the gradient motion equation for cyclone as (from equation 4.2.5)

$$C = -\frac{rf}{2} + \sqrt{\frac{f^2r^2}{4} - r(-f.V_g)}$$

$$\left[C + \frac{fr}{2}\right]^2 = \frac{f^2r^2}{4} + frV_g \tag{4.2.8}$$

$$V_g = \frac{C^2}{fr} + C = C(1 + \frac{C}{fr})$$

In normal flow around a low, c is cyclonic and therefore positive. Both 'r' and "f" are also positive in N.H. Thus in a 'low' $V_g$ is greater than 'C' and the geostrophic value is an over estimate of the gradient motion.

And around a high pressure both c and $V_g$ are negative. So taking the modulus

$$|V_g| = |C| - \left|\frac{C^2}{fr}\right| \tag{4.2.9}$$

Hence in a 'High', $V_g < C$, or the geostrophic value is an under estimate

## *Exercises*

### *Exercise 4.2.1*

a) Show in the case of anticyclonic motion, gradient wind balance is possible only when

$$\alpha\frac{\partial p}{\partial n} \le \frac{f^2 r}{4}$$

b) Calculate the maximum possible pressure gradient and wind speed in an anticyclone at a distance of 500 km from the centre at 14.5$^0$ latitude.

c) Determine the ratio of the maximum possible motion of the normal anticyclonic gradient wind speed to the geostrophic wind speed for the same pressure gradient.

### *Solution*

a) Balance of forces under anticyclone in N.H is ::

$$f.C = \alpha\frac{\partial p}{\partial n} + \frac{C^2}{r} \tag{1}$$

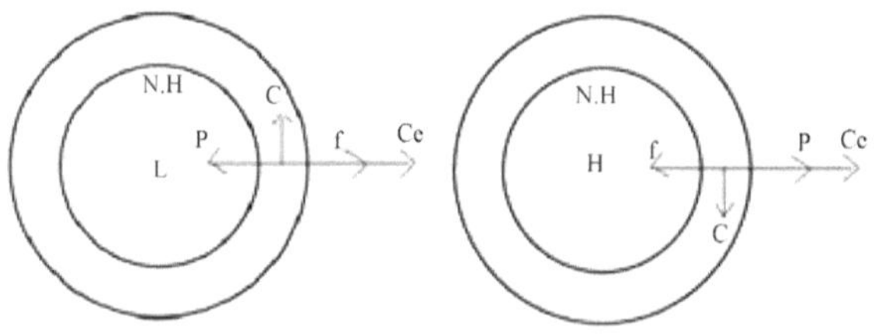


(a) cyclone (b) Anti cyclone

Balance of forces in gradient motion for a cyclone (a) and anticyclone (b). P denotes the pressure gradient, f denotes Coriolis and $C_e$ the Centrifugal force for Northern Hemisphere

Solving the flow for anticyclone (from equation1) is:

$$C^2 - rfc + r\alpha\frac{\partial p}{\partial n} = 0 \tag{2}$$

solving this quadratic equation, we get two roots $C = \frac{rf}{2} \pm \sqrt{\frac{f^2r^2}{4} - r\alpha\frac{\partial p}{\partial n}}$ (3)

To know which root is valid in (3) put the condition when $\alpha\partial p/\partial n = 0$, C=0 when the radical is taken as negative. So the valid root for anticyclone is:

$$C = \frac{rf}{2} - \sqrt{\frac{f^2r^2}{4} - r\alpha\frac{\partial p}{\partial n}} \qquad (4)$$

From the equation 4, it is obvious that if $r\alpha\frac{\partial p}{\partial n} \geq \frac{f^2r^2}{4}$ the roots will be imaginary. Therefore the pressure gradient in an anticyclone theoretically can not exceed the Coriolis force. Thus in order to maintain a real value of 'C' for a current around a high pressure, it is necessary that

$$r\alpha\frac{\partial p}{\partial n} \leq \frac{f^2r^2}{4} \text{ or } \alpha\left|\frac{\partial p}{\partial n}\right| \approx \frac{rf^2}{4} \qquad (5)$$

b) Given r = 500 × $10^3$m, $\rho_a$ = 1.3 kg $m^{-3}$; f = 2 × 7.3 × $10^{-5}$ sin14.5

Substitute all these values in equation (5) then :

$$\left|\frac{\partial p}{\partial n}\right| = \frac{rf^2\rho_a}{4} = \frac{500\times10^3\times1.3\times\left(2\times7.3\times10^{-5}\times\sin 14.5\right)^2}{4}$$

$$= 21.7\times10^{-5}\,Nm^{-2} = 21.7\times10^{-5}\,Pa/m = 0.22\times10^{-3}\,Pa/m = 0.22\,Pa/km$$

$$(\because 1mb = 100Nm^{-2} = 100\,Pascals)$$

Gradient wind speed from equation (4):

$$C = \frac{2\times7.3\times10^{-5}\times\sin 14.5\times500\times10^3}{2} - \left[\frac{\left(2\times7.3\times10^{-5}\times\sin 14.5\right)^2\left(500\times10^3\right)^2}{4} - \frac{500\times10^3}{1.3}\times21.7\times10^{-5}\right]^{1/2}$$

= -- 5.85 m/s, the negative sign indicates the anticyclone

c) Substituting (5) in equation (4) we get: $C = \frac{rf}{2} - \sqrt{\frac{r^2f^2}{4} - \frac{r^2f^2}{4}} = \frac{rf}{2}$ (6)

The balance of forces for geostrophic wind is: $f.v_g = \alpha\frac{\partial p}{\partial n}$ ..........(7)

Substituting equation (5) in (7) we get: $v_g = \frac{\alpha}{f}\frac{\partial p}{\partial n} = \frac{1}{f}\frac{f^2r}{4} = \frac{fr}{4}$ .....(8)

The ratio from equation (6) and (8), we get: $\frac{C}{v_g} = \frac{rf}{2} \div \frac{rf}{4} = 2$

### *Exercise 4.2.2*

i) Show that to maintain a real value of current around a high pressure, the condition necessary is $\left|\frac{\partial p}{\partial n}\right| < \frac{rf^2}{4\propto}$

ii) Show that the maximum possible ratio of the normal anticyclonic gradient wind speed to the geostrophic wind speed for the same pressure gradient is 2.

iii) Calculate the maximum possible pressure gradient in an anticyclone at a distance of 500km from the center at latitude $30^0$N(Take $\rho_a = 1.3$ kgm$^{-3}$ and $\Omega = 7.3 \times 10^{-5}$ radians /sec)

## *Solution*

i) Equation for anticyclonic motion (equation 4 of Exercise 4.2.1)

$$C = \frac{rf}{2} - \sqrt{\frac{f^2r^2}{4} - r\alpha\frac{\partial p}{\partial n}} \qquad \text{(P1)}$$

If $r\alpha\frac{\partial p}{\partial n} \geq \frac{f^2r^2}{4}$ the roots will be imaginary. So inorder to maintain a real value of C for current around high pressure, it is necessary that

$$r\alpha\frac{\partial p}{\partial n} \leq \frac{f^2r^2}{4}$$

$$or \left|\frac{\partial p}{\partial n}\right| \leq \frac{rf^2}{4\alpha} \quad \text{.........(P2)}$$

So the condition for maximum possible pressure gradient in an anticyclone is

$$\left|\frac{\partial p}{\partial n}\right| = \frac{rf^2\rho}{4} = \frac{rf^2}{4\alpha} \text{.............(P3)}$$

ii) From equations (P1) &(P2) above maximum possible pressure condition in anticyclone in question(i) we can write:

$$C = \frac{rf}{2} - \sqrt{\frac{f^2r^2}{4} - \frac{f^2r^2}{4}} = \frac{rf}{2} \text{.......(P4)}$$

we know the geostrophic motion from equation (4.1) relation as:

$$\left|-\alpha\frac{\partial p}{\partial n}\right| = f.v_g \text{.......(P5)}$$

So from equations (P3) &(P5) we can write:

$$\alpha\left|\frac{\partial p}{\partial n}\right| = f.v_g = \frac{rf^2}{4} \; or \; .v_g = \frac{rf}{4} \text{.....(P6)}$$

From equations (P4) &(P6): $\frac{c}{vg} = \frac{rf}{2} \div \frac{rf}{4} = 2$

iii) Calculate the maximum possible pressure gradient in an anticyclone at a distance of 500 km from the center at latitude 30°N (Take $\rho_a$ = 1.3 kgm$^{-3}$ and $\Omega = 7.3\times10^{-5}$ rad/sec).

Solution: From equation ($p_3$) of the problem 4.2.2: $\left|\frac{\partial p}{\partial n}\right| = \frac{1}{\rho}\frac{rf^2}{4}$

$$\frac{\partial p}{\partial n} = \frac{500\times10^3\times1.3\times2\times(7.3\times10^{-5}\times\frac{1}{2})^2}{4} = 0.87\times10^{-3}\,Pascals/meter = 0.87\times10^{-3}\times10^3\,Pa/km$$

= 0.87 Pa/km = 0.87 N/m$^2$

# 5

# Slope of the Sea Surface - Margule's Equation

## 5. Two Layer approximation-Margule's equation

The general distribution of density in the oceans is like a two layer ocean as shown in Fig.5.1 below. A layer of more or less uniform density(nearly vertical) near the surface called as isopycnal layer and a deeper layer of higher density below with sharp increase of density is called as pycnocline. The curve of density profile can be approximated distinctly by showing the stratification of two Layer Ocean as shown by the thick solid line in Fig.5.1. Here the pycnocline acts as a discontinuity or front between the two different water masses above and below.

A general derivation of the relationship between the inclination of a discontinuity between layers of differing density and velocity are given in Fig.5.2 below

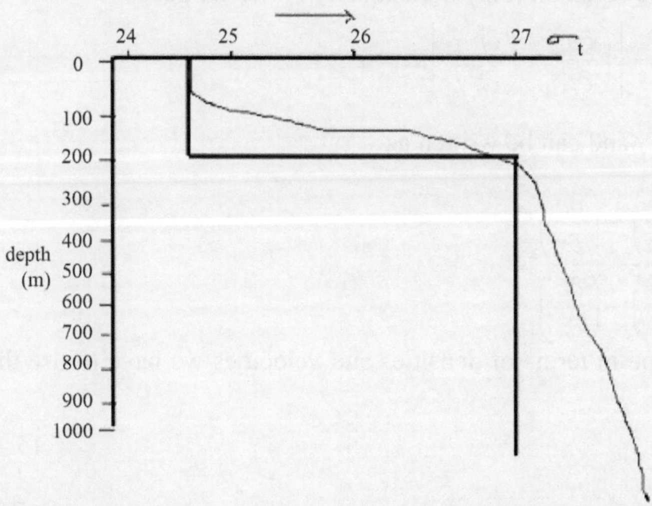


**Fig. 5.1.** Vertical profile of density showing the three layers of mixed, thermocline and stratified layers

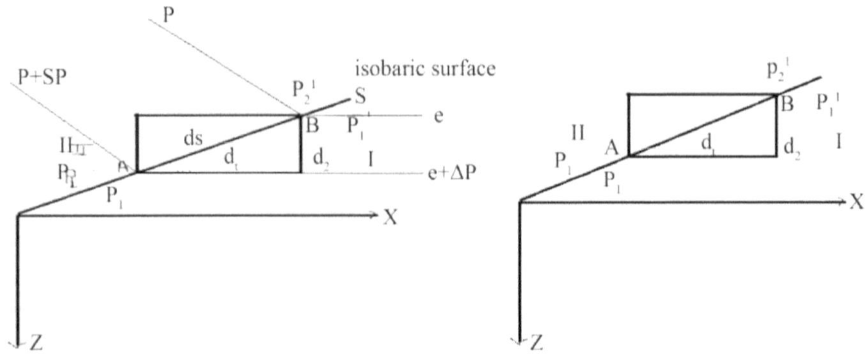


**Fig. 5.2.** The inclination of a discontinuity between layers of differing density

In the above Figure 5.2, the y-axis is oriented along the isobaric surface (P = constant) which is the discontinuity surface(S) such that AB is a small length element 'ds' and 'z' is pointing downward along depth of the ocean.

At the points A and B as pressure is uniform (taking $p_1' = p_2'$) and using Taylor's expansion we can write:

$$p_1 + \left(\frac{\partial p}{\partial x}\right)_1 dx + \left(\frac{\partial p}{\partial z}\right)_1 dz = p_1' \tag{5.1}$$

$$p_2 + \left(\frac{\partial p}{\partial x}\right)_2 dx + \left(\frac{\partial p}{\partial z}\right)_2 dz = p_2' \tag{5.2}$$

After subtracting the equation (5.2) from equation (5.1) we get as:

$$\left[\left(\frac{\partial p}{\partial x}\right)_1 - \left(\frac{\partial p}{\partial x}\right)_2\right] dx + \left[\left(\frac{\partial p}{\partial z}\right)_1 - \left(\frac{\partial p}{\partial z}\right)_2\right] dz = 0$$

Let the slope $\frac{dz}{dx} = \gamma$ and can be written as

$$\tan \gamma = \frac{dz}{dx} = -\frac{\left[\left(\frac{\partial p}{\partial x}\right)_1 - \left(\frac{\partial p}{\partial x}\right)_2\right]}{\left[\left(\frac{\partial p}{\partial z}\right)_1 - \left(\frac{\partial p}{\partial z}\right)_2\right]} \tag{5.3}$$

To express this slope in terms of densities and velocities we have to use the equation of motion.

$$\frac{du}{dt} = -\alpha \frac{\partial p}{\partial x} + fv \tag{5.4}$$

$$\frac{dw}{dt} = -\alpha \frac{\partial p}{\partial z} + g \tag{5.5}$$

Let $\frac{du}{dt} = b_x$ and $\frac{dw}{dt} = b_z$ and we know $\alpha = \frac{1}{\rho}$. Substituting and rewriting

$$b_x = -\frac{1}{\rho}\frac{\partial p}{\partial x} + fv \tag{5.6}$$

$$b_z = -\frac{1}{\rho}\frac{\partial p}{\partial z} + g \tag{5.7}$$

From equation (5.6) we can write as:

$$\frac{\partial p}{\partial x} = f\rho v - \rho b_x$$

For the two layers the equations can be written as

$$\left(\frac{\partial p}{\partial x}\right)_1 = f(\rho v)_1 - (\rho b_x)_1$$

$$\left(\frac{\partial p}{\partial x}\right)_2 = f(\rho v)_2 - (\rho b_x)_2$$

Subtracting one from the other we get

$$\left(\frac{\partial p}{\partial x}\right)_1 - \left(\frac{\partial p}{\partial x}\right)_2 = f\{[(\rho v)_1 - (\rho v)_2] - [(\rho b_x)_1 - (\rho b_x)_2]\} \tag{5.8}$$

Similarly we can write for $b_z$ using equation (5.7) as

$$\left(\frac{\partial p}{\partial z}\right)_1 - \left(\frac{\partial p}{\partial z}\right)_2 = g\{[(\rho)_1 - (\rho)_2] - [(\rho b_z)_1 - (\rho b_z)_2]\} \tag{5.9}$$

Hence from equation (5.3) and equations (5.8) and (5.9), it follows:

$$\tan\gamma = \frac{dz}{dx} = -\frac{f[(\rho v)_1 - (\rho v)_2] - [(\rho b_x)_1 - (\rho b_x)_2]}{g[(\rho)_1 - (\rho)_2] - [(\rho b_z)_1 - (\rho b_z)_2]}$$

Under non accelerated frictionless uniform motion we know

$$b_x = \frac{du}{dt} = b_z = \frac{dw}{dt} = 0$$

$$\therefore \tan\gamma = -\frac{f}{g}\left[\frac{\rho_1 v_1 - \rho_2 v_2}{\rho_1 - \rho_2}\right] \tag{5.9a}$$

Figure 5.3 shows schematically isobaric surfaces and the inclination of density discontinuity between two different water bodies for Northern Hemisphere. The Figure 5.3 shows the isobaric surfaces and currents in a two layer ocean where a less dense water mass $\rho_1$ is separated from a denser water mass $\rho_2$ such that $\gamma$ is the angle or slope of discontinuity with the level surface. The angles $\beta_1$ and $\beta_2$ are the angles of the sea surfaces with the level surface in the rarer and denser waters respectively corresponding to the geostrophic velocities $v_1$ and $v_2$.

With regard to the Margules's relation at the Equator, with a typical dynamic sea level difference of $\Delta\eta \cong -60$ cm in the Equatorial Pacific, we get a thermocline depth difference as:

$\Delta h \cong -200$ cm $\Delta\eta \cong +120$ m

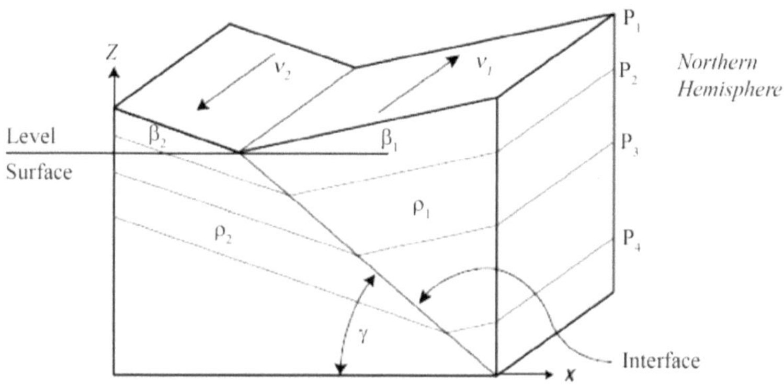


**Fig.5.3.** Slopes $\beta_1$ and $\beta_2$ of the sea surface and the slope of the interface between two homogeneous, moving layers, with densities $\rho_1$ and $\rho_2$ in the northern hemisphere.

## 5.1. Thermal wind

The variation of geostrophic velocity in the vertical is known as thermal wind. While the velocity is parallel to isobars in case of geostrophic motion, it is parallel to isotherms in case of thermal wind. Originally this is derived in meteorology.

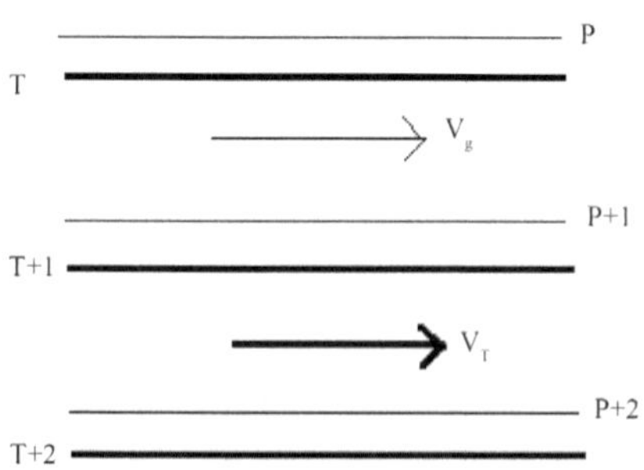


**Fig. 5.4.** Relation between geostrophic and thermal winds

Fig.5.4 shows the geostrophic and thermal winds in a Barotrophic mode. This implies that the temperature differences in the horizontal could lead to vertical variations in the geostrophic velocity.

Taking the x and y components of equations of motion:

$$\frac{du}{dt} = -\alpha\frac{\partial p}{\partial x} + fv \tag{5.10}$$

$$\frac{dv}{dt} = -\alpha\frac{\partial p}{\partial y} - fu \tag{5.11}$$

Under nonaccelerated frictionless conditions, the above equations changes as

$$\frac{\partial p}{\partial x} = f\rho v \; and \; \frac{\partial p}{\partial y} = -f\rho u \tag{5.12}$$

On differentiating equations (5.12) both sides with respect to "z" we get:

$$\frac{\partial}{\partial z}\left(\frac{\partial p}{\partial x}\right) = \frac{\partial}{\partial z}(f\rho v) \tag{5.13}$$

As thermal wind is no more parallel to isobars, Change the differential order to equation (5.13),

$$\frac{\partial}{\partial x}\left(\frac{\partial p}{\partial z}\right) = \frac{\partial}{\partial z}(f\rho v) \tag{5.14}$$

We know the hydrostatic equation

$$\left(\frac{\partial p}{\partial z}\right) = -\rho g \tag{5.15}$$

Substituting equation(5.15) in (5.14) we get:

$$\frac{\partial}{\partial x}(-\rho g) = \frac{\partial}{\partial z}(f\rho v)$$

$$\text{or } \frac{\partial}{\partial z}(f\rho v) = -g\frac{\partial \rho}{\partial x} \tag{5.16}$$

Similarly the other y-component equation of (5.12) can be analyzed: $\frac{\partial p}{\partial y} = -f\rho u$ and get:

$$\frac{\partial}{\partial z}(f\rho u) = g\frac{\partial \rho}{\partial y} \quad ................(5.17)$$

These equations (5.16) & (5.17) are called thermal wind equations applied to oceans. These say that from the density field we can determine the vertical variation of velocity (or velocity shear $\frac{dv}{dz}$).

In the N.H, "Buy's Ballots Law" states that when we look in the direction of flow, low pressures always will be in the left hand side. But in oceanography when this law is applied, it states that "Low topographies are always in the left hand side when one looks in the direction of flow".

Low topography implies low volume in the oceans. Low volume implies high density.

Equations (5.16) & (5.17) can be written as:

$$-g\frac{\partial \rho}{\partial x} = f\rho\frac{\partial v}{\partial z} \quad \text{.....................} (5.18)$$

$$g\frac{\partial \rho}{\partial y} = f\rho\frac{\partial u}{\partial z} \quad \text{.....................}(5.19)$$

To examine equation (5.18), the vertical variation of meridional velocity ($\frac{\partial v}{\partial z}$) can be understood from horizontal variation of density.

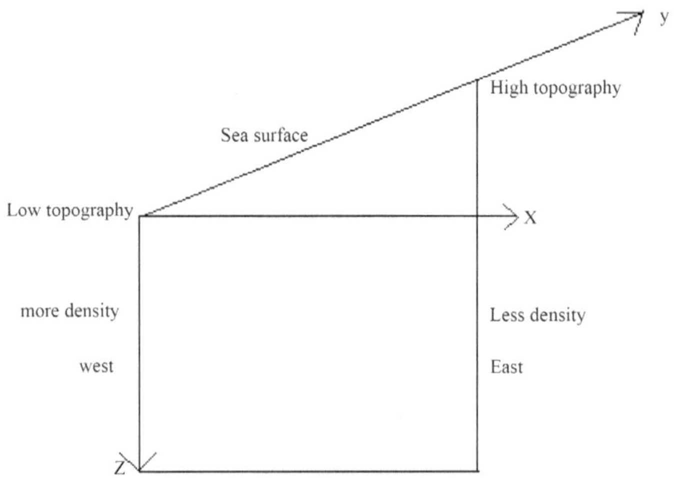


**Fig.5.5.** Sea surface topography and the density variation

In the above Fig.5.5, let the sea surface topography is increasing from west to east. Then the density decreases from west to east. This means the lateral density gradient ($\frac{\partial \rho}{\partial x}$) decreases eastward ($\frac{\partial \rho}{\partial x} < 0$). This means ρ is negative, $\frac{\partial v}{\partial z}$ is positive ($\frac{\partial v}{\partial z} > 0$). This implies meridional velocity(v) should increase as we go upward the slope (northward).

With regard to equation (5.19), as ρ decreases in the negative meridional direction (equatorward) (-y), equation (5.19) can be applied. Then $\frac{\partial \rho}{\partial y} > 0$ and so $\frac{\partial u}{\partial z} > 0$ which means eastward velocity increases with depth.

## References

Margules M. 1906. Uber Temperaturschichtung in stationarbewegter und ruhender Luft. Meteorologische Zeitschrift 241–244.

## *Exercises*

### *Exercise 5.1*

Find the angle of interface in a two layer system in which the surface is level and the densities of the upper layer is ($\rho_1$) and lower layer is ($\rho_2$) such that $\rho_2 > \rho_1$. Also estimate the geostrophic velocity ($V_2$) in the lower layer.

### *Solution*

The pressure gradient in the system =

$$\frac{\delta P}{\delta n} = \frac{P_B - P_A}{\delta n} = \frac{g}{\delta n}\left[\{\rho_2(Z_2 + \delta Z) + \rho_1 Z_1\} - \{\rho_2 Z_2 + \rho_1(Z_1 + \delta Z)\}\right] = \frac{g\delta Z}{\delta n}(\rho_2 - \rho_1) \tag{5.1.1}$$

We know for geostrophic motion: $\frac{1}{\rho_2}\frac{\delta P}{\delta n} = fV_2;\ so\ \frac{\delta P}{\delta n} = f\rho_2 V_2$ (5.1.2)

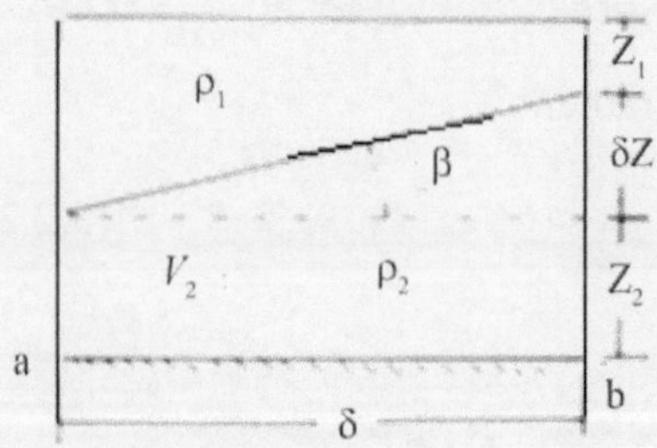


We also know $\frac{\delta Z}{\delta n} = \tan\beta$ (5.1.3)

So Substituting Equations (5.1.2) and (5.1.3) in equation (5.1.1) we can write:

$$V_2 = \frac{g}{f\rho_2}(\rho_2 - \rho_1)\tan\beta$$

### *Exercise 5.2*

i) Using the principle used in Exercise (5.1), calculate the slope of the interface in terms of sea surface (Margule's equation).ii) calculate tan $\beta_1$ and tan $\beta_2$ if f ~ $10^{-4}$rad/sec, g ~$10^3$cm/sec$^2$, and $V_1 = V_2$~ 100 cm/sec,

### *Solution*

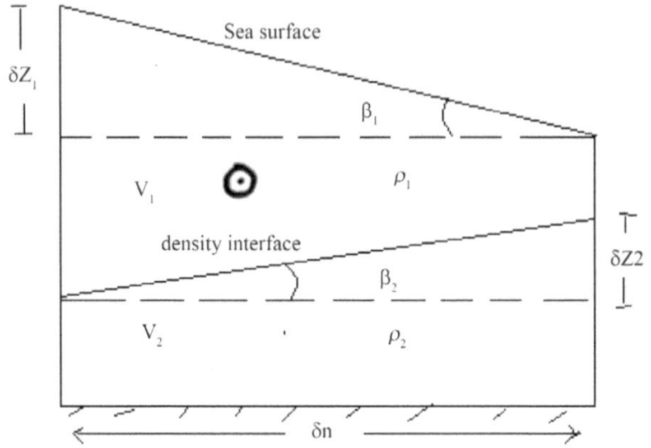


We know for the lower layer:

$$\frac{\delta P_2}{\delta n} = \frac{P_B - P_A}{\delta n} = \frac{g}{\delta n}\left[\{\rho_2(Z_2+\delta Z_2)+\rho_1 Z_1\} - \{\rho_1(Z_1+\delta Z_2)+\rho_1\delta Z_1+\rho_2 Z_2\}\right]$$

$$\Rightarrow f\rho_2 V_2 = \frac{g\delta Z_2}{\delta n}(\rho_2-\rho_1) - \frac{g\rho_1\delta Z_1}{\delta n} \qquad (5.2.1)$$

Since pressure gradient balances Coriolis for geostrophic motion: $\frac{\delta P_2}{\delta n} = f\rho_2 V_2$

Similarly for the upper layer, the pressure gradient =

$$\frac{1}{\rho_1}\frac{\delta P_1}{\delta n} = fV_1 = -g\frac{\delta Z_1}{\delta n} \qquad (5.2.2)$$

Substituting (5.2.2) in (5.2.1):

$$f\rho_2 V_2 = \frac{g\delta Z_2}{\delta n}(\rho_2-\rho_1) - \frac{g\rho_1\delta Z_1}{\delta n} = \frac{g\delta Z_2}{\delta n}(\rho_2-\rho_1) + f\rho_1 V_1$$

$$\frac{\delta Z_2}{\delta n} = \tan\beta_2 = \frac{f(\rho_2 V_2 - \rho_1 V_1)}{g(\rho_2-\rho_1)}$$

This is called Margule's equation.

### *Exercise 5.3*

Assume a two layer fluid with a northern flow in the upper layer of 19.6 cm $s^{-1}$, no flow in the lower layer, a density difference of 0.002 gm $cm^{-3}$ between layers at the latitude 43.3° N (take surface density as 1.025 gm $cm^{-3}$) a) draw a schematic figure of this setup, b) Estimate the east-west slope of the interface c) Estimate the east-west slope of the sea surface. Express the slope in degrees

### Solution

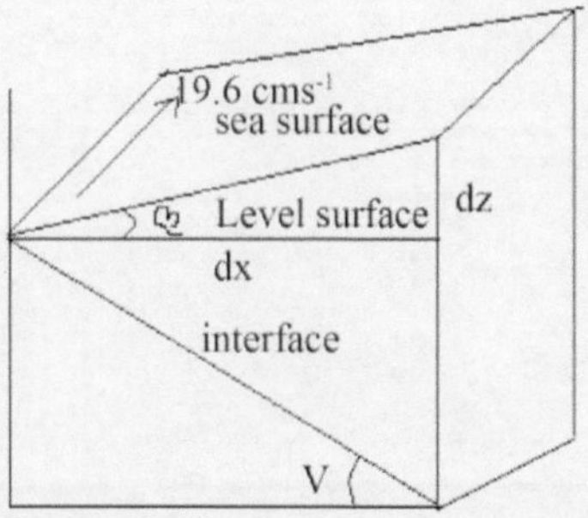


We know from Margules equation (5.9a): $\therefore \tan\gamma = -\frac{f}{g}\left[\frac{\rho_1 v_1 - \rho_2 v_2}{\rho_1 - \rho_2}\right]$

Given $V_1$ = 19.6 cm $s^{-1}$, $V_2$ =0; $\rho_1 - \rho_2$ = 0.002 g $cm^{-3}$; take $\rho_2$ = 1025g $cm^{-3}$. Then substituting in the above formula

$$\therefore \tan\gamma = \frac{f}{g}\left[\frac{\rho_1 v_1 - \rho_2 v_2}{\rho_2 - \rho_1}\right] = \frac{2\times 7.3\times 10^{-5}\times \sin 43.3}{980cms^{-2}}\left[\frac{1.025gcm^{-3}\times 19.6cms^{-1}}{0.002gcm^{-3}}\right]$$

$= 0.10263\times 10^{-2} = 0.00103$

Slope of the interface = 0.001;

$\tan\gamma = 0.001 \rightarrow \gamma = \tan^{-1}(0.001) = 0.057 \approx 0^0\ 3'26''$

$$\frac{du}{dt} = -\alpha\frac{\partial p}{\partial x} + fv$$

we know from equation of motion:

$$-\alpha\frac{\partial p}{\partial x} + fv = 0 \Rightarrow -gdz = -fvdx \Rightarrow \frac{dz}{dx} = \frac{fv}{g}$$

under non-accelerated conditions:

$$\Rightarrow \frac{dz}{dx} = \frac{fv}{g} = \frac{2\times 7.3\times 10^{-5}\sin 43.3\times 19.6cms^{-1}}{980cms^{-2}} = 0.2\times 10^{-5}$$

Slope of the sea surface = $2 \times 10^{-6}$

$\tan\beta = 2 \times 10^{-6} \rightarrow \beta = \tan^{-1}(2 \times 10^{-6}) = 1.14 \approx 0^0\ 0'\ 0.41''$

### Exercise 5.4

In the ocean, two homogeneous bodies of water flow geostrophically in the same direction with velocities of 40 cm $s^{-1}$ and 20 cm $s^{-1}$ respectively. The faster water has a density of 1.025 g $cm^{-3}$. The slower has a density of 1.030 g cm$^{-3}$. Relative to the level surface find the angle of interface between the two bodies giving a suitable figure (take f = $10^{-4}$ ).

### *Solution*

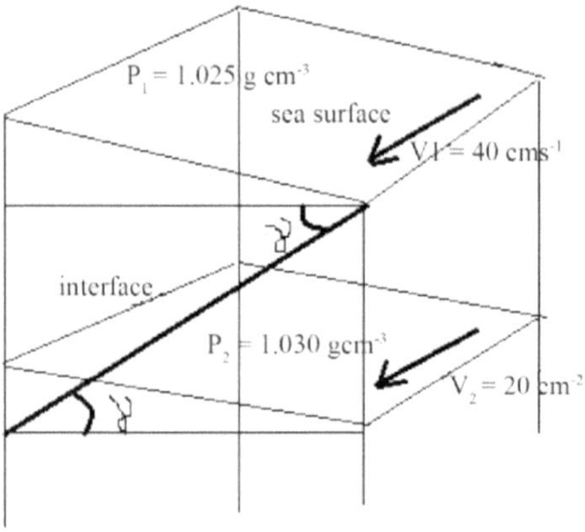


We know from Margule's equation: $\therefore \tan\gamma = \frac{f}{g}\left[\frac{\rho_1 v_1 - \rho_2 v_2}{\rho_2 - \rho_1}\right]$

In this, values given are f = $10^{-4}$; $V_1$ = 40 cm s$^{-1}$; $V_2$ = 20 cm s$^{-1}$; $\rho_1$ =1.025 g cm$^{-3}$; $\rho_2$ = 1.030 g cm$^{-3}$

Substituting these values in the above equation:

$$\tan\gamma = \frac{10^{-4}s^{-1}\left(1.025\times 40 - 1.030\times 20\right)gcm^{-3}cms^{-1}}{980cms^{-2}(1.030-1.025)} = \frac{20.4\times 10^{-4}}{4.9} = 4.16\times 10^{-4}$$

$\tan\gamma = 4.16 \times 10^{-4} \rightarrow \gamma = \tan^{-1}(4.16 \times 10^{-4}) = 5.72 \approx 0^0\ 0'\ 20.63''$

### *Exercise 5.5*

In a homogeneous ocean (i.e. $\rho$ = constant) with a sea surface tilt of θ what is the bottom pressure difference between stations a and b, which are separated laterally by a finite distance dx. Also find the angle, θ

### *Solution*

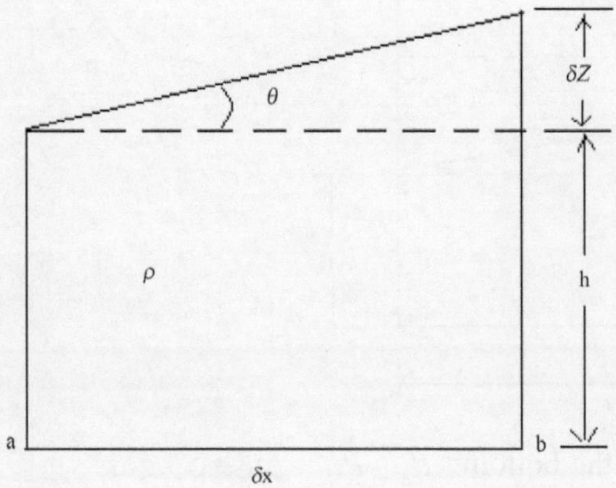


Consider a uniform ocean (i.e. ρ = constant) with surface tilt (θ ) as shown in Figure.

The bottom pressure difference is: dP= $P_b$ --$P_a$

So pressure gradient force: $-\frac{\delta P}{\delta x} = -\left[\frac{P_b - P_a}{\delta x}\right] = -\frac{g}{\delta x}\left[\rho(h+\delta z) - \rho h\right] = -\rho g \tan\theta$

So the bottom pressure difference = $\delta P = \rho g \delta z = \rho g \tan\theta \delta x$

Slope of the sea surface = $\tan\theta = \frac{1}{\rho g}\frac{\delta P}{\delta x} = \left[\frac{P_b - P_a}{\delta x}\right]$

### *Exercise 5.6*

What is the pressure difference and angle of the interface in a two layer ocean, with no surface tilt, in which a homogeneous upper layer $\rho_1$ overlies a homogeneous lower layer $\rho_2$.

### *Solution*

Consider a two layer ocean, without any surface tilt in which a homogeneous upper layer $\rho_1$ overlies a homogeneous lower layer $\rho_2$, with interface tilt of θ as shown in Figure.

The pressure gradient :

$$-\frac{\delta P}{\delta x} = -\left[\frac{P_b - P_a}{\delta x}\right] = -\frac{g}{\delta x}\left[\{\rho_1(h_1 + \delta z_1) + \rho_2 h_2\} - \{\rho_2(h_2 + \delta z_1) + \rho_1 h_1\}\right]$$

$$= -(\rho_2 - \rho_1) g \tan\theta$$

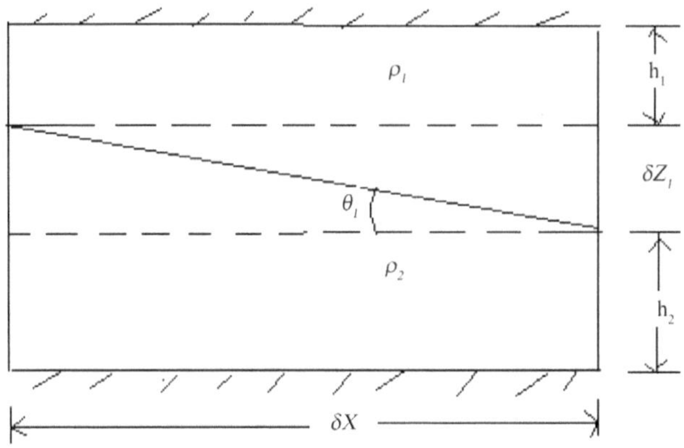


Hence the pressure difference at the bottom= $P_b - Pa = (\rho_2 - \rho_1)\, g\delta z$

The slope of the interface = $\tan\theta = \dfrac{1}{(\rho_2 - \rho_1)g}\dfrac{\delta P}{\delta x}$

## Exercise 5.7

(a) A useful version of the hydrostatic law for two layers of constant-density seawater is: $\nabla P_2 - \nabla P_1 = (\rho_2 - \rho_1)\nabla h$

In this formula the gradients are in the horizontal plane, the pressures are p1 and p2 in layers 1 and 2, the densities are $\rho_1$ and $\rho_2$ ($\rho_2 > \rho_1$; layer 1 is above the layer 2), g is acceleration due to gravity, and h is the height of interface between the layers. Derive the formula assuming that the pressure field is continuous across the interface.

(b) Derive the alternate thermal-wind formula ($V_T$) using your result from (a), or otherwise.

## Solution

Solution for (a) is given in previous Exercise 5.6 as: $P_b - Pa = (\rho_2 - \rho_1)\, g\delta z$

$\rightarrow \nabla P_2 - \nabla P_1 = (\rho_2 - \rho_1)\nabla h$

b) The variation of geostrophic velocity in the vertical is known as thermal wind. While the velocity is parallel to isobars in case of geostrophic motion, it is parallel to isotherms in case of thermal wind.

The Fig. 5.4 shows the geostrophic and thermal winds in a barotropic mode. This implies that the temperature differences in the horizontal could lead to vertical variations in the geostrophic velocity.

Taking the x and y components of equations of motion:

$$\frac{du}{dt} = -\alpha\frac{\partial p}{\partial x} + fv; \frac{dv}{dt} = -\alpha\frac{\partial p}{\partial y} - fu$$

Under non accelerated friction less conditions the above equations turn out to be:

$$\frac{\partial p}{\partial x} = f\rho v \text{..............(1)}$$

$$\frac{\partial p}{\partial y} = -f\rho u \text{............(2)}$$

On differentiating equation (1) both sides with respect to 'z'

$$\frac{\partial}{\partial z}(\frac{\partial p}{\partial x}) = \frac{\partial}{\partial z}(f\rho v).or$$

$$\frac{\partial}{\partial x}(\frac{\partial p}{\partial z}) = \frac{\partial}{\partial z}(f\rho v)$$

We know hydrostatic equation: $\frac{\partial p}{\partial z} = -\rho g$ substituting this in the above equation:

$$\frac{\partial}{\partial x}(-\rho g) = \frac{\partial}{\partial z}(f\rho v)or$$

$$\frac{\partial}{\partial z}(f\rho v) = -g\frac{\partial \rho}{\partial x} \text{...........(3)}$$

Similarly we can write: $\frac{\partial}{\partial z}(f\rho v) = -g\frac{\partial \rho}{\partial x}$ (4)

These equations (3) and (4) are called thermal wind equations applied to oceans. These equations implies that from the density field we can determine the vertical variation of velocity (velocity shear $\frac{\partial v}{\partial z}$)

In the northern hemisphere "Buys Ballots Law" states that when we look in the direction of flow, low pressures always will be in the left hand side. But in oceanography when this law is applied, it states that "Low topographies" are always in the left hand side when one looks in the direction of flow. Low topography implies low volume in the oceans. Low volume implies high density.

Equations (3) and (4) can be written as:

$$-g\frac{\partial\rho}{\partial x}=f\rho\frac{\partial v}{\partial z} \qquad \text{..........(5)}$$

$$g\frac{\partial\rho}{\partial y}=f\rho\frac{\partial u}{\partial z} \qquad \text{..............(6)}$$

To examine equation (5) the vertical variation of meridional velocity ($\frac{\partial v}{\partial z}$) can be understood from horizontal variation of density ($\frac{\partial\rho}{\partial x}$).

In the below figure, let the sea surface topography is increasing from west to east. Then the density decreases from west to east. This means the lateral density gradient ($\frac{\partial\rho}{\partial x}$) decrease from west to eastward ($\frac{\partial\rho}{\partial x}<0$).

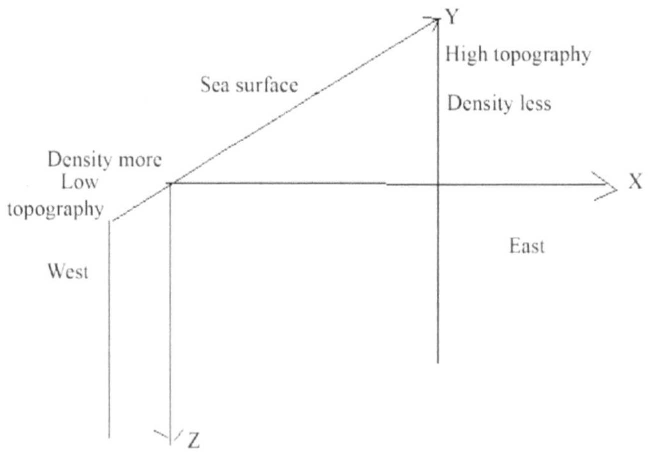


In the equation (5) if $\frac{\partial\rho}{\partial x}<0$, which means ρ is negative, means $\frac{\partial v}{\partial z}$ is positive. This means $\frac{\partial v}{\partial z}>0$.

This implies meridional velocity (v) should increase as we go upward (north ward).

Whereas if ρ decreases in the negative meridional direction (equatorward), equation (6) can be applied. Then $\frac{\partial\rho}{\partial y}>0$ and so $\frac{\partial u}{\partial z}>0$ which means eastward velocity increases with depth

### *Exercise 5.8*

Below is a schematic representation of the geometry and pressure field in a longitudinal section of the Strait of Bosporus. The ocean surface elevation at, $dz_1$ above the mean sea level (at z = 0) coincides with the atmospheric pressure ($P_a$) isobar. The slopes of this and the isobars at depth (i.e. $P_1$, $P_2$, $P_3$, etc.) are grossly exaggerated compared to the slope of the bold, dashed interface between the low density ($\rho_B$) Black Sea and the high density ($\rho_M$) Sea of Marmara waters. (a) Calculate the PGF at depths z = 0, $-z_1$, $z_2$, and $-z_3$ respectively.(b) Estimate the slopes of isobars $P_a$, $P_3$, and $P_6$ respectively, assuming $\rho_M$ =1.025 gm/cm³ and $\rho_B$ = 1.012 gm/cm³.

### *Solution*

We know from the previous exercise (5.6): $-\frac{\delta P}{\delta x} = -\left[\frac{P_b - P_a}{\delta x}\right]$

i) Pressure gradient at $Z_0 = -\frac{\delta P}{\delta x} = -\left[\frac{P_b - P_a}{\delta x}\right] = -\frac{g}{L}[\rho_B \delta z_1 - 0] = -\frac{g\rho_B \delta z_1}{L}$

ii) Pressure gradient at $Z_1$ =

$$-\frac{\delta P}{\delta x} = -\left[\frac{P_b - P_a}{\delta x}\right] = -\frac{g}{L}[\rho_B(z_1 + \delta z_1) - \rho_B z_1] = -\frac{g\rho_B \delta z_1}{L}$$

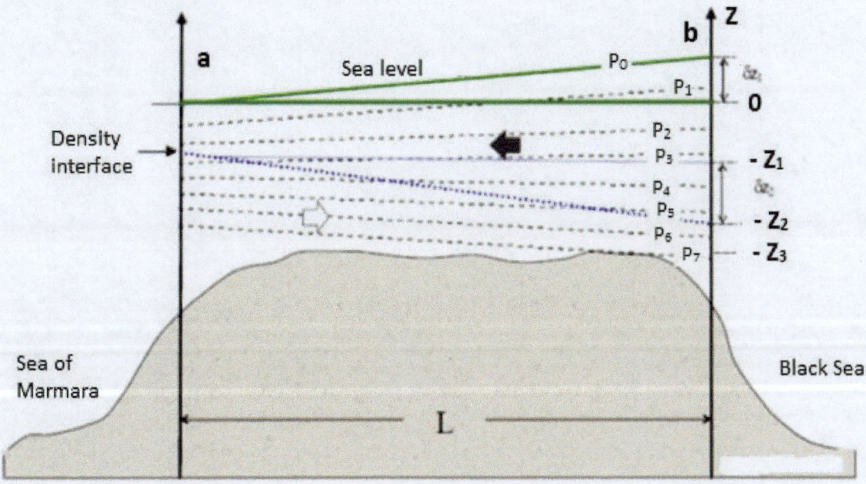


iii) Pressure gradient at $Z_2 = -\frac{}{\delta x} = -\left[\frac{P_b \quad P_a}{\delta x}\right]$

$$= -\frac{g}{L}[\rho_B(\delta z_2 + z_1 + \delta z_1) - (\rho_M \delta z_2 + \rho_B z_1] = \frac{g}{L}[\delta z_2(\rho_M - \rho_B) - \rho_B \delta z_1]$$

iv) Pressure gradient at $Z_3$ =

$$\frac{\delta P}{\delta x} = -\frac{g}{L}[\rho_B \delta z_2 - \rho_M(\delta z_2 + z_3 - z_1)] = \frac{g}{L}[\delta z_2(\rho_M - \rho_B) + (z_3 - z_1)\rho_M]$$

### *Slopes*

i) Slope of $P_A = \alpha dp = gdz \Rightarrow \frac{dz}{dx_{Pa}} = -\frac{dp}{\rho gdx} = -\frac{P_1 - P_a}{\rho_B gL} = \frac{1}{gL}\left[\frac{P_a - P_1}{\rho_B}\right]$

ii) Slope of P3 $= \frac{dz}{dx_{P3}} = -\frac{dp}{\rho gdx} = -\frac{P_3 - P_a}{\rho_B gL} = \frac{1}{gL}\left[\frac{P_a - P_3}{\rho_B}\right]$

iii) Slope of P6 $= \frac{dz}{dx_{P6}} = -\frac{P_6 - P_a}{\rho gL} = \frac{1}{gL}\left[\frac{P_a - P_6}{\rho_B - \rho_M}\right]$

# 6

# Ekman Circulation

## 6. The History of Wind driven Ocean Circulation

The following scientists greatly contributed to the wind driven circulation. Out of all Ekman, Sverdrup, Stommel and Munk are most important and ever remembered for their work. So in the following pages discussion is mostly concentrated on these four scientists theories.

| Name | Year | contribution |
|---|---|---|
| Fridtjof Nansen | 1897 | Suggested wind generates currents at an angle of 20-40$^0$ to the right in N.H due to Coriolis force |
| Vagan Wlafrid Ekman | 1902 &1905 | Suggested wind generates surface current at 45$^0$ to the right & Ekman spiral |
| Harald Sverdrup | 1947 | Worked mostly on the circulation of Eastern Pacific& gyres |
| Henry Stommel | 1948 | Studied extensively on eastern and western boundary currents and suggested westward intensification |
| Walter Munk | 1950 | Total interpretation & comparison of the theoretical circulation with that of the real ocean |
| Kirk Bryan | 1963 | Numerical models of Oceanic circulation was developed |
| Bert Semtner & Robert Chervin | 1992 | Global and eddy resolving realistic model of the ocean's circulation |

Earlier people were under the impression that if wind blows on a body in a fluid on the earth's surface, the body will move in the direction of the wind. But this concept was proved wrong by Nansen (1902) while studying the drift of ice berg due to wind during Fram expedition (1893-96). He found that the drift of ice with respect to the wind did not follow the wind direction, but deviated from it by about 20 to 40$^o$ to the right. He explained that this deviation may be due to earth's rotation.

Later his student Ekman investigated this problem and explained mathematically. This finally laid the foundation for one of the most remarkable theoretical developments in dynamic oceanography.

## 6.1. Pure drift currents by Ekman:

Pure drift currents are produced by the drag of the wind passing over homogeneous ocean. So Ekman considered the velocity into two parts. One is geostrophic velocity ($v_g$) and the other is velocity due to friction (it is called Ekman velocity, $v_E$).

$\therefore f.v = f(v_g + v_E)$

we know the equation of motion with friction as

$$\frac{du}{dt} = -\alpha\frac{\partial p}{\partial x} + f.v + \frac{A_z}{\rho}.\frac{\partial^2 u}{\partial z^2}$$

for non-accelerated uniform motion

$$f.v = \alpha\frac{\partial p}{\partial x} - \frac{A_z}{\rho}.\frac{\partial^2 u}{\partial z^2} \; or \; f.(v_g + v_E) = \alpha\frac{\partial p}{\partial x} - \frac{A_z}{\rho}.\frac{\partial^2 u}{\partial z^2}$$

We know the geostrophic balance $f.v_g = \alpha\frac{\partial p}{\partial x}$

$$\therefore \; f.(v_E) = -\frac{A_z}{\rho}.\frac{\partial^2 u}{\partial z^2} \quad \text{.................................(6.1)}$$

Similarly x component of equation can be written as

$$f.(u_E) = \frac{A_z}{\rho}.\frac{\partial^2 v}{\partial z^2} \qquad (6.2)$$

## 6.2. Assumptions of Ekman

To solve the frictional equations (6.1 & 6.2) of motion, Ekman made some assumptions. They are:

i) They are no boundaries to oceans (no lateral friction)
ii) Infinitely deep water ( so u = v = 0 at bottom)
iii) Homogeneous water (ρ = constant)
iv) No slope at the surface ($\frac{\partial p}{\partial x} = 0$)
v) A steady wind blowing over the sea surface for a long time in the direction of y component ( i.e. wind stress in x direction ,$\tau_x = 0$)
vi) Eddy viscosity coefficient ($A_z$) is constant.
vii) The flow is nonaccelerated ( i.e. $\frac{du}{dt} = \frac{dv}{dt} = 0$) and horizontal.

### 6.3. Ekman's Derivation

The equations (6.1) and (6.2) can be written after omitting the suffixes as:

$$.\frac{\partial^2 u}{\partial z^2} = -\frac{f\rho}{A}v \text{ and, } .\frac{\partial^2 v}{\partial z^2} = \frac{f\rho}{A}u, \quad Let\ \frac{f\rho}{A} = a^2$$

Substituting $i^2$ for –ve for former and multiplying both sides by i to the later we can write as:

$$\frac{\partial^2 u}{\partial z^2} + i\frac{\partial^2 v}{\partial z^2} = ia^2(u+iv)\, or\, \frac{\partial^2}{\partial z^2}(u+iv) = ia^2(u+iv)$$

Let $(u+iv) = W$

$$\frac{\partial^2}{\partial z^2}(W) = ia^2(W)$$

This is a second order differential equation and the solution is:

$\therefore\ W = Ae^{az\sqrt{i}} + Be^{-az\sqrt{i}}$ where $A = C_1\, e^{ip}$ and $B = C_2\, e^{-iq}$ are complex constants and p and q are also constants in the hyperbolic function and $C_1$ and $C_2$ are linear constants. So substituting them

$$\therefore\ W = C_1 e^{ip} e^{az\sqrt{i}} + C_2 e^{-iq} e^{-az\sqrt{i}}$$

we know $e^{ip} = \cos p + i \sin p$ and $e^{-iq} = \cos q - i \sin q$ substituting and simplifying using the relation $\sqrt{i} = \frac{1+i}{\sqrt{2}}$ we get

$$W = u + iv = (C_1 \cos p + iC_1 \sin p)e^{\frac{az}{\sqrt{2}}}.e^{i\frac{az}{\sqrt{2}}} + (C_2 \cos q - iC_2 \sin q)e^{-\frac{az}{\sqrt{2}}}.e^{-i\frac{az}{\sqrt{2}}}$$

On expanding and separating the real and imaginary parts we get

$$u = C_1 e^{az/\sqrt{2}} \cos\left(\frac{az}{\sqrt{2}} + p\right) + C_2 e^{-az/\sqrt{2}} \cos\left(\frac{az}{\sqrt{2}} + q\right)$$

similarly

$$v = C_1 e^{\frac{az}{\sqrt{2}}} \sin\left(\frac{az}{\sqrt{2}} + p\right) - C_2 e^{\frac{-az}{\sqrt{2}}} \sin\left(\frac{az}{\sqrt{2}} + q\right)$$

The four constants $C_1$, p, $C_2$ and q have to be determined from the boundary conditions.

i) the lower boundary condition is that the velocity becomes zero (u = v = 0) for infinitely deep water ( z = ∞). It is obvious that $C_1 = 0$ in order to satisfy this condition. Then the equations reduces to

$$u = C_2 e^{\frac{-az}{\sqrt{2}}} \cos\left(\frac{az}{\sqrt{2}} + q\right) \tag{6.3}$$

$$v = C_2 e^{\frac{-az}{\sqrt{2}}} \sin\left(\frac{az}{\sqrt{2}} + q\right) \tag{6.4}$$

ii) At the upper boundary a constant wind stress is applied to the sea surface (z = 0 ) along y direction. So the wind stress along x direction

$$\tau_x = -A\left(\frac{du}{dz}\right)_{z=0} = 0 \; and \; \tau_y = -A\left(\frac{dv}{dz}\right)_{z=0}$$

Differentiating equation (6.3) with respect to z and applying the boundary condition of $\tau_x$ on z = 0 we get

$$\left(\frac{du}{dz}\right)_{z=0} = -\frac{a}{\sqrt{2}} C_2 \left(\sin q + \cos q\right) = 0$$

which means ($\sin q + \cos q$) = 0, $\sin q = -\cos q$ or $\tan q = -1$ or $q = -\frac{\pi}{4}$

Similarly differentiating equation (6.4) and applying the boundary condition of $\tau_y$ on z = 0 we get

$$\left(\frac{dv}{dz}\right)_{z=0} = -\frac{a}{\sqrt{2}} C_2 \left(\cos q - \sin q\right) = -\frac{\tau_y}{A}$$

or

$$\frac{a}{\sqrt{2}} C_2 \left[\cos\left(-\frac{\pi}{4}\right) - \sin\left(-\frac{\pi}{4}\right)\right] = -\frac{\tau_y}{A}$$

on simplifying we get $C_2 = \frac{\tau_y}{Aa}$ where we put $a = \frac{f\rho}{A}$

Substituting all the known constants and rewriting finally we can write

$$u = V_0 e^{\frac{-\pi z}{D}} \cos\left(45 - \frac{\pi z}{D}\right) \tag{6.5}$$

$$v = V_0 e^{\frac{-\pi z}{D}} \sin\left(45 - \frac{\pi z}{D}\right) \tag{6.6}$$

Here $V_0 = \frac{\tau_y}{Aa} = \frac{\pi \tau_y \sqrt{2}}{Df\rho}$ and $D = \pi\sqrt{\frac{2A}{f\rho}} = \pi\sqrt{\frac{A}{\rho\Omega\sin\varphi}}$ (6.7)

These are called Ekman equations. These indicate that the surface current is deflected 45° right hand side to the direction of wind and the current vector decreases continuously with depth. This decrease continues up to a depth of frictional influence where the current speed will be $e^{-\pi}\left(\frac{1}{23}\right)$ times of the current and it is just opposite to the surface current. This is shown through a hodograph in Fig.6.1. The arrows show the velocity and direction of current vectors at depths of equal intervals. When projected on a horizontal plane the end points of the current vectors form a logarithmic spiral called 'Ekman Spiral' indicated by the dashed curve on the horizontal plate in the Fig. 6.1.

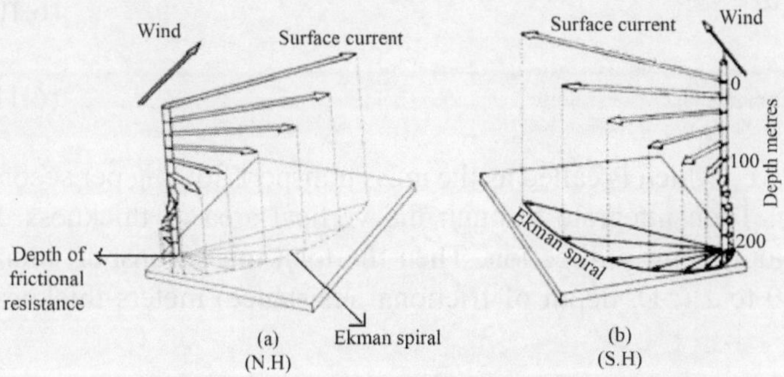


**Fig.6.1.** Hodograph of Ekman spiral for Northern Hemisphere **(a)** and for Southern Hemisphere **(b)**

## 6.4. Ekman Net transport and Ekman Pumping

Alternative way of deriving Ekman transport using Equations of motion with friction can be written as:

$$\frac{du}{dt} = -\alpha\frac{\partial p}{\partial x} + fv + \alpha\frac{\partial \tau_x}{\partial z}$$

$$\frac{dv}{dt} = -\alpha\frac{\partial p}{\partial y} - fu + \alpha\frac{\partial \tau_y}{\partial z}$$

As per the analogy and assumptions of Ekman, the equations are as below.

Pressure gradient is balanced by geostrophic component under non accelerated motion. So

$$0 = -\alpha\frac{\partial p}{\partial x} + f\left[v_g + v_E\right] + \alpha\frac{\partial \tau_x}{\partial z}$$

$$0 = -\alpha\frac{\partial p}{\partial y} - f\left[u_g + u_E\right] + \alpha\frac{\partial \tau_y}{\partial z}$$

As per the geostrophic balance:

$$f\left[v_g\right] = \alpha\frac{\partial p}{\partial x}; f\left[u_g\right] = \alpha\frac{\partial p}{\partial y}$$

Hence Ekman's frictional equations are:

$$0 = f\left[v_E\right] + \alpha\frac{\partial \tau_x}{\partial z} \tag{6.8}$$

$$0 = -f\left[u_E\right] + \alpha\frac{\partial \tau_y}{\partial z} \tag{6.9}$$

The Ekman mass transport can be computed using equations (6.8) & (6.9)as

$$fv_E = -\frac{1}{\rho}\frac{\partial \tau_x}{\partial z}, \ \rho f v_E dz = -d\tau_x \tag{6.10}$$

$$fu_E = \frac{1}{\rho}\frac{\partial \tau_y}{\partial z}; \ \rho f u_E \, dz = d\tau_y \tag{6.11}$$

Now put $\rho v_E$ dz = $M_{yE}$ which is called as the mass transport flowing per second in the y-direction. Then integrate through the vertical area of thickness dz and one meter width along y-direction. Then the total mass transport along y-direction from 0 to z (=D, depth of frictional resistance) meters thickness can be written as:

$$M_{yE} = \int_z^0 \rho v_E dz \tag{6.12}$$

Similarly the Ekman mass transport per unit width in the x-direction as

$$Mx_E = \int_z^0 \rho u_E dz \tag{6.13}$$

Then combining the equations (6.10) & (6.12) can be written as

$$fMy_E = f\int_D^0 \rho v_E dz = -\int_D^0 d\tau_x = -\tau_{xo} + \tau_{xD} \tag{6.14}$$

Where z=0 is at the surface and z=D is at the frictional resistance. Then the current speed beyond the frictional resistance ($e^{-\pi} V_0 = 0.002V_0$) is negligibly small according to Ekman.

Similarly the mass transport in x direction is from equations (6.11) and (6.13) can be written as:

$$fM_{xE} = f\int_D^0 \rho u_E dz = -\int_D^0 d\tau_y = \tau_{yo} - \tau_{yD} \tag{6.15}$$

In the equations (6.14) & (6.15), $\tau_{xD}$ & $\tau_{yD}$ will vanish because there is no wind stress at the depth of frictional resistance.

$$fM_{yE} = -\tau_{x0} \tag{6.16}$$

$$fM_{xE} = \tau_{y0} \tag{6.17}$$

Since Ekman has taken a homogeneous ocean (ρ = constant) we can take ρ out of integration. Taking $Q_{xE}$ and $Q_{yE}$ as the volume transports we can write equations (6.14) & (6.15) as:

$$M_{yE} = \int_D^0 \rho v_E dz = \rho \int_D^0 v_E dz = \rho Q_{yE} \tag{6.18}$$

$$M_{xE} = \int_D^0 \rho u_E dz = \rho \int_D^0 u_E dz = \rho Q_{xE} \tag{6.19}$$

So from equations (6.16) and (6.18) we can write:

$f M_{yE} = f r Q_{yE} = - \tau_{x0}$

$$\text{Or } fQ_{yE} = -\frac{\tau_{x0}}{\rho} = -\alpha \tau_{x0} \tag{6.20}$$

Similarly, $f M_{xE} = f \rho Q_{xE} = - \tau_{y0}$

$$\text{Or } fQ_{xE} = \frac{\tau_{y0}}{\rho} = \alpha \tau_{y0} \tag{6.21}$$

Equations (6.20) and (6.21) are called volume transports and are measured in sverdrups ($1SV = 10^6 \, m^3 s^{-1}$).

As Ekman assumed wind along y direction, wind stress along x direction is zero. So according to equation (6.20)

$\tau_{x0} = 0 \Rightarrow Q_{yE} = 0$

According to equation (6.21), as wind stress exists in y direction,

$$Q_{xE} = \frac{\tau_{y0}}{f\rho} = \frac{\alpha}{f} \tau_{y0} \tag{6.22}$$

Equation (6.22) indicates the volume transport in x direction is related to the wind stress in the y-direction. This means the net transport within the layer between o to D is perpendicular to the wind stress in the Northern Hemisphere as shown in Fig.6.2 and vice-versa.

Thus the net Ekman mass transport is 90° to the direction of wind such that it is right hand side in the N.H and left hand side in the S.H as shown in Fig.6.2. As a result of this horizontal motion vertical motions develop to fill the hollow space. These vertical motions are due to Ekman Pumping.

## 6.5. Practical computation of Ekman Mass transport

Ekman mass transport $Q_{xE} = \dfrac{\tau}{\rho_w f} = \dfrac{\rho_a C_D u_a^2}{\rho_w f} m^2 s^{-1}$ ; But technically this is not a transport because it is in $m^2\ s^{-1}$. In order to get the real transport ($m^3\ s^{-1}$) we have to multiply by width 'W' (sum up horizontally along a section). Hence

$$\text{Mass transport} = U_{EK} = \frac{\tau}{\rho_w f} \times W = \frac{\rho_a C_D u_a^2}{\rho_w f} \times W\ m^3 s^{-1} \tag{6.22a}$$

But mass transport in the whole stretch (Length) is

$$M_{xW} = \frac{\tau}{f} = \left( \frac{\rho_a C_D u_a^2}{f} \times W \right) kgs^{-1} \tag{6.22b}$$

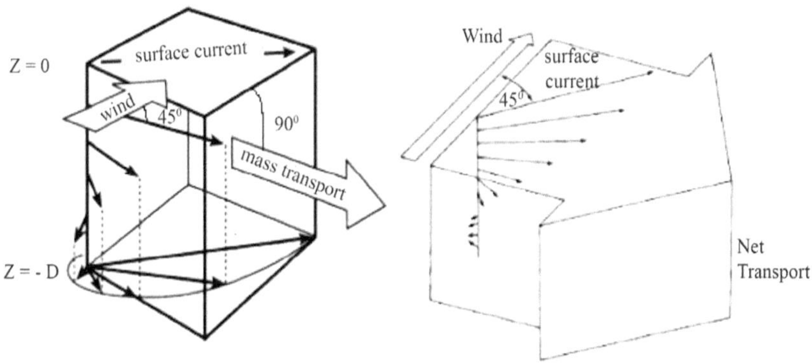


**Fig. 6.2.** Ekman net transport. Net transport is 90° to the wind (Pond and Pickard 1983& Neumann and Pierson 1966)

## 6.6. Interpretation of Ekman's results

i) At the sea surface ( where z = 0 ) the solutions of equations 6.5 and 6.6 become: $u = \pm V_o \text{ Cos } 45°$, $v = V_o \text{ Sin}45°$ (6.23)

which means that the surface current flows at an angle of 45° to the direction of wind speed such that right hand side of it in Northern Hemisphere(N.H) and left hand side of it in Southern Hemisphere (S.H)

ii) Below the surface, where z is no longer zero, the total current speed would be: $V_z = \sqrt{u^2 + v^2} = V_0 e^{\frac{-\pi z}{D}} \left[ \sin^2\left(\frac{\pi z}{D} + \frac{\pi}{4}\right) + \cos^2\left(\frac{\pi z}{D} + \frac{\pi}{4}\right) \right] = \pm V_0 e^{-\pi z/D}$. (6.24)

here +ve sign used for N.H and negative sign for S.H.). This also implies as z increases current speed decreases as z is in exponential function. That is, as z increases more in the negative z direction while the direction of velocity changes clockwise in the N.H and anticlockwise in the S.H.

iii) The direction of flow becomes opposite to that of the surface current at a depth z = -D where the speed falls as:

$$V_D = V_0.\exp\left(\pi\frac{-D}{D}\right) = V_0.\exp(-\pi) = 0.04.V_0 \quad (6.25)$$

such that $V_0$ is surface current. This implies that the current at the depth of frictional resistance is $e^{-\pi}$ (0.04) times of the surface current and it is just opposite to the surface current.

The depth $D_E$ is usually arbitrarily taken as the effective depth of the wind driven current or the Ekman layer. When viewed in a plane, the tips of the arrows of the current vector form a decreasing spiral called Ekman spiral as shown in Fig.6.1

## 6.7. Computation of Eddy coefficient of Viscosity

We know from equation 6.7, D = $\pi\sqrt{\frac{2A}{f\rho}}, \therefore \frac{D^2}{\pi^2} = \frac{2A}{f\rho}$

or $A = \frac{D^2 f\rho}{2\pi^2}$ kgm$^{-1}$ s$^{-1}$ (6.26)

If D = $10^2$ meters, f = $10^{-4}$ and ρ = 1025 kg/m$^3$=$10^3$, A will be equal to 50 kg m$^{-1}$S$^{-1}$

## 6.8. Estimation of wind factor and depth of frictional resistance

The ratio between the surface wind drift current and wind speed is known as 'wind factor' $\left(\frac{V_0}{W}\right)$. In order to obtain numerical relations between the surface current $V_0$, wind speed W and depth of frictional resistance D, Ekman used some experimental observations.

We know the surface current equation from equation (6.7) : $V_0 = \frac{\tau_y}{Aa} = \frac{\pi\tau_y\sqrt{2}}{Df\rho}$

We also know $\tau_y = \rho_a C_d W^2$ (6.26a)

where $\rho_a$ = density of air = 1.3 kg m$^{-3}$, $C_d$ = drag coefficient = $1.4\times10^{-3}$ for wind speeds less than 4 m/s and $2.4\times10^{-3}$ for wind speeds more than 4 m/s., W = wind speed in m/s. using these values for W < 4 m/s, $\tau_y = 1.8\times10^{-3}W^2$.

Substituting these values, we can write $V_0 = \frac{0.79\times10^{-5}W^2}{D.f}$ m/s, here ρ = density of water taken as 1025 kg m$^{-3}$

According to Ekman the wind factor was $\frac{V_0}{W} = \frac{0.0127}{\sqrt{\sin\phi}}$. This is applicable ±10$^0$ latitude from the equator. Substituting the value of $V_0$ in the wind factor comes to

$$\frac{V_0}{W} = \frac{0.79 \times 10^{-5} W}{D.f} = \frac{0.0127}{\sqrt{\sin \phi}}$$

$$\therefore D = \frac{4.3W}{\sqrt{\sin \phi}} \tag{6.27}$$

Therefore if we know 'W' at latitude φ we can calculate $V_0$ from equation (6.7) and the velocity at any depth below the sea surface from equation (6.24) and D from equation (6.27).

The variation of depth of frictional resistance with latitude and wind is as follows:

| | φ | $10^0$ | $45^0$ | $80^0$ | A($m^2$/s) |
|---|---|---|---|---|---|
| | $\frac{V_0}{W}$ | 0.03 | 0.05 | 0.013 | |
| W=10m/s | D (m) | 100 | 50 | 45 | 0.014 |
| W=20m/s | D(m) | 200 | 100 | 90 | 0.055 |

## 6.9. Ekman Depth

$D_{EK} = \pi \sqrt{\frac{2A}{f\rho}}$ *meters*; If Eddy viscosity (A) is about 0.05 $m^2s^{-1}$, Ekman layer depth is 20 to 60 m for latitudes $80^0$ to $10^0$.

## References

Academia: https://www.academia.edu/39969151/Chapter_III_Equations_of_ Motion_ in _ oceanography

Bryan, K.,1963: A numerical investigation of a nonlinear model of a wind driven ocean. J.Atmos.Sci, 20, 594-606.

Ekman,V. Walfrid. 1902. Om jordrotationensinverkanpii vindstrommari hafvet. Nyt. Msg. f. Naturvid,,v. 40, 1902. Kristiania.

Ekman V.W. 1905. On the influence of the Earth's rotation on ocean currents. Arkiv for Matematik, Astronomi, och Fysik: 2 (11).

Munk, W. H., and G. F. Carrier, 1950: The wind-driven circulation in ocean basins of various shapes. *Tel/us,* 2,158-167.

Nansen,Fridjof (1897): Farthest North, Volumes I & II.London, Archibald Constable & Co.

Neumann G., and W.J. Pierson. 1966. Principles of Physical Oceanography. New Jersey:Prentice-Hall.

Open University 1989. Ocean Circulation. Oxford: Pergamon Press,Oxford,.

Pond S & G.L.Pickard, 1983: Introductory Dynamical Oceanography,5th edition, Pergamon Press, 329pp

Semtner A. J., Jr., and R. M. Chervin, 1992: Ocean general circulation from a global eddy-resolving model. *J. Geophys. Res.,* 97, 5493-5550.

Stommel H. 1948. The westward intensification of wind-driven ocean currents. Transactions, American Geophysical Union 29 (2): 202–206.

Sverdrup, H. U., 1947: Wind driven currents in a baroclinic ocean; with applications to the equatorial currents of the eastern Pacific. Proc. Natl. Acad. Sci. USA, 33, 318–326, doi:10.1073/ pnas.33.11.318.

## Exercises

### *Exercise 6.1*

Calculate the current produced by a westerly wind of speed 10ms $^{-1}$ at $30^0$ N at the surface and half the depth of frictional resistance. Given the drag coefficient ($C_d$) = 1.4 × $10^{-3}$, density of air $\rho_a$ = 1.3 kg m $^{-3}$ and eddy coefficient of viscosity ($A_z$) = $10^{-1}$m $^2$ s$^{-1}$

### *Solution*

We know the equation for wind stress as: $\tau_y = \rho_a C_d W^2$ , Given $\rho_a$ = density of air = 1.3 kg m$^{-3}$, $C_d$ = drag coefficient = 1.4×$10^{-3}$, W= wind speed=10 m/s

i) Current produced by west wind at surface:

$\tau_y = \rho_a C_d W^2$= 1.3 × 1.4 × $10^{-3}$ ×$10^2$ = 182 × $10^{-3}$ = 0.182 kg m$^{-1}$s$^{-2}$ = 0.182 Pa

(Note: $10^3$mb = 1 bar = $10^5$Pascals = $10^6$ dynes/cm$^2$ )

We know from equation (6.7) $V_0 = \frac{\tau_y}{Aa} = \frac{\pi\tau_y\sqrt{2}}{Df\rho_w}$ and D = $\pi\sqrt{\frac{2A}{f}}$

$$V_0 = \frac{\tau_y}{Aa} = \frac{\sqrt{2}\rho_a C_d W^2}{\sqrt{\frac{2A}{f}}f\rho_w} = \frac{\tau_y}{\rho_w\sqrt{fA}}$$

$$\frac{\sqrt{2\times1.3\times1.4\times10^{-3}\times10^{2}}}{\sqrt{2\times{}^{1}\!/_{10}}\times\frac{1}{2\times7.3\times10^{-5}\times\text{Sin}\,30}\times1025\times2\times7.3\times10^{-5}\times\text{Sin}\,30} = 0.06m/s$$

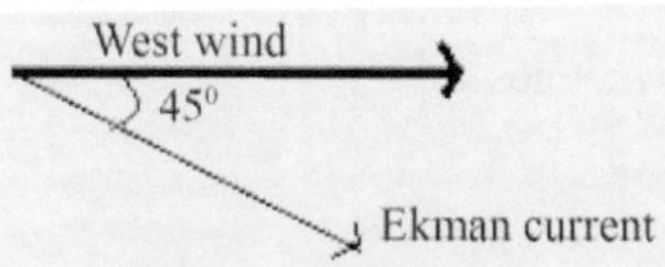


Substituting all the values and simplifying we get =0.06 m/s=6 cm/s. Since wind is west wind, the current at the surface should be $45^0$ right hand side in N.H to its direction which means $V_0$ will be from NW to SE.

ii) Current at half the depth of frictional resistance (D/2):

From equations (6.5) $u = V_0 e^{-\frac{\pi z}{D}} \cos\left(45 - \frac{\pi z}{D}\right)$, At half the depth (z = D/2)

$$u = V_0 e^{-\frac{\pi z}{D}} \cos\left(45 - \frac{\pi D/2}{D}\right),\ u = V_0 e^{-\frac{\pi z}{D}} \cos\left(45 - \frac{\pi}{2}\right)$$

Similarly from equation (6.6) we get $v = V_0 e^{-\frac{\pi z}{D}} \sin\left(45 - \frac{\pi}{2}\right)$,

According to equation 6.24,

$$V_z = \sqrt{u^2 + v^2} = V_0 e^{-\frac{\pi z}{D}} \left[\sin^2\left(\frac{\pi z}{D} + \frac{\pi}{4}\right) + \cos^2\left(\frac{\pi z}{D} + \frac{\pi}{4}\right)\right] = \pm V_0 e^{-\pi z/D}$$

$$\mathrm{v}_{D/2} = \pm V_0 e^{-\pi/2} = \frac{0.06}{4.8} = 0.01 m/s = 1 cm/s$$

The direction of this current at D/2 is perpendicular to west wind. So it is towards south.

### *Exercise 6.2*

Suppose in a certain region the geostrophic motion is westerly at 10m/s. Compute the net cross isobaric mass transport in the Ekman layer. Take $\rho = 1025 \text{ kg m}^{-3}$, A(Eddy coefficient) = $5 \text{ m}^2/\text{s}$ and $f = 10^{-4}/\text{s}$.

### *Solution*

The net cross isobaric Ekman Mass transport in the Ekman layer 0 to D is

$$M_y = \int_0^D \rho v dz = \int_0^D \rho u_g \left[e^{-az} Sin az\right] dz$$

Put $I = \int \sin az . e^{-az} dz$

$$M_y = \rho u_g I$$

$$= -\frac{1}{a} sin az e^{-az} + \left[cos az \frac{e^{-az}}{-a} - \int \frac{e^{-az}}{-a} sin az.dz\right]$$

$$\therefore 2I = -\frac{1}{a} e^{-az}\left[\sin az + \cos az\right] \text{ or } I = -\frac{1}{2a} e^{-az}\left[\sin az + \cos az\right]$$

$$M_y = -\frac{\rho u_g}{2a} e^{-az}\left[\sin az + \cos az\right], \text{ putting Z = D}$$

$$M_y = -\frac{\rho u_g}{2a}\left\{e^{-aD}\left[\sin aD + \cos aD\right] - e^0 (0+1)\right\}$$

Taking D=100m from table or compute from equation (6.18)

$$M_y = -\frac{\rho u_g}{2a}\left\{e^{-aD}\left[\sin aD + \cos aD\right] - e^0 (0+1)\right\}$$

we know $\frac{f\rho}{A} = a^2$, $a = \sqrt{\frac{10^{-4} \times 1025}{5}} = 0.0205$

Substituting, $\frac{1025 \times 10}{2 \times 0.0205}\left[e^{-0.0205\times100}(\{\sin(0.0205\times100)+\cos(0.0205\times100)-e^0(0+1)\right]$

250000{7.767[1.035-1]} = 67961.25 = 67.96125 × $10^3$ kg $m^{-1}s^{-2}$

*Exercise 6.3*

**Suppose that we measure a steady, ocean surface speed of 10 cm s -1 and a wind stress of 1 dyn cm-2. Assume that the water density, ρ = 1.0 gm cm-3, and the Coriolis parameter f = 10- 4 s-I. What is the value of the coefficient of eddy viscosity in cgs and mks units?**

*Solution*

We know the relation from equation (6.7):

Here $V_0 = \frac{\tau_y}{Aa} = \frac{\pi\tau_y\sqrt{2}}{Df\rho_w}$ and $D = \pi\sqrt{\frac{2A}{f\rho_w}}$

We can write: by substituting the given values: $V_0 = \frac{\pi\tau_y\sqrt{2}}{Df\rho}$

$$10\,cms^{-1} = \frac{\pi \times 1\text{dyn}\;cm^{-2} \times \sqrt{2}}{D \times 10^{-4}\,s^{-1} \times 1gmcm^{-3}}$$

$$\therefore D = \frac{\pi \times 1\text{dyn}\;cm^{-2} \times \sqrt{2}}{10\,cms^{-1} \times 10^{-4}\,s^{-1} \times 1gmcm^{-3}} = 4443\,cm \approx 44.43m$$

$$D = \pi\sqrt{\frac{2A}{f\rho}}$$

$$\rightarrow 4443cm = \pi\sqrt{\frac{2A}{f\rho}} \rightarrow A = \frac{(4443cm)^2 \times 10^{-4}s^{-1} \times 1gmcm^{-3}}{2\pi^2} = \frac{1974 \times 10^4 \times 10^{-4}s^{-1}}{19.74} = 1 \times$$

$\times 10^2\,gcm^{-1}s^{-1}$

$\therefore A = 100\,gcm^{-1}s^{-1}$ in cgs units;

A in MKS units $\frac{1 \times 10^2}{10^3} \times 10^2\,s^{-1} = 10\,kgm^{-1}s^{-1}$

### *Exercise 6.4*

To what zonally integrated sea level difference does it correspond in the equatorial Pacific for a surface Ekman frictional layer depth of 100m with a homogeneous density of $\rho_0$ = 1025 kg m$^{-3}$ with a balancing trade wind magnitude of $u_{10}$ = -5m/s. Take $\rho_a$ = 1kgm$^{-3}$;g = 10 m$^2$ s$^{-1}$; $C_d$ = 2 × 10$^{-3}$ and width of the Equatorial Pacific is 8000 km.

***Solution:***

The zonal momentum equation applicable in the case of the wind stress induced flow is (ignoring geostrophic effect) (from equation (8.2): $\alpha\frac{\partial p}{\partial x} = f.v + \alpha\frac{\partial \tau_x}{\partial z}$

$$-\frac{1}{\rho_0}\frac{\partial p}{\partial x} + \frac{1}{\rho_0}\frac{\partial \tau_x}{\partial z} = 0 \rightarrow -g\frac{\partial \eta}{\partial x} + \frac{\partial}{\partial z}\left(\frac{\tau_x}{\rho_0}\right) = 0$$

on integrating within the frictional surface layer(H):

$$gH\frac{\partial \eta}{\partial x} = \frac{\tau_x}{\rho_0} \rightarrow \Delta\eta = \frac{\tau_x \times \Delta x}{\rho_0 \times gH}$$

$$= \frac{\rho_a \times C_d \times u_{10}^2 \times \Delta x}{\rho_0 \times gH} = \frac{1kgm^{-3} \times 2\times10^{-3} \times \left(-5ms^{-1}\right)^2 \times 8\times10^6 m}{1025\,kg\,m^{-3} \times 10ms^{-2} \times 100m} = 400\times10^{-3} = 0.4\,m = 40\,cm$$

### *Exercise 6.5*

If a northerly wind of -10 m/s blows along California coast and that the Ekman theory holds at L=*100km* off the coast, Calculate the Ekman volume transport at the distance across the section width W=100 km within this coastal box. Take $\rho_a$= 1kgm$^{-3}$; g = 10 m$^2$ s$^{-1}$; $C_d$= 2 ×10$^{-3}$; $\rho_0$= 1000 kg m$^{-3}$; $f_0$= 10$^{-4}$ s$^{-1}$

### *Solution*

We know from equation 8.2: $\alpha\frac{\partial p}{\partial x} = f.v + \alpha\frac{\partial \tau_x}{\partial z}$
Chapter 8

$$-\frac{1}{\rho_0}\frac{\partial p}{\partial x} + \frac{1}{\rho_0}\frac{\partial \tau_x}{\partial z} = 0 \rightarrow -g\frac{\partial \eta}{\partial x} + \frac{\partial}{\partial z}\left(\frac{\tau_x}{\rho_0}\right) = 0$$

on integrating within the frictional surface layer(H):

We can get Ekman volume transport

$$T_E = \frac{\rho_a \times C_d \times u_{10}^2 \times W}{\rho_0 \times f_0} = \frac{1kgm^{-3} \times 2\times10^{-3} \times \left(-10ms^{-1}\right)^2 \times 10^5 m}{1000\,kg\,m^{-3} \times 10^{-4} s^{-1}}$$

$$= 2\times10^5 m^3 s^{-1} = 0.2\times10^6 m^3 s^{-1} = 0.2\,Sv$$

### Exercise 6.6

(1) Off the coast of Rhode Island a 4-knot wind blows from the west. (a) To what depth does the surface Ekman current extend?(b) What is the speed of the surface wind-generated Ekman current?

(2) Plot Ekman depth DE versus φ (latitude!) from 10° N - 50° N for wind speeds of 5, 10 and 20ms$^{-1}$ wind speeds. Put all three plots on the same graph. What do the graphs show you?

### Solution

$$u = V_0 e^{-\frac{\pi z}{D}} \cos\left(45 - \frac{\pi z}{D}\right) V_0 = \frac{\tau_y}{Aa} = \frac{\pi \tau_y \sqrt{2}}{Df\rho}; \tau_y = \rho_a C_d W^2; D_{EK} = \pi\sqrt{\frac{2A}{f\rho}} \text{ meters}$$

A= 0.01 or 0.05 m$^2$s$^{-1}$

where $\rho_a$ = density of air = 1.3 kg m$^{-3}$, $C_d$ = drag coefficient = 1.4×10$^{-3}$ for wind speeds less than 4 m/s and 2.4×10$^{-3}$ for wind speeds more than 4 m/s., W = wind speed in m/s. Using these values for W < 4 m/s, $\tau_y$ = 1.8×10$^{-3}$W$^2$. ρ = density of water taken as 1025 kg m$^{-3}$ or 1.0 g cm$^{-3}$.

Latitude of Rhode island 41.58$^0$ N; w = wind speed = 4kts = 2 ms$^{-1}$

a 1)

$$D_{EK} = \pi\sqrt{\frac{2A}{f\rho}} = \pi\sqrt{\frac{2\times 0.05m^2s^{-1}}{2\times 7.3\times \sin\left(41.58^0\right)s^{-1}\times 1025kgm^{-3}}}$$

$$= \pi\sqrt{\frac{0.1}{9.7\times 1025kgm^{-3}}} = \sqrt{0.01} = 0.1m$$

2) $$V_0 = \frac{\sqrt{2}\pi\rho_a C_d w^2}{Df\rho} = \frac{\sqrt{2}\pi\times 1.3kgm^{-3}\times 1.4\times 10^{-3}\times\left(2ms^{-1}\right)^2}{0.1m\times 2\times 7.3\times \sin\left(41.58^0\right)s^{-1}\times 1025kgm^{-3}} = 0.01ms^{-1}$$

b) Ekman depth $D_E$ versus φ :

From equation 6.27 we know: $\therefore D_E = \frac{4.3W}{\sqrt{\sin\phi}}$

The variation of depth of frictional resistance with latitude and wind is as follows:

| | φ | 10$^0$ | 30$^0$ | 45$^0$ | 50$^0$ | 80$^0$ | A(m$^2$/s) |
|---|---|---|---|---|---|---|---|
| W= 5 m/s | D(m) | 52 | 30 | 26 | 25 | 22 | 0.014 |
| W=10m/s | D (m) | 100 | 60 | 50 | 49 | 45 | 0.014 |
| W=20m/s | D(m) | 200 | 121 | 100 | 98 | 90 | 0.055 |

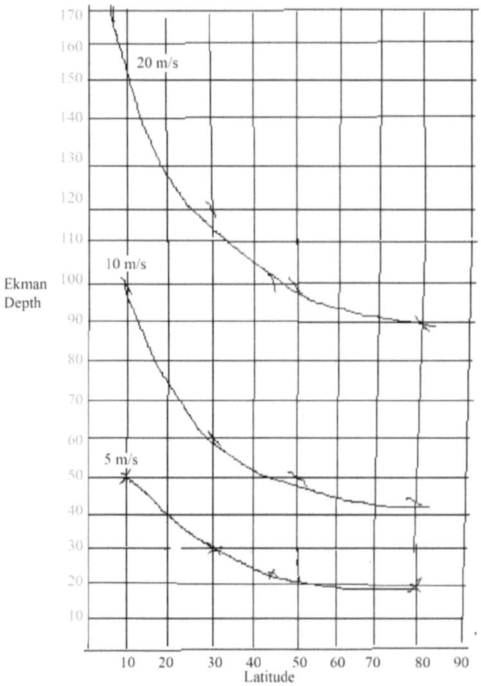


The graph shows with increase of latitude Ekman depth decreases for a particular wind and also with increase of wind Ekman depth increases at particular latitude.

## *Exercise 6.7*

During much of the year, a steady wind blows from the south along the northern New Jersey($\phi \approx 40^0$ N) coast which is oriented north-south at 40° N. Assume Ekman motion and $\rho_\omega = 1.0$ g cm$^{-3}$.(a) If the wind generates a surface stress of 2 dynes cm$^{-2}$, what is the Ekman transport along a 1-km stretch of the beach (Take $f = 10^{-4}$).

### *Solution*

We know from equation (6.22 a and b) as follows: $U_{EK} =$

$$\frac{\tau}{\rho_w f} \times W = \frac{\rho_a C_D u_a^2}{\rho_w f} \times W\ cm^3 s^{-1}$$

$$U_{EK} = \frac{\tau}{\rho_w f} \times W m^3 s^{-1} = \frac{2\, dynes cm^{-2} \times 10^5 cm}{1 gcm^{-3} \times 2 \times 7.3 \times sin 40 \times 10^{-5} s^{-1}} = \frac{2 gcms^{-2} cm^{-2} \times 10^5 cm}{1 gcm^{-3} \times 0.94 \times 10^{-5} s^{-1}}$$

$$= 2.13 \times 10^{10} cm^3 s^{-1} = 0.002 Sv \left(\because 10^{12} cm^3 s^{-1} = 1 Sv\right)$$

# 7

# Upwelling and Down Welling

Wind stress at the sea surface not only causes horizontal movement of water (Ekman Net transport), but also leads to vertical motion (Ekman Pumping). This horizontal movement of water leads to the occurrence of upwelling and down welling throughout the oceans both at coastal boundaries as well as in the open ocean. As divergence of surface water causes rising, convergence of the surface water causes sinking. This rising and sinking of water are called upwelling and down welling. While coastal upwelling occurs due to Ekman pumping, open ocean upwelling causes due to divergences caused due to currents or cyclonic motions in the oceans. Vertical motions have special significance because of their significant effects on biological productivity on the oceanic environment, though their speeds are much less compared to the horizontal currents.

According to Smith (1964), upwelling is defined as 'an ascending motion of some minimum duration and extent, by which water from subsurface layers is brought to the surface layer and is removed from that area by horizontal flow.

Though upwelling can occur anywhere in the oceans, but it is maximum and significant in the eastern shores (west coasts of continents) of the oceans.

## 7.1. Effects of Upwelling

i) The temperature decreases as the cooler subsurface water rises to the sea surface.

ii) As the subsurface water is denser than the surface water, increase of density results.

iii) Increase in the concentration of nutrients like phosphates, nitrates and silicates as subsurface water is not utilized due to less population.

iv) Decrease in the concentration of dissolved oxygen

v) Decrease of the thickness of the mixed layer or in other words cropping up of thermocline to naviface as shown in Fig.7.1a

vi) The mean sea level is lowered in the upwelling zone as cold subsurface water occupies lesser volume than the warm surface water.

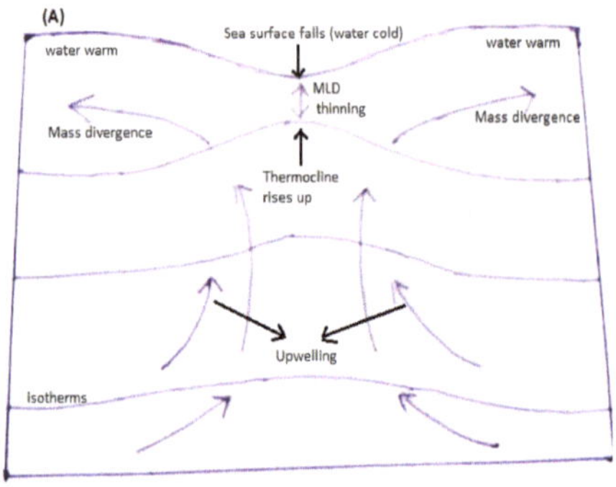


**Fig.7.1a.** Upwelling

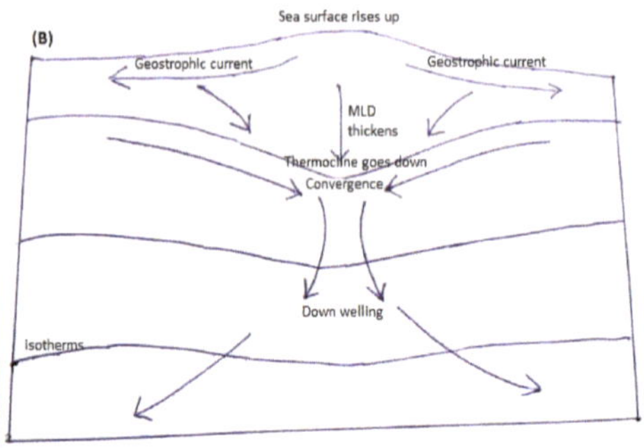


**Fig. 7.1b.** Down welling

## 7.2. Dynamics of Upwelling

Consider a reduced gravity ocean on f-plane (f > 0) bounded by a vertical wall and subjected to the surface stress acting with the wall on its left. The lower layer is infinitely deep and motionless. In the absence of longshore variations, the equations of motion are:

$$\frac{du}{dt} = -\alpha\frac{\partial p}{\partial x} + fv \tag{7.1}$$

This equation on expansion can be written as :

$$\frac{\partial u}{\partial t} + u\frac{\partial u}{\partial x} + v\frac{\partial u}{\partial y} + w\frac{\partial u}{\partial z} = -\alpha\frac{\partial p}{\partial x} + fv \tag{7.2}$$

Taking $u\frac{\partial u}{\partial x} + v\frac{\partial u}{\partial y} + w\frac{\partial u}{\partial z} = 0$, Equation (7.2) becomes $\frac{\partial u}{\partial t} - fv = -\alpha\frac{\partial p}{\partial x}$ (7.3)

We know $\alpha\partial p = g\partial z = g\partial h$,

So Equation (7.3) can be written as

$$\frac{\partial u}{\partial t} - fv = -g\frac{\partial z}{\partial x} = -g\frac{\partial h}{\partial x} \tag{7.4}$$

Similarly the y component can be written as: $\frac{\partial v}{\partial t} + u\frac{\partial v}{\partial x} + v\frac{\partial v}{\partial y} + w\frac{\partial v}{\partial z} = -\alpha\frac{\partial p}{\partial x} - fu$

$$\frac{\partial v}{\partial t} + fu = -g\frac{\partial z}{\partial y} = -g\frac{\partial \zeta}{\partial y} \tag{7.5}$$

$\frac{\partial v}{\partial t} = 0$, then equation (7.5) will become under non accelerated condition:

$$fu = -g\frac{\partial \zeta}{\partial y} \tag{7.6}$$

Multiplying both sides by ρ and integrating equation (7.6) in a thick ness of 'dz' we get,

$$\int_0^h \rho u dz = \frac{1}{f}\int_0^h \rho g \frac{\partial \zeta}{\partial y} dz = -\frac{\rho g h}{f}\frac{\partial \zeta}{\partial y} \tag{7.7}$$

From Ekman mass transport (equation 6.17) we know

$$M_x = \frac{\tau_{yo}}{f} = \int_0^h \rho u dz = \frac{\rho g h}{f}\frac{\partial \zeta}{\partial y} \tag{7.8}$$

So equation (7.8) can be written as:

$$g\frac{\partial \zeta}{\partial y} = \frac{\tau_{yo}}{\rho h} \tag{7.9}$$

Substituting equation (7.9) in equation (7.5)

$$\frac{\partial v}{\partial t} + fu = \frac{\tau_{yo}}{\rho h} \tag{7.10}$$

Equation of continuity can be written as

$$\frac{\partial h}{\partial t} + \frac{\partial}{\partial x}(hu) = 0 \tag{7.11}$$

The required equations are (7.4), (7.10) &(7.11) where x is offshore coordinate and $\tau_{yo}$ is long shore meridional wind stress at sea surface. As these equations are nonlinear, they are to be linearized to find the solution. So the following assumptions are taken.

Assumptions:

i) The wind stress τ and in turn, the ocean's reaction are weak.

ii) h = H ± η where H is the depth of the undisturbed surface layer and η is the disturbed sea surface as shown in Fig.7.2.

As in upwelling sea level falls, η is taken as negative

$$\therefore h = H - \eta \tag{7.12}$$

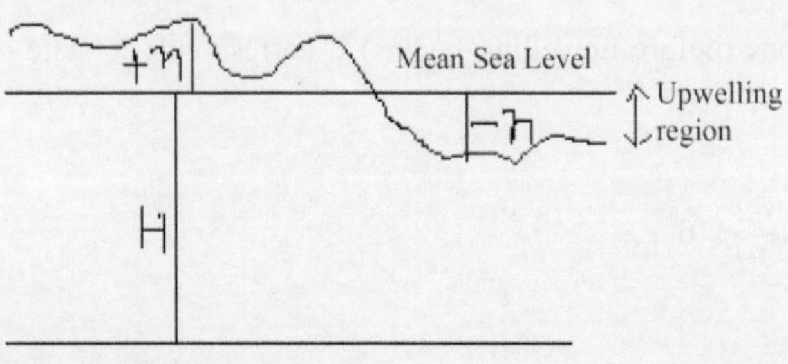


**Fig.7.2.** Fall of sea level (-η) is taken here as sea level falls in upwelling area)

So substituting equation (7.12) in (7.4) and writing:

$$\frac{\partial u}{\partial t} - fv = -g\frac{\partial}{\partial x}(H-\eta) = -g\frac{\partial H}{\partial x} + g\frac{\partial \eta}{\partial x} = g\frac{\partial \eta}{\partial x}$$

Since H is constant due to undisturbed sea surface $\frac{\partial H}{\partial x}$ =0

So $\frac{\partial u}{\partial t} - fv = g\frac{\partial \eta}{\partial x}$ (7.13)

Substituting equation (7.12) in equation (7.10):

$$\frac{\partial v}{\partial t} + fu = \frac{\tau_{yo}}{\rho}\left[\frac{1}{h}\right] = \frac{\tau_{yo}}{\rho}[H-\eta]^{-1}$$

Expanding and neglecting higher order terms in the inverse

$$\frac{\partial v}{\partial t} + fu = \frac{\tau_{yo}}{\rho}[H-\eta]^{-1} = \frac{\tau_{yo}}{\rho}\left[H^{-1} - (-1)c_1H^{-2}(-\eta^{-1}) + (-1)c_2H^{-3}(-\eta^{-2}) + \dots\dots\right]$$

Neglecting the 2[nd] term onwards in the brackets as they are negligible small:

$$\frac{\partial v}{\partial t} + fu = \frac{\tau_{yo}}{\rho H} \quad (7.14)$$

Substituting equation (7.12) in (7.11):

$$\frac{\partial h}{\partial t} + \frac{\partial}{\partial x}(hu) = 0$$

$\frac{\partial}{\partial t}(H-\eta) + \frac{\partial}{\partial x}(H-\eta)u = 0$ on expanding

$$\frac{\partial h}{\partial t} - \frac{\partial \eta}{\partial t} + u\frac{\partial H}{\partial x} + H\frac{\partial u}{\partial x} - u\frac{\partial \eta}{\partial x} - \eta\frac{\partial u}{\partial x} = 0 \quad (7.15)$$

Applying the following boundary conditions to equation (7.15):

i) 'u' vanishes at the coast (x = 0)
ii) 'η' vanishes at far offshore beyond upwelling zone (x → ∞)
iii) 'H' is constant
iv) At initial conditions (before upwelling begins) it corresponds to state of rest, which means

 (u = v = η = 0)

Hence $\frac{\partial h}{\partial t} = \frac{\partial H}{\partial x} = \frac{\partial \eta}{\partial x} = \frac{\partial u}{\partial x} = 0$

Then equation 7.15 becomes

$$-\frac{\partial \eta}{\partial t}+H\frac{\partial u}{\partial x}=0 \tag{7.16}$$

Thus the final equations to be solved are (7.13), (7.14) and (7.16). Kojo Yoshida (1967) solved these equations - considering them as periodic in time.

$$\{\frac{\partial u}{\partial t} fv = g\frac{\partial n}{\partial x} \quad ....(7.13) \quad \frac{\partial v}{\partial t} + fu = \frac{\partial n}{\partial t} \quad ...(7.14) \quad -\frac{\delta n}{\delta t} = H\frac{\partial u}{\partial x} = 0 \quad ...(7.16)$$

According to Yoshida (1967) the solutions of these equations are:

$$u = u_0 (x) \sin \omega t \tag{7.17}$$

$$v = v_0 (x) \cos \omega t \tag{7.18}$$

$$\eta = \eta_0 (x) \cos \omega t \tag{7.19}$$

$$\tau_{y0} = \tau_0 \sin \omega t \tag{7.20}$$

where $\tau_0$ = constant due to uniform wind stress

Differentiate equation (7.17) with respect to (w.r.t) 't':

$$\frac{\partial u}{\partial t} = u_0(\omega)\cos \omega t \tag{7.21}$$

Differentiate equation ((7.19) with respect to (w.r.t) 'x':

$$\frac{\partial \eta}{\partial x} = \frac{\partial \eta_0}{\partial x}\cos \omega t \tag{7.22}$$

Substitute equations (7.21), (7.22) & (7.18) in (7.13)

$$u_0(\omega)\cos \omega t - fv_0 \cos \omega t = g\frac{\partial}{\partial}\cos \omega t$$

$$u_0(\omega) - fv_0 = g\frac{\partial \eta_0}{\partial x} \tag{7.23}$$

Differentiate equation (7.18) with respect to (w.r.t) 't':

$$\frac{\partial v}{\partial t} = v_0(-\omega)\sin \omega t \tag{7.24}$$

Substitute equations (7.24), (7.17) & (7.20) in (7.14):

$$-v_0(\omega)\sin \omega t + fu_0 \sin \omega t = \frac{1}{\rho H}[\tau_0]\sin \omega t$$

$$-v_0(\omega) + fu_0 = \frac{\tau_0}{\rho H} \tag{7.25}$$

Differentiate equation (7.19) with respect to (w.r.t) ‘t’:

$$\frac{\partial \eta}{\partial t} = \eta_0(-\omega)\sin \omega t \tag{7.26}$$

Differentiate equation (7.17) with respect to (w.r.t) ‘x’:

$$\frac{\partial u}{\partial x} = \frac{\partial u_0}{\partial x}\sin \omega t \tag{7.27}$$

Substitute equations (7.26), (7.27) in (7.16):

$$-(-\omega\eta_0)\sin \omega t + H\frac{\partial u_0}{\partial x}\sin \omega t = 0$$

$$(\omega\eta_0) + H\frac{\partial u_0}{\partial x} = 0 \tag{7.28}$$

Now the equations to be solved are (7.23), (7.25)& (7.28):

$$u_0(\omega) - fv_o = g\frac{\partial \eta_0}{\partial x}\ (7.23);\ -v_0(\omega) + fu_o = \frac{\tau_0}{\rho H}\ (7.25);\ (\omega\eta_o) + H\frac{\partial u_0}{\partial x} = 0\ (7.28)$$

To solve these equations, first differentiate equation (7.28) with respect to ‘x’ and then substitute in (7.23). Then solve the equations (7.23) and (7.25) for $u_0$& $v_0$. Then we can get:

$$u_0 = \frac{1}{(f^2-\omega^2)}\left[\frac{\tau_0 f}{\rho H} + gH\frac{\partial^2 u_0}{\partial x^2}\right] \tag{7.29}$$

$$v_0 = \frac{1}{(f^2-\omega^2)}\left[\frac{\tau_0 \omega}{\rho H} + \frac{gH}{\omega}\frac{\partial^2 u_0}{\partial x^2}\right] \tag{7.30}$$

Further solving we can get (by substituting 7.29 & 7.30 in 7.17 & 7.18):

$$u = \frac{f\tau_0}{\rho H(f^2-\omega^2)}\left[1 - e^{\frac{-x}{R_\omega}}\right]\sin \omega t \tag{7.31}$$

$$v = \frac{\omega\tau_0}{\rho H(f^2-\omega^2)}\left[1 - \frac{f^2}{\omega^2}e^{\frac{-x}{R_\omega}}\right]\cos \omega t \tag{7.32}$$

$$\eta = \frac{-fR_\omega\tau_0}{\rho g H\omega}\left[e^{\frac{-x}{R_\omega}}\right]\cos \omega t \tag{7.33}$$

Where $R_\omega$ is a modified deformation radius

$$R_\omega = \left[\frac{gH}{(f^2-\omega^2)}\right]^{1/2} \tag{7.34}$$

Results:

i) From equation (7.34), we can understand that the upwelling or downwelling signal is trapped along the coast with in a distance of $R_\omega$

ii) Far offshore ($x \rightarrow \infty$) the Ekman transports are given by the equations (7.29)& (7.30)

iii) In equation 7.29, as ($x \rightarrow \infty$), ($\frac{\partial^2 u_0}{\partial x^2} \rightarrow 0$), and when ($\omega$<<f). $\omega$ can be very small and can be ignored in the inertial frequency ($f^2 - \omega^2$), but not in the harmonic mode

So $u_0 = \frac{\tau_0}{\rho f H} \text{Sin}\,\omega t$

iv) With the same analogy, Equation 7.30 becomes

$v_o = 0$

Cases:

i) When ($\omega$<<f), it is called sub-inertial frequency. At long periods such as months, the distance $R_\omega$ (equation 7.34) becomes:

$$R_\omega = \left[\frac{gH}{(f^2)}\right]^{1/2} = \frac{\sqrt{gh}}{f}$$

which is nothing but Rossby radius of deformation

ii) When ($\omega$>>f), it is called supra-inertial frequency, then the radius $R_\omega$ becomes imaginery. This means that the solution does not decay away from the coast but continuously oscillates as a wave. Physically the ocean's response is not trapped.

## Alternative discussion on upwelling

We know the equations of motion with friction as:

$$\frac{du}{dt} = -\alpha\frac{\partial p}{\partial x} + fv + \alpha\frac{\partial \tau_x}{\partial z}, \frac{dv}{dt} = -\alpha\frac{\partial p}{\partial y} - fu + \alpha\frac{\partial \tau_y}{\partial z} \quad (7.35)$$

these can be written as

$$-fv = -g\frac{\partial \zeta}{\partial x} + A_x\frac{\partial^2 u}{\partial x^2} + A_z\frac{\partial^2 u}{\partial z^2}; fu = -g\frac{\partial \zeta}{\partial y} + Ax\frac{\partial^2 v}{\partial x^2} + Az\frac{\partial^2 v}{\partial z^2} \quad (7.36)$$

The equation of continuity can be written as (Laplace equation,): $\frac{\partial u}{\partial x} + \frac{\partial w}{\partial z} = 0$ ….(7.37)

On scaling we know the depth of frictional resistance from Ekman circulation equation (6.26) can be written as

$$\therefore D_x = \left[\frac{A_x}{f}\right]^{\frac{1}{2}}, \quad D_y = \left[\frac{A_y}{f}\right]^{\frac{1}{2}} \tag{7.38}$$

Where $D_x$ is horizontal mixing length and $D_y$ is the Ekman layer depth.

Hidaka (1954) assumed for upwelling current as:

i) Water surface has no slope parallel to the coast so that $-g\frac{\partial \zeta}{\partial y} = 0$ (7.39)
ii) Infinitely deep water
iii) Constant wind stress acting over the areas extending from the coast to a distance $2\pi D_x$

The most significant features of the solution are:

a) the offshore mass transport takes place in a surface layer of thickness $D_y$
b) upwelling takes place within a zone adjacent to the coast of width $D_x$
c) the ratio of vertical horizontal velocity to that of $D_y$ to $D_x$ is

$$\frac{w}{u} = \frac{D_y}{D_x} = \left[\frac{A_y}{A_x}\right]^{1/2} \tag{7.40}$$

If $\frac{A_y}{A_x}$ is taken as $10^{-6}$ and u = 1 cm/sec then w = $10^{-3}$ cm/sec, then $\frac{D_x}{D_y} = 10^3$ which means if $D_y$ is taken as 20 m then $D_x$ = 20 km.

## 7.3. Steady state homogeneous water model (Garvine model)

Garvine (1971) retained the term $\frac{\partial \zeta}{\partial y}$ and considered water of uniform depth H which was large compared to $D_y$. He also included a finite $\frac{\partial \zeta}{\partial y}$ term which leads to the following arguments;

i) beyond a narrow coastal boundary layer, the frictional terms on the right hand side of equation (7.6) can be neglected, then

   $fu = -g\frac{\partial \zeta}{\partial y}$ ...........(7.41)

ii) Outside the coastal boundary layer the net transport offshore across any vertical section parallel to the coat is zero so that $M_x = \int_{-H}^{0} \rho u dz = 0$ (7.42)

iii) In the surface layer the offshore Ekman transport from equation (6.17) is

$$M_x = \frac{\tau_{sy}}{f} \tag{7.43}$$

iv) The transport due to geostrophic flow though out the depth is from equation (7.41) & (7.42) and (7.43) can be written as

$$M_x = \frac{\tau_{sy}}{f} = \int_{-H}^{0} -\rho \frac{g}{f}\frac{\partial \zeta}{\partial y} dz = \frac{\rho g H}{f}\frac{\partial \zeta}{\partial y} \tag{7.44}$$

$$\therefore \frac{\partial \zeta}{\partial y} = \frac{\tau_{sy}}{\rho g H} \tag{7.45}$$

v) Further analysis of Garvine showed $M_y = \int_{-H}^{0} \rho v dz = 0$ (7.46)

This means that below the surface current parallel to the coast in the direction of wind, there must be a compensating under current flowing in the opposite direction.

## Example

i) if $\tau_{sy} = 0.2\ Nm^{-2}$, $H = 100m$, $g = 9.8\ m\ s^{-2}$, $\rho = 1025\ kg\ m^{-3}$ then the longshore slope from equation (7.45), $\therefore \frac{\partial \zeta}{\partial y} = \frac{\tau_{sy}}{\rho g H} = \frac{0.2Nm^{-2}}{1025 \times 9.8 \times 100m} = 2 \times 10^{-7}$ = 2 cm in 100 km

ii) At latitude $30^0$ from equation (7.41), $u = -\frac{g}{f}\frac{\partial \zeta}{\partial y} = -\frac{10}{2 \times 7.29 \times 10^{-5} \times \frac{1}{2}} \times 2 \times 10^{-7}$ = 2.7 cm/s

iii) If the offshore Ekman transport is confined to the upper 20m, the average offshore velocity in the layer would be about 11 cm/s

## 7.4. Estimation of Vertical upwelling velocity

We know the equation of continuity as: $\frac{\partial u}{\partial x} + \frac{\partial v}{\partial y} + \frac{\partial w}{\partial z} = 0$ (7.47)

As the meridional velocity is uniform (v = constant) parallel to the coast, $\frac{\partial v}{\partial y} = 0$

$$\therefore \frac{\partial u}{\partial x} = -\frac{\partial w}{\partial z} \ or \ \therefore du = -\frac{dw}{dz} dx \tag{7.48}$$

The velocity across the vertical column at a distance 'L' from the coast can be obtained by integrating equation (7.48):

$$u_L = \int du_L = -\int_0^L \frac{dw}{dz} dx \tag{7.49}$$

The mass transport in the layer of thickness ( z = H to 0) from equation (7.49) can be written as:

$$M_x = \int_H^0 \rho_w u_L dz = -\int_H^0\int_0^L \rho_w \frac{dw}{dz} dxdz = -\int_H^0\int_0^L \rho_w dwdx = -\int_0^L \rho_w w_H dx = -\rho_w w_H L$$

Hence the vertical upwelling velocity is $w_H = -\dfrac{M_x}{\rho_w L}$ (7.50)

From equations (7.50) and (7.43) and (6.26a) of Ekman circulation we can write the vertical upwelling velocity as:

$$w_H = -\frac{M_x}{\rho_w L} = -\frac{\tau_y}{f\rho_w L} = \frac{-\rho_a C_D W^2}{f\rho_w L} \tag{7.51}$$

where $\rho_w$ is the density of sea water. The negative sign indicates the rising of water from below in the case of upwelling. Note here that at z = H, w = $w_H$ and at z=0, w=0 because vertical upwelling velocity at surface is zero.

## 7.5. Practical computation of Ekman Mass transport & upwelling

We know Ekman mass transport from equation (7.8) as ; $M_x = \dfrac{\tau_y}{f} = \dfrac{\rho_a C_D u_a^2}{f}$, but its units are kgm$^{-1}$ s$^{-1}$. So it should be converted to volume units by dividing with sea water density like this:

Ekman transport from equation 6.21 and section 6.5 $M_x = \dfrac{\tau}{\rho_w f} = \dfrac{\rho_a C_D u_a^2}{\rho_w f} m^2 s^{-1}$ ; But technically this is not a transport because it is in m$^2$s$^{-1}$. In order to get the real transport (m$^3$ s$^{-1}$) we have to multiply by width 'W' (sum up horizontally along a section). Hence

$$T = \frac{}{\rho_w f}\times W = \frac{{}_a C_D u_a}{\rho_w f}\times W\, m^3 s^{\ 1} \tag{7.52a}$$

But mass transport in the whole stretch (Length) is

$$M_{xW} = \frac{\tau\times W}{f} = \left(\frac{\rho_a C_D u_a^2}{f}\times W\right) kgs^{-1} \tag{7.52b}$$

Average Upwelling velocity = $W_h = -\dfrac{T_E}{WL}$ (7.53a)

Thus the actual upwelling velocity in proper units should be

$$w_H = \frac{M_x}{\rho_w L} = \frac{\tau_y}{\rho_w Lf} = \frac{\rho_a c_D u_a^2}{\rho_w Lf} \tag{7.53b}$$

For example for a wind speed of 10 m/s, width of ocean (W)=100 km and Length of travel is 100km, we can get Ekman volume transport and upwelling velocity as:

$$T_E = \frac{\tau \times W}{\rho_0 \times f_0} = \frac{\rho_a \times C_d \times u_{10}^2 \times W}{\rho_0 \times f_0} = \frac{1kgm^{-3} \times 2\times10^{-3} \times \left(-10ms^{-1}\right)^2 \times 10^5 m}{1000\,kg\,m^{-3} \times 10^{-4} s^{-1}} =$$

$2 \times 10^5\ m^3\ s^{-1} = 0.2 \times 10^6\ m^3\ s^{-1} = 0.2$ Sv

Average upwelling rate at the base of the Ekman layer :$W(-h_E)$:

$$= -\frac{T_E}{WL} = \frac{2\times10^5 m^3 s^{-1}}{100\,km \times 100km} = \frac{2\times10^5 m^3 s^{-1}}{10^5 m \times 10^5 m} = 2\times10^{-5} ms^{-1} = 2m/day$$

## 7.6. Upwelling along West coast of India

Coastal upwelling occurs along the coast during the southwest monsoon season (June to September) between 7 °N and 15 °N and it was observed to be a combined process of wind-driven and Kelvin waves Shetye 1985; Shankar & Shetye,1997; Haugen et al 2002; McCreary et al 1993; Rao et al 2010). The strengthening of the alongshore wind stress enhances upwelling and results in lower SST over the shelf(Darbyshire,1967; Banse,1959; Sharma, 1966; Johannessen et al 1978; Lathipha & Murthy,1978; Maheshwaran et al 1999; Sanilkumar et al 2004).

Various studies have hypothesized that bottom signature of upwelling were caused by a northward propagating Kelvin wave9-13 propagating from the Bay of Bengal around Sri Lanka and northward along the southwest coast of India. Gopalakrishna, *et al.*(2008) using XBT data in the Lakshadweep sea proposed that, in the South Eastern Arabian Sea (SEAS), upwelling is influenced mainly by the remote forcing from the propagating coastal Kelvin waves originating from south Sri Lanka which was forced by the alongshore wind stress off the southwest coast of India and not due to the equatorial forcing.

According to Fernandez et al (2019) in majority of the eastern Arabian Sea, the wind is southwesterly and it turns northerly along the west coast of India with a wind stress of the order of 0.02 Pa. The drastically plunging to lesser values of SST is observed to occur from early June to September. Rao et al (2006) explained in detail the existence of a mini cool pool in June caused by strong upliftment of the thermocline accompanied by enhanced blooming of chlorophyll. Ganer, *et al* (2009) simulated the cool pool at the southern tip of India from a thermo-dynamical model and reported the extensive cooling from July to September, they concluded that the intense cyclonic gyre of currents supported the cooling.

### 7.6.1. Onset of Upwelling along the Southwest Coast of India

According to Fernandez et al (2019) the ascending motion of isotherms along the west coast of India sets up by late February. For this study the 23 °C isotherm (referred as D23) is used as proxy to determine the upliftment of the thermocline as shown in Fig.7.3. It was noticed at the southern tip of India (8.25 °N), the D23 isotherm marked in black is undulated from Cape Comorin (77.5 °E) offshore to

11 °N in the northern most transect. The monthly rising of 23 $^0$C isotherm (D23) at 8.25 $^0$N latitude starting at about 120m in February reaches to naviface by 2nd week of July and continuing at this level throughout August it starts descending downward in October (Fig.7.6 ). However the steady upliftment is not noticed and the wave like nature of these undulations can signify as the onset of upwelling as reported by earlier researchers. Beyond 11 °N no undulations are observed.

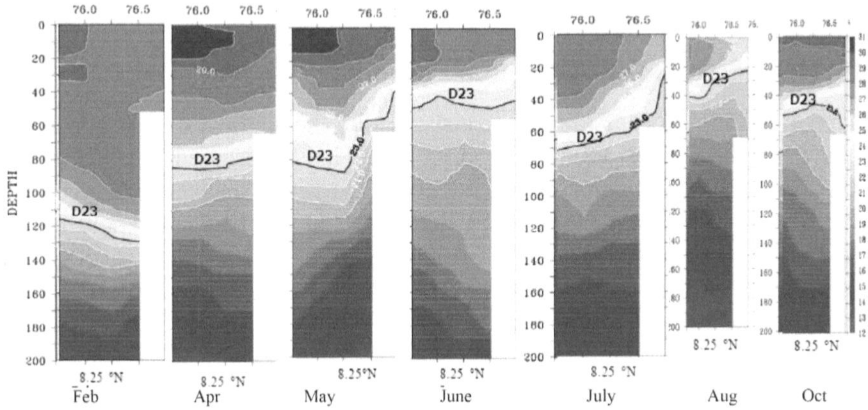


**Fig. 7.3.** Monthly rising of 23 $^0$C(D23) isotherm, an indicator of upwelling along West Coast of India (off 8.25$^0$N)(Fernandez et al,2019)

## References


Banse, K.(1959): On upwelling and bottom trawling off the southwest coast of India. J. Mar. Biol. Assoc. India, 1959, 1, 33– 49.

Darbyshire, M.(1967): The surface waters off the coast of Kerala, South west India. Deep Sea Res. 14, 1967, 3, 295-320.

Fernandez D.R, A. Raghunadha Rao, and T.Y. Niji (2019):Onset and Evolution of Upweling along the West Coast of India, Def. SCI. J., Vol. 69, No. 2, March2019

Ganer, D.W.; Deo, A.A. & Gnanaseelan, C. Variablilty of mini cold pool off the southern tip of India as revealed from a thermodynamic upper ocean model. Meteorol. Atmos. Phys., 2009, 104(303), 229-238. doi: 10.1007/s00703-009-0029-4

Garvine RW (1971) A simple model of coastal upwelling systems. J Phys Oceangr 1:169–179.

Gopalakrishna, V.V.; Rao, R.R.; Nisha, K.; Girishkumar, M.S.; Pankajakshan, T.; Ravichandran, M.; Johnson, Z.; Girish, K.; Aneeshkumar, N.; Srinath, M.; Rajesh, S. & Rajan, C.K. Observed anomalous upwelling in the Lakshadweep Sea during summer monsoon season of 2005. J. Geophys. Res, 2008, 113. doi: 10.1029/2007JC004240

Haugen, V.E.; Johannessen, O.M. & Evensen, G.(2002): Mesoscale modeling study of the oceanographic conditions off the southwest coast of India. In Proceedings of the Indian Academy of Sciences (Earth and Planetary Sciences), 2002, 111, 321-337. doi: 10.1007/BF02701978

Hidaka K (1954) A contribution to the theory of upwelling and coastal currents. Trans Am Geophys Union 35:431–444.

Johannessen, O.M.; Subbaraju, G. & Blindheim, J. (1978): Seasonal variation of oceanographic conditions off the southwest coast of India during 1971-1975. Fisk. Dir. Skr. Ser. Hav, 1978, 18, 247-261.

Lathipha, P.N. & Murthy, A.V.S.(1978): Studies of upwelling along the west coast of India using geo-potential anomaly. Indian J. Mar. Sci., 1978, 7(4), 219-221.

Maheshwaran, P.A., Rajesh, G., Revichandran, C. and K.K.C.Nair (1999): Upwelling and associated hydrography along the west coast of India during southwest monsoon. In Proceedings of PORSEC 1999, 2, pp. 873-878.

McCreary, J.P.; Kundu, P.K. & Molinari, R.L. (1993):. A numerical investigation of dynamics, thermodynamics and mixed layer processes in the Indian Ocean. Progress Oceanogra., 1993, 31, 181-244. doi: 10.1016/0079-6611(93)90002-U

Rao, R.R.; Kumar, G.; Ravichandran, M.; Rao, A.R.; Gopalakrishna, V.V. & Thadathil, P,(2010): Interannual variability of Kelvin wave propagation in the wave guides of the equatorial Indian Ocean, the coastal Bay of Bengal and the southeastern Arabian Sea during 1993–2006. Deep Sea Res. Part I: Oceanogra. Res. Papers, 2010, 57(1), 1-13. doi: 10.1016/j.dsr.2009.10.008

Rao, R.R.; Girish Kumar, M.S.; Ravichandran, M.; Samal, B.K. & Sreedevi, N. Observed minic cool pool off the southern tip of India and its intrusion into the south central Bay of Bengal during summer monsoon season. Geophys. Res. Lett., 2006, 33, 6607. doi: 10.1029/2006GL026086

Sanilkumar, K.V.; Unni, V.K. & James, V.V.(2004): Upwelling characteristics of the southwest coast of India during 2003. In Proceedings of METOC, 2004, pp. 137-143

Shankar, D. & Shetye, S.R.(1997): On the dynamics of the Lakshadweep high and low in the southeastern Arabian Sea. J.Geophys. Res., 1997, 102, 12551-12562. doi: 10.1029/97JC00465

Sharma, G.S.(1966): Thermocline as an indicator of upwelling. J. Mar. biol. Ass. India, 1966, 8(1), 8-19.

Shetye, S.R.; Shenoi, S.S.C.; Antony, M.K. & Krishna Kumar, V.(1985): Monthly mean wind stress along the coast of the north Indian Ocean. In Proceedings of the Indian Academy of Sciences (Earth and Planetary Sciences), 1985, 94(2), 129-137. doi: 10.1007/BF02871945

Yoshida K (1967) Circulation in the eastern tropical oceans with special reference to upwelling and undercurrents. Jap J Geophys 4:1–75

## Exercises

### *Exercise 7.1*

(a) Off the coast of Puri ($19.8^{0}$^0 N), east coast of India, a 4-knot wind blows from the west. (1) What is the speed of the surface wind-generated Ekman current? (2) To what depth does the surface Ekman current extend?

(b) During much of the year, a steady wind blows from the south along the Puri coast which is oriented north-south. Assume Ekman motion and ρ = 1.0 g cm--3.

(1) If the wind generates a surface stress of 2 dynes cm—2 during pre-monsoon season, what is the Ekman transport along a 1-km stretch of the beach?

(2) If the average width of the continental shelf in this region is 20 km, what would be the magnitude and direction of the vertical velocity of upwelling induced by the Ekman flow over the shelf? Assume that the upwelling velocity is constant over the entire shelf.

### Solution

We know the drag coefficient ($C_d$) = 1.4 × $10^{-3}$, density of air $\rho_a$ = 1.3 kg $m^{-3}$ and eddy coefficient of viscosity ($A_z$) = $10^{-1} m^2 s^{-1}$

Also we know: $\tau_y = \rho_a C_d W^2 = 1.3$ kg $m^{-3} \times 1.4 \times 10^{-3} \times (2ms^{-1})^2 = 7.28 \times 10^{-3}$ $kgm^{-1} s^{-2}$

We know Depth of frictional resistance as:

$$f = 2\Omega \sin\Phi = 2\times 7.3\times 10^{-5} \sin(19.8) = 4.9\times 10^{-5} s^{-1}$$

$$D = \pi\sqrt{\frac{2A}{f\rho_w}} = \pi\sqrt{\frac{2\times 0.5kgm^{-1}s^{-1}}{4.9\times 10^{-5}s^{-1}\times 10^3 \text{kgm}^{-3}}} = 0.2041\times 10^2 m = 20.4m$$

Surface current:

$$V_0 = \frac{\tau_y}{Aa} = \frac{\pi\tau_y\sqrt{2}}{Df\rho_w} = \frac{\pi\times 7.28\times 10^{-3}kgm^{-1}s^{-2}\times\sqrt{2}}{20.4m\times 4.9\times 10^{-5}s^{-1}\times 10^3 kgm^{-3}} \approx \frac{32.34}{99.96}\times 10^{-1}ms^{-1} = 0.32\times 10^{-1} = 0.032ms^{-1} = 3.2cms^{-1}$$

b)1. From equation (7.52) Ekman transport = $T_{Ek} = \dfrac{\tau}{\rho_w f}\times W$

Given values are $\tau$ = 2 dynes $cm^{-2}$; W = 1 km = $10^3$ m;

$$T_{Ek} = \frac{\tau}{\rho_w f}\times W = \frac{2gcms^{-2}cm^{-2}\times 10^5 cm}{1gcm^{-3}\times 4.9\times 10^{-5}s^{-1}} = 0.41\times 10^{10}cm^3s^{-1}$$ = 0.0041Sv(∵1 Sv = $10^{12}$ $cm^3s^{-1}$)

2. We know upwelling velocity from equation (7.53) as

$$W_{-HE} = \frac{T_{Ek}}{WL} = \frac{0.41\times 10^{10}cm^3s^{-1}}{10^5 cm\times 20\times 10^5 cm} \approx 0.02cm\,s^{-1}$$

### Exercise 7.2

What is the depth of frictional resistance (D) for a current velocity of 10 cm $s^{-1}$. Find the volume transport over a width of 200 km (take $f$=$10^{-4}$ $s^{-1}$ and eddy coefficient of viscosity (A)=50 $kgm^{-1}$ $s^{-1}$.

### Solution

Taking the standard values as: $\rho_a = 1kgm^{-3}; water\ density, \rho_w = 10^3 kgm^{-3}$; and given

$f = 10^{-4}s^{-1} V_0 = 10$ cm $s^{-1} = 10^{-1}$m $s^{-1}$ and eddy coefficient of viscosity (A) = 50 $kgm^{-1}s^{-1}$)

width W= 200 km = $2\times10^5 m$

From exercise 6.3. We can Write: $D = \pi\sqrt{\dfrac{2A}{f\rho_w}} = \pi\sqrt{\dfrac{A}{\rho\Omega\sin\varphi}}$ (1)

For calculation of Depth of frictional resistance, D, by substituting the above values in equation (1):

$$D = \pi\sqrt{\frac{2A}{f\rho_w}} = \pi\sqrt{\frac{2\times50\ kgm^{-1}s^{-1}}{10^{-4}s^{-1}\times10^3\ kgm^{-3}}} = \pi\sqrt{1000\,m^2} = 99.35\,m$$

Ekman volume transport as given in Exercise 7.1 = $V_0 = \dfrac{\tau_y}{Aa} = \dfrac{\pi_y\ \tau_y\sqrt{2}}{D\ f\ \rho_a}$ (2)

$$T_{Ek} = \frac{\tau}{\rho_w f}\times W \quad (3)$$

From equations (3), (2) and (1), we can write:

$$T_{Ek} = \frac{\tau}{\rho_w f}\times W = \frac{V_0 Df\rho_a}{\rho_w f\pi\sqrt{2}} = \frac{V_0 D\rho_a}{\rho_w\pi\sqrt{2}}$$

$$\therefore T_{Ek} = \frac{V_0 D\rho_a}{\rho_w\pi\sqrt{2}} = \frac{10^{-1}\text{m } s^{-1}\times99.35m\times1kgm^{-3}}{10^3\ kgm^{-3}\times\pi\times\sqrt{2}} = 447.4m^3s^{-1}$$

### *Exercise 7.3*

What is Ekman mass & volume transports? Explain with units.

### *Solution*

Ekman transport $M_x = \dfrac{\tau}{\rho_w f} = \dfrac{\rho_a C_D u_a^{\ 2}}{\rho_w f}\, m^2s^{-1}$; But technically this is not a transport because it is in square meters per second ($m^2 s^{-1}$). In order to get the real transport ($m^3\ s^{-1}$) we have to multiply by width 'W' ( sum up horizontally along a section). Hence $T_{EK}$ =

$$\frac{\tau}{\rho_w f}\times W = \frac{\rho_a C_D u_a^{\ 2}}{\rho_w f}\times W\ m^3s^{-1}$$

Ekman transport for $\tau$ = 0.1 N $m^{-2}$, $\rho_w$ = 1025(=$10^3$), f=$10^{-4}$ and whole width of ocean=5000km =

$$\frac{\tau}{\rho_w f}\times W = \frac{0.1Nm^{-2}\times5000\times10^3 m}{10^3\ kgm^{-3}\times10^{-4}} = \frac{0.1kgm^{-1}s^{-2}\times5000\times10^3 m}{10^3\ kgm^{-3}\times10^{-4}s^{-1}} = 5\times10^6 m^3s^{-1} = 5sv\ = 5$$

Sverdrups ( since 1 Sv = $10^6\ m^3 s^{-1}$)

### *Exercise 7.4*

Calculate the following for a wind of 10 m/s at $30^0$ latitude in N.H at a distance of 10 km from the coast. i) Ekman surface velocity ($V_0$) per meter of coast, ii)

Ekman depth ($D_E$), iii) velocity at Ekman depth ($V_D$), iv) Slope of the sea surface in the upwelling zone,v) Ekman mass transport ($M_x$) and vi) vertical upwelling velocity($W_h$ ) (Take , Cd=2.4, .A=50 kgm$^{-1}$ s$^{-1}$)

### Solution

i) We know Surface velocity ($V_0$ ) from equation (6.7)

ii) Ekman depth ($D_E$) from equation (6.7)

iii) Velocity at Ekman depth ($V_D$ ) from equation (6.25)

iv) Slope of the sea surface from equation (7.45)

v) We know from equation (6.17) $M_x = \frac{\tau_y}{f}$ and wind stress $\tau_y = \rho_a c_D w^2$

vi) We know the vertical velocity from equation (7.50) as:

I) $$V_0 = \frac{\tau_y}{LAa} = \frac{\tau_y}{Lf\rho_w} = \frac{\rho_a c_D u^2}{Lf\rho_w} = \frac{1.3kgm^{-3} \times 2.4 \times 10^{-3} \times 10^2 (m/s)^2}{10 \times 10^3 m \times 2 \times 7.3 \times 10^{-5} \times \sin 30 s^{-1} \times 1025 kgm^{-3}}$$

$$= \frac{3.12}{0.5} = 6.24 ms^{-1}$$

II) $$D_E = D_E = \pi\sqrt{\frac{2A}{f\rho_w}} = \pi\left[\frac{2 \times 50 kgm^{-1}s^{-1}}{2 \times 7.3 \times 10^{-5} \times \sin 30 s^{-1} \times 1025 kgm^{-3}}\right]^{\frac{1}{2}} = 3.632m$$

III) $V_D = e^{-\pi} V_0 = 0.04V_0 = 0.04 \times 6.24 = 0.25$ m/s

IV) $$\text{Slope} = \therefore \frac{\partial \zeta}{\partial y} = \frac{\tau_{sy}}{\rho_w g D_E} = \frac{1.3kgm^{-3} \times 2.4 \times 10^{-3} \times 10^2 (m/s)^2}{3.632m \times 10 ms^{-2} \times 1025 kgm^{-3}} = 8 \times 10^{-6}$$

V) Mass transport: from eqn (7.52):

$$M_x = \frac{\tau_y}{f_w} = \frac{\rho_a c_D u^2 \times W}{f_w \times \rho_w} = \frac{1.3kgm^{-3} \times 2.4 \times 10^{-3} \times 10^2 (m/s)^2 \times 10^4 m}{1025 kgm^{-3} \times 2 \times 7.3 \times 10^{-5} \times \sin 30 s^{-1}} = 41697 m^3 s^{-1}$$

VI) Vertical velocity, from eqn (7.53):

$$\frac{M_x}{WL} = \frac{41697 m^3 s^{-1}}{10^4 m \times 10^4 m} = 0.0042 ms^{-1} = 0.42 cms^{-1}$$

### Exercise 7.5

A steady wind is blowing on a rectangular ocean basin 8000 km wide. At 25°N, the wind blows from the east at a speed of $u_{10}$=8 m/s. At 40°N, the wind blows from the west at a speed of $u_{10}$=8 m/s. What is the rate of wind-driven net Ekman mass transport convergence between 25°N and 40°N (in kg/s)? What is the distance of the convergence point (P) from 25°N and 40°N latitude? What is the time of convergence in a square meter stretch in seconds? What is the average Ekman pumping (in cm/s) between 25°N and 40°N? If you consider the coast is on the left side and ocean is on the right side in Northern Hemisphere at what latitude region you expect upwelling and down welling [Take 1 degree of latitude = 111.12 km, $\rho_a$ = 1.3 kgm$^{-3}$, $C_d$ = 2.4 × 10$^{-3}$, $\rho_w$ = 1025 kg m$^{-3}$.

### *Solution*

Ekman mass transport $M_x = \dfrac{\tau}{\rho_w f} = \dfrac{\rho_a C_D u_a^{\;2}}{\rho_w f} m^2 s^{-1}$; But technically this is not a transport because it is in $m^2s^{-1}$. In order to get the real transport ($m^3\ s^{-1}$) we have to multiply by width 'W' (sum up horizontally along a section). Hence $T_{EK}$ = $\dfrac{\tau}{\rho_w f} \times W = \dfrac{\rho_a C_D u_a^{\;2}}{\rho_w f} \times W\, m^3 s^{-1}$.

i) We know from the previous chapter - 6 of equation (6.22b) Ekman mass transport for the whole length of the basin as: $M_{xW} = \dfrac{\tau \times W}{f} = \dfrac{\rho_a C_D u_a^{\;2} \times W}{f} kgs^{-1}$; But technically this is not a transport because it is in kg $s^{-1}$. In order to get the real transport ($m^3\ s^{-1}$) we have to divide by sea water density $\rho_w$ (sum up horizontally along a section). Hence $T_{EK} = \dfrac{\tau}{\rho_w f} \times W = \dfrac{\rho_a C_D u_a^{\;2}}{\rho_w f} \times W\, m^3 s^{-1}$ and wind stress $\tau_y$ $= \rho_a c_D W^2$

$$M_{Ex25} = \frac{\tau_y \times W}{f} = \frac{\rho_a c_D u^2 \times W}{f} = \frac{1.3kgm^{-3} \times 2.4 \times 10^{-3} \times 8^2 (m/s)^2 \times 8 \times 10^6 m}{2. \times 7.3 \times 10^{-5} \times \sin 25 s^{-1}}$$

$$= 2589 \times 10^8 kgs^{-1}$$

$$M_{x40} = \frac{\tau_y \times W}{f} = \frac{\rho_a c_D u^2 \times W}{f} = \frac{1.3kgm^{-3} \times 2.4 \times 10^{-3} \times 8^2 (m/s)^2 \times 8 \times 10^6 m}{2. \times 7.3 \times 10^{-5} \times \sin 40 rad/s}$$

$$= 170.24 \times 10^8 kgs^{-1}$$

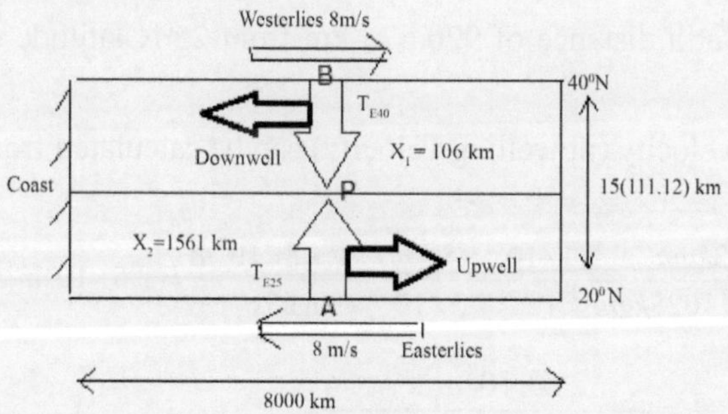


Volume transports:

$$T_{E25} = \frac{\tau_y \times W}{f \times \rho_w} = \frac{\rho_a c_D u^2 \times W}{f \rho_w} = \frac{1.3kgm^{-3} \times 2.4 \times 10^{-3} \times 8^2 (m/s)^2 \times 8 \times 10^6 m}{1025 kgm^{-3} \times 2. \times 7.3 \times 10^{-5} \times \sin 25 s^{-1}}$$

$$= 2.5 \times 10^8 m^3 s^{-1}$$

$$T_{E40} = \frac{\tau_y \times W}{f \rho_w} = \frac{\rho_a c_D u^2 \times W}{f \rho_w} = \frac{1.3kgm^{-3} \times 2.4 \times 10^{-3} \times 8^2 (m/s)^2 \times 8 \times 10^6 m}{1025 kgm^{-3} \times 2. \times 7.3 \times 10^{-5} \times \sin 40 rad/s}$$

$$= 0.17 \times 10^8 m^3 s^{-1}$$

Rate of flow in 1 $m^2$ stretch:

Along AP the flow = $T_{E25}$ per $m^2$ = 2.5 × $10^8$ $ms^{-1}$;

Similarly along BP the flow = $T_{E40}$ per $m^2$ = 0.17 ×$10^8$ $ms^{-1}$

Let both the transports meet at point 'P' and BP = $x_1$ and AP = $x_2$ such that $x_1 + x_2$ = 111.12 × 15 × $10^3$m

We know, distance = velocity rate × time. Let both the flows meet after time 't', then

The distance BP = $x_1 = u_{40}$ × t = 0.17 × $10^8$ ($ms^{-1}$) t (s); similarly distance AP = 2.5 × $10^8$ ($ms^{-1}$) × t(s)

then the total distance AP + BP = 2.5 × $10^8$ t + 0.17 × $10^8$ t = 2.67 t =111.12 × 15 × $10^3$ m

$$\therefore t\left(in 1m^2\right) = \frac{111.12\times 15\times 10^3}{2.67\times 10^8} = 624.27\times 10^{-5}\ s$$

Then distance AP = $x_2$ = velocity × time = $u_{25}$ × t = 2.5 × $10^8$ $ms^{-1}$ × 624.27 × $10^{-5}$ s ≈ 1561 × $10^3$ m = 1561 km

Similarly the distance BP = $x_1 = u_{40}$ × t = 0.17 × $10^8$ $ms^{-1}$ × 624.27 × $10^{-5}$ s ≈106 × $10^3$ m = 106 km

This means the rate of convergences are: $u_{25}$ = 2.5 × $10^8$ ms-1 = 90 × $10^7$ km $hr^{-1}$, and $u_{40}$ = 0.17 × $10^8$ $ms^{-1}$ = 6.12 × $10^7$ km $hr^{-1}$

The convergence point 'P' is at a distance of 920.676 km from 25$^0$N latitude & 746.21 km from 40$^0$N.

iii) Average Ekman pumping velocity (upwelling velocity) can be calculated from equation (7.53):

$$T_{E25} = \frac{\tau_y \times W}{f\times\rho_w} = \frac{\rho_a c_D u^2\times W}{f\rho_w} = \frac{1.3kgm^{-3}\times 2.4\times 10^{-3}\times 8^2 (m/s)^2\times 8\times 10^6 m}{1025kgm^{-3}\times 2.\times 7.3\times 10^{-5}\times \sin 25s^{-1}} = 2.5\times 10^8 m^3 s^{-1}$$

$$W_{H25} = -\frac{T_{E25}}{WL} = \frac{2.5\times 10^8 m^3 s^{-1}}{8000\,km\times x_2 km} = \frac{2.5\times 10^8 m^3 s^{-1}}{8\times 10^6 m\times 1561\times 10^3 m}$$

$$= -2\times 10^{-5} ms^{-1} = -1.72 m/day$$

$$W_{H40} = +\frac{T_{E40}}{WL} = \frac{0.17\times 10^8 m^3 s^{-1}}{8000\,km\times x_1 km} = \frac{0.17\times 10^8 m^3 s^{-1}}{8\times 10^6 m\times 106\times 10^3 m}$$

$$= +2\times 10^{-5} ms^{-1} = +1.72 m/day$$

If the coast is in the west side as shown in the Figure, at 25$^0$N easterlies and 40$^0$N Westerlies prevail. Easterlies generate Ekman mass transport ($T_E$) in the positive

Y direction and so to the coast upwelling ($-W_h$) occurs. Similarly the Westerlies generate Ekman mass transport ($T_E$) in the negative y direction and so to the coast downwelling ($+W_h$) occurs. Note as Z is positive downwards in the ocean, upwelling is a rising motion. So it is negative Z whereas downwelling is in the positive Z direction. Here the upwelling and downwelling velocities in both the latitudes is same, which is 1.72m/day

### *Exercise 7.6*

Calculate the net vertical rate of velocity between the latitudes of $10^0$N and $40^0$ N of easterlies velocity of 5 m/s and westerlies velocity of 10 m/s respectively in the subtropical North Atlantic such that the coast is left hand side. Explain whether this net vertical motion is upwelling or downwelling (Take $C_D = 2 \times 10^{-3}$, $\rho_a = 1$ kg/m³, $\rho_w = 1025$ kg/m³ and distance between the two latitudes (L) is 3000 km)

### *Solution*

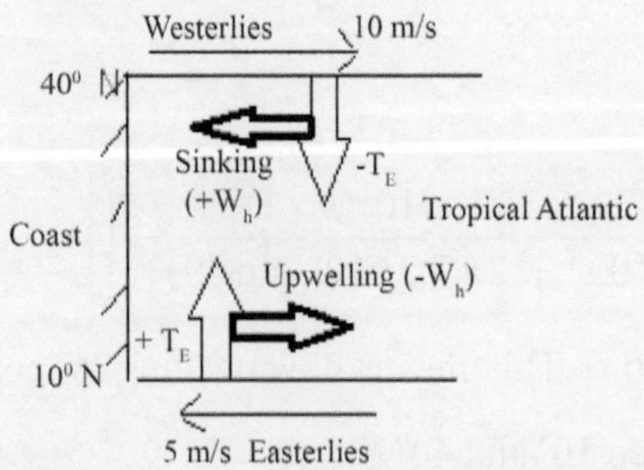


For the tropical Atlantic ocean as shown in the Figure coast is west side, at 10 ⁰N easterlies and 40 ⁰N westerlies prevail. Easterlies generate Ekman mass transport ($T_E$) in the positive Y direction and so to the American coast upwelling ($-W_h$) occurs. Similarly the westerlies generate Ekman mass transport ($T_E$) in the negative y direction and so to the coast downwelling ($-W_h$) occurs. Note as Z is positive downwards, upwelling is rising so it is negative Z and downwelling is in the positive Z direction.

Upwelling at 10⁰N (from eqn 7.53) is: $\dfrac{T_E}{WL} = \dfrac{\rho_a C_D u_a^{\,2}}{\rho_w f \times W \times L} \times W\, m^3 s^{-1} = \dfrac{\rho_a C_D u_a^{\,2}}{\rho_w f \times L} =$

$$W_{h10} = -\frac{1.3kgm^{-3} \times 2.4 \times 10^{-3}\left(5ms^{-1}\right)^2}{1025\,kgm^{-3} \times 2 \times 7.3 \times 10^{-5}\, sin10s^{-1} \times 3 \times 10^6 m} = -100051308 \times 10^{-14}\, ms^{-1}$$

Downwelling at 40⁰N

$$W_{h40} = +\frac{1.3kgm^{-3} \times 2.4\times10^{-3} \times \left(10ms^{-1}\right)^2}{1025\,kgm^{-3} \times 2\times7.3\times10^{-5} sin40s^{-1} \times 3\times10^6 m}$$

$= +108115601\times10^{-14} ms^{-1}$

The difference in upwelling = $10^{-14}$ (108115601-100051308)$ms^{-1}$ = 8064293 × $10^{-14}$ $ms^{-1}$ = 6.968 × $10^{-1}$ $\frac{\text{cm}}{day}$ ≈ 7 cm/10 days

## Another method

From the previous exercise: At 10 $^0$N upwelling and at 40 $^0$N down welling occurs. So we have to find the difference:

$$w_H = \frac{M_x}{\rho_w L} = \frac{\tau_y}{\rho_w Lf} = \frac{\rho_a c_D w^2}{\rho_w Lf} = \frac{\rho_a c_D}{\rho_w L}\left[-\frac{(w_{10})^2}{f_{10}} + \frac{W_{40}^2}{f_{40}}\right]$$

Note here negative sign is upwelling and positive sign is down welling direction

$$\frac{\rho_a c_D}{\rho_w L}\left[-\frac{W_{10}^2}{f_{10}} + \frac{W_{40}^2}{f_{40}}\right]$$

$$= \left[\frac{1.3kgm^{-3}\times2.4\times10^{-3}}{1025\times3000\times10^3}\left[\frac{-5^2(m/s)^2}{2\times7.3\times10^{-5}\sin10s^{-1}} + \frac{10^2(m/s)^2}{2\times7.3\times10^{-5}\sin40s^{-1}}\right]\right]$$

$= 82\ \times10^{-9} ms^{-1} = 0.71\frac{cm}{day} = 7.1cm/10days$ .This means downwelling is more by 7.1 cm in 10days than upwelling between $10^0$ and $40^0$ N.

### *Exercise 7.7*

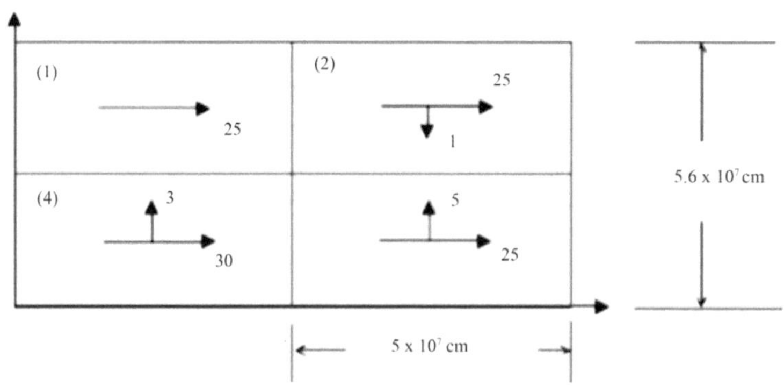


**Figure:** Horizontal current components, averaged for four 5-degree squares.

The mean horizontal current components (cm/sec) in the upper 50 m of an ocean have been averaged over five-degree squares and are shown in the figure above.

Estimate the vertical velocity at 50 m in each of the five-degree squares.

***Solution***

We know the vertical velocity from equation (7.50) as:

$$w_H = -\frac{M_x}{\rho_w L}; M_x = \frac{\tau_y}{f}; So\, w_H = -\frac{\tau_y}{f\rho_w L} \tag{1}$$

We also know: $V_0 = \frac{}{\sqrt{fA}} \Rightarrow \tau_y = V_0 \rho_w \sqrt{fA}$ (2)

We further know : $D = \pi\sqrt{\frac{2A}{f}}$ ); This can be written as: $\sqrt{A} = \frac{D\sqrt{f}}{\pi\sqrt{2}}$ (3)

Substituting eqn (3) in eqn (2): $\Rightarrow \tau_y = V_0 \rho_w \sqrt{f}\frac{D\sqrt{f}}{\pi\sqrt{2}}$ (4)

Substituting eqn (4) in eqn (1): $w_H = \frac{V_0 \rho_w f D}{f\rho_w L\pi\sqrt{2}} = \frac{V_0 D}{L\pi\sqrt{2}} ms^{-1}$ (5)

The negative sign in equation (1) is ignored as it only indicates that the flow is upward.

i) For horizontal component in box 1, the velocity given is $V_0 = 25$ cms$^{-1}$= 25 × 10$^{-2}$ms$^{-1}$

D = 50 m; L = $\frac{5.6\times10^7 cm}{2}$ = 2.8 10$^7$ cm = 2.8 × 10$^5$ m

Substituting the given items in equation (5)

$$w_{H25} = -\frac{V_0 D}{\sqrt{2}\pi L} = -\frac{25\times10^{-2} ms^{-1}}{2.8\times10^5 m}\times\left[\frac{50m}{\pi}\frac{1}{\sqrt{2}}\right] = -100.48\times10^{-7} ms^{-1}$$

The minus sign indicates the water is rising up to the surface.

ii) In Box 2, for the current between 25 & 1 ms$^{-1}$;

For 1 cms$^{-1}$ = 1 × 10$^{-2}$ms$^{-1}$; L= 5 × 10$^7$ cm = 5 × 10$^5$ m

$$\text{So } w_{H1} = -\frac{V_0 D}{\sqrt{2}\pi L} = \frac{1\times10^{-2} ms^{-1}}{5\times10^5 m}\times\left[\frac{50m}{\pi}\frac{1}{\sqrt{2}}\right] = -2.25\times10^{-7} ms^{-1}$$

As both are upwelled flows, the resultant is summation of both upwelled waters due to 25 and 1ms$^{-1}$.

So for box 2: $w_{H25} + w_{H1}$ = 100.48 × 10$^{-7}$ ms$^{-1}$ + (2.25 × 10$^{-7}$ ms$^{-1}$) = 102.73 × 10$^{-7}$ ms$^{-1}$

iii) In box 3: the current is : between 25 & 5 ms$^{-1}$; L = 5 × 10$^7$ cm = 5 × 10$^5$ m

$$w_{H5} = -\frac{V_0 D}{\sqrt{2}\pi L} = -\frac{5\times10^{-2} ms^{-1}}{5\times10^5 m}\times\left[\frac{50m}{\pi}\frac{1}{\sqrt{2}}\right] = -11.25\times10^{-7} ms^{-1}$$

For box 3: $w_{H25} + w_{H5} = 100.48 \times 10^{-7}\ ms^{-1} + (11.25 \times 10^{-7}\ ms^{-1}) = 111.73 \times 10^{-7}\ ms^{-1}$

For box 4: the current between 30 & 3 $ms^{-1}$;D = 50 m; L = $2.8 \times 10^7$ cm = $2.8 \times 10^5$ m

Substituting in equation (5): for 30:

$$w_{H30} = -\frac{V_0 D}{\sqrt{2}\pi L} = -\frac{30\times10^{-2} ms^{-1}}{2.8\times10^5 m}\times\left[\frac{50m}{\pi}\frac{1}{\sqrt{2}}\right] = -120.58\times10^{-7} ms^{-1}$$

For 3 $ms^{-1}$; D = 50 m; L = $5 \times 10^7$ cm = $5 \times 10^5$ m

$$w_{H3} = -\frac{V_0 D}{\sqrt{2}\pi L} = -\frac{3\times10^{-2} ms^{-1}}{5\times10^5 m}\times\left[\frac{50m}{\pi}\frac{1}{\sqrt{2}}\right] = -6.75\times10^{-7} ms^{-1}$$

For box 4: the resultant vertical velocity is:

$w_{H30} + w_{H3} = 120.58 \times 10^{-7}\ ms^{-1} + (6.75 \times 10^{-7}\ ms^{-1}) = 127.33 \times 10^{-7}\ ms^{-1}$

## *Exercise 7.8*

If the easterlies blow at 5 m/s all along the equatorial band and the Ekman relation is valid at $5^0$ N and S respectively, calculate the average poleward Ekman mass transport over a basin of width 6000 km. Also calculate the average upwelling rate in the equatorial band between both these latitudes. Assume the standard values of densities of water and air and other constants.(Take $1^0$ latitude=100 km in the upwelling zone, f = $10^{-5}$ $S^{-1}$, $\rho_a$ =1.3 kg $m^{-3}$, $\rho_w$ = 1025 kg $m^{-3}$, $C_D = 2 \times 10^{-3}$):

## *Solution*

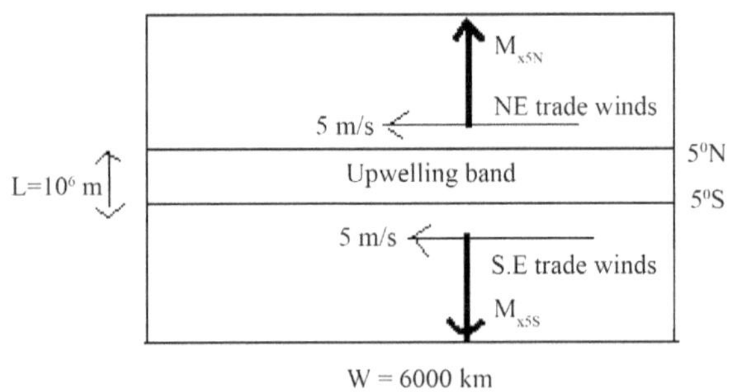


According to Equation (7.52) Ekman volume transport $T_E = \frac{\rho_a C_D u_a^2}{\rho_w f} \times W\ m^3 s^{-1}$

Hence $T_{E5N} = \frac{\tau_{y5N} \times W}{\rho_w f_{5N}} = \frac{\rho_a c_D u^2 \times W}{\rho_w f_{5N}} m^3 s^{-1}$ and upwelling $= W_{h5N} = \frac{T_{E5N}}{WL} = \frac{\rho_a c_D u^2}{\rho_w f_{5N} L}$

Given: Width of the basin = W =6000 km = $6 \times 10^6$ m; L = $5^0 + 5^0 = 10^0 = 10 \times 100 \times 10^3$ m

The volume transport for the whole width (W = 6000km) of the ocean:

Solution:

$$T_{E5N} = \frac{\tau_{y5N} \times w}{\rho_w f_{5N}} = \frac{\rho_a c_D u^2 \times w}{\rho_w f_{5N}} = \frac{1.3kgm^{-3} \times 2.4\times10^{-3} \times 5^2 (m/s)^2 \times 6000\times10^3 m}{2\times7.3\times10^{-5} s^{-1} \times \sin 5 \times 1025 kgm^{-3}}$$

$$= \frac{468\times10^3}{1304\times10^{-5}} = 0.36\times10^8 m^3 s^{-1} = 36Sv\left(\because 1Sv = 10^6 m^3 s^{-1}\right)$$

Suffix 5N denotes the value at $5^0$ N. Similarly same value we will get for $5^0$S.This means, the transports are

$T_{E5N}$ = 36Sv = 36 × $10^6$ $m^3$ $s^{-1}$ and $Q_{x5S}$ = 36Sv = 36 × $10^6$ $m^3$ $s^{-1}$

Upwelling = $W_h = \frac{T_{E5N}}{WL} = \frac{0.36\times10^8 m^3 s^{-1}}{6\times10^6 m \times 10^6 m} = 0.6\times10^{-4} ms^{-1} = 0.6\times10^{-2} cms^{-1}$

### Exercise 7.9

Calculate the Ekman volume transport across a section of width W = 100 km along California coast blowing a northerly wind of 10 m/s Assume that the influence of Ekman theory holds upto L = 100 km. Also calculate the average upwelling rate within the influence of the coastal box. Take f = $10^{-4}$ $S^{-1}$, $\rho_a$ =1.0 kg $m^{-3}$, $\rho_w$ = 1000 kg $m^{-3}$, $C_D$ = 2 × $10^{-3}$):

### Solution

Like the previous exercise:

$$Q_{x5N} = \frac{\tau_{y5N} \times W}{\rho_w f_{5N}} = \frac{\rho_a c_D u^2 \times W}{\rho_w f_{5N}} = \frac{1.0kgm^{-3} \times 2\times10^{-3} \times 10^2 (m/s)^2 \times 100\times10^3 m}{10^{-4} s^{-1} \times 1000 kgm^{-3}}$$

$$= 2\times10^5 m^3 s^{-1} = 0.2\times10^6 m^3 s^{-1} = 0.2Sv\left(\because 1Sv = 10^6 m^3 s^{-1}\right)$$

$$w_H = -\frac{\rho_a c_D u^2}{\rho_w f_{5N} L} = \frac{1.0kgm^{-3} \times 2.\times10^{-3} \times 10^2 (m/s)^2}{10^{-4} s^{-1} \times 1000 kgm^{-3} \times 10^5 m}$$

$$= \frac{2\times10^{-1}}{10^4} = 2\times10^{-5} ms^{-1}$$

### Exercise 7.10

If the Kuroshio Extension velocity is 10 cm/sec averaged from top to bottom, and the current width is 100km, assuming a depth of 5 km, what is its volume transport? Express transport in Sverdrups. What is its approximate mass transport assuming a constant density? (take $\rho_a$ = 1 kg$m^{-3}$, $\rho_w$ = $10^3$ kg$m^{-3}$,$C_d$ = 2 × $10^{-3}$, f =

$10^{-4}\ s^{-1}$)

### Solution

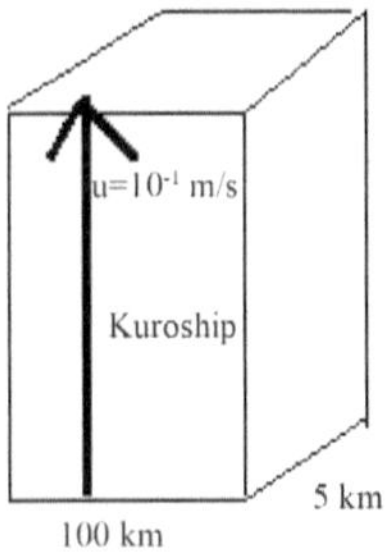


From equation (7.52) the volume transport is $T_{EK} = Q = \frac{\tau}{\rho_w f} \times W = \frac{\rho_a C_D u_a^{\ 2}}{\rho_w f} \times W\ m^3 s^{-1}$. But mass transport in the whole stretch (Length) is

$$M_{xW} = \frac{\tau \times W}{f} = \left( \frac{\rho_a C_D u_a^{\ 2}}{f} \times W \right) kgs^{-1}$$

The volume transport is =

$$Q_{x5N} = \frac{\tau_y}{\rho_w f} \times W = \frac{\rho_a c_D u^2 \times W}{\rho_w f} = \frac{1.0 kgm^{-3} \times 2 \times 10^{-3} \times 10^{-2} (m/s)^2 \times 100 \times 10^3 m}{10^{-4} s^{-1} \times 1000 kgm^{-3}}$$

$$= 2 \times 10 m^3 s^{-1} = 20 m^3 s^{-1} = 2 \times 10^{-5} Sv \left( \because 1Sv = 10^6 m^3 s^{-1} \right)$$

Mass transport =

$$M_x = -\frac{\rho_a c_D u^2}{f} = \frac{1.0 kgm^{-3} \times 2. \times 10^{-3} \times 10^{-2} (m/s)^2}{10^{-4} s^{-1}} = 2 \times 10^{-1} kgm^{-1} s^{-1} = 0.2 kgm^{-1} s^{-1}$$

Its appropriate mass transport = $M_{xW} = \frac{\tau \times W}{f} = \left( \frac{\rho_a C_D u_a^{\ 2}}{f} \times W \right)$ kgs$^{-1}$ = $2 \times 10^{-1} \times 10^5$ m = $2 \times 10^4$ kgs$^{-1}$

### Exercise 7.11

If the Ekman transport at 30$^0$N in the Atlantic is -2 Sv southward and the Ekman transport at 20$^0$N is -5 Sv southward, what is the average vertical velocity at the base of the Ekman layers between 20$^0$N and 30$^0$N? Take that the Atlantic is 5000 km wide (W) and 1$^0$ latitude is 100 km long (L).

### Solution

We know ; $M_{x30} = -\frac{\rho_a c_D u_{30}^{\ 2}}{f_{30}}; Q_{x30} = \frac{\tau_{y30} W}{\rho_w f_{30}} = \frac{\rho_a c_D u_{30}^{\ 2} W}{f_{30} \rho_w}; w_{h30} = \frac{M_{x30}}{\rho_w L} = \frac{\rho_a c_D u_{30}^{\ 2}}{f_{30} \rho_w L}$

$$\therefore w_{h30} = \frac{Q_{h30}}{WL} \;\&\; \therefore w_{h20} = \frac{Q_{h20}}{WL};\therefore w_{h30} + w_{h20} = \frac{1}{WL}\left[Q_{h30} + Q_{h20}\right]$$

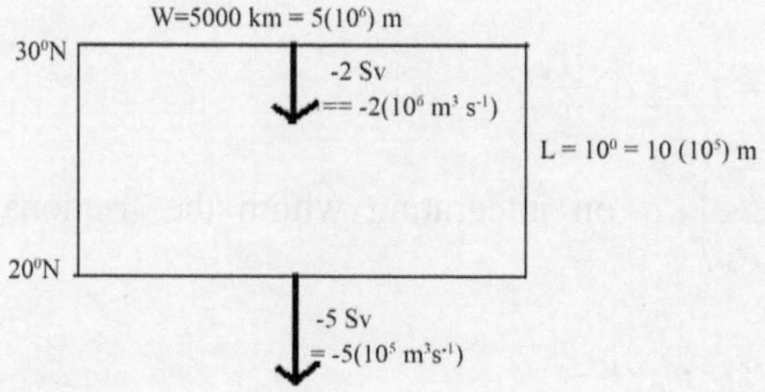


Given

$Q_{x30}$ = – 2Sv = – 2 × $10^6$ $m^3$ $s^{-1}$; $Q_{x20}$ = – 5Sv = – 5 × $10^6$ $m^3$ $s^{-1}$ Here negative sign indicates southward (equatorward) transport.

Putting width (W) = 5 × $10^6$m and Length between two latitudes ( L) = $10^6$ m, Net or average vertical velocity ($w_h$) i s=

$$\therefore w_{h30} + w_{h20} = -\frac{1}{WL}\left[Q_{h30} + Q_{h20}\right]$$

$$= -\frac{1}{5\times10^6\times10^6 m^2}\left[(2+5)10^6 m^3 s^{-1}\right] = -1.4\times10^{-6} m/s$$

Minus sign indicates rising up (upwelling) of water from subsurface

### *Exercise 7.12*

Calculate the zonal sea level gradient in the Equatorial Pacific of width 8000 km where the trade wind velocity has – 5m/s and the surface frictional layer of depth 100 m. (take $\rho_a$ = 1 kgm$^{-3}$, $\rho_w$ = 1025 kgm$^{-3}$,$C_d$ = 2 × $10^{-3}$)

### *Solution*

Given W = 8 ×$10^6$ m and $D_E$ = 100 m, u = 5 ms$^{-1}$. Slope of the sea surface from equation (7.45)

$$\therefore \frac{\partial\zeta}{\partial y} = \frac{\tau_{sy}W}{\rho_w g D_E} = \frac{\rho_a c_D u^2 \times W}{\rho_w \times g \times D_E}$$

$$= \frac{1.kgm^{-3}\times2.0\times10^{-3}\times(-5)^2(m/s)^2\times8\times10^3\times10^3 m}{10^2 m\times10ms^{-2}\times1025kgm^{-3}} = 0.39m = 39cm$$

### *Exercise 7.13*

If a northerly wind of -10 m/s blows along California coast and that the Ekman theory holds at L=*100km* off the coast, Calculate the Ekmman volume transport at the distance across the section width W=100 km and depth –$h_E$ and the average upwelling rate at the basis of the Ekman Layer W($-h_E$) within this coastal box. Assume Take $\rho_a = 1 kgm^{-3}$; $g = 10\ m^2\ s^{-1}$; $C_d = 2 \times 10^{-3}$; $\rho_0 = 1000 kgm^{-3}$; $f_0 = 10^{-4}\ s^{-1}$

### Solution

We know from equation 8.2: $\alpha \frac{\partial p}{\partial x} = f.v + \alpha \frac{\partial \tau_x}{\partial z}$

$-\frac{1}{\rho_0}\frac{\partial p}{\partial x} + \frac{1}{\rho_0}\frac{\partial \tau_x}{\partial z} = 0 \rightarrow -g\frac{\partial \eta}{\partial x} + \frac{\partial}{\partial z}\left(\frac{\tau_x}{\rho_0}\right) = 0$ on integrating within the frictional surface layer(H):

$$gH\frac{\partial \eta}{\partial x} = \frac{\tau_x}{\rho_0} \rightarrow \Delta\eta = \frac{\tau_x \times \Delta x}{\rho_0 \times gH} = \frac{\rho_a \times C_d \times u_{10}^2 \times W}{\rho_0 \times gH}$$

Similarly we can get Ekman volume transport:

$$T_E = \frac{\rho_a \times C_d \times u_{10}^2 \times W}{\rho_0 \times f_0} = \frac{1kgm^{-3} \times 2 \times 10^{-3} \times \left(-10ms^{-1}\right)^2 \times 10^5 m}{1000\,kg\,m^{-3} \times 10^{-4} s^{-1}}$$

$$= 2 \times 10^5 m^3 s^{-1} = 0.2 \times 10^6 m^3 s^{-1} = 0.2\,Sv$$

Average upwelling rate at the basis of the Ekman layer :W($-h_E$):

$$= -\frac{T_E}{WL} = \frac{2 \times 10^5 m^3 s^{-1}}{100\,km \times 100 km} = \frac{2 \times 10^5 m^3 s^{-1}}{10^5 m \times 10^5 m} = 2 \times 10^{-5} ms^{-1} = 2m/day$$

### Exercise 7.14

Calculate the average poleward Ekman mass transport for the easterly wind velocity of 5m/s all along the equatorial band over a basin width of 6000 km at $5^0$ N and S. Also calculate the average upwelling rate in the equatorial region between these two latitudes. Take $f_0 = \pm\ 10^{-5}\ s^{-1}$ at $5^0$ latitude. Take $\rho_a = 1\ kgm^{-3}$; $g = 10\ m^2\ s^{-1}$; $C_d = 2 \times 10^{-3}$; $\rho_0 = 1000\ kgm^{-3}$

### Solution

Ekman mass transport $M_x = \frac{\tau}{\rho_w f} = \frac{\rho_a C_D u_a^2}{\rho_w f} m^2 s^{-1}$; But technically this is not a transport because it is in. In order to get the real transport ($m^3\ s^{-1}$) we have to multiply by width 'W' (sum up horizontally along a section). Hence $U_{EK}$ =

$$\frac{\tau}{\rho_w f} \times W = \frac{\rho_a C_D u_a^2}{\rho_w f} \times W\ m^3 s^{-1}$$

We know the Ekman mass transport as:

$$M_x = -\frac{\rho_a c_D u^2}{\rho_0 f} = \frac{1.0kgm^{-3} \times 2.\times 10^{-3} \times 5^2 (m/s)^2}{1000kgm^{-3} \times 10^{-5} s^{-1}} = 5m^2 s^{-1}$$

$U_{EK} = \frac{\tau}{\rho_w f} \times W = \frac{\rho_a C_D u_a^2}{\rho_w f} \times$ W $m^3\ s^{-1}$ = $5m^2\ s^{-1} \times 6000$ km = $3 \times 10^7\ m^3\ s^{-1}$

# 8

# Sverdrup Circulation

## 8.1. Sverdrup's equation

Harald Ulrik Sverdrup (1888-1957), a Norwegian who began his career studying meteorology with Vilhelm Bjerknes in Oslo, was appointed director of Scripps Oceanographic Institution in 1936. His 1947 paper "Wind-driven currents in a baroclinic ocean", which showed the link between meridional currents and the curl of the wind stress, began the modern era of dynamical oceanography and initiated large-scale modeling of ocean circulation.

The problem the oceanographer's could not understand for long is if wind is to generate the currents and circulation, how some permanent strong currents like equatorial counter and under currents and western boundary currents are developed? Why eastern boundary currents are not as strong as western boundary currents?

Ekman's initial work (1905) on wind driven currents, intended to explain current flow at an angle to the wind direction. Sverdrup (1947) demonstrated another surprising relationship between wind stress and ocean circulation as shown in the following pages.

Sverdrup's aim was to determine current flow in response to wind stress and horizontal pressure gradients. Unlike Ekman, Sverdrup was not interested in determining how flow varied with depth; instead, he derived an equation for the net flow resulting from wind stress.

The three oceanographers, Sverdrup, Stommel and Munk, though discussed the theory of wind driven circulation in general, their main concern was North Atlantic and North Pacific circulations. The salient features of North Atlantic circulation are as below.

In the North Atlantic, the circulation is characterized by two cells, usually called 'gyres'. The smallest one, the subpolar gyre, is formed by currents south of Greenland and in the Labrador Sea such as the Irminger Current, the West Greenland Current and the Labrador Current. The largest one, the subtropical gyre, is formed by currents around the Sargasso Sea, such as the North Equatorial Current to the south. Near the western boundary, the latter current joins part of the

South Equatorial current and part of the combined currents flows northward as the Antilles Current, while the other part flows into the Gulf of Mexico. The latter water eventually 'escapes' between Florida and Cuba as the Florida Current.

A merger of the Florida and Antilles current leaves the east coast of the United States at Cape Hatteras and is then called the Gulf Stream. The Gulf Stream flows northeastwards towards the Grand Banks of Newfoundland near (40 $^0$N, 50$^0$W). The current that exists north-eastward of the Gulf Stream is called the North Atlantic Current. This current bifurcates into a part that contributes to the flows in the Norwegian Sea and a part that deflects southward as the Canary current.

Characteristics of the flow in the North Atlantic are strong boundary currents (Florida Current and Gulf Stream) which exist at the western side of the basin. The Florida Current is situated above the continental slope, but the Gulf Stream is situated in the open ocean. The horizontal velocities in the Gulf Stream (up to 2.5 $ms^{-1}$) are among the largest measured in the ocean. Its mean velocity is about 1.5 $ms^{-1}$ and its average width is about 115 km. The average volume transport of the Florida Current is 30 Sv (1 Sv = $10^6 m^3 s^{-1}$). This transport increases towards Cape Hatteras up to 150 Sv near 65$^0$W, but decreases again to 35 Sv near 40$^0$W. In comparison, the average surface current velocity in the subtropical gyre is about 0.1 $ms^{-1}$.

Sverdrup's (1947) theory of wind driven circulation began with the non-accelerated frictional equations of motion (8.1) integrated up to a depth of no motion (e.g., 1000 m) where the wind stress ($\tau_x$,$\tau_y$) was assumed as zero.

$$\frac{du}{dt} = -\alpha\frac{\partial p}{\partial x} + f.v + \alpha\frac{\partial \tau_x}{\partial z} = 0 \text{ and } \frac{dv}{dt} = -\alpha\frac{\partial p}{\partial y} + f.v + \alpha\frac{\partial \tau_y}{\partial z} = 0 \tag{8.1}$$

To derive the relationship, Sverdrup assumed that the flow is non accelerated (stationary), the lateral friction and molecular viscosity are small, the nonlinear terms such as $u\frac{\partial u}{\partial x}$ are small and that turbulence near the sea surface can be described using a vertical eddy viscosity. He also assumed that wind driven circulation vanishes at the depth of frictional resistance (- D). With these assumptions the horizontal components of equations of motion (8.1) become:

$$\alpha\frac{\partial p}{\partial x} = f.v + \alpha\frac{\partial \tau_x}{\partial z} \text{ and } \alpha\frac{\partial p}{\partial y} = -f.u + \alpha\frac{\partial \tau_y}{\partial z} \tag{8.2}$$

One may note that while Ekman ignored the horizontal pressure gradient terms, Sverdrup included them but he ignored horizontal friction terms. Sverdrup wanted to see the total net transport in the entire layer upto the depth of frictional resistance. So he integrated the equations (8.2) for the whole column between the limits – D to zero (sea surface).

$$\int_{-D}^{0}\frac{\partial p}{\partial x}dz=\int_{-D}^{0}f.\rho.v+\int_{\tau_D}^{\tau_0}d\tau_x,\rightarrow\int_{-D}^{0}\frac{\partial p}{\partial x}dz=f.M_y+\tau_{x0}$$

$$\int_{-D}^{0}\frac{\partial p}{\partial y}dz=-\int_{-D}^{0}f.\rho.u.dz+\int_{\tau_0}^{\tau_0}d\tau_y\rightarrow\int_{-D}^{0}\frac{\partial p}{\partial y}dz=-f.M_x+\tau_{y0} \tag{8.3}$$

To solve these two equations (8.3), differentiate the first one with respect to y and the second one with respect to x and equate as both on the left hand side are equal.

$$\int_{-D}^{0}\frac{\partial}{\partial y}\frac{\partial p}{\partial x}dz=\int_{-D}^{0}\frac{\partial}{\partial y}f.M_y+\int_{-D}^{0}\frac{\partial}{\partial y}\tau_{x0},\int_{-D}^{0}\frac{\partial}{\partial x}\frac{\partial p}{\partial y}dz=-\int_{-D}^{0}\frac{\partial}{\partial x}(f.M_x)+\int_{\tau_0}^{\tau_0}\frac{\partial}{\partial x}(\tau_{y0})$$

Further these equations can be written as;

$$\int_{-D}^{0}\frac{\partial^2 p}{\partial x\partial y}dz=f.\frac{\partial M_y}{\partial y}+M_y\frac{\partial f}{\partial y}+\frac{\partial \tau_{x0}}{\partial y} \tag{8.4}$$

$$\int_{-D}^{0}\frac{\partial^2 p}{\partial x\partial y}dz=-f.\frac{\partial M_x}{\partial x}-M_x\frac{\partial f}{\partial x}+\frac{\partial \tau_{y0}}{\partial x} \tag{8.5}$$

As equation (8.4) is equal to (8.5), we can re arrange and write

$$f\left[\frac{\partial M_x}{\partial x}+\frac{\partial M_y}{\partial y}\right]+M_x\frac{\partial f}{\partial x}+M_y\frac{\partial f}{\partial y}=\left[\frac{\partial \tau_{y0}}{\partial x}-\frac{\partial \tau_{x0}}{\partial y}\right]\text{.....(8.6)}$$

In equation (8.6), the first term on the left hand side is equation of continuity which is zero and the second term $\frac{\partial f}{\partial x}$ is variation of Coriolis parameter along x direction (zonal) is zero. Let $\frac{\partial f}{\partial y}=\beta$. Then we can write equation (8.6) as

$$\beta M_y=\left[\frac{\partial \tau_{y0}}{\partial x}-\frac{\partial \tau_{x0}}{\partial y}\right] \tag{8.7}$$

We know $\nabla=\left[i\frac{\partial}{\partial x}+j\frac{\partial}{\partial y}\right]$ and $\tau_o = i\tau_{xo}+j\tau_{yo}$

$$\therefore \beta M_y=\nabla_z\times\tau_0 \tag{8.8}$$

This is called Sverdrup's equation. The most interesting aspect of this equation is that the net meridional transport depends not on the absolute value of the wind stress but on the curl (shear), $(\nabla_z\times\tau_0)$, of the wind stress. That is the tendency of the wind to cause rotation or its ability to supply relative vorticity to the ocean. Thus Sverdrup showed that the net amount of water transported meridionally is directly proportional to the shear of the wind stress. Thus the Sverdrup balance, or Sverdrup relation, is a theoretical relationship between the wind stress exerted on the surface of the open ocean and the vertically integrated meridional (north-south) transport of ocean water such that the positive wind stress curl is northward mas transport and the negative wind stress curl is `southward mass transport

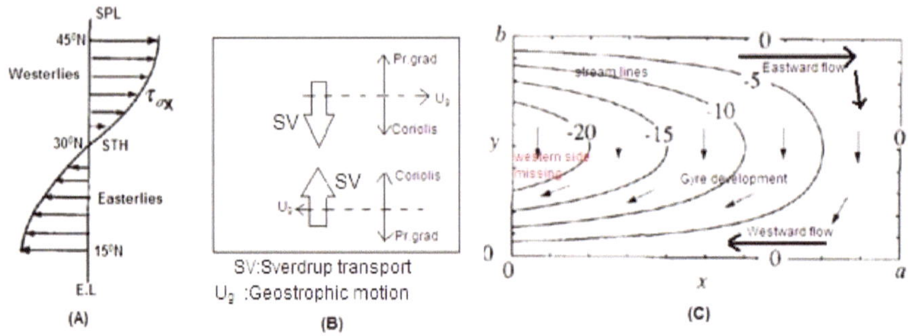


**Fig. 8.1:** Convergence, pileup and development of 'gyre' in midlatitudes according to Sverdrup. **Fig.8.1A**: Wind pattern in the equatorial and subtropical latitudes. **Fig. 8.1B:** Balance of forces (pressure gradient and Coriolis f) and directions of geostrophic velocity (due to easterlies and westerlies. **Fig. 8.1C:** clockwise gyre development in Northern Hemisphere) by joining both eastward and westward flows of (thick black arrows in Fig 8.1C). Note the gyre is incomplete in the western boundary (Ob) of the oceans.

Thus as per the above Fig 8.1, the Ekman Net transport in the northern half-basin is southward whereas that of the southern half-basin it is northward. Water piles up (converge) at mid-latitudes (STH, $30^0$ N) develop a high pressure ridge due to rising of free surface level. The induced eastward geostrophic flow in the northern half-basin requires a low pressure.

Since the general circulation is very large scale and not filled with ever-increasing vortices, the response is to move with latitude rather than to spin up locally. This response to Ekman pumping or suction is called the "Sverdrup transport"(Ekman Net Transport in Fig.8.1). Sverdrup transport is equatorward in subtropical regions of Ekman pumping and poleward in subpolar regions of Ekman suction (Fig.8.1).

It can be easily explained with potential vorticity concept. We know Sverdrup transport in relation to potential vorticity as

$$Q = \frac{f + \zeta}{H}$$

If there is Ekman convergence (pumping downwards), then H decreases. This must result in a decrease of the numerator ($f + \zeta$). Since we know from observations and scaling that relative vorticity does not spin up in the large-scale circulation, f must change, which means that the latitude must change. This means $Q \propto \frac{f}{H}$. If there is a decrease in H, then there is a decrease in latitude - water moves towards the equator. Hence

Sverdrup transport = meridional flow due to Ekman pumping/suction

The pattern of flow suggested by Sverdrup is as shown below in the Fig.8.1. This pattern comes due to the geostrophic velocity ($u_g$) generated due to the balance of pressure gradient $\alpha \frac{\partial p}{\partial n}$) and Coriolis ($fu_g$) forces (Fig. 8.1B). Also it is clear that at 30$^0$ latitude (STH), the shear is greatest and at 15$^0$ and 45$^0$ it is least. The areas where the shear is least, the flow is zonal. The pattern of flow shown by Sverdrup in Fig.8.1C is the 'semi-gyre'. According to Fig.8.1C, the Gyre developed because $u_g$ in the westerlies zone is eastward and $u_g$ in the easterlies zone is westward. So combining both eastward and westward flows (Fig.8.1C) a semi- clock wise 'gyre' comes in the entire ocean comprising of low and middle latitudes. The gyre is incomplete because he considered only the eastern boundary of the ocean. Western Boundary part was studied by Stommel (1948).

Thus the Sverdrup transport is the net Meridional transport diagnosed in both the subtropical and subpolar gyres, resulting from planetary vorticity changes that balance Ekman pumping or Ekman suction. The meridional transport $M_y$ as shown in equation (8.8) is the Sverdrup transport.

## 8.2. Application of Sverdrup equation to trade wind region

Sverdrup applied these equations to the trade wind region in lower latitudes where $\tau_y$ and $\frac{\partial}{\partial x}\left(\tau_y\right)$ are negligibly small and $\tau_x$ variations with x are averaged out taking $f = 2\Omega \sin\phi$ and $dy = R\, d\phi$ where R is radius of the earth, then

We can write $$\beta = \frac{\partial f}{\partial y} = \frac{d}{dy}(2\Omega \sin\phi) = \frac{d(2\Omega \sin\phi)}{Rd\phi} = \frac{2\Omega}{R}\frac{d}{d\phi}(\sin\phi) = \frac{2\Omega \cos\phi}{R} \tag{8.9}$$

Taking Sverdrup's equation (8.8): $$M_y = \frac{\left(\frac{\partial \tau_y}{\partial x} - \frac{\partial \tau_x}{\partial y}\right)}{\frac{\partial f}{\partial y}} \quad \text{........(8.10)}$$

we also know $$\left[\frac{\partial M_x}{\partial x} + \frac{\partial M_y}{\partial y}\right] = 0 \tag{8.11}$$

$$\therefore \quad \left[\frac{\partial M_x}{\partial x} = -\frac{\partial M_y}{\partial y}\right] \tag{8.12}$$

So differentiating equation (8.10) with respect to 'y' and then equate it to $\left[\frac{\partial M_x}{\partial x}\right]$

So $$\left[\frac{\partial M_x}{\partial x}\right] = -\frac{\partial}{\partial y}\frac{\left(\frac{\partial \tau_y}{\partial x}-\frac{\partial \tau_x}{\partial y}\right)}{\frac{\partial f}{\partial y}} = -\frac{\frac{\partial f}{\partial y}\left(-\frac{\partial^2 \tau_x}{\partial y^2}\right)-\left(\frac{\partial \tau_y}{\partial x}-\frac{\partial \tau_x}{\partial y}\right)\frac{\partial^2 f}{\partial y^2}}{\left(\frac{\partial f}{\partial y}\right)^2} \tag{8.13}$$

Over much of the open ocean, especially in the tropics, the wind is zonal and $\frac{\partial \tau_y}{\partial x}$ is negligibly small and from equation (8.9) we can write

$$\frac{\partial^2 f}{\partial y^2} = \frac{\partial}{\partial y}\left(\frac{\partial f}{\partial y}\right) = \frac{d}{R d\phi}\left(\frac{2\Omega\cos\phi}{R}\right) = \frac{2\Omega}{R^2}(-\sin\phi) = -\frac{2\Omega\sin\phi}{R^2} \tag{8.14}$$

Substituting all these values in equation (8.13) and simplifying we get

$$\frac{\partial M_x}{\partial x} = -\frac{\left[-\left\{\frac{2\Omega\cos\phi}{R}\left(-\frac{\partial^2 \tau_x}{\partial y^2}\right)+\left(\frac{\partial \tau_x}{\partial y}\right)\frac{2\Omega\sin\phi}{R^2}\right\}\right]}{\left(\frac{2\Omega\cos\phi}{R}\right)^2}$$

$$\frac{\partial M_x}{\partial x} = \frac{1}{2\Omega\cos\phi}\left[R\left(\frac{\partial^2 \tau_x}{\partial y^2}\right)+\frac{\partial \tau_x}{\partial y}\tan\phi\right] = \text{K (say)} \tag{8.15}$$

Then $dM_x = Kdx$ or $M_x = \int K dx = K.x$

Sverdrup integrated this equation (8.15) from a north-south eastern boundary at x = 0, assuming no flow into the boundary. This requires $M_x = 0$ at x = 0. Then

$$M_x = \frac{x}{2\Omega\cos\phi}\left[R\left(\frac{\partial^2 \tau_x}{\partial y^2}\right)+\frac{\partial \tau_x}{\partial y}\tan\phi\right] \tag{8.16}$$

Where x is the distance from the eastern boundary of the ocean basin as shown in Fig.8.2

Similarly $M_y$ can be obtained from equation (8.12) taking

$$\frac{\partial M_y}{\partial y} = -\frac{\partial M_x}{\partial x} = -\mu\frac{1}{2\Omega\cos\phi}\left[R\left(\frac{\partial^2 \tau_x}{\partial y^2}\right)+\frac{\partial \tau_x}{\partial y}\tan\phi\right]$$

$$\therefore \int dM_y = -\int\frac{1}{2\Omega\cos\phi}\left[R\left(\frac{\partial^2 \tau_x}{\partial y^2}\right)+\frac{\partial \tau_x}{\partial y}\tan\phi\right]dy$$

$$\therefore M_y = -\frac{1}{2\Omega\cos\phi}\left[R\frac{\partial \tau_x}{\partial y}+0\right] = -\frac{R}{2\Omega\cos\phi}\left[\frac{\partial \tau_x}{\partial y}\right] \tag{8.17}$$

Here 'x' is the distance from a north south coast line in the eastern boundary of the ocean (i.e coast is east side and ocean is west side as shown in the Fig 8.2). So from the coast west side into the ocean to a point 'P' is in the negative x-direction.

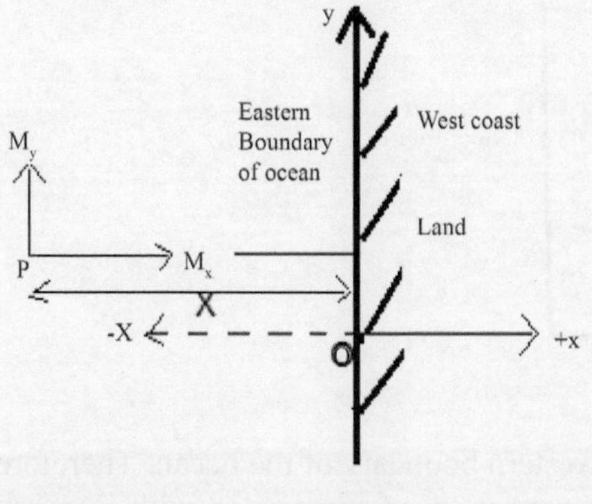


**Fig. 8.2.** Eastern boundary of the ocean according to Sverdrup

From the above equation (8.16) of $M_x$, it is apparent that in trade wind and equatorial zones of the ocean the important term is

$$\frac{\partial^2 \tau_x}{\partial y^2}$$

Using the above dynamics a simple ocean basin streamline pattern for the Atlantic Ocean (Fig.8.1C) was not closed on the western boundary. According to Sverdrup, in Figure 8.1C, each streamline drawn by Sverdrup taking the following constants. The boundaries are $x = 0$ and $x = a$, would roughly correspond to 75 $^0$W and 10 $^0$W whereas $y = 0$ and $y = b$ would correspond, to 15$^0$N and 50$^0$N. Labels on the streamlines are values of the transport stream function in units of $10^6$ m$^3$ s$^{-1}$. Note the deficiency that the streamlines is labeled with the value of $\psi$ for the typical values,$\tau_0 = 1$ dyne cm$^{-2}$, $\rho_0 = 1$gm cm$^{-3}$, $\beta = 2 \times 10^{-8}$ km$^{-1}$ s$^{-1}$, a = 6000 km, b = 4000 km. The units of $\psi$ are m$^3$ s$^{-1}$,. The value $1 \times 10^6$ m$^3$ s$^{-1}$ is called a "Sverdrup". This simple model yields a total anticyclonic circulation of about $25 \times 10^6$ m$^3$ s$^{-1}$ which is 25 Sverdrups.

The asymmetry in this flow pattern is a direct result of the circulation imparted by the wind stress distribution combined with the effects of earth rotation. The southward horizontal transport in the gyre is computed using the equation

$$\int_A^B M_y \, dx = -25 \times 10^6 \text{ m}^3 \text{ s}^{-1} = 25\text{Sv}$$

The minus sign indicates the flow of 25Sv is southward across the transect A-B in the gyre as shown in the Fig.8.3.

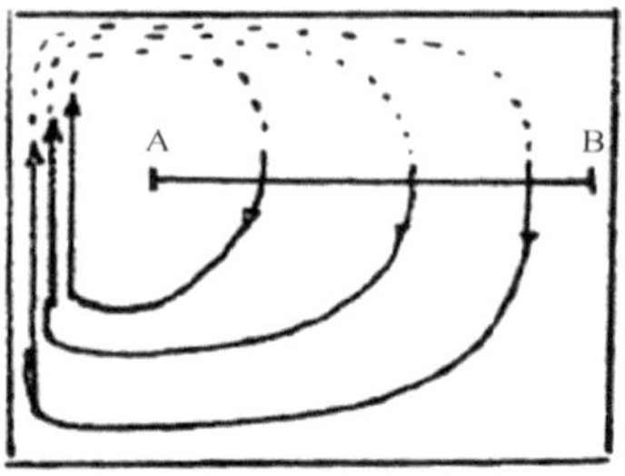


**Fig. 8.3.** Transport in a subtropical gyre.

This same transport must move up the western boundary of the ocean. Therefore, one would expect a large difference in current speeds associated with Sverdrup and western boundary current respectively. If the transect length from A to B = 3500 km and a depth of 1000 m, then $u_{Sverdrup} = 1\text{cm}^{-1}$ and $u_{western\ boundary} = 10\ \text{cm}^{-1}$. The real Gulf Stream is concentrated in a smaller cross-sectional area and exhibits maximum speeds of about 200 $\text{cm}^{-1}$.

Sverdrup's solution of equation (8.16) also explains the existence of the equatorial current system. Comparing the solutions of Ekman and Sverdrup while Ekman concentrated in the current variation with depth, Sverdrup studied the effect of coastal boundary in the east (Fig.8.2), which is more realistic than Ekman as Ekman considered horizontally infinite boundary less ocean.

All of the meridional flow is returned through western boundary currents according to Ekman. Therefore, subtropical gyres must be anticyclonic and subpolar gyres must be cyclonic as mentioned by Neumann and W.J. Pierson (1966)

A global map of the Sverdrup transport integrated from the eastern to the western boundary in three dimensional view and the size of the integral at the western boundary gives the transport of western boundary current since Sverdrup's model must be closed with a narrow boundary current that has at least one additional physical mechanism beyond those in the Sverdrup balance (a shift in latitude because of water column stretching driven by Ekman transport convergence). Also the mid latitude westerlies and low latitude trade winds generate the Ekman transport at the surface creating Ekman pumping and suction and Sverdrup transport results below thermocline in Northern Hemisphere.

To test the theory of Sverdrup, Reid (1948) compared transports calculated from known winds in the eastern tropical Pacific with transports calculated from hydrogrphic data collected by the Carnegie and Bushnell in October and November 1928, 1929 and 1939 between 34$^0$N and 10$^0$S and between 80$^0$W and 160$^0$W.

Note in Fig 8.1C the gyre is not closed in the western side of the oceans. In other words the information on the western side is missing in the Sverdrup's theory.

This is the deficiency in Sverdrup's theory. Also one can see that the subtropical gyre extends further north in the North Atlantic than the North Pacific, particularly during winter. This means that warm water is carried further north in the North Atlantic than in the North Pacific by the wind driven circulation. This would lead to more evaporation, higher salinity, greater density, etc., in the North Atlantic compared to the North Pacific. The reason why the subtropical gyre extends further north in the North Atlantic is because the maximum westerly winds is further north. It could be due to positions of mountain ranges or continents.

One interesting rationalization provided by Sverdrup theory is that it accounts for the Equatorial countercurrents observed in the tropical oceans i.e. currents that flow in a direction opposite to the prevailing winds. As can be seen in Fig.8.1B, Sverdrup balance implies meridional flow and convergence away from the Doldrums and subtropical Highs. This convergence of these Sv flows (Fig.8.1B) drives an eastward flow (Fig.8.1C) just north of the equator, even though the winds are Easterlies (Fig.8.1A) here without obeying Ekman physics between $15^0$ and $30^0$ N.

## 8.3. Constraints of Sverdrup's Circulation

Due to its extreme simplifications, the Sverdrup theory poses a series of issues: It is known to fail in representing the western side of gyres and subpolar gyres, which highly limits its domain of validity. Also it tends to largely underestimate (typically by a factor ~3) the magnitude of transports of both subtropical and subpolar gyres. Most importantly, it does not predict any return flow for the gyres, which is however, required by the continuity. In particular, it does not predict whether the return flow must occur in the eastern or western boundary: the integration of the Sverdrup relation from either the eastern or western boundary gives identical results in the interior.

Although we get the right sign and order of magnitude for the basin integrated gyral circulation, a gyre of negative vorticity will have positive values of its stream function, and vice versa for a cyclonic, largely under estimated it, which is a typical bias of Sverdrup theory. In the Pacific, the higher basin width by 25% mechanically induces a stronger basin integrated Sverdrup transport by 25%.

## 8.4. Fofonoff's Analysis of Interior Transport

In one further important simplified approach to large-scale ocean circulation is due to Nicholas Fofonoff, in 1954. He showed that circulation can arise as a free, unforced mode. The idea is that a very small amount of wind, with very little friction anywhere in the system, could set up such a circulation. In the Fofonoff circulation, there is no Sverdrup interior with flow moving northward or southward. The interior flow is exactly zonal (east-west). This is because there is no wind input of vorticity, so flow cannot change latitude

In order to analyze the vertically integrated equations of motion, let us start with the non-accelerated equations of motion with vertical friction terms as considered by Sverdrup in equations (8.2), rewritten as:

$$-f\rho v = -\frac{\partial p}{\partial x} + \frac{\partial \tau_x}{\partial z}$$

$$f\rho u = -\frac{\partial p}{\partial y} + \frac{\partial \tau_y}{\partial z}$$

Expressing in terms of mass transports and On integrating these equations from ocean bottom (D) to the sea surface (η), we get:

$$\begin{aligned} -\int_D^{\eta} f\rho v dz &= -\int_D^{\eta} \frac{\partial p}{\partial x} dz + \int_D^{\eta} \partial \tau_x \\ \int_D^{\eta} f\rho u dz &= -\int_D^{\eta} \frac{\partial p}{\partial y} dz + \int_D^{\eta} \partial \tau_y \end{aligned} \tag{8.18}$$

The whole exercise of these equations to understand the transport in the subsurface layer due to the effect of wind stress at the surface lies on the evaluation of the first term on the right hand side. Fofonoff (1962) did extensive analysis. His analysis is as below:

**Evaluation of** $\int_D^{\eta} \frac{\partial p}{\partial x} dz$ :

Put $E_p = \int_D^{\eta} p dz$ ……………………..(8.19)

where $E_P$ is the potential energy of a column of water of unit horizontal area, relative to the bottom of the ocean. Then we can write:

$$E_p = \int_D^{\eta} \frac{p\alpha}{g\alpha} g dz = \frac{1}{g}\int_D^{\eta} (p\alpha)\rho g dz = \frac{1}{g}\int_D^{\eta} (p\alpha)(-dp) = -\frac{1}{g}\int_D^{\eta} (p\alpha)(dp) = \frac{1}{g}\int_{\eta}^{D} (p\alpha)(dp)$$

$$E_P = \frac{1}{g}\int_{\eta}^{D} (p\alpha)(dp) = \frac{1}{g}\int_{\eta}^{D} (p\alpha_{35,0,p})(dp) + \frac{1}{g}\int_{\eta}^{D} (p\delta)(dp) = E_p^0 + \chi \tag{8.20}$$

Here $\chi$ is anomaly of potential energy and $E_p^0$ is the potential energy which is a function of bottom pressure $P_D$ only.

$$\therefore \frac{\partial E_p}{\partial x} = \frac{\partial E_P^0}{\partial x} + \frac{\partial \chi}{\partial x} \tag{8.21}$$

From equation (8.19) we can write:

$$\frac{\partial}{\partial x}(E_p) = \frac{\partial}{\partial x}\int_D^{\eta} p dz = \int_D^{\eta} \frac{\partial p}{\partial x} dz + p_{\eta}\frac{\partial \eta}{\partial x} - p_D\frac{\partial D}{\partial x} \tag{8.22}$$

As η and D vary with x the last two terms in equation (8.22) appear. But it may be noted that $P_\eta$ is pressure at sea surface is zero. So the second term in the right hand side of equation (8.22) vanishes. Hence equation (8.22) can be written as:

$$\int_D^\eta \frac{\partial p}{\partial x} dz = \frac{\partial}{\partial x}\left(E_p\right) + p_D \frac{\partial D}{\partial x} \tag{8.23}$$

Substituting (8.21) in (8.23) we can get

$$\int_D^\eta \frac{\partial p}{\partial x} dz = \frac{\partial E_P^0}{\partial x} + \frac{\partial \chi}{\partial x} + p_D \frac{\partial D}{\partial x} \tag{8.24}$$

Fofonoff combined these two terms ($\frac{\partial E_P^0}{\partial x} + p_D \frac{\partial D}{\partial x}$) and expressed as

$$\int_D^\eta \frac{\partial p}{\partial x} dz = \frac{\partial \chi}{\partial x} + \frac{p_D \alpha_D}{g} \frac{\partial P_D}{\partial x} \tag{8.25}$$

Where $\frac{\partial P_D}{\partial x}$ is a component of pressure gradient along a level surface at the bottom and $\alpha_{35,0,P_D}$ is assumed to be equal to the specific volume at the bottom, $\alpha_D$.

Substituting equation (8.25) in equations (8.18) we get:

$$-\int_D^\eta f \rho v dz = -\frac{\partial \chi}{\partial x} - \frac{p_D \alpha_D}{g} \frac{\partial P_D}{\partial x} + \tau_{x\eta} - \tau_{xD}$$

$$\int_D^\eta f \rho u dz = -\frac{\partial \chi}{\partial y} - \frac{p_D \alpha_D}{g} \frac{\partial P_D}{\partial y} + \tau_{y\eta} - \tau_{yD} \tag{8.26}$$

## i) First terms on the right hand side of the equations (8.26)

The first terms on the R.H.S are associated with variations in density and so these can be interpreted as the baroclinic part of the geostrophic velocity. So this is like balance between pressure gradient of mass transport to the Coriolis part of the transport

$$\therefore -fM_{yc} = -\frac{\partial \chi}{\partial x} \text{ and } fM_{xc} = -\frac{\partial \chi}{\partial y} \tag{8.27}$$

## ii) Second terms on the right hand side of the equations (8.26)

The mass transports near the bottom are assumed as geostrophic as velocities near the bottom are small and tangential stresses near the bottom are neglected. So the pressure gradient at the bottom is equal to the Coriolis at bottom.

$$\therefore fu_D = -\alpha_D \frac{\partial P_D}{\partial y} \text{ and } fv_D = -\alpha_D \frac{\partial P_D}{\partial x} \tag{8.28}$$

So the term $-\dfrac{p_D \alpha_D}{g}\dfrac{\partial P_D}{\partial x}$ can be written, substituting eqn (8.28) as

$$\frac{fp_D v_D}{g} \tag{8.29}$$

This equation (8.29) can be related to $fM_{yD}$ because $M_{yD} = \int_D^\eta \rho v_D dz = v_D \int_D^\eta \rho dz$ as $v_D$ does not vary with depth. Taking ρdz = -dp/g, this becomes

$$v_D \int_D^\eta \rho dz = -v_D \int_D^\eta \frac{dp}{g} dz = -\frac{v_D}{g}\left(p_\eta - p_D\right) = \frac{v_D p_D}{g} (\because p_\eta = 0) \tag{8.30}$$

So from 8.29 & 8.30 we can write $fM_{yD} = \dfrac{fp_D v_D}{g}$ (8.31)

Similarly y component of the second term can be written as

$$fM_{xD} = \frac{fp_D u_D}{g} \tag{8.32}$$

### iii) Third terms on the right hand side of the equations (8.26)

According to Ekman these terms can be written from equation (6.17) of Ekman circulation as: $M_{xDE} = \dfrac{\tau_{yD}}{f}$ and $M_{yDE} = -\dfrac{\tau_{xD}}{f}$ (8.33)

So finally equations (8.26) for total transport can be written using equations (8.27), (8.31) and (8.33) as:

$$-fM_y = -fM_{yc} - fM_{yD} - fM_{yDE}$$

$$fM_x = fM_{xc} + fM_{xD} + fM_{xDE}$$

Which means

$M_y = M_{yc} + M_{yD} + M_{yDE}$ ................(8.34)

$M_x = M_{xc} + M_{xD} + M_{xDE}$

Thus the total transport in the interior of the ocean can be considered to be made up of three types of transports (baroclinic, barotropic and Ekman types). While both the baroclinic ($M_{yc}$) and barotropic ($M_{yD}$) transports are geostrophic, the Ekman transport ($M_{yDE}$) is non geostrophic slope current due to the influence of wind stresses at the sea surface.

Though the mass continuity equation (8.11) requires that the divergence of sum of the three types of transports in equation (8.34) is zero (or equal to rate at which water is added at the surface), this need not hold good for each type of transport separately. So Fofonoff (1954) interprets the equation (8.26) following the approach of Sverdrup by cross differentiation and adding both the equations as :

$$f\left[\frac{\partial M_{xg}}{\partial x}+\frac{\partial M_{yg}}{\partial y}\right]+\beta M_{yg}=0 \tag{8.35}$$

$$f\left[\frac{\partial M_{xE}}{\partial x}+\frac{\partial M_{yE}}{\partial y}\right]+\beta M_{yE}=\frac{\partial \tau_{My}}{\partial x}-\frac{\partial \tau_{Mx}}{\partial y} \tag{8.36}$$

$$f\left[\frac{\partial M_{xD}}{\partial x}+\frac{\partial M_{yD}}{\partial y}\right]+\beta M_{yD}=-\rho' f\left[u_D\frac{\partial D}{\partial x}+v_D\frac{\partial D}{\partial y}\right] \tag{8.37}$$

Where $\rho'=\rho_D\left[1+\frac{p_D}{\alpha_D}\left(\frac{\partial \alpha}{\partial p}\right)_D\right]$

Combining all these equations of (8.35) to (8.37) we can write

$$\beta M_y=\nabla\times\tau_\eta-\nabla\times\tau_D-\rho' f\left[u_D\frac{\partial D}{\partial x}+v_D\frac{\partial D}{\partial y}\right] \tag{8.38}$$

Here $\chi$ terms cancel each other while cross differentiating both the equations of (8.26). If the barotropic deep flow is assumed as zero and Tangential stresses near the bottom are also assumed as zero, equation (8.38) reduces to the Sverdrup's equation (8.8).

According to Figure 8.6 an inertial boundary layer at the west diverts the flow northward and an eastward jet is formed such that the latter feeds into an inertial boundary layer on the east that supplies the steady westward flow of the interior.

## References


Dijkstra, H.A 2008. Dynamical Oceanography, ISBN: 978-3-540-76375-8, Springer-Verlag Berlin Heidelberg, 407pp

Dijkstra, H. A., 2005. Nonlinear Physical Oceanography: A Dynamical Systems Approach to the Large Scale Ocean Circulation and El Ni˜no, 2nd Revised and Enlarged edition. Springer, New York, 532 pp.

Ekman, V. Walfrid. 1905. "On the influence of the earth's rotation on ocean currents." Ark. f. Mat. Astr. och Fysik. K. Sv. Vet. Ak., Stockholm, 1905–06, v. 2, no. 11, 1905.

Fofonoff, N. P., 1954. Steady flow in a frictionless homogeneous ocean. J. Mar. Res., 13, 254-262.

Fofonoff, 1962.Dynamicsofoceancurrents,InTheSea,Vo.1:PhysicalOceanography,M.N.Hill,ed., Interscince, New York, 323-395.

Hidaka, K.1940. "Absolute evaluation of ocean currents in dynamical calculations" . *Imperial Acad. Japan, Proc.*, v. 16, p. 391–393, 1940. Tokyo

Lynne D Talley, 2011. Descriptive physical oceanography: an introduction. Academic press,.

Mellor, G.L 1996. Introduction to Physical Oceanography,193pp; ISBN 1-56396-210- I Springer-Verlag New York.

Munk, W. H., 1950: On the wind-driven ocean circulation. *J. Meteorol.,7, 79-93.*

Neumann G., and W.J. Pierson. 1966. Principles of Physical Oceanography, 2nd Edition. New Jersey:Prentice-Hall, Englewood cliffs.

Pickard, G. L. and W. J. Emery, 1990: *Descriptive Physical Oceanography, 5th edition.* Pergamon Press, Oxford, UK.

Pond, E. and G. L. Pickard, 1983: *Introduction to Dynamical Oceanography, 5th edition.* Pergamon Press, Oxford, UK.329pp

Reid R. O., 1948: The Equatorial Currents of the Eastern Pacific as Maintained by the Stress of the Wind. Jl. of Marine Res., 7, pp. 74-99.

Robert Stewart on line text book 'Introduction to Physical Oceanoraphy. https://www.uv.es/hegigui/Kasper/por%20Robert%20H%20Stewart.pdf

Stommel, H., 1948: The westward intensification of wind-driven ocean currents. *Trans., Am. Geophys. Union* 29, 202-206.

Stommel, H., 1977: *The Gulf Stream.* Univ. California Press, Los Angeles, USA.

Sverdrup, H. U., 1947: Wind driven currents in a baroclinic ocean; with applications to the equatorial currents of the eastern Pacific. Proc. Natl. Acad. Sci. USA, 33, 318–326, doi:10.1073/ pnas.33.11.318.

Sverdrup, H. U., and W. H. Munk, 1947: Wind, sea and swell: Theory of relations for forecasting. *U.S. Hydrographic Office, Publication* 301,44 pp.

Tolmazin, D., 1985: *Elements of Dynamic Oceangography.* Allen and Unwin Inc Winchester MA, 181 pp.

Vallis, G. K., 2006: *Atmospheric and Oceanic Fluid Dynamics: Fundamentals and Large-Scale Circulation.* Cambridge University Press, Cambridge, U.K.

## Exercises

### *Exercise 8.1*

Show that Sverdrup transport is equal to the sum of geostrophic and Ekman mass transports

### *Solution*

Taking the equation of continuity : $\frac{\partial u}{\partial x}+\frac{\partial v}{\partial y}+\frac{\partial w}{\partial z}=0$ (1)

We know the velocities in geostrophic motion as:

$$fu=-\frac{1}{\rho}\frac{\partial p}{\partial y} \tag{2}$$

and $fv=\frac{1}{\rho}\frac{\partial p}{\partial x}$ (3)

Differentiating equation (2) w.r.t "x" :

$$\frac{\partial}{\partial x}(fu)=\frac{\partial}{\partial x}\left[-\frac{1}{\rho}\frac{\partial p}{\partial y}\right]\rightarrow f\frac{\partial u}{\partial x}+u\frac{\partial f}{\partial x}=-\frac{1}{\rho}\frac{\partial^2 p}{\partial x\partial y} \tag{4}$$

Similarly differentiating equation (3) w.r.t "y":

$$\frac{\partial}{\partial y}(fv)=\frac{\partial}{\partial y}\left[\frac{1}{\rho}\frac{\partial p}{\partial x}\right]\rightarrow f\frac{\partial v}{\partial y}+v\frac{\partial f}{\partial y}=\frac{1}{\rho}\frac{\partial^2 p}{\partial x\partial y} \tag{5}$$

Adding equations (4) and (5): $f\left[\frac{\partial u}{\partial x}+\frac{\partial v}{\partial y}\right]+v\beta=0$ (6)

Note here: $\frac{\partial f}{\partial x}=0$ , since variation Coriolis parameter along the latitude is zero and $\frac{\partial f}{\partial y}=\beta$

From equation (1) we can write: $\frac{\partial u}{\partial x}+\frac{\partial v}{\partial y}=-\frac{\partial w}{\partial z}$ (7)

Substituting equation (7) in equation (6) we can write:

$$f\left[-\frac{\partial w}{\partial z}\right]+v\beta=0\rightarrow v\beta=f\frac{\partial w}{\partial z} \quad (8)$$

We know from chapter-6 of Ekman equation (6.8):

$$fv_E+\alpha\frac{\partial\tau_x}{\partial z}=0\rightarrow v_E dz=-\frac{1}{\rho f}d\tau_x \quad (9)$$

Similarly x-component: $fu_E-\alpha\frac{\partial\tau_y}{\partial z}=0\rightarrow u_E dz=\frac{1}{\rho f}d\tau_y$ (10)

Integrating equation (9) in the Ekman layer:

$$v_E\int dz=-\frac{1}{\rho f}\int d\tau_x\rightarrow v_E=-\frac{\tau_{x0}}{\rho f} \quad (11)$$

Similarly integrating equation (10): $u_E\int dz=\frac{1}{\rho f}\int d\tau_y\rightarrow u_E=\frac{\tau_{y0}}{\rho f}$ (12)

Integrating eqution (8): from the bottom of upper ocean ($D_w$) to the bottom of the Ekman layer ($D_E$):

$$\beta\int_{D_w}^{D_E}v\,dz=f\int_{W_{Dw}}^{w_E}dw\rightarrow\beta\int_{D_w}^{D_E}v\,dz=f\left[w_E-w_{Dw}\right] \quad (13)$$

From equation (1), we can write for Ekman: $\frac{\partial u_E}{\partial x}+\frac{\partial v_E}{\partial y}=-\frac{\partial w_E}{\partial z}$ (14)

Substituting equations (10) and (11) in (14):

$$\frac{\partial}{\partial x}\left(\frac{\tau_{y0}}{\rho f}\right)+\frac{\partial}{\partial y}\left(-\frac{\tau_{x0}}{\rho f}\right)=-\frac{\partial w_E}{\partial z} \quad (15)$$

Substituting equation (15) in equation (13):

$$\int_{D_w}^{D_E}v\,dz=f\left[w_E-w_{Dw}\right]=\frac{f}{\beta}\left[\frac{\partial}{\partial x}\left(\frac{\tau_{y0}}{\rho f}\right)+\frac{\partial}{\partial y}\left(-\frac{\tau_{x0}}{\rho f}\right)\right]-\frac{f}{\beta}w_{Dw}=\frac{1}{\rho\beta}\left[\frac{\partial\tau_{y0}}{\partial x}-\frac{\partial\tau_{x0}}{\partial x}\right]-\frac{\tau_{x0}}{\rho f}$$

$$\int_{D_w}^{D_E}v\,dz=\frac{1}{\beta\rho}\left[\frac{\partial\tau_{y0}}{\partial x}-\frac{\partial\tau_{x0}}{\partial x}\right]-\frac{\tau_{x0}}{\rho f}$$

$$\frac{1}{\beta\rho}\left[\frac{\partial\tau_{y0}}{\partial x}-\frac{\partial\tau_{x0}}{\partial x}\right]= \quad \int_{D_w}^{D_E} v\,dz \quad +\frac{\tau_{x0}}{\rho f} \tag{16}$$

*Sverdrup* *geostrophic* *Ekman*

Hence from equation (16), it is proved that Sverdrup transport is equal to the sum of geostrophic and Ekman mass transports

### Exercise 8.2

Estimate Sverdrup, Ekman and Geostrophic transports between the trade winds and subtropical westerlies of tropical oceans. Show that Sverdrup transport is ten times more than the Ekman transport in the midlatitude gyres. Estimate the heat transport associated with the whole gyre. (Take $C_P = 4.1 \times 10^3$ jkg$^{-1}$ k$^{-1}$, $\tau_{westerlies}$ = 0.05Pa; $\tau_{easterlies}$ = – 0.05Pa; $\beta = 2 \times 10^{-11}$ m$^{-1}$ s$^{-1}$; f = $10^{-4}$ and the meridional North-South length scale of the gyre is 1000 km and find the total Sverdrup volume transport in the Pacific ocean basin width of 12000 km and the gyre width of 1000km.

### Solution

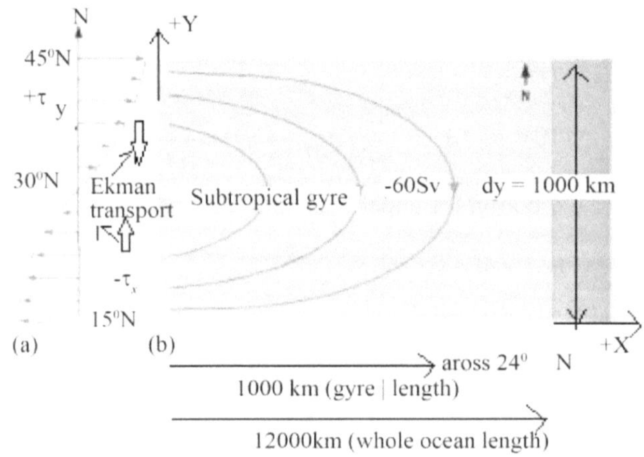


i) Sverdrup transport $= \beta M_y = \nabla\times\tau_0 = \left[\frac{\partial\tau_{y0}}{\partial x}-\frac{\partial\tau_{x0}}{\partial y}\right]$

As the wind is zonal, wind stress in the y direction ($\partial\tau_{y0}$) vanishes.

ii) So curl of wind stress=

$$\nabla\times\tau_0=\left[-\frac{\partial\tau_{x0}}{\partial y}\right]=-\left[\frac{0.05-(-0.05)}{1000\times10^3 m}\right]=-\frac{0.1Pa}{10^6 m}=-10^7\,kgm^{-2}s^{-2}=-10^7\,Nm^{-3}$$

iii) Sverdrup's meridional transport per unit distance in the 'x' direction taking $\beta = 2 \times 10^{-11}$ m$^{-1}$ s$^{-1}$

$$M_y = \frac{1}{\beta}(\nabla \times \tau_0) = -\frac{1}{\beta}\left[\frac{\partial \tau_{x0}}{\partial y}\right] = -\frac{10^{-7}\,kgm^{-2}s^{-2}}{2\times 10^{-11}\,m^{-1}s^{-1}} = -5000kgm^{-1}s^{-1}$$

Because it has minus sign the Sverdrup transport is southward at 5000 $kgs^{-1}$ per meter along west to east width.

iv) Ekman mass transport in the trade wind region taking $\tau_{easterlies}$ = – 0.05Pa; f = $10^{-4}$

$$M_{yEK} = -\left[\frac{\tau_0}{f}\right] = -\frac{(-0.05)kgm^{-1}s^{-2}}{10^{-4}\,s^{-1}} = 500kgm^{-1}s^{-1}$$

As the Ekman mass transport (product of volume and density) is positive, it is northward in $kgs^{-1}$ per meter of zonal distance.

v) Comparison of Ekman & Sverdrup transports: 500$kgm^{-1}$ $s^{-1}$ × 10 = 5000$kgm^{-1}$ $s^{-1}$. This means Sverdrup transport is ten times more than that of Ekman.

vi) Note that the total Sverdrup mass transport in the y-direction per unit horizontal distance x is $M_{yTot}$ which is the sum of Ekman (wind driven) and geostrophic (density driven) (See from Exercise 8.1)

$\therefore M_{yTot} = M_{yE} + M_{yG} \rightarrow \therefore M_{yG} = M_{yTot} - M_{yE} \rightarrow M_{yG} = -5000 - 500 = -5500kgm^{-1}\,s^{-1}$

The negative sign of the geostrophic transport indicates, it is southward (negative y direction)

**For Pacific Ocean**

vii) As these transports are per unit width in the east-west direction assuming the wind stress is uniform across the whole width of 12000km, the total southward Sverdrup mass transport in Pacific =

$$M_{ST} = -\frac{10^{-7}\,kgm^{-2}s^{-2}}{2\times 10^{-11}\,m^{-1}s^{-1}} = -5000kgm^{-1}s^{-1}\times 12000\times 10^{3}m = -60\times 10^{9}\,kgs^{-1}$$

viii) Total southward Sverdrup Volume transport = (Mass divided by density) =

$$Q_{ST} = \frac{-60\times 10^{9}\,kgs^{-1}}{10^{3}\,kgm^{-3}} = -60\times 10^{6}\,m^{3}s^{-1} = -60sv$$

ix) Ekman transport in Pacific = one tenth of Sverdrup transport= 60Sv/10= 6 Sv (from point V)

x) Geostrophic interior flow in Pacific= Total(Sverdrup) -Ekman= - 60 –( 6) = -66Sv (southward)

xi) To conserve this Sverdrup southward flow (-60Sv) at eastern boundary a western boundary mass flow of (+60 Sv) has to be developed (Kuroshio) northward.

xii) Estimation of heat transport in the entire gyre taking $T_G$, $T_{Kuroshio}$ & $T_{Ekman}$ as 15$^0$, 18$^0$& 22$^0$ C respectively:

Heat transport in the whole gyre =

$$H_f = \rho C_p \int QTdxdz = \rho C_p \{Q_G T_G + Q_E T_E + Q_{EK} T_{EK}\} dxdz$$

$$H_f = 10^3 m \times 4.1 \times 10^3 jkg^{-1}k^{-1}[-66\times15 + 6\times22 + 60\times18] kgm^{-1}s^{-1}\ {}^0c \times 12000\times1000m$$

$$= 10922.4\times10^{12} Joule/second \approx 10.9\times10^{15} watts \approx 11 petawatts$$

Note while -66Sv is transported southward in the eastern basin, +66 Sv(+60+6) transported northward through western basin.

To know the annual mean oceanic heat budget across 24$^0$ N in Pacific, take the width of 24$^0$ N as 1000 km instead of 12000 km, then

$$H_f = 10^3 m \times 4.1 \times 10^3 jkg^{-1}k^{-1}[-66\times15 + 6\times22 + 60\times18] kgm^{-1}s^{-1}\ {}^0c \times 1000\times1000m$$

$$\approx 910\times10^{12} Joule/second \approx 0.91\times10^{15} watts \approx 0.91 petawatts$$

Or $\dfrac{11 peta\,watts}{12} = 0.91\, peta\,watts$.

### Exercise 8.3

If the zonal winds in Atlantic at 10$^0$,40$^0$ and 75$^0$ N latitudes are --5m/s, 10 m/s and 0 m/s respectively and the Atlantic basin width is 6000 km, compute the Sverdrup's interior meridional transport between 30$^0$ N and 60$^0$ N. (b) Also calculate the same with the same data for Pacific ocean taking a Pacific width of 8000 km. (Take: $c_D$ = $2\times10^{-3}$, $\rho_a = 1 kgm^{-3}$, $\rho_w = 1000\ kgm^{-3}$, $\beta(30^o N) = 2\times10^{-11}\ m^{-1}\ s^{-1}$, $\beta(60^o N) = 10^{-11}\ m^{-1}\ s^{-1}$

### Solution

We know sverdrup's equation as:

$$\beta M_y = \left[\frac{\partial \tau_{y0}}{\partial x} - \frac{\partial \tau_{x0}}{\partial y}\right]$$

As wind is zonal, stress along y component is zero

$$\therefore \frac{\partial \tau_{y0}}{\partial x} = 0 \Rightarrow M_y = -\frac{1}{\beta}\left[\frac{\partial \tau_{x0}}{\partial y}\right]$$

According to Ekman: $M_{x30} = -\dfrac{\rho_a c_D u_{30}{}^2}{f_{30}}$ ; but this is not real mass transport but so volume transport is to be considered in place of $M_x$. So

$$Q_{x30} = \frac{\tau_{y30} W}{\rho_w f_{30}} = \frac{\rho_a c_D u_{30}{}^2 W}{f_{30}\rho_w}$$ ; Applying Munk formula to this:

$$\Rightarrow M_y = -\frac{1}{\beta}\left[\frac{\partial \tau_{x0}}{\partial y}\right] = -\frac{1}{\beta}\left[\frac{\tau_{x30} - \tau_{x60}}{30^0}\right] = -\frac{1}{\beta}\left[\frac{\rho_a c_D}{30^0}\left[u_{30}{}^2 - (-u_{60})^2\right]\right]$$

Hence $Q_{x10-40} = -\frac{W}{\beta \rho_w} \frac{\rho_a c_D}{\Delta y}\left[ u_{10}^{\ 2} - u_{40}^{\ 2} \right]$

$$= -\frac{6\times10^6 m \times 1 kgm^{-3} \times 2\times10^{-3}}{2\times10^{-11} m^{-1} s^{-1} \times 10^3 \times 30\times10^5 m}\left[ (-5)^2 - (10)^2 m^2 s^{-2} \right] = 15\times10^6 m^3 s^{-1} = 15 sv$$

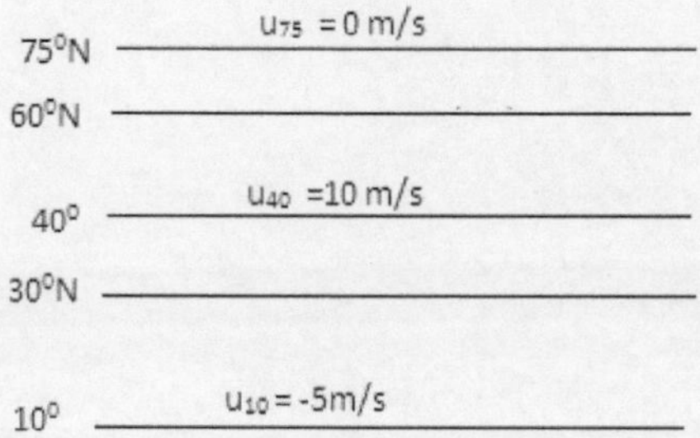


Assuming a uniform increase of transport between 10⁰ and 40⁰ N (30⁰), the transport is 15 sv. This means per 1⁰ the transport is 15/30=0.5 sv. Hence for 20⁰(between 10-30⁰ N), the transport is 0.5 × 20 = 10sv.

Similarly between 40 -75: $Q_{x40-75} = -\frac{W}{\beta \rho_w} \frac{\rho_a c_D}{\Delta y}\left[ u_{40}^{\ 2} - u_{75}^{\ 2} \right]$

$$= -\frac{6\times10^6 m \times 1 kgm^{-3} \times 2\times10^{-3}}{10^{-11} m^{-1} s^{-1} \times 10^3 \times 35\times10^5 m}\left[ (10)^2 - (0)^2 m^2 s^{-2} \right] = -34.3\times10^6 m^3 s^{-1} = -34.3 sv$$

For 35⁰ (40⁰-75⁰) the decrease of transport is 34.3 sv. For 1⁰, the decrease is 34.3/35=0.98sv. For 20⁰ (40⁰-60⁰), the decrease is 0.98(20) = 19.6sv. This means the net decrease of transport is 34.3-19.6=14.7Sv.

Thus for 60⁰N: the Sverdrup transport is -14.7 sv

## Similarly for Pacific Ocean

$$= -\frac{8\times10^6 m \times 1 kgm^{-3} \times 2\times10^{-3}}{2\times10^{-11} m^{-1} s^{-1} \times 10^3 \times 30\times10^5 m}\left[ (-5)^2 - (10)^2 m^2 s^{-2} \right] = 60\times10^6 m^3 s^{-1} = 60 sv$$

Assuming a uniform increase of transport between 10⁰ and 40⁰ N (30⁰), the transport is 60 sv. This means per 1⁰ the transport is 60/30=2 sv. Hence for 20⁰(between 0-30⁰ N), the transport is 2 × 20 = 40sv.

Similarly between 40 -75: $Q_{x40-75} = -\frac{W}{\beta \rho_w} \frac{\rho_a c_D}{\Delta y}\left[ u_{40}^{\ 2} - u_{75}^{\ 2} \right]$

$$= -\frac{8\times10^6 m \times 1 kgm^{-3} \times 2\times10^{-3}}{10^{-11} m^{-1} s^{-1} \times 10^3 \times 35\times10^5 m}\left[ (10)^2 - (0)^2 m^2 s^{-2} \right] = -45.71\times10^6 m^3 s^{-1} = -45.7 sv$$

Assuming a uniform decrease, for $35^0$ the decrease of transport is 45.7 sv. For $1^0$, the decrease is 45.7/35=1.31sv. For 20 (40-60), the decrease is 1.31(20) = 26.1sv. This means the net decrease of transport is 45.7-26.1=19.6.

Thus for $60^0$N: the Sverdrup transport is -19.6 sv

# 9

# Vorticity theorem

## 9.1. Vorticity equation

While circulation is a measure of tendency of fluid rotation, vorticity is the same property of fluid for an infinitely small area.

To obtain the vorticity theorem, consider the frictionless horizontal equations of motion as:

$$\frac{\partial u}{\partial t}+u\frac{\partial u}{\partial x}+v\frac{\partial u}{\partial y}=fv-\alpha\frac{\partial p}{\partial x} \tag{9.1}$$

$$\frac{\partial v}{\partial t}+u\frac{\partial v}{\partial x}+v\frac{\partial v}{\partial y}=-fu-\alpha\frac{\partial p}{\partial y} \tag{9.2}$$

and we know $\zeta=\frac{\partial v}{\partial x}-\frac{\partial u}{\partial y}$ (9.3)

Take partial derivative with respect to (w.r.t) y to equation 9.1 and differentiate w.r.t ' x' to equation 9.2 and subtract 9.1 from 9.2. and rearrange, then we get

$$\frac{\partial}{\partial t}\left[\frac{\partial v}{\partial x}-\frac{\partial u}{\partial y}\right]+u\frac{\partial}{\partial x}\left[\frac{\partial v}{\partial x}-\frac{\partial u}{\partial y}\right]+v\frac{\partial}{\partial y}\left[\frac{\partial v}{\partial x}-\frac{\partial u}{\partial y}\right]+\frac{\partial u}{\partial x}\left[\frac{\partial v}{\partial x}-\frac{\partial u}{\partial y}\right]+\frac{\partial v}{\partial y}\left[\frac{\partial v}{\partial x}-\frac{\partial u}{\partial y}\right]$$

$$=-f\left[\frac{\partial u}{\partial x}+\frac{\partial v}{\partial y}\right]-\left[u\frac{\partial f}{\partial x}+v\frac{\partial f}{\partial y}\right]-\alpha\left[\frac{\partial^2 p}{\partial x\partial y}-\frac{\partial^2 p}{\partial x\partial y}\right] \tag{9.4}$$

$$\frac{\partial}{\partial t}(\zeta)+u\frac{\partial \zeta}{\partial x}+v\frac{\partial \zeta}{\partial y}+\zeta(D)=-f.D-\left[\frac{\partial f}{\partial t}+u\frac{\partial f}{\partial x}+v\frac{\partial f}{\partial y}\right] \tag{9.5}$$

since adding (- $\frac{\partial f}{\partial t}$) as variation of 'f' with 't' is zero

$\frac{\partial f}{\partial x}=0$ Also since and $v\frac{\partial f}{\partial y}=\frac{dy}{dt}.\frac{df}{dy}=\frac{df}{dt}$

It can be further written as

$$\frac{d}{dt}(\zeta)+\frac{d}{dt}(f)=-D(f+\zeta)$$

$$\frac{d}{dt}(\zeta+f)=-D(f+\zeta) \tag{9.6}$$

I II

This equation (9.6) is called vorticity equation. The term (I) on the left hand side (L.H.S) is called the rate of change of absolute vorticity. This equation says that the absolute vorticity of a parcel of air can change only through the contribution of the IInd term on the right hand side.

## 9.2. Potential vorticity equation

For the case of incompressible barotropic fluid, the potential vorticity equation can be derived using the continuity equation:

$$\therefore\left(\frac{\partial u}{\partial x}+\frac{\partial v}{\partial y}+\frac{\partial w}{\partial z}\right)=0 \text{ which means}\therefore\left(\frac{\partial u}{\partial x}+\frac{\partial v}{\partial y}\right)=-\frac{\partial w}{\partial z} \tag{9.7}$$

The vorticity equation (9.6) can be written with the help of equation (9.7) as

$$\frac{d}{dt}(\zeta+f)=(f+\zeta)\frac{\partial w}{\partial z} \tag{9.8}$$

In the case of barotropic condition, the thermal wind vanishes and so the geostrophic wind is independent of height. Then integrating vertically from z=$D_1$ to $D_2$ equation (9.8) can be written as

$\frac{d}{dt}(\zeta+f)dz=(f+\zeta)dw$ on integration gives rise to

$$\frac{d}{dt}(\zeta+f)_g[Z]_{D_1}^{D_2}=(f+\zeta)_g W_{wD_1}^{wD_2}$$

$$\frac{d}{dt}(\zeta+f)_g[D_2-D_1]=(f+\zeta)_g[W_{D2}-W_{D1}] \tag{9.9}$$

The suffix 'g' denotes under geostrophic condition. But since W = $\frac{dz}{dt}$ we can further write $W_{D2}$ = $\frac{dD}{dt}$ and $W_{D1}$ = $\frac{dD_1}{dt}$ and let H =$D_2$ –$D_1$ is the depth of the column we can write (9.9) as

$$\frac{d}{dt}(\zeta+f)_g[D_2-D_1]=(f+\zeta)_g\left[\frac{dD_2}{dt}-\frac{dD_1}{dt}\right]=(f+\zeta)\frac{d}{dt}(D_2-D_1)$$

H. $\frac{d}{dt}(\zeta+f)=(f+\zeta)\frac{d}{dt}(H)$ re arranging the terms

$(\frac{1}{\zeta+f})\frac{d}{dt}(\zeta+f)=\frac{1}{H}\frac{d}{dt}(H)$ which gives rise to

$$\frac{d}{dt}[\ln\left(\frac{\zeta+f}{H}\right)] = 0$$

or $$\frac{d}{dt}\left(\frac{\zeta+f}{H}\right) = 0 \tag{9.10}$$

This equation (9.10) is the potential vorticity equation.

If the flow is purely horizontal (w = o), equation (9.8) reduces to

$$\frac{d}{dt}(\zeta+f) = 0 \tag{9.11}$$

This equation (9.11) is called the barotropic vorticity equation, which states that the absolute vorticity is conserved which means ($\zeta$ + f) = constant.

That is in the case of a ring of water parcel revolving round the earth, the absolute vorticity (f + $\zeta$) is constant. Which means as 'f' increases, the relative vorticity ($\zeta$) should decrease and vice versa in order that (f + $\zeta$)to be constant. So if water moves pole ward, 'f' increases and so $\zeta$ decreases. Which means pole ward moving water must lose cyclonic curvature and tend toward anti-cyclonic motion. The converse is true for the water moving equator ward.

## Conservation of Oceanic Vorticity on the Earth

A *resting fluid* on the Earth will be in solid body rotation and will have a well-defined vorticity relative to an inertial frame of reference. We call this *planetary vorticity*, which is 2 times the local overhead rotation rate, or f = 2 Ω sin ϕ! For example, the planetary vorticity of a fluid column at the pole is f = 2 Ω; at 30° N, f = Ω, and at the equator f = 0.

However a fluid moving relative to the Earth could have an additional component of vorticity relative to the Earth - called *relative vorticity* $\zeta$. The sum of planetary and relative vorticity is called the *absolute vorticity* (AV) or AV = f + $\zeta$, relative to an inertial frame of reference.

The ratio of absolute vorticity and the height of the water column H on a rotating Earth is called the *potential vorticity* (PV) according to $PV = \dfrac{f+\zeta}{H}$

It can be shown that under many important circumstances, the PV of a frictionless fluid column is conserved along the trajectory according to

$$\frac{d}{dt}\left[\frac{f+\zeta}{H}\right] = 0 \; or \; PV = \frac{f+\zeta}{H} = constant$$

Consider the implications of the conservation of PV for the following simple cases.

Case I: Fig.9.1shows the qualitative effects of stretching and shrinking (or squashing) a frictionless water column in an f = constant environment.

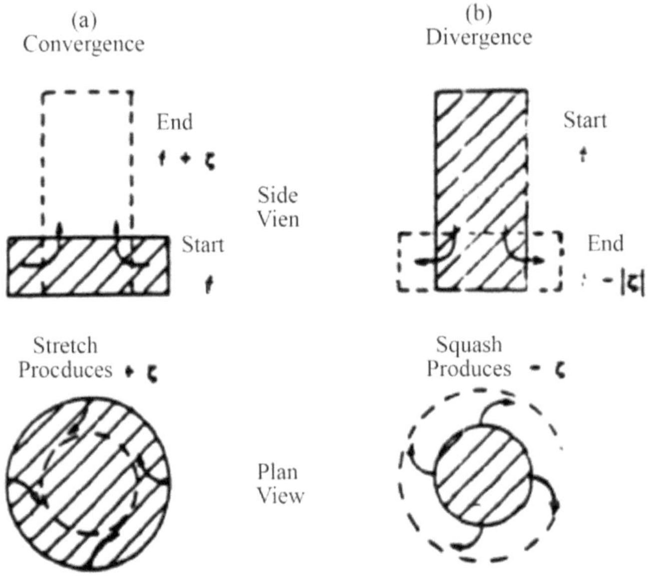


**Fig. 9.1.** Relative vorticity change due to water column stretching

Quantitatively (and symbolically), determine the amount and sign of the relative vorticity ζ that is produced, due to shrinking and/or stretching of the water column, with $f = f_0$ = a constant.

For *stretching*: Determine after an initially motion-free ($\zeta = 0$) fluid column is stretched to height H1 > Ho, where Ho is the fluid column height at t = 0.

To start, you know that PV conservation demands that $\frac{f+\zeta}{H} = constant$ following a water column for all time.

Thus at t = 0, the constant is: constant = $\frac{f_0}{H_0}$.

At some future time t = t1, PV conservation demands that: $\frac{f_0+\zeta_1}{H_1} = \frac{f_0}{H_0}$

Thus $\zeta_1 = f_0\left[\frac{H_1}{H_0} - 1\right]$

Case II: Fig.9.2 shows qualitatively that positive ζ is produced when a constant depth, frictionless water column is displaced from a poleward latitude to an equatorward latitude due to changes in f and vice versa.

Determination of the quantitative ζ changes (symbolically) that are illustrated in Fig.9.2.

As f + ζ = const, when f = const decreases, ζ should increase and vice versa.

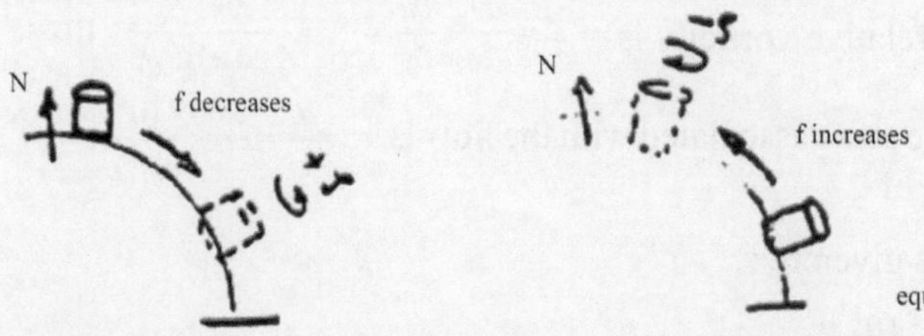


**Fig. 9.2.** Relative vorticity change due to meridional motion.

Case III: Fig.9.3 shows qualitatively that for a situation in which changes in relative vorticity are not allowed (i.e., ζ production is zero), equatorward movement produces water column height H decreases and vice versa.

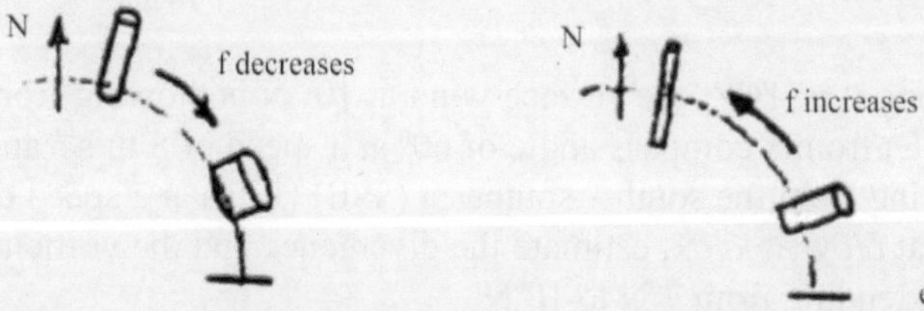


**Fig. 9.3.** Water column stretching due to meridional motion with $\frac{d\zeta}{dt} = 0$

Show how PV conservation produces the ζ changes illustrated in Fig.9.3.Use the approach illustrated in the Cases I and II analysis above.

As ζ is negligibly small when compared to *f*, we can take $\frac{f}{H} = const$. Then as *f* decreases, H decreases and as *f* increases, H should increase in order to be constant.

## Exercises

### *Exercise 9.1*

Consider a flow having a width of 200 km directed northward with a maximum velocity of 2 $ms^{-1}$ near the latitude 27$^0$N.

a. Calculate the value of the Coriolis parameter $f_0$ of this flow.
b. Determine a typical value of the relative vorticity of the flow.
c. Provide estimates of the time scales $\tau_f$, $\tau_a$ and the Rossby number ε

### *Solution*

a) The Coriolis parameter is given by f = 2Ω sinΦ. Where Ω = 7.3 × $10^{-5}$ $s^{-1}$ is the angular velocity due to the Earth's rotation and Φ is the latitude.

$\therefore$ f = 2Ω sinΦ = 2 × 7.3 × $10^{-5}$ $s^{-1}$ × sin 27 = 6.63 × $10^{-5}$ $s^{-1}$

b) The equation for relative vorticity is = $\zeta = \frac{\partial v}{\partial x} - \frac{\partial u}{\partial y} = \frac{U}{L} = \frac{2ms^{-1}}{200 \times 10^3 m} = 10^{-5} s^{-1}$

c) The advective time scale associated with the flow is $\tau_a = \frac{L}{U} = \frac{200 \times 10^3 m}{2ms^{-1}} = 10^5 s$, which is inverse of ζ.

The inertial time scale is given by:

$$\tau_f = \frac{1}{f} = \frac{1}{6.63 \times 10^{-5} s^{-1}} = \frac{10^5}{6.63} = 15082.9563s = 0.15082 \times 10^5 s = 1.5 \times 10^4 s$$

The Rossby number ε is the ratio of

$$\varepsilon = \frac{\tau_f}{\tau_a} = \frac{1/f}{L/U} = \frac{U}{fL} = \frac{2ms^{-1}}{6.63 \times 10^{-5} s^{-1} \times 200 \times 10^3 m} = \frac{1.5 \times 10^4 s}{10^5 s} = 0.15$$

### *Exercise 9.2*

At a certain location along the ITCZ, the surface wind at $10^0$ N is blowing from the east—northeast (ENE) from a compass angle of $60^0$ at a speed of 8 m $s^{-1}$ and the wind at $7^0$ N is blowing from the south—southeast (SSE) ($150^0$) at a speed of 5 m $s^{-1}$. (a) Assuming that ∂/∂y >> ∂/∂x, estimate the divergence and the vorticity averaged over the belt extending from $7^0$N to $10^0$N.

### *Solution*

### *Divergence*

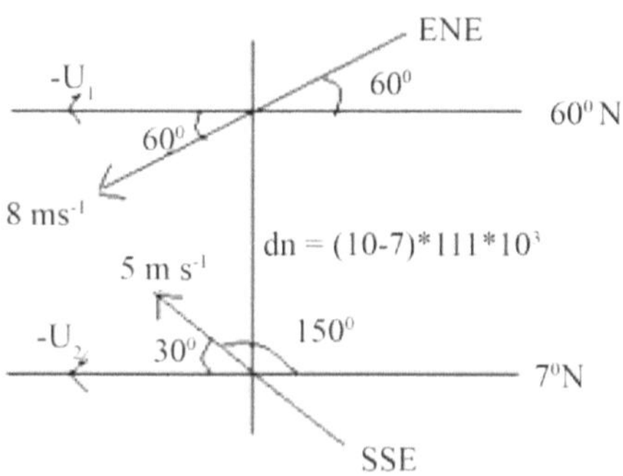


$\frac{\partial u}{\partial x} + \frac{\partial v}{\partial y} = \nabla.V$ ; The zonal wind in the layer between 7 and $10^0$ N is:

$$\frac{\partial U}{\partial n} = \frac{-U_1 + (-U_2)}{\Delta N} = \frac{-8\cos 60 - 5\cos(180-150)}{(10-7) \times 111 \times 10^3} = 0.025 \times 10^{-3} = 2.5 \times 10^{-5} s^{-1}$$

Vorticity: $\zeta = \frac{\partial v}{\partial x} - \frac{\partial u}{\partial y} = \nabla_Z \times V = \frac{5\cos 60}{5\cos 30} - \frac{8\cos 30}{8\cos 60} = -1.15 s^{-1}$

# 10

# Stommel's Circulation

## 10.1. Stommel's theory

Sverdrup's contribution could not take into account of the western boundary and so could not explain about the circulation what happened along the west coast of the ocean. Thus his circulation, as shown in Fig.8.1c in the chapter 8, is incomplete. This problem was solved by an American Oceanographer, Henry Stommel, in 1948.

Stommel's theory of wind driven circulation uses the same basic equations but integrates up to the bottom of the ocean and allows for bottom friction that is a simple linear function of velocity. The friction at the top of the ocean is just the applied zonal wind stress.

Unlike Sverdrup, Stommel included friction in his calculations and he worked out the flow that would result when the wind stress and frictional forces balanced (i.e. when there was a steady state with no acceleration or deceleration) for each of the three situations described above.

Figure 10.1 summarizes the results of Stommel's calculations. Stommel considered the effect of a symmetrical gyral wind field on a rectangular ocean and calculated the effect when the earth is rotating and Coriolis parameter varies with latitude. The Fig.10.1A represents the pattern of wind field and the Fig.10.1B shows the sea-surface topography that result, due to this wind field.

The striking difference in the stream line pattern of Fig.10.1B is that the stream lines are crowded in the west side whereas they are wide spread in the eastern side of the ocean. Stommel (1948) attributed this pattern as due to the variation of the Coriolis force. In order to have equilibrium, the sea surface would adjust to have balance with the horizontal pressure gradient forces.

Thus Stommel's result showed that the intensification of western boundary currents (Fig.10.1B) is due to increase of Coriolis force with increase of latitude in the west and decrease of the same in the east.

Because the Sverdrup's balance applies to the whole ocean basin, the return flow must be in a narrow, swift meridional jet where the potential vorticity balance is different from the Sverdrup's balance. Stommel included the dissipation of potential vorticity, and showed that the returning flow must be along the western boundary. His potential vorticity balance is change in planetary vorticity balanced by bottom friction. Stommel's idealized circulation resembles the westward intensification of Gulf Stream and Kuroshio in subtropical gyres .

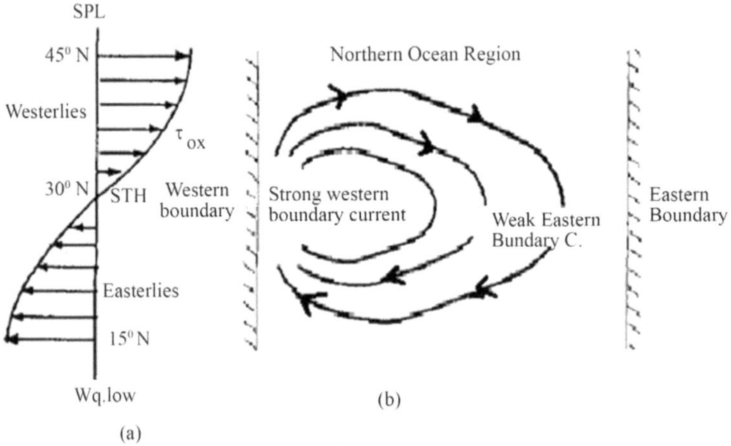


**Fig.10.1.** Stream function flow in a rectangular basin as calculated by Stommel (1948). **(A):** Wind stress in the Northern Hemisphere. **(B):** Stream line pattern resulted due to the wind stress.

Intensification of western boundary currents may also be explained in terms of vorticity balance. It is more convenient to work with vorticity than with linear current flow, because horizontal pressure gradient forces which lead to complications need not be considered.

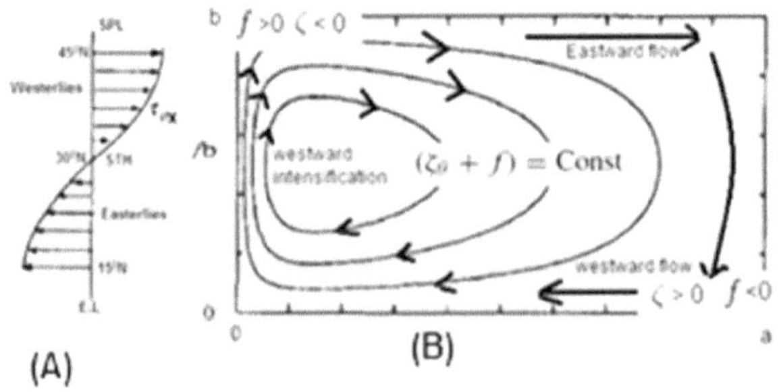


**Fig.10.2. (A).** Low latitude Easterlies and middle latitude Westerlies. **(C)** Development of gyre and eastern and western boundary currents

The most obvious factor affecting the relative vorticity ($\zeta$)of the water in the gyre is the wind. The wind field is symmetrical (Fig.10.2A) and acts to supply negative (clockwise) vorticity in the Northern Ocean gyre (say North Atlantic gyre).To put things in perspective, the flow of the Amazon river as it runs in to the sea is about 0.17(~0.2) Sverdrups when it is compared with the transport of the subtropical gyre of the north Pacific (a branch of it is Gulf stream the biggest river in oceans), it is about 50 Sv or 250 times that of the Amazon, the biggest river on Earth!

The next factor to be considered is the change in latitude or the variation of Coriolis parameter, $f$ which is called planetary vorticity. Water moving northwards on the western side of the gyre is moving into regions of larger positive planetary vorticity ($f > 0$) and hence acquires negative relative vorticity ($\zeta < 0$) since the absolute vorticity ($\zeta + f$) is constant (Fig.10.2B). Similarly water moving southwards on the eastern side of the gyre is moving into regions of smaller positive planetary vorticity ($f < 0$) and so gains positive relative vorticity ($\zeta > 0$). However, as more water moves northwards than it moves southwards, the net change in relative vorticity of water due to the change in latitude in the entire gyre is zero.

If we consider the potential vorticity ($\frac{\zeta + f}{H}$), the thickness of the column (H) also influences apart from absolute vorticity ($\zeta + f$). This means if H is small potential vorticity is more and vice versa. In other words the stretching (height of water column, H) is in the denominator since making a column taller makes it spin faster and vice versa.

Then what is acting to counteract the negative (clockwise) tendency supplied by the wind stress, so that the gyre does not speed-up indefinitely is friction. We know that wind driven circulation is confined to the upper layers and is not frictionally bound to the sea floor, but there will be significant friction with the coastal boundaries as a result of horizontal eddies. If we can calculate the vorticity balance using the appropriate value of eddy coefficient of viscosity ($A_h$), it is clear that frictional forces should be large to provide sufficient positive relative vorticity to balance the negative relative vorticity provided by the wind stress.

So western boundary currents are warm, deep, narrow, and fast flowing currents and occur on the west side of ocean basins. They are important in climate control by bringing warm water from the equator to poleward. Examples include the Gulf Stream, The Aghulas current and the Kuroshio current, the Brazil and Somali currents etc.

## 10.2. Reason for the Westward intensification of Western Boundary Currents

The net Sverdrup transport added up across the ocean (along a parallel of latitude) must be returned in a narrow flow somewhere. The transport of this narrow flow is the same and in the opposite direction to the transport across the whole ocean width; therefore the velocities are much larger than in the ocean interior.

The return flow must always be on the western side, regardless of northern or southern hemisphere. The reason for it is like this. Consider the subtropical gyre, where Sverdrup transport is equatorward (Fig.10.2B) in the eastern boundary. The wind through friction is putting "negative" vorticity into the subtropical ocean, through downward Ekman pumping that send the water columns towards lower rotation, e.g. towards the equator. For a steady state general circulation this vorticity must be removed somewhere. It is removed through friction in a narrow boundary current in the western boundary of the ocean which can be seen by crowding of streamlines in Fig.10.2B. Why "narrow"? Because friction acts best on high velocity, which can be achieved only in a narrow flow, and because the return flow has to be narrow enough to escape forcing by the Ekman pumping. A frictional boundary current has zero velocity along the boundary, and strong velocity offshore. If the boundary current for the northern hemisphere subtropical gyre is on the west, then the relative vorticity of the boundary current is positive. Thus the frictional boundary current puts positive vorticity into the ocean, which balances the negative vorticity put in by the Ekman pumping. This exact argument works for the low pressure regions as well (subpolar gyre, with northward Sverdrup transport and southward western boundary current). It also works in the southern hemisphere.

## 10.3. Summary of Stommel's Calculations

Henry Stommel (1948) showed that the north-south gradient of the strength of the Coriolis force (the "beta effect") was responsible for the observed fact that the return flow of the slow interior gyre circulations is concentrated in fast moving western boundary currents, such as the Gulf Stream and the Kuroshio Current, a process known as westward intensification. As a result, these western boundary currents have a larger and steadier transport than the corresponding eastern boundary currents, such as the California Current and Canary Current, on the eastern side of the ocean basins. In subtropical latitudes, the western boundary currents are important in transporting the excess heat in the low latitudes towards the poles.

The summary of the results of Stommel (1948) can be stated as below. The direction of Sverdrup flow (equatorward or poleward) is dictated by the Ekman pumping as the flow is geostrophic. That is, an equatorward flow from higher latitudes means that the sea surface is high in the west and low in the east and vice versa as the force balance is between pressure gradient force and Coriolis force. Thus the small terms in the momentum equation that is neglected in geostrophic balance are important when thinking about the very slow evolution of the flow. Note in Fig.10.2B the increase of sea surface height is from Eastern boundary to Western boundary. Also Left hand (West) side the lines are closer than right hand (East) side (Fig.10.1b) implies the topography is steeper in west and so currents are faster in west than in the East.

The resulting strong, narrow western boundary currents (crowding of lines) are called westward intensification, as shown in Fig.10.2B, is important for returning all of the interior transport back to the latitude where it started for continuity reasons. The Gulf Stream, Kuroshio, Oyashio, Labrador Current, etc are examples of such strong western boundary currents in wind-driven circulation. Because of various other factors, including just inertia, the western boundary currents are generally stronger than is required to simply the return of the interior flow. This results in an overshoot of the western boundary current at the latitude where it should end if it were very frictional. Thus the Kuroshio and Gulf Stream, for instance, separate and flow far out to sea before finally dying out.

Stommel's frictional solution is somewhat unrealistic since the friction is between the boundary current and the ocean bottom. This means that the wind-driven flow according to him will reach to the ocean bottom which is unrealistic as wind driven circulation is limited to the mixed layer zone. However, with stratification, it is not at all obvious that the circulation reaches so deep. A subsequent study by Walter Munk (1950) avoids this restriction and still yields westward intensification, which can be seen in the next chapter.

Although Fofonoff's (1954) solution appears to be very artificial, it is one of the survivors of the earlier theories. The strongly nonlinear version of Stommel's model leads to a solution that looks remarkably like that of Fofonoff' (Veronis, 1966; Niiler, 1966). The recirculation region just south of the Gulf Stream after the latter has separated from the coast has the appearance of a local inertial circulation. Thus, it is likely that some version of the latter will be part of any successful model of large-scale ocean circulation.

## 10.4. Stommel's derivation

Stommel's Theory of wind driven circulation uses the same basic equations but integrates to the bottom of the ocean and allows for bottom friction that is a simple linear function of velocity $F_x = -R_u$ and $F_y = -R_v$. The friction at the top of the ocean is just the applied zonal wind stress $\tau_x$. The starting equation of the wind driven circulation (Mellor, 1996) is

$$\psi \frac{\tau_0}{\rho_0} \sin \frac{\pi y}{b} = \frac{a\pi}{b\beta}\left(1 - \frac{x}{a}\right) \quad 10.1$$

It is noted that the deficiency in equation (10.1) is that streamlines do not close on the western boundary

In order to repair this deficiency, let us reexamine the basic assumptions that have been made. First of all bottom friction terms were ignored in the equation (10.2 a,b)

$$-fM_y = -gH\frac{\partial \eta}{\partial x} + \frac{\tau_{0x}}{\rho_0};\ fM_x = -gH\frac{\partial \eta}{\partial y} + \frac{\tau_{0y}}{\rho_0} \quad (10.2a,b)$$

Thus, upon integrating $\frac{\partial \tau_z}{\partial z}$ in (10.2a) from z =-*H* to z = 0, we could have obtained $\frac{\tau_{0x}}{\rho_0} - \frac{\tau_{Hx}}{\rho_0}$ where $-\tau_{Hx}$ is the bottom friction, and, similarly, $\frac{\tau_{0y}}{\rho_0} - \frac{\tau_{Hy}}{\rho_0}$ in (10.2b). Bottom friction indeed was the dissipative ingredient according to Stommel (1948). Another means of obtaining closed streamlines, while at the same time enforcing a no slip condition on the western boundary, is to include lateral friction; which was later Munk (1950) took this approach. Recall that the terms were eliminated due to the fact that the horizontal scale, *L,* for the variation of stress was assumed to be much greater than the vertical scale, *H*. However, it now appears that, near the western boundary, the horizontal scale is small.

$$-fv = -g\frac{\partial \eta}{\partial x} + \frac{1}{\rho_0}\frac{\partial \tau_{zx}}{\partial z}; fu = -g\frac{\partial \eta}{\partial y} + \frac{1}{\rho_0}\frac{\partial \tau_{zy}}{\partial z} \tag{10.3 a,b}$$

Thus the terms, $\frac{\partial \tau_{yx}}{\partial y} + \frac{\partial \tau_{xx}}{\partial x}$ in equation (10.3a) and $\frac{\partial \tau_{yy}}{\partial y} + \frac{\partial \tau_{xy}}{\partial x}$ in (10.3b) can be reinstated

$$-fv = -g\frac{\partial \eta}{\partial x} + \frac{1}{\rho_0}\frac{\partial \tau_{zx}}{\partial z} + \frac{\partial \tau_{yx}}{\partial y} + \frac{\partial \tau_{xx}}{\partial x}; fu = -g\frac{\partial \eta}{\partial y} + \frac{1}{\rho_0}\frac{\partial \tau_{zy}}{\partial z} + \frac{\partial \tau_{yy}}{\partial y} + \frac{\partial \tau_{xy}}{\partial x} \tag{10.4 a,b}$$

On integrating (10.4 a,b) after reinstating all terms from $z$ = -*H* to $z$ = 0 and we obtain

$$-fM_y = -gH\frac{\partial \eta}{\partial x} + \frac{\tau_{0x}}{\rho_0} - \frac{\tau_{Hx}}{\rho_0} + \frac{\partial \tau_{yx}}{\partial y} + \frac{\partial \tau_{xx}}{\partial x} \tag{10.4a}$$

$$fM_x = -gH\frac{\partial \eta}{\partial y} + \frac{\tau_{0y}}{\rho_0} - \frac{\tau_{Hy}}{\rho_0} + \frac{\partial \tau_{yy}}{\partial y} + \frac{\partial \tau_{xy}}{\partial x} \tag{10.4b}$$

Where $(\tau_{xx}, \tau_{yx}, \tau_{xy}, \tau_{yy}) = \int(\tau_{xx}, \tau_{yx}, \tau_{xy}, \tau_{yy})dz$

Stommel (1948) introduced a model for bottom friction such that

$$(\tau_{Hx}, \tau_{Hy}) = -R(M_x, M_y) \tag{10.5}$$

where *R* is an empirical constant. Alternatively, Munk (1950), in analogy to the stress matrix, assumed that

$$\begin{bmatrix} \tau_{xx} & \tau_{yx} \\ \tau_{xy} & \tau_{yy} \end{bmatrix} = A\begin{bmatrix} 2\frac{\partial M_x}{\partial x} & \frac{\partial M_x}{\partial y} + \frac{\partial M_y}{\partial x} \\ \frac{\partial M_x}{\partial x} + \frac{\partial M_y}{\partial y} & 2\frac{\partial M_y}{\partial y} \end{bmatrix} \tag{10.6}$$

where the lateral viscosity, *A,* must be empirically determined. Therefore, insert (10.5) and (10.6) into (10.4a,b) and obtain

$$-fM_y = -gH\frac{\partial \eta}{\partial x} + \frac{\tau_{0x}}{\rho_0} + RM_x + A\left[\frac{\partial^2 M_x}{\partial x^2} + \frac{\partial^2 M_x}{\partial y^2}\right] \tag{10.7a}$$

$$fM_x = -gH\frac{\partial \eta}{\partial x} + \frac{\tau_{0y}}{\rho_0} + RM_y + A\left[\frac{\partial^2 M_y}{\partial x^2} + \frac{\partial^2 M_y}{\partial y^2}\right] \tag{10.7b}$$

where the continuity equation $\frac{\partial M_x}{\partial x} + \frac{\partial M_y}{\partial y} = 0$ has been used to simplify the terms in the square brackets and form in the lines of, $\frac{\partial}{\partial x}\left(\frac{fM_x}{H}\right) + \frac{\partial}{\partial y}\left(\frac{fM_y}{H}\right)$ and obtain

$$M_x\frac{\partial}{\partial x}\left(\frac{f}{H}\right) + M_y\frac{\partial}{\partial y}\left(\frac{f}{H}\right) = \frac{\partial}{\partial x}\left(\frac{\tau_{0y}}{\rho_0 H}\right) - \frac{\partial}{\partial y}\left(\frac{\tau_{0x}}{\rho_0 H}\right) + R\left[\frac{\partial}{\partial x}\left(\frac{M_y}{H}\right) - \frac{\partial}{\partial y}\left(\frac{M_x}{H}\right)\right]$$
$$+A\left[\frac{\partial}{\partial x}\frac{1}{H}\left\{\frac{\partial^2 M_y}{\partial x^2} + \frac{\partial^2 M_y}{\partial y^2}\right\} - \frac{\partial}{\partial y}\frac{1}{H}\left\{\frac{\partial^2 M_x}{\partial x^2} + \frac{\partial^2 M_x}{\partial y^2}\right\}\right] \tag{10.8}$$

For a flat bottomed ocean and a purely zonal wind, (10.8) simplifies to

$$\beta M_y = -\frac{\partial}{\partial y}\left(\frac{\tau_{0x}}{\rho_0}\right) + R\left(\frac{\partial M_y}{\partial x} - \frac{\partial M_x}{\partial y}\right) + A\left[\frac{\partial^3 M_y}{\partial x^3} + \frac{\partial^3 M_y}{\partial x \partial y^2} - \frac{\partial^3 M_x}{\partial y \partial x^2} - \frac{\partial^3 M_x}{\partial y^3}\right]$$

It is again convenient to use the stream function defined by $M_x = \frac{\partial \psi}{\partial y}, M_y = -\frac{\partial \psi}{\partial x}$. Thus,

$$A\left[\frac{\partial^4 \psi}{\partial x^4} + 2\frac{\partial^4 \psi}{\partial x^2 \partial y^2} + \frac{\partial^4 \psi}{\partial y^4}\right] - R\left[\frac{\partial^2 \psi}{\partial x^2} + \frac{\partial^2 \psi}{\partial y^2}\right] - \beta\frac{\partial \psi}{\partial x} = -\frac{\partial}{\partial y}\left(\frac{\tau_{0x}}{\rho_0}\right) \tag{10.9}$$

## 10.4. Stommel's Model

Stommel included bottom friction in his model but not lateral friction so that after non-dimensionalizing *x* and *y* according to (10.10)

$$\tilde{x}, \tilde{y} \cong \left(\frac{x}{a}, \frac{y}{a}\right) \tag{10.10}$$

and setting A = 0, equation (10.9) may be written as

$$\varepsilon\left[\frac{\partial^2 \psi}{\partial \tilde{x}^2} + \frac{\partial^2 \psi}{\partial \tilde{y}^2}\right] + \frac{\partial \psi}{\partial \tilde{x}} = \frac{1}{\beta}\frac{\partial}{\partial \tilde{y}}\left(\frac{\tau_{0x}}{\rho_0}\right) \tag{10.11}$$

Where $\varepsilon = \frac{R}{a\beta}$ (10.12)

For a given wind stress, it is possible to integrate (10.11) and the resulting solution may be simplified for a flow morphology like the Gulf Stream where we shall see that ε must be a small number. For small ε, we expect a rapid change of ψ with respect to $\overline{x}$ near the western part of the basin, so that the product of the first term in the square brackets of (10.11) and ε is of the same order as other terms. Since we do not anticipate a similarly rapid change with respect to $\overline{y}$, the second term in the square brackets should be much smaller than the first term. Therefore this term can be neglected, so that

$$\varepsilon\left[\frac{\partial^2\psi}{\partial\tilde{x}^2}\right]+\frac{\partial\psi}{\partial\tilde{x}}=\frac{1}{\beta}\frac{\partial}{\partial\tilde{y}}\left(\frac{\tau_{0x}}{\rho_0}\right) \tag{10.13}$$

Using the simplified wind stress given by (10.14) so that the right side of (10.13) is a function of $\bar{y}$ and not a function of $\bar{x}$. We therefore assume a solution of the form

$$\tau_{0x}=\tau_{0x}\,(y),\ \tau_{0y}=0 \tag{10.14}$$

$$\bar{\psi}=\frac{1}{\beta}\frac{\partial}{\partial\tilde{y}}\left(\frac{\tau_{0x}}{\rho_0}\right)F(\bar{x}) \tag{10.15}$$

Where

$$\varepsilon\left[\frac{\partial^2 F}{\partial\tilde{x}^2}\right]+\frac{\partial F}{\partial\tilde{x}}=1 \tag{10.16}$$

The particular solution is simply $\bar{x}$ which may be added to the homogeneous solution, $A+Be^{-\frac{\tilde{x}}{\varepsilon}}$ Then the boundary conditions, (0,y) = ψ (1,y) = 0, determine the constants, A and $B$, so that. $F$ = $\bar{x}-(1-e^{-\frac{\tilde{x}}{\varepsilon}})/(1-e^{-\frac{1}{\varepsilon}})$ For small, this result can be simplified since $e^{-\frac{1}{\varepsilon}}$ = O. In place of Eqn (10.15), we can therefore write

$$\psi=\frac{1}{\beta}\frac{\partial}{\partial\tilde{y}}\left(\frac{\tau_{0x}}{\rho_0}\right)(e^{-\frac{\tilde{x}}{\varepsilon}}+\bar{x}-1) \tag{10.17a}$$

Now note that

$$\frac{\partial^2\psi}{\partial\tilde{y}^2}\approx order\ of\,\frac{\tau_0}{\rho_0\beta}$$

Whereas

$$\frac{\partial^2\psi}{\partial\tilde{x}^2}\approx order\ of\,\varepsilon^{-2}\frac{\tau_0}{\rho_0\beta}$$

the former term is therefore much smaller than the latter term as was assumed in obtaining equation (10.13). When the streamlines from Eqn(10.17a) are plotted it comes like the Fig. 10.3 together with the northward transport for $y = b/2$ obtained from

$$M_y=\frac{1}{a\beta}\frac{\partial}{\partial\tilde{y}}\left(\frac{\tau_{0x}}{\rho_0}\right)\left[\frac{1}{\varepsilon}e^{-\frac{\tilde{x}}{\varepsilon}}-1\right] \tag{10.17b}$$

for the values ε = 0.050 and 0.025. The wind stress given by (10.18) can be used

$$\tau_{0x}=-\tau_0 cos\frac{\pi y}{b} \tag{10.18}$$

and the following constant values are also used

$\tau_0 = 1$ dyne cm$^{-2}$; $\beta = 2\times10^{-8}$ km$^{-1}$ s$^{-1}$; a = 6000 km; b = 4000 km

In this figure and in equation (10.17b) note how differentiation has increased the magnitude of the velocities in the western boundary region by the factor, $\epsilon^{-1}$; however, the distance, over which the velocity is large, decreases by a factor,ε, so that the total flow in the western boundary remains constant as ε varies.

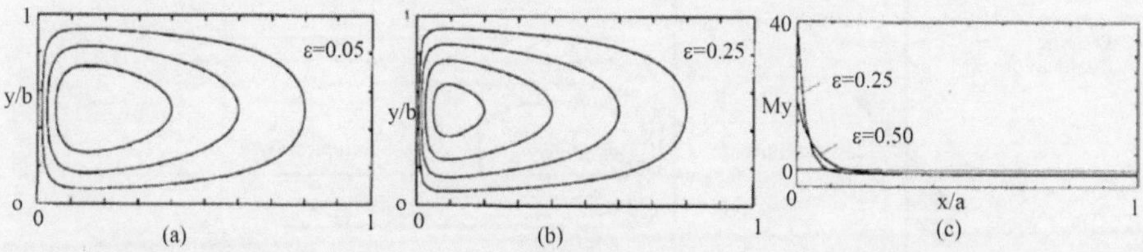


**Fig. 10.3.** The two left panels (A&B) are Stommel's solutions for ε = 0.050 and 0.025 calculated from equations (10.17) and 10.18. The panel (C) is $M_y\, at\, y = \frac{b}{2}$ for the two values of ε. The units of $M_y$ are $m^2\, s^{-1}$. Note that the integral of $M_y$ with respect to x is zero.

## 10.6. Sverdrup's Calculations with assumed standard values in the North Pacific

The calculation of order of magnitude in the tropical Northern Pacific Ocean taking the Easterlies (trade winds) with maximum values of wind stress + 0.05Pa in the tropical regions and westerlies in the middle latitudes with -0.05Pa (see Fig.10.4) and the meridional (north south) length scale between these wind systems (Easterlies and Westerlies) is 1000 km gave rise to -60Sv. In a closed basin like North Pacific, all this southward transport in the eastern branch has to be balanced by a northward flow in the western side of the ocean for continuity reasons and to complete the gyre. In other words to conserve this southward mass of the Sverdrup Volume flow (= - 60Sv) must be balanced by the western boundary current "Kuroshio" with a transport of +60 Sv north ward! That is the Western Boundary Current (Kuroshio) in Pacific Ocean and refer to as westward intensification. (Refer Exercise 8.2 of chapter 8 of Sverdrup circulation for details).

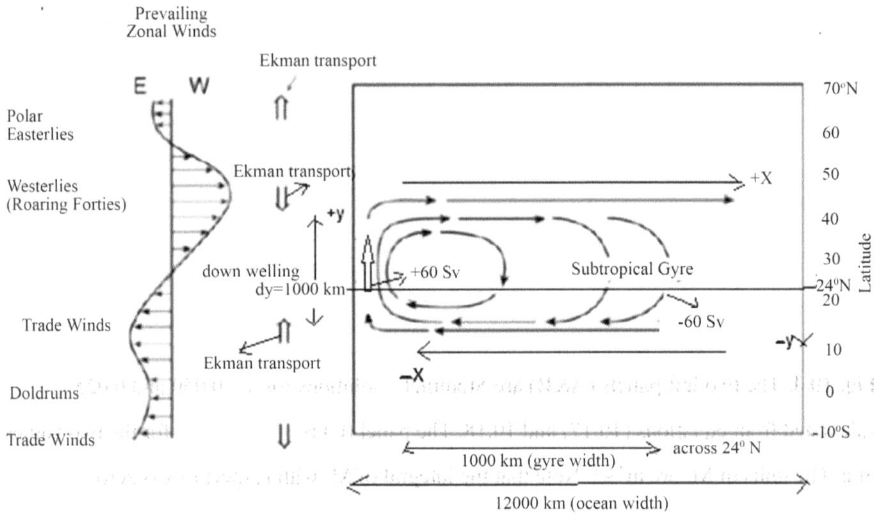


**Fig. 10.4.** Estimation for Sverdrup calculation of transports in the tropical pacific

## References

Fofonoff, N. P., 1954. Steady flow in a frictionless homogeneous ocean. J. Mar. Res., 13, 254-262.

Mellor, G.L (1996): Introduction to Physical Oceanography,193pp; ISBN 1-56396-210- I Springer-Verlag New York.

Munk, W. H., 1950: On the wind-driven ocean circulation. *J. Meteorol.,7, 79-93.*

Niiler, P., 1966: On the theory of wind-driven ocean circulation. DeepSea Res., 13, 597-606.

Pond, E. and G. L. Pickard, 1983: *Introduction to Dynamical Oceanography, 5th edition.* Pergamom Press, Oxford, UK.329pp

Stommel, H., 1948: The westward intensification of wind-driven ocean currents. *Trans., Am. Geophys. Union* 29, 202-206.

Stommel, H., 1977: *The Gulf Stream.* Univ. California Press, Los Angeles, USA.

Veronis, G., 1966. Wind-driven ocean circulationepart II.Numerical solution of the nonlinear problem. Deep-Sea Res. 13, 30e55.

## Exercises

### *Exercise10.1*

If the Ekman transport southward across 20°N in the Atlantic is 5 Sv and the Sverdrup transport southward across the same latitude is 50 Sv, what is the transport of the western boundary current? What is the geostrophic interior transport?

### *Solution*

To conserve this Sverdrup southward flow (-50Sv) at eastern boundary a western boundary mass flow of (+50 Sv) has to be developed (Gulf Stream) northward at 20°N which is the western boundary current.

As the total transport (Sverdrup transport) = Ekman transport+ Geostrophic transport

So Geostrophic interior transport = Sverdrup transport – Ekman transport

= -50 sv - (-5sv) = -45 sv

The negative sign is given as the transport is southward (-y direction)

### *Exercise10.2*

In this exercise, we consider the dominant balances in the horizontal momentum equations for the Gulf Stream. Characteristic length scales for the Gulf Stream are Lx = 50 km and Ly = 500 km. The characteristic velocity in east-west direction is about U = 0.1 m/s and that in north-south direction V = 1.0 m/s. Assume furthermore that the Gulf Stream is a steady current and take $A_H = 10^5\ m^2s^{-1}$ and $A_V = 0.1\ m^2s^{-1}$. (a) Show that in the meridional direction the Gulf Stream is not quite in geostrophic balance. Which other terms contribute to this momentum balance? (b) Show that the zonal momentum equation is dominantly geostrophic.

### *Solution*

a) The meridional momentum balance in Cartesian coordinates is given by:

$$\frac{dv}{dt} = -\frac{1}{\rho}\frac{\partial p}{\partial y} - fu + A_H \nabla_H^2 v + A_v \frac{\partial^2 v}{\partial z^2}$$

In order to be in geostrophic, pressure gradient should balance the Coriolis and frictional terms must be negligible. Hence the ratio of horizontal frictional and Coriolis terms in the characteristic scales, taking $f = 10^{-4} s^{-1}$ is

Given Lx = 50 km; $A_H = 10^5\ m^2s^{-1}$; U = 0.1 m/s; V = 1.0 m/s

$$\frac{A_H V}{L_x^2 f U} = \frac{10^5\, m^2 s^{-1} \times 1 ms^{-1}}{(50 \times 10^3\, m)^2 \times 10^{-4} s^{-1} \times 0.1 ms^{-1}} = \frac{100}{25} = 4$$

As this value is very high frictional terms can not be neglected in the meridional direction. So it is not geostrophic.

b) The equation of motion in the zonal direction with frictional terms is

$$\frac{du}{dt} = -\frac{1}{\rho}\frac{\partial p}{\partial x} + fv + A_H \nabla_H^2 u + A_v \frac{\partial^2 u}{\partial z^2}$$

In order to be geostrophic the same condition mentioned above holds good and friction must be negligible. Hence the ratio of horizontal frictional and Coriolis terms in the characteristic scales, taking $f = 10^{-4} s^{-1}$ is

Given Lx = 50 km; $A_H = 10^5\ m^2s^{-1}$; U = 0.1 m/s; V = 1.0 m/s,

$$\frac{A_H U}{L_x^2 fV} = \frac{10^5 m^2 s^{-1} \times 0.1 ms^{-1}}{(50 \times 10^3 m)^2 \times 10^{-4} s^{-1} \times 1 ms^{-1}} = \frac{1}{25} = 0.04$$

Advection and vertical frictional terms will be even smaller. Hence the zonal flow is in geostrophic balance.

# 11

# Munk's Circulation

## 11.1. Munk's derivation

Munk (1950) built upon Sverdrup's theory, using both the vertical and lateral eddy viscosity. Munk combined the basic features contributed by Ekman, Sverdrup and Stommel to provide the first comprehensive solution of the wind driven circulation, using the real wind field Munk included the effect of eddy viscosity in both the horizontal and the vertical dimensions (both $A_h$ and $A_z$). Instead of using the curl of the hypothetically varying wind, he used the curl that results from the real winds of Pacific and Atlantic Oceans.

Munk's theory of wind-driven circulation starts with the same basic equations that Sverdrup (1947) and Stommel (1948) started with and adds a lateral (north/south) eddy diffusion.

Sverdrup's Original Equations are:

$$\frac{\partial p}{\partial x} = f\rho v + \frac{\partial \tau_x}{\partial z}$$

$$\frac{\partial p}{\partial y} = -f\rho u + \frac{\partial \tau_y}{\partial z}$$

Munk's Starting Equations:

$$\frac{\partial p}{\partial x} = f\rho v + \frac{\partial}{\partial z}\left(A_z \frac{\partial u}{\partial z}\right) + A_h \frac{\partial^2 u}{\partial x^2} + A_h \frac{\partial^2 u}{\partial y^2}$$

$$\frac{\partial p}{\partial y} = -f\rho u + \frac{\partial}{\partial z}\left(A_z \frac{\partial v}{\partial z}\right) + A_h \frac{\partial^2 v}{\partial x^2} + A_h \frac{\partial^2 v}{\partial y^2}$$

Munk replaced u and v velocity components in the starting equations and then followed Sversrup's approach of taking x and y derivatives of respective equation and then subtracting one equation from another to get:

$A_h \nabla^4 \psi - \beta \frac{\partial \psi}{\partial x} = -\nabla_z \times \tau_w$ or this can be written as

$$\beta M_y = \nabla \times \tau - A_h \nabla^4 \psi \quad (11.1)$$

Where $\frac{\partial f}{\partial y} = \beta$, τ is wind stress and ψ is stream function

In this equation (11.1), the first term ($A_h \nabla^4\psi$) is the loss of vorticity due to lateral friction which is large close to lateral boundaries such as continents. The remaining two terms ($\beta M_y = \nabla \times \tau$) is Sverdrup Balance which expresses the conservation of potential vorticity (in the absence of frictional loss of vorticity)

Equation (11.1) is called Munk's equation. Munk (1950) assumed that the currents went to zero near the bottom or that the bottom was level and the bottom stress was level and the bottom stress was negligible because the currents are very small in deep water. Munk (1950) also assumed that in the west the friction terms associated with horizontal shears in the currents are important. The vanishing of curl $\tau$ at certain latitudes breaks the flow into gyres.

In principle, the wind driven circulation theory presented here is applicable to the major sub-tropical regions between$10^0$ N (or S) and $50^0$ N (or S) in the Atlantic, Pacific and perhaps the Indian Oceans where major continents block the zonal flow. Walter Munk has applied a modified version of Stommel's model with more realistic wind stress forcing on more realistic model ocean basins including the tropics between $10^0$ N to $10^0$ S as shown in Fig.11.1.

Thus in many ways, Munk's model of ocean circulation (Fig.11.1) reproduces the circulation pattern in the real ocean. This means that the oceanic processes considered by Munk (1950) are important. They are flow of water into different latitudes (i.e. regions in which the Coriolis force is different) and horizontal and vertical frictional forces which are important in determining the large scale current patterns that result from the wind in the real oceans.

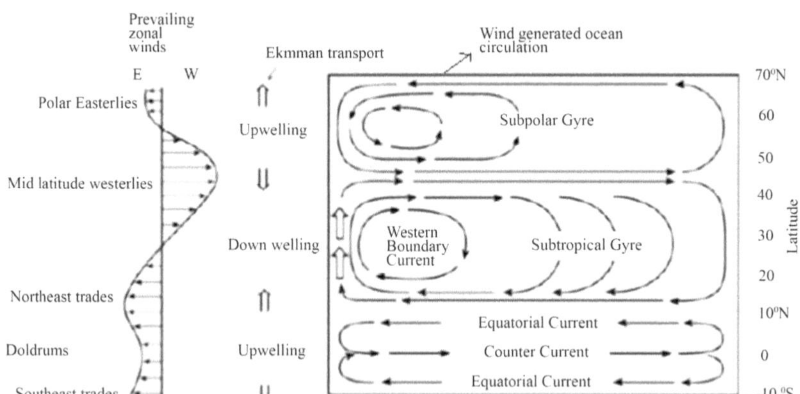


**Fig. 11.1:** Schematic configuration of current gyres driven by wind in Pacific by Munk (1950)

Fig.11.1 shows the pattern of ocean circulation that Munk calculated in North Pacific Ocean. If this pattern is compared with the global surface circulation of any Northern Oceans (like North Atlantic or North Pacific), it bears a strong resemblance to the actual circulation like gyres, North Equatorial current, Equatorial counter current, westward intensification etc.

Models of ocean circulation are being improved and refined all the time using the equation of motion. In this regard Fofonoff's (1954) analysis of interior transport (section 8.4) also is worth to be noted. Later Munk and Carrier (1950) solved for a triangular ocean which represents, more nearly, the shape of the North Pacific which results the same physics of Munk as described above.

Thus Sverdrup (1947) showed that the circulation in the upper kilometer or so of the ocean is directly related to the curl of the wind stress if the Coriolis force varies with latitude. Stommel (1948) showed that the circulation in oceanic gyres is asymmetric also because the Coriolis force varies with latitude. Walter Munk (1950) added eddy viscosity and calculated the circulation of the upper layers of the Pacific. Together the three oceanographers laid the foundations for a complete understanding of a general wind driven circulation of the oceans.

## References

Fofonoff, N. P., 1954. Steady flow in a frictionless homogeneous ocean. J. Mar. Res., 13, 254-262.

Lynne D Talley,2011. Descriptive physical oceanography: an introduction. Academic press,.

Munk, W. H., 1950: On the wind-driven ocean circulation. J. Meteor., 7, 80–93, doi:10.1175/1520-0469(1950)007,0080: OTWDOC.2.0.CO;2.

Munk, W. H., and Carrier, G. F., 1950: The wind driven ocean circulation in ocean basins of various shapes. Tellus, 2.MIT Open course ware: https://ocw.mit.edu/resources/res-12-000-evolution-of-physical-oceanography-spring-2007/part-1/

Stommel H. 1948. The westward intensification of wind-driven ocean currents. Transactions, American Geophysical Union 29 (2): 202–206.

Sverdrup, H. U., 1947: Wind driven currents in a baroclinic ocean; with applications to the equatorial currents of the eastern Pacific. Proc. Natl. Acad. Sci. USA, 33, 318–326, doi:10.1073/ pnas.33.11.318

# 12

# Water Masses and Thermohaline Circulation

## 12.1. Introduction

As water consists of two positively charged hydrogen ions and a single negatively charged oxygen ion, water is arranged as a polar molecule having positive and negative sides. This molecular polarity leads to water's high dielectric constant (ability to withstand or balance an electric field). Water is able to dissolve many substances because the polar water molecules align to shield each ion, resisting the recombination of the ions. The ocean's salty character is due to the abundance of dissolved ions. The polar nature of the water molecule causes it to form polymer-like chains of up to eight molecules. Approximately 90% of the water molecules are found in these chains. Energy is required to produce these chains, which is related to water's heat capacity. Water has the highest heat capacity of all liquids except ammonia. This high heat capacity is the primary reason the ocean is so important in the world climate system. Unlike the land and atmosphere, the ocean stores large amounts of heat energy it receives from the sun. This heat is carried by ocean currents, exporting or importing heat to various regions. Approximately 90% of the anthropogenic heating associated with global climate change is stored in the oceans, because water is such an effective heat reservoir.

### 12.1.1. The wind stress

The ocean waters are driven by wind and temperature-salinity (thermohaline) forces. The magnitude and direction of the shear stress which is exerted by the atmospheric winds on the ocean surface depends on the wind velocities. If **U** = (U, V ) indicates the wind velocity at 10 m height, then a semi-empirical bulk relation provides the wind stress $\tau$ (in Pa = $Nm^{-2}$) as

$$\tau = \rho_a C_D |U| U,$$

where $\rho_a = 1.2\ kgm^{-3}$ is the density of air and $C_D \approx 10^{-3}$ is a friction coefficient that depends on |U|

## 12.2. The heat budget of Oceans

Motion of the atmosphere is driven by the uneven distribution of solar energy. Solar heating is greatest in equatorial regions and causes water in the oceans to evaporate and the moist air to rise. The warm, humid air forms a belt of equatorial clouds and heavy rainfall. It is bordered in the middle latitudes by high-pressure zones that are cloud-free and contain dry descending air.

The total net heat flux $Q_{oa}$ in $Wm^{-2}$ which is positive when heat is transported from the atmosphere into the ocean) is the sum of several components, i.e.,

$$Q_{oa} = Q_H + Q_E + Q_{LW} + Q_{SW}$$

In this expression, $Q_H$ is the sensible heat flux, $Q_E$ is the latent heat flux, $Q_{LW}$ is the longwave heat flux and $Q_{SW}$ is the shortwave heat flux. The magnitudes of these fluxes can also be estimated by bulk formulae. For example, the sensible heat flux $Q_H$ is the flux of heat arising through random motion near the ocean-atmosphere interface (a 'diffusive flux') and it is proportional to the difference between the sea-surface temperature ($T_S$) and the atmospheric temperature ($T_a$) at a height ( $Z_a$) just above this interface. The bulk formula estimating this flux is given by

$$Q_H = \rho_a C_{Pa} C_H |U| (T_s - (T_a + \gamma_a z_a))$$

where $\rho_a$ ($kgm^{-3}$) and $C_{pa}$ ($Jkg^{-1}K^{-1}$) are the density and heat capacity of air, $|U|$ ($ms^{-1}$) is the wind speed, $\gamma_a$ ($Km^{-1}$) the (dry or moist) atmospheric 'lapse rate' and $C_H$ is a dimensionless semi-empirical transfer coefficient.

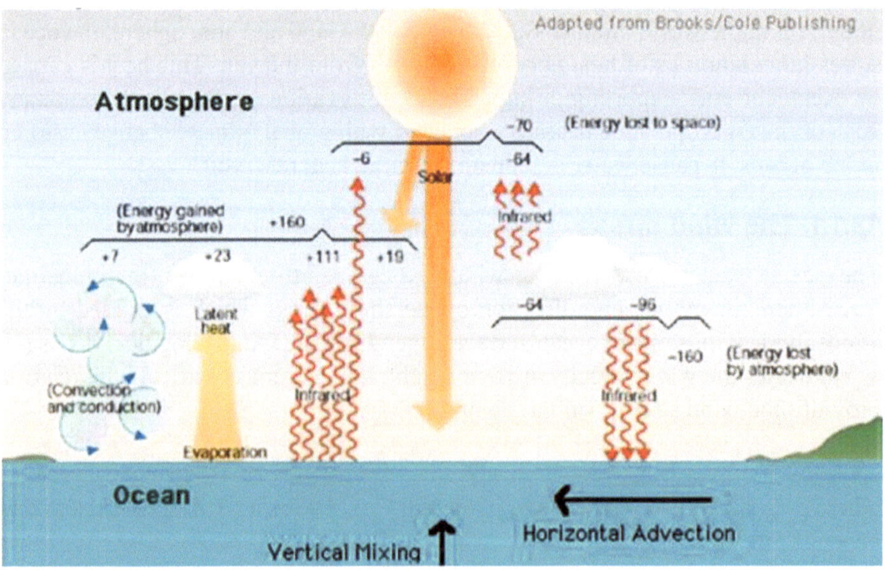


**Fig. 12.1.** Heat budget of oceans (Roxy, 28328964)

Figure 12.1 shows the heat budget of oceans. While energy gained by atmosphere is + 160 units, its loss to Oceans through longwave radiation is -160 units which is balanced and the energy lost to space once for all is -70 units. The different ways ocean loses to atmosphere are through conduction and convection (-7 units), latent heat (-23 units), long wave radiation (-111 units). Atmosphere also gains energy from sun through short wave radiation as +19 units. Thus the total gain of energy to atmosphere is +160 units. Atmosphere in turn loses energy (-160 units) through long wave radiation towards oceans (-96 units) as well as to space (-64 units). Ocean is also directly losing -6 units to space. Thus space gain is 70 units.

The evaporation E ($ms^{-1}$) over the ocean can be estimated by a bulk formula of the form

$$E = C_E |U|(q_s - q_a)$$

where $C_E$ is a semi-empirical exchange coefficient, $q_a$ the relative humidity in the atmosphere and $q_s$ the saturation relative humidity at the ocean-atmosphere boundary. The latent heat flux $Q_E$ is related to the evaporation E through

$$QL = \rho_a L_f E,$$

where $L_f = 2.5 \times 10^6$ $Jkg^{-1}$ is the latent heat of evaporation.

## 12.3. Water characteristics of North Indian Ocean

Ocean temperatures vary with depth and with latitude. In most cases, the ocean is stratified into warm surface water and cooler (denser) deep water. Except in Polar Regions, ocean water becomes markedly cooler with depth. Surface waters are warmest near the equator, but the temperature at 1 km depth varies little with latitude. The average temperature of the global ocean is 3.6°C. The region of maximum gradient of temperature in the vertical is called thermocline which is the region between warm, light surface water and cold, dense deep water. Mixed layer is the region where temperature mixed and homogeneous to a certain level. The SST and precipitation over the Indian Ocean is shown in Fig.12.2 a and b respectively during northern summer (June-September). Note in Fig.12.2a the SST values in Arabian Sea (AS) are relatively lower than Bay of Bengal (BOB) while the precipitation is opposite.

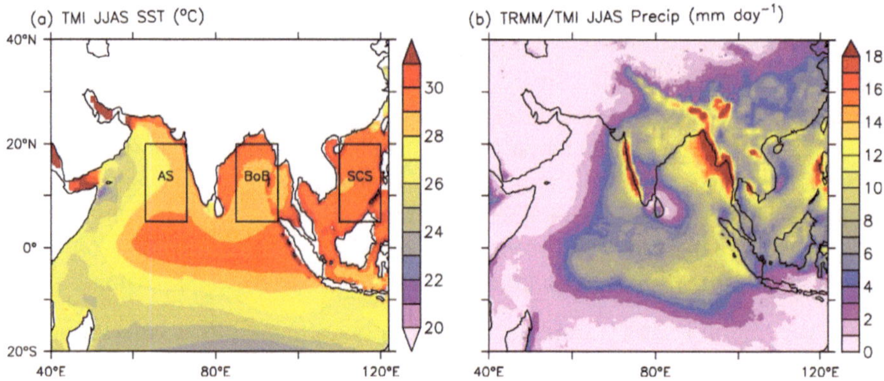


**Fig.12.2.** SST**(a)** and precipitation **(b)** in North Indian Ocean during northern summer monsoon (Roxy, 28328964)

Since the specific heat of ocean water is much larger than that of air, the top 2.5 m of the ocean holds as much heat as the entire atmosphere above it. Thus the heat required to change a mixed layer of 25 m by 1 °C would be sufficient to raise the temperature of the atmosphere by 10 °C. The depth of the mixed layer is thus very important for determining the temperature range in oceanic and coastal regions. The heat stored within the oceanic mixed layer provides a source for heat that drives global variability such as El Niño. The mixed layer is also important as its depth determines the average level of light seen by marine organisms. In very deep mixed layers, the tiny marine plants known as phytoplankton are unable to get enough light to maintain their metabolism.

Salinity in a given region of the ocean can change due to river discharge, precipitation, freezing/thawing, and evaporation. However, the overall amount of salt in the oceans remains relatively constant. So salinity is considered as a conservative property in the oceans. In subtropical regions, the salinity of surface water is high because intense evaporation leaves the water rich in salts that cannot evaporate. At high latitudes where the temperature is lower, the evaporation rate is much lower and fresh rainwater makes the surface waters low in salinity. With regard to density the ocean can be divided into two layers. They are: (1) a thin upper layer of warm, well-stirred water and (2) a thick mass of deeper, colder water that is relatively calm and slow-moving. For the normal range of salinity in the ocean, the maximum density occurs at the freezing point, which is depressed to well below 0°C. As seawater is heated, molecular activity increases and thermal expansion occurs which reduces the density.

Concentration of oxygen in near-surface waters is large (1 in Fig. 12.3) because of exchange with the oxygen rich atmosphere at the sea surface and because of downward mixing by waves and eddies. Concentrations of $O_2$ just below the thermocline are low because oxidation of sinking organic particles consumes $O_2$.

This causes the oxygen-minimum layer (2 in Fig. 12.3), in which $O_2$ concentrations may reach zero. The $O_2$ minimum layer generally occurs at depths of 300 to 800 meters (stramma et al 2008)

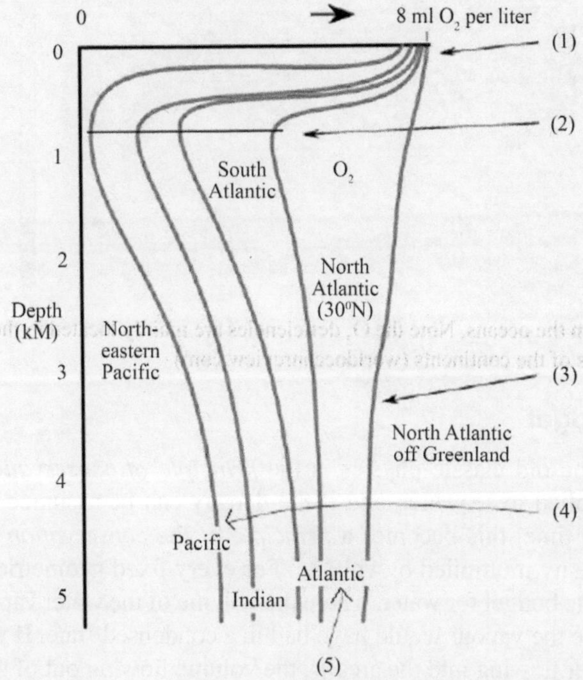


**Fig.12.3.** Vertical variation of oxygen in the oceans and $O_2$ minimum layer (Dietrich 1963, Chesler 1990)

Concentrations of $O_2$ in abyssal waters are less than those in surface waters (3 in Fig.12.3) because oxidation of sinking organic particles consumes $O_2$.Concentrations of $O_2$ in deep waters of the Pacific are less than that of Atlantic and Indian Ocean (4 of Fig. 12.3) because Pacific and Indian deep waters have had more time to lose $O_2$ to oxidation of organic matter. Concentrations of $O_2$ in deep waters of the Atlantic are typically greater than that of Pacific and Indian (5 of Fig. 12.3) because Atlantic deep waters have more recently left the surface and so had less time to lose $O_2$ to organic matter.

Marine regions with oxygen deficiencies are mainly located in the mid-latitudes on the west sides of the continents (Fig. 12.4). There is very little mixing here of the warm surface waters with the cold deep waters, so not much of oxygen penetrates to greater depths. In addition, high bio-productivity and the resulting large amounts of sinking of biomass here lead to strong oxygen consumption at depth, especially between 100 and 1000 meters

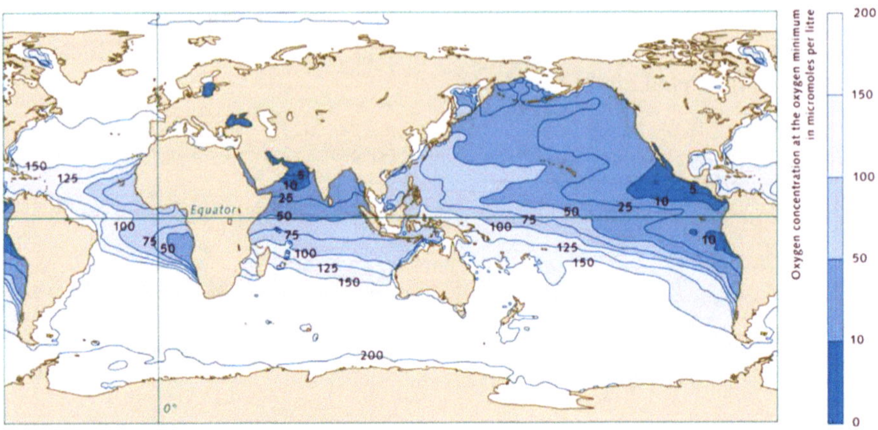


**Fig. 12.4.** $O_2$ minimum layer in the oceans. Note the $O_2$ deficiencies are mainly located in the mid-latitudes on the west sides of the continents (worldoceanreview.com)

## 12.4. The volume Budget

An important axiom in the old classic physics is the *principle of conservation of mass*. When this is applied to seawater, and if the density can be assumed to be practically constant in time, this becomes a *principle of the conservation of volume*, since mass is density multiplied by volume. For every fixed geometrical area we can set up a volume budget for water, where the volume of the water vapor is calculated as the volume the vapour would have had in a condensed state. If we call the volume of seawater flowing into the area $V_i$, the volume flowing out of the area $V_0$ the water volume supplied by rivers *R*, the evaporated volume *E*, and the precipitated volume *P*, then the volume budget is

$$V_i + R + P - V_0 - E = Fh \tag{12.1}$$

where *F* is the surface area and *h* is the average increase of the water level. If there is no change of the water level ($h = 0$), we get

$$V_0 - V_i = + R + P - E \tag{12.2}$$

This can be interpreted to mean that the net volume of seawater leaving an area must be the same as the net volume of freshwater entering it.

## 12.5. The Salt Budget

As earlier mentioned, we can assume with a high degree of accuracy that the amount of salts in the oceans is constant. Salts are not created nor used. We will now look at the simplified case where neither the water level nor the average salinity changes with time in a specific area. An equal amount of salts must be brought into the area as taken out of it, and the salt budget for this special condition yields

$$S_i \rho_i V_i = S_o \rho_o V_o \qquad (12.3)$$

where $S_i$ and $\rho_i$ are salinity and density of the entering volume of sea water. $S_0$ and $\rho_0$ are the corresponding values for the leaving volume of sea water. Because there is little variation in ρ in comparison with S, we can reduce the equation by ρ and obtain

$$S_i V_i = S_o V_o \qquad (12.4)$$

In some cases, when *h*=0 and we know *R+P-E* and have measured $S_0$ and $S_i$ equations (12.2) and (12.4) can be combined to express $V_0$ and $V_i$ as functions of the known quantities. These are the so-called *Knudsen relations*. Since it is easier to record salinity than volume transport, these are quite handy for estimating transports in areas where the inward and outward currents have clearly defined salinities.

## 12.6. Water masses

There are two basic circulation systems in the oceans. One is wind driven at the sea surface and the other is density driven circulation in the deep ocean. While only about 10% of the ocean volume is involved in wind-driven surface currents, the other 90% circulates due to density differences in water masses. Water masses are identified by their temperature, salinity, and other properties such as nutrients or oxygen content. All water masses gain their particular characteristics because of interaction with the surface during their development. Water properties, such as T and S, are formed only when the water is at the surface or in the mixed layer. The characteristics of water masses are obtained by different processes at the time of formation of water masses are different inputs of freshwater patterns of precipitation, evaporation, temperature regimes, brine rejection etc.

Water with a specific temperature and salinity will result if a body of water is subject to specific meteorological influences over a significant period, during which it remains in the mixed surface layer. Further, if the water eventually becomes isolated from the atmosphere, it sinks down from the surface. All these conditions are generally satisfied in the convergence regions.

The formation and geographical distribution of surface water masses is strongly influenced by the pattern of surface currents. The thickness of surface water masses generally include from the top of the mixed layer to upper part of permanent thermocline. In the equatorial zone, a shallow water mass forms as density is low due to high temperature and lower salinity due to higher precipitation.

Whereas the water masses that form in the subtropical zones are very thick as subtropical zones are convergence zones and so the sea surface is raised (High) and the thermocline is depressed. So a thick water mass forms in the subtropical regions.

Water from a particular region has a particular temperature associated with a particular salinity, and the relationship changes little as the water moves through the deep ocean because the process of movement is very slow. Once water masses sink, their temperature and salinity are modified only by mixing with other water masses (diffusive and turbulent heat exchange).

### 12.6.1. Temperature –Salinity diagram (T-S diagram)

Helland-Hansen pointed out in 1916 when in a given area the temperatures and corresponding salinities of the sub surface water are plotted against each other the points generally fall on a well-defined curve called TS-curve, showing the temperature salinity association of that water. The TS-diagram containing salinity on the x-axis, temperature on the y-axis and $\sigma_t$ lines over printed on it is called TS-diagram as shown in Fig.12.5 Montgomery used specific volume anomaly lines or thermosteric anomaly lines in place of $\sigma_t$ lines and called it as TSV-diagram.

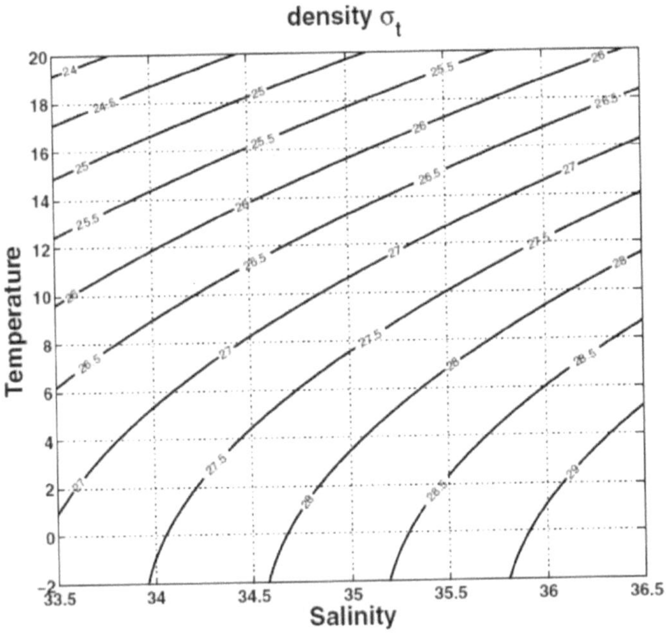


**Fig.12.5.** T-S diagram

### 12.6.2. Uses of T-S diagram

The T-S diagram has become one of the valuable tools in physical oceanography. By means of this diagram characteristic features of the temperature salinity distribution are conveniently represented and anomalies in the distribution are easily recognized. The diagram is also useful in correcting the data. More important use is water masses can be classified on the basis of temperature-salinity characteristics. But density cannot be used for classification because two water masses of different temperatures and salinities may have the same density. In recent years the amount of mixing between one water mass to the other is also being estimated with the help of T-S diagram.

A water mass is defined by a portion of T-S curve on the T-S diagram which has a range of temperature and a range of salinity of a huge body of water of almost homogeneous characteristics whereas a water type is represented by a point on the T-S diagram which has a single value of temperature and a single value of salinity and has no thickness.

Because the number of possible combinations of temperature and salinity is limited, it follows that only a reasonably small number of water masses are formed in the oceans. However, the density of a water mass alone is not sufficient for its identification because various combinations of temperature and salinity can produce the same density.

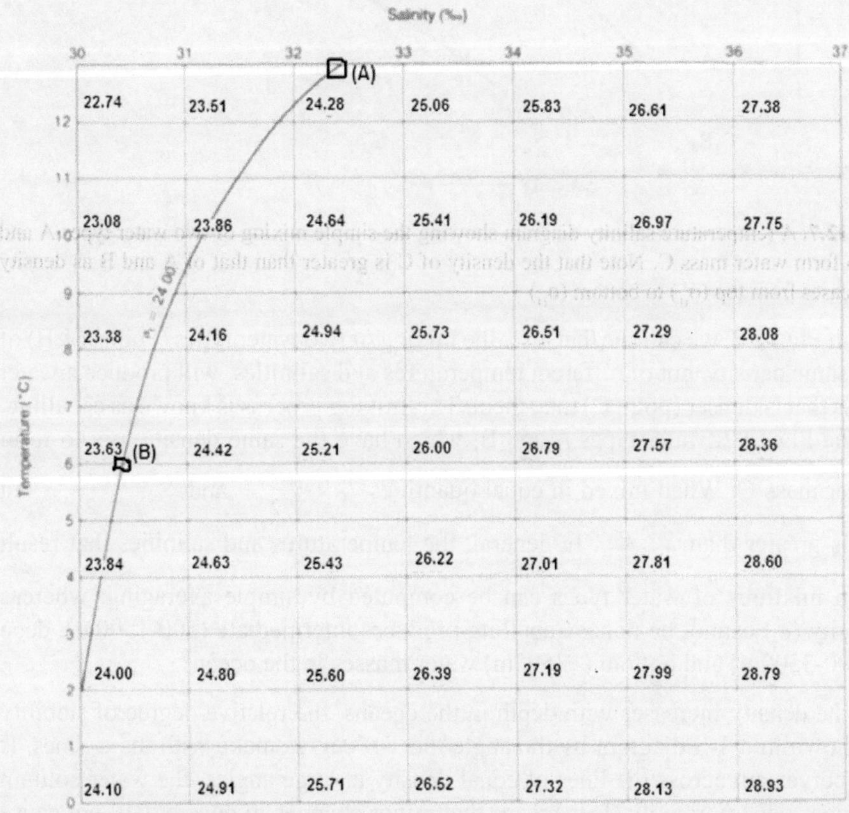


**Fig. 12.6.** T-S diagram (chegg.com)

If we plot the density for a series of combinations of temperature and salinity on a T-S diagram as in Fig. 12.6, the isopycnals are curved. For example, water with a salinity of 32‰ at a temperature of 10 °C has $\sigma_t$ of 24.64. When this $\sigma_t$ value is plotted on a TS-diagram (as shown in Fig. 12.6), it is the intersecting point of (10 °C, 32 ‰). For example, note the $\sigma_t = 24$ line crossing points of t and S in the Fig. 12.6. Note this $\sigma_t = 24$ line crosses the 12 °C temperature line at a salinity of about 31.85 ‰.

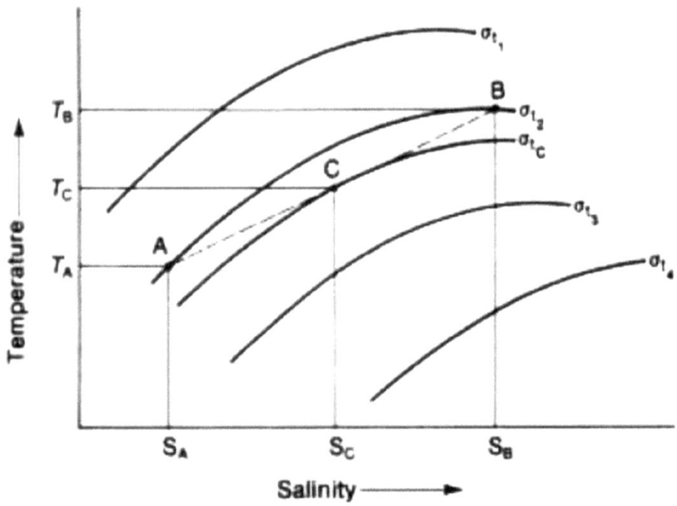


**Fig.12.7.** A temperature salinity diagram showing the simple mixing of two water types A and B to form water mass C. Note that the density of C is greater than that of A and B as density increases from top ($\sigma_{t1}$) to bottom ($\sigma_{t4}$)

From Fig.12.7, we can see that after the mixing of two water types (Point A &B) of the same density, but of different temperatures and salinities, will produce a water mass that is denser (point C) than them. This mixing process is known as caballing. In the Fig.12.7, water types A and B, which have the same density mix to form water mass C. When mixed in equal quantities $T_C = \frac{T_A + T_B}{2}$ and $S_C = \frac{S_A + S_B}{2}$, but $\sigma_{tC}$ is greater than $\frac{\sigma_{tA} + \sigma_{tB}}{2}$. In general, the temperatures and salinities that result from mixtures of water types can be computed by simple averaging, whereas density ($\sigma_t$) cannot be done. Caballing produces intermediate (500-1500m), deep (1500-3500m) and bottom (>3500m) water masses in the oceans.

As the density increases with depth in the oceans, the relative degree of stability is shown in a T-S diagram by the angle the T-S curves make with the $\sigma_t$ lines. If the curves cut across the lines of equal density at large angles, the water column changes density rapidly. This means that minor changes in density will not cause water to sink or rise a great distance, and that the water column is therefore very stable. If the curves are at very small angles to the density lines, then the density changes only slightly with increasing depth, so that minor changes in temperature and salinity will cause water to move to different depths rapidly.

Also a T-S diagram in a well-known oceanic area can be used to correct or find the missing data points as the incorrect data appear as isolated points on the T-S diagram outside of the area and can easily be found.

### 12.6.3. Mixing of water masses

#### 12.6.3.1. Simple mixing between two water types

The simplest mixing that can occur between two water types is the mixing of two equal volumes. For example, if you mix equal volumes of water type X and water type Y, the resulting density of the mixture would be a simple average of both. The average density of X + Y = (density X + density Y)/2. This can be illustrated on a TDS diagram as a point on the straight line between the two water types. In Fig. 12.8, the resultant mixture of two equal volumes would be exactly halfway along the line connecting the two water types.

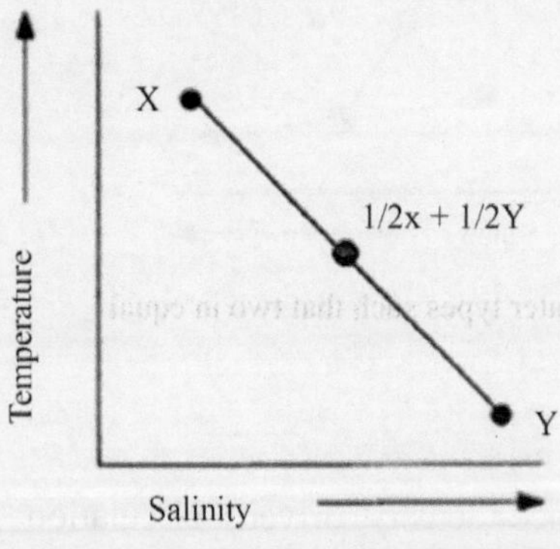


**Fig. 12.8.** Simple mixing of water masses

Thus in general mixing of two water masses show a straight line on a T-S diagram, though it is always not true. For example see Fig.12.7 for the mixing of two water masses A and B and the mixture is C on a straight line.

#### 12.6.3.2. Mixing of unequal volumes of two water types

For example 80mL of water type X mixes with 160mL of water type Y, then the resultant mixture will be $\frac{80}{240}X + \frac{160}{240}Y = \frac{1}{3}X + \frac{2}{3}Y$

Hence the mixing in this case is also along the same straight line between X and Y as shown above Fig.12.8 with ratio 1/3 and 2/3

### 12.6.3.3. Mixing of three water types in equal proportions

Suppose three water types (X, Y, Z) mix with different sigma-t's (σt) in equal proportions. Then to predict the mixture's final density by plotting on a T-S diagram, consider it in two parts. First assume Z mixes with Y in equal proportions as shown in Fig. 12.9. Then the final water type 'X' mixes with the resultant point 1/3 the way up the line from Z+Y---- > to X.

For example, if 80ml of three different water samples X,Yand Z are mixed, then the ratios of each in the mixture is $\frac{80}{240}X + \frac{80}{240}Y + \frac{80}{240}Z = \frac{1}{3}X + \frac{1}{3}Y + \frac{1}{3}Z$

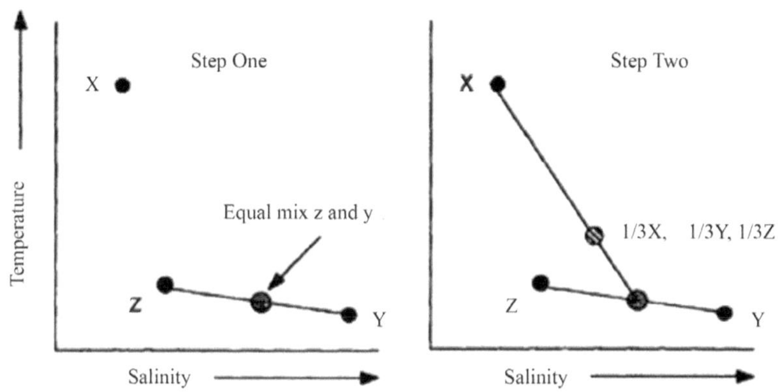


**Fig. 12.9.** Steps in determining the mixing of three water types such that two in equal proportions

## 12.7. Thermohaline circulation

Thermohaline (temperature-salinity and hence density driven) circulation is completely separate from that of the surface or wind driven circulation. While surface circulation is largely in an east-west direction moving warm water towards the poles, the thermohaline circulation is sinking down and moving north-south direction by returning cooler water along the meridians towards the equator (See Fig.12.8.6 in Exercise 12.8.4). The cold water eventually returns to the sea surface to be reheated and returned to the poles through the chosen depths or $\sigma_\theta$ surfaces. The velocity of thermohaline currents is very slow, about 1 cm/ sec, whereas the surface currents are 10 to 20 times faster. Using the concept of residence time (the average time that a water particle resides in an area before being recycled), the water mass replacement at a location is understood. It was found that about 500-1000 years would be required to replace all the deep water in the Atlantic Ocean.

Most water involved in deep-ocean currents (thermohaline circulation) originates in high latitudes *at the surface*. In these regions, surface water becomes cold and its salinity increases as sea ice forms. When this surface water becomes dense enough, it sinks, initiating deep-ocean currents. Once this water sinks, it is removed from the physical processes that increased its density in the first place, so its temperature and salinity remain largely unchanged for the duration it spends in the deep ocean. Thus, temperature salinity (T- S) diagram can be used to identify deep-water masses based on their characteristic temperature, salinity, and resulting density.

Thus the thermo-haline circulation is driven by density differences between ocean water masses caused by temperature and salinity variations. Variations in temperature and salinity are controlled, in turn, primarily by balances between solar heating and radiative cooling, and between solar evaporation and precipitation, respectively. Wind driven currents dominate water motions in the upper layer of the oceans above the pycnocline. Below the pycnocline, currents are driven by density differences between water masses. Density differences cause water masses to sink or rise to the appropriate density level. In areas where water masses sink or rise, the density distribution with depth, and hence the pressure distribution, is different from that in the surrounding areas. Thus horizontal pressure gradients are formed. Once a water mass has moved vertically to reach its density equilibrium level, it flows horizontally in response to the gradient. Thus thermohaline circulation (THC) is a part of the large-scale ocean circulation that is driven by global density gradients created by surface heat and freshwater fluxes. The word thermo-haline derives from the words: thermo-referring to temperature and haline-referring to salt content. These two words together determine the density of sea water.

The thermohaline circulation (THC) is well developed and well-studied in the North Atlantic Ocean. The reason for this is the North Atlantic is much more saline and hence, denser. North Atlantic is more saline because it is warmer due to the mighty ocean currents (such as Gulf Stream) bring warm water from low latitudes. It begins on the surface of the ocean near the pole in the North Atlantic as shown in Fig.12.10. Here, the water is chilled by arctic temperatures, and the winds. It also gets saltier because when sea ice forms, the salt does not freeze and is left behind in the surrounding water. The cold water is now denser, due to the added salts, and sinks toward the ocean bottom (See Fig. 12.8.6 in Exercise 12.8.4). Surface water moves in to replace the sinking water, thus creating a current. This deep water moves south, between the continents, past the equator, and down to the ends of Africa and South America. The current travels around the edge of Antarctica, where the water cools and sinks again, as it does in the North Atlantic. Thus, the conveyor belt gets "recharged." As it moves around Antarctica, two sections split off the conveyor and turn northward. One section moves into the Indian Ocean, the other into the Pacific Ocean. These two sections that split off warm up and become less dense as they travel northward toward the equator, so that they rise to the surface (upwelling). They then loop back southward and westward to the South Atlantic, eventually returning to the North Atlantic, where the cycle begins again.

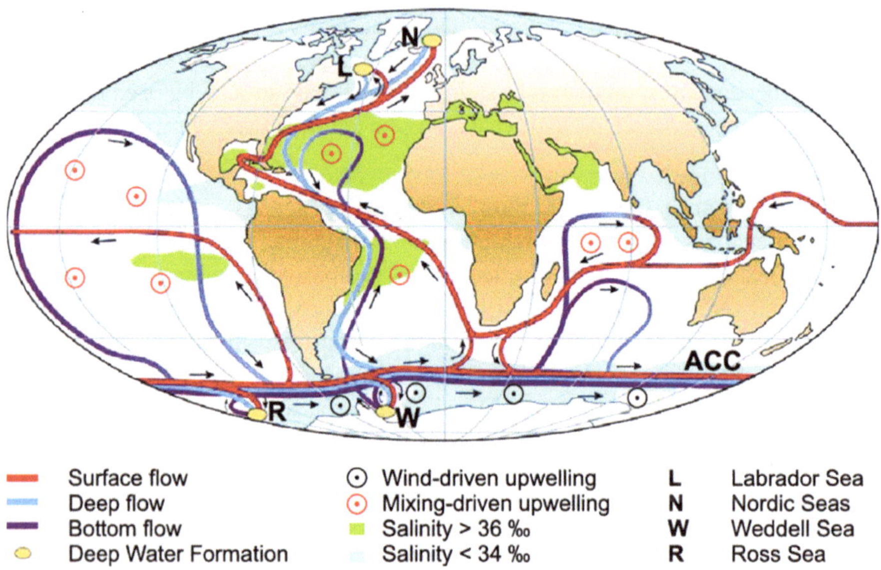


**Fig.12.10.** Thermohaline circulation. The circles denote the sinking areas. Blue denotes cold water flow after sinking and red denotes the warm water return flow (Broecker, 1991). https://www.researchgate.net/figure/Simplified-Conceptual-Model-of-the-Thermohaline-Circulation-figure-from-Kuhlbrodt-et_fig3_283802113

Cold water in polar zones sink relatively rapidly over a small area, while warm water in temperate and tropical zones rise more gradually across a much larger area. It then slowly returns poleward near the surface to repeat the cycle. The continual diffuse upwelling of deep water maintains the existence of the permanent thermocline found everywhere at low and mid-latitudes. This slow upward movement is estimated to be about 1 cm/day over most of the ocean.

According to Fig.12.10, the cold water in Nordic Seas near Green land (N) and Labrador (L) in Arctic Sea and Weddell (W) and Ross(R) seas in Antarctic sink at N, L,W and R along the route shown by blue light curve and returns back from surface along the thick red curve. That is why this is called conveyor circulation. As deep circulation is controlled by temperature and salinity characteristics this is also called thermohaline circulation.

### 12.7.1. Method of measuring Thermohaline circulation

Thermohaline circulation is difficult to study, and most of our knowledge on this came from studies of deep ocean water masses and modeling studies.

The most common method of measuring the flow in the deep ocean is to track parcels of water containing molecules not normally found in the ocean like tritium. The distribution of trace molecules is calculated from the concentration of the molecules in water samples collected on hydrographic sections. The distribution of tritium in the north Atlantic collected during 1972-73 by Geosecs program is an example of it.

If we date a water parcel from the time that it leaves the surface and sink into the time that it leaves the surface and sink into the deep ocean again it takes several thousands of years. It is understood from the dating that the youngest water is in the deep North Atlantic and the oldest water is the deep northern Pacific where its age is estimated to be 1000 years.

It is also understood from dating that the CFC, the chemical tritium and $C^{14}$ that were released through atmospheric atomic bomb test in 1950s and 1960s entered the deep ocean in the northern Atlantic and are still moving southward slowly. These pollutants just crossed the equator in the Atlantic and have not yet reached the northern Pacific.

### 12.7.2. Climate Change Concerns and Thermohaline Circulation

The following are related to carbon cycle influence on global warming.

1. The temperature of bottom water formation determines how much $CO_2$ is dissolved in deep ocean water
2. The rate of overturn of the oceans determines the "burial rate" of C from the atmosphere
3. Organic C accumulates in sediments, depending on the $O_2$ content of deep ocean
4. Organic C in sediments is reduced to $CH_4$ (methane gas)
5. Methane gas migrates upward and can be trapped as frozen "gas hydrates" near the ocean floor

If sea level falls, it gives a negative feedback and if deep water warms, it gives a positive feedback for thermohaline circulation. Both effects liberate gas hydrates ($CH_4$), which combines with $O_2$ to form $CO_2$, ultimately reaching the atmosphere. With regard to Deep Ocean Circulation the ocean has an enormous capacity to absorb and release greenhouse gases and so the rate at which temperature and composition of seawater circulating through the deep ocean is vitally important in assessing long term climate change.

### 12.7.3. Whether the thermohaline circulation will collapse?

To some extent the THC appears to be self-sustaining. However, if some unexpected event occurs to break this self-sustaining chain of processes, then there is a chance for this circulation to break down (i.e., over several decades) and later may remain in a state of reduced-circulation for several centuries.

Some simple models of the world's oceans simulated a rapid break down of the THC, when the density of the water in the North Atlantic Ocean is lowered due to addition of fresh water (rain) due to global warming. Increased greenhouse gas concentrations will not only warming the North Atlantic but also increase rainfall and so it can be argued that global warming may cause a rapid collapse of the thermohaline circulation.

Nevertheless, observations and palaeo-climatic studies indicated that the THC was affected to a large extent both in the recent and distant past.

### 12.7.4. What would happen to our climate if the thermohaline circulation is collapsed?

Vellinga and Wood (2007) provide estimates of the response of pre-industrial surface climate variables if the thermohaline circulation in the Atlantic Ocean collapse. For this they have used HadCM3, an ocean-atmosphere general circulation model that is run without flux adjustments. In this model a temporary collapse was forced by applying a strong initial freshening to the top layers of the North Atlantic. In the first five decades after the collapse surface air temperature response is dominated by cooling of much of the Northern Hemisphere (locally up to 8 $^0$C, 1–2 $^0$C on an average) and weak warming of the Southern Hemisphere (locally up to 1 $^0$C, 0.2 $^0$C on average). Response is strongest around the North Atlantic but significant changes occur over the entire globe and highlight rapid teleconnections. Precipitation is reduced over large parts of the Northern Hemisphere. A southward shift of the Intertropical Convergence Zone over the Atlantic and eastern Pacific creates changes in precipitation that are particularly large in South America and Africa. Colder and drier conditions in much of the Northern Hemisphere reduce soil moisture and net primary productivity of the terrestrial vegetation. This is only partly compensated by more productivity in the Southern Hemisphere. The total global net primary productivity by the vegetation decreases by 5%. It should be noted, however, that in this version of the model the vegetation distribution cannot change, and atmospheric carbon levels are also fixed. After about 100 years the model's thermohaline circulation has largely recovered, and most climatic anomalies disappear (Vellinga and Richard, 2002; Wood et al 2003).

### 12.7.5. Dynamics of Thermohaline circulation

The deep flow is sluggish with very long timescales. It will therefore be in geostrophic, hydrostatic, and thermal wind balance. We know from quasi-geostrophic vorticity considerations that if columns of fluid are stretched or squashed then meridional motion results.

The Fig.12.11 represents the ocean as two layers: an upper layer largely driven by the winds with localized regions while the lower layer is formed by air-sea interaction. These localized sources, S, are in high latitudes such as the northern North Atlantic and the Antarctic. The return flow consists of weak upwelling everywhere else in the ocean. Since the upper layer is on contact with the atmosphere, its properties are constantly forced to some specified surface values. If the deep water is not going to eventually fill up the basin, it must be returned to the surface values. If this is done by downward turbulent diffusion from the upper to lower layer, a balance can be reached when the weak upwelling balances this vertical diffusion. This is the basic interior balance envisioned by Munk (1966).

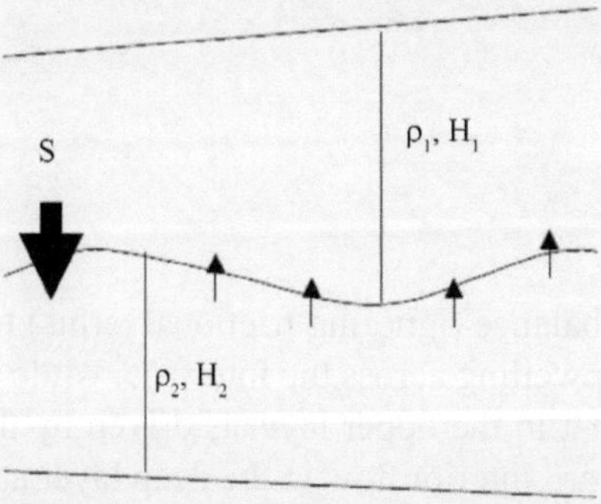


**Fig. 12.11.** A two layers ocean of upper wind driven and lower thermohaline type (Munk 1966)

The source/sink flows represent water mass formation and Stommel and Arons (1960) developed a global model for the horizontal circulation driven by these transformations. The two high latitude sources are each of the order of 15-20 Sv., and a deep circulation of one sign develops in the lower layer will be exactly opposite to the one that develops in the upper layer in response to these sources.

The model is essentially one in which the potential vorticity of each layer is conserved. The conservation of Potential Vorticity in each layer is given by the following:

$$\frac{\mathrm{d}}{\mathrm{dt}}\left[\frac{f+\zeta_1}{H_1}\right] = -\frac{r_1}{H_1}\zeta_1 + \frac{1}{\rho_1 H_1} curl\overline{F}$$

$$\frac{\mathrm{d}}{\mathrm{dt}}\left[\frac{f+\zeta_2}{H_2}\right] = -\frac{r_2}{H_2}\zeta_2$$

where F is the body force (i.e. wind stress) at the upper surface. Now in the interior, away from any sources, there is an upward velocity of the interface, which is taken as a constant = $w_0$: $H_1 = H^0_1 - w_0 t$, $H_2 = H^0_2 + w_0 t$. If these are inserted into the above, we can obtain a pair of equations for the linearized vorticity of the layer [this comes like that of the wind-driven circulation]. These are:

$$\beta v_1 + \frac{f}{H_1^0} w_0 = -r_1\zeta_1 + \frac{1}{\rho_1} curl\overline{F}$$

$$\beta v_2 - \frac{f}{H_2^0} w_0 = -r_2\zeta_2$$

These equations should be integrated over the layers using the definition of wind stress at the upper surface. Here one may note that while the current in each layer is uniform with depth, the friction parameter is not uniform. It is strong near the surface and bottom and sometimes near the interface also. The integration results an average of the friction parameter in each layer. It is assumed that the variable topography of the internal interface and bottom can be ignored (i.e. flat). Then the following is the result:

$$\beta \int v_1 dz = -fw_0 - \overline{r_1} \int \zeta_1 dz_1 + \frac{1}{\rho_1} curl\overline{F}$$

$$\beta \int v_2 dz = +fw_0 - \overline{r_2} \int \zeta_2 dz_2$$

This is something like the Sverdrup interior balance (ignoring frictional terms) for the upper layer, which is augmented by an upwelling across the interface, while in the deeper layer, a circulation opposite to that in the upper layer is driven by the upwelling. This upwelling results in a poleward interior flow in the deep layer and an equatorward flow in the upper layer driven by the thermohaline mass exchange. What emerges from the above balance is the importance of western boundary currents in redistribution of the required transport for the interior. Though the Stommel and Arons (1960) model has not been verified in detail for the interior of the ocean, its inferences about the location and direction of abyssal flows has been one of the few theoretical predictions made before the observations of strong, deep western boundary currents were made.

Note that the upper layer circulation (opposite to that plotted in the Fig. 12.10) will be such as to increase northward transports in Gulf Stream over wind-driven dynamics, but will decrease transport of the Brazil Current, because the two flows are opposed. The actual flows in the western boundary currents are, much like those for the wind-driven circulation, dependent crucially on the nature of the frictional damping with Stommel (1961) and Munk (1966) type boundary layers.

This thermohaline circulation is an internal mode which vanishes if we average vertically over both layers. It is ultimately tied to the interior balance reflected by Munk (1966) as the cold water injected into the deep layer must be "converted" to warm water in the upper layer by diffusive processes. This is often overlooked, but it is the central part of the theory. Numerical work with more complicated thermohaline models shows that the strength of the overturning circulation is proportional to the amount of small-scale vertical diffusion of heat/salt/density in the ocean.

To know the contribution of each of the element of heat/salt/density on the strength of the vertical velocity, Warren (1981) estimated that for a lower layer thickness of 3 km, with total source strength of 34 Sv. of deep water formation, mean vertical velocities will be approximately 1.5 x $10^{-7}$m/s, horizontal interior flows of the order of $10^{-3}$ m/s and western boundary transports of the order of 10 Sv. [=10 $^{7}$m $^{3}$/s].

### 12.7.6. Abyssal Mixing

If the vertical flux of salinity is modeled, (for example) in the deep water as a balance between vertical advection and vertical mixing, the latter is modeled as an eddy- diffusivity so that small-scale turbulent fluxes of salinity are given by the equation:

$$-\rho < w'S' > = \rho\kappa \frac{\partial < S >}{\partial z} = \rho w < S(z) >$$

Where κ is the vertical eddy diffusivity. If typical values are given for the vertical variation of salinity, w' is the estimate for the vertical velocity and an estimate of the vertical diffusivity required for an overall balance of the thermohaline circulation comes as: $\kappa \approx 10^{-4}m^2/s$, which is one of the canonical numbers in oceanography.

### References


Broecker, W.S., 1991. The Great Ocean conveyor. Oceanography 4 (2), 79–89

Dietrich, G., K. Kalle, W. Krauss and G. Siedler, 1980: Oceanography. John Wiley & Sons, 626 pp.

Munk, W. H. (1950): On the wind-driven ocean circulation. Journal of Meteorology, 7, 79-93.

Munk,W., 1966. Abyssal recipes. Deep-Sea Res. 13, 707-730.

Roxy, 28328964: coursehero.com/file/28328964/roxy-lecture-physical-oceanographypdf/)

Stommel, H. (1948): The westward intensification of wind-driven currents. Transactions, American Geophysical Union, 29, 202-206

Stommel, H.M., Arons, A., 1960. On the abyssal circulation of the World Ocean d II. An idealized model of the circulation pattern and amplitude in oceanic basins. Deep-Sea Res. 6, 217-233.

Stommel, H.M., 1961. Thermohaline convection with two stable regimes of flow. Tellus 13, 224-230.

Stramma L.G, C.Johnson and M.Volker ( 2008): Expanding Oxygen –minimum zones in the Tropical Oceans. Science Vol.320,Issue 5876,pp 655-658. DOI: 10.1126/science.1153847

Vellinga, M and A.W.Richard (2002): Global climatic impacts of a collapse of the Atlantic Thermohaline Circulation. Climatic Change. 54(3): 251–267. doi:10.1023/A:1016168827653. S2CID 153075940.

Vellinga M and R. A. Wood (2007): Impacts of thermohaline circulation shutdown in the twenty-first century. Climatic Change, DOI: 10.1007/s10584-006-9146-y.

Warren, B.A., 1981. Deep circulation of the world ocean. In:Warren, B.A., Wunsch, C. (Eds.), Evolution of Physical Oceanography. MIT Press, Cambridge, MA, pp. 6-41.

Wood R. A., M. Vellinga and R. Thorpe (2003): Global Warming and thermohaline circulation stability. Phil. Trans. R. Soc. Lond. A, 361, 1961-1975.

## Exercises

### *Exercise 12.1*

Determine density ($\sigma_t$ & $\rho$) for the following temperature and salinity values: i) $15^0$ c, 35g/kg; ii)$20^0$ c, 32g/kg; iii) $5^0$ c, 32g/kg iv) Why density or $\sigma_t$ lines are curved on a T-S diagram?

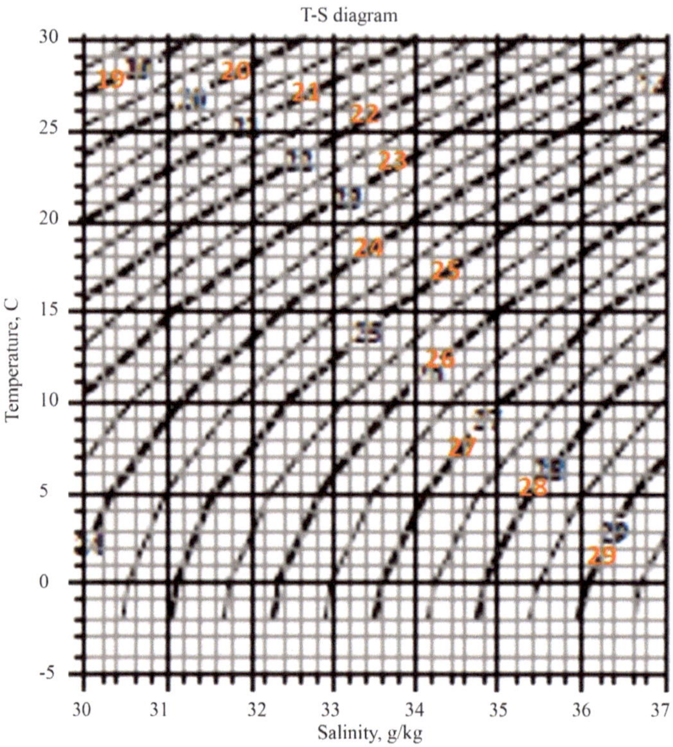


**Fi.g. 12.1.1** T-S diagram: With the courtesy of Richard Strickland, Lecture 19, oceanography 101, university of Washington, 2006.

### *Solution*

Using the T-S diagram above in Fig. 12.1.1, read directly

i) For $15^0$ c, 35g/kg, $\sigma_t = 26$, since $\sigma_t = (\rho_{s,t,p}\text{-}1)10^3$; $\rho_{s,t,p} = 1.026$ g $cm^3$

ii) For $20^0$ c, 32g/kg, $\sigma_t = 22.5$, $\rho_{s,t,p} = 1.0225$ g $cm^3$

iii) For$5^0$ c, 32g/kg, $\sigma_t = 25.3$, $\rho_{s,t,p} = 1.0253$ g $cm^3$

iv) In general, when two water types mix, resulting temperatures and salinities may be obtained by simply averaging, but resulting densities may not. The effect of temperature on density is different than the effect of salinity on density. That is why density is a curved line on a T-S diagram. In other words two water masses can have the same density but when they mix their density will be different. For example the two water masses A and B in the Fig. 12.1.2 have the same ($\sigma_t$ = 24) but their mixture (c) has different $\sigma_t$ (>24).

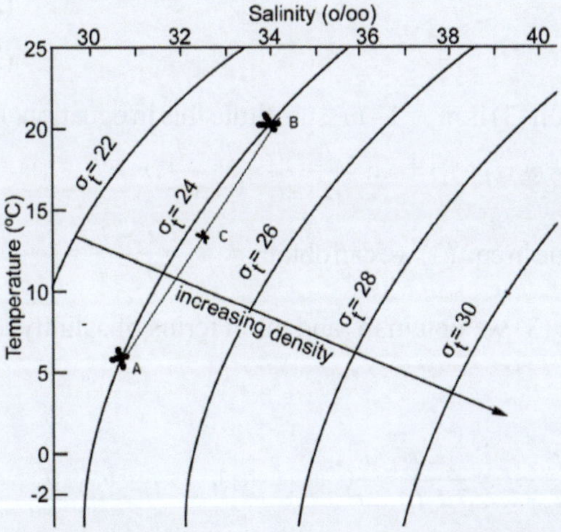


**Fig. 12.1.2** (MS20 water mass Lab version 3/8/2005).

## *Exercise 12.2*

If two water types $I(T_1,S_1)$ and $II(T_2,S_2)$ are mixing linearly along a straight line, Find the values of the resultant mixture 'R'$(T_R,S_R)$.

## *Solution*

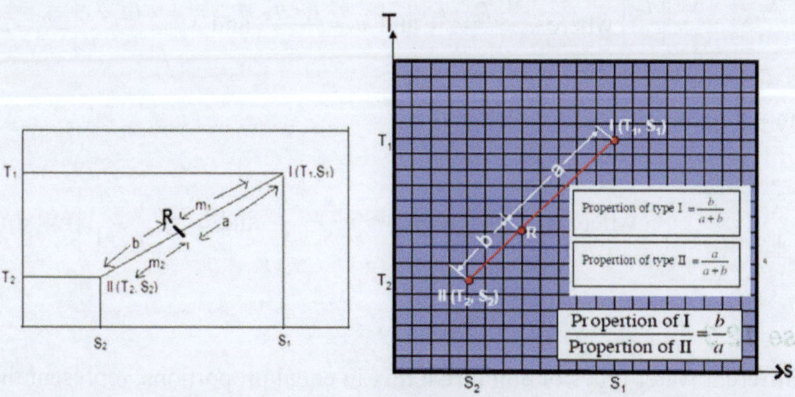


**Fig.12.2** Linear mixing of two Water masses: (With the courtesy of C. Chen, General Physical Oceanography, MAR 555; School for Marine Sciences and Technology(SMAST); Umass-Dartmouth)

(http://fvcom.smast.umassd.eduCoursesMAR555Lectures_pdfMAR555_Lec_11.pdf)

We can write from the above Fig.12.2 :

$T_R = T_1 m_1 + T_2 m_2$ (1)

$S_R = S_1 m_1 + S_2 m_2$ (2)

where $m_1 + m_2 = 1$ (3)

Solving the three equations: equation (3) is $m_1 = 1 - m_2$ substitute this in equation(1):

$T_R = T_1(1 - m_2) + T_2 m_2; \Rightarrow T_R - T_1 = m_2(T_2 - T_1) \Rightarrow m_2 = \frac{T_R - T_1}{T_2 - T_1} = \frac{T_R - T_1}{\Delta}; put(T_2 - T_1) = \Delta$

In the same way substituting $m_2$ value in eqn (3) we can obtain $\Rightarrow m_1 = \frac{T_2 - T_R}{T_2 - T_1} = \frac{T_2 - T_R}{\Delta}$

Similarly solving equation (2) and (3) we obtain $m_1$ and $m_2$ in terms of salinity as:

$m_1 = \frac{S_2 - S_R}{S_2 - S_1}$ and $m_2 = \frac{S_R - S_1}{S_2 - S_1}$

$\frac{\text{Proportion I}}{\text{Proportion II}} = \frac{m_1}{m_2} = \frac{T_2 - T_R}{T_1 - T_2} \div \frac{T_R - T_1}{T_1 - T_2} = \frac{T_2 - T_R}{T_R - T_1} = \frac{T_R - T_2}{T_1 - T_R} = \frac{b}{a} (\because \text{in the graph } T_R - T_2 = b; T_1 - T_R = a)$

Percentage of I $= \frac{m_1}{m_1 + m_2} \times 100$ ; Percentage of II $= \frac{m_2}{m_1 + m_2} \times 100$

## Alternative solution

The solution in the matrix notation is:

For temperature:

$$\begin{bmatrix} m_1 \\ m_2 \end{bmatrix} = \begin{bmatrix} \frac{1}{\Delta_T} & -\frac{T_2}{\Delta_T} \\ -\frac{1}{\Delta_T} & \frac{T_1}{\Delta_T} \end{bmatrix} \begin{bmatrix} T_R \\ 1 \end{bmatrix} \text{ where } m_1 = \frac{T_R - T_2}{\Delta} \text{ and } m_2 = \frac{T_1 - T_R}{\Delta} \text{ and } \Delta_T = \begin{bmatrix} T_1 & T_2 \\ 1 & 1 \end{bmatrix} = T_1 - T_2$$

Similarly for salinity:

$$\begin{bmatrix} m_1 \\ m_2 \end{bmatrix} = \begin{bmatrix} -\frac{1}{\Delta_S} & -\frac{S_2}{\Delta_S} \\ \frac{1}{\Delta_S} & \frac{-S_1}{\Delta_S} \end{bmatrix} \begin{bmatrix} S_R \\ 1 \end{bmatrix} \text{ where } m_1 = \frac{S_2 - S_R}{S_1 - S_2} \text{ and } m_2 = \frac{S_R - S_1}{S_1 - S_2} \text{ and } \Delta_S = \begin{bmatrix} S_1 & S_2 \\ 1 & 1 \end{bmatrix} = S_1 - S_2$$

### *Exercise 12.3*

If four different water types of 80ml each mix in equal proportions, represent the mixture in the following Fig.12.3.

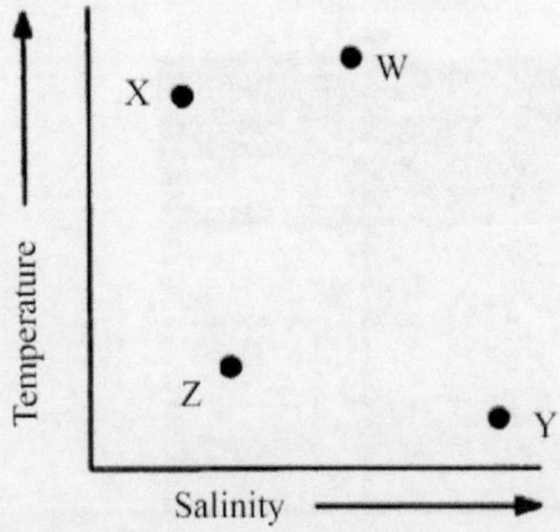


**Fig 12.3.** Four water types in a T-S diagram

### *Solution*

As each water type of 80ml mixed. Then the mixture has the proportions as:

$$P = \frac{80}{320}X + \frac{80}{320}Y + \frac{80}{320}Z + \frac{80}{320}W = \frac{1}{4}X + \frac{1}{4}Y + \frac{1}{4}Z + \frac{1}{4}W$$

This means the mixture should have each $\frac{1}{4}$ share. This can be represented as in Fig. 12.3.1 like this

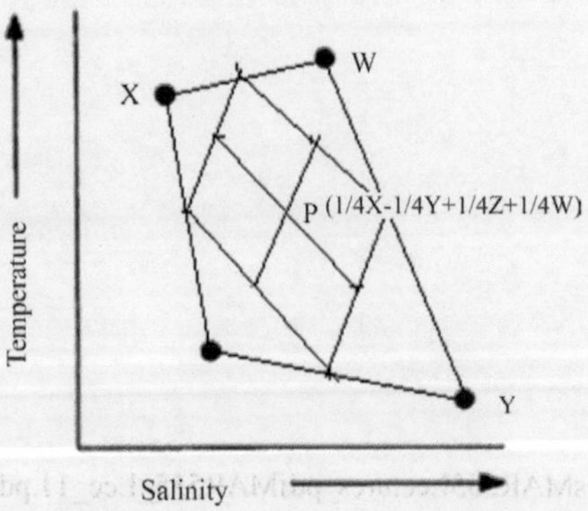


**Fig. 12.3.1.** Representation of mixing of four different water types

### *Exercise 12.4*

The mixing and flow of different water masses in the North Atlantic Central water in the subtropical gyre region is as shown in Figure below. The salinity and temperature values vertically for surface, intermediate and deep waters are given as (36.6,20), (36.2,16), (35.8,12) respectively.(i) Draw the T-S graph if it is a straight line mixing. (ii) Find the values of salinity and temperature at the points A, B,C and D indicated in the Fig.12.4.

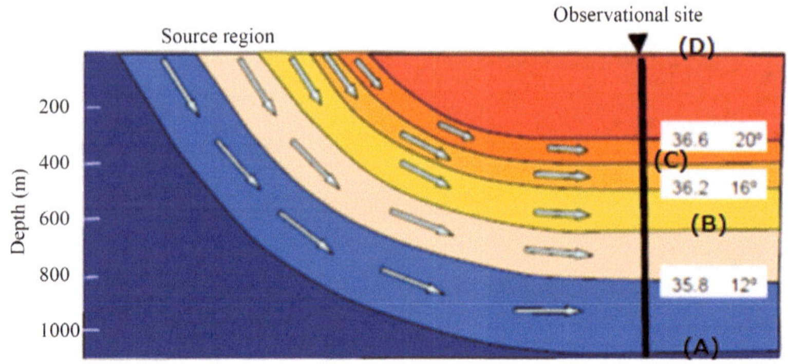


**Fig.12.4** (httpfvcom.smast.umassd.eduCoursesMAR555Lectures_pdfMAR555_Lec_11.pdf)

## *Solution*

i) The T-S diagram for the above Fig.12.4 in case of a straight line mixing is as shown in Fig.12.4.1 below:

ii) The values of salinity and temperature at A, B and C are: A(35.6,10), B(36.0,14), C(36.4,18), D(36.6,20)

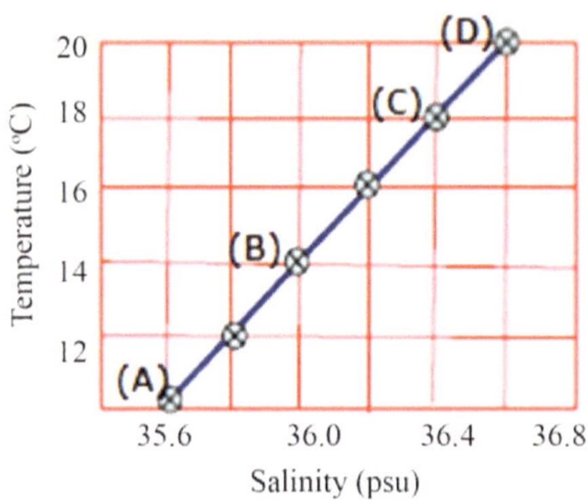


**Fig.12.4.1.** (httpfvcom.smast.umassd.eduCoursesMAR555Lectures_pdfMAR555_Lec_11.pdf)

### *Exercise 12.5*

### *Triangular Mixing*

Find the values of the proportions $m_1$,$m_2$ and $m_3$ of the mixture of the three water types ($T_1$,$S_1$),($T_2$,$S_2$) and ($T_3$,$S_3$) when they mix linearly.

### *Solution*

If more than two water types are mixing in a locality, the resultant mixture obeys the principle of triangular mixing as shown in Fig.12.5. If the three water types

$(T_1,S_1)$,$(T_2,S_2)$ and $(T_3,S_3)$ are mixing linearly, the resultant mixture will have the characteristics as:

$$T = T_1m_1 + T_2m_2 + T_3m_3 \tag{1}$$

and $$S = S_1m_1 + S_2m_2 + S_3m_3 \tag{2}$$

where $m_1$,$m_2$,$m_3$ are the proportions of mixing such that $m_1 + m_2 + m_3 = 1$ (3)

Substituting equation (3) in equation (1) we can write: $T = T_1\{1 - m_2 - m_3\} + T_2m_2 + T_3m_3$

$$\Rightarrow T = T_1 + m_2(T_2 - T_1) + m_3(T_3 - T_1) \Rightarrow T - T_1 = m_2(T_2 - T_1) + m_3(T_3 - T_1) \tag{4}$$

In the same way substituting eqn (3) in eqn (2) we can write for salinity as:

$$S - S_1 = m_2(S_2 - S_1) + m_3(S_3 - S_1) \tag{5}$$

Eqn (4) further can be written as: $$m_2 = \frac{T - T_1}{(T_2 - T_1)} - \frac{(T_3 - T_1)}{(T_2 - T_1)}m_3 \tag{6}$$

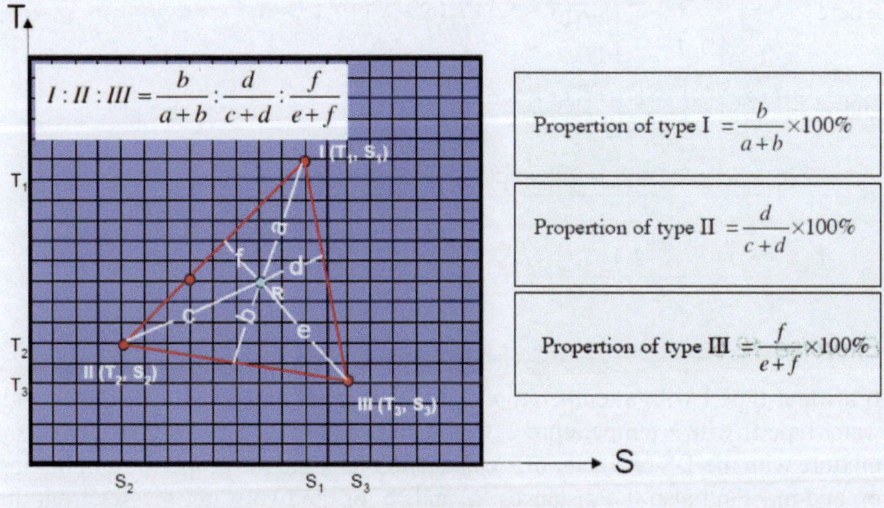


**Fig.12.5.** Triangular mixing of Water masses: With the courtesy of C. Chen, General Physical Oceanography, MAR 555; School for Marine Sciences and Technology(SMAST); Umass-Dartmouth.
(httpfvcom.smast.umassd.eduCoursesMAR555Lectures_pdfMAR555_Lec_11.pdf)

Similarly eqn (5) can be written as $$m_2 = \frac{S - S_1}{(S_2 - S_1)} - \frac{(S_3 - S_1)}{(S_2 - S_1)}m_3 \tag{7}$$

Equating and Solving eqns (6) and (7) we can write: $m_3 = \frac{T(S_1 - S_2) - S(T_1 - T_2) + T_1S_2 - T_2S_1}{\Delta}$

Where $\Delta = T_1(S_2 - S_3) + T_2(S_3 - S_1) + T_3(S_1 - S_2)$

In the same way solving $m_1$ and $m_2$ we get: $m_1 = \dfrac{T(S_2 - S_3) - S(T_2 - T_3) + T_2S_3 - T_3S_2}{\Delta}$

$$m_2 = \frac{-T(S_1 - S_3) + S(T_1 - T_3) - T_1S_3 + T_3S_1}{\Delta}$$

### *Alternative Solution*

The solution for the proportion of mixing in this case is:

$$\begin{bmatrix} m_1 \\ m_2 \\ m_3 \end{bmatrix} = \begin{bmatrix} \dfrac{S_2 - S_3}{\Delta} & -\dfrac{T_2 - T_3}{\Delta} & \dfrac{T_2S_3 - T_3S_2}{\Delta} \\ -\dfrac{S_1 - S_3}{\Delta} & \dfrac{T_1 - T_3}{\Delta} & \dfrac{-T_1S_3 - T_3S_1}{\Delta} \\ \dfrac{S_1 - S_2}{\Delta} & -\dfrac{T_1 - T_2}{\Delta} & \dfrac{T_1S_2 - T_2S_1}{\Delta} \end{bmatrix} \begin{bmatrix} T \\ S \\ 1 \end{bmatrix}$$

where $= \Delta = \begin{bmatrix} T_1 & T_2 & T_3 \\ S_1 & S_2 & S_3 \\ 1 & 1 & 1 \end{bmatrix} = T_1(S_2 - S_3) + T_2(S_3 - S_1) + T_3(S_1 - S_2)$

and $m_1 = \dfrac{T(S_2 - S_3) - S(T_2 - T_3) + T_2S_3 - T_3S_2}{\Delta}$

$$m_2 = \frac{-T(S_1 - S_3) + S(T_1 - T_3) - T_1S_3 + T_3S_1}{\Delta}$$

$$m_3 = \frac{T(S_1 - S_2) - S(T_1 - T_2) + T_1S_2 - T_2S_1}{\Delta}$$

### *Exercise 12.6*

If a water type I with a temperature of $5^0$ c and a salinity of 35.5 ‰ mixes with water type II with a temperature $2^0$ C and a salinity of 34.5 ‰ to give a resultant mixture with the T-S character of $3^0$c and 34.85 ‰. Find the proportions of mixing $m_1$ and $m_2$. Find also the distances 'a' and 'b' of the two water masses from the mixture in the T-S graph.

### *Solution*

Given I: $T_1 = 5^0$ c ; $S_1 = 35.5$ for II: $T_2 = {=}2^0$ C ; $S_2 = 34.5$; $T_R = {=}3^0$ C ; $S_R = 34.85$

We know $m_1 = \dfrac{T_2 - T_R}{T_2 - T_1} = \dfrac{2-3}{2-5} = \dfrac{-1}{-3} = \dfrac{1}{3}$ ; $m_2 = \dfrac{T_R - T_1}{T_2 - T_1} = \dfrac{3-5}{2-5} = \dfrac{2}{3}$

This means 1/3 rd of type I and 2/3 of type II got mixed-up.

$$\frac{\text{Proportion I}}{\text{Proportion II}} = \frac{b}{a} = \frac{T_R - T_2}{T_1 - T_R} = \frac{3-2}{5-3} = \frac{1}{2}$$

This means b =1/2 of a.

### *Exercise 12.7*

If three water types A($5^0$ c,35.5 ‰), B($2^0$ c, 34.5 ‰) and C(1 $^0$c,35.25 ‰) got mixed-up in proportions 38%, 46% and 16% respectively, find the T-S values of the resultant mixture.

### *Solution*

Given A($T_1$ =$5^0$ c, $S_1$ = 35.5 ‰), B($T_2$ =$2^0$ c, $S_2$ = 34.5 ‰) and C($T_3$=1 $^0$c, $S_3$= 35.25 ‰). Let T and S are for resultant mixture.

We can write from Exercise 12.6 as:

Substituting in triangular mixing equation: $m_1 = \dfrac{T(S_2 - S_3) - S(T_2 - T_3) + T_2S_3 - T_3S_2}{\Delta}$

Where = $\Delta = T_1(S_2 - S_3) + T_2(S_3 - S_1) + T_3(S_1 - S_2)$

$$\Delta = 5(34.5 - 35.25) + 2(35.25 - 35.5) + 1(35.5 - 34.5) = -1$$

$$\frac{38}{100} = \frac{T(34.5 - 35.25) - S(2-1) + 2\times 35.25 - 1\times 34.5}{-1} = -0.75T - S + 36 \qquad (1)$$

Similarly, $m_2 = \dfrac{-T(S_1 - S_3) + S(T_1 - T_3) - T_1S_3 + T_3S_1}{\Delta} =$

$$\frac{46}{100} = \frac{-T(35.5 - 35.25) + S(5-1) - 5\times 35.25 + 1\times 35.5}{-1} = 0.25T - 4S + 140.75 \qquad (2)$$

Solving equation (1) & (2): 75T+100S= 3562 & 25T-400S = -14029 ; T =219/325 = $0.67^0$ C

Substituting T=0.67 in equation (1) and solving we get, S = 35.11

### *Exercise 12.8*

If a Water mass A (with volume $V_A$ and density ) is mixed with water mass B (with volume $V_B$ and density $\rho_B$) to give a water mass with a volume V, a temperature T, a salinity S, and a density ρ, calculate the ratios of $V_A$ /V and $V_B$ /V

If A water mass A, with a temperature $T_A$ = $5^0$C and a salinity $S_A$ = 35.5 ppt, is mixed with a water mass B, with a temperature $T_B$ = $2^0$C and a salinity $S_B$ = 34.5 ppt. The resulting water has a temperature of $3^0$C and a salinity of 34.85 ppt. Calculate the volume ratio of the water masses $V_A$/ V and $V_B$/ V and the percentages of each water mass in the mixture.

### *Solution*

Conservation of total mass, internal energy and salt imply that

$$\rho_A V_A + \rho_B V_B = \rho_V$$

$\rho_A V_A C_p T_A + \rho_B V_B C_p T_B = \rho v C_p T$

$\rho_A V_A S_A + V_B \rho_B S_B = \rho v S$

Here Cp is the heat capacity of water, which we assume to be independent of temperature so that it cancels out. Furthermore, $\frac{\rho_A}{\rho} \approx 1$ and $\frac{\rho_B}{\rho} \approx 1$ are reasonable approximations because absolute density variations in the ocean are small. Hence

$V_A T_A + (V -- V_A) T_B = V\,T$

$V_A S_A + (V --V_A) S_2 = SV$

Dividing by V and putting $\alpha = V_A / V$ gives

$\alpha T_A + (1 - \alpha) T_B = T \Rightarrow \alpha = \frac{T - T_B}{T_A - T_B}$

Similarly, $\alpha S_A + (1 - \alpha) S_B = S \Rightarrow \alpha = \frac{S - S_B}{S_A - S_B}$ With the values of the temperatures and salinities as above we find

For temperature

$\alpha = \frac{V_A}{V} = \frac{T - T_B}{T_A - T_B} = \frac{3-2}{5-2} = \frac{1}{3} = 0.33$, Similarly for salinity: $\alpha = \frac{V_A}{V} = \frac{S - S_B}{S_A - S_B} = \frac{34.85 - 34.5}{35.5 - 34.5} = 0.35$

; Average α=0.34=34% $\left(\frac{0.33 + 0.35}{2}\right)$

and similarly the average value of $V_B / V$ between temperature and salinity for $1 - \alpha = V_B / V = 0.76 = 76\%$ (1-0.34 = 0.76)

This means 34% of A and 76% of B is in the resultant mixture.

## *Exercise 12.9*

A water mass A with potential temperature $\theta_A = 2^0$C and a salinity $S_A = 34.04$ ppt, is mixed with a water mass B, having a potential temperature $\theta_B = 8.5^0$C and a salinity $S_B = 36.0$ ppt in equal proportions. Find the values of the resultant mixture and mention the interesting part in it.

## *Solution*

Let $T_2 = \theta_B = 8.5^oC$, $T_1 = \theta_A = 2^oC$ and $S_A = S_1$ and $S_B = S_2$

we know $m_1 = \frac{T_2 - T_R}{T_2 - T_1} = \frac{8.5 - T_R}{8.5 - 2} = \frac{8.5 - T_R}{6.5}; m_2 = \frac{T_R - T_1}{T_2 - T_1} = \frac{T_R - 2}{8.5 - 2} = \frac{T_R - 2}{6.5}$;

As both are in equal proportions: $m_1 = m_2$; $\frac{8.5 - T_R}{6.5} = \frac{T_R - 2}{6.5} \Rightarrow T_R = 5.25$ ;

Similarly for salinity: $m_1 = \frac{S_2 - S_R}{S_2 - S_1} = \frac{36 - S_R}{36 - 34.04} = \frac{36 - S_R}{1.96}$ ;

$$m_2 = \frac{S_R - S_1}{S_2 - S_1} = \frac{S_R - 34.04}{36 - 34.04} = \frac{S_R - 34.04}{1.96}$$

$$m_1 = m_2 \Rightarrow \frac{36 - S_R}{1.96} = \frac{S_R - 34.04}{1.96} \Rightarrow S_R = 35.02$$

The resulting mixture is (5.25, 35.02). The interesting part is simply the averages of temperatures and salinities if two water masses mix in equal proportions.

### Exercise 12.10

If the temperature difference of the northward flowing warm water and southward flowing cooler water of Gulf Stream is $5^0$ c, estimate the meridional heat transport associated with Gulf Stream in North Atlantic.

### Solution

Heat transport : $Q_H$ = (volume transport × temperature difference) = UDL ×$\rho C_p$ $\Delta T$ = $50 \times 10^6 \times 1025 \times 4.3 \times 10^3 \times 5 = 1101875 \times 10^9 = 1.1 \times 10^{15}$ watts = 1.1 peta watts (Where U is Velocity, D = thickness and L = Length of Gulfstream. Values are taken from the text.)

### Exercise 12.11

The salinities and temperatures of two sea water samples A and B are like this: $S_A$ =33.2‰ ,$T_A$ = $4^0$ C and $S_B$ = 34.8 ‰ and $T_B$ = 1°C. Mark them in a TS-diagram. (a)What is the temperature and salinity of a mixture if 10% of A and 90% of B is mixed?(b)What is the temperature and salinity of a mixture if 73% of A and 27% of B is mixed?

### Solution

Given a =10%, b = 90°; $S_A^1 = \dfrac{b}{a+b}$ *and* $S_B^1 = \dfrac{a}{a+b}$

a) $S'_A = 33.2\left(\dfrac{90}{90+10}\right) = 29.88$ , $S'_B = 34.8\left(\dfrac{10}{90+10}\right) = 3.48$

   $S_C = S'_A + S'_B$ =29.88+3.48 = 33.36, similarly $T_C$ = 3.7

b) $S_C = S'_A + S'_B$ =8.96+25.4 = 34.36, similarly $T_C$ = 1.81 (Here a = 73% and b = 27%)

### Exercise 12.12

i) What is the value of $\sigma_t$ for A (34.21‰,-0.5°C) &B (34.5‰, 1.5°C). ii) Find S,T, & $\sigma_t$ of the mixture, if both mix in the ratio 10:90.iii) Whether the mixture is denser or lighter with respect to the original

## *Solution*

i) From the TS-plot (Fig.12.12) $\sigma_{tA}$, =27.5 and $\sigma_{tB}$ = 27.6

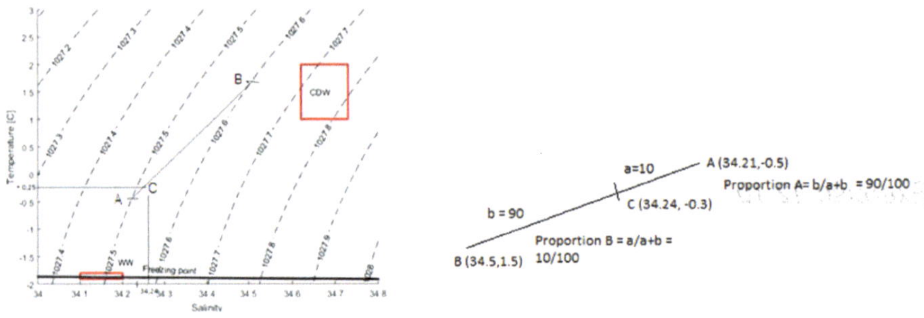


**Fig.12.12**

ii) The mixture = $S_A^{'} = 34.21 \times \dfrac{90}{100} = 30.79; \; T_A^{'} = -0.5 \times \dfrac{90}{100} = -0.45$

$$S_B^{'} = 34.5 \times \frac{10}{100} = 3.45; \; T_B^{'} = 1.5 \times \frac{10}{100} = 0.15$$

The mixture $S_C = S_A' + S_B'$ = 34.24‰ and $T_C = T_A' + T_B'$ = – 0.30; , $\sigma_{tC}$ = 27.52 (From T-S figure above)

iii)The mixture (C ) is denser than A and lighter than B as $\sigma_{tC}$ =27.52 is in between $\sigma_{tA}$ = 27.5 and $\sigma_{tB}$ = 27.6

## *Exercise 12.13*

If a third water mass'D' (34.51‰, -0.5°C) joins with A (34.21‰,-0.5°C) & B (34.5‰,1.5°C) in exercise 12.12 and mixes in equal proportions, what is the resultant mixture ?

## *Solution*

1/3 A + 1/3B + 1/3D = $\frac{1}{3}(A + b + d)$

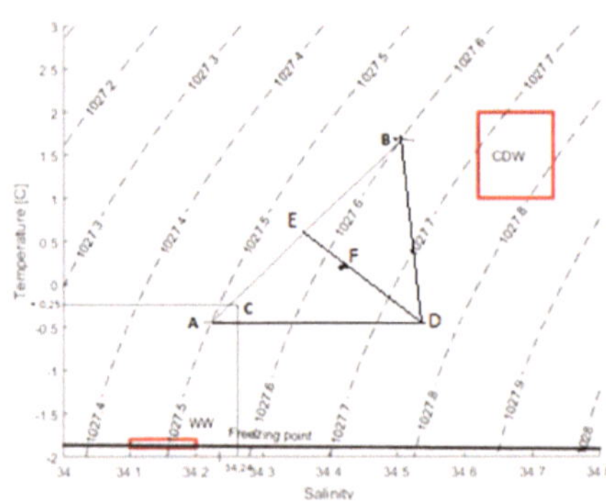


**Fig.12.13**

First A & B joins in equal proportions as shown in Fig.12.13: So (A+B)/2: from figure 12.13, E is midpoint of AB and 'F' is the mixture between D and E which is the centroid of ABD. So EF will be 1/3 of DE.

$$S_F = \frac{34.21 + 34.5 + 34.51}{3} = 34.41;\ T_F = \frac{-0.5 + 1.5 - 0.5}{3} = 0.25$$

### *Another method*

First take the average of A & B then share between E and D in 1:3 ratio

$$S_E = \frac{34.21 + 34.5}{2} = 34.4\ \upsilon T_E = \frac{-0.5 + 1.5}{2} = 0.5$$

Hence Between DE if we mix, D(34.51,-0.5) and E(34.355,0.5)in 1/3 proportion: EF:FD=1:3

$$S'_E = 34.355 \times \frac{3}{4} = 25.766;\ S'_D = 34.51 \times \frac{1}{4} = 8.63;\ T'_E = 0.5 \times \frac{3}{4} = 0.375;\ T'_D = -0.5 \times \frac{1}{4} = -0.125$$

So the mixture $S_F = S_E' + S_D' = 34.4‰$ and $T_F = T_E' + T_D' = 0.25$;, $\sigma_{tF} = 27.75$ (From T-S figure above)

### *Exercise 12.14*

Prepare a T-S diagram for the ranges of temperatures and salinities 0 to 13$^0$ C and 30 to 37 ‰ and $\sigma_t = 24$ to 28

### *Solution*

Take a graph paper. Plot temperature on the y-axis with the scale 1cm=1$^0$c and salinity on x-axis with the scale 1cm = 1 ‰. Then for example you want to draw $\sigma_t$=24 line. Already $\sigma_t$ values are plotted on the T-S diagram. So estimate the place of points at every grid of temperature and salinity where $\sigma_t$ = 24 comes. After marking the points of $\sigma_t$ = 24 at every grid, join the points with a smooth curve. In other words, put the points on each of the horizontal line estimating where $\sigma_t$ = 24 comes. Then join all these points of $\sigma_t$ = 24 with a smooth curve. Like this draw $\sigma_t$ = 25, 26, 27 and 28 curves in Fig.12.14. For your understanding $\sigma_t$ = 24 curve was already drawn in Fig.12.14.

### *Exercise 12.15*

Let 'A'& 'B' are the two water types on the T-S graph shown in (Fig.12.14). (a) What are the values of T,S and of these two water types A & B. (b) What are the ranges of temperature and salinity of the water mass 'AB' (c) If these two water types mixed in equal proportions (50% ratio) what are the values of temperature, salinity and $\sigma_t$ of the mixture. (d) Indicate whether this mixture is stable or not relative to the water mass AB.

## *Solution*

(a) A( 13°C, 32.5 ‰); B ( 6°C, 30.5 ‰); $\sigma_t = 24$; $\rho = 1.024$

(b) Range of temperature = 7°C, Range of salinity = 2‰

(c) Values of the mixture : $T_R = \frac{13+6}{2} = 9.5^0C$; $S_R = \frac{32.5+30.5}{2} = 31.5‰$

(d) From the figure 12.14, it is evident that the density of the mixture C is more than 24. So it is stable.

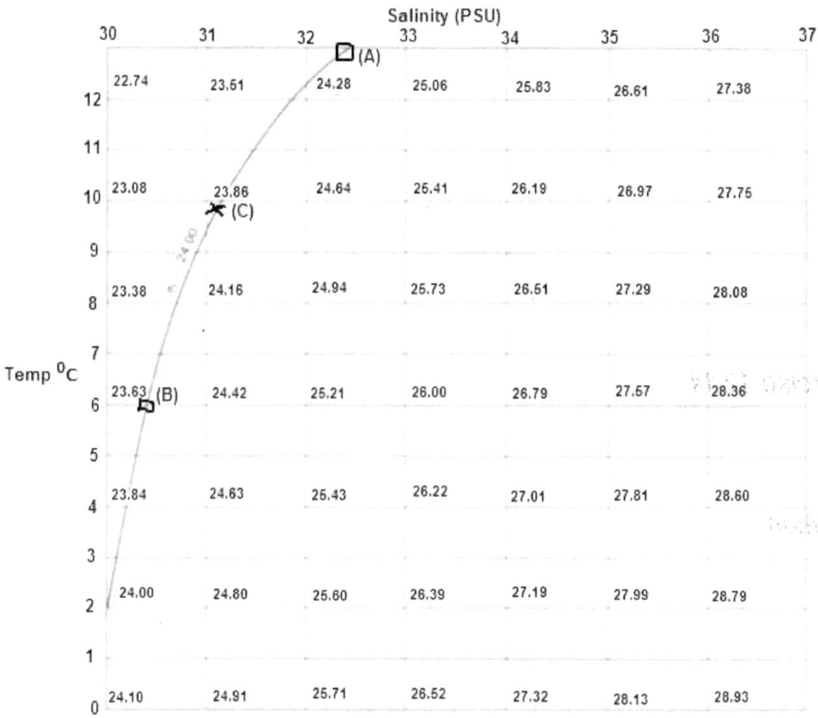


**Fig.12.14. T-S diagram**

## *Exercise 12.16*

The following table gives the temperature and salinity data at a oceanographic station. Plot the data on the T-S diagram given below and connect them with straight lines

| Depth (m) | 0 | 10 | 20 | 50 | 100 | 200 | 300 | 400 |
|---|---|---|---|---|---|---|---|---|
| Temp °C | 12 | 12 | 12 | 10 | 8 | 7 | 6 | 5 |
| Sal (‰) | 33.7 | 33.7 | 33.8 | 33.9 | 33.7 | 34.1 | 34.2 | 34.2 |

1. Draw the contours of $\sigma_t = 25$ to $\sigma_t = 28$ in the given T-S graph in Exercise 12.14.

2. Plot the above data on the $\sigma_t$ = 25 to 28 contoured T-S diagram and join the points with straight lines
3. What is the relative stability of the water column at this station? How deep is the mixed layer and in what portion of the curve is the water most unstable (portion below the mixed layer)?
4. In what portion of the curve is the water most stable?

## *Exercise 12.17*

The following data is from North Atlantic Ocean at 20° N latitude. (i) Plot the data on a blank T-S diagram given below in Fig.12.17.1 and draw T-S curve for this data and indicate the depths of different water masses. (ii) List the different water masses and write their temperature and salinity ranges in this area. (iii) Find the sigma-t (density 0 for these values of temp and salinity from T-S graph you have drawn. (iv) Using the ranges of values of temp (T) and salinity (S) for surface, intermediate, deep and bottom water masses that occur in North Atlantic name and indicate them on the T-S diagram.

| Dep (m) | 100 | 200 | 400 | 500 | 600 | 850 | 950 | 1200 | 1500 | 2000 | 2200 | 2500 | 3000 | 4000 | 5000 |
|---|---|---|---|---|---|---|---|---|---|---|---|---|---|---|---|
| Tem °C | 16.0 | 13.0 | 11.0 | 9.0 | 8.0 | 13.0 | 12.5 | 11.0 | 4.8 | 4.0 | 3.5 | 2.0 | 0 | -1.9 | -2.0 |
| sal ‰ | 36.1 | 35.8 | 35.5 | 35.3 | 35.0 | 37.3 | 37.1 | 36.7 | 34.8 | 35.0 | 34.9 | 34.8 | 34.7 | 34.6 | 34.6 |

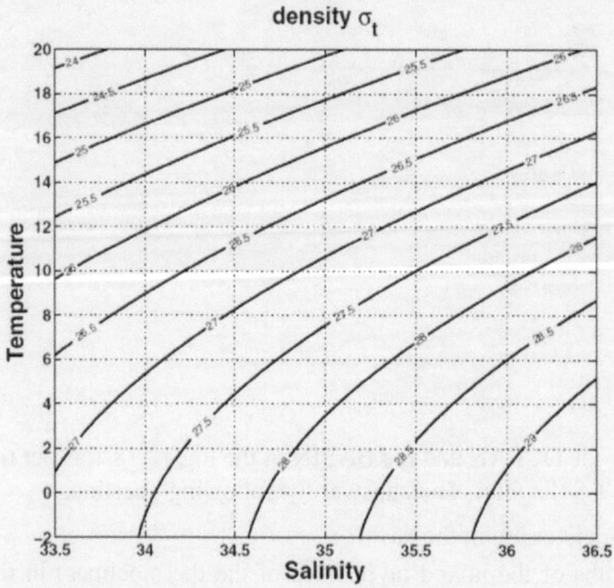


Fig. 12.17.1

## *Solution*

ii) Ranges of temp and salinity of different water masses in North Atlantic

| Water mass | Temp $^0$C | Salinity (ppt) |
|---|---|---|
| Antarctic Intermediate water (AIW) | 3-7 | 33.8-34.7 |
| North Atlantic Central water (NCW) | 8-19 | 35.1-36.5 |
| Antarctic Bottom water (ABW) | -0.4 | 34.7 |
| North Atlantic Deep Water (NADW) | 2-4 | 34.8-35.1 |
| North Atlantic Intermediate Water (NAIW) | 3-9 | 34.0-35.3 |

iii) Indicating on the T-S diagram Fig.12.17.2

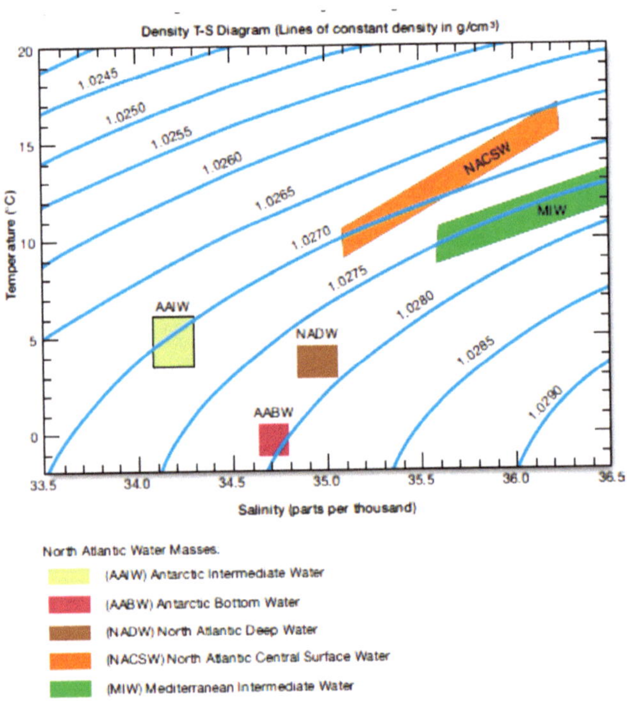


**Fig. 12.17.2**

## *Exercise 12.18*

Draw the T-S curves for the OCEAN and the BASIN of the Fig.12.18.1 as per the data given below in the T-S diagram. Then answer the following questions:

1. How many thermoclines are in the basin?
2. What are the depths of the mixed layers (top of the thermoclines) in the ocean and in the basin?
3. Is there a halocline in the ocean plot? If so, at what depth does it start?

4. Why there is a higher salinity at the surface waters of the basin compared with those of the ocean?

| Ocean | depth | 0 | 50 | 100 | 150 | 300 | 800 | 1000 | 1500 | 2000 |
|---|---|---|---|---|---|---|---|---|---|---|
| | temp | 20.1 | 20.0 | 13 | 8.6 | 7.3 | 5.5 | 4.2 | 3.0 | 2.3 |
| | sal | 33.8 | 34.0 | 34.0 | 33.4 | 33.5 | 33.8 | 34.0 | 34.4 | 34.5 |
| Basin | depth | 0 | 50 | 100 | 150 | 300 | 800 | 1000 | 1500 | 2000 |
| | temp | 22.0 | 21.9 | 21.5 | 16.0 | 15.2 | 7.4 | 6.0 | 5.0 | 4.0 |
| | sal | 35.3 | 35.8 | 36.0 | 35.4 | 34.1 | 34.8 | 35.6 | 35.8 | 36.0 |

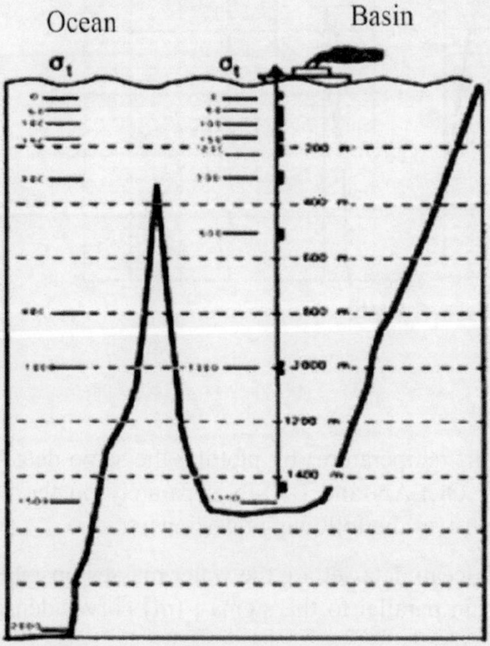


**Fig. 12.18.1** Ocean and Basin topography

## *Solution*

The T-S curves for salinity and temperature are separately given in the Figure 12.18.2 The following are the observations on seeing the plots.

1. Basin has two thermoclines
2. The thickness of Mixed layers in both the ocean and Basin is 100m
3. Yes, the halocline starts at 100m
4. The reason for the higher salinity in surface waters of the basin may be due to more temperature and so more evaporation and also enclosed. Whereas ocean is open to a large extent and so mixing and ocean currents reduce the salinity.

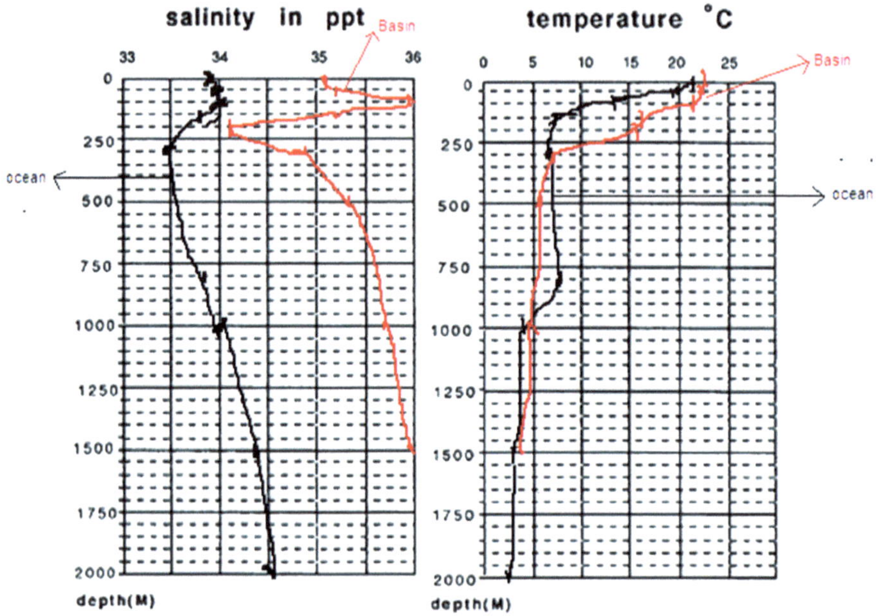


**Fig. 12.18.2**

## *Exercise 12.19*

Draw the T-S curves (salinity versus temperature) by plotting the same data set given in question 12.18 above for OCEAN and BASIN separately on the T-S diagrams. After plotting this data, answer the following questions:

i) Between what depths in the ocean data set are the water masses unstable? (Hint: where does the plot run parallel to the sigma-t ($\sigma t$) curves/density curves?)

ii) In the basin is there any depth where the water mass is unstable? Explain how you know this by looking at the T-S diagram. (hint: in a normal basin profile, water masses increase sequentially in density from the surface to the bottom)

iii) Determine sigma-t ($\sigma t$) values for the ocean and basin temperature and salinity data. This can be done by inspection of your T-S profile. Write this sigma-t ($\sigma t$) values directly on the topography of Basin and Ocean figure of 12.18.1 in the spaces provided at the correct water depths. In the same figure, indicate with an arrow the direction of flow of water basing on density structure.

## *Solution*

The T-S profiles (red curves) are drawn on a T-S diagram (Fig.12.19.1) for Ocean and Basin separately.

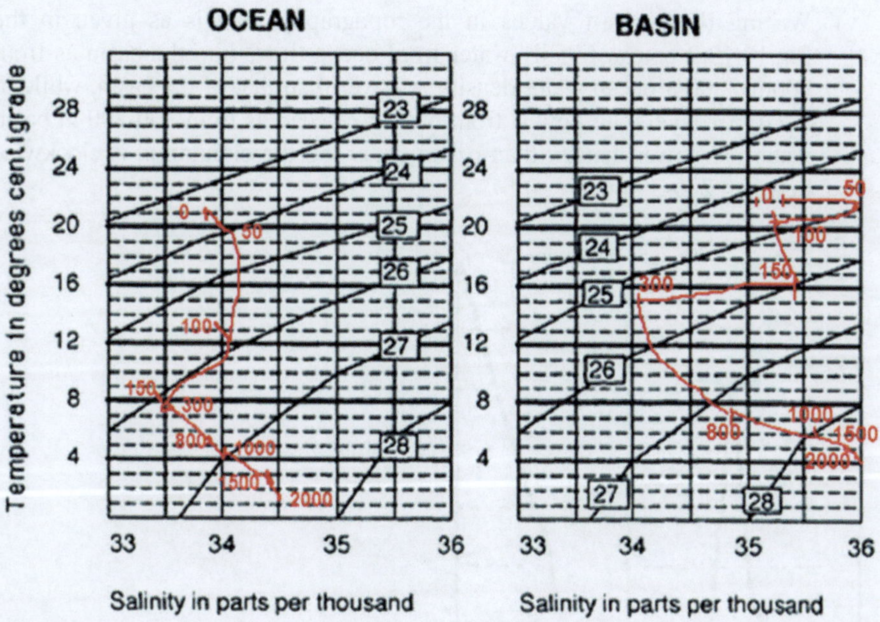


**Fig. 12.19.1**

i) In the Ocean between about 120 to 150 m the water mass is unstable as the profile is parallel to density surface at σt=26

ii) In the basin between 50 to 100m the water mass is unstable as the profile is parallel to density profile at σt=25. However, by looking at the sigma-t values in the basin at 150m, σt=26 and at 300m σt=25.3. This means denser water (150m) is above the less denser water (300m). So this layer between 150-300 m is unstable.

iii) Sigma-t values for the given data are as given in the table below. These are directly read from the profiles figures (12.19.1) drawn above

| Ocean | Depth (m) | 0 | 50 | 100 | 150 | 300 | 800 | 1000 | 1500 | 2000 |
|---|---|---|---|---|---|---|---|---|---|---|
| | Temp $^0$C | 20.1 | 20.0 | 13 | 8.6 | 7.3 | 5.5 | 4.2 | 3.0 | 2.3 |
| | Sal (ppt) | 33.8 | 34.0 | 34.0 | 33.4 | 33.5 | 33.8 | 34.0 | 34.4 | 34.5 |
| | | 23.5 | 24 | 25.8 | 26.0 | 26.2 | 26.8 | 27.3 | 27.4 | 27.5 |

| Basin | Depth (m) | 0 | 50 | 100 | 150 | 300 | 800 | 1000 | 1500 | 2000 |
|---|---|---|---|---|---|---|---|---|---|---|
| | Temp °C | 22.0 | 21.9 | 21.5 | 16.0 | 15.2 | 7.4 | 6.0 | 5.0 | 4.0 |
| | Sal (ppt) | 35.3 | 35.8 | 36.0 | 35.4 | 34.1 | 34.8 | 35.6 | 35.8 | 36.0 |
| | | 24.1 | 24.8 | 24.9 | 26.0 | 25.3 | 27.2 | 28.0 | 28.4 | 28.5 |

v) Writing the sigma-t values in the topography map is as given in the Fig.12.19.2 below. Finally water from ocean flows into the basin as from 800m-2000m ocean water density is lower than that of the basin, while in the upper layers the flow is from basin to Ocean as from 200-300 m basin water has lower density than ocean water. So the circulation is clockwise between 200-300m.

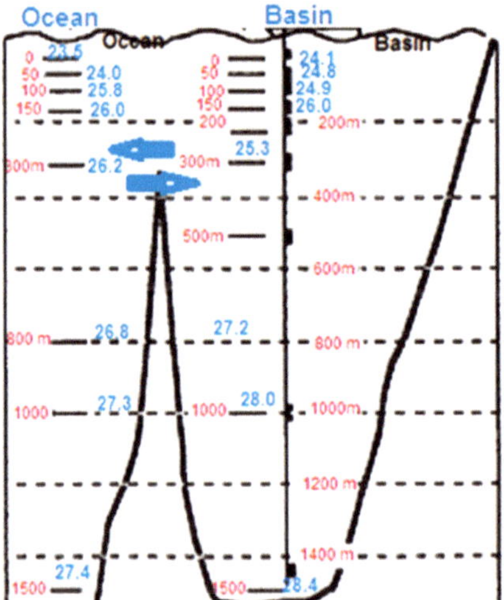


**Fig. 12.19.2.** Ocean-basin topography.The red numbers are depth (m) and the blue values are sigma-t. The blue thick arrow indicates the direction of flow.

## *Exercise 12.20*

1. Write down the Temp-Salinity ranges of the following water masses of Atalntic Ocean from your class notes:

   Antarctic Intermediate Water (AIW), **North and South Atlantic Central Waters (NSCW)** (given for your reference), Antarctic Bottom Water (ABW), North Atlantic Deep Water (NADW), North Atlantic Intermediate Water (NAIW).

2. Find the sigma-t values of the above water masses as given in the table below 'A' of Fig.12.20.1 using the T-S diagram and identify these water masses basing on the ranges of temp and salinity in table A using your class notes.
3. Basing on the sigma-t values you obtained from T-S graph (in Exercise 12.20-2) using the stability criteria (such that denser water is below and lighter water is above), demarcate the above water masses (in Exercise 12.20- 1) or fill-up the question marks in the tableA and the figure 12.20.1.

| Water mass | Salinity | Temp | σt |
|---|---|---|---|
| ? | 34.8 | 7 | ? |
| ? | 36.0 | 15.0 | ? |
| ? | 34.7 | 0 | ? |
| ? | 33.8 | 4.0 | ? |
| ? | 34.9 | 4.0 | ? |

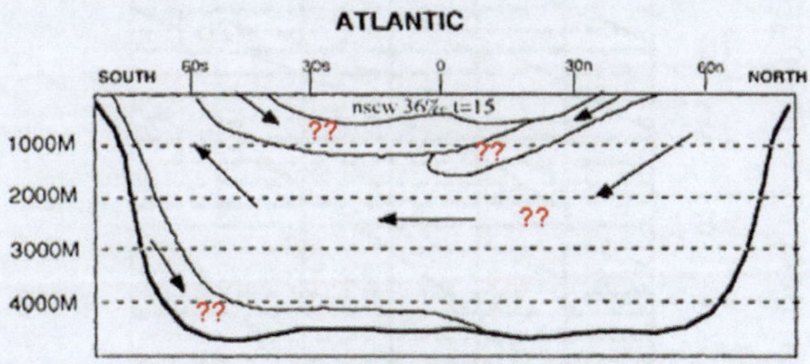


**Table A & Fig.12.20.1.** Fill the question marks for Atlantic Ocean

## *Solution*

1) Atlantic water mass T-S ranges according to class notes:

| Water mass | Temp $^0$C | Salinity ppt |
|---|---|---|
| Antarctic Intermediate water (AIW) | 3-7 | 33.8-34.7 |
| North & South Atlantic Central water (NSCW) | 8-19 | 35.1-36.5 |
| Antarctic Bottom water (ABW) | -0.4 | 34.7 |
| North Atlantic Deep Water (NADW) | 2-4 | 34.8-35.1 |
| North Atlantic Intermediate Water (NAIW) | 3-9 | 34.0-35.3 |

2) Identification of water masses and sigma-t for the question marks

| Water mass | Salinity ppt | Temp $^0$C | Sigma-t |
|---|---|---|---|
| AIW | 34.8 | 7 | 27.2 |
| NSCW | 36.0 | 15.0 | 26.7 |
| ABW | 34.7 | 0 | 27.8 |
| NAIW | 33.8 | 4.0 | 26.9 |
| NADW | 34.9 | 4.0 | 27.7 |

2) Finding sigma-t from T-S graph (Fig. 12.20.2)

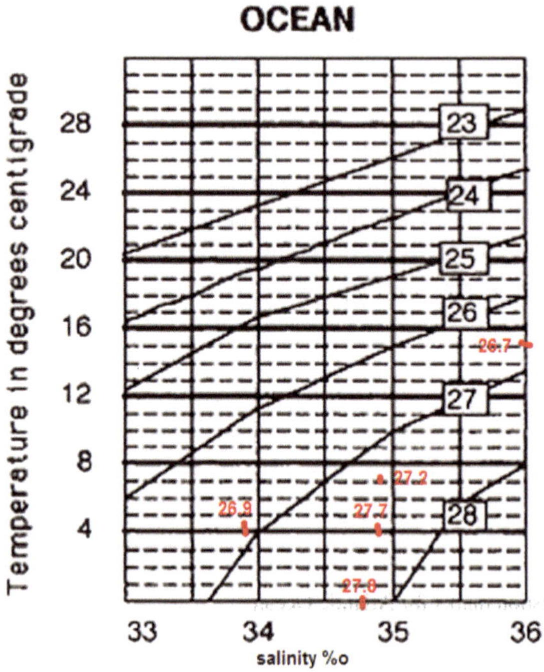


**Fig. 12.20.2.** Water masses of Atlantic ocean from North to south

2) Identification of the water masses of Atlantic (Fig.12.20.3) basing on ranges of Temp-salinity and sigma-t and stability criteria:

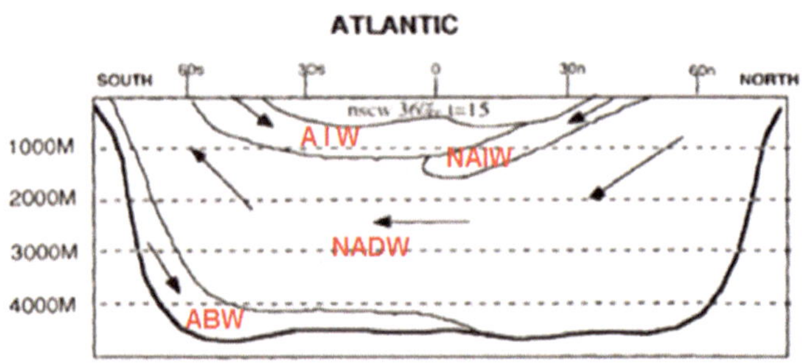


**Fig. 12.20.3**

## *Exercise 12.21*

a.) Write down the Temp-Salinity ranges of the following water masses of Pacific Ocean from your class notes:

**North Pacific Intermediate Water (NPIW),** North Atlantic Deep Water (NADW), Pacific Equatorial Water Mass (PEWM), North Pacific Deep Water (NPDW), North and South Pacific Central Waters (NSPCW) and Antarctic Intermediate Water (AIW)

b) Find the sigma-t values of the above water masses for the table given below 'B'(at the question marks) using the T-S diagram and identify these water masses basing on the ranges of temp and salinity in table B using your class notes.

c) Basing on the sigma-t values you obtained from T-S graph (in question b) using the stability criteria (such that denser water is below and lighter water is above), demarcate the above water masses (in Question a) in the figure below or fill-up the question marks in the table and the figure below (Fig. 12.21.1).

| Water mass | Salinity (PSU) | Temp (°C) | Sigma-t |
|---|---|---|---|
| ? | 33.8 | 4.0 | ? |
| ? | 35.0 | 15.0 | ? |
| ? | 34.3 | 2.0 | ? |
| ? | 34.9 | 12.0 | ? |
| ? | 34.5 | 5.0 | ? |
| NPIW (given) | 34.3 | 6.0 | 27.25 |

For this T-S ranges find the question marks in Pacific

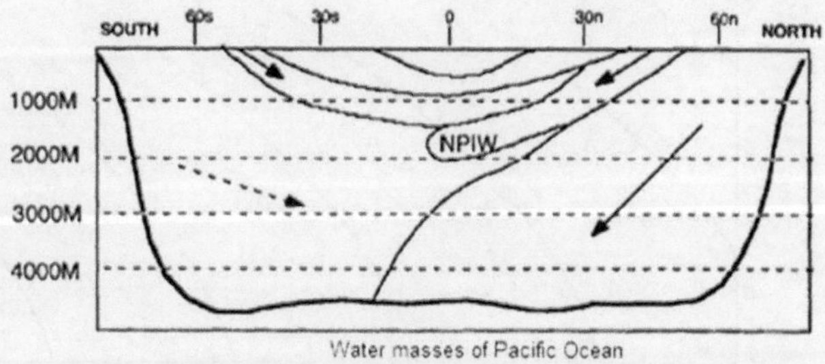


**Table 'B' and Figure 12.21.1.** Pacific Ocean for filling-up the density and water masses

## *Solution*

Pacific Water masses and T-S ranges according to class notes

| Water mass | Temp °C | Salinity psu |
|---|---|---|
| Antarctic Intermediate water (AIW) | 2-10 | 34.0-34.5 |
| North & South Pacific Central water (NSPCW) | 6-24 | 34.4-36.4 |
| Pacific Equatorial Water Mass (PEWM) | 7-23 | 34.5-36.0 |
| North Pacific Deep Water (NPDW) | 2-4 | 34.8-35.1 |
| North Pacific Intermediate Water (NPIW) | 3-9 | 34.0-35.3 |
| North Atlantic Deep water (NADW) | 2.2-3.5 | 34.9-34.97 |

| Water mass | Salinity (PSU) | Temp (°C) | Sigma-t | Number as per stability |
|---|---|---|---|---|
| AIW | 33.8 | 4.0 | 27.0 | 3 |
| PEWM | 35.0 | 15.0 | 26.3 | 1 |
| NADW | 34.3 | 2.0 | 27.6 | 6 |
| NSPCW | 34.9 | 12.0 | 26.6 | 2 |
| NPDW | 34.5 | 5.0 | 27.4 | 5 |
| NPIW (given) | 34.3 | 6.0 | 27.25 | 4 |

In the Fig.12.21.2 Sigma-t was estimated using T-S graph. In the Fig.12.21.3 water masses were indicated in the Basin.

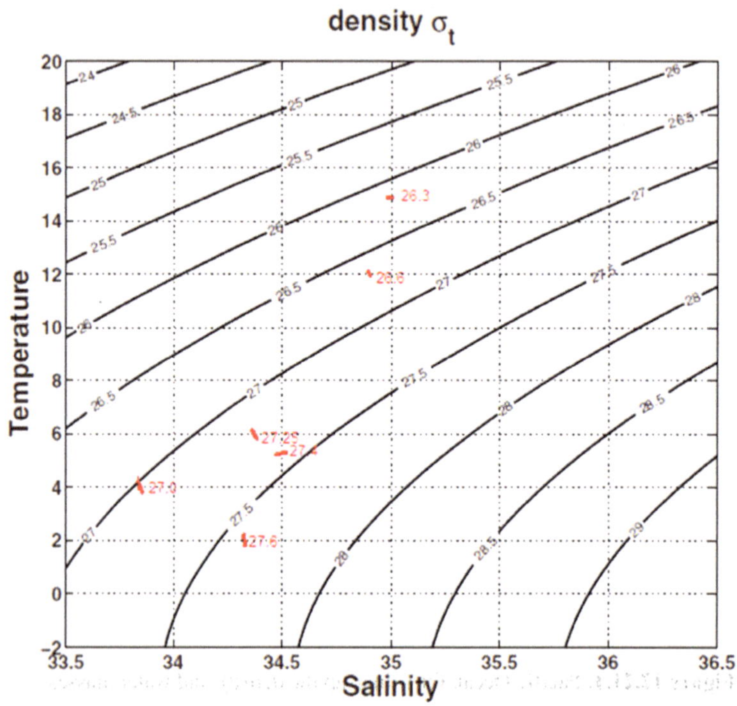


Fig. 12.21.2

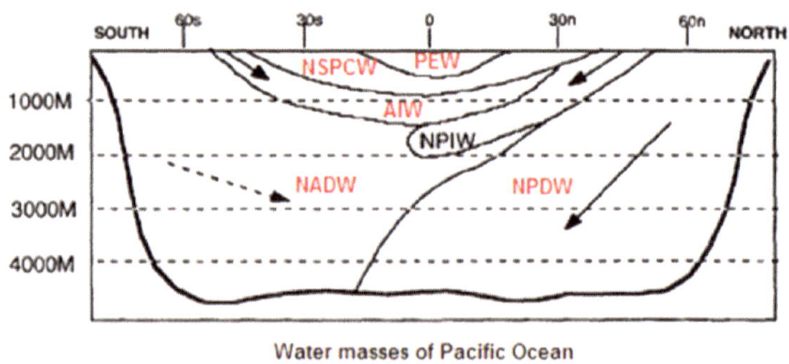


Water masses of Pacific Ocean

Fig. 12.21.3

## *Exercise 12.21.3*

What are the temperature ($^0$C), salinity(‰) and density (g/mL) ranges of the following water masses ?

Atlantic Antarctic Bottom Water (AABW); Atlantic Antarctic intermediate water(AAIW) , North Atlantic Central Surface Water(NACSW), Mediterranean intermediate Water(MIW), North Atlantic Deep Water (NADW)

## *Solution*

**Table 1:** Use these ranges of temperature and salinity and density for the next home work exercise

| Water mass name | Temp range (°C) | Salinity range (‰) | Density (g/ml) |
|---|---|---|---|
| Atlantic Antarctic Bottom Water (AABW) | 0 | 34.6-34.8 | 1.0275-1.0280 |
| Atlantic Antarctic Intermediate Water (AAIW) | 3.0-6.0 | 34.1-34.3 | 1.0270-1.0275 |
| North Atlantic Central Surface Water (NACSW) | 9.0-17.0 | 35.1-36.3 | 1.0265-1.0270 |
| Mediterranean Intermediate Water (MIW) | 9.0-14.0 | 35.6-36.5 | 1.0275-1.0280 |
| North Atlantic Deep Water (NADW) | 3.0-6.0 | 34.1-34.4 | 1.0275-1.0280 |

***Home work Exercise***: Similarly workout this problem

i) Plot the following temperature salinity values (Table 2) on a T-S diagram and find the density. Basing on the ranges of temp, salinity and density given in table 1of Exercise 12.22, find the name of water masses and fill the table 2.

ii) Demarcate these water masses on the Fig.12.22.1 and mention the dominant water mass

iii) Mention the name of each water mass on the Atlantic Ocean map (Fig.12.22.2) at the numbers 1 to 5.

**Table 2:** Fill the density and water masses for these ranges of temp and salinity

| Depth (meters) | Temperature (°C) | Salinity (%) | Density(g/mL) | Water Mass Name |
|---|---|---|---|---|
| 100 | 15.0 | 36.0 | | |
| 500 | 4.0 | 34.2 | | |
| 1000 | 10.0 | 35.8 | | |
| 2000 | 4.0 | 34.9 | | |
| 4000 | 0.0 | 34.7 | | |
| | | | | |

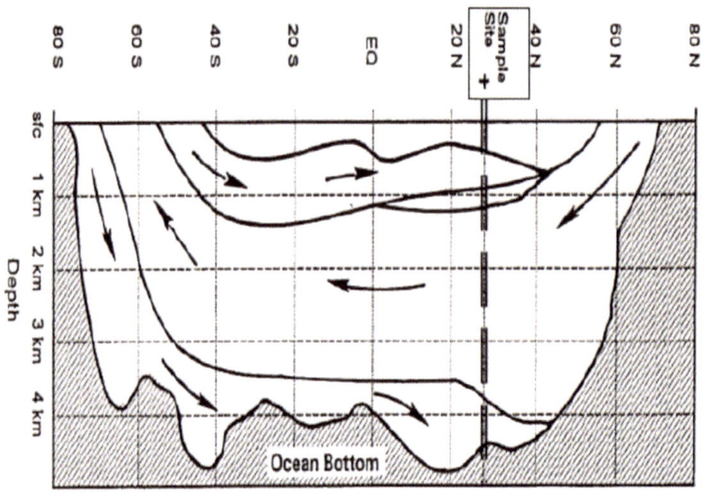


The dominant water mass in this region is the ______________________

**Fig.12.22.1.** Demarcate the different water masses in this figure and mention the dominant water mass

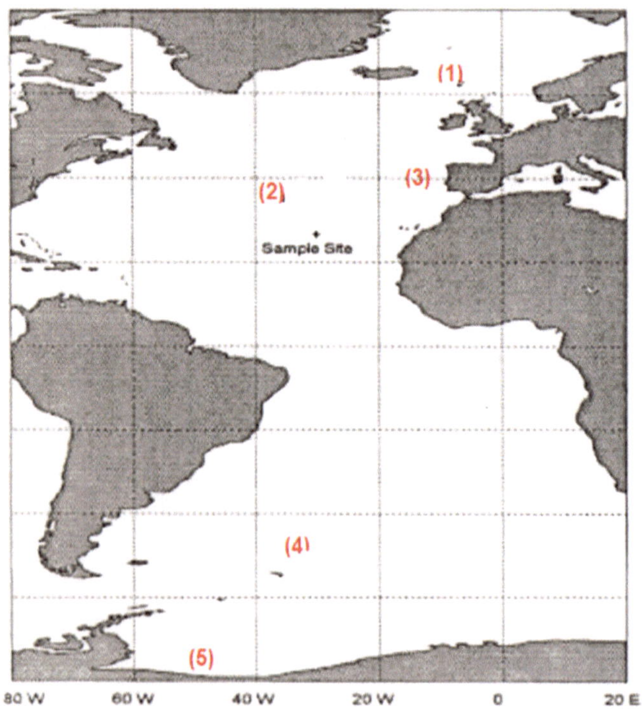


**Fig. 12.22.2.** Map for showing the location of each water mass. Mention which water mass exists at what number between 1 to 5

## *Exercise 12.23*

A water mass A with a temperature $T_A = 5^0$ C and a salinity $S_A = 35.5$ ppt is mixed with a water mass B, with a temperature $T_B = 2\ ^0$C and a salinity $S_B = 34.5$ ppt. The resulting water has a temperature of 3 $^0$C and a salinity of 34.85 ppt. Determine the percentage density ratios of the water masses A and B in the mixture.(Hint use T-S diagram and find ratios of density)

***Solution:*** Values of density marked as red points in T-S diagram (Fig.12.23) are given in table below.

| Water mass | Temp | sal | Sigma-t | density |
|---|---|---|---|---|
| A | 5 | 35.5 | 28.25 | 1.02825 |
| B | 2 | 34.5 | 27.75 | 1.02775 |
| Mixer | 3 | 34.85 | 28 | 1.0280 |

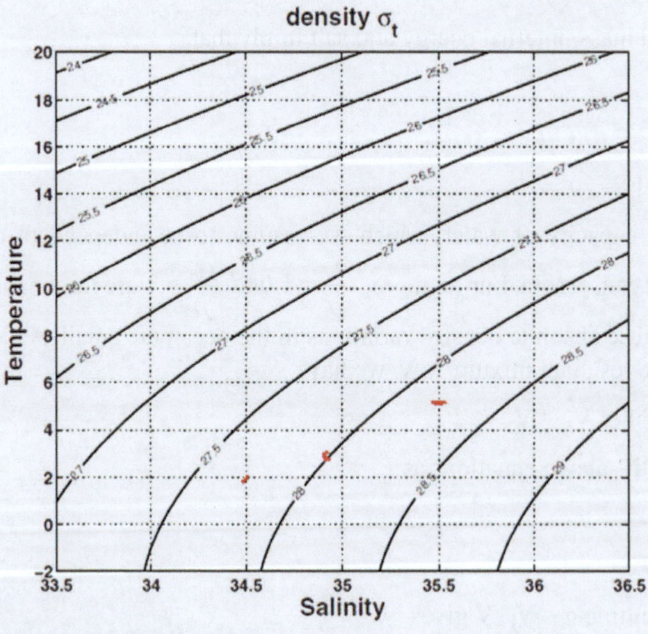


**Fig. 12.23**

Let x is the fraction of density proportional to the mixture. Then we can write:

$$28.25x + 27.75x = 28 \Longrightarrow 56x = 28 \Longrightarrow \therefore x = \frac{28}{56} = \frac{1}{2}$$

Percentage of density of A in mixer $= \frac{28.25}{2 \times 28} \times 100 = 50.44\%$; Percentage of density of B in the mixer $= \frac{27.75}{2 \times 28} \times 100 = 49.5\%$

Hence water mass A is 50.44% and water mass B is 49.5% in the mixture.

### *Exercise 12.24*

A water mass A with a temperature $T_A$ =5° C and a salinity $S_A$ =35.5 ppt is mixed with a water mass B, with a temperature $T_B$ = 2 °C and a salinity $S_B$ = 34.5 ppt. The resulting water has a temperature of 3 °C and a salinity of 34.85 ppt. Determine the percentage volume ratio of the water masses A and B in the mixture. a) Calculate the volume ratio of the water masses A and B in the mixture.(b) Determine this ratio graphically using a T-S graph.

### *Solution*

a. Calculation of the volume ratio of the water masses A and B in the mixture.

Let the Water mass A (with volume $V_A$ and density $\rho_A$) is mixed with water mass B (with volume $V_B$ and density $\rho_B$)) to give a water mass with a volume V , a temperature T, a salinity S, and a density ρ.

Conservation of total mass, internal energy and salt imply that:

$$\rho_A V_A + \rho_B V_B = \rho V$$

$$\rho_A V_A C_p T_A + \rho_B V_B C_p T_B = \rho V C_p T$$

$$\rho_A V_A S_A + \rho_B V_B S_B = \rho V S$$

Here Cp is the heat capacity of water, which we assume to be independent of temperature so that it cancels out. Furthermore, $\frac{\rho_A}{\rho} \approx 1$ *and* $\frac{\rho_B}{\rho} \approx 1$ are reasonable approximations because absolute density variations in the ocean are small. As $V_A$ and $V_B$ are mixed the resultant mixture is V. we have

$V_A + V_B = V$; This implies $V_B = V - V_A$

Hence we can write the above equations as:

$$V_A T_A + (V - V_A) T_B = VT$$

$$V_A S_A + (V - V_A) S_2 = VS$$

Dividing by V and putting $\alpha = V_A / V$ gives

$$\alpha T_A + (1-\alpha) T_B = T \Rightarrow \alpha_T = \frac{T - T_B}{T_A - T_B} \quad (1)$$

$$\alpha S_A + (1-\alpha) S_B = S \Rightarrow \alpha_S = \frac{S - S_B}{S_A - S_B} \quad (2)$$

Substituting the values of $T_A$ , $T_B$ and T in the above equation(1),

$$\alpha_T = \frac{T - T_B}{T_A - T_B} = \frac{3-2}{5-2} = \frac{1}{3} = 0.33$$

Similarly substituting the values of $S_A$ , $S_B$ and S in the above equation(2),

$$\alpha_S = \frac{S - S_B}{S_A - S_B} = \frac{34.85 - 34.5}{35.5 - 34.5} = 0.35$$

Finally taking the average of $\alpha_T$ and $\alpha_S$ can be taken as α, Which comes to

$$\alpha = \frac{\alpha_T + \alpha_S}{2} = \frac{0.33 + 0.35}{2} = 0.34$$

As we have taken $\alpha = \frac{V_A}{V} = 0.34$

We know $V_B = V - V_A$. This means $\frac{V_B}{V} = 1 - \frac{V_A}{V} = 1 - \alpha = 1 - 0.34 = 0.76$

Hence the ratios of water masses A and B with the mixture are: 0.34 and 0.76

b. How can one determine this ratio graphically using a T-S diagram?

By equating both expressions for α, we see that there is a linear relationship between T and S, as

$$\frac{T - T_B}{T_A - T_B} = \frac{S - S_B}{S_A - S_B} \rightarrow T = T_B + (T_A - T_B)\,\frac{S - S_B}{S_A - S_B}$$

For the values of $T_A$, $T_B$ and $S_A$, $S_B$ above, this line is plotted in Fig. 12.24. The distances a and b in this figure represent the ratio's of volumes as a : b = 2 : 1, i.e., a = 2b. This can easily be seen by realizing that the sine of the angle between the line and the horizontal axis can be expressed as

$$\frac{T - T_B}{b} = \frac{T_A - T}{a} \rightarrow \frac{a}{b} = \frac{T_A - T}{T - T_B} \approx 2$$

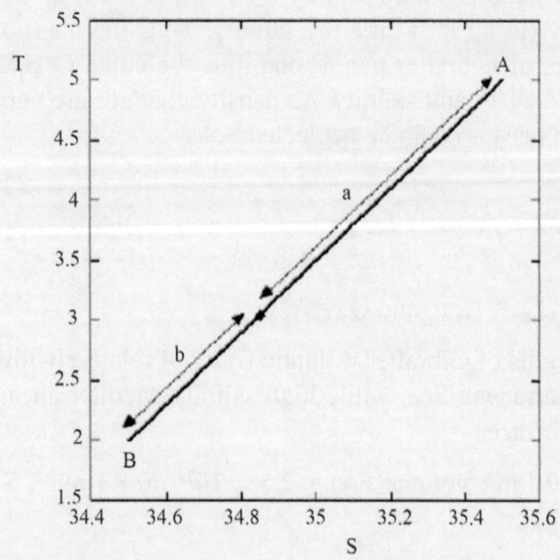


**Fig. 12.24**

### *Exercise 12.25*

The $Q_1$ and $Q_2$ are the inflow and outflow respectively of the Mediterranean Sea through Gibralter straight (Fig. 12.25). The inflowing Atlantic water has a salinity of 36.2 ‰ whereas the outflowing Mediterranean Seawater has a salinity of 38.4 ‰. The surface area of the Sea is about 2.5 x $10^{12}$ $m^2$and about 1 m $yr^{-1}$of fresh water is evaporated into the atmosphere. Ignoring smaller contributions of fresh water from small rivers and the Black Sea, estimate for$Q_1$ and $Q_2$ and mention which is more·

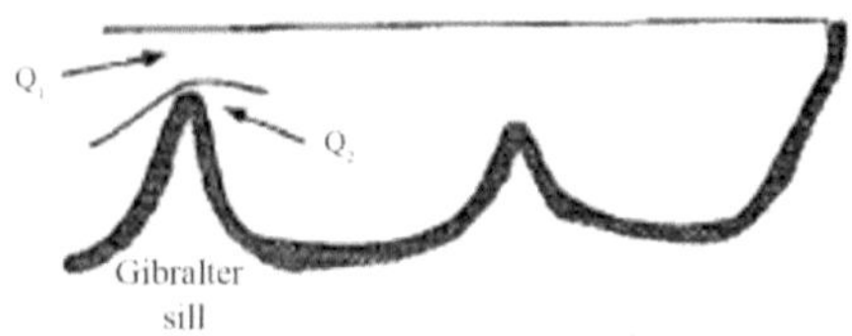


**Fig. 12.25.** Gibralter Sill

### *Solution*

Let A is the area of the basin. Let water enters the channel at a rate of $Q_1$ $m^3$ $s^{-1}$ and leaves it at a rate of $Q_2$ $m^3$ $s^{-1}$. Water also can enter and leave through precipitation and evaporation in the basin. So let 'F' is the net rate of loss through evaporation. Then the volume of loss through evaporation is 'FA'. Then we can write:

$$Q_1 - FA = Q_2 \tag{1}$$

Because salinity is measured in parts per thousand by weight, the mass of salt in kg of sea water is the density (in kg $m^{-3}$) time the salinity. Thus the mass of salt transported through straights of Gibralter per second must be equal $Q_1 \rho_1 S_1 = Q_2 \rho_2 S_2$ where ρ and S are the density and salinity. As density changes are very small compared to salinity in the oceans ρ can be neglected.So

$$Q_1 S_1 = Q_2 S_2 \tag{2}$$

From (1) and (2) we can write: $$FA = Q_2\left(\frac{S_2}{S_1} - 1\right) \text{ and } Q_2 = \frac{FA}{\left(\frac{S_2}{S_1} - 1\right)} \text{ or } Q_1 = \frac{FA}{\left(1 - \frac{S_1}{S_2}\right)} \tag{3}$$

If $u_1$ and $u_2$are the velocities then $Q_1 = A_1 u_1$ and $Q_2 = A_2 u_2$

In the above problem at the straights of Gibralter, Atlantic water of relatively low salinity flow($Q_1$) into the Mediterranean Sea, while high salinity Mediterranean water flows ($Q_2$) out below the surface.

Given: F = 1myr$^{-1}$, A = 2.5 × $10^{12}$ $m^2$; volume(FA) = 2.5 × $10^{12}$ $m^2$ × 1myr$^{-1}$, $S_1$ =36.2 ‰, $S_2$= 38.4‰

From equation (3): $2.5 \times 10^{12}\ m^2 \times 1 myr^{-1} = Q_2\left(\frac{38.4}{36.2} - 1\right)$

$$\Rightarrow 2.5 \times 10^{12}\ m^2 \times \frac{1}{3600 \times 24 \times 365} = Q_2\left(\frac{2.2}{36.2}\right)$$ ; $Q_2 = 41.13 \times 10^{12} = 4.113 \times 10^{13} m^3 yr^{-1}$

$$Q_2 = 2.5 \times 10^{12} m^2 \times \frac{1}{3600 \times 24 \times 365} \times \frac{36.2}{2.2} = 13044255 \times 10^{12} \times 10^{-13}$$

$= 13044255 \times 10^{-1} = 1304425.5 m^3\ s^{-1} = 1.3 \times 10^6 = 1.3$ Sv

Again from eqn (3): $Q_1 = \dfrac{2.5 \times 10^{12} m^2 \times 1m\ yr^{-1}}{(1 - \frac{36.2}{38.4})} = 43.63 \times 10^{12} = 4.363 \times 10^{13}\ m^3 yr^{-1}$

$= 1383498 m^3 s^{-1} = 1.38$Sv

This means both $Q_1$ and $Q_2$ are almost same, though $Q_1$(the flow from Atlantic into Mediterranean) is little more by $0.25 m^3 yr^{-1}$ (0.08 Sv).

### *Exercise 12.26*

Estimate the volume transport (in Sv) of the Gulfstream which has a width of 100 km, depth of 500 m and a mean speed of 1 m $s^{-1}$

### *Solution*

Volume transport: Q = UDL = $1ms^{-1} \times 500m \times 100 \times 10^3 = 50 \times 10^6 m^3 = 50Sv$

### *Exercise 12.27*

The data given below are some in-situ measurements of the temperature and salinity at a certain location in the Atlantic. Investigate whether the water column is statically stable when a linear equation of state is assumed. (a) Calculate $\frac{\rho}{\rho_0}$ for all the depths (b) Determine the potential temperature as a function of depth for these measurements. (C) Is the water column statically stable?

| Depth (m) | Temp ($^0$C) | Salinity (ppt) |
|---|---|---|
| 0 | 18.909 | 32.574 |
| 1000 | 2.697 | 34.410 |
| 2000 | 1.868 | 34.600 |
| 3000 | 1.528 | 34.661 |
| 4000 | 1.456 | 34.679 |
| 5000 | 1.503 | 34.686 |
| 5460 | 1.547 | 34.686 |

### *Solution*

The linear equation of state is given as: $\rho = \rho_0 [1 - \alpha_T (T - T_0) + \alpha_S (S - S_0)]$

Where $\rho_0$, $T_0$ and $S_0$ are density, temperature and salinity at reference surface. Here surface values are taken as reference values. We know $\alpha_T = 10^{-4} K^{-1}$ and $\alpha_S = 7.6 \times 10^{-4}$. From the equation of state:

$$\frac{\rho}{\rho_0}=\left[1-\alpha_T\left(T-T_0\right)+\alpha_S\left(S-S_0\right)\right]$$

Calculating $\frac{\rho}{\rho_0}$ for all the depths

| Depth (m) | Temp ($^0$C) | Salinity (ppt) | $\frac{\rho}{\rho_0}$ |
|---|---|---|---|
| 0 | 18.909 | 32.574 | 1 |
| 1000 | 2.697 | 34.410 | 1.003017 |
| 2000 | 1.868 | 34.600 | 1.003244 |
| 3000 | 1.528 | 34.661 | 1.003324 |
| 4000 | 1.456 | 34.679 | 1.003345 |
| 5000 | 1.503 | 34.686 | 1.003346 |
| 5460 | 1.547 | 34.686 | 1.003343 |

a) On examining the $\frac{\rho}{\rho_0}$ values in the above table, When a linear equation of state is used, the water column does not appear to be statically stable between 5000 and 5460 m as the density decreases downward

b) The potential temperature is the in-situ temperature corrected for the adiabatic temperature gradient, Γ. where, Γ ($^0$C m$^{-1}$), and a good approximation is given by: $\vartheta = T - \Gamma dZ$, such that T is given for different depths as following: (See Exercise 12.8.43)

| 0-1000m | Very small (negligible |
|---|---|
| 1000-5000m | $\Gamma = 0.14$ ($^0$C km$^{-1}$) |
| 5000-9000m | $\Gamma = 0.19$ ($^0$C km$^{-1}$) |

For example, calculation of potential temperature is as below: $\vartheta_{2000}$ = 1.868-0.14 = 1.728

$$\vartheta_{3000}=1.528-\left\{0.14\times\frac{(3000-1000)}{1000}\right\}=1.528-0.28=1.248$$

c) Please note as 0-1000, Γ is negligible the depth difference is calculated from 1000. Also note salinity has not much influence on potential temperature.

Now calculate $\frac{\rho_\vartheta}{\rho_{\vartheta 0}}$ using potential temperature;

$$\frac{\rho_\vartheta}{\rho_{\vartheta 0}}=\left[1-\alpha_T\left(\vartheta-\vartheta_0\right)+\alpha_S\left(S-S_0\right)\right]$$

$$\frac{\rho_\vartheta}{\rho_{\vartheta 0}}=\left[1-10^{-4}\left(\vartheta-\vartheta_0\right)+7.6\times10^{-4}\left(S-S_0\right)\right]$$

$$\frac{\rho_\vartheta}{\rho_{\vartheta 0}}=\left[1-10^{-4}\left(2.697-18.909\right)+7.6\times10^{-4}\left(34.41-32.574\right)\right]=1.003017$$

| Depth (m) | Temp ($^0$C) | Salinity (ppt) | Potential Temp $^0$C ($\vartheta$) | $\frac{\rho_\vartheta}{\rho_{90}}$ |
|---|---|---|---|---|
| 0 | 18.909 | 32.574 | 18.909 | 1 |
| 1000 | 2.697 | 34.410 | 2.697 | 1.003017 |
| 2000 | 1.868 | 34.600 | 1.728 | 1.003258 |
| 3000 | 1.528 | 34.661 | 1.248 | 1.003352 |
| 4000 | 1.456 | 34.679 | 1.036 | 1.003387 |
| 5000 | 1.503 | 34.686 | 0.943 | 1.003402 |
| 5460 | 1.547 | 34.686 | 0.900 | 1.003406 |

The water column is statically stable as the relative potential density $\frac{\rho_\vartheta}{\rho_{90}}$ continuously increased

### *Exercise 12.28*

The average wind speed over the region of the Antarctic Circumpolar Current is about 10 ms$^{-1}$.(a) Give an estimate of the mean wind-stress at the ocean-atmosphere interface in this region.(b) If a unidirectional wind blows over an initially motionless water column of 10 m thickness, then calculate the flow velocity in this layer after 3 hours.

### *Solution*

Given that U = 10 ms $^{-1}$, Cd = 10 $^{-3}$, $\rho_a$ = 1.2 kg/m$^3$, We know the wind stress equation as :

a) Wind stress for 10 m/s wind = $\tau = \rho_a C_d U^2$= 1.2 kgm$^{-3}$ × 1.2 × 10$^{-3}$ × 10$^2$ m$^2$ s$^{-2}$ = 1.44 × 10$^{-1}$ = 0.14 kgm$^{-1}$ s$^{-2}$ = 0. 14Nm$^{-2}$ = 0.14 Pa

1N= 1kg.m s$^{-2}$ = 10$^5$ dynes

1 Pascal = 1N/m$^2$ =1 kg. m$^{--1}$ s$^{-2}$ = 1 J m$^{--3}$

b) The wind stress $\tau$ will transfer momentum to the water column of depth h according to the balance

$$\rho h \frac{du}{dt} = \tau$$

where u is the (uniform) velocity of the layer. From this equation we estimate on integration

$$u(t) \approx u(0) + \frac{\tau}{\rho h} t$$

and find with t* = 3 hours, taking $\tau$ = 0.12 Pa, $\rho_w$ = 1025(=10$^3$),h=10m

$$u_t = 0 + \frac{0.12 kgm^{-1}s^{-2}}{1000 kgm^{-3} \times 10m} \times 3 \times 60 \times 60 s = 0.1296 m/s$$

that u(t*) = 0.13 ms$^{-1}$.

### *Exercise 12.29*

Consider a layer of seawater (heat capacity $Cp = 4.2 \times 10^3$ $Jkg^{-1}K^{-1}$, density $\rho = 1.027 \times 10^3$ $kgm^{-3}$) with a thickness of 1 m and an area of $10^4$ $m^2$. Assume that a constant heat flux $Q = 400$ $Wm^{-2}$ is applied over the area, for example due to a warmer atmosphere.

a. Determine how much energy is needed to raise the temperature of the water by $1^0$C.
b. How long does it take before the temperature of the water layer has increased by $1^0$C? (Note: During a seasonal cycle, there is heat uptake in an ocean region whereas during the winter the opposite occurs).
c. Assume that the upper 100 m of ocean water is in contact with the atmosphere and that the seasonal temperature change in the water is about $10^0$C. Calculate the amount of energy which is stored (released) by the water layer.

### *Solution*

a) The mass m of the water layer is $m = \rho V = 1.027 \times 10^3$ $kgm^{-3} \times 1m \times 10^4$ $m^2$ $= 1.027 \times 10^7$ kg. The energy E needed to raise the temperature by $\Delta T = 1$ $^0$C is given by:

$E = C_p\, m\, \Delta T = 4.2 \times 10^3$ J $kg^{-1}$ $k^{-1} \times 1.027 \times 10^7$ kg $\times 1^0$ C $= 4.31 \times 10^{10}$ J = 43.1 GJ

b) The mean temperature in the layer of depth h will change according to the balance

$$\rho h C_p \frac{dT}{dt} = Q$$

On integration, we can write: $\int dT = \frac{Q}{\rho h C_p}\int dt \Rightarrow T(t) = T(0) + \frac{Q}{\rho h C_p} t$

$$T(t) - T(0) = \frac{Q}{\rho h C_p} t = \frac{400wm^{-2}}{1.027 \times 10^3 kgm^{-3} \times 1m \times 4.2 \times 10^3 Jkg^{-1}k^{-1}} \times t = 92.73 \times 10^{-6}$$

Since $T(t) - T(0) = 1^0$ K,

$$t = \frac{\rho h C_p}{Q} \times 1k = \frac{1.027 \times 10^3 kgm^{-3} \times 1m \times 4.2 \times 10^3 Jkg^{-1}k^{-1}}{400wm^{-2}} \times 1k = 1.078 \times 10^4 S = 1.08 \times 10^4 S$$

c) Now h =100m, area = $10^4$ m. So volume = $10^4 \times 10^2 = 10^6$ $m^3$. For seasonal energy change is for $10^0$ C

$E = \rho V C_p \Delta T = 1.027 \times 10^3$ $kgm^{-3} \times 10^6$ $m^3 \times 4.2 \times 10^3$ J $kg^{-1}$ $k^{-1} \times 10k = 4.31 \times 10^{13}$ J

Note since it is change in temperature, we need not have to convert $^0$ C to K, because the difference will be same.

### Exercise 12.30

Warm western boundary currents such as the Gulf Stream play an important role in the climate system because they contribute to the meridional heat transport. The Gulf Stream is about 100 km wide, has depth of 500 m and has a mean speed of 1 $ms^{-1}$. (a) Provide an estimate of the volume transport of the Gulf Stream (in Sv). (b) The meridional heat transport due to the Gulf Stream depends on the temperature difference between the warm water flowing northward and the cooler water flowing southward. Assuming this temperature difference as 5 $^0$C and of the meridional heat transport associated with the Gulf Stream., estimating the meridional heat transport.

### Solution

The volume transport $V_G$ of the Gulf Stream can be estimated by

$V_G$= UDL

where L is its width, D its depth and U its average speed. Given L=100km=$10^5$ m, D = 500 m, U= 1m/s

$V_G = 1ms^{-1} \times 500m \times 10^5\ m = 5 \times 10^7\ m^3\ s^{-1} = 50 \times 10^6\ m^3\ s^{-1} = 50\ Sv$

(b) The meridional heat transport $\varphi_H$ is given by

$\varphi_H = \rho C_p\ U\Delta TDL$

$= 1.027 \times 10^3\ kgm^{-3} \times 4.2 \times 10^3\ Jkg^{-1}\ k^{-1} \times 1ms^{-1} \times 5k \times 500m \times 10^5\ m = 107.8 \times 10^{13}$ $J \approx 1.1 \times 10^{15}\ J \approx 1.1pw$

### Exercise 12.31

Consider a flow over a sill such as that in the Gibraltar Strait. West of the sill, the water has a salinity $S_a$ = 36.2 ppt and east of the sill (in the Mediterranean basin), the salinity is $S_b$ = 38.4 ppt. The volume flux of the surface transport of the Atlantic to Mediterranean is $Q_a$, and the reverse transport at depth has a volume flux of $Q_b$ (both in Sv).

(a). Derive that

$Q_a + Q_b = 0$

$Q_a S_a + Q_b S_b = (E - P)AS_0$

where $S_0$ is a mean salinity in the Mediterranean basin. Assume that the difference between evaporation E and precipitation P is about 1 m/year over the Mediterranean basin. This basin has an area of about A = 2.5 × $10^{12}$ $m^2$.

b. Determine $Q_a$ and $Q_b$ for the Mediterranean outflow and compare the result to values in the literature.

### *Solution*

The above equations are based on the peculiar yet often used approximation that the effect of E-P may be represented by a `virtual salt flux' $F_s = (E–P)AS_0$, where $S_0$ is a reference salinity. In this approximation there is no freshwater transport through the ocean surface. Hence the volume change of the Mediterranean is determined solely by the transports through the Gribraltar Strait. When $Q_a$, is the volume flux of surface water (positive eastward) and $Q_b$, that of the flow at depth (also positive eastward) then it follows by continuity that $Q_a + Q_b = 0$. The salt leaving the Mediterranean basin is $– Q_bS_b$ while the salt entering the basin is $Q_a$ + $(A–P)AS_0$,. A balance of salt over the Mediterranean basin then gives

$Q_a S_a + Q_b S_b = (E–P)AS_0$

As $Q_a + Q_b = 0 \Rightarrow Q_a = –Q_b$. So substituting this in the above equation

$Q_a S_a – Q_a S_b = (E – P)AS_0$

$$Q_a = \frac{(E-P)AS_0}{S_a - S_b}$$

And hence we find that

$$Q_a = -Q_b = \frac{2.5\times10^{12}\times35\times1\times\dfrac{1}{3600\times24\times365}}{2.2} \approx 1.3Sv$$

Which is in good agreement with observations

## ***Heat and Energy***

### *Exercise 12.32*

a) Calculate the energy flux at the surface of the sun and the power of sun in watts /m$^2$;( b) Calculate the energy flux received by the earth in the outer surface from the sun in watts /m$^2$(Take $\sigma = 5.67\times10^{-8} wm^{-2}k^{-4}$ and Sun's temperature ($T_S$) as 5800 K, Sun's radius (R)6.9 x $10^5$ km, the distance from sun to earth (d) 150 x $10^6$ km) c) also compare the temperature of sun and earth

### *Solution*

a) According to Stafan Boltzman law: the energy flux of the sun $E_S$ is:

$E_S = \sigma T_S^4 = 5.67\times10^{-8} wm^{-2}k^{-4}\times(5800K)^4 = 6.4\times10^7 wm^{-2}$

Power of Sun ($P_S$)= $6.4\times10^7 wm^{-2}\times4\times\pi\times(6.9\times10^5\times10^3 m)^2 \approx 3.95\times10^{26} w$

The intensity of solar radiation is called irradiance and is measured in the units of power per unit area (w m$^{--2}$). However, the total amount of solar radiation energy is called irradiation and is measured in the units of energy per unit area (J m $^{--2}$)

b) Energy received by the earth in the outer atmosphere (solar irradiance) or solar constant (S):

$$\frac{P_S}{4\pi d^2} = \frac{3.95\times10^{26} w}{4\pi(150\times10^6\times10^3 m)^2} = 1397 wm^{-2}$$

(Denoting Ps as the power of the sun, 'S' as the power of the sun on earth, R as the radius of the sun and d as the distance between the earth and the sun. Then: $Ps = 6.31\text{x}10^7 \text{ x } 4\pi R^2 = 3.85\text{x}10^{26}W$; $S = 3.85\text{x}10^{26}/4\pi d^2 = 1370 W/m^2$. This is the solar constant, $1370W/m^2$)

Due to the fact that the earth rotates around the sun and its distance depends on the time of the year, the actual direct solar incidence that Earth receives varies from 1,412 $W/m^2$ in January to 1,321 $W/m^2$ in July.

How to calculate approximately the total amount of solar energy received by the Earth? The Earth intercepts parallel rays of the Sun in an area equal to the disk of Earth radius (Fig.12.32). The area of the disk is $\pi R^2$, where R=6400km. Recall that the solar constant is the amount of energy incident on a unit area. The total amount of solar energy intercepted by the Earth is $S\pi R^2$. Here the solar constant (S) $=1337W/m^2$. Hence total amount of solar energy received by the earth is

$$S\pi R^2 = 1337 Wm^{-2}\times\pi\times(6.4\times10^6 m)^2 = 172044\times10^{12} w = 1.72\times10^{17} w$$

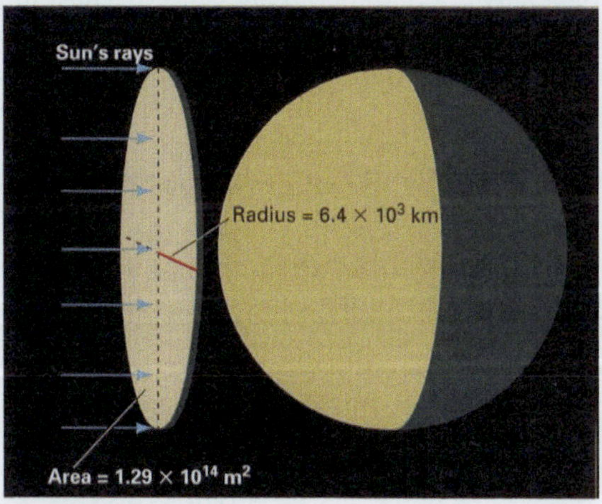


**Fig. 12.32.**

c) Applying Stefan-Boltzmann Law To Sun and Earth:

For **Sun:**

$E_s = 5.67\times10^{-8}\text{ wm}^{-2}\text{ K}^{-4}\times(6000K)^4 = 73{,}483{,}200\text{ wm}^{-2}$

For **Earth:**

$E_e = 5.67\times10^{-8}\text{ wm}^{-2}\text{ K}^{-4}\times(300K)^4 = 459\text{ wm}^{-2}$

This means Sun emits about 160,000 times more radiation per unit area than the Earth because Sun's temperature is about 20 times higher (300K $\times$ 20 = 6000K) than Earth's temperature. $20^4 = 160{,}000$

### *Exercise 12.33*

a) What is astronomical unit? (b) What is solar constant? C) How much is the solar constant or solar irradiance in calories and watts? (d)What is the solar incidence or irradiance at 1.5 AU?

### *Solution*

a) The **astronomical unit** (symbol: **AU**) is a unit of length, roughly the distance from Earth to the Sun and equal to about 150 million kilometres (93 million miles). However, that distance varies as Earth orbits the Sun, from a maximum (aphelion) to a minimum (perihelion) and back again once a year. Originally conceived as the average of Earth's aphelion and perihelion, since 2012 it has been defined as exactly $1.495978707\times10^{11}$ m. The astronomical unit is used primarily for measuring distances within the Solar System or around other stars. It is also a fundamental component in the definition of another unit of astronomical length called the parsec.

b) At the outer limit of the earth's atmosphere the intensity of solar energy incident on a surface at right angles to the solar beam is nearly equal to 2 gm.cal.cm$^{-2}$ min$^{-1}$(langleys per day). This is called solar constant. The earth intercepts a circular sample of this radiation which has an area of $\pi R^2$. The earth's rotation distributes this sample of energy over the spherical surface of the earth, which has an area of $4\pi R^2$. As a result the average input of solar radiation at the top of the atmosphere is $\pi R^2/4\pi R^2 = 0.5$ gm.cal.cm$^{-2}$ min$^{-1}$ (ly/day).

c) The solar constant is the amount of solar radiation incident per minute on a unit surface placed in the outer atmosphere at a distance of 1 astronomical unit (AU) from the Sun. It is presently calculated as being 1366 W/m$^2$. This is also called solar irradiance. It is also equal to 2 calcm$^{-2}$ min$^{-1}$( ly /min).

d) The formula to calculate specific solar incidence or solar irradiance for a specific distance in terms of AU's is: $E = S\times\left(\frac{1}{\text{Distance in A.U}}\right)^2 wm^{-2}$
WhereE:solar Incidence in W/m$^2$.; **S**: Solar Constant at 1 AU (1366 W/m$^2$)

e) The solar incidence for 1.5 A.U is 1366wm$^{-2}$ $\times\left[\frac{1}{1.5}\right]^2 = 607$ W/m$^2$

### *Exercise 12.34*

If α is the Earth's mean planetary albedo, show that the solar constant $S_\alpha = \sigma T_{\alpha E}{}^4 = \frac{S_0}{4}(1-\alpha)$, where $S_0$ is the solar constant without albedo.

### *Solution*

First let us denote the solar radiative flux at the top of the Earth's atmosphere by $S_0$ (for solar constant) and the albedo of the earth by α. Then let us estimate the total amount of radiation absorbed by the Earth. Let Earth's radius is R then the area of the disc intercepted is $\pi R^2$. Then

the heat absorbed by Earth = $(1-\alpha)\,\pi R^2 S_o$

The total heat radiated from the earth according to Stefan–Boltzman law on whole of the earth surface is = $(4\pi R^2)\,\sigma T^4_E$ Where $T_E$ temperature of the earth.

As heat absorbed by earth is equal to that radiated:

$(4\pi R^2)\,\sigma T^4_E = (1-\alpha)\,\pi R^2 S_o$

Hence $\sigma T_E{}^4 = \frac{S_0}{4}(1-\alpha)$

### *Exercise 12.35*

a) Using the table.1 of hydrologic cycle, calculate the heat lost from oceans per day in joules (taking latent heat of evaporation as $2.25\times10^6$J/kg)

b) What is the residence time of water in oceans, on land and atmosphere?

**Table 1.**

| S.No | Stored in different stages | | Annual movements (per annum) | |
|---|---|---|---|---|
| 1. | In atmosphere | $13\times10^{15}$ kg | Precipitation on land | $100\times10^{15}$ kg |
| 2. | On land | $38000\times10^{15}$ kg | Evaporation from land | $64\times10^{15}$ kg |
| 3 | In the Oceans | $1322000\times10^{15}$ kg | Blows over land | $36\times10^{15}$ kg |
| 4 | Total world water supply | $1360000\times10^{15}$ kg | Run off through streams | $36\times10^{15}$ kg |
| 5 | | | Evaporation from oceans | $336\times10^{15}$ kg |
| 6 | | | Precipitation in oceans | $300\times10^{15}$ kg |

**Table 2.** Water on land ($\times10^{15}$ kg)

| Rivers and streams | 1 |
|---|---|
| Freshwater lakes | 125 |
| Salt lakes and inland seas | 104 |
| Total surface water | 230 |
| Glaciers and ice caps | 29300 |
| Soil moisture and seepage | 70 |
| Ground water | 8400 |
| Total on land | 38000 |

### Solution

The residence time is the average length of time that a water molecule resides or is stored in any particular stage of the hydrologic cycle. The value of residence time can be estimated from table 1. It is calculated by dividing the amount of water in that part of the cycle by the amount that enters and leaves it in unit time.

a) Mass of water evaporates from oceans=$336 \times 10^{15}$ kg per annum

- Mass of water evaporates per day =$336\times10^{15}$ kg $\div$365days = $920\times10^{12}$ kg / day
- For 1 kg, latent heat of evaporation is = $2.25\times10^{6}$ J

  For $920\times10^{15}$ kg latent heat of evaporation is= ?

  = $2.25\times10^{6}\times920\times10^{15}$ = $2.07\times10^{21}$ J/d

b) **Residence time (RT) in oceans** : Water in oceans = 97%

- Total world water in the hydrologic cycle= $1360\times10^{18}$ kg. The evaporation from oceans is 336 $\times10^{18}$ kg per year. So residence time in oceans = $1360\times10^{18}\div336\times10^{18}$ =4 yrs

**Residence time on land**

- Water on land = 3% of $1360\times10^{18}$ =$38\times10^{18}$ kg. Falls on land = $100\times10^{15}$kg

  So R.T on land = $(38\times10^{18})\div(100\times10^{15})$=$0.38\times10^{3}$ = $3.8\times10^{2}$yrs =380 years

**Residence time in Atmosphere**

- Atmosphere contains : $13\times10^{15}$kg
- Moving in the atmosphere= Evap from land($64\times10^{15}$) + Evap from oceans ($336\times10^{15}$) = 400 $\times10^{15}$ kg/year
- So RT in atmosphere = $(13\times10^{15})\div(400\times10^{15})$ = $3.25\times10^{-2}$yrs=1.08 days

Thus residence time of water in the atmosphere is very small as it stays only 1.08 days.

### Exercise 12.36

(a) The sun (radius $R_S$) radiates energy uniformly in all directions at a temperature $T_S$. If a spherical planet of radius R is at a distance d from the sun, how much energy does it intercept in terms of $T_S$, $R_S$, R, d and the Stephan-Boltzmann constant ?

(b) Suppose the planet is perfectly heat conducting and is black so that it is uniform temperature. If the planet radiates away the same amount of energy that it receives from space, then what must be its temperature ($T_E$) in terms of the variables in part (a)? Now compute this temperature assuming the planet is the Earth using $T_S = 5800°K$ ; d = 150 x 10 $^6$ km and $R_S = 6.9 \times 10^5$ km ; R = 6371 km

(c) Suppose 1/2 of the solar heat flux is reflected from the Earth and 1/2 is absorbed and then re radiated. Then what would be the Earth's temperature, $T_E$?

(d) Explain why the Earth's surface is warmer than the temperature in part (c).

(e) Suppose only 40% of the radiation radiated by the Earth in case (c) can escape. What is the temperature at the surface necessary for a radiation balance? Suppose by adding $CO_2$ to the atmosphere, the window opening decreases by 2% so that only 39% of the radiation can escapes. What is $T_E$ under that scenario?

## *Solution*

Sun's power = $\sigma T_S^4 \times 4\pi R_S^2$ ;

a) Sun power on earth or (Earth's interception of Sun's power) or Solar incidence or Solar constant = $E_I = \dfrac{\sigma T_S^4 \times 4\pi R_S^2}{4\pi d^2} = \dfrac{\sigma T_S^4 \times R_S^2}{d^2}$

b) The temperature of the planet ($T_E$)is:

$$\sigma T_E^4 = \frac{\sigma T_S^4 \times R_S^2}{d^2} \Rightarrow T_E^4 = \frac{T_S^4 \times R_S^2}{d^2} \Rightarrow T_E = \left[\frac{T_S^4 \times R_S^2}{d^2}\right]^{1/4} = T_S \times \left[\frac{R_S}{d}\right]^{1/2}$$

If $T_S = 5800°K$ ; d = 150 x 10 $^6$ km; $R_S = 6.9 \times 10^5$ km ; R = 6371 km

$$T_E = T_S\left[\frac{R_S}{d}\right]^{1/2} = 5800K\left[\frac{6.9\times10^8 m}{150\times10^9 m}\right]^{1/2} = 393.4K$$

Other way of calculation of $T_E$:

$$E_I = \frac{\sigma T_S^4 \times R_S^2}{d^2}$$

$$= \frac{5.67\times10^{-8} wm^{-2}k^{-4} \times (5800K)^4 \times (6.9\times10^8 m)^2}{(150\times10^9 m)^2} = 1397wm^{-2}$$

Earth's energy = $\sigma T_E^4 = 1397wm^{-2} \Rightarrow T_E^4 = \dfrac{1397wm^{-2}}{5.67\times10^{-8} wm^{-2}K^{-4}} \Rightarrow T_E = 396K$

a) If half of the solar flux is reflected, the amount of flux absorbed by earth $= \dfrac{S}{4}(1-\alpha) = \sigma T_E^4$ where, S = solar constant, and α = earth's mean planetary albedo.

i.e. $\dfrac{1397wm^{-2}}{4}(1-50\%) \approx 175wm^{-2}$

This means $\sigma T_E^4 = 175wm^{-2} \Rightarrow T_E^4 = \dfrac{175wm^{-2}}{5.67\times10^{-8} wm^{-2}k^{-4}} \Rightarrow T_E = 235K$

b) The difference is 396-235k =161 K. This shows how much influence the reflection (albedo) makes

c) If 40% is radiated by the earth means 60% reflected:

$$\frac{1397wm^{-2}}{4}(1-60\%) \approx 140wm^{-2}$$

Then the temperature of the earth is:

$$\sigma T_E^{\ 4} = 140wm^{-2} \Rightarrow T_E^{\ 4} = \frac{140wm^{-2}}{5.67\times 10^{-8}wm^{-2}k^{-4}} \Rightarrow T_E = 223K$$

d) If 39% escapes from the earth, : $\frac{1397wm^{-2}}{4}(1-39\%) \approx 213wm^{-2}$

Then the temperature of the earth is:

$$\sigma T_E^{\ 4} = 213wm^{-2} \Rightarrow T_E^{\ 4} = \frac{213wm^{-2}}{5.67\times 10^{-8}wm^{-2}k^{-4}} \Rightarrow T_E = 248K$$

## *Exercise 12.37*

(a) Determine how much energy is needed to raise the temperature of the water by $1^0$C in a layer of seawater with a thickness of 1 m and an area of $10^4$m $^2$ .(Take the heat capacity Cp= $4.2 \times 10^3$Jkg$^{-1}$K$^{-1}$, density $\rho_w = 1.027 \times 10^3$kgm$^{-3}$). (b)How much time does it take to raise the above layer of ocean (mentioned in question 1) to raise by $1^0$ C, if a constant heat flux of Q = 400 Wm$^{-2}$ is applied.

## *Solution*

(a) The energy E needed to raise the temperature by ΔT degrees is given by

E = Cp m ΔT

Let the mass m of the water layer is m = $\rho_w$ V = $1.027 \times 10^3$ kg m $^{-3}$(1m × $10^4$ m$^2$) = $1.027 \times 10^7$kg.

With ΔT = $1^0$ C = 274K, we can find E = $4.2 \times 10^3$Jkg$^{-1}$K$^{-1}$ × $1.027 \times 10^7$ kg × 274K =1181.87 × $10^{10}$J = 11818.7 GJ.

(b) Q = 400 w/m$^2$ = 400 J/s/m$^2$

The mean temperature in the layer of depth h will change according to the equation:

$$\rho h C_p \frac{dT}{dt} = Q$$

Where T is time and 't' is temperature

$$T(t) \approx T(0) + \frac{Q}{\rho h C_p} t$$

Or

$$dT = \frac{Q}{\rho h C_p} dt$$

$$\Rightarrow T = \frac{Q}{\rho h C_p} t + const(T_0) = \frac{400\, J\, s^{-1} m^{-2} \times 274 K}{1.027 \times 10^3\, kg\, m^{-3} \times 1m \times 4.2 \times 10^3\, Jkg^{-1} K^{-1}}$$

$$= 25409.18 \times 10^{-6} = 0.0254\, seconds\,(\text{Here } T_0 = 0)$$

### Exercise 12.38

Calculate the amount of energy stored or released by the ocean layer of thickness 100 m and a seasonal change in temperature as $10^0$ C which is in contact with the atmosphere. Assume a uniform dimensions of the cubical volume of the layer. Take the heat capacity Cp = $4.2 \times 10^3 Jkg^{-1}K^{-1}$, density $\rho_w = 1.025 \times 10^3 kgm^{-3}$).

### Solution

The energy stored is given by the equation: $E = \rho_w V C_p \Delta T$ where $\rho_w$ = water density, v = volume = $h^3 = (10^2)^3$, $C_p$ = specific heat at constant pressure, $\Delta T$ is the change in temperature = $10^0$ C = $10^0$ K (since it is difference)

### Exercise 12.39

Suppose the heat flux at the surface of the ocean is 1/4 ly/min. What is the change in temperature if this heating continues through 12 hours and is distributed through the upper 10 meters and upper 100 meters?

### Solution

**Heat Flux (Q)** - The amount of heat energy passing through a unit area in a unit time. Heat flux is measured in units of cal/cm²/sec (or langleys / sec or ly/sec). The effect of heat flux on the temperature changes $\Delta T$ in a laterally infinite slab of material with thickness $\Delta h$, density ρ specific heat at constant pressure Cp over a time increment $\Delta t$, according to the equation:

$$\Delta T = \frac{Q_{in} - Q_{out}}{\Delta h} \frac{1}{\rho C_p} \Delta t$$

where Qin and Qout are total incoming and outgoing heat fluxes respectively in Fig.12.39 and $C_p = 0.95 \pm 0.05 ca\lg m^{-1} C^{-1}; \rho = 1.025 gcm^{-3}$

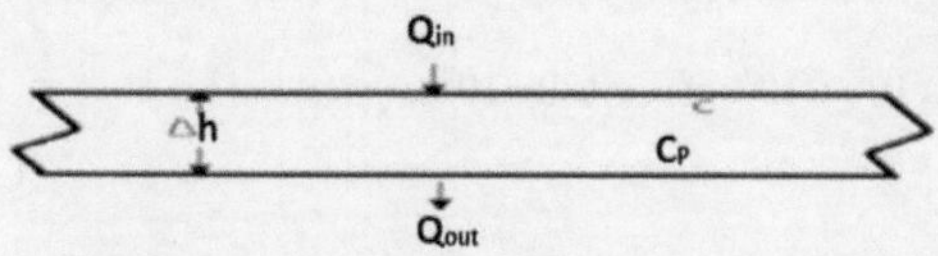


**Fig. 12.39.** A laterally semi-infinite slab of material with incoming and outgoing heat fluxes.

Given heat flux at ocean surface ($Q_{in}$ -- $Q_{out}$) : ¼ ly/min= 0.25 cal cm $^{-2}$ min$^{-1}$ = 0.25/60 cal cm $^{-2}$/ second

i) $\Delta h = 10$ m = $10^3$cm; $\Delta T$ = 12 hours = 12 × 60 × 60 seconds

we know $C_p$ ℵ0.95 $0.05 cal g m^{ℵ} C$ ; $\rho$ $1.025 gcm$

Substituting in the above formula:

$$\Delta T = \frac{0.25 calcm^{-2} \times 12 \times 60 \times 60 s}{60 s \times 1.025 gcm^{-3} \times 10^3 cm \times 0.95 ca \lg^{-1} C^{-1}} = 0.18^0 C$$

The change in temperature is 0.18 $^0$c

ii) $\Delta h = 100$ m=$10^4$cm: then

$$\Delta T = \frac{0.25 calcm^{-2} \times 12 \times 60 \times 60 s}{60 s \times 1.025 gcm^{-3} \times 10^4 cm \times 0.95 ca \lg^{-1} C^{-1}} = 0.018^0 C$$

## Exercise 12.40

a) If the following figure (12.40) is a model ocean and currents uniformly present in the whole ocean till bottom, what is the total kinetic energy KE of the system expressed in ergs and joules in this model ocean b) How much heat energy (in calories) would be released if all the system's kinetic energy were converted to heat c) How many minutes of sunlight at 0.25 ly/min would supply this much heat to the ocean surface?

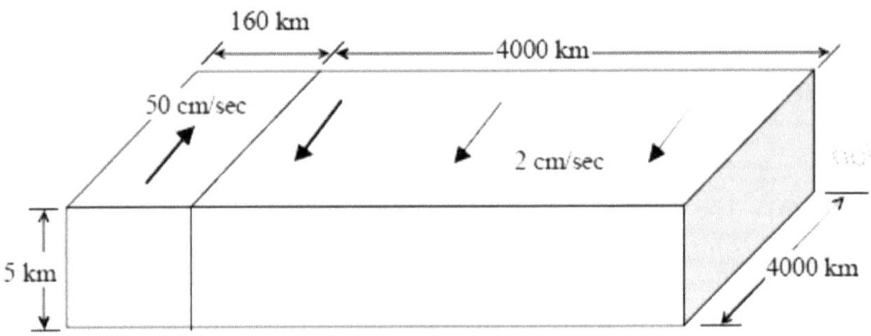


**Fig. 12.40.**

### *Solution*

a) I) K.E$_1$ = $\frac{1}{2} \times \rho_w \times V_1 \times u_1^2$

$$= \frac{1}{2} \times 10^3 \times 4 \times 10^6 \times 4 \times 10^6 \times 5 \times 10^3 \times (2 \times 10^{-2} ms^{-1})^2 \times 3 = 480 \times 10^{14} kgm^2 s^{-2}$$

$$= 480 \times 10^{14} joules = 480 \times 10^{14} \times 10^7 ergs = 480 \times 10^{21} ergs$$

K.E$_2$

$\frac{1}{2} \times \rho_w \times V_2 \times u_2^2$

$$= \frac{1}{2} \times 10^3 kgm^{-3} \times 160 \times 10^3 m \times 5 \times 10^3 m \times 4 \times 10^6 m \times (50 \times 10^{-2} ms^{-1})^2 = 120 \times 10^8 kgm^2 s^{-2}$$

$$= 120 \times 10^8 joules = 120 \times 10^8 \times 10^7 ergs = 120 \times 10^{15} ergs$$

Total K.E = K.E1 -- K.E2 = $(480\times10^{6}-120)\times10^{8}\approx5\times10^{17}\,Joules\approx5\times10^{24}\,ergs$

Note as the currents are in opposite direction the difference in K.E is the total K.E.

b) The value in heat :1 calorie = 4.186 Joules. So

$$5\times10^{17}\,Joules=\frac{5\times10^{17}}{4.186}=119\times10^{15}\,calories$$

c) 0.25 ly/min = 0.25 cal/cm²/min:

So the number of minutes = $\dfrac{119\times10^{15}\,cal}{0.25cal}=4.76\times10^{17}$ minutes

### *Exercise 12.41*

From January to June (6 months, winter to summer) a location in Bay of Bengal receives 150 wm$^{-2}$ of heat flux at the sea surface. Assuming a constant mixed layer depth of 50m in this period, how much is the change in SST in this period from winter to summer( Take $\rho_w$ =1025 kg m$^{-3}$, $C_p$ = 3850 Joules/ kg $^0$C)

### *Solution*

Given: $Q_{net}$ = 150 w m$^{-2}$, thickness of MLD (D) = 50m, change in time $\Delta$t = 6months=184 × 86400 seconds

SST tendency equation is: $\dfrac{dT_m}{dt}=\dfrac{d_{SST}}{dt}=\dfrac{Q_{net}}{\rho C\,H}$ ; So

$$\Delta_{SST}=\frac{Q_{net}}{\rho C_p D}dt=\frac{150\,wm^{-2}}{1025\,kgm^{-3}\times3850Jouleskg^{-1}C^{-1}\times50m}\times(184\times86400\,s\,)=12.08^{0}C$$

## Potential temperature calculation

### *Exercise 12.42*

Consider a water parcel moving downwards in the ocean to deeper layers from surface at a level where pressure is 500 bar (i.e. about to a depth of 5000 m). Estimate the adiabatic temperature gradient and the potential temperature at 5000m if the initial temperature of the water parcel is T = 288 K.

### *Solution*

Given $\delta$ p = 500 ×10$^5$ Pa ; then adiabatically moving parcel will have a change in temperature by $\delta$T which is given by:

$$\delta T=\left(\frac{\partial T}{\partial p}\right)_{\eta S}\delta p=\Gamma\delta p=\frac{T\alpha_T}{\rho C_p}\delta p$$

$$\delta T = \frac{T\alpha_T}{C_p}\delta p = \frac{T\alpha_T}{C_p}g\delta Z = \frac{288 \times 2.2 \times 10}{4200} \times 9.81 \times 5000 = 0.74K$$

In above calculations, $\alpha_T = 2.2 \times 10^{-4} K^{-1}$ and $C_p = 4200 J\, Kg^{-1} K^{-1}$. It is very informative to know that $\Gamma$ can also be estimated from $\frac{\delta T}{\delta p} = \frac{0.74(K)}{5000\ (db)} = 0.148 K / 1000db$.

So the potential temperature gradient upto 5000 km is $\frac{\delta T}{\delta P} = \frac{\delta T}{\delta Z} =$ 0.148 per kilometer and potential temperature ($\theta$) at 5000 m = T- $\delta$T = 288K- 0.74K = 287.26

### Exercise 12.43

A water parcel with an in-situ temperature of T = 10°C and a salinity of S = 35 ppt moves adiabatically and without salinity changes from a depth of 2 km to a depth of 6 km. What is the potential temperature of the parcel at 5 km and 6 km depths? Determine the density difference of this parcel between 5km and 6 km, taking the reference density as $\sigma_0 = 26.98$.

### Solution

The equation used for calculation of $\theta = T + \Gamma$, where T is surface insitu temperature and $\Gamma$ is adiabatic temperature gradient ($\frac{\partial T}{\partial z}$) which can be obtained from the table below (See exercise 12.42 and Exercise 12.27 for details):

| | |
|---|---|
| 0-1000m | Very small (negligible |
| 1000-5000m | $\Gamma = 0.14\ (0_C\ km^{-1})$ |
| 5000-9000m | $\Gamma = 0.19\ (0_C\ km^{-1})$ |

i) From the above table upto 5 km, value of $\Gamma$ is 0.14 per kilometer. Hence $\theta$ = 10 + (0.14 × 5) = 10.95 °C

ii) At 6 km depth: $\theta$ = 10 + (0.19 × 5) + 0.19 × 1 = 10.95 + 0.19 = 11.14 °C

iii) Density can be calculated using the equation:

$\rho = \rho_0[1 - \alpha_T(T - T_0) + \alpha_S(S - S_0)]$ where $\alpha_T = 10^{-4}K$ and $\alpha_S = 7.6 \times 10^{-4}$

At 5 km: $\rho_5 = 1.02698\left[1 - 10^{-4}(10.95 - 10) + 7.6 \times 10^{-4}(35 - 35)\right] = 1.01671$

At 6 km: $\rho_6 = 1.02698\left[1 - 10^{-4}(11.14 - 10) + 7.6 \times 10^{-4}(35 - 35)\right] = 1.01671$

The potential density between 5 & 6 km is same. There is no variation.

$\sigma_\theta = (\rho_\theta - 1)\ 10^3$

| Depth | Temp(°C) | °K (insitu) | Salinity | Density ($\rho_{STP}$) | Pot temp ($\theta$) | Pot.density ($\rho_\theta$) | $\sigma_\theta$ |
|---|---|---|---|---|---|---|---|
| 0 | 10.00 | 283.00 | 35.00 | 1.02698 | 10 | 1.02698 | 26.98 |
| 1000 | | | | | | | |
| 2000 | | | | | | | |

| | | | | | | | |
|---|---|---|---|---|---|---|---|
| 3000 | | | | | | | |
| 4000 | | | | | | | |
| 5000 | | | | | 10.95 | 1.01671 | 16.71 |
| 6000 | | | | | 11.14 | 1.01671 | 16.71 |

## *Exercise 12.44*

Calculate potential temperature (θ), potential density ($\rho_\theta$) and $\sigma_\theta$ for the following data and find its stability

| Depth | 0 | 1000 | 2000 | 3000 | 4000 | 5000 | 5460 |
|---|---|---|---|---|---|---|---|
| Temp(⁰C) | 18.909 | 2.697 | 1.868 | 1.528 | 1.456 | 1.503 | 1.547 |
| Salinity | 32.574 | 34.410 | 34.600 | 34.661 | 34.679 | 34.686 | 34.688 |

### *Solution*

First convert temperature (⁰C) to degrees Kelvin. Calculate density using the following equation and note to know the stability find whether vertical distribution of density is in the decreasing or increasing order if it is increasing it is stable and if it is decreasing it is unstable with reference to the overlying layers.

Density can be calculated from the equation:

$\rho = \rho_0[1 - \alpha_T(T - T_0) + \alpha_S(S - S_0)]$ where $\alpha_T = 10^{-4}\ K^{-1}$ and and $\rho_0$, $T_0$, $S_0$ are the reference values taken at sea surface (0 meters). Remember this temperature (T) is referred to Kelvin temperatures and potential density is density calculated using potential temperature$(\theta)$ in place of T in the above equation. The equation used for conversion is :

$\sigma_\theta = (\rho_\theta - 1)\ 10^3$

| Depth | Temp(⁰C) | ⁰K (insitu) | Salinity | Density ($\rho_{STP}$) | Pot temp (θ) | Pot.density ($\rho_\theta$) | $\sigma_\theta$ |
|---|---|---|---|---|---|---|---|
| 0 | 18.909 | 291.909 | 32.574 | 1.02322 | 18.909 | 1.02322 | 23.22 |
| 1000 | 2.697 | 275.697 | 34.410 | 1.00022 | 19.049 | 1.02302 | 23.02 |
| 2000 | 1.868 | 274.868 | 34.600 | 1.00150 | 19.189 | 1.02298 | 22.98 |
| 3000 | 1.528 | 274.528 | 34.661 | 1.00152 | 19.329 | 1.02296 | 22.96 |
| 4000 | 1.456 | 274.456 | 34.679 | 1.00153 | 19.469 | 1.02295 | 22.95 |
| 5000 | 1.503 | 274.503 | 34.686 | 1.00153 | 19.609 | 1.02293 | 22.93 |
| 5460 | 1.547 | 274.547 | 34.688 | 1.00152 | 19.67 | 1.02292 | 22.92 |

From the above table by observing potential density, it is continuously decreasing with depth. This means upper layers are denser than the layers below. This implies it is highly unstable and the upper layers try to go down while the lower layers try to come down.

## Exercise 12.45

Imagine a 4000m deep column of seawater that has uniform potential density (i.e., the same at all depths) and is resting. The (in-situ) temperature decreases by 2°C for every 1000m depth increase. The surface temperature and salinity are 10°C and 35.00ppm, respectively. What are the temperature, salinity, and pressure at the bottom? (Assume seawater is incompressible, and make a sensible assumption for the equation of state.)

## Solution

Given surface temperature and salinity are: 10°C, 35‰; the temperature gradient given is: dt/dz=2° C/1km. From Knauss tables or the T-S diagram for 10°C, 35%; sigma-t value is 27.

i) So at the depth of 4 km, the temperature value is = [10° C -- (2° C × 4)] = 2° C.

ii) The pressure = P = h ρ g = 4000m × 1025 kg $m^{-3}$ × 9.8 $ms^{-2}$

= 4 × $10^3$ × $10^3$ × 10 = 4 × $10^7$kg$m^{-1}$ $s^{-2}$ = 4 × $10^7$ N$m^{-2}$ or at 4000 m, the sea pressure is 4000 db.

iii) Using the relation in Exercise 12.8.43, Potential temperature = θ = T + Γ = 10° C + 0.14 × 4 = 10.56°C

iv) We know $\alpha dp = gdz \Rightarrow \rho = \dfrac{dp}{gdz} = \dfrac{4\times10^7 kgm^{-1}s^{-2}}{10ms^{-2}\times4000m} = 10^3 kgm^{-3}$

v) Find salinity from the equation: $\rho = \rho_0\left[1-\alpha_T(T-T_0)+\alpha_S(S-S_0)\right]$ where $\alpha_T = 10^{-4}K$ and $\alpha_S = 7.6\times10^{-4}$

$1000 = 1027\left[1-10^{-4}(10.56-10)+7.6\times10^{-4}(S-35)\right] \Rightarrow s-35 = 34.47$ ‰. As the surface salinity is 35, the value at 4000 m is 34.47‰

## Exercise 12.46

a) It is estimated that the flow southward of North Atlantic Deep water is 6 × $10^6$ $m^3$ $s^{-1}$ . If the volume of the Atlantic Ocean is 3.24 × $10^{17}$ $m^3$, what would be the length of time required to circulate the entire Atlantic through the deep water?

b) If the oceans are at least 3 × $10^9$ years old and we assume that the mixing rate has been constant, how many times has the ocean been "stirred" through the deep water?

## *Solution*

a) Length of time =

$$\frac{Volume}{flow} = \frac{3.24 \times 10^{17} m^3}{6 \times 10^6 m^3 s^{-1}} = 0.54 \times 10^{11} s = \frac{0.54 \times 10^{11} S}{60 \times 60 \times 24 \times 360 S} \approx 1736\, years$$

b) Number of times stirred = $\dfrac{3 \times 10^9\, years}{1736\, years} = 1728110.6\, times$

# 13

# Small Amplitude Ocean Waves

## 13.1. Basics of waves

The wind blows over the water, changing its surface into ripples and then to waves. Thus most of the normal waves are wind-driven. As waves grow in height, the wind pushes them along faster and higher. Sometimes waves can become unexpectedly strong and destructive. As waves enter shallow water, they become taller and slow down, eventually break on the shore. Waves cut the rocks and make them erode faster and create beaches by transporting sand towards the shore. Waves are only oscillations on the water surface and don't transport much of mass. Waves are only movement of energy along air-sea interface.

Waves are defined as successive crests with intervening troughs advance in an undulatory motion.

Waves are considered under two categories. They are forced waves and free waves. A forced wave propagates as long as the generating force acts and once the generating force is removed the wave stops whereas a free wave continues to propagate even after the removal of the generating force.

Waves can be distinguished basing on the shape of the profile. If the profile of a wave moves relative to the medium, it is a progressive wave. If the profile doesn't move, but merely oscillates in one place relative to the medium, it is a standing wave. A wave in which the particle motion is entirely parallel to the direction in which the wave moves is called a longitudinal wave. One in which the motion is entirely perpendicular is called a transverse wave. Ocean surface waves belong to neither of these classes as these are theoretical profiles and as ocean waves are formed due to combination of several profiles and bands called spectrum.

As the wind begins to blow, the waves generate within a few minutes and grow steadily in amplitude to a saturation level that depends on the strength of the wind. When the wind stops blowing, the waves gradually decay.

To describe ocean waves, we use a right-handed, Cartesian coordinate system in which the z-axis points upward. The x- and y-axes point in horizontal directions at right angles. In the state of rest, the ocean surface coincides with z = 0. When waves are present, the surface is located at $z = \eta(x, y, t)$, where t is time. The ocean bottom is at, and it is located at z = - h, where h is a constant equal to the depth of the ocean as shown in Fig.13.1.

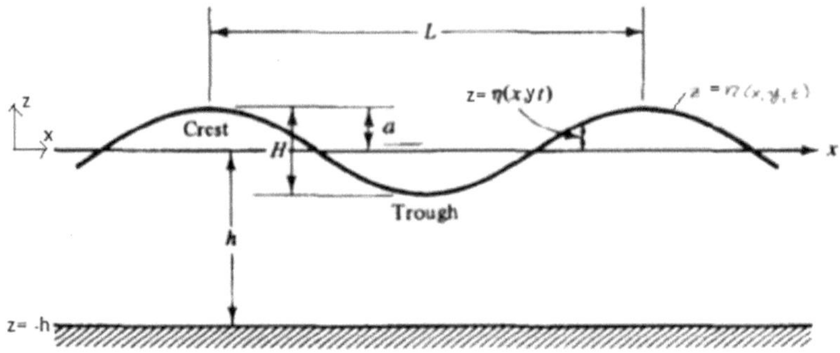


**Fig. 13.1** Wave profile with reference to sea surface and bottom

While the crest of the wave is the highest point of water above the mean water level, the trough is the lowest point below the water level. In the case of sinusoidal wave the crest and the trough are displaced symmetrically from the mean level and so the height of the wave is twice the amplitude or crest height as shown in Fig.13.1. The wavelength, L, is the horizontal distance between two crests or troughs and the period,T, is the time interval between the passage of two successive crests past a fixed point. The wave velocity or Celerity, C, is the ratio of wave Length to time period (L/T).

### 13.1.1 Classification of waves

Waves are classified according to their wave periods as shown in the table 13.1. The different forces that cause different types of waves and their restoring forces are indicated in Fig.13.2 which contains very smallest period "capillary" waves (with period less than 0.01 sec) to "ocean swells" (period about 15 sec) and tsunamis (period about 20 min). The amount of energy in a particular wave class is determined by the typical wave height in that class. The other higher order waves in this category are ultra- gravity waves, gravity waves, infra-gravity waves, tidal and trans- tidal waves.

**Table:13.1:** Classification of waves basing on the period.

| S.No. | Wave name | Period |
|---|---|---|
| 1. | Capillary waves | < 0.01 sec |
| 2. | Ultra gravity waves | 0.01sec-1.0sec |
| 3. | Gravity waves | 1sec-30sec |
| 4. | Infra Gravity waves | 30sec-5 min |
| 5. | Long waves | 5 min-12 hrs |
| 6. | Tidal waves | 12 hrs-24 hrs |
| 7. | Trans tidal waves | > 24 hrs. |

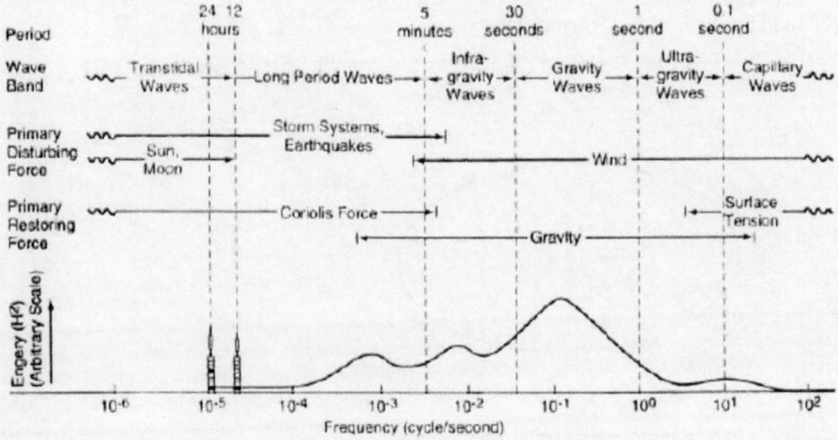


**Fig. 13.2.** Wave spectrum basing on period (frequency) (SPM, 1984)

## 13.1.2. Wave generation by wind

In the Ocean waves are generated due to various forces. The normal day to day waves are generated due to wind and so they are called wind waves. The restoring force for *small amplitude* waves is either surface tension or gravity and so they are also called *gravity waves*. But for the *finite or nonlinear* waves the restoring force is gravity.

The region of blowing wind is called *generating area.* The generating area is highly chaotic and confusing and contains high energy. The development of waves ultimately depends on three factors *speed, duration* and *fetch* of the wind. The duration of time that the wind acts on the water surface is called the wind duration. The distance over which the wind blows is called the fetch. A minimum speed, duration and fetch are necessary for the development of wavelets. When these minimum conditions over steady state reach then the sea is called *fully arisen sea* (Fig.13.3).

The first wavelets appear in the fully arisen sea are *capillary waves* which are only a few centimeters in length and a few meters in height. The capillary waves, though small, wrinkle the sea surface. This wrinkling is restricted by the surface tension. However, if the wind continues to blow, the capillary waves grow in size and become gravity waves.

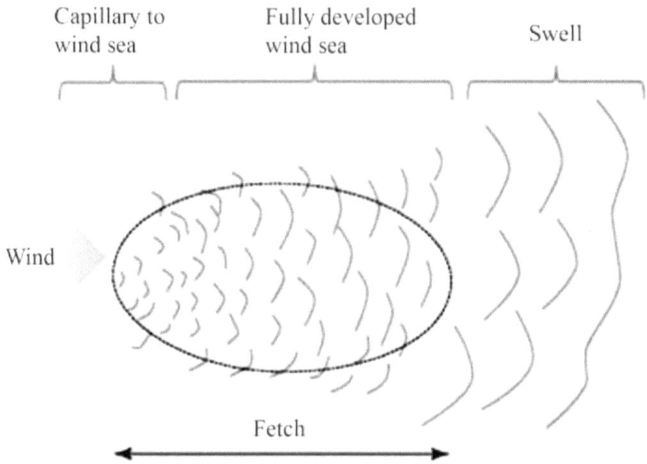


**Fig.13.3.** Fully developed (arisen) sea

Waves of different lengths run at different speeds, so that waves of different sizes combine and recombine in a constantly changing interference pattern. Once waves have escaped from the wind generating area that made them, they can run for several days with very little loss of energy. That is how we are always able to see waves on the coast even if wind speeds are zero sometimes at the place of observation.

## Interference

When sea waves from separate generating areas move out and propagate as swells, they invariably coincide called interference. Thus when swells from two different storm areas collide, the interference pattern results. There are three kinds of interference. They are constructive, destructive and mixed. Constructive interference occurs when wave trains having the same wavelength come together in phase, meaning crest to crest and trough to trough coincide. If the displacements from each wave are added together, the interference pattern results in a wave with the same wavelength as the two overlapping wave systems, but with a wave height equal to the sum of the individual wave heights. Destructive interference occurs when wave trains having the same wavelength come together out of phase, meaning the crest from one wave coincide with the trough from a second wave. If the waves have identical heights, the sum of the crest of one and the trough of other is zero, so the energy of these waves cancels each other. Generally several swell systems interact, creating complex wave patterns. It is more likely that the two swells consist of waves of various heights and lengths that come together with a mixture of constructive and destructive interference. The waves always tend to run in groups due to interference. The grouping of high and low waves is called *surf beat*. The velocity with which these groups of waves move called *group velocity*. As waves grow in height they also grow in length in the early part of their history. Even after the height is stopped growing, the length continues to grow as it propagates in the ocean. Thus the steepness is more during early part of the wave formation than the later part. Observations show that the wave steepness varies from 0.1 to 0.008 from the formation of wave to its later part. However, as the wave enters the shallow water, Stokes put a maximum possible wave steepness (H/L) at 1/7 = 0.143.

### 13.1.2.1. Seas and Swells

Waves that are generated in the generating area are called *seas*. All the seas that are generated in the generating area gets nullified sometimes and some resulting waves will come out of this generating area and can travel to far off distances up to the coasts. These long period waves that come out of the generating areas and reach the coasts are called *swells* (Fig.13.3).

*Sea* waves have usually shorter periods (higher frequency) than *swell waves* (Fig.13.4). Generally 10 seconds (0.1 cycles per second) can be taken as the demarcation period from sea to swell. *Sea* is shorter in length, steeper, and more rugged and confused in state than the swell.

Further transformation is as the wave enters into the shallow water the transformations that occur are wave reflection, refraction and diffraction and then wave breaking. As the waves enter into shallow water, the wave length and velocity decreases and period remains same and wave height enormously increases as the wave energy is conserved and so the waves break at the shore.

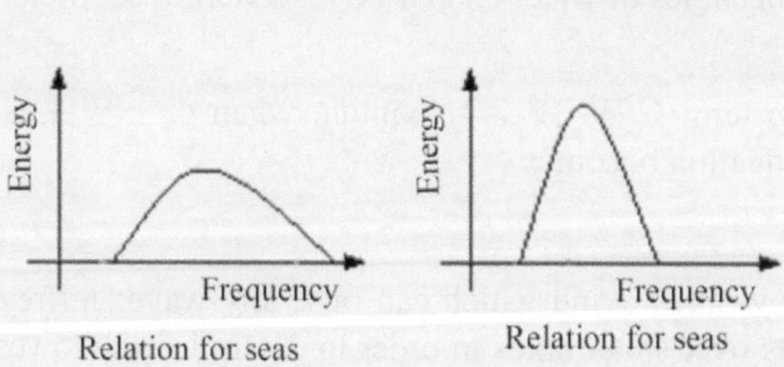


**Fig. 13.4.** Frequency & energy contained in the case of sea and swell

### 13.1.2.2. Theories of wind wave generation

Actually the transfer of energy from wind to waves, and the resulting wave growth, is not completely understood. Many theories are in vogue to explain the wave generation by wind. The wave may get energy from the direct push of the wind on their upwind faces if they are remaining slower than the wind. The following are the wind wave generation mechanism theories.

### 13.1.2.2a.The sheltering theory of Jeffrey's

The sheltering theory of Jeffreys (1925) assumes that the air flow is laminar over the wind ward slope of the wave and turbulent on the lee side slope that the lee side of the wave form is sheltered by the crest and that low velocities are found there. The normal pressure is thus said to differ between the windward and leeward faces (Fig.13.5). If the wind velocity exceeds the wave velocity, energy is transferred from the air to the water. Tangential friction is ignored in the theory. One important parameter in the analysis is the "sheltering coefficient-S" which is related to the proportion of the wind ward slope of the wave which offers resistance to the wind. The value of S is given as 0.27.

Jeffrey's theory accounts reasonably satisfactory for the initial generation of waves, when their steepness is very low, but as the waves increase in steepness the value of 'S' appears too great.

If a strong wind blows against an obstacle the wind exerts a pressure on the wind ward side of the obstacle but on the leeward side there will be suction. Similarly when the wind blows over a sequence of waves the pressure of the wind will be greater on the windward slope than on the leeward slope that is sheltered by the crest (Fig.13.5). This condition can prevail only if the waves travel at a velocity smaller than the speed of the wind. On the basis of these arguments, Jeffreys (1925) found that wave can grow only if

$C(W-C)^2 \geq \frac{4\gamma g(\rho-\rho"}{S\rho'}$ where W is wind velocity, C wave velocity; $\gamma$=kinematic viscosity of water; ρ and ρ' = densities of water and air ; S = sheltering coefficient (nondimensional).

At a given wind velocity the term $C(W-C)^2$ is maximum when C = W/3. So substituting for C the above equation becomes

$$(W)^3 \geq \frac{27\gamma g(\rho-\rho")}{S\rho'}$$

This equation determines the weakest wind which can raise any wave. Jeffreys conducted wind measurements over small lakes in order to determine the lowest velocity of the wind at which small waves appear. He found that only at a speed of the wind a little greater than one meter per second distinct waves seem to appear. Thus substituting this value of 'W' in the above equation the sheltering coefficient w found to be about 0.27 which is related to the proportion of the windward slope of the wave offering resistance to the wind. This value seems to have been too great. This theory also cannot account for the original generation of waves from a flat surface, although it does seem to apply better in the early stages of wave generation than when the waves become large.

The sheltering coefficient $S$ was calculated by determining the rate of energy loss due to molecular viscosity and the minimum wind $W_{min}$ that could maintain waves against this loss. This value was compared with observations of the minimum winds required to sustain waves. The calculated value of $S$ varies as $W^3 min$. The choice of $W_{min}$ is thus critical. Jeffreys chose $W_{min}$ equal to 1.1 m/s, giving $S$ =0.3.

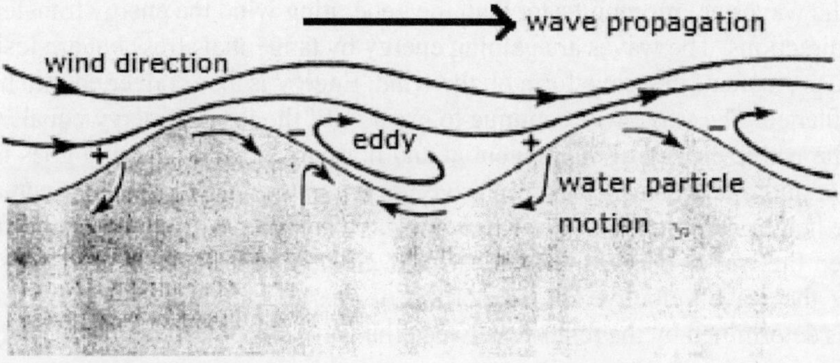


**Fig. 13.5.** Jeffreys sheltering theory. Note ( +) indicates higher pressure, minus ( - ) indicates lower pressure due to sheltering of the crest. Circular arrows indicate the formation of eddies due to pressure difference. Small arrows indicate the water particle movement. Large wavy arrows indicate the wind flow

### 13.1.2.2b. Sverdrup & Munk theory

The next great advance in the theory of wave generation in deep water is given by H.U.Sverdrup and W.H.Munk (1947). The theory is based on rough turbulent flow which occurs when the wind speed exceeds 7m/s above this velocity ripples of small dimension are formed on the wave surface and these alter the type of surface from hydrodynamically smooth to rough. The air flow is considered to be similar to that suggested by Jeffreys, as eddies form in the lee side of the wave. The sheltering coefficient used is about 1/20 of Jeffreys, that is 0.013.

According to this theory waves grow because energy is transferred from the wind to waves. This transfer takes place by two processes namely, transfer by pressure difference and transfer by tangential stress of the wind (Fig.13.6). The energy which is imparted to the wave is used in part to increase its height and in part to increase its velocity of progress. When the wave velocity is less than the wind velocity, wave receives energy from both the pressure difference and the tangential stress. If, however, the wave velocity exceeds the wind velocity the wave still gains energy by tangential stress but it loses energy through pressure difference because it has to overcome the air resistance.

It is suggested that the tangential stress of the wind, which is an important factor in this theory varies with the square of the wind velocity above a certain value. The waves acquire energy from the wind by tangential stress when the water particles move in the same direction as the wind, and lose energy when the particles move against the wind, but because of mass transport, there is a net gain of energy. The wave velocity is much greater than the velocity of the particles within the wave, and it is thus possible for the wave velocity to exceed that of the generating wind. The limit would be reached when the particle velocity equaled the wind speed. When the waves are moving faster than the generating wind the energy transfer is in two directions. The waves are gaining energy by tangential stress but are losing it by the pressure of the wave form on the wind. Energy is also converted into heat by turbulence. The wave will continue to grow until the loss of energy equals the gain. The whole energy of the tangential and normal stress of the wind goes into the wave and causes an increase of wave height and wave length. The increase in wave height is more important when the wave is moving lower than the wind, but when the wave speed exceeds the wind velocity most of the energy goes to increase the wave velocity. The proportion of energy increasing the height and length is determined by the relation between the wave steepness and the wave age.

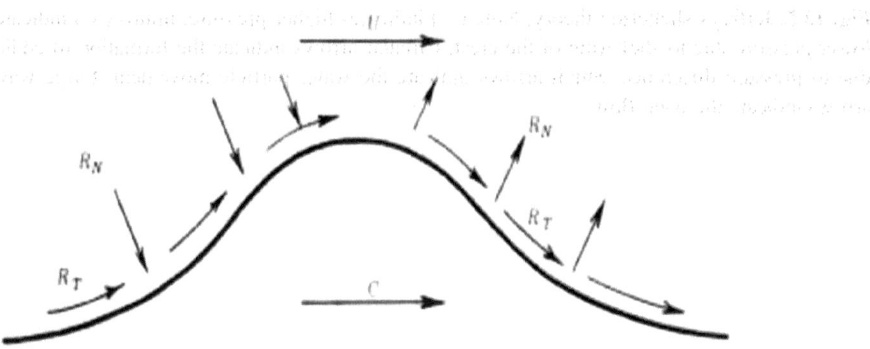


**Fig.13.6.** Wave grows and propagates due to tangential ( $R_T$ ) and normal stresses ($R_N$ ) caused by wind (U)

The average rate at which energy is transferred to a wave by normal stress ($R_N$) is equal to (Bretschneider 1965)

$R_N = \frac{1}{L}\int_0^L p_z w_0 dx$ where $w_0$ = - kac Cos k (x-ct) is the vertical component of the particle velocity at the surface and $p_z$ is the normal pressure (surface tension) acting on the sea surface.

The average rate at which energy is transmitted to the waves by tangential stress is equal to

$R_T = \frac{1}{L}\int_0^L \tau u_0 dx$ where $u_0$ = kac Sin k (x-ct) denotes the horizontal component of particle velocity at the sea surface and $\tau = \rho\ C_d\ U^2$ is the wind stress and 'U' is the wind speed and ρ is air density. Here $C_d$ is the drag coefficient. Various experiments and observations have been made leading to different values of $C_d$ as a function of wind speed. However, a number of authors agreed to put at $1.25\times10^{-3}$ for wind speeds less than 6 m/s and $2.6\times10^{-3}$for wind speeds greater than 6 m/s.

According to the above the energy of waves can increase only if the rate at which energy is added by both normal and tangential stresses of the wind ($R_N + R_T$), exceeds the rate at which energy is dissipated by viscosity. The energy added by the wind goes into building the wave height and increasing the wave speed.

That is $R_H + R_C = R_N \pm R_T$ where $R_H$ is the portion of energy transformed into wave height and $R_c$ is that portion of energy transformed into wave speed.

### 13.1.2.2c. O. M. Phillisps theory

One of the most promising theories was put forward by O.M.Phillips in1957. He assumes that a turbulent wind at a given moment starts to blow over a surface previously at rest. The pressure on the surface fluctuates in a distribution which is a stationary random function of position and time. The study aims to discover the properties of the surface displacement at subsequent periods.

The random distribution assumed by Phillips involves a type of resonance. The turbulent nature of the wind is an essential factor in the growth of waves and causes random stresses on the water surface. These include both normal and tangential stresses. It is true that winds blowing over water do not consist of streams of air in steady and uniform motion but rather of an irregular series of 'puffs' and 'lulls' carrying eddies and swirls distributed in a disordered manner. Eddies, in the air stream, are carried by the wind and change as they move, so that the stress distribution moves across the surface with a certain velocity dependent on the wind speed. This convection velocity of the stress fluctuations is defined as the velocity of the frame of reference, $U_c$, in which their frequency scale is least, or their time scale is greatest. This velocity is equal to the wind velocity as measured from a ship.

The fluctuating pressure upon the water surface is held responsible for the birth and early growth of the waves, an analysis of the surface should show a wide variety of wave numbers and frequencies in both two dimensional space and time. The components of the pressure fluctuations acting on the surface generate small forced oscillations. These fluctuations affect the component of the amplitude on the surface. If the pressure distribution includes components whose wave numbers and frequencies coincide with possible dimensions of free surface waves a type of resonance is setup, and the continued presence of these particular frequencies in the pressure distribution will generate surface waves whose amplitude will increase. Thus it is asserted that the frame of reference is not changing in character but is moving with the convection velocity, and if one of the wave frequencies generated is moving with the same velocity, then the two move together and growth can continue. This, however, ignores the evolution of the stress pattern and so is an over simplification. If the pressure pattern is changing slowly the growth of the wave lets is reduced.

At first the surface may be assumed to have no effect on the pressure distribution, but this will not last for long. Small ripples will soon form and these have a considerable effect on the vertical velocity distribution. The long gravity waves do not appear to influence the pressure distribution. The viscosity of the water is ignored, which is probably justified for all but the shortest waves. The motion is then irrotational. The results apply when the mean square slope is small so that the surface boundary conditions can be considered linear. At later stages of growth the non- linear effects may become important.

In the initial generation of waves the minimum velocity is given by the equation $C_{\min} = \left(4gT_s\frac{1}{\rho}\right)^{\frac{1}{4}}$ where ρ is the water density, $T_s$ is the surface tension at the interface and $C_{min}$ is the minimum velocity. The wave length of the critical waves is about:

$L_{cr} = 2\pi\left(T_s\frac{1}{\rho g}\right)^{\frac{1}{2}}$ For water if ρ = 1 gm/cm$^3$ $T_s$ = 73 gm/sec$^2$ and g = 980 cm/sec$^2$, $C_{min}$ = 23 cm/sec and $L_{cr}$ = 1.7 cm.

According to Phillips theory of 1957, the main problem of wave generation is related to the distribution of pressure on the moving, random water surface under the influence of a turbulent wind. The pressure fluctuations are of two kinds. One is produced by the turbulent eddies in the wind and the other is induced by the air flow over the irregular surface. The total pressure pattern is the sum of these two. The first, the turbulent pressures, provide energy input over a wide spectral range, while the second, the directly induced pressure, provides a selective feed-back which allows certain components to grow rapidly.

### 13.1.2.2d. Miles' theory

Miles (1957) assumed that the air was inviscid and incompressible and, in the absence of waves, can be described by a prescribed mean shear flow which varies only with height above the surface. It was assumed that there is a feedback from the waves such that wave-induced air pressure perturbations are two-dimensional and sufficiently small that they can be neglected in the nonlinear processes of the equations of motion. As turbulence in the air is neglected, the resulting theory is often termed as *quasi-laminar.*

The growth rate predictions of Miles were of the same order of magnitude as observations but too low. The determination of the growth rate by this theory is, however, critically dependent on the shape of the near surface atmospheric boundary layer and the surface roughness.

The neglect of turbulence within the Miles Theory is often raised as a shortcoming of the theory and further experimental corroboration of the theory is still required. Despite these concerns, the theory has proved successful in operational wave models such as WAM (WAMDI, 1988).

## 13.2. Coastal Processes

The areas, most directly affected by the forces of the sea are the beach topography as shown in Fig.13.7, the gulfs, and the near-shore zone regions that experience the full impact of the sea's energy. There are two general types of dynamic beach responses to wave motion:i) Response to normal condition and ii) Response to storm condition. Normal condition prevails most of the time, and the wave energy is easily dissipated by the beach's natural defence mechanisms. During stormy conditions when large amount of wave energy is available, large section of beach and dune (Fig.13.7) will be affected. However, under normal days alternate erosion and accretion may be seasonal on some beaches such that the winter storm waves erode the beach, and the summer waves rebuild it.

Longshore drift is the movement of sediments along the coast parallel to the shoreline where waves are strong, the coast will be eroded and sediment is carried away and where they are weak, sediment is deposited. Longshore drift can have undesirable effects for humans, such as beach erosion. To prevent this, sea defences such as groynes or gabions may be constructed.

At any given instant of time the sea state is never a single harmonic oscillation, but appears to the observer as a chaotic process. Therefore ways have to be found to describe the wave conditions in terms of measurable statistical quantities. Two approaches are used:

a) Determination of significant wave parameters (description in the time domain)

b) Determination of the wave spectrum (description in the frequency domain)

The combination of many measurements has resulted in wave parameter estimates for fully developed seas. A fully developed sea is one for which the fetch and duration are not limiting i.e., no further growth occurs, as momentum and energy loss through breaking balances input from the wind.

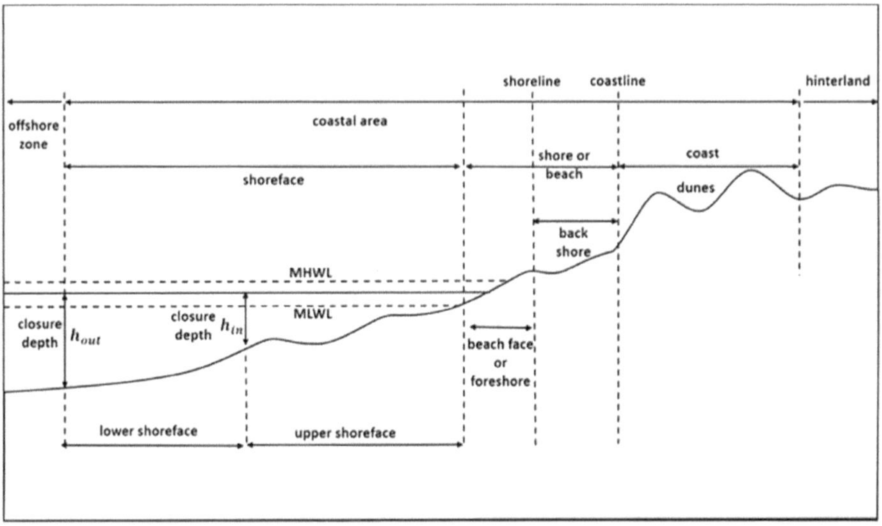


**Fig. 13.7.** The beach topography and shore area that are subjected to the impact of waves

For many marine applications, for example the routing of ships or the design of platforms, only the highest waves are of interest. The quantity significant wave height has therefore been introduced. It is defined as either $H_{1/3}$ or $H_{1/10}$, ie as the average of the 1/3 or 1/10 highest waves over an observation period. (The use of $H_{1/3}$ is more common than the use of $H_{1/10}$). From observations, the largest wave height $H_{max}$ is related to the significant wave height by (SPM, 1984)

$$\frac{H_{max}}{H_{1/3}} = 1.45 \; and \; \frac{H_{max}}{H_{1/10}} = 1.3$$

Measured maximum wave heights depend to some degree on the length of the observation period; different values for $H_{max}$ are found from a 10 minute time series compared to a 3 hour time series.

### 13.2.1. Statistical description of waves

The relations between largest wave height and significant wave height indicate that the sea state has certain statistical properties. A statistical description is based on the representation of the wave field through the energy spectrum. For a given frequency, wave energy is proportional to the square of the wave amplitude. An energy spectrum shows wave energy as a function of wave frequency (Fig.13.8). A single harmonic wave has a 'monochromatic' spectrum:

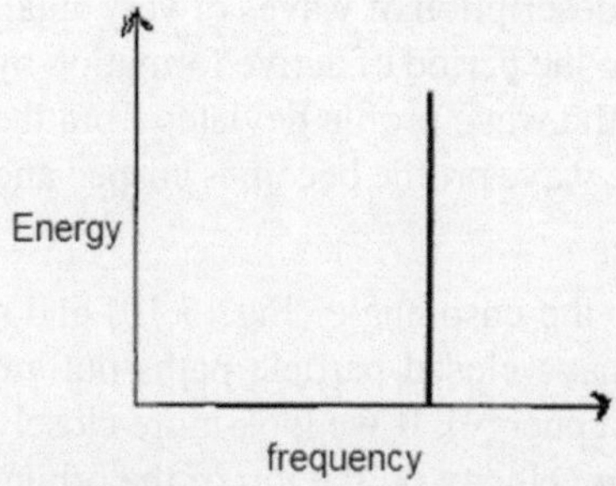


**Fig. 13.8.** Relation between wave frequency and wave energy

Statistical description assumes that waves of all frequencies and corresponding wavelengths are present. It does not attempt to describe the form of the sea surface but concentrates on wave energy. In a wind sea with random distribution of wave energy over all wave frequencies, the theoretical form of the energy spectrum is that of a Gaussian or 'normal' distribution. Note that the energy is spread in the case of seas whereas when only swell is present the energy is concentrated near the swell frequency and the spectrum is much narrower as shown in Fig.13.4.

In real oceanic situations energy is not randomly distributed but cascades from the shortest possible waves stirred up by the wind to the longer wavelengths. As a result the shape of ocean wave spectra depends strongly on the wind speed. Figure 13.9 shows the observed energy (proportional to square of wave height) spectra for a fully developed sea at various wind speeds. Early measurements showed that, for a given wind speed, the spectrum evolved to the point where it became essentially invariable but that the total energy in the spectrum increased with increasing wind speed and the peak frequency shifted toward the lower frequency (longer period) waves (Figure 13.9).

Note that in Fig (13.9) as the wind speed decreases from top curve (20 m/s) to the bottom curve (10 m/s), the wave energy and period decreases. The center joining curve of the peaks (that contain most of the energy) shows the shift towards lower periods as the wind speed decreases. Note that the spectrum has a normal distribution only for very low wind speeds; as the wind speed increases, waves of short period are still present but most of the energy is found in longer period waves. The spectrum then drops off rapidly at longer periods. That is, it is "skewed" towards the longer periods.

Finally wave decay occurs during the breaking phase on the beach, which occurs when the particle velocity becomes larger than the phase velocity (wave speed). In this state the waves transport both energy and mass towards the beach. While much of the energy is dissipated in mechanical reworking of the beach, the mass moved by the waves has to be returned to the sea. This occurs at regular intervals along the beach in so-called rip currents, strong seaward movement of the water along the sea floor ("undertow").

The harmonic oscillation concept gives a good description of waves of very small steepness. On the approach of a beach, or during the period of active formation by wind, wave steepness is not small enough and the wave profile deviates from the harmonic profile. As the steepness increases, the wave profile becomes cusped and eventually the waves will break.

The limiting value before breaking is 120° for the cusp angle (Fig.13.10) and a steepness = H/L= 1/7. Cusped waves do not have closed particle paths but are associated with a net transport of water (not just energy). If we look more closely into the equations for $u$ and $w$, we see that the $u$ velocity on the top of the orbital motion of particle is slightly larger than the velocity at the bottom of the orbital velocity of the particle as the orbits exponentially decrease below. As a result each particle moves forward a very small distance during each orbit. Therefore, there is a very small net mass transport of water in the direction of wave called Stokes drift (Fig. 13.11).

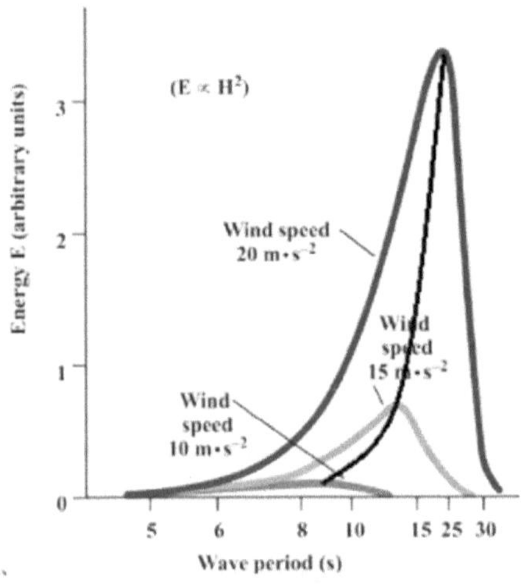


**Fig.13.9.** Wave energy (square of wave height) as a function of wave period (frequency) for fully developed seas under different wind speeds.

**Fig. 13.10.** The angle of the cusped wave before breaking

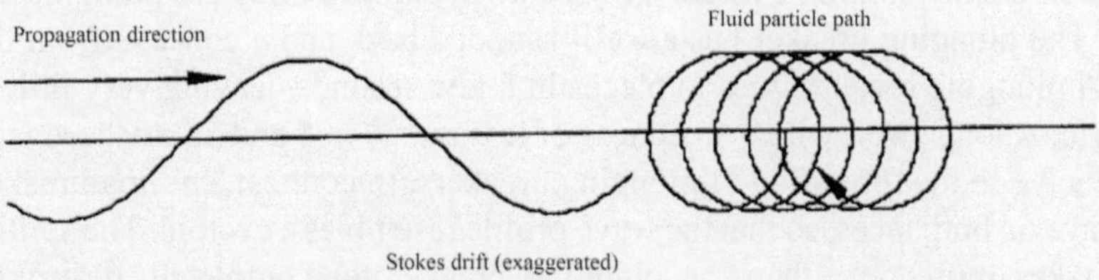


**Fig. 13.11.** Orbital motion of particles and the stokes drift

### 13.2.2. Wave transformation and breaking

There are two main sets of conditions in which waves break. The first is when wind blows over the waves in deep water and the second is when waves arrive at the shore over a sloping bottom. The long period swell waves or Stokes waves of finite amplitude finally run ashore and break up into surf when the steepness exceeds one by seven (1/7). In an unbroken wave the orbital velocity of water particles at the crest is less than that of the wave. However, if the velocity of water particles at the crest exceeds the wave velocity it will break away from the wave profile. As the waves enter into shallow water (the ratio of water depth to wave length, d/L, the relative depth, is less than 1/20), the wave length and velocity decreases and period remains same and wave height enormously increases as the wave energy is conserved.

When the wave enters shallow water over a shoaling bottom their wave length is reduced and their height is increased with a resulting increase in steepness. When the waves reach a depth of 1 to 1.5 times of their height, generally the particle velocity at the crest is equal to the wave velocity and so they break. The exact ratio of wave height to water depth (H/d) at which the wave breaks depends on the deep water steepness of the waves and the slope of the beach. The types of breakers described are spilling, plunging, surging and collapsing as shown in Fig.13.12.

In swell waves on a gently sloping bottom, the typical relation of wave height to water depth at the outer edge of the breakers is 0.78 for short period steep waves its value is 0.6. So storm waves break farther in the sea than the long swell waves of the same height at the coast.

The spilling type of breaking occurs in deep water or over gentle bed slope (m = 1:50). This involves gradual release of energy and is characterized by appearance of foam on forward side of the crest. Plunging and collapsing breakers occur on moderately steep slopes. In plunging, the water mass plunges and falls on forward face of the crest while in collapsing type the breaking is sudden, the crest form is steep and the foam is seen on lower side of the forward face. The last type of breaking i.e. surging, takes place when the seabed slope is steep with m = 1:10 or so. In this case the entire water mass gets piled up and the foam is seen on beach face.

Breakers on evenly sloping beaches may be of two kinds. They are plunging and spilling. The plunging breaker has a well- rounded back and a concave front like a hook. A plunging breaker goes to pieces in a few seconds, leaving very little of wave form. A long swell with a steepness of less than 0.005 and an offshore wind are conducive to the formation of plunging breakers. In contrast, spilling breakers are concave on both faces, so that the wave profile resembles a cycloid. The spilling breaker takes more time than the plunging breaker for completely destroying the wave shape. Waves with steepness of more than 0.01 often produce spilling breakers, particularly, when there is an offshore wind. Instead some breakers come out of the mixture of both kinds of plunging and spilling and break more than once. Such breakers are called surging breakers which do not actually break but surge up steep braches. This breaker may break three or four times before the wave form is completely destroyed. The collapsing breaker also is another variety of the mixture of these two main categories.

Two parameters, viz. Galvin's (β) and Irribarren's (or Surf Similarity) ($N_1$) are often used to distinguish between different types of breaking. If $H_0$ is the deep water height, $L_0$ is corresponding wavelength and m is the seabed slope then,

$$\beta = \frac{H_0 / L_0}{m^2} \; and \; N_1 = (\beta)^{-\frac{1}{2}} = \frac{m}{\sqrt{H_0 / L_0}}$$

Where $H_0$ is deep water wave height and $L_0$ is deep water wave length. Figure 13.12 shows different values of β and $N_1$ exhibited in various breaking modes.

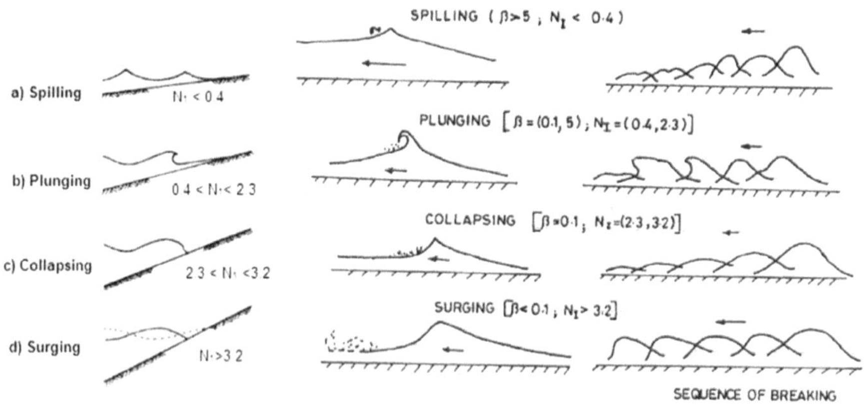


**Fig. 13.12. Left:** Classification of breaking waves as a function of beach steepness and wave steepness offshore. **Right:** Condition of breaking waves by different authors. (After Horikawa, 1988)

Wave shoaling, refraction and diffraction transform the waves from deep water to the point where they break and then the wave height begins to decrease markedly, because of energy dissipation. The sudden decrease in the wave height is used to define the breaking point and determines the breaking parameters (Hb, db and xb). Breaking waves may be classified as one of four types as shown in Figure 13.12. The type can be approximately determined by the value of the surf similarity parameter (or Irribarren Number), $N_1$.

$N_1 = \frac{tan\beta}{\sqrt{\frac{H_b}{L_b}}} = \frac{m}{\sqrt{\frac{H_b}{L_b}}}$ where $L_b$ is the wavelength at breaking and m is the beach slope.

Spilling breakers occur when $N_1 < 0.4$; plunging breakers when $0.4 \leq N_1 \leq 2.3$ and collapsing breaker $2.3 \leq N_1 \leq 3.2$ and Surging breaker when $N_1 > 3.2$ as shown in Fig. 3.13 below. Further details may be found in Horikawa (1988). Using the following formulae the breaker type, using Irribarren number $N_1$, can be calculated.

The boundary between spilling and plunging breakers occurs at a higher ratio between open sea wave steepness and wave length for a steeply sloping beach than for a gently sloping beach. To produce spilling breakers on a steep beach, the deep sea wave steepness must be relatively high, and conversely to produce plunging breakers on a gently sloping beach, the deep sea wave steepness must be relatively low. The diagram in Fig.13.13 shows a relation between plunging, surging and spilling breakers. It may be noted here that the gentler sloping beaches do not have surging breakers. The surging and collapsing breakers are only a mixture of the original plunging and spilling breakers. The plunging breaker collapses and the spilling breaker surges depending on the wind and slope of the beach. Collapsing breakers are rare and occur where the seafloor has a steep slope and the lower part of the wave is slowed such that foam and bubbles are formed with little splash.

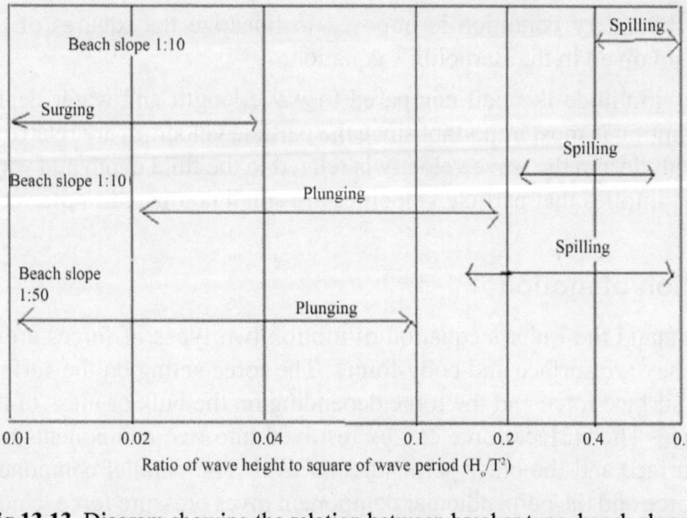


**Fig.13.13.** Diagram showing the relation between breaker type, beach slope and steepness of Deep water waves

## 13.3. Small amplitude waves

### 13.3.1. Derivation involved in small amplitude wave theory

The small amplitude theory for a two dimensional free, periodic gravity wave is developed by linearizing the equation that defines the free surface boundary condition. Along with this using the bottom boundary condition, a periodic velocity potential is sought that satisfies the requirements for irrotational flow. This velocity potential is then used to derive the equations for various wave characteristics like wave celerity, particle and group velocity, pressure, energy etc.

### Assumptions involved in small amplitude wave theory

i) The water is homogeneous, incompressible and surface tension forces are negligible. Thus there are no internal gravity or pressure waves affecting the flow, and the surface waves being considered are larger than that size where capillary effects are important (i.e. wave lengths greater than 3 cm.)

ii) Flow is considered irrotational and so velocity potential exists which satisfy the Laplace equation. Thus there is no shear stress at the air-sea interface or at the sea bottom and so the fluid slips freely at the bottom.

iii) The bottom is considered impermeable and horizontal and so doesn't move. So the bottom cannot add or remove energy to the propagating wave. A separate bottom boundary condition is defined.

iv) The pressure along the air-sea interface is considered constant. Thus no wind pressures exist and hydrostatic pressure differences due to surface elevation differences are negligible.

v) A surface boundary condition is imposed to linearize the squares of the velocities involved in the Bernoulli's equation.

vi) The wave amplitude is small compared to wave length and water depth. This assumption is most important since the particle velocities are related to wave amplitude and the wave velocity is related to the fluid depth and wave length, this implies that particle velocities are small compared to the wave velocity.

### Euler's Equation of motion

In order to understand the Euler's equation of motion two types of forces are to be considered. They are surface and body forces. The force acting on the surface of an element is surface force and the force depending on the bulk or mass of the fluid is body force. The surface force can be resolved into two components one parallel to the surface and the other perpendicular to it. The parallel component gives the shear force and the perpendicular component gives pressure force. Under frictionless conditions, the shear force can be neglected and so the only surface force remains is pressure force. Gravity is considered as the principal body force in the present study.

The equation of motion can be written as:

$$X-\frac{1}{\rho}\frac{\partial p}{\partial x}=\frac{du}{dt}$$

$$Y-\frac{1}{\rho}\frac{\partial p}{\partial y}=\frac{dv}{dt} \qquad (13.1)$$

$$Z-\frac{1}{\rho}\frac{\partial p}{\partial z}=\frac{d\omega}{dt}$$

Where X, Y and Z are the body forces per unit mass in the respective directions. Please note here while X and Y are Coriolis accelerations Z is acceleration due to gravity. These equations were first published by Euler in 1775 and so are known as Euler's equation of motion.

## Existence of velocity potential

The velocity potential or potential function ϕ is a function of space and time, the derivative of which in a given coordinate direction is the force in that direction. The velocity potential by definition is related to the three components of velocity as

$$u=-\frac{\partial\phi}{\partial x}, v=-\frac{\partial\phi}{\partial y}, w=-\frac{\partial\phi}{\partial z} \qquad (13.2)$$

If a velocity potential exists the flow must be irrotational, that is the vorticity is zero ($\zeta = 0$ ).This can be shown by obtaining derivatives of the three velocities as shown below:

$ie. \frac{\partial u}{\partial z}=\frac{\partial}{\partial z}\left(-\frac{\partial\phi}{\partial x}\right)=-\frac{\partial^2\phi}{\partial x\partial z}$ Similarly, $\frac{\partial w}{\partial x}=\frac{\partial}{\partial x}\left(-\frac{\partial\phi}{\partial z}\right)=-\frac{\partial^2\phi}{\partial x\partial z}$

This means $\frac{\partial u}{\partial z}=\frac{\partial w}{\partial x}$ (13.3)

this implies $\frac{\partial w}{\partial x}-\frac{\partial u}{\partial z}=0 \Rightarrow \zeta=0$

## 13.3.2. The equation of Continuity

An important restriction upon the velocity of a fluid may be obtained by kinematic methods applied to the law of conservation of mass.

Consider an infinitesimal rectangular volume fixed in space with sides of lengths δx, δy, δz in a moving fluid having velocities u,v,w respectively as shown in Fig.13.14. Consider first the flow parallel to the x-axis. As 'u' is x- component of velocity and ρ is density of the fluid the mass flow through the left face is ρu.

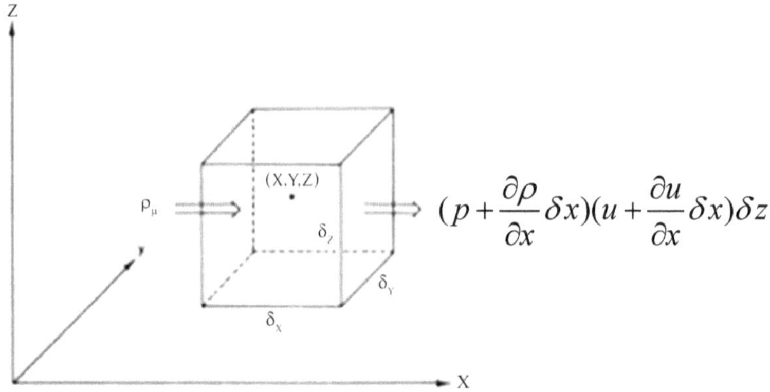


**Fig. 13.14.** The cube for derivation of equation of continuity

The mass flow into the volume per unit time is $(\rho u)\delta y\delta z$

The mass flow out of the volume per unit time on the right face is

$(\rho+\frac{\partial\rho}{\partial x}\delta x)(u+\frac{\partial u}{\partial x}\delta x)\delta y\delta z$ (This comes from Taylor's series)

So the net flow out of the volume in the x-direction is

$$(\rho u\delta y\delta z)-\left[(\rho+\frac{\partial\rho}{\partial x}\delta x)(u+\frac{\partial u}{\partial x}\delta x)\right]\delta y\delta z$$

$$-\left[\frac{\partial}{\partial x}(\rho u)+\frac{\partial\rho}{\partial x}\frac{\partial u}{\partial x}\delta x\right]$$

As the volume of the rectangular element considered is infinitesimal, δx is very small and so higher order term (second term) is negligible.

Therefore, the net flow in the x-direction is = $-\left[\frac{\partial}{\partial x}(\rho u)\right]\delta x\delta y\delta z$

Taking the mass flow in all the three component directions, the total flow out would be

$$-\left[\frac{\partial}{\partial x}(\rho u)+\frac{\partial}{\partial y}(\rho v)+\frac{\partial}{\partial z}(\rho w)\right]\delta x\delta y\delta z$$

where 'v' and 'w' are the velocity components in the y and z directions respectively.

We shall compute the rate at which mass enters the box as a result of fluid motion and, since mass must be conserved, we shall equate this to the rate of change with time of the mass contained within the cube.

In other words, if mass is neither created nor destroyed and if the sum is not zero, it must be balanced by a change of density (rate of change of density $\frac{\partial\rho}{\partial t}$ )

$$\therefore\left(\frac{\partial\rho}{\partial T}\right)\delta X\delta Y\delta Z=-\left[\frac{\partial}{\partial X}(\rho U)+\frac{\partial}{\partial y}(\rho V)+\frac{\partial}{\partial z}(\rho W)\right]\delta x\delta y\delta z$$

$$-\frac{\partial\rho}{\partial t}=\frac{\partial}{\partial x}(\rho U)+\frac{\partial}{\partial y}(\rho V)+\frac{\partial}{\partial z}(\rho W)=\nabla.\rho\bar{V}\;.............(13.4)$$

$$\frac{1}{\rho}\frac{d\rho}{dt}+\left[\frac{\partial U}{\partial X}+\frac{\partial V}{\partial Y}+\frac{\partial W}{\partial Z}\right]=0.......(13.5)$$

This equation (13.5) is called the equation of continuity. The first term is the fractional rate of change of density and the second term is the fractional rate of change of volume.

If the fluid is incompressible, the density is uniform (constant), then $\frac{d\rho}{dt}=0$

$$\therefore\frac{1}{\rho}\frac{d\rho}{dt}=0$$

Then the equation of continuity is $\frac{\partial u}{\partial x}+\frac{\partial v}{\partial y}+\frac{\partial w}{\partial z}=\nabla.\bar{V}=0..........(13.6)$

$$where\,\bar{V}=iu+jv+kw$$

Equation (13.5) is frequently considered convenient as it is nothing but horizontal three dimensional divergence.

### 13.3.3. Laplace Equation

For treatment of wave equation only two dimensional motion in the x-z plane will be considered. For compressible two dimensional motion in the x, z plane the equation of continuity as per equation (13.6) is:

$$\frac{\partial u}{\partial x}+\frac{\partial w}{\partial z}=0 \tag{13.7}$$

If in addition the flow is irrotational, the velocity potential exists so by substituting equation (13.2) in equation (13.7), we can write:

$$\frac{\partial^2\phi}{\partial x^2}+\frac{\partial^2\phi}{\partial z^2}=0 \tag{13.8}$$

This equation (13.8) is known as the Laplace equation of continuity. Any solution for the water wave motion must obey this continuity relationship.

### 13.3.4. Bernoulli's equation

Considering the Euler's equation of motion (13.1) in x-z coordinate system;

$$\begin{aligned}x-\frac{1}{\rho}\frac{\partial p}{\partial x}&=\frac{du}{dt}=\frac{\partial u}{\partial t}+u\frac{\partial u}{\partial x}+w\frac{\partial u}{\partial z}\\z-\frac{1}{\rho}\frac{\partial p}{\partial z}&=\frac{dw}{dt}=\frac{\partial w}{\partial t}+u\frac{\partial w}{\partial x}+w\frac{\partial w}{\partial z}\end{aligned} \tag{13.9}$$

The body force per unit mass (z) is restricted to acceleration due to gravity only here. So the other body force x = 0.

$\therefore$ x = 0 and z = - g (13.10)

We can write 'g' in a derivable form as $z=-g=-\frac{\partial}{\partial z}(gz)$ (13.11)

We can also write using the concept of φ: $-\frac{\partial u}{\partial t}=\frac{\partial^2\phi}{\partial x\partial t}$ and $-\frac{\partial w}{\partial t}=\frac{\partial^2\phi}{\partial z\partial t}$ (13.12)

Substituting equations (13.10), (13.11), (13.3) and (13.12) in (13.9) we get

$$-\frac{1}{\rho}\frac{\partial p}{\partial x}=-\frac{\partial^2\phi}{\partial x\partial t}+u\frac{\partial u}{\partial x}+w\frac{\partial w}{\partial x}$$
$$-\frac{\partial}{\partial z}(gz)-\frac{1}{\rho}\frac{\partial p}{\partial z}=-\frac{\partial^2\phi}{\partial z\partial t}+u\frac{\partial w}{\partial x}+w\frac{\partial w}{\partial z} \quad (13.13)$$

Re writing again using $u\frac{\partial u}{\partial x}=\frac{1}{2}\frac{\partial}{\partial x}(u^2)$ and $w\frac{\partial w}{\partial z}=\frac{1}{2}\frac{\partial}{\partial z}(w^2)$ & $\frac{1}{\rho}\frac{\partial p}{\partial x}=\frac{\partial}{\partial x}\left(\frac{p}{\rho}\right)$ and $\frac{1}{\rho}\frac{\partial p}{\partial z}=\frac{\partial}{\partial z}\left(\frac{p}{\rho}\right)$, Equations (13.13) turns out to be:

$$\frac{\partial}{\partial x}\left[-\frac{\partial\phi}{\partial t}+\frac{1}{2}\left(u^2+w^2\right)+\frac{p}{\rho}\right]=0$$
$$\frac{\partial}{\partial z}\left[-\frac{\partial\phi}{\partial t}+\frac{1}{2}\left(u^2+w^2\right)+\frac{p}{\rho}+gz\right]=0 \quad (13.14)$$

On integrating (13.14) we will get

$$\left[-\frac{\partial\phi}{\partial t}+\frac{1}{2}\left(u^2+w^2\right)+\frac{p}{\rho}\right]=F(x,t)$$
$$\left[-\frac{\partial\phi}{\partial t}+\frac{1}{2}\left(u^2+w^2\right)+\frac{p}{\rho}+gz\right]=F(z,t) \quad (13.15)$$

To find the constants F(x,t) and F(z,t) subtracting the equations of (13.15) one from the other then

F(z,t) – F(x,t) = gz or F(x,t) = F(z,t)-gz .....(13.16)

Replacing the constant of 13.16 in (13.15) we get only one equation i.e.,

$$\left[-\frac{\partial\phi}{\partial t}+\frac{1}{2}\left(u^2+w^2\right)+\frac{p}{\rho}+gz\right]=F(z,t) \quad (13.17)$$

Since the fluid motion is affected only by pressure gradients the constant F(z, t) is not a variable constant throughout the fluid at any time 't' and so can be taken as zero

$$\left[-\frac{\partial\phi}{\partial t}+\frac{1}{2}\left(u^2+w^2\right)+\frac{p}{\rho}+gz\right]=0 \quad (13.18)$$

This equation (13.18) is called generalized Bernoulli's equation

For steady flows which do not change with time at a fixed locality $\frac{\partial \phi}{\partial t} = 0$, So equation (13.17) reduces to $\left[\frac{1}{2}\left(u^2 + w^2\right) + \frac{p}{\rho} + gz\right] = constant$ ∴ (13.19)

This equation (13.19) is called the steady state Bernoulli's equation.

## Boundary conditions

Kinematic Boundary condition: At any boundary whether it is fixed (sea bottom) or free (the sea surface) the water is free to deform under the influence of any force. So certain physical conditions must be satisfied by the fluid velocities. These conditions on the water particle kinematics are called kinematic boundary condition. The physical interpretation here is at any surface or interface, there cannot be any flow across the surface, because it is a barrier.

Bottom Boundary condition (BBC): The lower boundary is described as z = - d for a two dimensional wave where the origin is located at the still water level and 'h' represents the depth. If the bottom is impermeable, the condition holds good is as below as the bottom does not move with time.

From equation (13.2) we know $w = -\frac{\partial \phi}{\partial z}$. This must be equal to zero on z = - d as vertical velocity ceases at the sea bottom.

$$\therefore w = -\frac{\partial \varphi}{\partial z} = 0 \quad (13.20)$$

Dynamic Free Surface Boundary Condition (DFSBC): This boundary condition can be easily fixed on a rigid surface but it cannot be fixed on a free surface like air-water interface because they cannot support variations in pressure. So a dynamic boundary condition is thus required on any free surface to prescribe the pressure distribution.

The condition for a dynamic free surface boundary is that the pressure on the free surface be uniform along the wave form which means it is constant and is equal to zero. So Bernoulli's equation (eqn.13.18) can be applied.

$$\left[-\frac{\partial \phi}{\partial t} + \frac{1}{2}\left(u^2 + w^2\right) + \frac{p}{\rho} + gz\right] = 0$$

As 'η' is the instantaneous elevation of free surface, here z = η because on the surface η is the surface of oscillation from the mean sea surface and the sea surface pressure starts from zero (p = 0). Small amplitude theory further based upon the assumption that all motions are so small that the above equation may be linearized. In particular this assumes that the squares of the velocity components $u^2$ and $w^2$ are negligible in compared to the remaining terms. Then the equation (13.18) reduces to give the boundary condition on the free surface as:

$$-\frac{\partial\varphi}{\partial t}+g(+\eta)=0$$

Or $\eta=\dfrac{1}{g}\dfrac{\partial\varphi}{\partial t}$ on z = 0 (13.21)

This means when there is no elevation also this boundary condition is valid but not η ≠ 0.

### 13.3.5. Wave Solutions

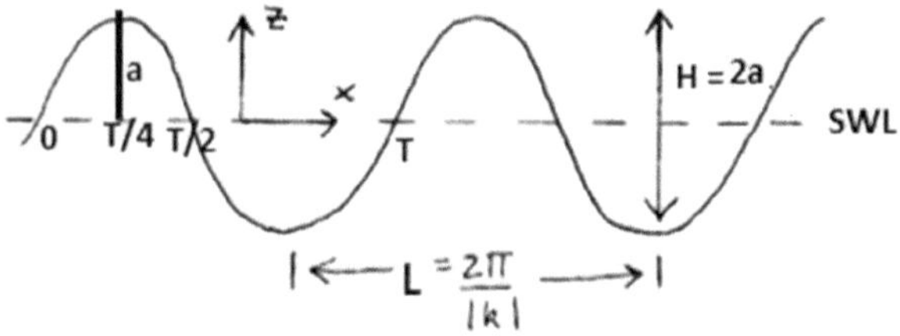


**Fig. 13.15.** Wave profile

As shown in Figure 13.15, the highest point of the wave is the *crest* and the lowest point is the *trough*. For linear or small-amplitude waves, the height of the crest above the still-water level (SWL) and the distance of the trough below the SWL are each equal to the wave amplitude *a*. Therefore *a* = *H/2*, where *H* = *the wave height*. The time interval between the passage of two successive wave crests or troughs at a given point is the *wave period T*. The *wavelength L* is the horizontal distance between two identical points on two successive wave crests or two successive wave troughs.

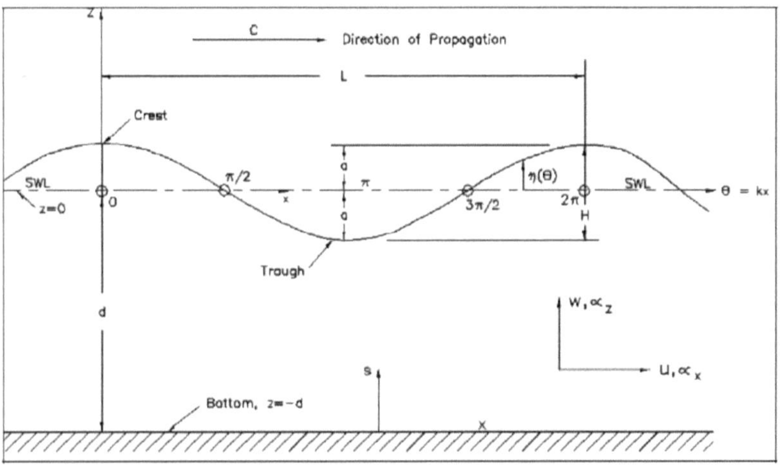


**Fig.13.16** A simple, progressive wave as it passes a fixed point

A progressive wave may be represented by the variables *x* (spatial) and *t* (temporal) or by their combination (phase), defined as $\theta = kx - \sigma t$, where *k* and $\sigma$ are wave number and frequency respectively. The values of $\theta$ vary between 0 and $2\pi$. A simple, periodic wave of permanent form propagating over a horizontal bottom may be completely characterized by the wave height *H* wavelength *L* and water depth *d* as shown in Fig.13.16.

Other wave parameters include $\sigma = 2\pi/T$ the *angular* or *radian frequency*, the *wave number* $k = 2\pi/L$, the *phase velocity* or *wave celerity* $C = L/T = \omega/k$, the *wave steepness* $\varepsilon = H/L$, the *relative depth* $d/L$, and the *relative wave height* $H/d$.

The governing second order differential equation for the fluid motion under a periodic two dimensional water wave is the Laplace equation (13.5), which holds throughout the fluid domain consisting of one wave.

$$\frac{\partial^2 \phi}{\partial x^2} + \frac{\partial^2 \phi}{\partial z^2} = 0$$

To solve this equation use the method of separation of variables. As the velocity potential (potential function) is a function of x, z and t, let suppose its solution is φ such that it is a function of x, z and t. Then we can write the velocity potential (φ) as:

$$\Phi\,(x,z,t) = X(x).Z(z).T(t) \qquad 13.22$$

Double differentiating (13.22) with respect to x and z and on substitution in it we can write

$$\frac{\partial^2 \varphi}{\partial x^2} + \frac{\partial^2 \varphi}{\partial z^2} = X''ZT + XZ''T = O$$

Dividing throughout by XZT and rewriting, we get

$$X''\,Z\,T = -\,X\,Z''\,T$$

$$\frac{X''}{X} = -\frac{Z''}{Z} = -K^2 \text{ (say)}$$

where K is some constant. Then we can write two differential equations as:

$$X'' + K^2 X = 0$$

$$Z'' - K^2 Z = 0$$

Then the general solution of the first one is X = A coskx + B sin kx and the second one is $Z = Ce^{kz} + De^{-kz}$. On substituting in( 13.22) and writing

$$\Phi(x,z,t) = (A \cos kx + B \sin kx).(Ce^{kz} + De^{-kz}).T(t)$$

Since the desired solutions are simple harmonic in time, it is reasonable to express T(t) as cos (σ t) or sin (σ t) where (σ = 2π/T)

$$\Phi(x,z,t) = (A \cos kx + B \sin kx).(Ce^{kz} + De^{-kz}).\,\mathrm{Cos}(\sigma t)$$

$$= (A \cos kx + B \sin kx).(Ce^{kz} + De^{-kz}).\,\mathrm{Sin}\,(\sigma t)$$

Expanding the terms and considering each one as solution, we can write as

$$\phi_1 = \left(Ce^{kz} + De^{-kz}\right)(A\cos kx.cos\sigma t)$$
$$\phi_2 = \left(Ce^{kz} + De^{-kz}\right)(B\sin kx.sin\sigma t)$$
$$\phi_3 = \left(Ce^{kz} + De^{-kz}\right)(B\sin kx.cos\sigma t)$$
$$\phi_4 = \left(Ce^{kz} + De^{-kz}\right)(A\cos kx.sin\sigma t) \tag{13.23}$$

These four separate elementary combinations of the terms which are periodic in x and t and themselves are solutions to Laplace's equation. To find the constants A,B,C,D, we have to use the boundary conditions. Let us consider first ϕ1 (eqn 13.23) and try to find the constants so that in a similar fashion rest other potential functions ϕ2, ϕ3 and ϕ4 can be determined.

For applying the bottom boundary condition (equation 13.20) to equation (13.23), differentiate equation 13.23 with respect to z then we get

$$\frac{\partial \phi_1}{\partial z} = k\left(Ce^{kz} + De^{-kz}\right)(A\cos kx.cos\sigma t)$$

According to equation 13.20 on z = - d, $\dfrac{\partial \phi_1}{\partial z} = 0$

$$\therefore \frac{\partial \phi_1}{\partial z} = k\left(Ce^{-kd} - De^{kd}\right)(A_1 \cos kx.cos\sigma t) = 0$$

This means that as A1 and k cannot be equal to zero, $\left(Ce^{-kd} - De^{kd}\right)$ must be equal to zero

$$\left(Ce^{-kd} - De^{kd}\right) = 0 \quad \text{or } C = De^{2kd}$$

Substituting this value of C in equation 13.23 after much manipulation we can write

$$\phi = (A\cos kx.cos\sigma t)\left(De^{\ kd}e^{kz} + De^{\ kz}\right)$$

$$A coskx.\cos \quad t\left[2\frac{De^{kd}e^{k(d+z)} \quad De^{kd}e^{-k(d+z)}}{}\right]$$

$$= 2ADe^{kd}coshk\left(d+z\right).\cos kx.\cos \sigma t \tag{13.24}$$

To find the constant A in equation 13.24, apply the surface boundary condition (equation 13.21). To do that differentiate equation 13.24 with respect to 't' then

$$\frac{\partial \phi_1}{\partial t} = -2A\sigma De^{kd}coshk\left(d+z\right).\cos kx.\sin \sigma t$$

According to equation (13.21) on Z = 0, $\dfrac{1}{g}\dfrac{\partial \phi_1}{\partial t} = \eta$

$$\therefore \frac{1}{g}\frac{\partial \phi_1}{\partial t} = \eta = -\frac{2A\sigma}{g} De^{kd} coshk\left(d\right).\cos kx.\sin \sigma t \tag{13.25}$$

$$\eta = a\sin\left(\sigma t - kx\right) \tag{13.25a}$$

We know from equation (13.25a) that η = a when cos kx.sin σt = 1. Using this condition, we can write equation 13.25 as:

$a = -\frac{2\sigma A}{g}$ D $e^{kd}$Coshk(d). Or 2 A D $e^{kd}$ = $-\frac{ag}{\sigma \cosh kd}$ substituting this value in equation (13.24)

$\Phi_1 = -\frac{ag}{\sigma \cosh kd}$ Cosh k(d +z). Cos kx. Cos σ t

Here the constant $k = \frac{2\pi}{L}$ is wave number and $\sigma = \frac{2\pi}{T}$ is wave radian frequency.

Similarly if all the other φ's in equation 13.23 are evaluated we get

$\Phi_2 = +\frac{ag}{\sigma \cosh kd}$ Cosh k(d +z). Sin kx. Sinσ t

$\Phi_3 = -\frac{ag}{\sigma \cosh kd}$ Cosh k(h +z). Sin kx. Cos σ t

$\Phi_4 = +\frac{ag}{\sigma \cosh kd}$ Cosh k(h +z). Cos kx. Sinσ t

As mentioned earlier these elementary solutions may be combined linearly and due to the linearity of Laplace equation the combination in various ways or with one another will also be a solution to the problem. Suppose the subtraction of $\varphi_1$ from $\varphi_2$ gives rise to:

$\Phi = (\varphi_2 - \varphi_1) = +\frac{ag}{\sigma \cosh kd}$ Cosh k(d +z).Cos (k x – σ t) (13.26a)

If $(\varphi_4 - \varphi_3)$ is used we can get

$\Phi = (\varphi_4 - \varphi_3) = +\frac{ag}{\sigma \cosh kd}$ Cosh k(d +z).Sin (k x – σ t) 13.26b

Either of the equations can be used for potential function representing the wave solution. The wave parameters of wave celerity, Length and period can be extracted from this potential function

Where K = $\frac{2\pi}{L}$ $and\ \sigma = \frac{2\pi}{T}$

**Wave celerity, length and period:**

The rate of rise of water surface can be written as: $w = \frac{d\eta}{dt} = \frac{\partial \eta}{\partial t} + u\frac{\partial \eta}{\partial x} + w\frac{\partial \eta}{\partial z}$

Since we are dealing with waves of very small amplitude, the surface slope elevation $\frac{\partial \eta}{\partial x}$ is negligibly small and so can be neglected. On Z = 0, $\frac{\partial \eta}{\partial z}$ does not exist and so

$$w = \frac{\partial \eta}{\partial t} \text{ on } z = 0$$

Relating this with the dynamic free surface boundary condition (equation 13.21) and the velocity potential (equation 13.2) we can write:

$\eta = \frac{1}{g}\frac{\partial \varphi}{\partial t} \Rightarrow \frac{\partial \eta}{\partial t} = \frac{1}{g}\frac{\partial^2 \varphi}{\partial t^2} = w$, on z = 0, We know w = $-\frac{\partial \phi}{\partial z}$ on z = 0

$$\therefore -\frac{\partial \phi}{\partial z} = \frac{1}{g}\frac{\partial^2 \varphi}{\partial t^2} \text{ on } z = 0 \tag{13.27}$$

To Apply equation (13.27) on to the equation (13.26) we have to partially double differentiate 13.26 with respect to 't' and then partially' differentiate 13.26 with respect to 'z' and then equate and simplify.

Double differentiation of 13.26 with respect to 't':

$$\frac{1}{g}\frac{\partial^2 \phi}{\partial t^2} = -a\sigma \frac{Coshk(d+z)}{Coshkd} Cos(kx - \sigma t) \tag{13.28}$$

Differentiating 13.26 with respect to 'z' gives rise to:

$$-\frac{\partial \phi}{\partial z} = -kag \frac{Sinhk(d+z)}{\sigma Coshkd} Cos(kx - \sigma t) \tag{13.29}$$

Equating (13.28) and (13.29) using the relation 13.27, on simplification we get:

$\sigma^2 = gk \tanh k(d+z)$ on z =0 this tends to

$$\sigma^2 = gk \tanh(kd) \Rightarrow gk \tanh(r) \tag{13.30}$$

Or $\left(\frac{2\pi}{T}\right)^2 = \frac{2\pi g}{L} \tanh\left(\frac{2\pi d}{L}\right)$

where kh = r is the asymptote of the hyperbolic wave profile.

We know the equation of free moving surface of wave profile as:

$$\eta = aSin(kx - \sigma t) = \frac{H}{2} Sin(\frac{2\pi}{L}x - \frac{2\pi}{T}t) \tag{13.30a}$$

where, a = amplitude which is half of the wave height, H. Let suppose Kx – σ t is constant, then Kdx – σ dt = 0 or Kdx = σ dt, $\Rightarrow \frac{dx}{dt} = \frac{\sigma}{k} = \frac{2\pi/T}{2\pi/L} = \frac{L}{T} = c \therefore c = \frac{\sigma}{k} \Rightarrow \sigma^2 = k^2c^2$, substituting the value of $\sigma^2$ from equation (13.30) we get

$$c^2 = \frac{g}{k}\tanh(kd)$$

$$\Rightarrow \frac{L^2}{T^2} = \frac{gL}{2\pi}\tanh(kd) \Rightarrow \frac{L}{T} = \frac{gT}{2\pi}\tanh(kd) = c \qquad 13.31$$

Similarly, $L = \frac{gT^2}{2\pi}\tanh(kd) = \frac{gT^2}{2\pi}\tanh\left(\frac{2\pi d}{L}\right)$

Use of this equation involves some difficulty since the unknown $L$ appears on both sides of the equation. Tabulated values of d/L and d/$L_0$ (SPM 1984) where $L_0$ is the deep water wavelength may be used to simplify the solution of this Equation. Eckart (1952) gives an approximate expression for the above Equation, which is correct to within about 10 percent. This expression is given by

$$L = \frac{gT^2}{2\pi}\tanh\left(\frac{4\pi^2 d}{T^2 g}\right)$$

This Equation explicitly gives $L$ in terms of wave period $T$ and is sufficiently accurate for many engineering calculations. The maximum error 10 percent occurs when d/L =1/2.

### 13.3.6. Classification of waves according to relative depth:

As the wave approaches from deep water to shallow water, the wave profile changes as shown in Fig.13.20. Ocean waves are generally classified according to a parameter called 'relative depth'. It is the ratio of the water depth (d) to the wave length (L). This ratio d/L is called the 'relative depth'.

The table 13.2 and Fig.13.17 gives Classifications made according to the magnitude of *d/L* and the resulting limiting values taken by the function *tanh (2πd/L)*. Note that as the argument of the hyperbolic tangent r = kd = 2πd/L gets large, then tanh (r) approaches 1, and for small values of kd, tanh (kd) tends to kd.

**Table 13.2.** Wave classification according to relative depth

| Range of d/L | Range of r = Kd=2πd/L | Types of waves |
|---|---|---|
| 0 to 1/20 | 0 to π/10 | Shallow water waves (Long waves) |
| 1/20 to ½ | π/10 to π | Intermediate waves |
| ½ to ∞ | π to ∞ | Deep water waves ( short waves) |

The simplifications which occur in the various wave equations arise by replacing the hyperbolic functions by their asymptotes for the particular range of relative depth are listed below in the table 13.3.

**Table 13.3.** Approximations for hyperbolic functions

| Function | Asymptotes (r = Kd) | |
|---|---|---|
| | Shallow water | Deep water |
| Sinh r | r | $\frac{e^r}{2}$ |
| Cosh r | 1 | $\frac{e^r}{2}$ |
| Tanh r | r | 1 |

The deep and shallow water expressions can be obtained by using the simplifications resulted due to the asymptotes as given in table 13.3 and Fig.13.19.

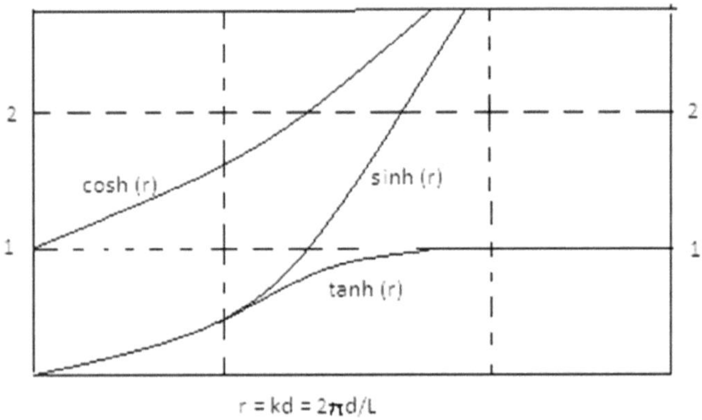


**Fig.13.17.** Asymptotic variation for shallow and deep water relative depths ( $r = kd = \frac{2\pi d}{L}$ )

**Shallow water wave equation:**

The shallow water condition from table 13.2 is tanh(kd) → kd, So the equation 13.31 turns out to

$$C^2 = gd \tag{13.32}$$

This equation implies that wave velocity is independent of wave length and depends only on the water depth.

**Deep water wave equation:**

The deep water condition from table 13.2 is tanh(kd) → 1 and so equation 13.31 turns out to be

$C_o^2 = g/k_o$ as we know $k_o = 2\pi/L_o$; $\quad C_0 = \frac{gT}{2\pi} \quad L_0 = \frac{gT^2}{2\pi} = C_o T \qquad$ 13.33

The suffix zero indicates the deep water conditions. The celerity and wave length for deep water conditions are therefore independent of water depth. One should note here that 'T' doesn't vary with the local depth of water.

If $g = 9.8\ m/s^2$ and $\pi = 3.14$ then $\frac{g}{2\pi} = 1.56$

then $C_0 = 1.56T$ and $L_0 = 1.56\ T^2$ (13.34)

Comparison of deep and shallow water wave velocities:

Shallow water wave velocity from equation (13.31)= $C = \frac{gT}{2\pi}\tanh(kd)$

Comparing equations (13.31) and (13.33): $\frac{C}{C_0} = \frac{L}{L_0} = \tanh(kd) = \tanh\left(\frac{2\pi d}{L}\right)$ (13.34a)

If both sides of Equation 13.34a are multiplied by d/L, it becomes:

$$\frac{d}{L_0} = \frac{d}{L}\tanh(kd) = \frac{d}{L}\tanh\left(\frac{2\pi d}{L}\right)$$

The terms $d/L_o$ and $d/L$ and other useful functions such as $kd = 2\pi d/L$ and *tanh (kd)* have been tabulated by Wiegel (1954) as a function of $d/Lo$ (see SPM 1984).

These functions simplify the solution of wave problems described by the linear theory.

### 13.3.7. Orbital motion of wave particles in a progressive wave

The velocity potential for a small amplitude progressive wave traveling in the positive x direction in an inviscid fluid is (Equation 13.26)

$$\Phi = (\varphi_2 - \varphi_1) = +\frac{ag}{\sigma \cosh kd}\ \text{Cosh}\ k(d+z).\text{Cos}\ (k\,x - \sigma\, t)$$

By definition of velocity potential, the horizontal and vertical components of velocity potential are given in equations 13.2 as

$$u = -\frac{\partial \phi}{\partial x}, w = -\frac{\partial \phi}{\partial z}$$

### Local fluid velocities

In wave force studies, it is often desirable to know the local fluid velocities and accelerations for various values of z and t during the passage of a wave. The horizontal component u and the vertical component w of the local fluid velocity are obtained in the following manner:

So differentiating partially equation 13.26b with respect to x and z we get 'u' and 'w' as

$$\Phi = (\varphi_4 - \varphi_3) = +\frac{ag}{\sigma \cosh kd}\ \text{Cosh}\ k(d+z).\text{Sin}\ (k\,x - \sigma\, t) \quad 13.26b$$

$$u = +\frac{agk}{\sigma \cosh kd}\ \text{Coshk}(d+z).\cos\ (k\,x - \sigma\, t) \quad (13.35)$$

$$\frac{\pi H}{T}\frac{Coshk(d+z)}{Coshkd}\cos(kx-\sigma t)(\because L=\frac{gT^2}{2\pi},K=\frac{2\pi}{L},\sigma=\frac{2\pi}{T},a=\frac{H}{2}and$$

$$\frac{agk}{\sigma}=\frac{\pi H}{T}=\frac{Hg}{2}\frac{2\pi}{L}\frac{T}{2\pi}=\frac{HgT}{2L}$$

$$=\text{u}=\frac{H}{2}\frac{gT}{L}\frac{\cosh\frac{2\pi}{L}(d+z)}{\cosh\frac{2\pi d}{L}}\cos\left(\frac{2\pi x}{L}-\frac{2\pi t}{T}\right) \quad (13.35)$$

$$u=\frac{H}{2}\frac{gT}{L}\frac{\cosh\frac{2\pi}{L}(d+z)}{\cosh\frac{2\pi d}{L}}\cos\theta,\ where;;\theta=\left(\frac{2\pi x}{L}-\frac{2\pi t}{T}\right)$$

Differentiating equation (13.26b) with respect to Z

$$\text{w}=\frac{agk}{\sigma}\frac{Sinhk(d+z)}{Coshkd}\sin(kx-\sigma t)=\frac{\pi H}{T}\frac{Sinhk(d+z)}{Coshkd}\sin(kx-\sigma t)$$

$$=\frac{H}{2}\frac{gT}{L}\frac{\sinh\frac{2\pi}{L}(d+z)}{\cosh\frac{2\pi d}{L}}\sin\left(\frac{2\pi x}{L}-\frac{2\pi t}{T}\right)=\rightarrow\frac{H}{2}\frac{gT}{L}\frac{\sinh\frac{2\pi}{L}(d+z)}{\cosh\frac{2\pi d}{L}}\sin\theta \quad (13.36)$$

These equations express the local fluid velocity components at any distance (z + d) above the bottom. The velocities are harmonic in both x and t. For a given value of the phase angle $\theta=\left(\frac{2\pi x}{L}-\frac{2\pi t}{T}\right)$ the hyperbolic functions of Z (i.e.cosh and sinh), result in an approximate exponential decay of the magnitude of velocity components with increasing distance below the free surface as shown in the Fig.13.18. The maximum positive horizontal velocity occurs when $\theta = 0, 2\pi$ etc. while the maximum horizontal velocity in the negative direction occurs when $\theta = \pi, 3\pi$ etc. On the other hand, the maximum positive vertical velocity occurs when $\theta = \frac{\pi}{2},\frac{5\pi}{2}$ etc. and the maximum vertical velocity in the negative direction occurs when $\theta=\frac{3\pi}{2},\frac{7\pi}{2}$ etc. As the depth $z=-\frac{L}{2}$ is reached, the velocities become negligibly small and below this depth $z=-\frac{L}{2}$ there is essentially no motion.

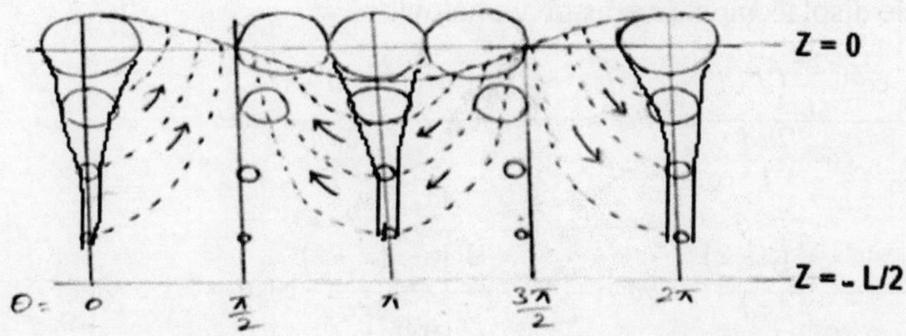


**Fig. 13.18.** Exponential decay of the water particle motions from surface to below. The dashed lines are the streamlines, i.e., lines everywhere parallel to the flow. The direction and length of the arrows indicate direction and velocity of orbital motion

## The particle displacements and paths of orbits: Displacement of individual water particles

Another important aspect of linear wave theory deals with the displacement of individual water particles within the wave. Water particles generally move in elliptical paths in shallow and intermediate (transitional) depth water and in circular paths in deep water (Fig. 13.20). If the mean particle position is considered to be at the center of the ellipse or circle, then vertical particle displacement with respect to the mean position cannot exceed one-half the wave height. Thus, since the wave height is assumed to be small, the displacement of any fluid particle from its mean position must be small. Integration of Equations (13.35) and (13.36) gives the horizontal and vertical particle displacements from the mean position, respectively (Figure 13.19). The horizontal and vertical particle displacements from the mean positions are denoted as $\zeta$ and $\epsilon$ respectively as shown in Fig.13.19.

The velocity field during the period "T' can be obtained by integrating the equations (13.35 and 13.36).

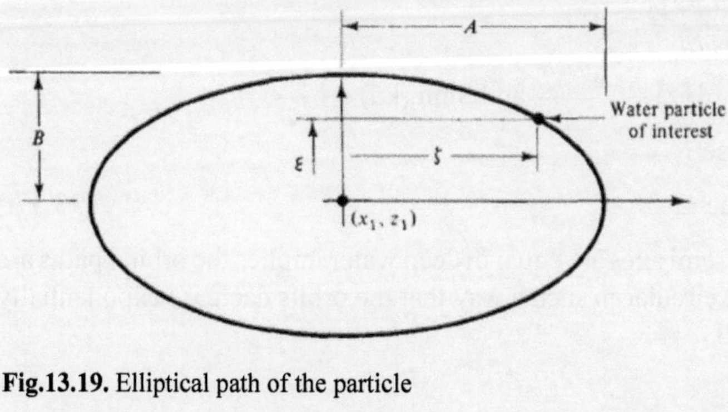


**Fig.13.19.** Elliptical path of the particle

The Fluid particle displacements are as given below:

$$\zeta = \int udt = \int \frac{H}{2}\frac{gT}{L}\frac{\cosh\frac{2\pi}{L}(d+z)}{\cosh\frac{2\pi d}{L}}\cos\theta dt = -\frac{H}{4\pi}\frac{gT^2}{L}\frac{\cosh\frac{2\pi}{L}(d+z)}{\cosh\frac{2\pi d}{L}}\sin\theta$$

$$\varepsilon = \int wdt = \int \frac{H}{2}\frac{gT}{L}\frac{\sinh\frac{2\pi}{L}(d+z)}{\cosh\frac{2\pi d}{L}}\sin\theta dt = \frac{H}{4\pi}\frac{gT^2}{L}\frac{\sinh\frac{2\pi}{L}(d+z)}{\cosh\frac{2\pi d}{L}}\cos\theta$$

We know from equation (13.30), $\sigma^2 = gk.\tanh(kd)$, on substituting we get

$$\therefore\ \zeta = a\frac{Coshk(d+z)}{Sinhkd}\ \sin(k\,x\text{-}\sigma t)$$

$$\varepsilon = a\frac{Sinhk(d+z)}{Sinhkd}\ \cos(k\,x\text{-}\sigma t)$$

Put $A = a\frac{Coshk(d+z)}{Sinhkd}$ and $B = a\frac{Sinhk(d+z)}{Sinhkd}$ 13.37

then we can write: $\frac{\zeta^2}{A^2} + \frac{\varepsilon^2}{B^2} = 1$

This is the equation of an ellipse with a major semi-axis of 'A' and a minor semi-axis of 'B'. Hence the particle paths, in general, are elliptical in shape. The specific form of the particle paths for shallow and deep water conditions can be determined by examining their axes A and B.

## Deep water case

$A = a\frac{Coshk(d+z)}{Sinhkd}$,

For deep water from table 13.2, Coshk(d+z) $\rightarrow \frac{e^{k(d+z)}}{2}$ and Sinh (kd) $\rightarrow \frac{e^{kd}}{2}$

On substitution, $A = ae^{kd}$

Similarly, $B = a\frac{Sinhk(d+z)}{Sinhkd}$

for deep water Sinh k(d+z) $\rightarrow \frac{e^{k(d+z)}}{2}$ and Sinh (kd) $\rightarrow \frac{e^{kd}}{2}$

On substitution, $B = ae^{kd}$

This means $A = B = ae^{kd}$ (13.37i)

The major and minor semi axes are equal in deep water implies the orbital paths are no more elliptical but circular in such a way that the orbits decrease exponentially as shown in Fig.13.20

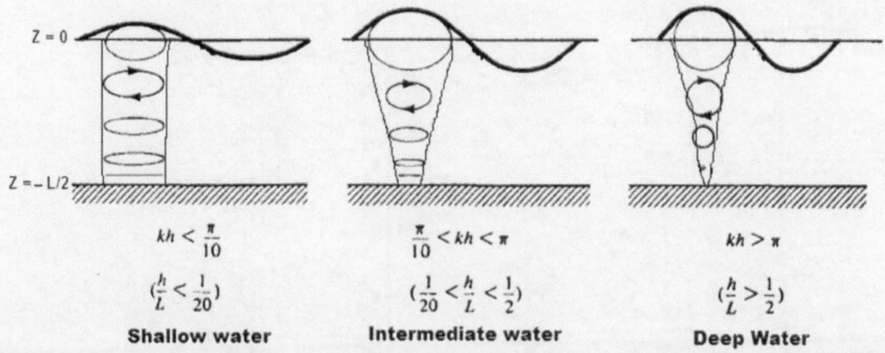


**Fig.13.20.** Orbital motion of water particle motion in shallow, intermediate and deep water waves. Note that at the bottom the particle orbit is a line in the case of shallow and intermediate water waves and a point (dot) in the case of deep water waves. (Varma, 2014) (https://www.ias.ac.in/article/fulltext/reso/019/11/1047-1057)

The individual water particles move in closed circular orbits with radii ($ae^{kd}$) decreasing exponentially with depth below the sea surface. The Fig.13.20 depicts the orbits, instantaneous orbital velocities and streamlines (dashed lines) in a deep water wave. Just ahead of the crest there is an area of convergence and just behind the crest an area of divergence. Hence the surface rises ahead of the crest and falls behind it. This qualitatively describes the progress of the wave.

## Shallow water case

The values of A and B for shallow water condition can be obtained by applying the condition of table 13.2 to the equations (13.37).

$$A = a\frac{Coshk(d+z)}{Sinhkd},$$

For shallow water from table 13.2 Cosh k(d+z) → 1 and Sinh (kd) → kd

$$\therefore A = \frac{a}{kd}$$

$B = a\frac{Sinhk(d+z)}{Sinhkd}$ for shallow water Sinhk(d+z) → k (d+z) and Sinh (kd) → kd

$$\therefore B = \frac{a(d+z)}{d}$$

This means the particle paths of shallow water waves are *elliptical* with a major semi-axis of A = a/kd and a minor semi-axis of B = a(d+z)/d as shown in Fig.13.19. The Fig 13.21 depicts the orbits, instantaneous orbital velocities and streamlines (dashed lines) in a shallow water wave. The same condition is applicable to Intermediate waves or transitional waves also (Fig. 13.20).

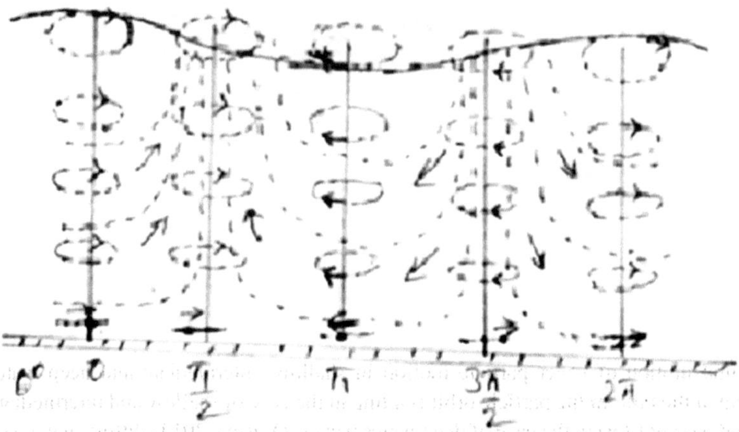

**Fig. 13.21.** Orbital motion of water particle motion in shallow water waves. Dashed curves are stream lines of water particle motion in shallow water waves.

**Table 13.4**. Differences between deep (short) and shallow (long) water waves

| S.No | Character of the wave | Surface waves (short waves) | Long waves (shallow water waves) |
|---|---|---|---|
| 1. | Velocity | i). Progressive, standing, free or forced waves<br>ii). Dependent on wave Length $C^2 = \frac{gL}{2\pi}$ | i). Progressive, standing, free or forced waves<br>ii). Dependent on water depth $C^2 = gd$ |
| 2. | Movement of water Particles in a Vertical plane | Orbital motion of water particles is circular. The Radii of theses circles decrease with depth. Motion is imperceptible at a depth equal to half of its wave length. | The motion is in ellipses. While the size of the ellipses decrease with depth and flattened, at a depth of half the wave length the ellipses become so flat like a line and so particles move back and forth. |
| 3. | Pressure distribution | Below the depth of perceptible motion of water particles the pressure is not influenced by the wave | The wave influences the pressure distribution in the same manner at all depths. |
| 4. | Influence of earth's rotation | Negligible | Cannot be neglected if the period of wave approaches the period of earth's rotation. |

## Local fluid velocities

In wave force studies, it is often desirable to know the local fluid velocities and accelerations for various values of z and t during the passage of a wave. The horizontal component u and the vertical component w of the local fluid velocity are given by the following equations:

$$u = \frac{H}{2}\frac{gT}{L}\frac{\cosh\frac{2\pi}{L}(d+z)}{\cosh\frac{2\pi d}{L}}\cos\left(\frac{2\pi x}{L} - \frac{2\pi t}{T}\right)$$

$$u = \frac{H}{2}\frac{gT}{L}\frac{\cosh\frac{2\pi}{L}(d+z)}{\cosh\frac{2\pi d}{L}}\cos\theta$$

Substitution of the equations for A and B of shallow water into Eqs. (13.35) and (13.36) respectively yields (after some algebraic manipulation) the following equations for water particle velocity in shallow water:

$$u = \frac{H}{2}\sqrt{\frac{g}{d}}\cos(kx - \sigma t) \tag{13.37a}$$

Similarly $w = \frac{H}{2}\frac{gT}{L}\frac{\sinh\frac{2\pi}{L}(d+z)}{\cosh\frac{2\pi d}{L}}\sin\left(\frac{2\pi x}{L} - \frac{2\pi t}{T}\right)$

Similarly 'w' can be written as:

$$w = \frac{\pi H}{T}\left(1 + \frac{z}{d}\right)\sin(kx - \sigma t) \tag{13.37b}$$

These equations express the local fluid velocity components at any distance (z + d) above the bottom. The velocities are harmonic in both x and t. For a given value of the phase angle

$$\theta = \left(\frac{2\pi x}{L} - \frac{2\pi t}{T}\right)$$

The local fluid particle accelerations are obtained from the above equations (13.37a &b) by differentiating each equation with respect to t. Thus,

$$a_x = \frac{du}{dt} = \frac{g\pi H}{L}\frac{\cosh\left[\frac{2\pi}{L}(d+Z)\right]}{\cosh\left(\frac{2\pi d}{L}\right)}\sin\left(\frac{2\pi x}{L} - \frac{2\pi t}{T}\right) \tag{13.37c}$$

$$a_z = \frac{dw}{dt} = -\frac{g\pi H}{L}\frac{\sinh\left[\frac{2\pi}{L}(d+Z)\right]}{\cosh\left(\frac{2\pi d}{L}\right)}\cos\left(\frac{2\pi x}{L} - \frac{2\pi t}{T}\right) \tag{13.37d}$$

Positive and negative values of the horizontal and vertical fluid accelerations for various values of $\theta = \left(\frac{2\pi x}{L} - \frac{2\pi t}{T}\right)$ are shown in the Figure 13.22. It is a sketch of the local fluid motion and indicates that the fluid under the crest moves in the direction of wave propagation and returns during passage of the trough. Linear theory does not predict any net mass transport; hence, the sketch shows only an oscillatory fluid motion.

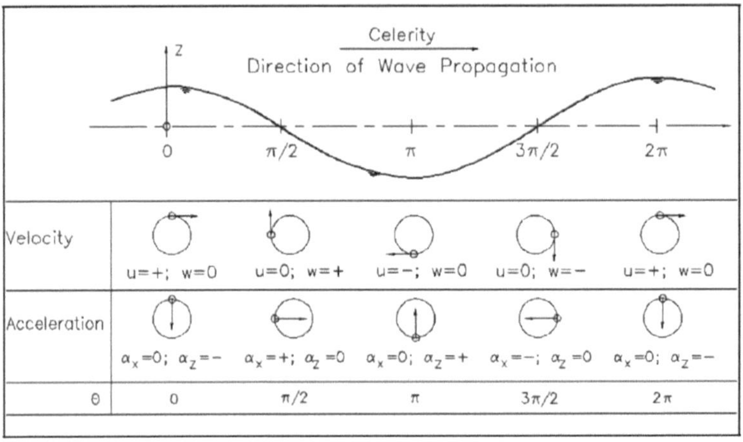


**Fig.13.22.** Local fluid velocities and accelerations (SPM,1984)

## Displacement of individual water particles

Another important aspect of linear wave theory deals with the displacement of individual water particles within the wave. Water particles generally move in elliptical paths in shallow or transitional depth water and in circular paths in deep water (Figure 13.20). If the mean particle position is considered to be at the center of the ellipse or circle, then vertical particle displacement with respect to the mean position cannot exceed one-half the wave height. Thus, since the wave height is assumed to be small, the displacement of any fluid particle from its mean position must be small. Integration of Equations (13.37a) and (13.37b) gives the horizontal and vertical particle displacements from the mean position, respectively (Figure 13.19) are given below.

The Fluid particle displacements are as given below:

$$\zeta = \int u dt = \int \frac{H}{2}\frac{gT}{L}\frac{\cosh\frac{2\pi}{L}(d+z)}{\cosh\frac{2\pi d}{L}}\cos\theta dt = -\frac{H}{4\pi}\frac{gT^2}{L}\frac{\cosh\frac{2\pi}{L}(d+z)}{\cosh\frac{2\pi d}{L}}\sin\theta \tag{13.37e}$$

$$\in = \int w dt = \int \frac{H}{2}\frac{gT}{L}\frac{\sinh\frac{2\pi}{L}(d+z)}{\cosh\frac{2\pi d}{L}}\sin\theta dt = +\frac{H}{4\pi}\frac{gT^2}{L}\frac{\sinh\frac{2\pi}{L}(d+z)}{\cosh\frac{2\pi d}{L}}\cos\theta \tag{13.37f}$$

### 13.3.8.Group velocity

Consider two waves moving in the positive x direction whose periods are different. As the wave advances from left to right, each wave moves through the group to die out at the front as new waves form at the rear. In this process, the distance travelled by each individual wave as it moves from the rear to front of the group is twice that travelled by the group as a whole. Hence, the wave speed is twice that of the group speed. Wave energy is contained within each group and advances with the group speed. Thus a wave profile is not a single wave but it is a composite of many waves. That means in the ocean a mixture of many waves together travel. This mixture of wave profile is called a spectrum or band as shown in Fig. 13.23 below. So a single wave has phase velocity (C) and the spectrum of waves has group velocity ($C_g$). This group velocity ($C_g$) in shallow water is half of the phase velocity (C).

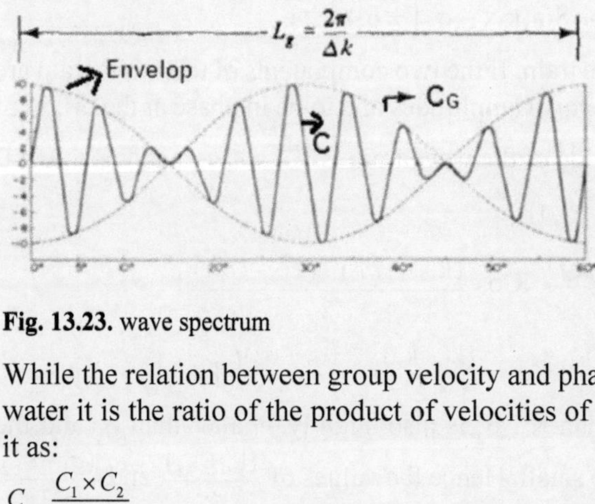


**Fig. 13.23.** wave spectrum

While the relation between group velocity and phase velocity is $C_g$ = ½C, in deep water it is the ratio of the product of velocities of individual waves to the sum of it as:

$$C_g = \frac{C_1 \times C_2}{C_1 + C_2}$$

Where $C_1$ and $C_2$ are phase velocities of two individual waves. From the above expression it is clear that the group velocity is half the phase velocity and $\frac{cg}{c}$ can be seen to have the value ½ in deep water and approach to a limiting value of unity in shallow water. As water becomes shallower, wave length becomes less important and wave speed depends only on water depth. As a result wave speed in shoaling water becomes closer to group velocity. Eventually at depths less than L/20, all waves approaching a particular depth travel at the same speed. So the group velocity and phase velocity at that depth are equal. The detailed derivation is as below.

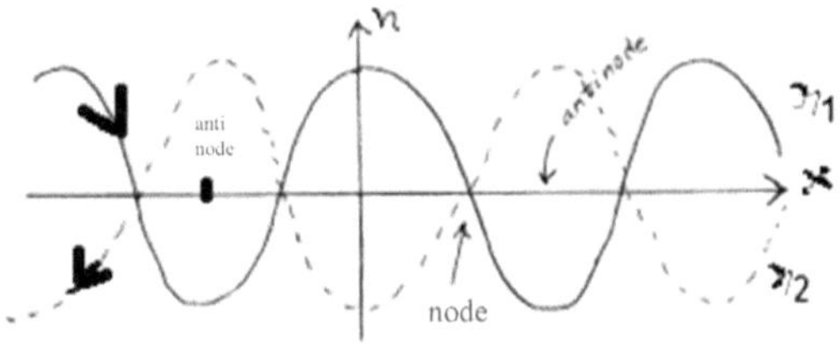


**Fig.13.24.** Wave groups produced by two harmonic waves of nearly identical frequencies.

The composite wave profile can be computed from the Fig.13.24 using equation (13.25a) as:

Then $\eta_1 = a_1 \text{Sin}(k_1x - \sigma_1 t + \delta_1)$ and $\eta_2 = a_2 \text{Sin}(k_2x - \sigma_2 t + \delta_2)$

Then the composite wave is $\eta_T = \eta_1 + \eta_2$

So $\eta_T = a_1 \text{Sin}(k_1x - \sigma_1 t + \delta_1) + a_2 \text{Sin}(k_2x - \sigma_2 t + \delta_2)$

This equation represents a wave train. If the two components of this wave train are assumed for simplicity to have equal amplitudes and to be in phase at the origin of (kx-t), then the average amplitude of the wave train is $\frac{a1 + a2}{2} = a$ and $\delta_1 = \delta_2 = 0$

$\eta_T = a\{\text{Sin}(k_1x - \sigma_1 t) + \text{Sin}(k_2x - \sigma_2 t)\}$

$$\eta_T = 2a\text{Sin}\{\frac{(k_1x - \sigma_1 t) + (k_2x - \sigma_2 t)}{2} .\text{Cos}\ \frac{(k_1x - \sigma_1 t) - (k_2x - \sigma_2 t)}{2}\}$$

$$= 2a\{\text{Sin}[\frac{(k_1 + k_2)x}{2} - \frac{(\sigma_1 + \sigma_2)t)}{2}].\text{Cos} - \frac{(k_1 - k_2)x}{2} - \frac{(\sigma_1 - \sigma_2)t}{2}\}$$

When $K_2$ is slightly smaller than $K_1$, $\sigma_2$ is also slightly smaller than $\sigma_1$ and the difference between each is also small. Hence the values of $\frac{(k_1 + k_2)}{2}$ and $\frac{(\sigma_1 + \sigma_2)}{2}$ are essentially the wave number and wave radian frequency of the composite wave. The cosine term represents a slowly varying term in the above equation which only modulates the sine term.

We can see that this composite wave takes the form of a series of Sine waves the amplitude of which varies slowly between 0 and 2a according to the Cosine factor. The points of zero amplitude of the wave envelope thus separate groups of individual waves. These nodal points are located by finding the zeros of the cosine factor

$\text{Cos}\ (2m+1)\frac{\pi}{2} = 0$ which implies $\eta_T = 0$

$\text{Cos}\ \frac{(k_1 - k_2)x}{2} - \frac{(\sigma_1 - \sigma_2)t}{2} = \text{Cos}\ (2m+1)\frac{\pi}{2} = 0$ which means

$$\frac{(k_1 - k_2)x}{2} - \frac{(\sigma_1 - \sigma_2)t}{2} = (2m+1)\frac{\pi}{2}$$

Or $x (k_1 - k_2) = (2m + 1) \pi + (\sigma_1 - \sigma_2) t$

Or $X_{node} = \frac{(\sigma_1 - \sigma_2)t}{(k_1 - k_2)} + \frac{(2m+1)\pi}{(k_1 - k_2)}$

Since the position of all nodes is a function of time, they are not stationary. At t = 0 there will be nodes at (putting m = 0,1,2,….)

$X = \frac{\pi}{(k_1 - k_2)}, \frac{3\pi}{(k_1 - k_2)}, \frac{5\pi}{(k_1 - k_2)}$ , ……..

Thus the distance between one node to another is $dx = \frac{2\pi}{(k_1 - k_2)} = \frac{L_1 L_2}{(L_2 - L_1)}$ { since $K_1 = \frac{2\pi}{L_1}$ and $K_2 = \frac{2\pi}{L_2}$ }.

The speed of propagation of the nodes associated with this wave group is given by

Group velocity $C_g = \frac{dXnode}{dt} = \frac{(\sigma_1 - \sigma_2)}{(k_1 - k_2)} = \frac{d\sigma}{dk}$

**Derivation of groupveloety:** $C_g = \frac{d\sigma}{dk} = \frac{d}{dk}(Ck)$ {since $C = \frac{\sigma}{k}$ }

So $C_g = \frac{d}{dk}(Ck) = \frac{d}{dk} k\sqrt{\frac{g}{k}\tanh kd} = \frac{d}{dk}\sqrt{gk\tanh kd} = \frac{d}{dk}[gk\tanh kd]^{-½}$

On differentiating

$= \frac{1}{2}[gk\tanh kd]^{-½}$ {gkd sech²(kd) + g tanh (kd) .1}

$= \frac{1}{2}\frac{g}{\sqrt{gk\tanh kd}}\{ kd.\frac{1}{Cosh^2 kd} + \frac{Sinh\, kd}{Cosh\ kd}\}$ on further simplification of dividing and subtracting by √k we can write

$= \frac{1}{2}\frac{g}{\frac{\sqrt{k}}{\sqrt{k}}\sqrt{gk\tanh kd}}\{ kd.\frac{1}{Cosh^2 kd} + \frac{Sinh\, kd}{Cosh\ kd}\}$

$= \frac{1}{2}\frac{g}{k\sqrt{\frac{g}{k}\tanh kd}}\{ kd.\frac{1}{Cosh^2 kd} + \frac{Sinh\, kd}{Cosh\ kd}\} = \frac{g}{2k}\frac{1}{C}\{\frac{kd + \frac{1}{2}Sinh\ kd}{Cosh^2 kd}\}$

On separating and simplifying within the brackets we can write

$= \frac{g}{2k}\frac{1}{C}\{\frac{2Sinh\ kd\ Cosh\ kd}{2Cosh^2 kd}\}\{1 + \frac{2kd}{Sinh\, 2kd}\} = \frac{g}{2k}\frac{1}{C}\tanh kd\ \{1 + \frac{2kd}{Sinh\, 2kd}\}$

We know $C^2 = \frac{g}{k}\tanh kd$ substituting this in the above equation we get

$C_g = \frac{C}{2}\{1 + \frac{2kd}{Sinh\, 2kd}\} = \frac{1}{2}\frac{L}{T}\left[1 + \frac{4\pi d / L}{Sinh\, ^{4\pi d}/_{L}}\right]$

In deep waters the term $\left[\dfrac{4\pi d/L}{Sinh{}^{4\pi d}\!/_{L}}\right] \to 0$

So $C_g = \frac{1}{2}\frac{L_0}{T}[1+0] = \frac{1}{2}C_0$

From the above expression it is clear that the group velocity is half the phase velocity in deep water. This means $\frac{\text{Cg}}{\text{C}}$ can be seen to have the value $\frac{1}{2}$ in deep water. Also the envelope 'A'( Fig.13.24) moves half the speed of 'C'

In shallow water the term $Sinh{}^{4\pi d}\!/_{L} \to 4\pi d/L$, Hence

$$C_g = \frac{1}{2}\frac{L_S}{T}\left[1+\frac{4\pi d/L}{{}^{4\pi d}\!/_{L}}\right] = \frac{1}{2}\frac{L_S}{T}(2) = \frac{L_S}{T} = C_S = \sqrt{gd}$$

Hence, the group and phase velocities are equal in shallow water as $\frac{C_g}{C_s}$ approach to a limiting value of unity in shallow water. Thus, in shallow water, because wave celerity is fully determined by the depth (d), all component waves in a wave train will travel at the same speed precluding the alternate reinforcing and canceling of components. In deep and intermediate (transitional) water, wave celerity depends on the wavelength; hence, slightly longer waves travel slightly faster and produce the small phase differences resulting in wave groups. These waves are said to be dispersive or propagating in a dispersive medium; i.e., in a medium where their celerity is dependent on wavelength.

Outside of shallow water, the phase velocity of gravity waves is greater than the group velocity; an observer that follows a group of waves at group velocity will see waves that originate at the rear of the group move forward through the group traveling at the phase velocity and disappear at the front of the wave group.

Group velocity is important because it is with this velocity that wave energy is propagated. Although mathematically the group velocity can be shown rigorously from the interference of two or more waves, the physical significance is not as obvious as it is in the method based on the consideration of wave energy. Therefore an additional explanation of group velocity is provided on wave energy and energy transmission.

### 13.3.9. Pressure of a progressive wave

We have seen that the integrated and linearized equation of motion for irrotational motion of an incompressible fluid can be written from Bernoullils equation (equation 13.15) as:

$$\left[-\frac{\partial \phi}{\partial t}+\frac{p}{\rho}+gz\right]=0 \qquad 13.38$$

The velocity potential for a progressive wave moving in the positive x direction (equation 13.26 ), can be differentiated with respect to 't' and substituting in the equation (13.38) we get:

$$\frac{p}{\rho}=g\frac{\cosh k(d+z)}{\cosh kd}a\sin(kx-\sigma t)-gz \qquad 13.39$$

Using the wave profile equation (13.25a), $\eta = a \sin(\sigma t - k x)$, equation 13.39 can be modified as:

$$p=\gamma\left[\eta\frac{\cosh k(d+z)}{\cosh kd}-z\right] \qquad 13.40$$

In equation (13.36), $\gamma = \rho g$ and z is negative downward from mean water level.

The quantity $\frac{\cosh k(d+z)}{\cosh kd}=K_p$ is called the pressure response factor and $K_p$ is less than unity for all depths below the mean water level. Equation (13.40) is not valid for all positive z, though it exists under the crest above the mean water level because of the boundary condition. If we recall that in free surface boundary condition we set p = 0 at the surface but in determining $\varphi$ we applied this surface boundary condition at z = 0, rather than z = $\eta$ at the peak level of the profile of the crest. Because of this inconsistency we will restrict our discussion of equation (13.40) to the region of negative z only.

Equation (13.40) can further be simplified as:

$$\frac{p}{\gamma}=K_p\eta-z \qquad \text{.......................} \ 13.41$$

where $\frac{p}{\gamma}$ is pressure head, $K_p$ is pressure response factor and z is elevation head. This can be readily interpreted physically in comparison with the familiar relationships of hydrostatics. The second term ($K_p\eta$) represents the negative one-to-one correspondence between pressure head ($\frac{p}{\gamma}$) and elevation head (z) associated with hydrostatic pressure distribution. The first term ($\frac{p}{\gamma}$) provides a positive or negative (depending on wave phase) deviation from the hydrostatic produced by the vertical accelerations accompanying the orbital motions.

Looking first at the limiting values of z, equation (13.40) turns out to:

i) at z = 0 and $\frac{p}{\gamma} = \eta$, the equation 13.41 is equal to:
   $\eta = \eta K_p \rightarrow \eta(1 - K_p) = 0$ which means $K_p = 1$
   It would thus be consistent to assume the pressure to be hydrostatic for positive z.

ii) at z = - d (below the mean water level) the equations 13.40 turns out to:

$$\frac{p}{\gamma} = \frac{\eta}{\cosh kd} + d \qquad 13.42$$

as $K_p = \frac{1}{\cosh kd} \because \cosh 0 \rightarrow 1, \cosh(z+d) \rightarrow \cosh(d-d) \rightarrow \cosh 0 \rightarrow 1$

Thus the hydrostatic bottom pressure as given by the equation (13.42) turns out to be:

$\frac{p}{\gamma} = \eta + d$ (as $\cosh kd \rightarrow 1$)

For wave phases at which η is negative (under the trough) the bottom pressure is thus greater than the hydrostatic pressure ( h – η) since $h - \frac{\eta}{\cosh kd} > h - \eta$

Conversely for positive 'η' (under the wave crest)

$h + \frac{\eta}{\cosh kd} < h + \eta$

Which means the bottom pressure is less than hydrostatic pressure under the wave crest.

The Fig.13.25 below represents these arguments of resulting vertical pressure distribution.

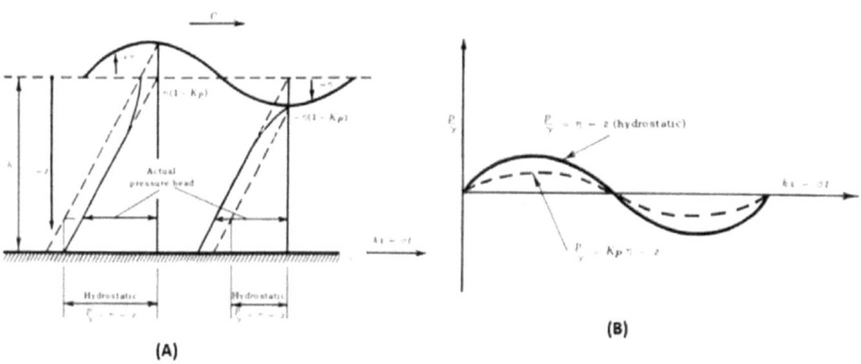


**Fig. 13.25** . Pressure distribution in a wave **A:** Vertical pressure distribution under crest and trough of a wave; **B:** Phase distribution of pressure head.

## Alternative explanation

Starting from equation (13.39), $\frac{p}{\rho} = g\frac{\cosh k(d+z)}{\cosh kd} a\sin(kx - \sigma t) - gz$

We know $\eta = \frac{H}{2}\sin(kx - \sigma t)$ and put $K_p = \frac{\cosh k(d+z)}{\cosh kd}$ = pressure response factor

Then we can write equation 13.39 as: $p = \rho g \eta K_p - \rho g Z = \rho g(\eta K_p - Z)$ (13.39a)

At Z = - d at sea bottom, the pressure response factor will become: $K_p = \dfrac{1}{\cosh kd}$

Hence at Z = - d, the equation 13.39a will become: $\dfrac{p}{\rho g} = (\eta K_p - Z) \Rightarrow \eta K_p = \dfrac{p}{\rho g} + Z)$

$$\eta = \frac{1}{K_p}(\frac{p}{\rho g} + Z)$$

$$\Rightarrow \eta = N\left[\frac{p + \rho g Z}{\rho g K_p}\right] \tag{13.39b}$$

where z is the depth below the SWL of the pressure gage, and N a correction factor equal to unity if the linear theory applies. Several empirical studies have found N to be a function of period, depth, wave amplitude, and other factors. In general, N decreases with decreasing period, being greater than 1.0 for long-period waves and less than 1.0 for short-period waves. Generally equation 13.39b is used to find the wave height with the help of a pressure gauge which gives wave pressure.

## 13.4. Wave Energy

The energy possessed by a wave is in two forms. One is kinetic energy and the other is potential energy. Kinetic energy is inherent in the orbital motion of the water particles and the potential energy is possessed by the particles when they are displaced from their mean position.

The total energy of a wave is partly potential and partly kinetic. The potential energy of a wave is equal to the work necessary to distort a horizontal sea surface into the wave profile which may be occurring beneath the sea surface. Thus the potential energy, resulting from the position of the wave and the displacement of the free surface from mean sea level and the kinetic energy due to the fact that the water particles throughout the fluid are moving are important. The total energy and its transmission are important in determining how waves change in propagating toward shore, the power required to generate waves, and the available power for wave energy extraction.

### 13.4.1. Potential energy

Potential energy occurring in waves is the result of displacing a mass from a position of equilibrium against a gravitational field. When water is at rest with a uniform free surface elevation, it can be shown readily that the potential energy is minimum. However, a displacement of an assemblage of particles resulting in the displacement of the free surface will require that work be done on the system and results in an increase in potential energy.

We will derive the average potential energy associated with a sinusoidal wave by considering a strip of unit depth in the y-direction in a two dimensional wave.

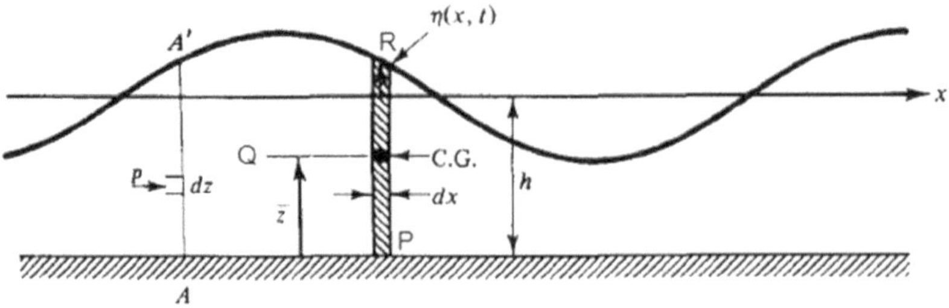


**Fig.13.26 .** The potential energy of a column of water in a section of wave

The potential energy of a column of water PR at height 'η' with width dx is the weight of the water column in the Fig.13.26.

∴ The area of the column = dx.η

If this area is multiplied by the thickness in y direction, then it gives the volume of the fluid within the strip (column). Since the thickness of the column in y direction is unity, the volume of the fluid in the column is dx.η. 1

Mass of the fluid = ρ. dx.η

Where ρ is the density of sea water. The centre of gravity of water in the strip is at 'Q' that is, at half the height ($\frac{\eta}{2}$).

The potential energy is = mgh = (ρ. dx.η).g. $\frac{\eta}{2} = \frac{1}{2}$ ρ.g. dx.$\eta^2$

The average P.E per unit horizontal area is found by integrating over a wave length.

$$\therefore \text{P.E} = \int_0^L \frac{\rho g}{2}\eta^2 dx$$

The horizontal area covered by the wave of length L and thickness unity in y direction is ( L × 1 = L) units. So the average potential energy per unit horizontal surface area is

$$\text{P.E} = \frac{g}{2L}\int_0^L \rho.\eta^2.dx$$

If the fluid has constant density, then

$$\text{P.E} = \frac{g\rho}{2L}\int_0^L \eta^2.dx$$

But we know η = a cos (kx – σ t) for progressive wave

$$\therefore \text{ P.E} = \frac{\rho g a^2}{2L}\int_0^L \cos^2(Kx-\sigma t)dx \text{ , putting L} = 2\pi/\text{K}$$

$$\therefore \text{ P.E} = \frac{\rho g a^2}{4\pi}\int_0^{2\pi/k} \cos^2(Kx-\sigma t)kdx$$

$$= \frac{\rho g a^2}{4\pi}\int_0^{2\pi/k}\frac{1+\cos 2(kx-\sigma t)}{2}kdx$$

$$= \frac{k.\rho g a^2}{2(4\pi)}\{\int_0^{2\pi/k}dx+\int_0^{2\pi/k}\frac{1+\cos 2(kx-\sigma t)}{2}kdx\}$$

$$= \frac{k.\rho g a^2}{8\pi)}\left\{[x]_0^{2\pi/k}+\left[\frac{Sin2(kx-\sigma t)}{2k}\right]_0^{2\pi/k}\right\} =$$

$$\frac{k.\rho g a^2}{8\pi)}\left\{\left(\frac{2\pi}{k}-0\right)+\frac{1}{2k}\left[\sin 2\left(k.\frac{2\pi}{k}-\sigma t\right)-\sin 2(k.0-\sigma t)\right]\right\}$$

$$= \frac{k.\rho g a^2}{8\pi)}\left\{\left(\frac{2\pi}{k}\right)+\frac{1}{2k}[-\sin 2\sigma t+\sin 2\sigma t]\right\} = \frac{k.\rho g a^2}{8\pi}.\frac{2\pi}{k} = \frac{\rho g a^2}{4} \qquad 13.43$$

### 13.4.2. Kinetic Energy

The kinetic energy of the wave is due to moving particles in the wave.

We know K.E = $\frac{1}{2}mv^2 = \frac{1}{2}\rho\left(u^2+v^2\right) = \int_0^L\int_0^{-d}\frac{1}{2}\left(u^2+w^2\right)dxdy$

We know the particle velocities as from equations (13.35&13.36)

$$u = +\frac{agk}{\sigma\cosh kd}\ \text{Coshk}(d+z).\text{Cos}\ (k\,x-\sigma\,t)$$

$$w = -\ \frac{agk}{\sigma}\frac{Sinhk(d+z)}{Coshkd}Sin(kx-\sigma t)$$

substituting and writing

$$K.E = \int_0^L\int_0^{-d}\frac{\rho}{2}\left[\frac{agk}{\sigma}\frac{1}{\cosh kd}\right]^2\left[Cosh^2k(d+z)Cos^2(kx-\sigma t)+Sinh^2k(d+z).Sin^2(kx-\sigma t)\right]dxdz$$

Using the trigonometric identities just as was done earlier, this can be re-casted as

$$K.E = \frac{\rho}{2}\left[\frac{agk}{\sigma}\frac{1}{\cosh kd}\right]^2\frac{1}{2}\int_0^L\int_0^{-d}\left[Cosh2k(d+z)+Cos2(kx-\sigma t)\right]dxdz$$

On integration and further simplification, it yields to : K.E = $\frac{1}{4}\rho g a^2$ 13.44

### Total Energy

The total energy per unit width of the crest in a wave is P.E + K.E (equations 13.43 + 13.44)

$$= \frac{1}{4}\rho g a^2+\frac{1}{4}\rho g a^2 = \frac{1}{2}\rho g a^2 = \frac{1}{8}\rho g H^2\ (\because a = \frac{H}{2}) \qquad 13.45$$

According to the Airy (1845) theory, if the potential energy is determined relative to Still water level (SWL), and all waves are propagated in the same direction, potential and kinetic energy components are equal, and the total wave energy in one wavelength per unit crest width is given by

$$E = \frac{1}{8}\rho g H^2 L \qquad 13.45a$$

Total average wave energy per unit surface area, termed the specific energy or energy density, is given by:

$$\bar{E} = \frac{E}{L} = \frac{1}{8}\rho g H^2 \qquad 13.45b$$

### 13.4.3. Energy flux

Small amplitude waves do not transmit mass as they propagate across a fluid, as the trajectories of the water particles are closed. However, all waves do transmit energy. The rate at which the energy is transferred is called the energy flux "Ғ" and for linear waves it is the rate at which work is being done by the fluid on one side of a vertical section. For the vertical section A in the Fig.13.27, the instantaneous rate at which work is being done by the dynamic pressure $P_D$ (= P + ρgz) per unit width in the direction of wave propagation is:

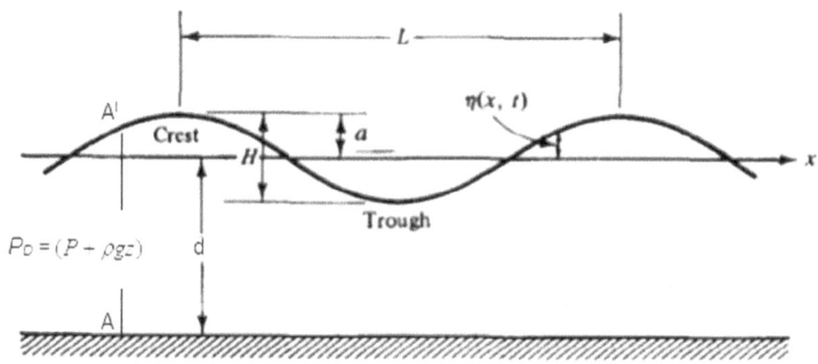


**Fig. 13.27.** Energy flux calculation (SPM 1984)

$$Ғ = \int_0^{-d} P_D.u dz$$

$$Ғ = \int_0^{-d} (P + \rho g z).u dz \qquad 13.46$$

The average energy flux is obtained by averaging over a wave period

$$Ғ = \frac{1}{T}\int_0^T \int_0^{-d} P_D.u dz dt \qquad 13.47$$

Substituting the values of P from equation (13.40), η from equation (13.25a) and u from equation (13.35) in 13.46 and then substituting in eqn (13.47) we get:

$$Ғ = \frac{1}{T}\int_0^T \int_0^{-d} \left[\rho g \eta \frac{Coshk(d+z)}{Coshkd}\right]\left[\frac{agk}{\sigma}\frac{Coshk(d+z)}{Coshkd} Sin(kx - \sigma t\right] dz dt$$

$$= \frac{1}{T}\int_{0}^{T}\int_{0}^{-d}\left[\rho g\eta\frac{Coshk(d+z)}{Coshkd}\right]\left[\eta\sigma\frac{Coshk(d+z)}{Sinhkd}\right]dzdt\left(\because \sigma^2 = gk\tanh kd\right) \qquad 13.48$$

To retain the terms upto the second order in wave height, it is only necessary to integrate up to the mean free surface, on integration the equation reduces to

$$Ꞙ = \frac{\rho g\sigma}{4k}\left(\frac{H}{2}\right)^2\left[\frac{2kd + Sinh2kd}{Sinh2kd}\right] = \left(\frac{1}{8}\rho gH^2\right)\left(\frac{\sigma}{k}\right)\left[\frac{1}{2}\left(1+\frac{2kd}{Sinh2kd}\right)\right]$$

$$= (E)(C)(N) = E\,C_g \qquad 13.49$$

Where CN = $C_g$ is the group velocity at which the energy is transmitted. Thus it is evident from group velocity

$$\frac{C_g}{C} = N = \left[\frac{1}{2}\left(1+\frac{2kd}{Sinh2kd}\right)\right] \qquad 13.50$$

The factor N has deep and shallow water asymptotes and has the values ½ and 1 respectively. The energy flux 'Ꞙ' has the unit of power and for this reason it is sometimes referred to as the 'wave power'.

Thus the energy flux is the rate at which the wave energy is transmitted in the direction of wave propagation. The rate at which the energy is propagated per unit length of the wave crest is called the wave power. So the energy flux has the unit of power.

$\therefore$ Ꞙ = E $C_g$ N

where Ꞙ is wave power, E is energy, $C_g$ is group velocity and N is the shoaling constant. In deep water N = ½ and in shallow water N = 1.

If a vertical plane is taken other than perpendicular to the direction of wave advance, P = E C sin φ where φ is the angle between the plane across which the energy is being transmitted and the direction of wave advance.

For deep and shallow water, equation (13.49) becomes:

$$F = \bar{P} = \frac{1}{2}\bar{E}_0 C_0 (deepwater)$$
$$F = \bar{P} = \bar{E}C_g = \bar{E}C(shallowwater) \qquad (13.50a)$$

An energy balance for a region through which waves are passing will reveal that, for steady state, the amount of energy entering the region will be equal to the amount leaving the region provided no energy is added or removed from the system. Therefore, when the waves are moving, their crests are parallel to the bottom contours,

$$E_0 N_0 C_0 = \bar{E}NC$$

$$\sin ce; N_0 = \frac{1}{2} \qquad (13.50b)$$

$$\frac{1}{2} E_0 C_0 = \bar{E}NC$$

This equation does not apply when the wave crests are not parallel to the bottom contours as some parts of the wave will be traveling at different speeds and the wave will be refracted.

### 13.4.4. Wave power for the spectrum

An alternate formula used for estimation of wave power from the spectrum is

$$P = \frac{\gamma g}{64\pi} H_S^2 T_Z \text{ watts} \qquad 13.51$$

Where $\gamma$ = Specific weight of sea water (10.05newtons /m$^3$), g = 9.8 m/s$^2$, $H_S$ = Significant wave height in meters, $T_Z$ = zero crossing period in seconds.

### 13.4.5. Table of wave characteristics

All the wave characteristics are summed up and presented in table 13.5

**Table 13.5.** Summation of all wave characteristics

| Relative Depth | Shallow Water | Transitional Water | Deep Water |
|---|---|---|---|
| | $\frac{d}{L} < \frac{1}{20}$ <br> $kd < \frac{\pi}{10}$ | $\frac{1}{20} < \frac{d}{L} < \frac{1}{2}$ <br> $\frac{\pi}{10} < kd < \frac{\pi}{2}$ | $\frac{d}{L} > \frac{1}{2}$ <br> $kd > \frac{\pi}{2}$ |
| 1. Wave profile | Same As > | $\eta = \frac{H}{2}\cos\left[\frac{2\pi x}{L} - \frac{2\pi t}{T}\right] = \frac{H}{2}\cos\theta$ | < Same As |
| 2. Wave celerity | $C = \frac{L}{T} = \sqrt{gd}$ | $C = \frac{L}{T} = \frac{gT}{2\pi}\tanh\left(\frac{2\pi d}{L}\right)$ | $C = C_0 = \frac{L}{T} = \frac{gT}{2\pi}$ |
| 3. Wavelength | $L = T\sqrt{gd} = CT$ | $L = \frac{gT^2}{2\pi}\tanh\left(\frac{2\pi d}{L}\right)$ | $L = L_0 = \frac{gT^2}{2\pi} = C_0 T$ |
| 4. Group velocity | $C_g = C = \sqrt{gd}$ | $C_g = nC = \frac{1}{2}\left[1 + \frac{4\pi d/L}{\sinh(4\pi d/L)}\right]C$ | $C_g = \frac{1}{2}C = \frac{gT}{4\pi}$ |
| 5. Water particle velocity <br> (a) Horizontal | $u = \frac{H}{2}\sqrt{\frac{g}{d}}\cos\theta$ | $u = \frac{H}{2}\frac{gT}{L}\frac{\cosh[2\pi(z+d)/L]}{\cosh(2\pi d/L)}\cos\theta$ | $u = \frac{\pi H}{T}e^{\left(\frac{2\pi z}{L}\right)}\cos\theta$ |
| (b) Vertical | $w = \frac{H\pi}{T}\left(1+\frac{z}{d}\right)\sin\theta$ | $w = \frac{H}{2}\frac{gT}{L}\frac{\sinh[2\pi(z+d)/L]}{\cosh(2\pi d/L)}\sin\theta$ | $w = \frac{\pi H}{T}e^{\left(\frac{2\pi z}{L}\right)}\sin\theta$ |
| 6. Water particle accelerations <br> (a) Horizontal | $a_x = \frac{H\pi}{T}\sqrt{\frac{g}{d}}\sin\theta$ | $a_x = \frac{g\pi H}{L}\frac{\cosh[2\pi(z+d)/L]}{\cosh(2\pi d/L)}\sin\theta$ | $a_x = 2H\left(\frac{\pi}{T}\right)^2 e^{\left(\frac{2\pi z}{L}\right)}\sin\theta$ |
| (b) Vertical | $a_z = -2H\left(\frac{\pi}{T}\right)^2\left(1+\frac{z}{d}\right)\cos\theta$ | $a_z = -\frac{g\pi H}{L}\frac{\sinh[2\pi(z+d)/L]}{\cosh(2\pi d/L)}\cos\theta$ | $a_z = -2H\left(\frac{\pi}{T}\right)^2 e^{\left(\frac{2\pi z}{L}\right)}\cos\theta$ |
| 7. Water particle displacements <br> (a) Horizontal | $\xi = -\frac{HT}{4\pi}\sqrt{\frac{g}{d}}\sin\theta$ | $\xi = -\frac{H}{2}\frac{\cosh[2\pi(z+d)/L]}{\sinh(2\pi d/L)}\sin\theta$ | $\xi = -\frac{H}{2}e^{\left(\frac{2\pi z}{L}\right)}\sin\theta$ |
| (b) Vertical | $\zeta = \frac{H}{2}\left(1+\frac{z}{d}\right)\cos\theta$ | $\zeta = \frac{H}{2}\frac{\sinh[2\pi(z+d)/L]}{\sinh(2\pi d/L)}\cos\theta$ | $\zeta = \frac{H}{2}e^{\left(\frac{2\pi z}{L}\right)}\cos\theta$ |
| 8. Subsurface pressure | $p = \rho g(\eta - z)$ | $p = \rho g\eta\frac{\cosh[2\pi(z+d)/L]}{\cosh(2\pi d/L)} - \rho gz$ | $p = \rho g\eta\, e^{\left(\frac{2\pi z}{L}\right)} - \rho gz$ |

## 13.5. Wave refraction

The process by which ocean water wave, moving in shallow water from deep water as it approaches the shore at an angle, tends to turn from its original direction. The part of the wave advancing in shallower water moves more slowly than the part still advancing in deeper water, causing the wave crests to bend toward parallel alignment with the shoreline. A consequence of this is that the wave fronts tend to become aligned with the depth contours. Currents also can result in refraction and this is termed as current refraction.

Wave celerity equation depends on the water depth in which the wave propagates. If the wave celerity decreases with depth, wavelength must also decrease proportionally. Variation in wave velocity occurs along the crest of a wave moving at an angle to underwater contours because the part of the wave in deeper water is moving faster than the part in shallower water. This variation causes the wave crest to bend toward alignment with the contours (see Fig.13.28). This bending effect, called refraction, depends on the relation of water depth to wavelength. It is analogous to refraction for other types of waves; i.e., light and sound.

Wave approach angle is the angle between the wave orthogonal and normal of the bottom contour or the angle between wave crest line and the bottom contour such that $d_1$ is deep water and $d_2$ is shallow water.

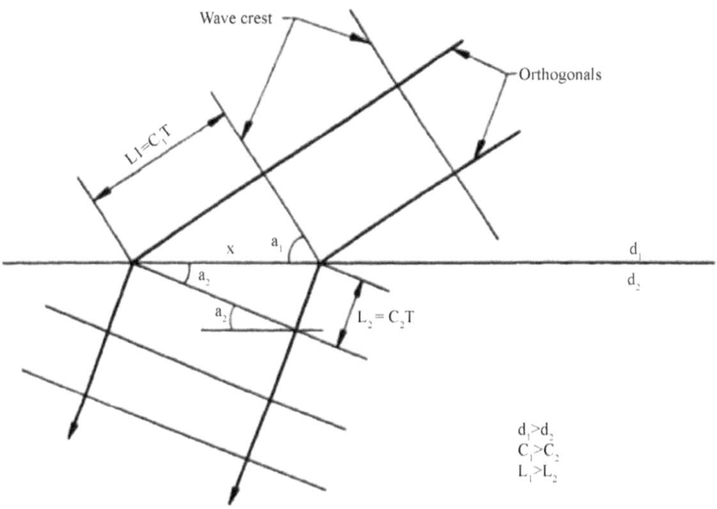


**Fig.13.28.** Wave refraction

Refraction has significant influence on the wave height and distribution of wave energy along the coast. The change in wave direction of different parts of the wave results in convergence or divergence of wave energy and materially affects the forces exerted by waves on structures. Refraction contributes to the alteration of bottom topography by its effect on the erosion and deposition of beach sediments.

The general assumptions in wave refraction studies are:

a) Wave energy between wave rays or orthogonals remains constant. (Orthogonals are lines drawn perpendicular to the wave crests, and extend in the direction of wave advance.)

b) Direction of wave advance is perpendicular to the wave crest; i.e., in the direction of the orthogonals.

c) Speed of a wave with a given period at a particular location depends only on the depth at that location.

d) Changes in bottom topography are gradual.

The change of direction of an orthogonal as it passes over relatively simple hydrography may be approximated by the Snell's Law:

$$Sin\alpha_2 = \left(\frac{C_2}{C_1}\right) Sin\alpha_1$$

Where $\alpha_1$ is the angle a wave crest (perpendicular to an orthogonal) makes with the bottom contour over which the wave is passing; $\alpha_2$ is a similar angle measured as the wave crest (or orthogonal) passes over the next bottom contour. $C_1$ = the wave velocity at the depth of the first contour; $C_2$ = the wave velocity at the depth of the second contour.

As the wave travels from deep ($d_1$) water to shallow water ($d_2$) as shown in Fig.13.28 the wave celerity and length will decrease. For an orthogonal spacing x and time interval T,

$$Sin\alpha_1 = \frac{C_1 T}{X} = \frac{L_1}{X}$$

$$Sin\alpha_2 = \frac{C_2 T}{X} = \frac{L_2}{X}$$

Hence, according to snell's law $\frac{Sin\alpha_1}{Sin\alpha_2} = \frac{C_1}{C_2} = \frac{L_1}{L_2}$

We know : $c^2 = \frac{g}{k}\tanh(kd) = \frac{gL}{2\pi}\tanh\left(\frac{2\pi d}{L}\right) ==> C = \frac{gT}{2\pi}\tanh\left(\frac{2\pi d}{L}\right)$

In water deeper than one-half the wavelength, the hyperbolic tangent function, $\tanh\left(\frac{2\pi d}{L}\right)$, in the formula becomes unity and so the wave celerity in deep water becomes: (See equations 13.33):

$$c_0^2 = \frac{gL}{2\pi} ==> C_0 = \frac{gT}{2\pi}$$

The subscript '0' of any parameter indicates that they are of the deep water.

In this equation, the velocity C does not depend on depth; therefore, in those regions deeper than one-half the wavelength (deep water), refraction by bathymetry will not be significant. Where the water depth is between one-half and one-twenty-fifth the wavelength (transitional water), and in the region where the water depth is less than one-twenty-fifth, the wavelength (shallow water), refraction effects may be significant. In transitional water, wave velocity must be computed from the above equation; in shallow water, $\tanh\left(\frac{2\pi d}{L}\right) \to \left(\frac{2\pi d}{L}\right)$ and so the equation reduces to

$$c^2 = gd => C = \sqrt{gd}$$

From the above two equations we can write: $\frac{C}{C_0} = \frac{gT}{2\pi}\tanh(kd) \div \frac{gT}{2\pi} = \tanh\left(\frac{2\pi d}{L}\right)$ (13.52)

Thus both the equations show the dependence of wave velocity on depth.

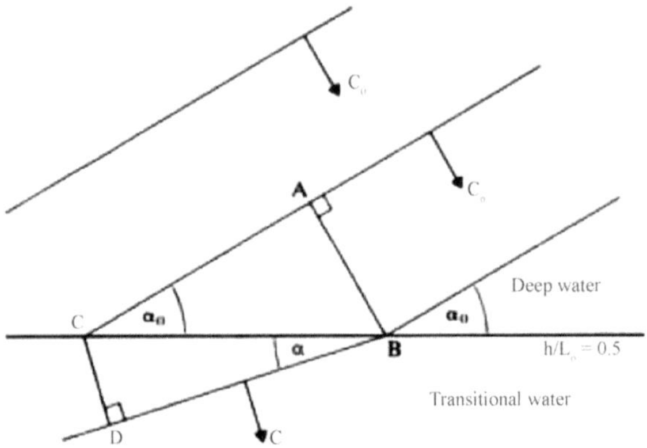


**Fig.13.29.** Wave refraction

Consider a deep water wave approaching the transitional depth limit ($d/L_0 = 0.5$), as shown in Fig.13.29. Let suppose a wave travelling from A to B in deep water travels a distance $L_0$ in one wave period, T. However, the wave travelling from C to D travels a smaller distance, L, at the same time as it is in the shallower depth region. Hence the new wave front is now BD, which has rotated with respect to AC.

If α represents the angle of the wave front to the depth contour, then we can write:

$$Sin\alpha = \frac{L}{BC} \qquad Sin\alpha_0 = \frac{L_0}{BC} \quad \text{or} \quad \frac{Sin\alpha}{Sin\alpha_0} = \frac{C}{C_0} = \frac{L}{L_0} = \tanh(kd) \text{ (it follows from equation 13.52)} \quad (13.53)$$

As $C < C_0$, α must be less than $\alpha_0$. This means shallow water angle is less than that in deep water. This means the wave ray is bent. This implies that as the wave approaches the shoreline from an oblique angle the wave fronts tend to align themselves with the underwater contours. Equation (13.53) is known as Snell's law, according to which sin(α/c) is constant along a wave ray.

From this Snell's law equation a template (or ray plotter) may be constructed which will show the angular changes in α that occurs as an orthogonal passes over a particular contour interval to construct the changed-direction orthogonal. In application to wave refraction problems, it is simplest to construct this template on a transparent material.

### 13.5.1. Shoaling

To a first approximation, the total energy in a wave per unit crest width may be written as:

$$E = \frac{\rho g H^2 L}{8}$$

It has been noted that not all the wave energy E is transmitted forward with the wave; only one-half is transmitted forward in deep water. The amount of energy transmitted forward for a given wave remains nearly constant as the wave moves from deep water to the breaker line if energy dissipation due to bottom friction ($K_f$ = 1) percolation, and reflected wave energy is negligible.

Consider first a wave front travelling parallel to the seabed contours when there is no refraction. Making the assumption that wave energy is transmitted shoreward without loss due to bed friction or turbulence, then from wave power equation (13.49):

$$P=\left(\frac{1}{8}\rho gH^2\right)\left[\frac{C}{2}\left(1+\frac{2kd}{Sinh2kd}\right)\right]=EC_g \text{ where } C_g=\left[\left(1+\frac{2kd}{Sinh2kd}\right)\right] \text{ is group velocity.}$$

In deep water (d/L > 0.5) the group velocity $C_g$ = C/2 and in shallow water $C_g$ = C. Hence in deep water wave energy is transmitted forward at only half the wave celerity. Hence from above we can write:

$$P=\left(\frac{1}{8}\rho gH^2\right)C_g \text{ and } P_0=\left(\frac{1}{8}\rho gH_0{}^2\right)C_{g0}$$

$$P=EC_g=E_0C_{g0}=\rightarrow P=P_0=\left(\frac{1}{8}\rho gH^2\right)C_g=\left(\frac{1}{8}\rho gH_0{}^2\right)C_{g0}$$

Since wave power remained constant as it approaches the coast:

$$\frac{P}{P_0}=\frac{\bar{E}C_g}{\bar{E}Cg_0}=\frac{\frac{1}{8}\rho gH^2}{\frac{1}{8}\rho gH_0{}^2}=\frac{H^2}{H_0^2}=1$$

$$\frac{H}{H_0}=\sqrt{\frac{C_{g0}}{C_g}}=K_S$$

Where $K_S$ is called shoaling coefficient. The shoaling coefficient shows that as the local group velocity $C_g$ decreases with water depth, the wave height increases. The variation of $K_S$ with $d/L_0$ is shown in Fig.13.30. We can get by substituting all the related expressions and evaluated from the equation of group velocity given above as;

$$K_S=\sqrt{\frac{C_{g0}}{C_g}}=\sqrt{\frac{C_0}{C}\times\frac{1}{\left\{1+\frac{2kd}{\sinh 2kd}\right\}}}\Rightarrow K_S=\sqrt{\frac{C_{g0}}{C_g}}=\frac{C}{C_0}\left[1+\frac{2kd}{\sinh 2kd}\right]^{-1/2} \quad (.13.54a)$$

Where $C_0$ is the wave speed in deep water and C that at depth d. we can relate these two by writing $C_0=\sigma/K_0$ and $C=\sigma/K$ where σ is the frequency of the waves which is same whether they are in deep or shallow water, giving $\sigma=\sqrt{gk\tanh kd}\rightarrow\sigma=\sqrt{gk_0}$

$$\frac{C_0}{C} = \frac{K}{K_0}$$

As in deep water, when d → ∞, tanh kd → 1. This gives

$$\frac{K}{K_0} = \frac{1}{\tanh kd}$$

$$K_S = \left[\tanh kd\left\{1 + \frac{2kd}{\sinh 2kd}\right\}\right]^{-1/2} \Longrightarrow K_S = \left[\tanh kd + kd\left\{1 - \tanh^2 kd\right\}\right]^{-1/2}$$

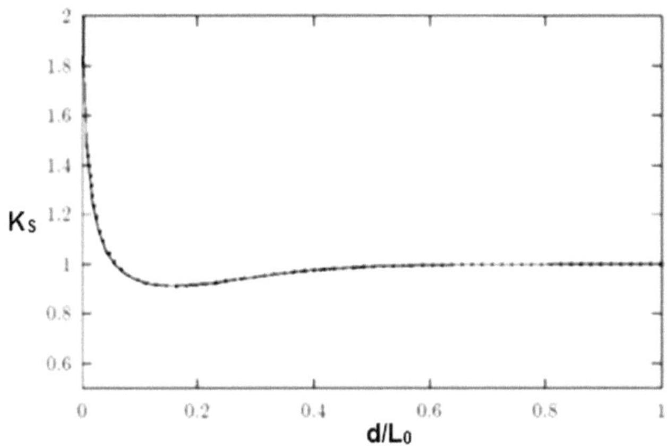


**Fig.13.30.** Variation of shoaling coefficient with depth

## 13.5.2. Bathymetry effects on refraction

Consider next a wave front travelling obliquely to the seabed contours as shown in Fig.13.31. As the wave rays bend, they may either converge or diverge as they travel shoreward.

Then at the contour $d/L_0 = 0.5$, $BC = \frac{b_0}{\cos\alpha_0} = \frac{b}{\cos\alpha} \Longrightarrow \frac{b}{b_0} = \frac{\cos\alpha}{\cos\alpha_0}$

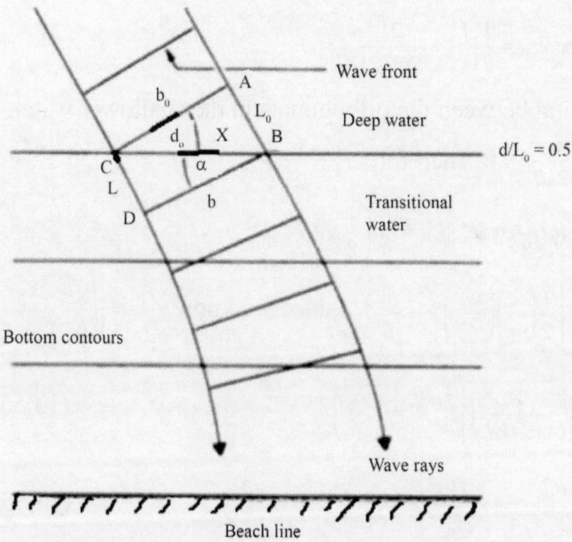


**Fig. 13.31.** Diverging wave rays over parallel contours

If we choose AC and BD as bo and b respectively as in the (Fig.13.31) and the orthogonal lengths are equal to AB= Lo and CD= L, and BC = X, then we can write:

$\frac{b_0}{\cos\alpha_0} = \frac{b}{\cos\alpha} = X$ where X is the bottom contour or bathymetry line. It follows from this as

$$\sqrt{\frac{b_0}{b}} = \sqrt{\frac{\cos\alpha_0}{\cos\alpha}} = K_R \tag{13.54b}$$

This $K_R$ is called refraction coefficient.

In refraction analyses, it is assumed that for a wave advancing toward shore, no energy flows laterally along a wave crest; i.e., the transmitted energy remains constant between orthogonals. In deep water the wave energy transmitted forward across a plane between two adjacent orthogonals (the average energy flux) (Fig.13.31) is

$$\overline{P}_0 = n_0 b_0 \overline{E}_0 C = \frac{1}{2} b_0 \left(\frac{1}{8}\rho g H_0^2\right) C_0$$

Where $b_0$ is the distance between the selected orthogonals in deep water. The subscript '0' refers to deep water conditions. This power may be equated to the energy transmitted forward between the same two orthogonals in shallow water

$$\bar{P} = nb\bar{E}C = nb\left(\frac{1}{8}\rho g H^2\right)C \text{ where } n = \frac{1}{2}\left[\left(1 + \frac{2kd}{Sinh2kd}\right)\right]$$

where b is the relative spacing between the orthogonals in the shallower water. In the case of deep water $\frac{2kd}{Sinh2kd} \to 0$ Therefore $\bar{P}_0 = \frac{1}{2}b_0\bar{E}_0C_0$

$$\bar{P}_0 = \bar{P} = \frac{1}{2}b_0\bar{E}_0C_0 = nb\bar{E}C \to \frac{1}{2}b_0\left(\frac{1}{8}\rho g H_0^2\right)C_0 = nb\left(\frac{1}{8}\rho g H^2\right)C$$

$$\frac{\bar{E}}{\bar{E}_0} = \frac{1}{2}\left(\frac{1}{n}\right)\left(\frac{b_0}{b}\right)\left(\frac{C_0}{C}\right) \to \frac{H}{H_0} = \left[\frac{1}{2}\left(\frac{1}{n}\right)\left(\frac{C_0}{C}\right)\right]^{1/2}\left(\frac{b_0}{b}\right)^{1/2} \text{ since we know } \frac{H}{H_0} = \sqrt{\frac{\bar{E}}{\bar{E}_0}}$$

$$K_S = \left[\frac{1}{2}\left(\frac{1}{n}\right)\left(\frac{C_0}{C}\right)\right]^{1/2} = \left[\frac{1}{2}\left(\frac{1}{n}\right)\left(\frac{1}{\tanh kd}\right)\right]^{1/2}; K_R = \left(\frac{b_0}{b}\right)^{1/2} \tag{13.55a}$$

$$\frac{P}{P_0} = 1 = \frac{\bar{E}bC_g}{\bar{E}b_0Cg_0} = \left(\frac{H}{H_0}\right)^2 \frac{\cos\alpha C_g}{\cos\alpha_0 Cg_0} = \to \frac{H}{H_0} = K_R K_S \tag{13.55b}$$

When the energy flux is conserved between the wave rays, then n C E b = const, where b is the distance between adjacent wave rays. Then we can also write

$$\frac{H_2}{H_1} = \sqrt{\frac{\eta_1 C_1}{C_1 C_2}}\sqrt{\frac{b_1}{b_2}}$$

$$\frac{H}{H_0} = \sqrt{\frac{\eta_0 C_0}{\eta C}}\sqrt{\frac{b_0}{b}} = \sqrt{\frac{1}{2}\frac{1}{\eta}\frac{1}{\tanh kd}}\sqrt{\frac{b_0}{b}} = K_s K_r$$

such that $K_s = \sqrt{\frac{1}{2}\frac{1}{\eta}\frac{1}{\tanh kd}}$ *and* $K_r = \sqrt{\frac{b_0}{b}}$ (13.55c)

The term $K_S$ is known as the shoaling coefficient and $K_R$ is called refraction coefficient. This shoaling coefficient is a function of wavelength and water depth. The values of $K_S$ are available in Shore Protection manual (SPM 1984).

Equation (13.55b) enables determination of wave heights in transitional or shallow water, knowing the deep water wave height when the relative spacing between orthogonals can be determined. In a more general case $K_R$ can be found from the refraction diagram directly by measuring b and $b_0$.

Various methods may be used for constructing refraction diagrams. The earliest approaches required the drawing of successive wave crests. Later approaches permitted the immediate construction of orthogonals, and also permitted moving from the shore to deep water (Arthur et at. 1952 and Kaplan 1952).

In general the seabed contours are neither straight nor parallel but curved (Fig.13.32). This causes a significant refraction effects such as convergence of orthogonals over the submerged ridge (Point 1), resulting in, higher wave heights, higher wave breaking and concentration of energy as such more erosion takes place here whereas over the submerged trough, (coves and Bays) (Point 2) the energy is dissipated due to diverging wave orthogonals resulting in lesser erosion.

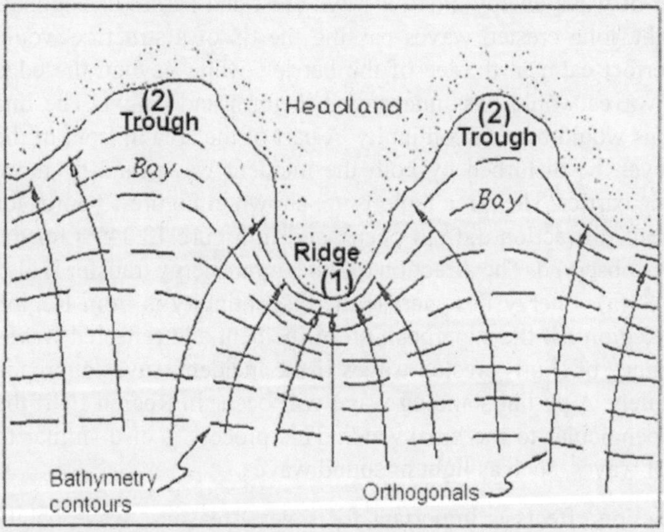


**Fig.13.32.** wave refraction in ridges and bays

## 13.6. Wave Diffraction

Diffraction of water waves is a phenomenon in which energy is transferred laterally along a wave crest. It is most noticeable where an otherwise regular train of waves is interrupted by a barrier such as a breakwater or small island as shown in Fig.13.33.

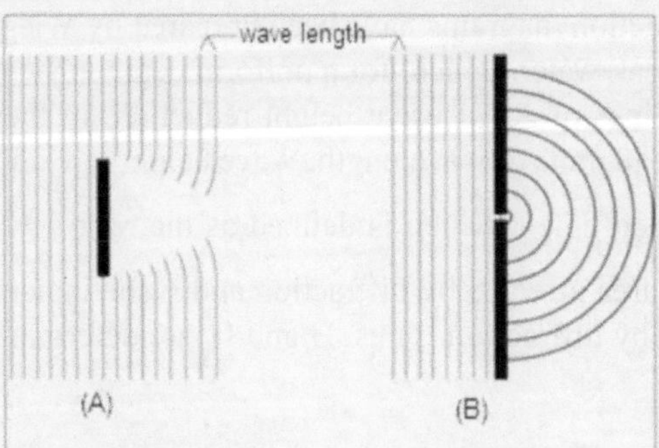


**Fig.13.33.** Diffraction of waves through a slit or orifice (small opening) of a break water. Note the obstacle is longer than the wavelength in (A) and the opening in the obstacle is shorter than the wavelength.

In the case of Fig.13.33A, the obstacle is longer than the wavelength so there is little diffraction whereas in the case of Fig.13.33B, the opening in the obstacle is shorter than the wavelength so there is lot of diffraction.

If the lateral transfer of wave energy along a wave crest and across orthogonals did not occur, straight, long-crested waves passing the tip of a structure would leave a region of perfect calm in the lee of the barrier, while beyond the edge of the structure the waves would pass unchanged in form and height. The line separating two regions would be a discontinuity. A part of the area in front of the barrier would, however, be disturbed by both the incident waves and by those waves reflected by the barrier. The three regions are shown in Figure 13.34(a) for the hypothetical case if diffraction did not occur, and in Figure 13.34(b) for the actual phenomenon as observed. The direction of the lateral energy transfer is also shown in Figure 13.34(a). Energy flow across the discontinuity is from Region II into Region I. In Region III, the superposition of incident and reflected waves results in the appearance of short- crested waves if the incident waves approach the breakwater obliquely. A partial standing wave will occur in Region III if the waves approach perpendicular to the breakwater. This process is also similar to that for other types of waves, such as light or sound waves.

Calculation of diffraction effects is important for several reasons. Wave height distribution in a harbor or sheltered bay is determined to some degree by the diffraction characteristics of both the natural and manmade structures affording protection from incident waves. Therefore, knowledge of the diffraction process is essential in planning such facilities. The proper design and location of harbor entrances to reduce such problems as silting and harbor resonance also require knowledge of the effects of wave diffraction. The prediction of wave heights near the shore is affected by diffraction caused by naturally occurring changes in hydrography.

For calculation of diffraction, diffraction diagrams have been prepared by Wiegel (1962) for waves passing single break water, which, for a uniform depth adjacent to an impervious structure, show lines of equal wave height reduction. In these diagrams, the graph coordinates are in units of wavelength. Wave height reduction is given in terms of a diffraction coefficient which is defined as the ratio $\left(\dfrac{H}{H_i}\right)$ where H is the wave height in the area affected by diffraction and is the incident wave height in the area unaffected by diffraction. Thus, H and $H_i$ are determined by $H = K' H_i$.

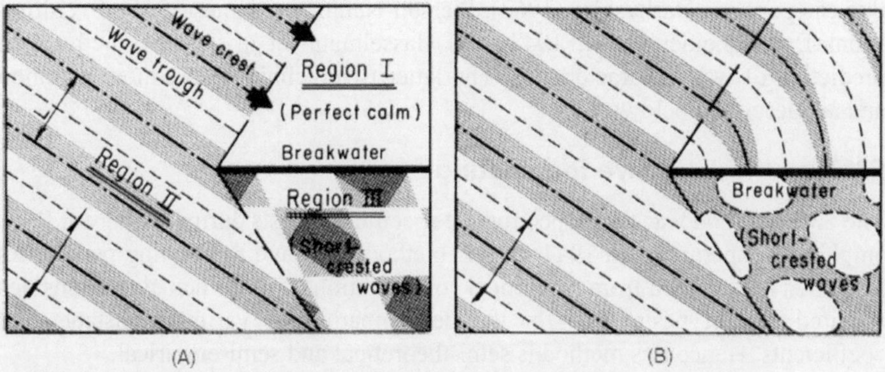


**Fig. 13.34.(A):** Without wave diffraction; (B): with wave diffraction (SPM,1984)

## 13.7. Wave Forecasting

The wave forecasting for practical purposes has gained importance during world war II for planning and landing operations of different Naval ships and submarines, particularly on the shallow beaches. It is also important from both the design and operational point of view.

The calculation of wave characteristics from the weather maps is known as hind-casting. This hind-casting technique is useful for obtaining statistical information on the possible wave conditions expected at a given location. This statistical information is also useful for choosing the equipment needed or for choosing the suitable time of the year for marine operations.

Wave forecasting may also be used to predict the wave dimensions at a given location on a day-to-day basis. This information can be used whether a dredge can operate, a mobile offshore oil platform can be moved, a barge can be towed etc.

### Forecasting Procedures

The wave forecasting techniques can be classified into two broad types viz., (i) the simplified or parametric and (ii) the elaborate or numerical. The former methods explicitly give wave height and period from the knowledge of wind-speed, fetch and duration while the later ones require numerical solution of the equation of wave growth. The numerical methods are far more accurate than the parametric and give information over a number of locations simultaneously. They however require a number of oceanographic and meteorological parameters. They are more justified when the wind speed varies considerably along with its direction in a given time duration and area. When the wind field can be assumed to be fairly stationary and when accurate and elaborate wind data are not available, simplified parametric wind-wave relationships, involving an empirical treatment, could be a workable alternative to the elaborate techniques. Common methods under this category are Darbyshire (1963), Pierson-Neumann- James(1955), Sverdrup-Munk-Bretschneider (1944, 1947) and Hasselmann et al, (1976), methods of prediction of wave characteristics. The latter two techniques are more common and are described below:

## SMB method of wave forecasting

The SMB method was developed for site specific forecasts during world war II for amphibious operations in 1944. In this method the wave forecasting parameters presented are evolved from theoretical considerations, but the actual relationships required certain basic data for the determination of various constants and coefficients. Hence this method is semi-theoretical and semi-empirical.

The SMB method entails two important parameters. They are significant wave height and significant wave period. These two are considered important because around which the wave energy is concentrated.

The significant wave height is the mean or average wave height of the highest one third of all the waves present in the wave train. The significant wave period represents the mean period of the significant wave height.

This method works on duration and fetch limited cases. In the duration limited case, the significant height ($H_{1/3}$) and significant wave period ($P_{1/3}$) depend on the wind duration. In fetch limited case, which is quasi-steady state, the wave dimensions depend upon the fetch. So before forecasting it is necessary to distinguish between these two cases and proceed accordingly. A third case of wind-speed limited also introduced under 'fully developed sea'.

The wave parameters required for SMB method can be obtained from dimensional considerations as detailed below:

The factors on which the wind wave parameters for deep water depend are the wind velocity U, the fetch 'F' and the wind duration 't'. The wave parameters required are wave height 'H' and wave period 'T' such that in deep water the wave length and celerity (velocity) are:

$$L = \frac{g}{2\pi} T^2 \ and \ C = \frac{g}{2\pi} \mathrm{T}$$

So 'C' and 'H' are functions as below:

$$\mathrm{C} = \mathrm{f}_1 \ (\mathrm{U,F,t,g}) \tag{13.56}$$

$$\mathrm{H} = \mathrm{f}_2 \ (\mathrm{U,F,t,g}) \tag{13.57}$$

The above two equations contain five variables and two dimensional units. Hence equation (13.56) can be written as:

$$0 = \mathrm{F}_1 \ [(\mathrm{C\ U^a\ g^b})(\mathrm{F\ U^c\ g^d})(\mathrm{t\ U^e\ g^f})]$$

Equating to unity the sum of the exponents for the corresponding dimensions we can get

$$0 = F_1\left[\left(\frac{C}{U}\right)\left(\frac{gF}{U^2}\right)\left(\frac{gt}{U}\right)\right]$$

Or

$$\frac{C}{U} = \psi_1\left[\left(\frac{gF}{U^2}\right),\left(\frac{gt}{U}\right)\right] \tag{13.58}$$

In a similar manner equation (13.57) can be written as:

$$\frac{gH}{U^2} = \psi_2\left[\left(\frac{gF}{U^2}\right),\left(\frac{gt}{U}\right)\right] \tag{13.59}$$

Equations (13.58) and (13.59) represent the wave generation parameters for deep water, based on dimensional considerations, $\psi_1$, and $\psi_2$ are functional relations that must be determined by the help of wave data.

The quantities $\frac{gH}{U^2}, \frac{C}{U}, \frac{gF}{U^2}$ *and* $\frac{gt}{U}$ respectively are the parameters of wave height, velocity, fetch and wind duration. The wave speed parameter $\left(\frac{C}{U}\right) = \left(\frac{gt}{2\pi U}\right)$ can also be used as as $\frac{gt}{2\pi U}$ as T can be easily measured than the wave speed. Using the above dimensionless parameters Bretschneider revised the original forecasting relations of Sverdrup and Munk utilizing much additional wave data. So these forecasting relationships have acquired the name SMB method.

The Sverdrup-Munk and Bretschneider (SMB) equations thus are based on dimensional analysis considerations. These equations are suggested for deep water (where depth may exceed about 90 m) are given below (SPM 1984). The wind of speed (u) blowing over fetch (F) will produce the $H_s$ and $T_s$ values according to the following equations:

$$\frac{gH_s}{u^2} = 0.283\tanh\left[0.0125\left(\frac{gF}{u^2}\right)^{0.42}\right] \tag{13.60}$$

$$\frac{gT_s}{u} = 2.4\pi\tanh\left[0.077\left(\frac{gF}{u^2}\right)^{0.25}\right] \tag{13.61}$$

The above $H_s$, $T_s$ values would occur only if the wind blows for a duration given in terms of fetch 'F' as follows:

$$\frac{gT_{min}}{u} = 68.8\left[\left(\frac{gF}{u^2}\right)^{0.67}\right] \tag{13.62}$$

If actual duration < $t_{min}$, then find F from equation (13.62) for the given t and then substitute the new F value in equation (13.60) and (13.61). This is duration limited sea (with fetch controlled by duration). If $\geq t_{min}$, the wave heights and periods are controlled by the given fetch.

A graphical representation of the above equation is known as SMB (Sverdrup, Munk, Bretschneider) curves for deep water. Fig. 13.35 shows the same. In this Fig. 13.35 there are five sets of lines along which one parameter is kept constant. Horizontal lines represent constant (corrected or adjusted) wind speed, $U_A$ , while the vertical ones show constant fetch length (F). These two belong to the input parameters along with the third one represented by the inclined set of lines for constant wind duration (t). The output of significant wave height ($H_S$) and associated wave period ($T_S$) can be seen in terms of corresponding isolines in the Fig. 13.35. In order to use these curves proceed horizontally from the given $U_A$ value and find intersection of this horizontal line (interpolated, if necessary) with the vertical line coming up from the given F value. Then find out $H_S$ and $T_S$ values by interpolation. Similarly proceed horizontally from given $U_A$ line and find its intersection with given duration line and get another set of $H_S$ and $T_S$ values. The smaller set of $H_S$ and $T_S$ values is the required answer. If these values are obtained on the basis of given fetch then the sea is called fetch controlled. Otherwise it becomes duration controlled. Alternatively we can get the required significant height and period after stopping at those values wherever the given F or t is reached first.

### *Example 13.1*

Obtain the values of the significant wave height and period in deep water generated by wind of (corrected) speed of 24.2 m/sec lasting for 3 hours over a fetch of 10 Km. State whether the sea is fetch or duration controlled?

### *Solution*

By following the above procedure we get the final values as:

Significant wave height = 1.25m

Significant wave period = 3.9 sec.

The sea is fetch limited

### *Example 13.2*

Obtain the changed values of the significant height and period in the above example, if the same wind has a fetch of 100 km instead of 10 km.

### *Solution*

Following the same procedure as above,
Significant wave height = 2.2 m
Significant wave period = 5.7sec
The sea is duration limited.

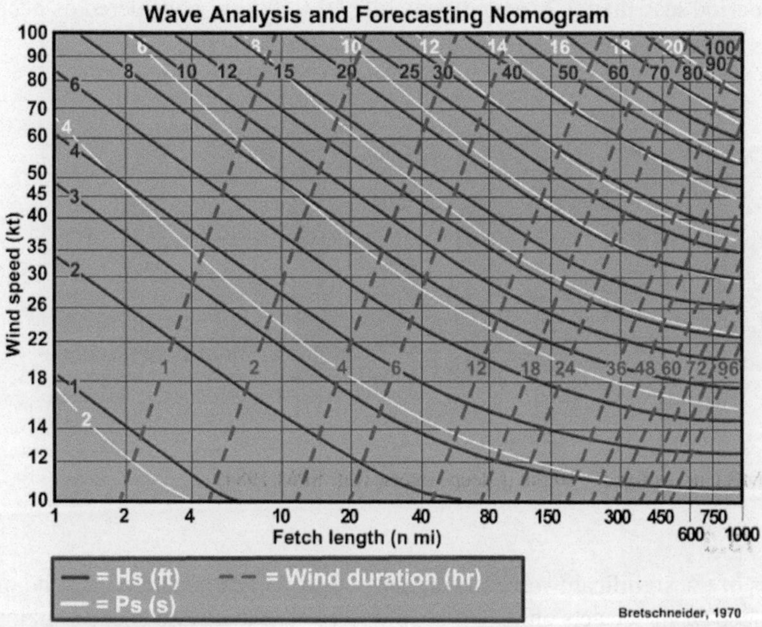


**Fig.13.35.** Forecasting wave height and period if wind speed and fetch and duration are known in deep water according to Bretschneider, (1970)

In shallower water of depth (d) the three equations equivalent to equation (13.60), (13.61) and (13.62) are, respectively

$$\frac{gH_s}{u^2} = 0.283\tanh\left[0.53\left(\frac{gd}{u^2}\right)^{0.75}\right]\tanh\left\{\frac{0.0125\left(\frac{gF}{u^2}\right)^{0.42}}{\tanh\left[0.53\left(\frac{gd}{u^2}\right)^{0.75}\right]}\right] \tag{13.63}$$

$$\frac{gT_s}{u} = 7.54\tanh\left[0.833\left(\frac{gd}{u^2}\right)^{0.375}\right]\tanh\left\{\frac{0.077\left(\frac{gF}{u^2}\right)^{0.25}}{\tanh\left[0.833\left(\frac{gd}{u^2}\right)^{0.375}\right]}\right] \ldots\ldots(13.64)$$

$$\frac{gT_{min}}{u} = 6.5882exp\left\{\left[0.016\left(ln\left(\frac{gF}{u^2}\right)\right)^2 - 0.3692ln\left(\frac{gF}{u^2}\right) + 2.2024\right]^{0.5} + 0.8798\,ln\left(\frac{gF}{u^2}\right)\right\} \tag{13.65}$$

These equations (each drawn for a separate water depth), are available in the graphical forms in the case of curves for the shallow water. Fig 13.36 shows an example corresponding to water of depth 10.5m. Waves in water depth of 10.5 m with wave period less than 3.7 seconds (i.e., $d/T^2 < 0.78$) are considered as deep water.

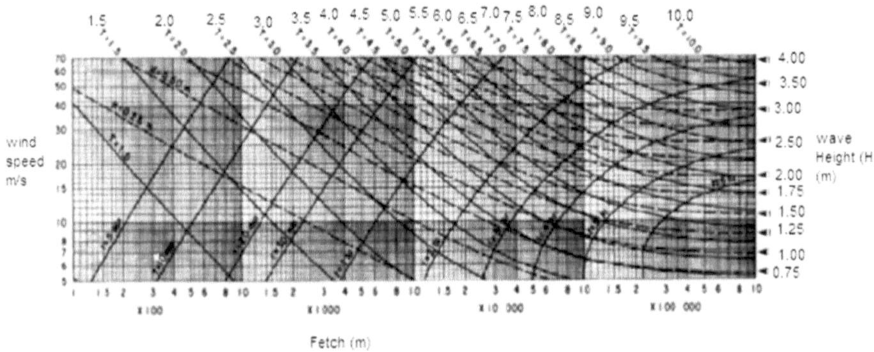


**Fig. 13.36.** SMB Curves for d = 10.5m ,( deep water)( Ref: SPM 1984)

### *Example 13.3*

What values of the significant wave height and period will be generated if wind of corrected speed 22 m/sec acts on 10.5m deep water and if this wind speed remains same over a fetch distance of 24.4 km and a duration of 5 hours?

### *Solution*

From Fig.13.36, one can easily find, Significant wave height = 1.5m

Significant Wave period = 4.4 sec.

### Wilson method

When the state of the sea is limited by fetch (or wind duration) if the fetch is unlimited, one would expect the spectrum. Then the significant wave height and period are the functions of fetch as well as the wind speed. The two variables may be combined by introducing the concept of 'dimensionless fetch' as defined by $\frac{gF}{U^2}$. The prediction equations may then be expressed in dimensionless form, giving the dimensionless significant wave height $\frac{gH_S}{U^2}$ and dimensionless wave velocity $\frac{C}{U}$. Where U is wind speed and $H_S$ is significant wave height. Keeping this concept in view, the following empirical forecasting equations were derived by Wilson in 1966.

$$\frac{C}{U_{10}} = 1.37\left[1-\left\{1+0.008F^{1/3}\right\}\right]^{-5}$$

$$\frac{gH_S}{U_{10}^2} = 0.30\left[1-\left\{1+0.004F^{1/2}\right\}\right]^{-2}$$

These equations are valid only for large values of F. For small values of F the equations given are

$H_S \propto F^{1/2}$ and $T_m \propto F^{1/3}$

Where $T_m$ is the period corresponding to the maximum frequency.

## Hasselmann method

Hasselmann et al (1976) developed a simplified parametric model of wave growth to obtain the Hs and Ts values for given quantities of u and F. These equations along with the one that gives the minimum duration necessary to produce these values of Hs and Ts are given below:

a) For Deep Water:

$$\frac{gH_s}{u^2} = 0.0016\sqrt{\frac{gF}{u^2}} \tag{13.66}$$

$$\frac{gT_s}{u} = 0.2857\left(\frac{gF}{u^2}\right)^{1/3} \tag{13.67}$$

$$\frac{gt_{min}}{u} = 68.8\left(\frac{gF}{u^2}\right)^{2/3} \tag{13.68}$$

b) For Shallow water:

$$\frac{gH_s}{u^2} = 0.283\tanh\left[0.53\left(\frac{gd}{u^2}\right)^{0.75}\right]\tanh\{\frac{0.00565\left(\frac{gF}{u^2}\right)^{\frac{1}{2}}}{\tanh\left[0.53\left(\frac{gd}{u^2}\right)^{0.75}\right]}] \tag{13.69}$$

$$\frac{gT_s}{u} = 7.54\tanh\left[0.833\left(\frac{gd}{u^2}\right)^{0.375}\right]\tanh\{\frac{0.0379\left(\frac{gF}{u^2}\right)^{0.33}}{\tanh\left[0.833\left(\frac{gd}{u^2}\right)^{0.375}\right]}] \tag{13.70}$$

$$\frac{gt_{min}}{u} = 537\left[\left(\frac{gF_s}{u}\right)^{\frac{7}{3}}\right] \tag{13.71}$$

The graphical forms of these equations are shown in Figures13.37 and 13.38.

### *Example 13.4*

For a wind of corrected speed 20 m/sec remaining constant over a fetch distance of 30 Km, obtain Hs and Ts values using Hasselmann technique, if (a) water is very deep and if (b) water depth is 2.86 m. Also obtain the minimum duration necessary to generate above values of Hs and Ts in deep water.

## *Solution*

(a) Deep water case:

$$\frac{gF}{U^2} = \frac{9.81 \times 30{,}000}{20^2} = 737.75$$

In Fig 13.37, proceeding vertically from this value and noting the point of intersection of this vertical line with the inclined one corresponding to the 'Deep water wave, we get,

$\frac{gH_s}{U^2} =$ 0.044. This gives Hs = 1.79 m.

In Fig 13.38 following the similar procedure, we get,

$\frac{gT_s}{U}$ = 2.6. This gives Ts = 5.4 s.

(b) Shallow water case:

$$\frac{gd}{U^2} = \frac{9.81 \times 2.86}{20^2} = 0.07$$

From Fig 13.37 proceeding vertically from $\frac{gF}{}$ = 737.75 and noting the point of intersection with the curve (imagined) corresponding to $\frac{gd}{U^2}$ = 0.02, This gives Hs = 0.82 m.

Similarly from Fig. 13.38, $\frac{gT_s}{U}$ = 1.8. This gives Ts = 3.67 s.

The minimum duration necessary to generate Hs = 1.79 m and Ts = 5.5 s in deep water is given by equation 13.62,i.e.,

$$\frac{gt_{min}}{U} = 68.8\left(\frac{gF}{U^2}\right)^{\frac{2}{3}}$$

Hence, $t_{min} = 68.8\left(\frac{9.81 \times 30000}{20^2}\right)^{\frac{2}{3}}\left(\frac{20}{9.81}\right)$ = 3.17 expressed in hours.

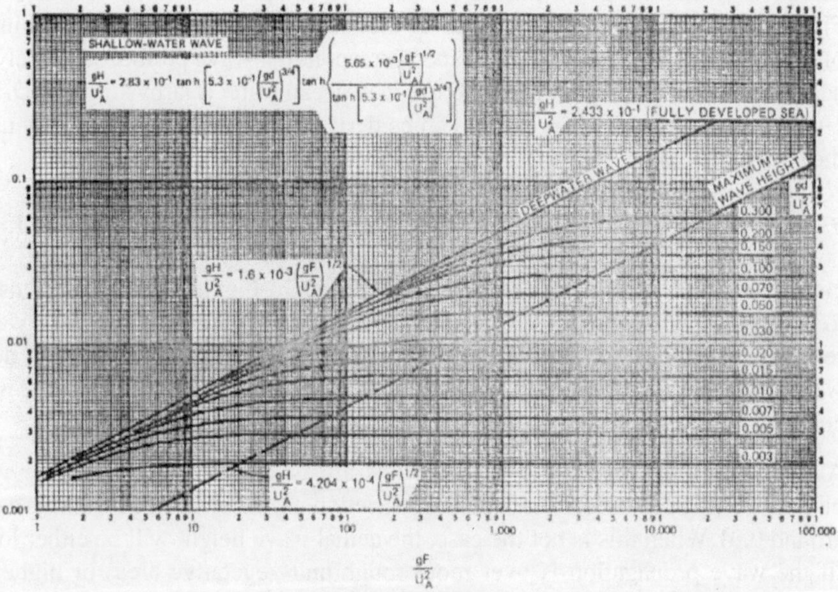


**Fig. 13.37.** Hasselmann Curves for H (Refer SPM 1984)

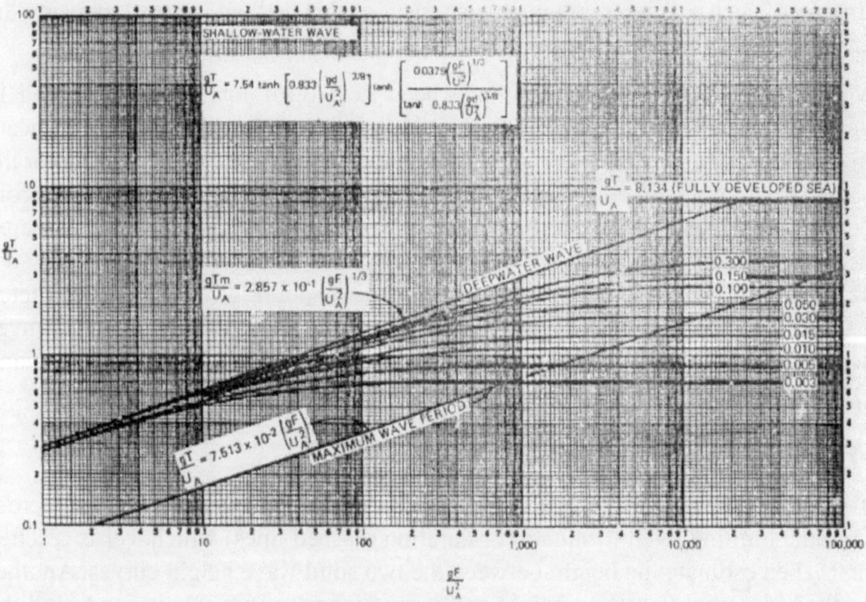


**Fig. 13.38.** Hasselmann Curves for T (Refer SPM 1984)

**Darbyshire and Draper's Technique:** Another widely used alternative wave prediction technique is that developed by Darbyshire and Draper (1963). This can be conveniently given in terms of the curves as in Fig 13.39.The curves in Figures 13.40 to 13.43 give the value of the maximum wave height in 10 minutes of the observation period, i.e., $H_{10}$, and corresponding wave period, separately for coastal waters (up to 45 m deep) and for deep ocean water (Darbyshire and Draper 1963). The significant wave height can be derived from this value of $H_{10}$ by using the equation:

$$H_s = \frac{H_{10}}{1.6} \tag{13.72}$$

Given the value of wind speed, fetch and duration, Fig 13.40& 13.41 is useful to know the wave height ($H_{10}$) and period (Ts) in coastal (shallow) waters respectively while Fig 13.42 & 13.43 help to know the same in oceanic (deep) waters. Application of these curves follow the same procedure as that for the SMB curves.

Above simplified wind-wave relationships are generally valid when the wind blows over a fetch area over which the friction factor is averaged and is taken around 0.01.When this is not the case, the actual wave height will be either lower (if the wave propagation is over more rough and vegetative area) or higher (if the waves move over smoother sea bed where the friction is less than the stress exerted by the wind). SPM (1984) gives a method to calculate the wave height at the end of such a different fetch characteristics. This method is based on using the same wind-wave relationships with an adjustment of the fetch value.

Most wave prediction schemes have been developed for waves generated in deep water. So Darbyshire and Draper(1963) deduced that there were significant differences in the properties of waves generated in deep and shallow waters for the same speed and fetch as for example given in Figures 13.40 to 13.43. Therefore he produced separate prediction schemes for the two cases. So he prepared nomograms that give prediction for significant wave height and zero upcrossing period ($T_Z$) in coastal waters (shallow waters) as functions of wind speed and fetch or duration. Nomograms also were prepared for prediction of deep water waves separately by PNJ (1955), Bretschneider (1970)(Fig.13.35) and Darbyshire and Draper(1963)(Fig.13.39).

### Wave prediction by nomograms

To predict the wave height, enter with wind speed at left hand side, move across until the limiting fetch (ordinate) or duration (dashed lines) whichever is reached first. Then estimate the height between the two solid wave height curves. Another method is to find the wave height and period for the given wind speed, find the meeting points of the curves of duration as well as fetch separately. Take the lower values of wave height and period out of the two sets obtained from these two curves.

### *Example 13.5: a) Fetch limited case*

Estimate the significant wave height if a wind speed of 30 m/s blows for 36 hours over an extent of 300 km in deep water.

### *Solution*

Starting with 30 m/s on left hand side ordinate of Fig.13.39 and proceed towards right hand side until either F = 300 km or t = 36 hrs whichever first is reached. When we proceed along 30 horizontal first it touches 300 km ordinate, as 36 hrs light green curve is far away. The point of intersection of 30 m/s and 300 km fetch is in between 8 and 10 m solid red curves. As the meeting point is close to 10 m curve but less than that, the height may be around 9 m.

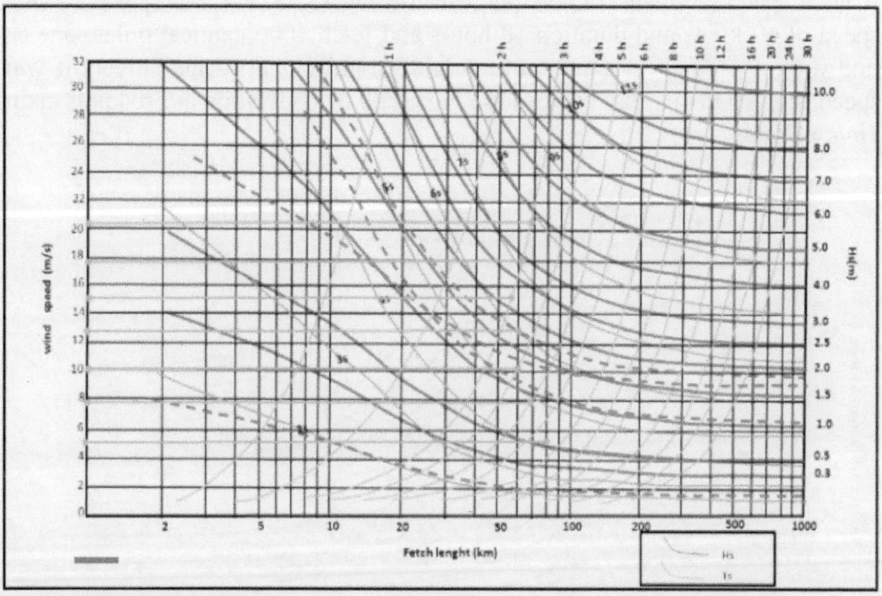


**Fig.13.39.** Forecasting wave height and period if wind speed, duration and fetch are known in deep water. The x-axis is fetch, y-axis is wind speed, Light green curves: duration of wind speed, orange red curves: wave height, Navy blue curves: wave period. Please note we have to find wave height and wave period from red and navy blue curves. (According to Darbyshire and Draper, 1963.

### *Example 13.6*

Similarly find the significant wave height and period if the wind speed is 30 m/s blown over a stretch of 1000 km for 21 hours.

### *Solution*

Taking the wind speed 30 m/s move horizontally in the duration limited nomogram (Fig.13.39). Then we can find that t = 21 hrs is reached first as 1000 km fetch is farther off. So the meeting point of 30 m/s and 21 seconds is on 11.5 seconds significant period (Navy blue curve). Hence significant wave period is $T_s = 11.5$ seconds.

From Fig.13.39 we can find significant wave height using wind speed 30 m/s and significant wave period is 11.5 seconds or duration of wind is 21 hours which is around 10m (red curve).

### *Example 13.7*

If wind speed is given in knots one can use Fig.13.35. For example for a wind speed of 60 knots and duration 18 hours and fetch 1000 nautical miles, one can find wave period as 14 s and wave height as 40ft (12 m). Alternatively if wind speed is given in m/s, still one can use Fig.13.39 by converting m/s to knots taking $1 \text{m/s} \approx 2 \text{knots}$.

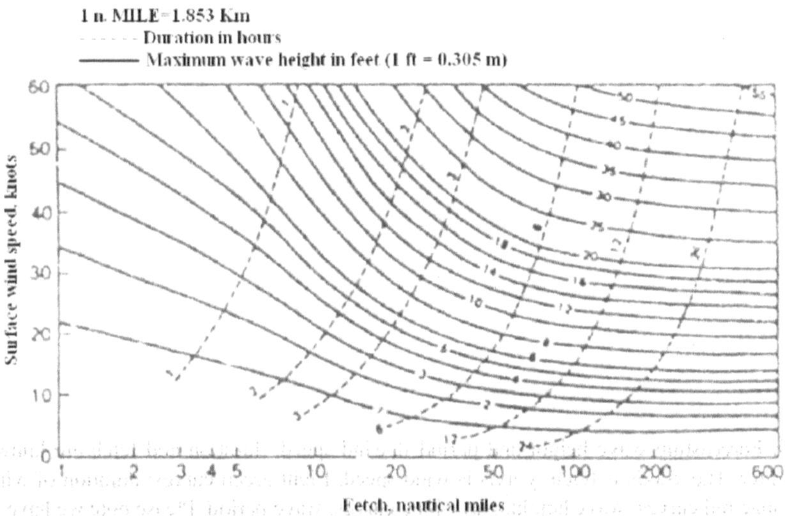


**Fig.13.40.** $H_{10}$ for coastal waters (After Draper and Darbyshire 1963; taken from Bretschneider, 1964) (Ref. Brebbia and Walker, 1979)

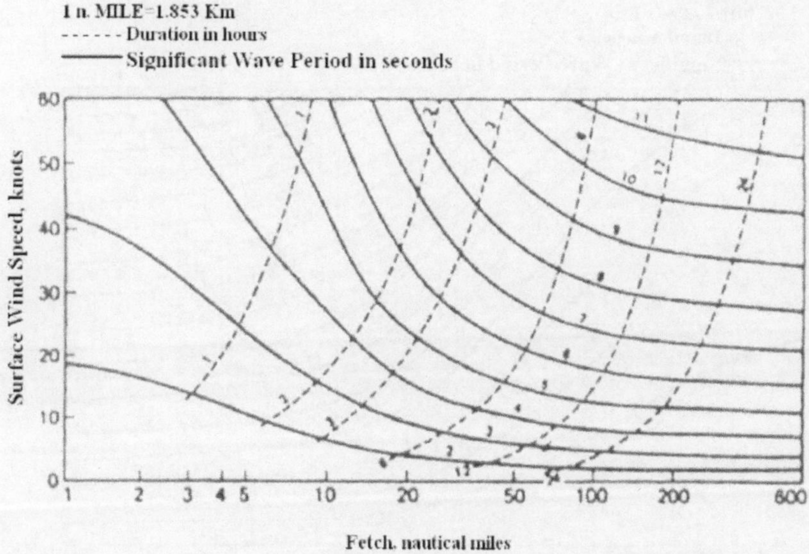


**Fig. 13.41.** $T_S$ for Coastal Waters (After Darbyshire and Draper 1963, taken from Bretschneider, 1964) (Ref. Brebbia and Walker, 1979)

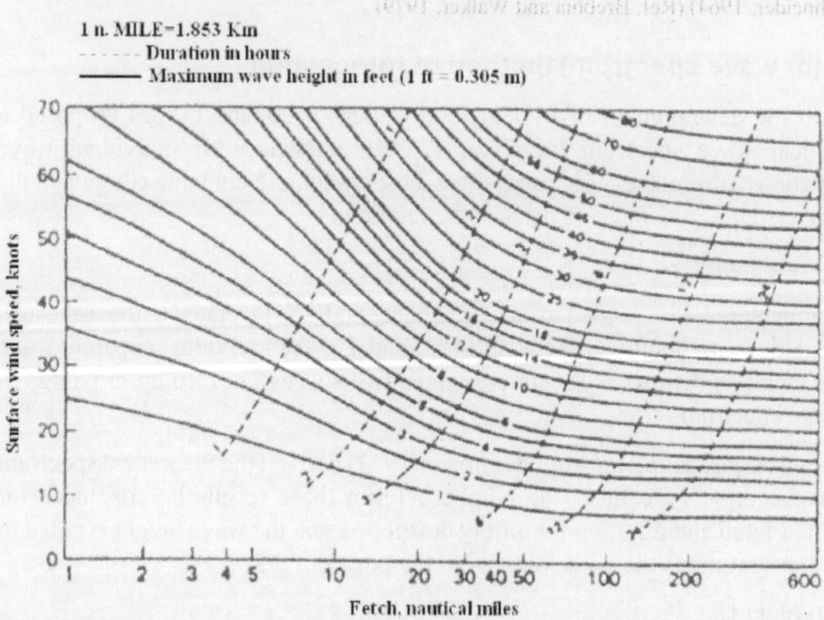


**Fig. 13.42.** H10 for Oceanic Waters (After Darbyshire and Draper 1963; taken from Bretschneider, 1964) (Ref. Brebbia and Walker, 1979)

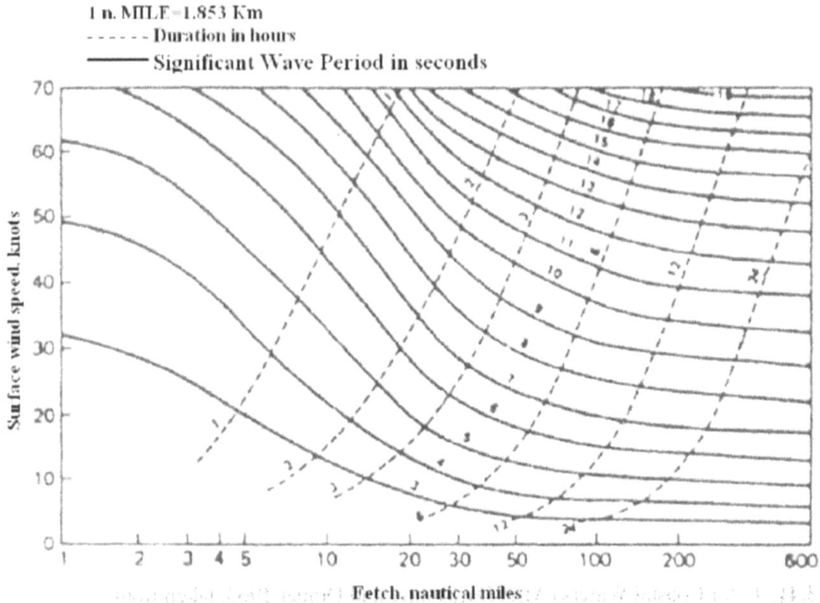


**Fig. 13.43.** TS for Oceanic Waters (After Darbyshire and Draper 1963; taken from Bretschneider, 1964) (Ref. Brebbia and Walker, 1979)

## PNJ (or wave spectrum) method of forecasting

Prior to the development of PNJ method (1958), Neumann (1952) proposed a theoretical wave spectrum based on a great abundance of individual wave observations. From the wind and wave observations, Neumann computed the parameters $H/T^2$ and T/U and derived the relation:

$$\frac{H}{gT^2} = Constant\, e^{-\{\frac{gT}{2\pi U}\}^2}$$

The parameter $H/T^2$ is related to wave steepness and T/U is related to the wave age. Where H is wave height, U the wind speed and T is the individual apparent wave period which is defined as the time between two successive zero up-crossings of the wave spectrum.

From this empirical relationship, Neumann(1953) derived the theoretical spectrum of wave energy 'E' called 'one number'. From these results he concludes for unlimited fetch and duration for a fully developed sea, the wave height is equal to some constant times the wind velocity to the power of 5/2.

$H = \text{constant } U^{5/2}$

The concept of the wave spectrum is certainly a great advance in trying to understand the nature of wave generation. It is shown that the waves generate from the high frequency side of the spectrum and with this concept the relationships for the young and the transient state of the sea can be derived. That is, these developed relations are called co-cumulative power spectra from which 'E' values can be predicted.

Longuet-Higgins (1952) later derived the theoretical wave distribution that is related to E-values as:

Average wave height data (in feet)

i) The most frequent wave height = 1.41 $\sqrt{E}$

ii) The average wave height (($\bar{H}$) = 1.772 $\sqrt{E}$

iii) Significant wave height ($H_S$) or ($H_{1/3}$) = 2.832 $\sqrt{E}$

iv) Average of the height of 1/10th of the highest waves ($H_{1/10}$) = 3.600 $\sqrt{E}$

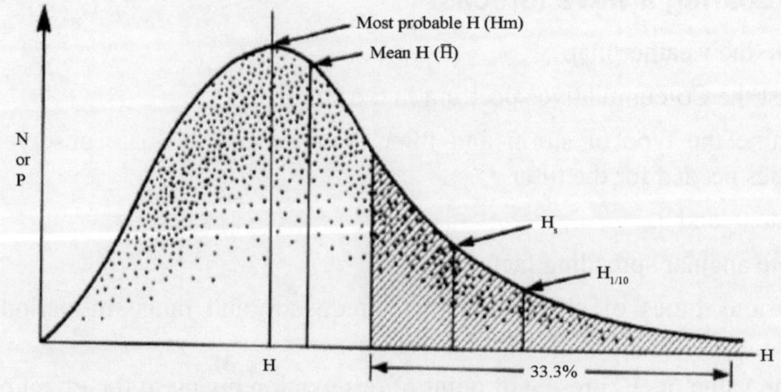


**Fig.13.44.** Statistical distribution of wave height ranges

The Figure 13.44 shows the distribution of wave heights in a histograph which is a graph of number of waves (or percentage of total waves) in each wave height range. The significant wave description is a simple and practical means of dealing with problems in wave forecasting. As the waves in the sea are very complex and made up of many variable heights and periods, one important wave height which represents the spectrum is important, that is significant wave height ($H_{1/3}$) as shown in the above figure 13.44.

The significant wave height is the average of the highest one third of the waves in a given wave train which is a statistical parameter used by both SMB and PNJ methods. Whereas the significant wave period is the average period of the highest one-third wave heights and is used only by SMB method while PNJ uses a mean apparent wave period and ranges in wave period.

Further the average wave height ($\bar{H}$), the significant wave height ($H_{1/3}$) and the average of the highest 10% of waves ($H_{1/10}$) are related as:

($\bar{H}$) = 0.625 $H_{1/3}$

$H_{1/10} = 2.03\ \bar{H} = 1.27\ H_{1/3}$

Longuet-Higgins found these relations and said about 16 or 17% of the waves will be higher than the significant wave height.

Material needed for wave forecasting according to PNJ:

1. Weather map and weather data
2. Co-cumulative spectra
3. Dispersion diagrams
4. Angular spreading factor diagrams
5. Height table

## Steps in preparing a wave forecast

1. Analyze the weather map
2. Forecast the Co-cumulative spectrum in the storm area
3. Determine the type of storm and filter required and evaluate observed quantities needed for the filter.
4. Evaluate $\theta_4$ and $\theta_3$
5. Evaluate angular spreading factor
6. For various times of observation find frequency and range of periods present.
7. Find the value of 'E' present at point of observation owing to the effect of dispersion by subtracting the value of 'E' at the upper frequency from the value of E at the lower frequency.
8. Multiply by angular spreading factor to find the forecast value of E.
9. From E compute the needed height data like significant, most frequent, average or $1/10^{th}$ of the height using Longuet-Higgins statistics.

## Calculation of 'One number ' E

To compute E, the value of the wave records is read off at a hundred or so equally spaced points with any arbitrary zero reference level. An average value is computed from these values and this average is subtracted from each of the original readings. Then each resultant number is squared and then the total is found. Then this total if multiplied by two is called the number 'E'.

Let suppose the wave data is $n_1$ $n_2$, $n_3$, and so on upto $n_i$. Then the average of these values is say $\bar{n}$. Then the difference of average from each of the wave data is $(n_i - \bar{n})$. Then the square of the difference is $(n_i - \bar{n})^2$. Then twice the sum of the squares of all values (variance) is "E"

$E=2\,[\sum_{i=1}^{i}(n_i-\bar{n})^2]$

For example computation of E is in the following way:

| Wave height | No. of waves ($n_i$) | Diff. with average | Square of the difference |
|---|---|---|---|
| 1 | $n_1$ | $n_1-\bar{n}$ | $(n_1-\bar{n})^2$ |
| 2 | $n_2$ | $n_2-\bar{n}$ | $(n_2-\bar{n})^2$ |
| 3 | $n_3$ | $n_3-\bar{n}$ | $(n_3-\bar{n})^2$ |
| Average | $\bar{n}$ | | $\sum_{i=1}^{i}(n_i-\bar{n})^2$ |

## Reliability and limitation in SMB and PNJ methods of wave forecasting

The value of the day-to-day forecasts has been subject to criticism by marine operators due to their consistencies and so they must be used with caution.

On comparing the two methods of SMB and PNJ, no method is appreciably better than the other. It is difficult to compare the results of the two methods of forecasting, as each set of forecasting curves is based largely upon its own set of original data and none of which can be considered to be adequate. There is no theoretical set of forecasting curves. These empirical and semi-empirical procedures rely on physical measurements. These physical measurements in the open ocean are few and unreliable. In fact there is not much difference between the spectrum method and significant method as far as the results are concerned.

Though at present considerable data has been obtained and used for the open ocean wave forecasting according to SMB method, still it is difficult to make reliable predictions for large fetches. The scatter of data is large and so it is difficult to see how predictions can be made more reliable. Additional information on the effect of air-sea temperature difference will improve the forecasts in future.

The relationships among winds and sea are not precise, especially when considering the moving fetches (moving storms) and varying wind fields. The numerical effects of these must be determined and incorporated in the wave forecasting data for a reliable forecast.

## Forecasting in cyclones

Above referred simple parametric forecasting models fail when wind conditions like, its speed, direction and profile rapidly change with time as in the case of the cyclones. It becomes difficult to forecast waves using simple equations in such situations. However, tropical cyclones exhibit relatively stable wind profile and hence they can be tackled by parametric modelling. For slowly moving cyclones, waves in deep water can be predicted by knowing (i) forward speed of the cyclone,

UF, (ii) radial distance from the eye of the cyclone to the point of maximum wind on isobar map & (iii) air pressure at the eye of the cyclone. At the point of maximum wind, the Hs and Ts values are given by, (SPM 1984):

$$H_s = 5.03\exp\left(\frac{R\Delta P}{4700}\right)\left\{1+\left[\frac{0.29\alpha U_F}{(U_R)^{\frac{1}{2}}}\right]\right\} \quad (13.73)$$

$$T_s = 8.60\exp\left(\frac{R\Delta P}{9400}\right)\left\{1+\left[\frac{0.145\alpha U_F}{(U_R)^{\frac{1}{2}}}\right]\right\} \quad (13.74)$$

Where : R = radius of maximum wind (km); ΔP = normal pressure (=760 mm of mercury) at cyclone center (mm); $U_F$ = wind speed along the cyclone forward direction; $U_R$ wind speed at R corresponding to maximum wind (at 10 m above MSL-sustained); $U_R = 0.865U_{max}$ if cyclone is stationary or $U_R = 0.865\ U_{max} + 0.5U_F$ if cyclone is moving.

Where $U_{max}$ = maximum gradient wind speed at 10 m above MSL

$= 0.447[14.5\ (\Delta P)^{1/2} - R(0.31\ f)]$ where f is Coriolis parameter, α = 1 if cyclone is slowly moving or α = f($U_F$, fetch) otherwise

Above equations give the values of Hs and Ts at the point of maximum wind. To find the significant wave height value at any other point, say $H_s'$, same Hs is required to be reduced by a reduction factor as given in SPM (1984) while corresponding $T_s'$ value is to be obtained by using the equation 13.75,

$$T_s' = \left(\frac{H_s'}{g}\right)^{\frac{1}{2}} \quad 13.75$$

## Numerical Wave Modelling

The numerical wave models deal with a spectrum of waves rather than unique wave height and period values of the above simplified schemes. They involve a detailed modeling of wave generation, propagation and dissipation mechanisms. They basically solve a differential wave energy balance equation given below in terms of a directional spectrum $G_\eta$:

$$\frac{\partial}{\partial t} G_\eta(f, \theta, \bar{x}, t) + \bar{C}_g(f, \theta).\nabla G_\eta(f, \theta, \bar{x}, t) = S \quad 13.76$$

Where $G_\eta$ (f, θ, $\bar{x}$, t) = Directional wave spectrum at wave frequency f and direction θ at given position $\bar{x}$, and time 't'. Note directional wave energy spectrum (Fig.13.45) gives energy of a wave component of certain frequency along a certain direction. $\bar{C}_g$ (f, θ) = Group velocity vector for wave frequency (f) and direction (θ). S = Source function = $S_{in} + S_{ds} + S_{nl}$ such that $S_{in}$ = wind energy input, $S_{ds}$ = wave energy dissipation in bottom friction and wave breaking; $S_{nl}$ = wave energy input being transferred from one wave frequency component to the other in a non-linear way.

Many numerical models employ a net source function (S) as shown in equation (13.76) rather than its separation into three parts as above. The source functions are based on some theoretical understanding and may require modifications based on measurements. The above governing differential equation, (13.76), is generally solved using finite difference schemes so as to obtain wave directional spectrum over a number of locations and over a series of time instants. This requires specification of initial temporal conditions and spatial boundary conditions. The directional spectrum may typically be resolved into finite number of frequencies and directions. Equation (13.76) is applicable for deep water and can be modified to account for shallow water effects like refraction and diffraction.

Resolution of the directional spectrum into discrete frequencies and directions is laborious and can be substituted by parameterizing it into assumed forms of wave spectrum and energy spreading function.

Actual waves at site may result from a combination of wind waves and swells arriving from a distant storm. In that case separate governing equations are required to be written.

There is a variety of numerical wave models used worldwide to obtain spatial wave forecasts with lead time of 6 to 72 hours (WMO, 1988). They can be classified as First Generation, Second Generation and Third Generation models- each indicating significant improvement in the wave modeling technique. The First Generation models, evolved in 1960s and 1970s, are the simplest. They assume growth of each wave spectral component independently. They are useful mainly in constant wind field. In the Second Generation model, the concept of a non-linear interaction between different wave components was introduced with simplified terms. These simplified terms are substituted by their exact solution in the Third Generation models. These can be usefully employed when the wind field is rapidly changing. In India, a Second Generation Model code named 'Dolphin' has been applied to make single location forecasts (Mandal, 1985).

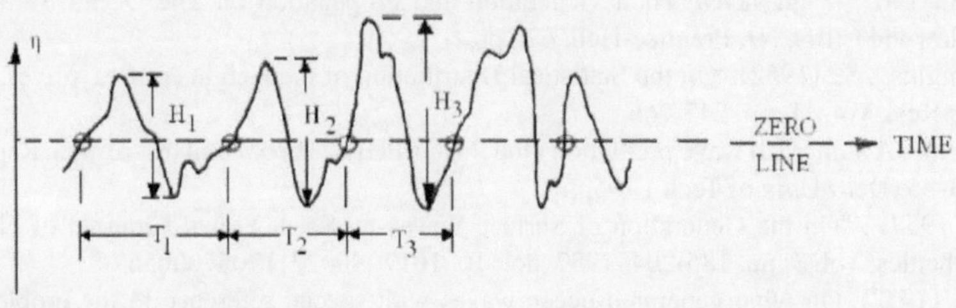


**Fig 13.45.** Individual Waves

## References

Airy, G.B., 1845, "Tides and waves", EncycLMetrop., London

Arthur, R.S., Munk, W.H., and Isaacs, J.D. (1952), "The Direct Construction of Wave Rays," Transactions, American Geophysical Union, Vol. 33, pp. 855–865.

ASN handout in academia.edu site: small amplitude waves https://berhampurunivrsitybhanjabihar.academia.edu/AyapillaMurty

Brebbia, C.A. and Walker, S. (1979). "Dynamic Analysis of Offshore Structures", Springer Verlag.

Bretschneider 1964: Generation of waves by wind state of the art, Presented at International Summer Course Lunteren, The Netherlands September 1-18, 1964, Netherlands University International Cooperation and North Atlantic Treaty Organization, NESCO Report SN-134-6

Bretschneider, C. L. 1969. "Wave Forecasting," Chapter 11 in Handbook of Ocean and Underwater Engineering, J. J. Myers et al., eds., McGraw-Hill Book Co., New York.

Darbyshire, M. (1963). "Wave Measurements made by the National Institute of Oceanography, " Proc. , 1961 Conf. on Ocean Wave Spectra, Prentice-Hall, pp. 111-132.

Darbyshire, M., & Draper, L. (1963). Forecasting wind-generated sea waves. Engineering, 195, 482-484.

Dean, R.G. and Dalrymple, R.A. (1984), Water Wave Mechanics for Engineers and Scientists, Prentice-Hall, Englewood CliVs, NJ.

Eckart C,. 1951, "Surface Waves on Water of variable Depth," SIO 51-12, Scripps Institute.of Oceanography

Fenton, John (2020): Coastal and ocean engineering, Institute of Hydraulic Engineering and Water Resources Management Vienna University of Technology, Karlsplatz 13/222, 1040 Vienna, Austria.

http://johndfenton.com/Lectures/Coastal-and-Ocean-Engineering/Coastal-and-Ocean.pdf

Hasselmann, K., D.B. Ross, P. Muller and W. Sell, 1976: A parametric wave prediction model. Journal of Physical Oceanography, 6(2):200–228, doi: 10 .1175/ 1520 0485(1976)006<0200:APWPM>2.0.CO;2.

Horikawa, K., Editor, (1988), Nearshore Dynamics and Coastal Processes—Theory; Measurement and Predictive Models, University of Tokyo Press, Tokyo.

Ippen, A.T. (1966), Estuary and Coastline Hydrodynamics, McGraw-Hill, New York.

Jeffreys, H., 1925.On the formation of water waves by wind, Proc. Roy.Soc. Lond. A, 107, 189-206

Kaplan, K.,1952 "Effective Height of Seawalls," Vol. 6, Bulletin No. 2, U.S. Army, Corps of Engineers, Beach Erosion Board, Washington, D.C., 1952.

Kinsman, B., 1965, WindWaves, Their Generation and Propagation on The Ocean Surface, Englewood Cliffs, NJ, Prentice-Hall, 676 pp.

Longuet-Higgins. , S., (1952):"On the Statistical Distribution of the Heights of Sea Waves," J. Mar: Res., Vol. 11, pp. 245-266.

Mandal ,S 1985: A numerical wave prediction model DOLPHIN. Theory and test results. Report No.3-85 (Delft Univ of Tech,) 195pp

Miles J.W (1957) : "On the Generation of Surface Waves by Shear Flows," Journal of Fluid Mechanics, Vol. 3, pp. 185-204, 1957.,doi: 10 .1017/ S0022112057000567

Neumann,G (1952): On wind generated ocean waves with special reference to the problems of wave forecasting. New York university, college of Engineering, Department of Meteorology and Oceanography. 136 pp, ONR contact Nonr-285 (05)

Neumann, G (1953): On ocean wave spectra and a new method of forecasting wind-generated sea, U.S.Beach Erosion Board, Technical Memorandum No. 43, 42pp

Neumann, G., and W. J. Pierson jr.,1966: Principles of physical oceanography, Prentice-Hall Inc., Englewood Cliffs (NJ).

P h i l l i p s , O. M-. (1957). "On the Generation of Waves by Turbulent Wind. " Journal of F l u i d Mechanics, Vol. 2, Part 5; pp. 417-445.

Pierson, W.J., JR., Neumann, G., and James, R.W.,1955 "Practical Methods for Observing and Forecasting Ocean Waves by Means of Wave Spectra and Statistics," Publication No. 603, U.S. Navy Hydrographic Office, Washington, D.C., 1955.

Sorensen, R.M. (1978), Basic Coastal Engineering, John Wiley, New York.

SPM (1984). "Shore Protection Manual", Coastal Engineering Research Centre, US Department of Army, Corps of Engineers, Waterways Experiment Station, Vicksburg, Mississippi. USA

Sverdrup, H.U. and Munk, W.H., 1944, "Wind waves and swell: Principles in forecasting", Hydrographic Office, U.S. Navy, Misc. 11, 275.

Sverdrup, H.U., and Munk, W.H., (1947) "Wind, Sea, and Swell: Theory of Relations for Forecasting," Publication No. 601, U.S. Navy Hydrographic Office, Washington, D.C., Mar. 1947., doi: 10 .5962/ bhl .title .38751.

Varma, 2014:, Finite amplitude ocean waves, Waves with Peaked Crests and Broad Troughs Resonance 1047- 1057.

WAMDI Group (Hasselmann, S., Hasselmann, K., Bauer, E., Janssen, P.A.E.M., Komen, G.J., Bertotti, L., Lionello, P., Guillaume, A., Cardone, V.C., Greenwood, J.A., Reistad, M., Zambresky, L. and Ewing, J.A.), 1988,"The WAM model - a third generation ocean wave prediction model", J. Phys. Oceanogr., 18, 1775-1810. U S Army Corps of Engineers, Coastal Engineering Manual, Part II, Chapter 1. pp.II-1.1-1.121, 2008.

Wiegel, R. L. 1954. Gravity Waves, Tables of Functions, Univ. of Calif., Council on Wave Res., The Engr.Found., Berkeley, CA.

Wiegel,R, L.1962, "Diffraction of Waves by a Semi-infinite Breakwater," J. Hydraulics Div., ASCE,Vol. 88, No. HY1, pp. 27-44, Jan. 1962.

Wilson, W.S. (1966), "A Method for Calculating and Plotting Surface Wave Rays," Technical Memorandum 17, U.S. Army Coastal Engineering Research Center, Washington, DC.

WMO, 1988 Guide to Wave Analysis and Forecasting. First edition (WMO-No. 702). Geneva

## Exercises

### *Exercise 13.1*

Find the wavelength, phase speed and group velocity for a wave of period 5 seconds and a basin depth of 1000 m. Prove that this is a deep water wave.

### *Solution*

We know the wave equation as

$$c^2 = \frac{g}{k}\tanh(kd) \Rightarrow \frac{L^2}{T^2} = \frac{gL}{2\pi}\tanh(kd) \Rightarrow \frac{L}{T} = C = \frac{gT}{2\pi}\tanh(kd) \qquad (1)$$

i) But in the case of deep water waves from equation (1),

$$\tanh(kd) \to 1$$

$$\Rightarrow C = \frac{gT}{2\pi} = \frac{9.8ms^{-2} \times T}{2\pi} = 1.56T = 1.56 \times 5s = 7.8ms^{-1}$$

ii) Similarly wavelength from equation (1) for deep water waves:

$$\tanh(kd) \to 1 \Rightarrow L = \frac{gT^2}{2\pi} = \frac{9.8ms^{-2} \times 25s^{-2}}{2\pi} = 1.56 \times 25s = 39m$$

iii) The condition for deep water waves is $\frac{d}{L} > \frac{1}{2} \Rightarrow \frac{1000m}{39m} = 25.6$ , this value is greater than 0.5. So it is deep water.

iv) The group velocity in the deep water is half of the phase velocity: $C_G = \frac{C}{2}$

$$\Rightarrow C_G = \frac{gT}{2\pi} = \frac{7.8ms^{-2}}{2} = 3.9ms^{-1}$$

## *Exercise 13.2*

Find the equations for wave velocity and length in the case of shallow waters

### *Solution*

The condition for shallow water waves is $\frac{d}{L} < \frac{1}{20} \Rightarrow \tanh\left(\frac{2\pi d}{L}\right) \to \frac{2\pi d}{L} \to kd$ ; hence the wave equation

$$c^2 = \frac{g}{k}\tanh(kd) \Rightarrow \frac{L^2}{T^2} = \frac{gL}{2\pi}\tanh(kd) \Rightarrow \frac{L}{T} = C = \frac{gT}{2\pi}\tanh(kd)$$

turns out to be:

$$c^2 = \frac{g}{k}(kd) \Rightarrow gd; L = \frac{gT^2}{2\pi}$$

## *Exercise 13.3*

A wave in water of 100 m depth has a period of 10 s and a height of 2 m. Determine the wave celerity, length, and steepness. What is the water particle speed at the wave crest?

### *Solution*

Assume that this is a deep water wave. Then, from Eq. $L_0 = \frac{gT^2}{2\pi} = \frac{9.8ms^{-2} \times 10^2 s^2}{2\pi} = 156m$

At 100 m, $d/L_0 = 0.641$,

From SPM tables given at the end for $d/L_0 = 0.641$, the value of $d/L = 0.641$, $\tanh(2\pi / \frac{d}{L}) = 0.9994$.

The wave celerity =

$$C = \frac{gT}{2\pi}\tanh\left(\frac{2\pi d}{L}\right) = \frac{9.8ms^{-2} \times 10 \times 0.9994}{2\pi} = 1.56 \times \times 10 \times 0.9994 = 15.6ms^{-1}$$

Or since the depth is greater than half of the calculated wave length, the wave is in deep water and the wave length is 156 m.

The deep water wave celerity Eq. is $C_0 = \frac{L_0}{T} = \frac{156m}{10s} = 15.6ms^{=1}$

and the steepness is $\frac{H_0}{L_0} = \frac{2m}{156m} = 0.013$

For deep water the particle orbits are circular having a diameter at the surface equal to the wave height. Since a particle completes one orbit in one wave period, the particle speed at the crest would be the orbit circumference divided by the period:

$$u_0 = \frac{2\pi \frac{H_0}{2}}{T} = \frac{\pi \times 2m}{10s} = 0.63ms^{=1}$$

Note that this particle velocity is much less than Co.

### Exercise 13.4

Consider the shallow water wave when it has propagated in to a near-shore depth of 2.3 m. Calculate the wave celerity and length.

### Solution

As this is a shallow water wave, wave velocity for shallow water is :

$$C = \sqrt{gd} = \sqrt{9.8ms^{-2}2.3m} == 4.75ms^{-1}$$

L = CT= 4.75ms$^{-1}$10s =47.5 m

Check whether this is shallow water or not: For shallow water wave d/L< 1/20 =0.50;

d/L =2.3m/47.5m = 0.048. As 0.048 is less than 0.5 it is shallow water.

### Exercise 13.5

Calculate the speed of wave in deep water (a) if its period is 20 seconds.( b) At what speed will a wave of wave length 312 m travel over deep water. (c ) At what speed will each of the waves in (a) and (b) travel in water of depth 12 meters. (d) What inference you will get from this example.

### Solution

Condition for shallow water waves: $\frac{d}{L} < \frac{1}{20} \Rightarrow \tanh\left(\frac{2\pi d}{L}\right) \rightarrow \frac{2\pi d}{L} \rightarrow kd$ ; $C^2 = gd$;

Condition for deep water waves: $\frac{d}{L} > \frac{1}{2}; C = \sqrt{\frac{gL}{2\pi}}$ ; $L = \frac{gT^2}{2\pi}$ ;$\frac{L}{T} = C = \frac{gT}{2\pi}$;

a) In the case of (a): T = 20 seconds, so L = 1.56 × 20$^2$ = 624 m

b) When d = 12 m, d/L = 12/624 = 1/52; 1/52< 1/20. So it comes under shallow water category. So shallow water wave velocity equation is to be used. Then the wave velocity according to shallow water equation is $C = \sqrt{gd} = \sqrt{9.8 \times 12}$

= 10.8 m/s

c) Similarly in case of (b), L = 312 m; so d/L = 12/312 = 1/26 which is also less than 1/20. So this also comes under shallow water category. So $C = \sqrt{gd} = \sqrt{9.8 \times 12}$ = 10.8 m/s

d) This example implies that in shallow water where d/L< 1/20 all waves travel at the same speed when they reach to that particular depth.

### *Exercise 13.6*

A generating area consists of 'seas' of wavelength 6m and 'swells' of wave period of 22 seconds i) calculate the speed of 'seas' and 'swells' ii) when these seas and swells propagate against a current of 5.0 m/s in a bay whether these can overcome and pass through it or not?

### *Solution*

**Seas:** Seas are short waves or deep water waves. So the equation is $C^2 = \frac{gL}{2\pi}$ = 1.56×6 or C= 3.06 m/s. But the deep water velocity in groups (group velocity) is $C_g$ = C/2 = 3.06/2 = 1.5 m/s

**Swells:** Swells are also generated in the generating area. So deep water wave equation used is $C = \frac{gT}{2\pi}$ = 1.56 × 22 = 34.32 m/s. The deep water group velocity is $C_g$ = C/2 = 34.32/2 = 17.16 m/s

**Propagation in the Bay against the current 5.0 m/s:** The group speed of seas is 1.5 m/s whereas the current speed is 5 m/s. So seas cannot overcome the current. Whereas the swells can overcome because their group speed is 17.16 m/s which is much greater than the current speed.

### *Exercise 13.7*

The ship Sagarkanya is a 146 m long Research Vessel of NIO. In one of her cruises in Indian ocean, it was moving at a reduced speed of 5.14 m/s and that the waves took 6.3 seconds to pass the ship from stern to bow. Then calculate

i) the apparent speed of overtaking waves

ii) the actual wave speed, wave length and steepness if the wave height reported by the ship was 34 m.

### *Solution*

i) the apparent speed of overtaking waves = Ship length/period of passing waves = 146/6.3 = 23.17 m/s

ii) Actual wave speed = Apparent speed of waves + The ships speed = 23.17 + 5.14 = 28.31 m/s

iii) Actual wave length = L = $C^2/1.56 = 28.31^2/1.56$ = 513.75 m (since in deep water $C^2 = \frac{gL}{2\pi}$ = 1.56L)

iv) Actual wave period = T = C/1.56 = 28.31/1.56 = 18.14 seconds (since $C = \frac{gT}{2\pi}$ = 1.56T)

v) The steep ness = H/L = 34/513.75 = 0.066 which is a very steep wave.

### Exercise 13.8

Calculate the speed of tsunami in deep water at a depth of 5 km.

### Solution

Tsunami is a shallow water wave. So shallow water wave equation is to be used in the deep ocean as: = $C = \sqrt{gd} = \sqrt{9.8 \times 5000}$ = 70 m/s = 252 km/hr.

### Exercise 13.9

Calculate the speed and time period of semidiurnal tides in the deep ocean at a depth of 10,000 m if their wave length is 31,590 km.

### Solution

As tides are shallow water waves: the equation used is $L = \frac{gT^2}{2\pi}$.

So the time period is $T = \sqrt{\frac{2\pi L}{g}} = \sqrt{\frac{2\pi \times 31590 \times 10^3 m}{9.8m}} = 45 \times 10^2\ seconds$

$$C = \frac{L}{T} = \frac{31590 \times 10^3 m}{45 \times 10^2} = 2527.2 km/hr = 702 ms^{-1}$$

### Exercise 13.10

If a wave with a period T=10 seconds is propagated shoreward over a uniformly sloping shelf from a depth d = 200 meters to a depth d = 3 meters, find the wave celerities and lengths corresponding to depths 200 and 3 meters respectively.

### Solution

Using the equation for deep water wave length:

i) $L_0 = 1.56T^2 = 1.56(10)^2 = 156m$

The condition for deep water waves is $\frac{d}{L} > \frac{1}{2} \Rightarrow \frac{200m}{156m} = 1.28$, this value is greater than 0.5. So it is deep water.

ii) Wave celerity at 200 m is : C = L/T = 156/10 =15.6 $ms^{-1}$

iii) For d =3 meters: According to Eckart (1952) we can write:

$$L = \frac{gT^2}{2\pi}\sqrt{\tanh(\frac{4\pi^2}{T^2}\frac{d}{g})}$$

This can be further written as $L = L_0\sqrt{\tanh\left(\frac{2\pi d}{L_0}\right)}$:

The wave length for the condition of 3 m water is: $L = \frac{gT^2}{2\pi}\tanh(kd)$

$$L = L_0\sqrt{\tanh\left(\frac{2\pi d}{L_0}\right)} = 156\sqrt{\tanh\left(\frac{2\pi 3}{156}\right)} = 156\sqrt{\tanh(0.1208)} = 156\sqrt{(0.1202)} = 54.1m$$

So the value of $\frac{d}{L} \Rightarrow \frac{3m}{54.1m} = 0.055$, this value is between 0.05 and 0.5. So it is intermediate water. The wave celerity equation for intermediate (3m) water is: $C = \frac{L}{T} = \frac{54.1}{10} = 5.41ms^{-1}$

### *Exercise 13.11*

For example, if $\alpha_0 = 30^0$ and the period and depth of the wave are such that $C/C_0 = 0.5$ then calculate α, and $K_R$

### *Solution*

From Snells Law:

$$Sin\alpha = \left(\frac{C}{C_0}\right)Sin\alpha_0 ==> \alpha = \sin^{-1}[(.0.5)\sin 30] = \sin^{-1}[(.0.5)0.5] = \sin^{-1}[(.0.25)] = 14.5^0$$

If $\alpha = 14.5^0$, $\cos(14.5^0) = 0.968$ and $\cos\alpha_0 = \cos(30^0) = 0.866$

From the above equation of refraction coefficient

$$K_R = \left(\frac{b_0}{b}\right)^{1/2} = \left(\frac{\cos\alpha_0}{\cos\alpha}\right)^{1/2} = \left(\frac{0.866}{0.968}\right)^{1/2} = 0.945$$

### *Exercise 13.12*

Waves of T = 9sec and H = 2.2 m are measured at a depth of 25m with an approach angle $\alpha = 30^0$ at this depth. Sea bed has a uniform slope. Find deep water wave height $H_0$ and deep water approach angle $\alpha_0$

### *Solution*

As T remains constant both in shallow and deep waters:

Using eqn: $L_0 = gT^2/2\pi = 1.56T^2 = 1.56 \times 9^2 = 126.36$ m

For d = 25m, $d/L_0 = 25/126.36 = 0.198$,

In the SPM (1984) tables, for $d/L_0$ = 0.198, d/L = 0.2234, n = 0.6700, $C/C_0$ = 0.5938; and $H/H_0$ =0.9176. From equation 13.55a.

$$K_S = \left[\frac{1}{2}\left(\frac{1}{n}\right)\left(\frac{C_0}{C}\right)\right]^{1/2} \rightarrow K_S = \left[\frac{1}{2}\left(\frac{1}{0.67}\right)\left(\frac{1}{0.5938}\right)\right]^{1/2} = 1.121$$

We know from eqn 13.55b: $\frac{H}{H_0} = K_R K_S \rightarrow 0.917 = K_S\ K_R = 1.121\ K_R$

So $K_R$ = 0.917/1.121 = 0.818

$$\frac{H}{H_0} = 0.9176 ==> \frac{2.2}{H_0} = 0.9176 ==> H_0 = 2.4m$$

Weknowfromequation13.54b: $\sqrt{\frac{\cos\alpha_0}{\cos\alpha}} = K_R ==> \sqrt{\frac{\cos\alpha_0}{\cos\alpha}} = 0.818 ==> \sqrt{\frac{\cos\alpha_0}{\cos 30^0}} = 0.818$

$$\cos\alpha_0 = 0.58 ==> \alpha_0 = 54.6^0$$

### *Exercise 13.13*

A wave of period T = 8 sec is approaching from deep water towards shore on straight and parallel bottom contours. The angle of incidence of the wave at d=15m water depth is $\alpha_{15} = 15^0$. If the wave height at d = 10m is $H_{10}$ = 2 m. Calculate wave height($H_{15}$) at d = 15m and wave height in deep water($H_0$), $K_S$ & $K_R$ at both the depths and the ratio of spacing between orthogonals between deep water($b_0$ ) and at 15m ($b_{15}$ ).

### *Solution*

Given T = 8S, d =15m, $\alpha_{15} = 15^0$

$L_0 = gT^2/2\pi = 1.56T^2 = 1.56 \times 8^2 = 99.8$ m ⟶ 100 m

i) At 10m: $\frac{d_{10}}{L_0} = \frac{10}{100} = 0.1$

From Table $C_1$ of SPM (1984) $\frac{d_{15}}{L_0} = 0.15$ for ; the values are:

$$\frac{d_{10}}{L_{10}} = 0.1410\ ;\ n_{10} = 0.8103;\ \frac{C_{10}}{C_0} = 0.5747\ ;\ \frac{H_{10}}{H_0} = 0.9327$$

$$\frac{H_{10}}{H_0} = \frac{2}{H_0} = 0.9327 ==> H_0 = \frac{2}{0.9327} = 2.14m$$

From equation (13.55a)

$$K_{S10} = \left[\frac{1}{2}\left(\frac{1}{n_{10}}\right)\left(\frac{C_0}{C_{10}}\right)\right]^{1/2} \rightarrow K_S = \left[\frac{1}{2}\left(\frac{1}{0.8103}\right)\left(\frac{1}{0.5747}\right)\right]^{1/2} = 1.036$$

$$\frac{H_{10}}{H_0} = K_{S10} \times K_{R10} = 1.036 \times K_{R10} = 0.9327 ==> K_{R10} = \frac{0.9327}{1.036} = 0.9$$

ii) At 15m: $\frac{d_{15}}{L_0} = \frac{15}{100} = 0.15$

From Table $C_1$ of SPM (1984) for $\frac{d_{15}}{L_0} = 0.15$; the values are:

$\frac{d_{15}}{L_{15}} = 0.1813$; $n_{15} = 0.7325$; $\frac{C_{15}}{C_0} = 0.5994$ ; $\frac{H_{15}}{H_0} = 0.9133$

From i) we calculated $H_0 = 2.14$ m. So

$$\frac{H_{15}}{H_0} = \frac{H_{15}}{2.14} = 0.9133 ==> H_{15} = 0.9133 \times 2.14 = 1.96$$

For calculation of $K_{R15}$: use the equation (13.54b): Given: $\alpha_{15} = 15^0$ ; $C_{15}/C_0$ =0.5994

$$\text{Sin}\,\alpha_{15} = \left(\frac{C_{15}}{C_0}\right)\text{Sin}\,\alpha_0 ==> \alpha_0 = \sin^{-1}\left[\frac{\sin 15}{0.5994}\right] = \sin^{-1}[0.4317] = 25.58^0$$

**Cos** $(\alpha_{15})$ = cos(15) =0.966; **Cos** $(\alpha_0)$ = cos(25.58) =0.902

We know :
$$K_{R15} = \left(\frac{b_0}{b_{15}}\right)^{1/2} = \left(\frac{\cos\alpha_0}{\cos\alpha_{15}}\right)^{1/2} = \left(\frac{0.902}{0.966}\right)^{1/2} = 0.966$$

$$==> \frac{b_0}{b_{15}} = 0.933$$

$$K_{S15} = \left[\frac{1}{2}\left(\frac{1}{n_{15}}\right)\left(\frac{C_0}{C_{15}}\right)\right]^{1/2} \to K_S = \left[\frac{1}{2}\left(\frac{1}{0.7325}\right)\left(\frac{1}{0.5994}\right)\right]^{1/2} = 1.07$$

## *Exercise 13.14*

(a) In deep water waves, what is the energy per square metre of a wave field made up of waves with an average amplitude of 1.3m ($\rho = 1.03\times10^3$ kgm$^{-3}$).(b)What would be the wave power in KW/m of crest length if the waves had a steepness of 0.04 ( 1 watt = 1JS$^{-1}$ and one kilowatt = $10^3$ watts). (c) Also calculate wave power in shallow water.

## *Solution*

a) Energy E = 1/8 $\rho g H^2$ = 1/8 ×9.8 ×$1.03\times10^3\times(2.6)^2 = 8.529\times10^3$ J/m$^2$

b) Wave power: P = group velocity ×wave energy per unit area

Given steepness = H/L = 0.04, we know H = 2.6; L = 2.6/0.04 = 65m

Phase velocity in deep water = C = $\sqrt{\frac{gL}{2\pi}}$ = $\sqrt{1.56\times65} = 10.07 m/s$

Group velocity in deep water = half of phase velocity; $C_g = C/2 = 10.07/2 = 5.035$ m /s

Wave power = E.C.n =$EC_g$ = 8.529 ×$10^3$×5.035 = 42.7×$10^3$ J/m/s = 42.7×$10^3$watts/m=42.7 Kw/m

c) wave power in shallow water = P = E.C.n = E.C as n = 1 in shallow water = 8.529 ×$10^3$×10.07 = 85.4 ×$10^3$J/m/s = 85.4 kw/m

### *Exercise 13.15*

A wave of period 10 seconds approaching the shore has a height of 1m in deep water. Calculate a) the wave speed and group speed in deep water b) the wave steepness in deep water c) the wave power per meter of crest in water of 2.5 m deep

### *Solution*

a) The deep water wave celerity, $C^2 = gL/2\pi$ or $C = gT/2\pi = 1.56\times T = 1.56\times 10 =$ 15.6 m/s

### *Exercise 13.16*

If a wave with a period T = 8 seconds, in a water depth d = 15 meters (49 feet), and a height H = 5.5 meters (18.0 feet), find the local horizontal and vertical velocities u and w, and accelerations $\alpha_x$ and $\alpha_z$ at an elevation Z = - 5 meters below the SWL when

$$\theta = \left(\frac{2\pi x}{L} - \frac{2\pi t}{T}\right) = \frac{\pi}{3} = 60^0$$

### *Solution*

First calculate $L_0$. Using $L_0 = 1.56T^2 = 1.56\,(8)^2 = 99.84$ meters

Calculate relative depth using: $\frac{d}{L} = \frac{15m}{99.84m} = 0.15$ .

From table c of shore protection manual (SPM) (1984). for $\frac{d}{L_0} = 0.15$, the value of $\frac{d}{L} = 0.1833$

As d = 15 m, $L = \frac{15m}{0.1833} = 81.83m$

This value is in between 0.05 to 0.5. So the equation for intermediate waves has to be used. So the equation for wave length is as below:

From table c of SPM (1984); Tanhkd = 0.8183,

$$L = \frac{gT^2}{2\pi}\tanh(kd) = \frac{9.8ms^{-2} \times (8s)^2}{2\pi}(0.8183) = 81.68m$$

{note tanh (kd) =0.8183 for d/L =0.1833 from SPM}.

Similarly From table c of SPM Cosh 2πd/L = 1.740

$$u=\frac{H}{2}\frac{gT}{L}\frac{\cosh\frac{2\pi}{L}(d+z)}{\cosh\frac{2\pi d}{L}}\sin\left(\frac{2\pi x}{L}-\frac{2\pi t}{T}\right)=\frac{5.5m}{2}\frac{9.8ms^{-2}\times 8s}{81.68m}\frac{\cosh\frac{2\pi}{81.68m}(15-5)m}{1.740}\sin(60^0)$$

$$2.64\times 0.75\times 0.86=1.71ms^{-1}$$

Alternative way;

$$u=\frac{H}{2}\frac{gT}{L}\frac{\cosh\frac{2\pi}{L}(d+z)}{\cosh\frac{2\pi d}{L}}\sin\left(\frac{2\pi x}{L}-\frac{2\pi t}{T}\right)=\frac{5.5m}{2}\frac{9.8ms^{-2}\times 8s}{12.47m}\frac{\cosh\frac{2\pi}{12.47}(15-5)m}{\cosh\frac{2\pi 15m}{12.47m}}\sin(\frac{\pi}{3})$$

$$=2.16\times 8\times\frac{77.13}{957.97}\times 0.86=1.19ms^{-1}$$

Similarly, $$w=-\frac{H}{2}\frac{gT}{L}\frac{\sinh\frac{2\pi}{L}(d+z)}{\cosh\frac{2\pi d}{L}}\cos\left(\frac{2\pi x}{L}-\frac{2\pi t}{T}\right)=$$

$$=-\frac{5.5m}{2}\frac{9.8ms^{-2}\times 8s}{12.47m}\frac{\sinh\frac{2\pi}{12.47}(15-5)m}{\cosh\frac{2\pi 15m}{12.47m}}\cos(\frac{\pi}{3})=-2.16\times 8\times\frac{77.13}{957.97}\times 0.5=-0.7ms^{-1}$$

The local fluid particle accelerations are obtained from equations (13.37 C and d)

$$a_X=\frac{du}{dt}=-\frac{g\pi H}{L}\frac{\cosh\left[\frac{2\pi}{L}(d+Z)\right]}{\cosh\left(\frac{2\pi d}{L}\right)}\cos\left(\frac{2\pi x}{L}-\frac{2\pi t}{T}\right)$$

$$a_Z=\frac{dw}{dt}=-\frac{g\pi H}{L}\frac{\sinh\left[\frac{2\pi}{L}(d+Z)\right]}{\cosh\left(\frac{2\pi d}{L}\right)}\sin\left(\frac{2\pi x}{L}-\frac{2\pi t}{T}\right)$$

$$a_x=-\frac{9.8ms^{-2}\times\pi\times 5.5m}{12.47m}\frac{\cosh\left(\frac{2\pi}{12.47}(15-5)m\right)}{\cosh\frac{2\pi 15m}{12.47m}}\sin\left(60^0\right)=-13.58\times\frac{77.13}{957.97}\times 0.86=-0.94ms^{-2}$$

$$a_Z=-\frac{9.8ms^{-2}\times\pi\times 5.5m}{12.47m}\frac{\sinh\left(\frac{2\pi}{12.47}(15-5)m\right)}{\cosh\frac{2\pi 15m}{12.47m}}\cos\left(60^0\right)=-13.58\times\frac{77.13}{957.97}\times 0.5=-0.55ms^{-2}$$

The horizontal and vertical particle velocities are given by: (equations 13.37e and 13.37f)

$$\zeta = \int u dt = \int \frac{H}{2}\frac{gT}{L}\frac{\cosh\frac{2\pi}{L}(d+z)}{\cosh\frac{2\pi d}{L}}\cos\theta dt = -\frac{H}{4\pi}\frac{gT^2}{L}\frac{\cosh\frac{2\pi}{L}(d+z)}{\cosh\frac{2\pi d}{L}}\sin\theta$$

$$\varepsilon = \int w dt = \int \frac{H}{2}\frac{gT}{L}\frac{\sinh\frac{2\pi}{L}(d+z)}{\cosh\frac{2\pi d}{L}}\sin\theta dt = +\frac{H}{4\pi}\frac{gT^2}{L}\frac{\sinh\frac{2\pi}{L}(d+z)}{\cosh\frac{2\pi d}{L}}\cos\theta$$

$$\zeta = -\frac{5.5\times 9.8\times 8^2}{4\pi\times 12.47}\frac{\text{Cosh}\frac{2\pi}{12.47}(15-5)}{\text{Cosh}\left(\frac{2\pi\times 15}{12.47}\right)}\sin\left(60^0\right) = 22.01\times\frac{77.13}{958}\times 0.866 = -1.53ms^{-1}$$

$$\varepsilon = -\frac{5.5\times 9.8\times 8^2}{4\pi\times 12.47}\frac{\sinh\frac{2\pi}{12.47}(15-5)}{\text{Cosh}\left(\frac{2\pi\times 15}{12.47}\right)}\cos\left(60^0\right) = 22.01\times\frac{77.13}{958}\times 0.5 = 0.89ms^{-1}$$

### Exercise 13.17

If a wave in a depth d = 12 meters (39.4 feet), height H = 3 meters (9.8 feet), and a period T = 10 seconds and a corresponding deep water height $H_0$ = 3.13 meters (10.27 feet), FIND

(a) the major and minor semi axes of the water particle velocities in shallow and deep water waves and the horizontal and vertical displacement of a water particle from its mean position when z = 0, and when z = -d.

(b) The maximum water particle displacement at an elevation z = 7.5 meters (-24.6 feet) when the wave is in infinitely deep water.

(c) For the deepwater conditions of (b) above, show that the particle displacements are small relative to the wave height when z = $-L_0/2$.

### Solution

First let us find the value of d/L and $d/L_0$ using table C1 of SPM (1984):

$L_o = 1.56T^2 = 1.56(10)^2 = 156$m, Given d = 12 m; so $\frac{d}{L_0} = \frac{12}{156} = 0.0769$

From table C1 of SPM: For $d/L_0$, $\sinh\left(\frac{2\pi d}{L}\right) = 0.8306; \tanh\left(\frac{2\pi d}{L}\right) = 0.6389$

a) We know the equation for major semi-axis of orbital velocities of particle as: $A = a\frac{\text{Cosh}\, k(d+z)}{\text{Sinh}\, kd}$ where 'a' is the amplitude.

When Z = 0, $A = \frac{H}{2}\frac{\text{Cosh}\, k(d+0)}{\text{Sinh}\, kd} = \frac{H}{2}\frac{1}{\tanh kd} = \frac{H}{2}\frac{1}{\tanh\left(\frac{2\pi d}{L}\right)} = \frac{3m}{2}\left[\frac{1}{0.6389}\right] = 2.348m$

Similarly When Z =0, the equation for minor semi-axis of orbital velocity of water particle is ; $B = a\dfrac{\text{Sinh}\,k(d+z)}{\text{Sinh}\,kd} = \dfrac{H}{2}\dfrac{\text{Sinh}\,k(d+0)}{\text{Sinh}\,kd} = \dfrac{H}{2} = \dfrac{3m}{2} = 1.5m$

When Z = -d, $A = \dfrac{H}{2}\dfrac{\text{Cosh}\,k(d+0)}{\text{Sinh}\,kd} = \dfrac{H}{2}\dfrac{1}{\sinh kd} = \dfrac{H}{2}\dfrac{1}{\sinh\left(\dfrac{2\pi d}{L}\right)} = \dfrac{3m}{2}\left[\dfrac{1}{0.8306}\right] = 1.81m$

When Z = -d, $B = a\dfrac{\text{Sinh}\,k(d-d)}{\text{Sinh}\,kd} = \dfrac{H}{2}\dfrac{\text{Sinh}\,k(0)}{\text{Sinh}\,kd} = 0m$

b) Orbital velocities in the case of deep water waves are : A = B = $ae^{kd}$

$$A = B = \frac{H}{2}e^{\frac{2\pi d}{L_0}} = \frac{3.13m}{2}[e]^{\frac{2\pi(-7.5m)}{156m}} = 1.57\times[e]^{-0.302} = 1.57\times 0.739 = 1.157 = 1.16$$

As the orbit is circle and its radius is 1.157, the maximum displacement is diameter = 1.16 × 2 = 2.3m

c) When $Z = -\dfrac{L_0}{2} = -\dfrac{156m}{2} = -78m$

$$A = B = \frac{H}{2}e^{\frac{2\pi d}{L_0}} = \frac{3.13m}{2}[e]^{\frac{2\pi(-78m)}{156m}} = 1.57\times[e]^{-3.14} = 1.57\times 0.04 = 0.0678 = 0.07$$

The maximum displacement of the water particle (diameter) is 0.07 × 2 = 0.14m which is far less than the deep water wave height 3.13m.

## Exercise 13.18

If an average maximum pressure p = 124 kilo newtons per square meter is measured by a subsurface pressure gage located in salt water 0.6 meter above the bed in water depth d = 12 meters. The average frequency f = 0.0666 cycles per second (hertz). Find the height of the wave H and the average frequency corresponds to the average wave amplitude assuming that linear theory applies.

***Solution***: Using the equation of wave pressure (equations 13.39b) (you can also see **Ayapilla Murty handout** in academia.edu site: https://berhampurunivrsitybhanjabihar.academia.edu/AyapillaMurty

$$\Rightarrow \eta = N\left[\frac{p+\rho g Z}{\rho g K_Z}\right]$$

The height of the wave under crest when pressure is maximum = $\eta = a = \dfrac{H}{2}$

$$\therefore \frac{H}{2} = \eta = N\left[\frac{p+\rho g Z}{\rho g K_Z}\right]$$

For linear theory N =1, ρ = 1.026kgm$^{-3}$

As the pressure gauge is 0.6 m above the sea bed and the depth of the sea is d= 12 m, d+Z = 12-0.6 = 11.4m, but as it is negative Z direction, d+Z = -- 11.4 m

$$T = \frac{1}{f} = \frac{1}{0.0666} = 15.01s$$

Lo = 1.56T$^2$ = 1.56(15)$^2$ = 352m

$$\frac{d}{L_0} = \frac{12}{352} = 0.034$$

***From table C1 of SPM (1984):*** for 0.034, $\frac{d}{L} = 0.07651$,

$$L = \frac{12}{0.07651} = 156.8m$$

Similarly $\cosh\frac{2\pi d}{L} = 1.1178$

We know the pressure response factor from the equations 13.40 as:

$$K_p = \frac{\cosh k(d+z)}{\cosh kd} = \frac{\cosh\frac{2\pi}{L}(d+z)}{\cosh\frac{2\pi}{L}d} = \frac{\cosh\frac{2\pi}{156.8}(12-11.4)}{1.1178} = \frac{1.0003}{1.1178} = 0.8949$$

Substituting these values in the pressure equation:3

$$H = 2N\left[\frac{p+\rho gZ}{\rho g K_p}\right] = 2\times 1\left[\frac{124+(1.026\times 9.8\times -11.4)}{1.026\times 9.8\times 0.8949}\right] = 2.08m$$

### *Exercise 13.19*

Show in the case of wave refraction $\frac{H}{H_0} = K_R K_S$ where H is wave height and $K_R$ & $K_S$ are refraction and shoaling coefficients.

### *Solution*

Refer Fig.13.35 and the derivation of equations 13.54(a ,b) and 13.55 (a,b) for full derivation

When the energy flux is conserved between the wave rays, then , where b is the distance between adjacent wave rays. Then we can also write (from equations 13.55c)

$$\frac{H_2}{H_1} = \sqrt{\frac{\eta_1 C_1}{C_1 C_2}}\sqrt{\frac{b_1}{b_2}}$$

$$\frac{H}{H_0} = \sqrt{\frac{\eta_0 C_0}{\eta C}}\sqrt{\frac{b_0}{b}} = \sqrt{\frac{1}{2}\frac{1}{\eta}\frac{1}{\tanh kd}}\sqrt{\frac{b_0}{b}} = K_s K_r$$

such that $K_s = \sqrt{\frac{1}{2}\frac{1}{\eta}\frac{1}{\tanh kd}}$ *and* $K_r = \sqrt{\frac{b_0}{b}}$

The term $K_S$ is known as the shoaling coefficient and $K_R$ is called refraction coefficient. This shoaling coefficient is a function of wavelength and water depth. The values of $K_S$ are available in Shore Protection manual (SPM1984).

For example, if $\alpha_0 = 30^0$ and the period and depth of the wave are such that $C/C_0 = 0.5$ then from Snells Law:

$$\text{Sin}\,\alpha = \left(\frac{C}{C_0}\right)\text{Sin}\,\alpha_0 ==> \alpha = \sin^{-1}[(.0.5)\sin 30] = \sin^{-1}[(.0.5)0.5] = \sin^{-1}[(.0.25)] = 14.5^0$$

If $\alpha = 14.5^0$, cos $(14.5^0) = 0.968$ and cos $\alpha_0$ = cos $(30^0) = 0.866$

From the above equation of refraction coefficient

$$K_R = \left(\frac{b_0}{b}\right)^{1/2} = \left(\frac{\cos\alpha_0}{\cos\alpha}\right)^{1/2} = \left(\frac{0.866}{0.968}\right)^{1/2} = 0.945$$

***Exercise 13.20:***

A wave in deep water has the following characteristics: $H_0 = 3$m, T = 8 sec and $\alpha_0 = 30^0$. Calculate $L_0$, $K_S$, $K_R$, H for 10m and 2 m.

*Solution*

***From the tables of SPM(1984):***

Using eqn: $L_0 = gT^2/2\pi = 1.56T^2 = 1.56 \times 8^2 = 99.8 = 100$ m

For d = 10m, $d/L_0 = 10/100 = 0.1$,

In the SPM tables, for $d/L_0 = 0.1$, d/L = 0.5747, n = 0.8103, $C/C_0 = 0.5747$; and $H/H_0 = 0.9327$

We know from eqn 13.55a: $K_S = \left[\frac{1}{2}\left(\frac{1}{n}\right)\left(\frac{C_0}{C}\right)\right]^{1/2} \rightarrow K_S = \left[\frac{1}{2}\left(\frac{1}{0.8103}\right)\left(\frac{1}{0.5747}\right)\right]^{1/2} = 1.03$

We know from eqn 13.55b $\frac{H}{H_0} = K_R K_S \rightarrow$ ➤ $0.9327 = K_S K_R = 1.03 K_R$

So $K_R = 0.9327/1.03 = 0.91$

As $H/H_0 = 0.9327$ ➤ H/3m ➤ $H = 0.9327 \times 3 = 2.79$ m

**In the same way for 2m can be calculated**

For d = 2m, $d/L_0 = 2/100 = 0.02$

: ***From the tables of SPM (1984)***, for $d/L_0 = 0.02$, d/L = 0.05763, n = 0.9588, $C/C_0 = 0.3327$; and $H/H_0 = 1.226$

Weknowfromeqn13.55a: $K_S = \left[\frac{1}{2}\left(\frac{1}{n}\right)\left(\frac{C_0}{C}\right)\right]^{1/2} \rightarrow K_S = \left[\frac{1}{2}\left(\frac{1}{0.9588}\right)\left(\frac{1}{0.3327}\right)\right]^{1/2} = 1.252$

We know from eqn 13.55b $\frac{H}{H_0} = K_R K_S$ ➔ $1.226 = K_S\,K_R = 1.252\,K_R$

So $K_R = 1.226/1.252 = 0.98$

As $H/H_0 = 1.226 = H/3\text{m}$ ➔ $H = 1.226 \times 3 = 3.68$ m

### Exercise 13.21

A deep water oscillatory wave with a wavelength L = 156 meters, a height H = 2 meters, and a celerity Co = 15.6 meters per second, moving shoreward with its crest parallel to the depth contours. Any effects due to reflection from the beach are negligible.

(a) Derive a relationship between the wave height in any depth of water and the wave height in deep water, assuming that wave energy flux per unit crest width is conserved as a wave moves from deep water into shoaling water.

(b) Calculate the wave height for the given wave when the depth is 3 meters.

(c) Determine the rate at which energy per unit crest width is transported toward the shoreline and the total energy per unit width delivered to the shore in 1 hour by the given waves.

### Solution

(a) Take the equation (13.50b) : $\frac{1}{2}E_0 C_0 = \bar{E}NC$ (13.50b)

When the wave does not refract, the wave crests are parallel to the bottom contours. Then energy of the wave in deep water without refraction can be written from equation 13.45 as:

$\bar{E}_0 = \frac{1}{8}\rho g H''^2_0$ and $\bar{E} = \frac{1}{8}\rho g H^2$ where H' represents the wave height in deep water if the wave is not refracted. Substituting these in equation (13.50b):

$$\frac{1}{2}C_0\frac{\rho g H''^2_0}{8} = NC\frac{\rho g H^2}{8} \quad (13.50c)$$

$$\frac{H^2}{\bar{H}_0^2} = \frac{C_0}{2NC} \text{ where } N = \left[\frac{1}{2}\left(1+\frac{2kd}{\text{Sinh}\,2kd}\right)\right] = \frac{1}{2}\left[1+\frac{4\pi d/L}{\text{Sinh}\left(4\pi d/L\right)}\right] \quad (13.50d)$$

We also know from equation (13.52):

$$\frac{C}{C_0} = \tanh(kd) = \tanh\left(\frac{2\pi d}{L}\right) \quad (13.50e)$$

Substituting (13.50d) and (13.50e) in (13.50c):

$$\frac{H}{H_0} = \sqrt{\frac{1}{\tanh(kd)}\left\{\frac{1}{1+\dfrac{(2kd)}{\sinh 2kd}}\right\}};\ \frac{H}{H_0} = \left[\frac{1}{\tanh\left(2\pi d/L\right)}\left\{\frac{1}{1+\dfrac{\left(4\pi d/L\right)}{\sinh\left(\left(4\pi d/L\right)\right)}}\right\}\right]^{1/2} = K_S \quad (13.50f)$$

where $K_S$ or $H/H_0$' is termed as the shoaling coefficient. Values of H/H' as a function of d/Lo and d/L have been tabulated in Tables C-1 and C-2 of SPM (1984).

(b) For the given wave, d/L = 3/156 = 0.01923. From Table C-1 of SPM or from an evaluation of equation (13.50f): $\dfrac{H}{H_0} = 1.237$

Therefore, As $H_0$ is given as 2 m; H = 1.237(2) = 2.474 m

(c) The rate at which energy is being transported toward shore is the wave energy flux (taking equation (13.50b)

$\overline{P} = \dfrac{1}{2}E_0 C_0 = \overline{E}NC$ ; Taking the left side relation:

$$\overline{P} = \frac{1}{2}E_0 C_0 = \frac{1}{2}\times\frac{\rho g H_0^2}{8}C_0 = \frac{1}{2}\times\frac{1..025\times10^3 kgm^{-3}\times 9.8kgm^{-3}\times(2m)^2\times 15.6ms^{-1}}{8}$$
$$= 39175.5kg^2m^{-3}s^{-1} = 39{,}175Nms^{-1}/m$$

= 39,175 $Nms^{-1}$ per meter of wave crest

The rate of energy transported per unit crest width in one hour = 39175 × 3600 = 14.1 × $10^7$ joules. This means 14.1 × $10^7$ joules of energy is released by waves in each meter of beach in one hour.

## *Exercise 13.22*

Find the local horizontal and vertical velocities $u$ and $w$, and accelerations "ax" and "az" at an elevation $z$ = -5 m below the SWL when $\theta = \left(\dfrac{2\pi x}{L} - \dfrac{2\pi t}{T}\right) = \dfrac{\pi}{3} = 60^0$ for a wave with a period $T$ = 8 sec, in a water depth $d$ = 15 m, and a height $H$ = 5.5 m

### *Solution*

Equations of local horizontal and vertical velocities are:

$$u = \frac{H}{2}\frac{gT}{L}\frac{\cosh\frac{2\pi}{L}(d+z)}{\cosh\frac{2\pi d}{L}}\cos\theta \quad w = -\frac{H}{2}\frac{gT}{L}\frac{\sinh\frac{2\pi}{L}(d+z)}{\cosh\frac{2\pi d}{L}}\sin\theta$$

A wave of period 10 seconds approaching the shore has a height of 1m in deep water. Calculate a) the wave speed and group speed in deep water b) the wave steepness in deep water c) the wave power per meter of crest in water of 2.5 m deep.

*Solution:*

a) The deep water wave celerity is

$$C^2 = \frac{gL}{2\pi} \; or \; C = \frac{gT}{2\pi} = 1.56 \times T = 1.56 \times 10 = 15.6 \; m / s^{-1}$$

**Table 13.6.** Wave table

| $d/L_0$ | tanh kd | d/L | Kd | Sinhkd | Cosh kd | N | $d/L_0$ | tanh kd | d/L | Kd | Sinhkd | Coshkd | N |
|---|---|---|---|---|---|---|---|---|---|---|---|---|---|
| 0.000 | 0.000 | 0.000 | 0.000 | 0.000 | 1.00 | 1.000 | 0.22 | 0.909 | 0.242 | 1.52 | 2.18 | 2.40 | 0.646 |
| .002 | 0.112 | 0.0179 | 0.112 | 0.113 | 1.01 | 0.996 | 0.23 | 0.918 | 0.251 | 1.57 | 2.31 | 2.52 | 0.635 |
| .004 | 0.158 | 0.0253 | 0.159 | 0.160 | 1.01 | 0.992 | 0.24 | 0.926 | 0.259 | 1.63 | 2.45 | 2.65 | 0.626 |
| .006 | 0.193 | 0.0311 | 0.195 | 0.197 | 1.02 | 0.998 | 0.25 | 0.933 | 0.268 | 1.68 | 2.60 | 2.78 | 0.616 |
| .008 | 0.222 | 0.0360 | 0.226 | 0.228 | 1.03 | 0.983 | 0.26 | 0.940 | 0.277 | 1.74 | 2.75 | 2.93 | 0.608 |
| .010 | 0.248 | 0.0403 | 0.253 | 0.256 | 1.03 | 0.979 | 0.27 | 0.956 | 0.285 | 1.79 | 2.92 | 3.09 | 0.599 |
| .015 | 0.302 | 0.0496 | 0.312 | 0.317 | 1.05 | 0.969 | 0.28 | 0.952 | 0.290 | 1.85 | 3.10 | 3.25 | 0.592 |
| .020 | 0.347 | 0.0576 | 0.362 | 0.370 | 1.07 | 0.959 | 0.29 | 0.957 | 0.303 | 1.90 | 3.28 | 3.43 | 0.585 |
| .025 | 0.386 | 0.0648 | 0.407 | 0.418 | 1.08 | 0.949 | 0.30 | 0.961 | 0.312 | 1.96 | 3.48 | 3.62 | 0.578 |
| .030 | 0.420 | 0.0713 | 0.448 | 0.463 | 1.10 | 0.939 | 0.31 | 0.965 | 0.321 | 2.02 | 3.69 | 3.83 | 0.571 |
| .035 | 0.452 | 0.0775 | 0.487 | 0.506 | 1.12 | 0.929 | 0.32 | 0.969 | 0.330 | 2.08 | 3.92 | 4.05 | 0.566 |
| .040 | 0.480 | 0.0833 | 0.523 | 0.548 | 1.14 | 0.919 | 0.33 | 0.972 | 0.339 | 2.13 | 4.16 | 4.28 | 0.560 |
| .045 | 0.507 | 0.0888 | 0.558 | 0.588 | 1.16 | 0.910 | 0.34 | 0.975 | 0.349 | 2.19 | 4.41 | 4.53 | 0.555 |
| .050 | 0.531 | 0.0942 | 0.592 | 0.627 | 1.18 | 0.900 | 0.35 | 0.978 | 0.358 | 2.25 | 4.68 | 4.79 | 0.550 |
| .055 | 0.554 | 0.0993 | 0.624 | 0.665 | 1.20 | 0.891 | 0.36 | 0.980 | 0.367 | 2.31 | 4.97 | 5.07 | 0.546 |
| .060 | 0.575 | 0.104 | 0.655 | 0.703 | 1.22 | 0.880 | 0.37 | 0.983 | 0.377 | 2.37 | 5.28 | 5.37 | 0.542 |
| .065 | 0.595 | 0.109 | 0.686 | 0.741 | 1.24 | 0.872 | 0.38 | 0.984 | 0.386 | 2.43 | 5.61 | 5.70 | 0.538 |
| .070 | 0.614 | 0.114 | 0.716 | 0.779 | 1.26 | 0.863 | 0.39 | 0.986 | 0.395 | 2.48 | 5.96 | 6.04 | 0.535 |
| .075 | 0.632 | 0.119 | 0.745 | 0.816 | 1.29 | 0.853 | 0.40 | 0.988 | 0.405 | 2.54 | 6.33 | 6.41 | 0.531 |
| .080 | 0.649 | 0.123 | 0.774 | 0.854 | 1.31 | 0.845 | 0.41 | 0.989 | 0.415 | 2.60 | 6.72 | 6.80 | 0.529 |
| .085 | 0.665 | 0.128 | 0.803 | 0.892 | 1.34 | 0.836 | 0.42 | 0.990 | 0.424 | 2.66 | 7.15 | 7.22 | 0.526 |
| .090 | 0.681 | 0.132 | 0.831 | 0.929 | 1.37 | 0.827 | 0.43 | 0.991 | 0.434 | 2.73 | 7.60 | 7.66 | 0.523 |
| .095 | 0.695 | 0.137 | 0.858 | 0.968 | 1.39 | 0.819 | 0.44 | 0.992 | 0.443 | 2.79 | 8.07 | 8.14 | 0.521 |
| .100 | 0.709 | 0.141 | 0.886 | 1.01 | 1.42 | 0.810 | 0.45 | 0.993 | 0.453 | 2.85 | 8.59 | 8.64 | 0.519 |
| .110 | 0.735 | 0.150 | 0.940 | 1.08 | 1.48 | 0.794 | 0.46 | 0.994 | 0.463 | 2.91 | 9.13 | 9.18 | 0.517 |
| .120 | 0.759 | 0.158 | 0.994 | 1.17 | 1.54 | 0.778 | 0.47 | 0.995 | 0.472 | 2.97 | 9.71 | 9.76 | 0.516 |

| $d/L_0$ | tanh kd | d/L | Kd | Sinhkd | Cosh kd | N | $d/L_0$ | tanh kd | d/L | Kd | Sinhkd | Coshkd | N |
|---|---|---|---|---|---|---|---|---|---|---|---|---|---|
| .130 | 0.780 | 0.167 | 1.05 | 1.25 | 1.60 | 0.762 | 0.48 | 0.995 | 0.482 | 3.03 | 10.3 | 10.4 | 0.514 |
| .140 | 0.800 | 0.175 | 1.10 | 1.33 | 1.67 | 0.747 | 0.49 | 0.996 | 0.492 | 3.09 | 11.0 | 11.0 | 0.513 |
| .150 | 0.818 | 0.183 | 1.15 | 1.42 | 1.74 | 0.733 | 0.50 | 0.996 | 0.502 | 3.15 | 11.7 | 11.7 | 0.512 |
| .160 | 0.835 | 0.192 | 1.20 | 1.52 | 1.82 | 0.718 | 0.75 | 1.000 | 0.746 | 4.69 | 54.5 | 54.5 | 0.501 |
| .170 | 0.850 | 0.200 | 1.26 | 1.61 | 1.90 | 0.705 | 1.0 | 1.000 | 0.981 | 6.16 | 269.5 | 269.5 | 0.500 |
| .180 | 0.864 | 0.208 | 1.31 | 1.72 | 1.99 | 0.692 | | | | | | | |
| .200 | 0.888 | 0.225 | 1.41 | 1.94 | 2.18 | 0.668 | | | | | | | |
| .210 | 0.899 | 0.234 | 1.47 | 2.05 | 2.28 | 0.656 | | | | | | | |

### *Exercise 13.24*

Given: T=10 sec, $H_0$=4 m, $\alpha_0 = 60°$ .Find: H and α at the depth of d =15.6 m. Check whether the wave is broken at this depth. Assume $\gamma = \frac{H_b}{d_b} = 0.78$

### *Solution*

Similar to the previous problem, calculate $L_0$, then find d/ $L_0$. Find d/L, tanh kd and η from wave table 13.6.

$$L_0 = \frac{gT^2}{2\pi} = \frac{9.8\times10^2}{2\pi} \approx 156m; \frac{d}{L_0} = \frac{15.6m}{156m} = 0.1;$$

From wave table 13.6 or SPM (1984) tables $\frac{d}{L_0} = 0.1, \frac{d}{L} = 0.14$ , $tan\,h\,kd = 0.71$ and η = 0.81 sinα = sin $\alpha_0$ tanh $\frac{2\pi d}{L}$ = sin (60) × 0.71 = 0.61 ➔ α = 17.9⁰

$$K_r = \sqrt{\frac{cos\alpha_0}{cos\alpha}} = \sqrt{\frac{cos60^0}{cos17.9^0}} = \sqrt{\frac{0.5}{0.95}} = \sqrt{0.53} = 0.73$$

$$\frac{H}{H_o} = K_s K_r = 0.93\times0.73$$

$H = H_b = H_o \times 0.87 \times 1.44 = 4.0 \times 0.93 \times 0.73 = 2.72$m

Now to know whether the wave has broken at the depth of 15.6 m, find $\gamma = \frac{H_b}{d_b}$

$$\gamma \aleph \frac{H_b}{d_b} \quad \frac{2.72m}{15.6m} \quad 0.17$$

As γ = 0.17 is less than the given value of 0.78, it is not broken.

### *Exercise 13.25*

The Figure below shows the passing of ocean waves through a gap of the break water in a harbour. The wave fronts are reaching the berthing place where ships anchor for loading and unloading operations in the harbour. (a)What is the kind of wave transformation takes place here, and (b) how can you improve the safety conditions in the berthing place due to the instability caused by these waves?

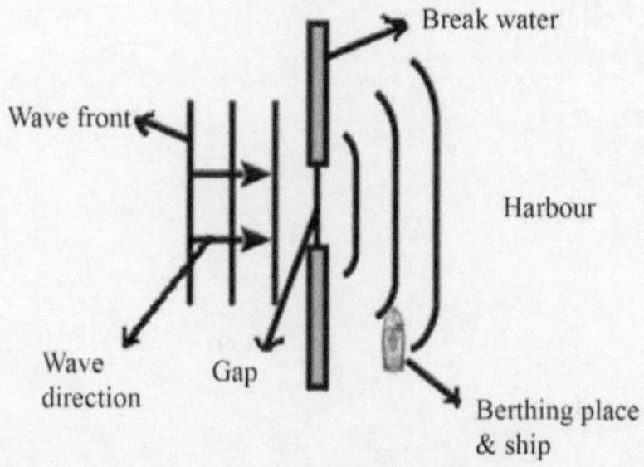


Or choose the correct answer from the following table:

| Choose alternative | Name of wave transformation | Solution for the problem |
|---|---|---|
| A | Reflection | Make the gap wider than the wave length |
| B | Refraction | Close the gap |
| C | Refraction | Make the gap smaller than the wave length |
| D | Diffraction | Make the gap wider than the wavelength |
| E | Diffraction | Make the gap smaller than the wave length |

### *Solution*

Answer is 'D'

### *Exercise 13.26*

For the exercise 13.25, assuming that the width of the gap is comparable to the wavelength of the wave, answer which one is correct.

| Choose | Speed | Wavelength |
|---|---|---|
| A | Decreases | Decreases |
| B | decreases | Remains constant |
| C | Remains constant | Decreases |
| D | Remains constant | Remains constant |

### *Solution*

Answer is 'D' A wave with a speed of 4.43 m/s in the harbor, passing through the gap of a break water is shown in the next exercise.

### *Exercise 13.27*

A wave with a wave length of 64.3m in 2 m in the harbor, is returning through the gap of break water shown below into deep water. If the depth of water near the gap in the break water is equal to of 10 meters, i) Find the wave speed, and period in the harbor ii) what is the speed and wave length at the time of crossing the gap of the break water. iii) What is wave number and group velocity?

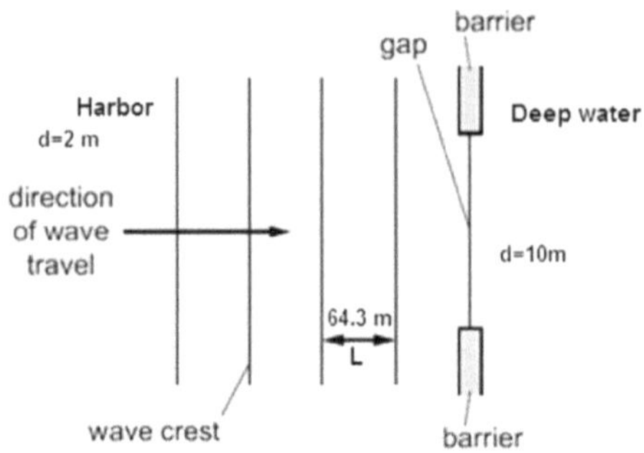


## *Solution*

Wave is returning from harbor (shallow) into deep water. Its speed in harbor can be found from the SPM tables. So $\frac{d}{L}$ in the harbor = $\frac{2}{64.3} = 0.0311$

From SPM tables for $\frac{d}{L} = 0.0311 \rightarrow \frac{d}{L_o} = 0.006$ . and $\tanh\frac{2\pi d}{L} = 0.1929$

We know the wave equation for speed is $c^2 = \frac{gL}{2\pi}tanh\frac{2\pi d}{L}$

i) Wave velocity in the harbor = $c_h^2 = \frac{g \times 64.3}{2\pi} \times 0.1929 = 19.35$ ➔ $c_h = 4.4ms^{-1}$

ii) Wave period we know L = cT ➔ $T = \frac{L}{c_h} = \frac{64.3}{4.4} = 14.61s$

iii) Wave velocity at the time of crossing the gap
$= c_0^2 = \frac{gL_0}{2\pi}tanh\frac{2\pi d}{L} = \frac{g \times 1666.7}{2\pi} \times 0.1929 = 501.5$ ➔ $c_o = 22.4ms^{-1}$

iv) As water depth at the gap is 10 m, $\frac{d}{L_o} = \frac{10}{L_o} = 0.006 \rightarrow L_o = \frac{10}{0.006} = 1666.7m$

v) From SPM tables, wave number K, for $\frac{d}{L} = 0.0311$ ➔ K = 0.9812

vi) From SPM tables, $\frac{c_G}{c} = 0.1905$ ➔ $c_G = c_o \times 0.1905 = 1666.7 \times 0.1905 = 317.5ms^{-1}$

## *Exercise 13.28*

Calculate the speed in meters per second for all and wavelength for b and c for deep water waves with the following characteristics: a) L=351 m, T=15 seconds, b) T=12 seconds, c) f = 0.125 waves per second.

### Solution

a) We know : $C = \frac{L}{T} = \frac{351m}{15s} = 23.4ms^{-1}$

b) We know from equation 13.34, $C_0 = 1.56T$ and $L_0 = 1.56\ T^2 \rightarrow Co = 1.56 \times 12 = 18.72\ ms^{-1}$

$L_0 = 1.56\ T^2 \rightarrow L_0 = 1.56 \times 12^2 = 224.64$ m

c) If f=0.125, $T = \frac{1}{f} = \frac{1}{0.125} = 8s \rightarrow C_0 = 1.56 \times 8 = 12.48ms^{-1}$ ; $L_0 = 1.56 \times 8^2 = 99.84$ m

### Exercise 13.29

When sea waves from separate generating areas move out and propagate as swells, they invariably coincide called interference. If the sea wave 'A' has a height of 1.5 m and sea wave 'B' has a height of 3.5 m, what would be the height of resulting wave from constructive and destructive interference?

### Solution

See section 13.2.2 for details of constructive and destructive interferences.

i) For constructive interference, crest of 'A' with crest of 'B' and trough of 'A' with trough of 'B' should coincide assuming their wavelengths are same.

   So height of the resultant wave for constructive interference = heights of A + B = 1.5 + 3.5 = 5 m

ii) For destructive interference, crest of A should coincide with trough of B as their wave lengths are different. So height of the resultant wave = +1.5- 3.5 = - 2.0 or -1.5+ 3.5 = + 2.0. So the answer is $\pm$ 2.0.

### Exercise 13.30

Describe the physical changes that take place when a deep water wave enters a shallow zone with regard to speed, length, height, steepness, period and energy.

### Solution

As the wave enters from deep to shallow water (d/L <1/2), the shoaling depth interfere with water particle movement at the base of the wave, so the wave speed decreases. As one wave slows down, the following wave form, which is still moving at its original speed, moves closer to the wave that is being slowed, causing a decrease in wavelength. As the energy can neither be created nor destroyed, energy should remains same but it will act to increase the wave height. The energy will also act to increase the steepness (H/L). However, as the wave steepness reaches the ratio 1:7, wave will break and the entire energy is dissipated into foam. The wave period remains constant.

### *Exercise 13.31*

i)Calculate ($\bar{H}$), ($H_{1/3}$), and ($H_{1/10}$) for the following data using cumulative frequency. ii) Also calculate 'One number ($\sqrt{E}$) and calculate average, significant and one-tenth wave heights using Longuet- Higgins statistics and compare both.

| Wave height | 1 | 2 | 3 | 4 | 5 | 6 | 7 | 8 |
|---|---|---|---|---|---|---|---|---|
| Number of waves | 7 | 33 | 30 | 20 | 7 | 3 | 1 | 1 |

### *Solution*

| S.No | Wave height (ft) | No. of waves | Cumulative value | Wave height × Number of waves | $H_{1/3}$ | $H_{1/10}$ |
|---|---|---|---|---|---|---|
| 1. | 1 | 7 | 102 | 1× 7 = 7 | | |
| 2. | 2 | 33 | 102-7=95 | 2×33 = 66 | | |
| 3. | 3 | 30 (28 + 2) | 95-33=62 | 3×30=90 | 3×2=6 | |
| 4. | 4 | 20 | 62-30=32 | 4×20=80 | 4×20=80 | |
| 5 | 5 | 7 | 32-20=12 | 5×7=35 | 5×7=35 | 5×5=25 |
| 6 | 6 | 3 | 12-7=5 | 6×3=18 | 6×3=18 | 6×3=18 |
| 7 | 7 | 1 | 5-3=2 | 7×1=7 | 7×1=7 | 7×1=7 |
| 8 | 8 | 1 | 2-1=1 | 8×1=8 | 8×1=8 | 8×1=8 |
| Total | | 102 | | 311 | 102/3=34 | |

Note: Always take from highest to lowest height of wave such that the number of waves is equal to the required. For example in the case of $H_{1/3}$, number of waves required is total number of waves divided by 3 (102/3=34). Taking from the highest height (8) move toward lower side in the descending order (8,7,6,5,4,3 etc). For these waves (column 2), take column three values against each wave height until 34 becomes. Similarly follow in the case of $H_{1/10}$

Average wave height = 311/102 = 3.0 ft (1)

Significant wave height $H_{1/3}$ (102/3 =34) = $\frac{6+80+35+18+7+8}{(2+20+7+3+1+1)} = \frac{154}{34}$ = 4.5 ft (2)

Average height of 1/10th of highest H1/10 = $\frac{25+18+7+8}{5+3+1+1)} = \frac{58}{10}$ = 5.8 ft (3)

According to Longuet-Higgins the average wave height data is :

iii) The most frequent wave height is $1.41\sqrt{E}$ (4)

iv) Average wave height is $1.77\sqrt{E}$ (5)

v) Significant wave height is $2.83\sqrt{E}$ (6)

vi) Average of the heights of the 1/10th highest waves is $3.60\sqrt{E}$ (7)

When the average wave height from equation (5) and equation (1) is equated:

$1.77\sqrt{E} = 3.0 \text{ ft} \rightarrow \sqrt{E} = \frac{3}{1.77} = 1.69$ (8)

Significant wave height from equations (6) and (8) is

$2.83\ \sqrt{E} = 2.83 \times 1.69 = 4.796 \approx 4.8$ ft (9)

Comparing equations (2) and (9), the difference is $4.8 - 4.5 = 0.3$ ft: (10)

Similarly computing average of the heights of the 1/10th highest waves from equations (7) and (8) is : 3.60 ft $\sqrt{E} = 3.6 \times 1.69 = 6.08 \approx 6.1$ (11)

Comparing equations (11) and (3), the difference is $6.1 - 5.8 = 0.3$ ft: (12)

Thus comparing equations (10) and (12), the difference is same and the agreement between the two ways of computing is almost same.

### *Exercise 13.32*

Using the same data of Exercise 13.31, find the number of waves in each of the following cases i) less than 2.4 ft; ii) $\leq 2$ ft; iii) $= 4.3$ ft; iv) $\geq 5$ft

### *Solution*

By examining the values of table of solution of Exercise 13.31, it is clear that :i) there should be at least 41 (>33+7) out of 102 waves, less than 2.4ft. ii) There are 40(33+7) waves which are less than or equal to 2 feet; iii) There are 20 waves which are equal to 4.3 ft; iv) There are 12 (7+3+1+1) waves greater than or equal to 5 ft.

### *Exercise 13.33*

Explain the graphical procedure of drawing or preparing the refraction diagram

### *Solution*

At first a template or ray plotter has to be made as shown below in Fig.35.1 using a transparent or perspex sheet or plate.

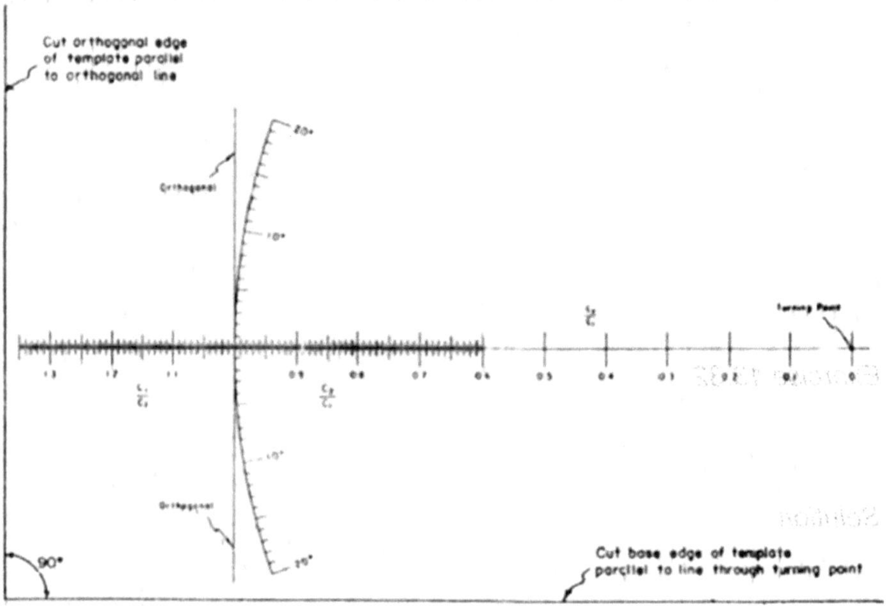


**Fig. 35.1.** Ray plotter or template

Using the ray plotter, the procedure to draw the refraction diagram is as below:

(i) Obtain the hydrographic chart of the area of interest.

(ii) Draw smooth bed contours at suitable intervals, like 2 m or so.

(iii) Select a range of wave periods (T) and wave directions (θ) and draw separate diagrams for each T and (θ) as below:

Obtain $C_0/C_1$ values, where $C_0$ is wave speed before refraction in deep water and $C_1$ is the wave speed in shallow water after refraction over bed contour. In open areas rays are drawn from deep to shallow water and in highly sheltered bays, it may become more useful to follow a reversed procedure to draw rays from the shallow to deep water.

As shown in Figure 35.2 draw orthogonals (in deep water) to some spacing and write the $C_0/C_1$ values obtained from the above table on the hydrographic chart at appropriate places within the region of the two underlying contours.

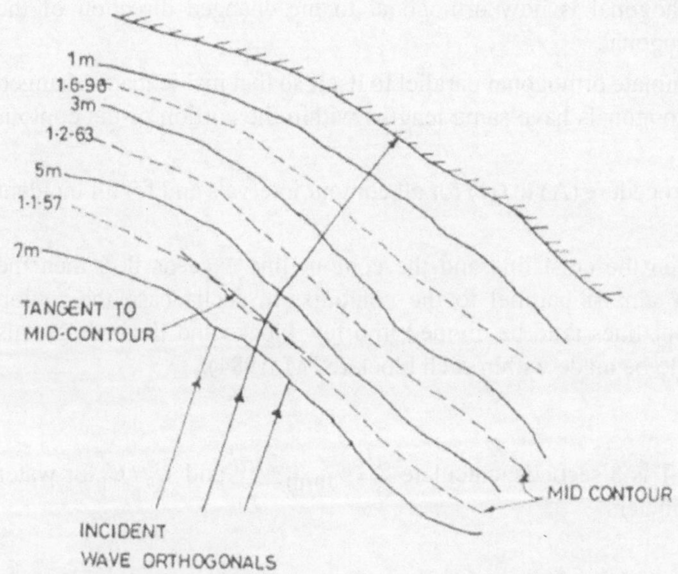


**Fig. 35.2.** Bathymetry map

If the angle between the underlying contour line and the crest line is less than 90$^0$, follow the procedure given below (Figure 35.2)

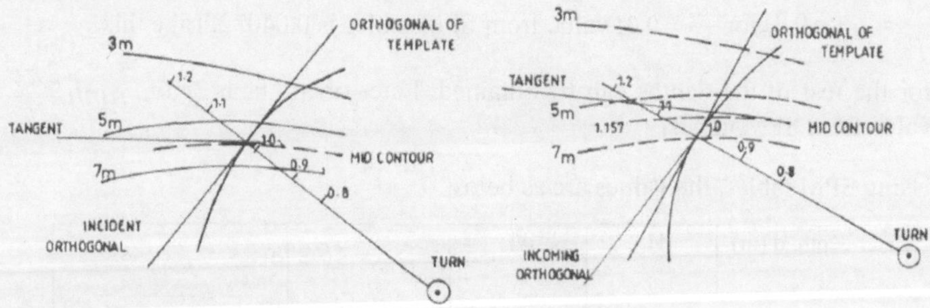


**Fig. 35.3.** Drawing the contours

(A) Referring to Figure 35.2 and 35.3 construct a mid-contour. Extend the incident orthogonal m upon it and draw tangent to the mid-contour at the intersection point.

(B) Place the refraction template (Figure 35.1) on the top of the hydrographic chart such that the template orthogonal coincides with the intersection point # at mark '1.0' on the ray plotter or template.

(C) Turn the template around its turning point until $C_0/C_1$ value of the contour interval read on the template crosses the tangent to the mid-contour. The template orthogonal is now orthogonal to the changed direction of the incident orthogonal.

(D) Move the template orthogonal parallel to itself so that incident and changed or turned orthogonals have same lengths within the portion of the contour intervals.

(E) Repeat the procedure (A) to (D) for all contour intervals and for all incident orthogonals.

If the angle between the crest line and the contour line exceeds $80^0$, then the orthogonals appear almost parallel to the contours. In sucha case the region between two contour lines is to be divided into few blocks and the orthogonals turning is required to be made within each block (SPM, 1984).

## Exercise 13.34

If the wave period T is 8 seconds, calculate $\frac{d}{L_0}$, $\tanh\frac{2\pi d}{L}$ and $C_0/C_1$ for water depths 1,3,5 and 7 meters

## Solution

The equation for deep water wave length is:

$$L_0 = \frac{gT^2}{2\pi} = \frac{g\times 8^2}{2\pi} = 99.82m \approx 100\,m$$

$\frac{d}{L_0} = \frac{1}{100} = 0.01$ for $\frac{d}{L_0} = 0.01$ value from SPM tables is 0.04032. Like this for the rest of the depths can be obtained. Later obtain tanh 2πd/L $tanh\frac{2\pi d}{L}$ value from SPM tables.

Using SPM tables, the values are as below.

| Water depth, d (m) | d/L$_0$ | d/L | $tanh\frac{2\pi d}{L}$ | $\frac{C_0}{C_1} = \frac{tanh\frac{2\pi d_1}{L}}{tanh\frac{2\pi d_0}{L}}$ |
|---|---|---|---|---|
| 1 | 0.01 | 0.04032 | 0.2480 | |
| | | | | $\frac{0.4205}{0.2480}$ 1.696 |
| 3 | 0.03 | 0.07135 | 0.4205 | |
| | | | | $\frac{0.5310}{0.4205} = 1.263$ |
| 5 | 0.05 | 0.09416 | 0.5310 | |

| 7 | 0.07 | 0.1139 | 0.6144 | $\frac{0.6144}{0.5310} = 1.157$ |
|---|---|---|---|---|

**Another method :** Using Fenton's (2020) equation: $k = \frac{\sigma^2}{g}\left[coth\left\{\left(\sigma\sqrt{\frac{d}{g}}\right)^{3/2}\right\}\right]^{2/3}$,

where $\sigma = \frac{2\pi}{T}$, calculate k and later $kd = \frac{2\pi d}{L}$. Then calculate $tanh\frac{2\pi d}{L}$ or 1,3,5 and 7 meters.

### *Exercise 13.35*

For a wave with a period of 5 seconds in a basin whose depth is 1000 m find the wave length, phase speed and group velocity for deep water waves. Verify that the wave is indeed a deep water wave.

### *Solution*

The equation for deep water wave length is: $L_0 = \frac{gT^2}{2\pi} = \frac{g \times 5^2}{2\pi} = 38.99 \approx 39\,m$

The wave velocity = $C_0 = \frac{gT}{2\pi} = \frac{9.8 \times 5}{2\pi} = 7.8ms^{-1}$

Verification for deep water wave is $\frac{d}{L}$: must be greater than ½,

$\frac{d}{L} = \frac{1000m}{39m} = 25.64$

As $\frac{d}{L} = 25.64$ is greater than ½, it is a deep water wave.

### *Exercise 13.36*

A wave has 3 m height and 7 seconds period in deep water. It travels towards shore over parallel bed contours. If its crest line makes an angle of $30^0$ with the bed contour of 10 m before refraction, calculate the wave height after crossing this contour line.

### *Solution*

$L_0 = \frac{gT^2}{2\pi} = 1.56T^2 = 1.56(7)^2 = 76.5\,m$

At 10 m depth $\frac{d}{L_0} = \frac{10}{76.5} = 0.1307$

From SPM tables, for $\frac{d}{L_0} = 0.1307$, $\frac{d}{L} = 0.1671$, (after interpretation)

For d=10 m, and $\frac{d}{L} = 0.1671 \rightarrow \frac{10}{L} = 0.1671 \rightarrow L = \frac{10}{0.1671} = 59.84m$

As $C = \frac{L}{T} = \frac{59.84m}{7s} = 8.55\,ms^{-1}$ Similarly $C_0 = \frac{L_0}{T} = \frac{76.5m}{7s} = 10.93\,ms^{-1}$

Now $\frac{C_0}{C} = \frac{\sin(\alpha_0)}{\sin(\alpha)} \rightarrow \frac{\sin(30)}{\sin(\alpha)} = \frac{10.93}{8.55} \approx 1.3 \rightarrow \sin(\alpha) = \frac{\sin(30)}{1.3} = 0.38 \rightarrow \alpha = 22.6^0$

$$K_r = \sqrt{\frac{b_0}{b}} = \left(\frac{\cos(\alpha_0)}{\cos(\alpha)}\right)^{1/2} = \left(\frac{\cos(30)}{\cos(22.6)}\right)^{1/2} = \left(\frac{0.87}{0.92}\right)^{1/2} = (0.95)^{1/2} = 0.97$$

we know $n = \frac{1}{2}\left[\frac{4\pi d/L}{Sinh(4\pi d/L)}\right]$; From SPM tables: for $\frac{d}{L_0} = 0.1307$, n=0.7616

Wave height after refraction,

$$H_r = K_s K_r H_0 = K_r H_0 \sqrt{\frac{C_0}{2nC}} = (0.97)(3)\sqrt{\frac{10.93}{2\times(0.7616)\times 8.55}} = 2.76\,m$$

### *Exercise 13.37*

Confused sea waves are striking approximately normally against a semi-infinite breakwater in 6 m deep water with Hs = 2.5 m and Ts = 10 sec. What are the values of Hs and Ts at a location 400 m behind and 400 m on the lee of the breakwater?

### *Solution*

$$L_0 = \frac{gT^2}{2\pi} = 1.56T^2 = 1.56(10)^2 = 156\,m$$

At 6 m depth $\frac{d}{L_0} = \frac{6}{156} = 0.03846$

From SPM tables, for $\frac{d}{L_0} = 0.03846, \frac{d}{L} = 0.0815 \rightarrow$

For d=6 m, and $\frac{d}{L} = 0.0815 \rightarrow \frac{6}{L} = 0.0815 \rightarrow L = \frac{6}{0.0815} = 73.62m$

$$\frac{x}{L} = \frac{400m}{73.62m} = 5.4 = \frac{y}{L}(\because x = y = 400m)$$

From Fig.7-61 of SPM (1984) vol-2, in $S_{max}$ = 10, for $\frac{x}{L} = \frac{y}{L} = 5.4$, the height ratio = 0.31 and period ratio = 0.87

Hence $H_s$ = (0.31) (2.5) = 0.78m and $T_s$ = (0.87) (10) = 8.7seconds

### *Exercise 13.38*

Assume that the sea is calm and suddenly a storm develops in the sea at a distance of 300 km from the coast. After how long swells of period 14 seconds develop in the coastal waters? If the period is 10 seconds, what happens? Assume the water depth in the coastal waters is 200m.

### Solution

For time period T= 14s, the wave length in deep water is

$$L_0 = \frac{gT^2}{2\pi} = 1.56T^2 = 1.56(14)^2 = 306\,m$$

This means $\frac{h}{L} = \frac{200m}{306m} = 0.65 > \frac{1}{3}$. As this is greater than $\frac{1}{2}$, they are deep water waves.

The group velocity is $C_g = \frac{C}{2} = \frac{1}{2}\left(\frac{L}{T}\right) == \frac{1}{2}\left(\frac{306}{14}\right) = 10.9\,ms^{-1} = \frac{1}{2}(1.56T) = 0.78T$

Reaching time of swell to the coast from deep water =

$$\Delta t = \frac{distance\ from\,coast}{group\,speed} = \frac{l}{C_g} = \frac{300km}{0.78T}$$

For T=14 s ➤ = $\Delta t = \frac{300km}{0.78T} = \frac{300 \times 10^3\,m}{0.78 \times 14ms^{-1}} = 27{,}472.5\,s \approx 7.6\,hours$

For T=10 s ➤ = $\Delta t = \frac{300km}{0.78T} = \frac{300 \times 10^3\,m}{0.78 \times 10ms^{-1}} = 38461.5\,s \approx 10.7\,hours$

### Exercise 13.39

What is the average energy *E* stored on a horizontal square metre of the water surface for a sinusoidal wave of height *H= 2m*

### Solution

The average energy E is: $\frac{1}{8}\rho g H^2$ where ρ = mass density of sea water ≈ 1020 $kgm^{-3}$

g = acceleration of gravity ≈ 9.8 $ms^{-2}$

$$E = \frac{1}{8}\rho g H^2 = \frac{1}{8}\left[1020\ \text{kgm}^{-3} \times 9.8\ \text{ms}^{-2} \times (2m)^2\right] = 4998kgs^{-2} \approx 5\,KW.s^{-2}$$

### Exercise 13.40

Calculate the wave power level (J) for a period of 10s and wave height 2meters (Take ρ = 1020 $kgm^{-3}$)

### Solution

The wave power level (energy transport per meter width of the wave front) is J = $C_g$ E:

In deep water the group velocity, $C_g$ is : $\frac{C_0}{2} = \frac{1}{2}\left(\frac{gT}{2\pi}\right) = \frac{gT}{4\pi}$

Hence $J = C_g E = \frac{gT}{4\pi}\left[\frac{1}{8}\rho g H^2\right] = \frac{\rho g^2 H^2 T}{32\pi} = K_j H^2 T$ ; where $K_j = \frac{\rho g^2}{32\pi} = 1kw/m^3 s$

$\therefore J = K_j H^2 T = 1\frac{kw}{m^3 s} \times (2m)^2 \times 10s = 40\,kw/m$

## Exercise 13.41

We know the paths of the fluid particles become circles with radius decreasing exponentially with depth in deep water. At what depth is the radius reduced to exp(-1) of the surface value (reduced to 37%)? How deep will a 10 m long wave reach? What about a 100 m long wave?

## Solution

We know from equations (13.37i): the radius of the orbital particle is $A = B = ae^{kd}$

a) Given $e^{kd} = e^{-1}$ ➔ $kd = -1$ ➔ $d = -\frac{1}{k} = -\frac{L}{2\pi} \approx 0.16L$

b) If L=10m, d = 0.16L = 0.16 × 10m = 1.6m

c) If L=100m, d = 0.16L = 0.16 × 100m = 16m

## Exercise 13.42

Sitting at east coast of visakhapatnam, it was noticed that waves coming in from a storm center in the Bay Bengal are typically 100 m long. 12 hours later, wave lengths have decreased to 50 m. Assuming that the waves are short compared to the depth, what are the distance to the storm center from the coast?

## Solution

As they are deep water waves, $L_0 = \frac{gT^2}{2\pi} = 1.56T^2 = 100m \rightarrow T^2 = \frac{100m}{1.56} \rightarrow T = 8s$

If T= 8s, the speed of the deep water wave is $C_0 = \frac{gT}{2\pi} = 1.56 \times 8 = 12.48\,ms^{-1}$

The wave travelled 12 hours with the speed 12.48 $ms^{-1}$,

So the distance travelled in 12 hours is = 12.48 × 12 × 3600 = 539136 m = 539km

Then after 12 hours its wave length is 50 m, still it is in deep water, so

$L_0 = \frac{gT^2}{2\pi} = 1.56T^2 = 50m \rightarrow T^2 = \frac{50m}{1.56} \rightarrow T = 5.66s$

If T= 5.66s, the speed of the deep water wave is $C_0 = \frac{gT}{2\pi} = 1.56 \times 5.66 = 8.83\,ms^{-1}$

With speed 12.48 $ms^{-1}$, the distance travelled is 539136m

With 8.83 $ms^{-1}$, the distance travelled is ? $\rightarrow \frac{539136 \times 8.83}{12.48} = 4892.4\,m$

So total distance travelled by wave is : 539136m+4892.4m = 544028.4m

Hence the distance of the storm is 544 km (approx.)

### *Exercise 13.43*

A storm was generated offshore of Visakhapatnam on the East coast of India. The energy spectrum of surface gravity waves was measured on October 1 at noon and then again on October 3 at noon. (A) If the peak energy occurs at a period of 12s on October 1, and 6 s on October 3, What is the wavelength for each of those waves? (B) What is the phase velocity and group velocity for each of the waves found in A? (C) Estimate *when* the storm occurred assuming 'D' be the offshore distance same for both cases.

### *Solution*

A) We know from equation (13.31) of text $L = \frac{gT^2}{2\pi}$ (13.43.1)

So for T = 12S, $L = \frac{gT^2}{2\pi} = 1.56\times(12)^2 = 224.64$m

For T = 6S, $L = \frac{gT^2}{2\pi} = 1.56\times(6)^2 = 56.16$m

B) The phase velocity for deep water waves is $C = \frac{gT}{2\pi} = 1.56\times12 = 18.72\,m/s$

For 12 s wave $C = \frac{gT}{2\pi} = 1.56\times12 = 18.72\,m/s$ (13.43.2)

For 6 s wave $C = \frac{gT}{2\pi} = 1.56\times6 = 9.36\,m/s$

The group velocity is half that of the phase velocities: 9.35 m/s and 4.68 m/s respectively for 12 and 6 S waves.

C) Given $D = C_1\, t_1 = C_2\, t_2$ (13.43.3)

Here $C_1$ and $C_2$ are the group velocities for 12 and 6 S respectively.

and $t_2 = t_1 + 48$ hours (13.43.4)

From equations (13.43.3) and (13.43.4) we can write: $C_1\, t_1 = C_2\,(t_1 + 48 \text{ hours})$

$(C_1 - C_2)t_1 = C_2$ (48 hours)

$$\therefore t_1 = \frac{C_2(48\,hours)}{(C_1 - C_2)} \quad (13.43.5)$$

Substituting the values of $C_1$ and $C_2$ in equation (13.43.5)

$$t_1 = \frac{C_2(48\,hours)}{(C_1 - C_2)} = \frac{9.36\times48}{18.72 - 9.36} = 48hours$$

From equation (13.43.3) = $C_1\, t_1 = 9.35 \times 48 \times 60 \times 60$S = 1615680 m ≈ 1616km

## Exercise 13.44

A straight coastline borders a uniformly sloping sea bed. Regular waves are observed to cross the 8m depth contour at an angle of 14° to the coastline-normal, with wave length 45m. Find:

(a) the wave period; (b) the wave length in deep water; (c) the direction in deep water.

## Solution

i) In shallow water ( In shore):

Given d = 8 m; $\theta = 14^0$; L = 45m

We know $\sigma^2 = gk \tanh kd$ (13.44.1)

and $k = \frac{2\pi}{L} = \frac{2\pi}{45m} = 0.1396\, m^{-1}$ (13.44.2)

Substituting (13.44.2) in (13.44.1) and using the constant values

$\sigma^2 = gk \tanh kd = 9.8ms^{-2} \times 0.1396m^{-1} \tanh (0.1396m^{-1} \times 8m)$

$\sigma = 1.051 \text{ rad s}^{-1}$

We know $\sigma = \frac{2\pi}{T} \rightarrow T = \frac{2\pi}{\sigma} = \frac{2\pi\, rad}{1.051\, rad\, s^{-1}} = 5.978\, s$

ii) In deep water: $L_0 = \frac{gT^2}{2\pi} = \frac{9.8 \times (5.978S)^2}{2\pi} = 55.8\, m$

iii) For the direction in deep water: use Snell's law

$(k \sin \theta)_0 = (k \sin \theta)_{8m} \longrightarrow k_0 \sin\theta_0 = k_{8m} \sin\theta_{8m}$ (13.44.3)

$k_0 = \frac{2\pi}{L_0} = \frac{2\pi}{55.8} = 0.1126\, m^{-1}$; $k_{8m} = 0.1396 \text{ m}^{-1}$ (13.44.4)

Substituting values of eqn (13.44.4) in eqn (13.44.3): $0.1126 \text{ m}^{-1} \sin\theta_0 = 0.1396 \text{ m}^{-1} \sin 14$

$\therefore \sin\theta_0 = \frac{0.1396\, m^{-1} \sin 14}{0.1126\, m^{-1}} = 1.228142 \rightarrow \theta_0 = 17.45^0$

Please note here d is water depth, L is wave length and suffix zero indicates for the deep water and H is wave height.

### *Exercise 13.45*

Waves propagate towards straight coastline that has a very gradual bed slope normal to the coast. In water depth of 20m, regular waves propagate at heading θ=40° relative to the bed slope.

(a)For a wave with period T = 8s and height 1.2m at 20m depth, calculate the wave heading and wave height at the 5m depth contour.

### *Solution*

Given T = 8 s at both the depths of 20 m and 5 m

We also know $\sigma = \frac{2\pi}{T} = \frac{2\pi\, rad}{8s} = 0.7854\, rad\, S^{-1}$

Dispersion:

We know $\sigma^2$ = gk tanh kd $\Rightarrow \frac{\sigma^2 d}{g} = kd \tanh kd$ (13.45.1)

$$\rightarrow kd = \frac{\frac{\sigma^2 d}{g}}{\tanh kd} \quad or \quad kd = \frac{1}{2}\left\{ kd + \frac{\frac{\sigma^2 d}{g}}{\tanh kd} \right\}$$

Wave number (k): according to Fenton & McKee (1990).

$$k = \frac{\sigma^2}{g}\left[ coth\left\{ \left( \sigma\sqrt{\frac{d}{g}} \right)^{3/2} \right\} \right]^{2/3} \tag{13.45.2}$$

$$k \tanh kd = \frac{\sigma^2}{g} = k_0 \tag{13.45.3}$$

Refraction (Snell's Law):

$(k \sin\theta)_{5m} = (k \sin\theta)_{20m}$ (13.45.4)

Shoaling:

From equation (13.55) of text we can write: $(H^2\, nc \cos\theta)_{5m} = (H^2\, nc \cos\theta)_{20m}$ (13.45.5)

For calculation of refraction and shoaling the following values are calculated and tabulated below at first.

| $\sigma = 0.7854$ rad $S^{-1}$ | d = 5m | d=20m |
|---|---|---|
| $\frac{\sigma^2 d}{g}$ | 0.3144 | 1.258 |
| Iteration $k = \frac{\sigma^2}{g}\left[coth\left\{\left(\sigma\sqrt{\frac{d}{g}}\right)^{3/2}\right\}\right]^{2/3}$ (13.47.2) | $Kd = \frac{1}{2}\left\{kd + \frac{\frac{\sigma^2 d}{g}}{\tanh kd}\right\}$ $= \frac{1}{2}\left\{kd + \frac{0.3144}{\tanh kd}\right\}$ | $kd \quad \frac{}{\tanh kd}$ $\frac{1.258}{\tanh kd}$ |
| Kd | 0.5918 | 1.416 |
| K | $k = 0.1184\ m^{-1}$ | $k_0 = 0.07080\ m^{-1}$ |
| $C = \frac{\sigma}{k}$ | 6.633 $ms^{-1}$ | 11.09$ms^{-1}$ |
| $n = \frac{1}{2}\left\{1 + \frac{2kd}{\sinh 2kd}\right\}$ | 0.8999 | 0.6674 |
| θ | -- | $40^0$ |
| H | - | 1.2m |

Note: K from eqn 13.45.2 can also be calculated. Also note from equation 13.45.3 can be obtained

Using the above table refraction and shoaling can be calculated from equations 13.45.4 and 13.45.5 as below:

Refraction: From eqn 13.45.4

$(k \text{ Sin } \theta)_{5m} = (k \text{ Sin } \theta)_{20m} \longrightarrow 0.1184 \text{ Sin}(\theta)_{5m} = 0.07080 \text{ Sin}(\theta)_{20m} \longrightarrow (\theta)_{5m} = 22.6^0$

Shoaling: from equation 13.45.5

$(H^2 \text{ nc } \cos\theta)_{5m} = (H^2 \text{ nc } \cos\theta)_{20m}$

$(H^2)5m \times 0.8999 \times 6.633 \times \cos 22.6^0 = (1.2^2 \times 0.6674 \times 11.09 \times \cos 40^0)_{20m} \longrightarrow (H)_{5m} = 1.217m$

### *Exercise 13.46*

If the wave period is 8 seconds, calculate kd and $\frac{C_0}{C}$ at 1 and 3 meters depths.

### *Solution*

First calculate $\sigma = \frac{2\pi}{T} = \frac{2\pi\, rad}{8s} = 0.7854\, rad\, S^{-1}$

Using Wave number (k) equation (13.45.2) of exersise 13.45 according to Fenton & McKee (1990)

$$k = \frac{\sigma^2}{g}\left[ coth\left\{ \left( \sigma\sqrt{\frac{d}{g}} \right)^{3/2} \right\} \right]^{2/3}$$

For d=1 m:

$$k = \frac{0.7854\,rad\,S^{-12}}{g}\left[ coth\left\{ \left( 0.7854\,rad\,S^{-1}\sqrt{\frac{1m}{g}} \right)^{3/2} \right\} \right]^{2/3}$$

$$k = \frac{(0.7854\,rad\,S^{-1})^2}{g}\left[ coth\left\{ \left( 0.7854\,rad\,S^{-1} \times 0.319 \right)^{3/2} \right\} \right]^{2/3}$$

$$k = \frac{(0.7854\,rad\,S^{-1})^2}{g}\left[ coth\left\{ (0.251)^{3/2} \right\} \right]^{2/3}$$

$$k = \frac{(0.7854\,rad\,S^{-1})^2}{g}\left[ coth\{0.126\} \right]^{2/3}$$

$$k = \frac{(0.7854\,rad\,S^{-1})^2}{g}\left[ coth\{0.126\} \right]^{2/3}$$

$$k = \frac{(0.7854\,rad\,S^{-1})^2}{g}[63.69]^{1/3} = \frac{(0.7854\,rad\,S^{-1})^2}{g} \times 3.99 = 0.3197$$

$kd_1 = 0.3197m^{-1} \times 1m = 0.3197$

Similarly for d=3 m

$$k = \frac{(0.7854\,rad\,S^{-1})^2}{g}\left[ coth\left\{ \left( 0.7854\,rad\,S^{-1}\sqrt{\frac{3m}{g}} \right)^{3/2} \right\} \right]^{2/3}$$

$$k = \frac{0.6169\,rad\,S^{-1}}{9.8}\left[ coth\left\{ \left( 0.7854\,rad\,S^{-1} \times 0.5533 \right)^{3/2} \right\} \right]^{2/3}$$

$0.0629\ [\coth\{\ (0.4345)^{3/2}\ \}]^{2/3} = 0.0629\ [\coth\{0.2864\}]^{2/3} = 0.0629\ [3.586]^{2/3} = 0.0629\ [12.86]^{1/3} = 0.0629 \times 2.343 = 0.1473m^{-1}$

$kd_3 = 0.1473m^{-1} \times 3m = 0.4419$

Calculation of $\frac{C_0}{C}$:

We know $C = \frac{gT}{2\pi} Tanh\, kd = 1.56 \times T \times \tanh kd$

and for deep water velocity as $C_0 = \frac{gT}{2\pi} = 1.56T$:

So $C_0 = 1.56T = 1.56 \times 8 = 12.48\ ms^{-1}$

$$C_{d1} = \frac{gT}{2\pi} Tanh\, kd_1 = 1.56 \times 8 \times \tanh(0.3197) = 3.859$$

$$\therefore \frac{C_o}{Cd_1} = \frac{12.48}{3.859} = 3.23$$

Similarly, $C_{d3} = \frac{gT}{2\pi} Tanh\, kd_3 = 1.56 \times 8 \times \tanh(0.4419) = 5.18 ms^{-1}$

$$\frac{C_0}{Cd_3} = \frac{12.48}{5.18} = 2.41$$

Refer Exercise 13.36 for another method of computation.

### *Exercise 13.47*

Describe with relevant equations on i) breaking criterion,ii) breaking condition in deep and shallow waters,iii) breaker height index, iv)breaker depth index and v) classification of breaker types basing on irribaren number and figures.

### *Solution*

i) **Breaking criterion**: $\left(-\right) \quad 0.14 \tanh(kd)$ (13.47.1)

Breaking occurs when $\frac{U_{max}}{C}$ exceeds a critical value

$$\frac{U_{max}}{C} = \frac{Agk/\sigma}{\sigma/k} = \frac{Agk^2}{\sigma^2} = \frac{Ak}{\tanh kd} = \frac{\pi H/L}{\tanh kd}$$

At breaking : $\frac{H}{L} = constant \times \tanh(kd)$

ii) **The breaking condition**: for deep water : $\left(\frac{H}{L}\right)_b = 0.14 \approx \frac{1}{7}$

iii) The breaking condition in shallow water:

$$\left(\frac{H}{L}\right)_b = 0.14 \approx \frac{1}{7} \text{ and } \left(\frac{H}{d}\right)_b = 0.88 \qquad (13.47.2)$$

Where H is wave height, d is water depth and L is wave length

iv) **Breaker height index**: $\Omega_b = \frac{H_b}{H_0}$, $\Omega_b = 0.56\left(\frac{H_0}{L_0}\right)^{-1/5}$ (13.47.3) where $H_b$ is wave height at breaking and $H_0$ is deep water wave height.

Deep water quantities extrapolated from height and period measured at one point:

$H_0$ from shoaling:

$(H^2 \text{ nc} \cos\theta)_0 = (H^2 \text{ nc} \cos\theta)$ measured (13.47.4)

$L_0$ from: $L_0 = \dfrac{gT^2}{2\pi}$

v) **Breaker Depth Index**: $\gamma_b = \left(\dfrac{H}{d}\right)_b$ (13.47.5)

where H and d are wave height and depth respectively at breaking.

On a mild slope: $\gamma_b = 0.78$

On a beach slope of m : $\gamma_b = \left\{b - a\dfrac{H_b}{gT^2}\right\} = \left\{b - a\dfrac{H_b}{2\pi L_0}\right\}$ (13.47.6)

Where $a = 43.8(1 - e^{-19m})$ and $b = \dfrac{1.56}{\left(1+e^{-19.5m}\right)}$

vi) **Types of Breakers:** See Figure 13.12 for details

The Irribarren Number (Aka surf-similarity parameter) is given as:

$$\xi = \frac{Beach\,slope}{\sqrt{wave\,steepness}} \quad (13.47.7)$$

The Irribarren number for deep water is given as: $\xi_0 = \dfrac{m}{\sqrt{H_0 / L_0}}$

The Irribarren number at breaking is given as: $\xi_b = \dfrac{m}{\sqrt{H_b / L_0}}$

The classification of breakers basing on irribarren number is as below:

| Irribarren number | Type of Breaker |
|---|---|
| $\xi_0 < 0.5$ | Spilling breaker |
| $0.5 < \xi_0 < 3.3$ | Plunging breaker |
| $3.3 < \xi_0$ | Surging or collapsing breaker |

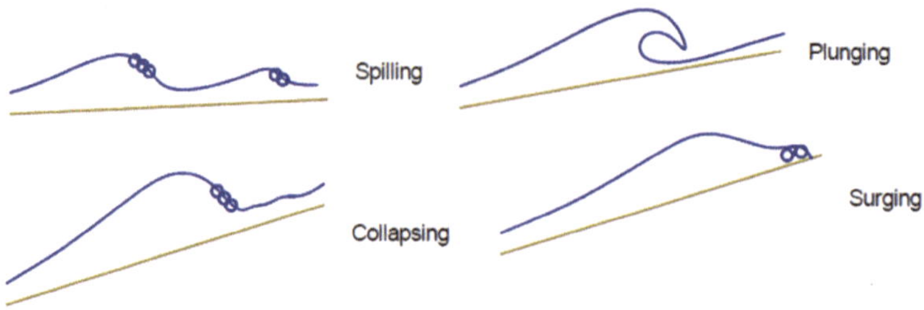


### *Exercise 13.48*

Waves propagate towards a long straight coastline that has a constant bed slope (m) of 1 in 100. Consider the *x*-axis to be normal to the coastline and the *y*-axis parallel to the coastline. Waves propagate at an angle θ to the *x*-axis. If a wave with period 7s and height 1.2m crosses the 36m depth contour at angle θ=22°,(i) determine the direction, height and power per metre width of wave crest at the 4m depth contour.(ii) Explain how height changes between these depths.

### *Solution*

Given wave period 7s and height 1.2m crosses the 36m depth contour at angle θ = 22°.

Referring Exercise 13.45:

Given T = 7 s at both the depths of 4m and 36 m.

We also know $\sigma = \dfrac{2\pi}{T} = \dfrac{2\pi\,rad}{7s} = 0.8976\,rad\,S^{-1}$

Dispersion:

We know from exercise (eqn 13.45.1), $\sigma^2 = gk \tanh kd \rightarrow \dfrac{\sigma^2 d}{g} = kd \tanh kd$

$$\rightarrow kd = \frac{\frac{\sigma^2 d}{g}}{\tanh kd} \quad or \quad kd = \frac{1}{2}\left\{kd + \frac{\frac{\sigma^2 d}{g}}{\tanh kd}\right\} \tag{13.48.1}$$

Refraction (Snell's Law): From exercise 13.45 of equation 13.45.4)

$$(k \operatorname{Sin} \theta)_{4m} = (k \operatorname{Sin} \theta)_{36m} \tag{13.48.2}$$

Shoaling:

From equation (13.45.5) we can write: $(H^2\, nc \cos\theta)_{4m} = (H^2\, nc \cos\theta)_{36m}$ (13.48.3)

Power:

$P = EC_g \qquad E = \frac{1}{8}\rho g H^2 \qquad C_g = nC$

$$P = \frac{1}{8}\rho g H^2 nC \tag{13.48.4}$$

For calculation of refraction and shoaling the following values are calculated and tabulated below at first.

| $\sigma = 0.8976$ rad $S^{-1}$ | d = 4m | d = 36m |
|---|---|---|
| $\frac{\sigma^2 d}{g}$ | 0.3285 | 2.957 |
| Iteration | $kd = \frac{1}{2}\left\{kd + \frac{\frac{\sigma^2 d}{g}}{\tanh kd}\right\}$ $= \frac{1}{2}\left\{kd + \frac{0.3285}{\tanh kd}\right\}$ | $kd \quad \frac{\overline{\phantom{x}}}{\tanh kd}$ $\frac{2.957}{\tanh kd}$ |
| Kd | 0.6065 | 2.973 |
| K | $k = 0.1516\ ms^{-1}$ | $k_0 = 0.08258\ m^{-1}$ |
| $C = \frac{\sigma}{k}$ | $5.921\ ms^{-1}$ | $10.87 ms^{-1}$ |
| $n = \frac{1}{2}\left\{1 + \frac{2kd}{\sinh 2kd}\right\}$ | 0.8956 | 0.5156 |
| θ | -- | $22^0$ |
| H | - | 1.2m |

Note: K from eqn 13.45.2 can also be calculated or From SPM tables can be obtained. . Also note k tanh kd = $k_0$ from equation 13.45.3

Using the above table refraction and shoaling can be calculated from equations 13.48.2 and 13.48.3 as below:

i) Refraction: direction at 4m depth contour (from equation 13.48.2): $(k \sin\theta)_{4m} = (k \sin\theta)_{36m} \longrightarrow 0.1516 \sin(\theta)_{4m} = 0.08258 \sin(22^0)_{36m} \longrightarrow (\theta)_{4m} = 11.77^0$

ii) Shoaling: height at 4m depth contour (from equation 13.48.3)::

$(H^2\ nc \cos\theta)_{4m} = (H^2\ nc \cos\theta)_{36m}$

$(H^2)_{4m} \times 0.8956 \times 5.921 \times \cos 11.77^0 = (1.2^2 \times 0.5156 \times 10.87 \times \cos 22^0)_{36m}$
$\longrightarrow (H)_{4m} = 1.201\text{m}$

iii) Power at 4m depth (from equation 13.48.4):

$$\frac{1}{8}\rho g H^2 nC = \left\{\frac{1}{8}\times 1.025\times 10^3\,kgm^{-3}\times(1.2m)^2\times\left(0.5156\times 10.87ms^{-1}\right)\right\}$$

$$= 9614W\,m^{-1}$$

## *Exercise 13.49*

In continuation of exercise 13.48, if a wave with period 7s and height 1.0m crosses the 4m depth contour at angle θ=0°, determine the breaking wave height and breaking depth from their corresponding indices and identify the type of breaker expected.

### *Solution*

We know from Breaker height index from exercise 13.47 of equation (13.47.3)

$$\Omega_b = \frac{H_b}{H_0} = 0.56\left(\frac{H_0}{L_0}\right)^{-1/5} \rightarrow H_b = 0.56H_0\left(\frac{H_0}{L_0}\right)^{-1/5}$$ and from equation (13.47.6)

$$\gamma_b = \left(\frac{H}{d}\right)_b = \left\{b - a\frac{H_b}{gT^2}\right\}$$ where $a = 43.8(1 - e^{-19m})$ and $b = \frac{1.56}{\left(1+e^{-19.5m}\right)}$

From shoaling we know from equation (13.48.3):

$(H^2\, nc \cos\theta)_0 = (H^2\, nc \cos\theta)$measured

As θ=0°, cos 0°= 1, So $(H^2\, nc)_0 = (H^2\, nc)_{4m}$ and given T = 7 seconds

| Depth 4m | Deep water |
|---|---|
| H=1.0 m | ? |
| n=0.8956 | n=0.5 |
| C=5.921m $s^{-1}$ | $C_0 = \frac{gT}{2\pi} = 1.56\times 7S = 10.93ms^{-1}$<br>$L_0 = \frac{gT^2}{2\pi} = 1.56\times 49S^2 = 76.5m\left(\because \frac{g}{2\pi} = 1.56\,ms^{-2}\right)$ |

## Calculation of Wave height in deep water

Shoaling :

$(H^2\, nc)_0 = (H^2\, nc)_{4m} \longrightarrow = (H^2)_0 \times 0.5 \times 10.93 = (1^2 \times 0.8956 \times 5.921)_{4m}$

$$H_0 = \frac{\left(1^2\times 0.8956\times 5.921\right)_{4m}}{0.5\times 10.93} = 0.9851m$$

A wave with period 7s and height 1.0m crosses the 4m depth contour at angle θ=0°. Determine the breaking wave height and breaking depth from their corresponding indices and identify the type of breaker expected.

It was given constant bed slope of 1 in 100. This means $m = \dfrac{1}{100} = 0.01$

**Breaking height:** $\Omega_b = \dfrac{H_b}{H_0} = 0.56\left(\dfrac{H_0}{L_0}\right)^{-1/5} \rightarrow H_b = 0.56H_0\left(\dfrac{H_0}{L_0}\right)^{-1/5}$

Substituting the values:

$$H_b = 0.56H_0\left(\frac{H_0}{L_0}\right)^{-1/5} = 0.56\times0.9851\left(\frac{0.9851}{76.51}\right)^{-1/5} = 1.317m$$

**Calculation of $\gamma_b$:**

$$\gamma_b = \left\{b - a\frac{H_b}{gT^2}\right\} = \left\{b - a\frac{H_b}{2\pi L_0}\right\}$$

Where a = 43.8(1 - $e^{-19m}$) and $b = \dfrac{1.56}{\left(1+e^{-19.5m}\right)}$

a = 43.8 (1-$e^{-19\times0.01}$) = 7.579

$$b = \frac{1.56}{\left(1+e^{-19.5\times0.01}\right)} = 0.8558$$

$$\gamma_b = \left\{0.8558 - 7.579\frac{1.317m}{9.8ms^{-2}\times7^2}\right\} = 0.8350$$

Calculation of $d_b$:

We know or $\gamma_b = \left(\dfrac{H}{d}\right)_b$ $or\ d_b = \dfrac{H_b}{\gamma_b} = \dfrac{1.317m}{0.8350} = 1.577m$

Calculation of breaker type:

The Irribarren number for deep water is given as: $\xi_0 = \dfrac{m}{\sqrt{H_0/L_0}} = \dfrac{0.01}{\sqrt{0.9851/76.5}} =$ 0.0881 ≈ 0.08

The Irribarren number at breaking is given as: $\xi_b = \dfrac{m}{\sqrt{H_b/L_0}} = \dfrac{0.01}{\sqrt{1.317/76.5}} = 0.076 \approx$ 0.08

In both cases as 0.08 is less than 0.5 , so it is spilling breaker.

### Exercise 13.50

In continuation of exercises 13.48 and 13.49, further along the coast, waves propagate over the outflow of a river. In water depth of 14m, measurements indicate a period of 7s and depth-averaged flow velocity of 0.8 ms$^{-1}$ against the wave direction. Determine the wavelength.

### Solution

Given d= 14m, $T_a$ = 7 s and U = - 0.8 $ms^{-1}$ (since depth averaged flow velocity is against the wave direction).

We know $\sigma_a = \frac{2\pi}{T_a} = \frac{2\pi}{7s} = 0.8976\,rad\,s^{-1}$

We also know $(\sigma_a - kU)^2 = \sigma_t^2 = gk\tanh kd \rightarrow k = \frac{(\sigma_a - kU)^2}{g\tanh kd}$

$\therefore k = \frac{(\sigma_a - kU)^2}{g\tanh kd} = \frac{(0.8976-(-0.8k))^2}{9.8\tanh(14k)} \rightarrow$ on solving k = 0.1087 $m^{-1}$

As $L = \frac{2\pi}{k} = \frac{2\pi}{0.1087m^{-1}} = 57.8m$

### Exercise 13.51

Explain diffraction and diffraction coefficient with illustration?

### Solution

Diffraction is the spreading of waves into a region of geometric shadow.

It occurs because there cannot be discontinuities at the boundary of the illuminated zone as shown in Fig.13.51.1.

Diffraction coefficient : $K_D(x) = \frac{H(x)}{H_0}$

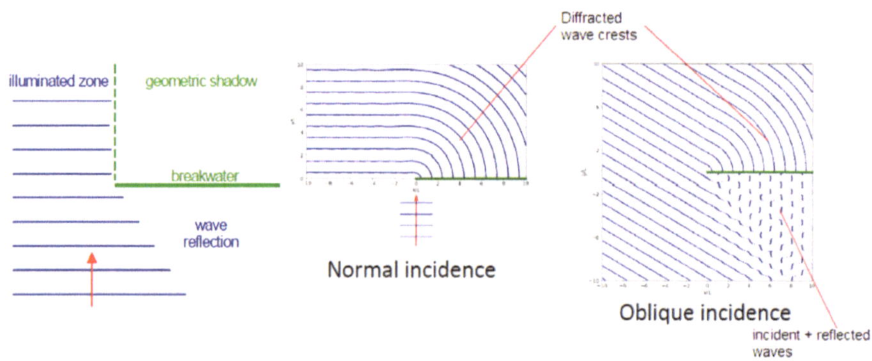


**Fig. 13.51.1.**

### Exercise 13.52

A harbor is to be protected by an L-shaped break water as sketched. Determine the length X of the outer arm necessary for the wave height at point P to be 0.3m when incident waves have a height of 3m and a period of 5 s. Assume the depth is everywhere uniform at 5m. Neglect reflections within the harbour.

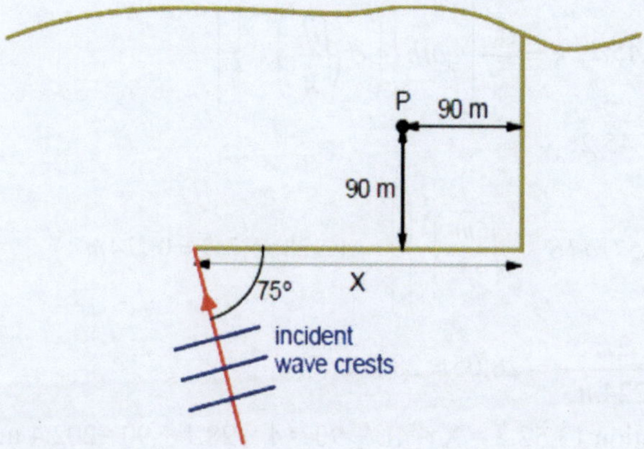


*Solution*

The value of diffraction coefficient K is $K = \dfrac{0.3}{3} = 0.1$ (13.52.1)

The diffraction diagram for the appropriate approach angle is as shown below

From the diffraction diagram for the value of 0.1, the value of $\dfrac{x}{L} = 4$. Hence x is:

$\dfrac{x}{L} = 4.0 \rightarrow x = 4L$

The total arm length in meters is X = x + 90 = 4L + 90: (13.52.2)

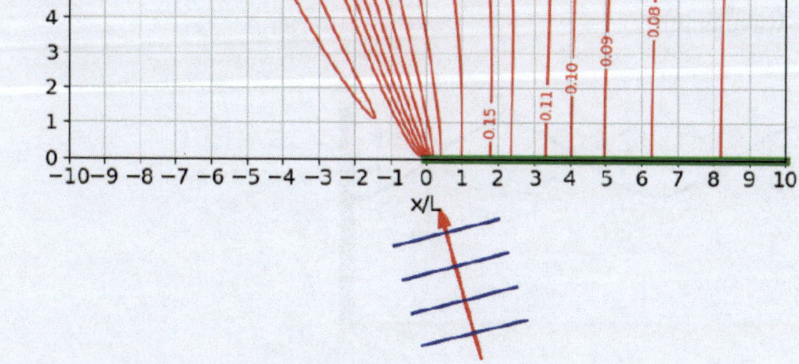


Given d = 5 m; T = 5 s and we know

$\sigma = \dfrac{2\pi}{T} = \dfrac{2\pi\, rad}{5s} = 1.257\, rad\, S^{-1}$

We know From equation 13.45.2: $k = \frac{\sigma^2}{g}\left[coth\left\{\left(\sigma\sqrt{\frac{d}{g}}\right)^{3/2}\right\}\right]^{2/3}$

Substituting the values in 13.45.2:

$$k = \frac{\left(1.257\, rad\, S^{-1}\right)^2}{9.8 m s^{-2}}\left[coth\left\{1.257\, rad\, S^{-1}\left(\sqrt{\frac{5m}{9.8}}\right)^{\frac{3}{2}}\right\}\right]^{\frac{2}{3}} = 0.128 \times 1.747 = 0.224 m^{-1}$$

We know $L = \frac{2\pi}{k} \rightarrow L = \frac{2\pi}{0.224 m^{-1}} = 28.05 \approx 28.1 m$

Hence arm length from equation 13.52.2 = X = 4L + 90 = 4 × 28.1 + 90 =202.4 m.

### *Exercise 13.53*

What is reflection of a wave, standing wave and seiching? Explain with relevant equations and figures.

### *Solution*

### Standing waves

Standing waves occur when two progressive waves moving in the opposite direction in a impermeable basin as shown in the figure:

The superposition of the two wave profiles is :

$\eta = A \cos(kx - \sigma t) + A \cos(-kx - \sigma t)$

$A[\cos(kx - \sigma t) + \cos(kx + \sigma t)] = 2A\cos kx \cos \sigma t$

This is the equation of a standing wave with twice the amplitude and nodes at every half wavelength.

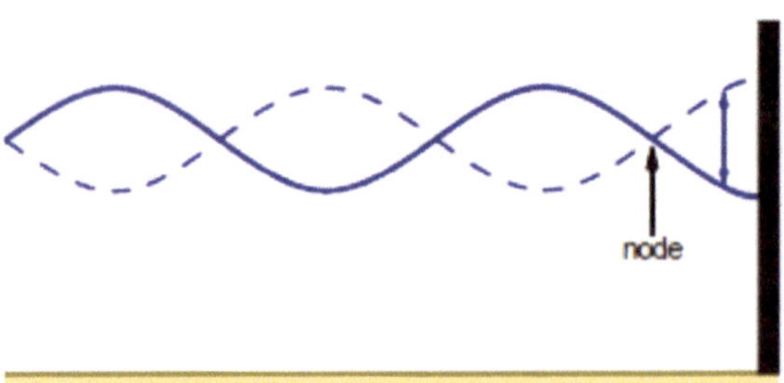


The wave velocity is given as:

$$u = \frac{2Agk}{\sigma}\frac{\cosh k(d+z)}{\cosh kd}\sin kx \sin \sigma t$$

Thus Reflection can be represented by the superposition of two equal and opposite progressive waves to form a standing wave. The standing wave has the same wavelength and frequency but twice the amplitude. There are surface nodes (i.e. zeroes of $\eta$) separated by half a wavelength, with velocity nodes intermediate between. The point of reflection corresponds to a point of zero velocity and double-amplitude displacement.

## Seiches

Seiches are resonant standing waves set up in enclosed basins (lakes, harbours, ...). The basin length (B) is a whole number of half-wavelengths: (L) of Seiches

$$B = n_S\left(\frac{1}{2}\right); L = \frac{2B}{n_S}; period = \frac{wave\,length}{Speed} \rightarrow T = \frac{2B}{n_S\sqrt{gd}}$$

Where $n_s$ is the node of seiche

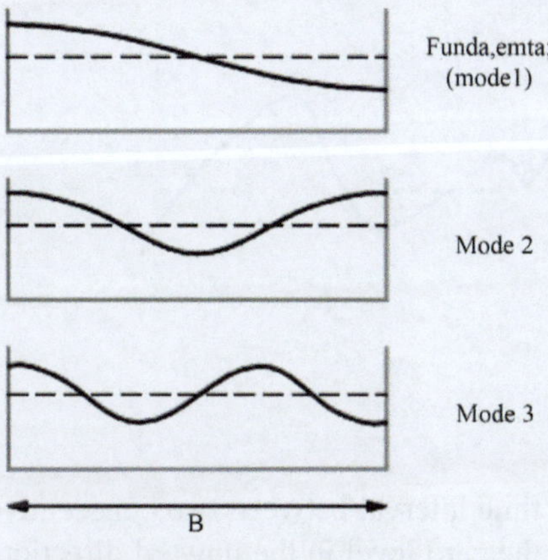


### *Exercise 13.54*

Lake Baikal in Siberia contains about one fifth of the world's fresh-water resources. It is 636km long, with an average depth of 744m. Calculate length, velocity and fundamental period of seiching.

### *Solution*

As per the above figure the fundamental mode of seiching is

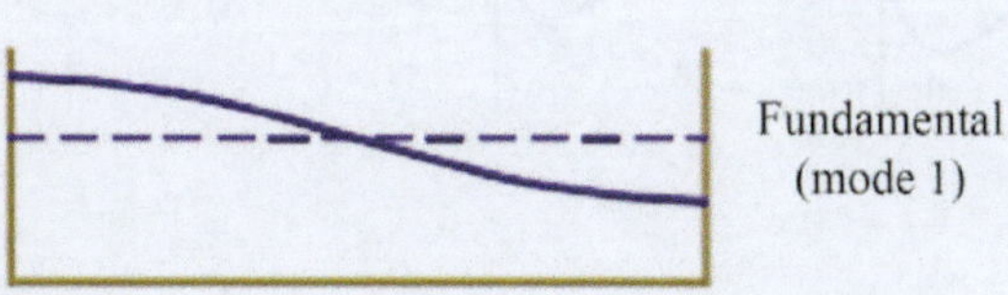

The wave length of the seiche is twice the length of the lake : So

L = 2 × 636 km = 1272 km = 1.272 × $10^6$ m

Seichevelocity:Shallowwaterwavevelocity $= C = \sqrt{gd} = \sqrt{9.8 \times 744m} = 85.4\,m\,s^{-1}$

Period of the seiche: $T = \frac{L}{C} = \frac{1.272 \times 10^6\,m}{85.4\,m\,s^{-1}} = 14894.6\,s \approx 4.1\,hours$

### *Exercise 13.55*

What is significant wave height explain with illustration.

### *Solution*

The *significant wave Height* is traditionally defined as the average of the highest one third of the individual wave heihgts (trough-to-crest heights *Hi*) (*i*=1,2,3,…), and is denoted by *H*1/3.

$$H_{1/3} = \frac{H_1 + H_2 + H_3 + \ldots\ldots + H_{N/3}}{N/3}$$

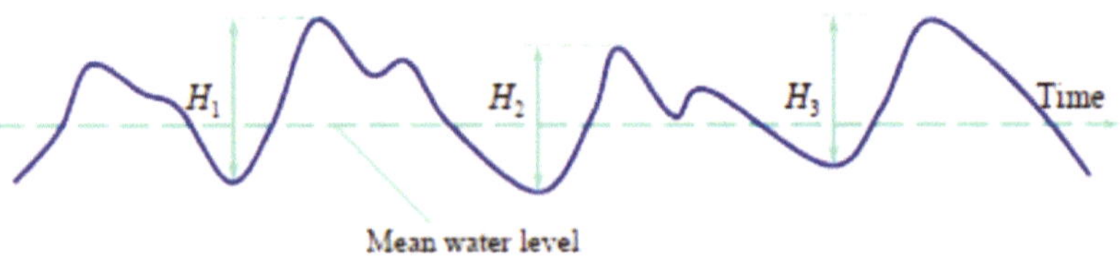


### *Exercise 13.56*

What is average zero up-crossing period?

The individual *zero up-cross time Ti* is the time interval between two consecutive instants where the wave elevation crosses the zero level in the upward direction. An average of these over a certain time provides a useful measure of the real-sea wave period.

$$T_Z = \frac{T_1 + T_2 + T_3 + \ldots\ldots + T_N}{N}$$

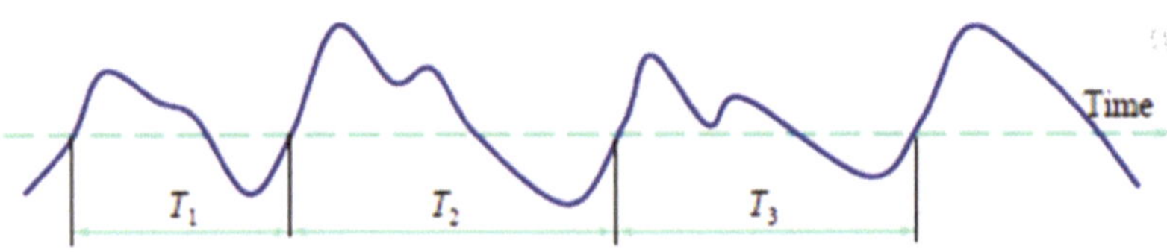

# 14

# Finite Amplitude Ocean Waves

## 14.1. Basics of Finite amplitude waves

Small amplitude wave theory was based on the premise that motions are sufficiently small to allow the free surface boundary condition to be linearized. Finite amplitude wave theories or Stokes' higher order theories are similar as per the concept in which additional higher order terms that were neglected in the Airy's theory are included. In general, the higher the order of the wave theory, higher will be the limiting wave height for which it would be valid. Shoaling computations are done using cnoidal theory, which best describes wave motion in relatively shallow waters. Although the Stokes waves satisfy the basic Laplace equation and the seabed boundary conditions, the free surface boundary conditions are not fully satisfied. The components of flow at the surface are not necessarily in accordance with the shape of the surface and its motions.

The basic feature of the finite amplitude waves is that the amplitude cannot be considered as small in comparison with either the wavelength or the depth of water column. Therefore, the higher order terms of Bernoullis equation (Equation 13.18 of chapter 13) cannot be neglected and thus, they are also known as nonlinear waves. A consequence of considering the full form of equation (13.18) is that the wave profiles are not sinusoidal.

One of the main characteristics of the finite amplitude waves is that they do not have sinusoidal shape and instead have peaked crests and broad troughs. Also, the upward amplitude, i.e., towards the crest is more than the downward amplitude, i.e., towards the trough. The retention of nonlinear terms makes the solution of the equation difficult. As a result of this, there is no unique theory that is applicable to all depth regions from deep to very shallow water. It must be noted that in the case of small amplitude waves, a single theory is applicable irrespective of water depth. On the other hand, different finite amplitude wave types are applicable to waves at different depths. For deep water locations, Stokes wave or trochoidal wave is applicable. If water depth is less, it is Cnoidal wave; and in very shallow waters, solitary wave is found to be suitable (see Table1). Another aspect is that different wave parameters gain importance in different depth regions. If the depth is more than half wavelength, then the important parameters are wavelength and wave steepness, given by *H/L*. In regions with depth between half and 1/20 of wavelength, *H/L* and water depth, *h* are important, and in still shallower regions, the depth of water and *H/h* become important parameters. Some basic features of these three types of finite amplitude waves are discussed in the following pages.

**Table 1.** Applicability of different finite ampliude wave theories. *L*, *H* and *h* are wavelength, wave height and water depth respectively. It must be noted that the depth limits are not exact and theories will overlap at the fringe regions.

| Water depth in relation to Wave length | Theories |
|---|---|
| $h > \frac{L}{10}$ | Stokes and Trochiodal |
| $\frac{L}{10} > h > \frac{L}{50}$ | Cnoidal |
| $h < \frac{L}{50}$ | Solitary |

It has already been noted that the Airy wave equations only strictly apply to waves of relatively small height in comparison to their wavelength and water depth. For steep waves and shallow water waves the profile becomes asymmetric with high crests and shallow troughs. For such waves, celerity and wavelength are affected by wave height and are better described by other wave theories. To categorize finite amplitude waves, three parameters are required. These are the wave height (H), the water depth (h) and wavelength (L). Using these parameters various non-dimensional parameters can be defined, namely relative depth (h/L), wave steepness (H/L) and wave height to water depth ratio (H/h). Another useful non-dimensional parameter is the Ursell number ($U_r = L^2H/d^3$), first introduced in 1953.

The first finite amplitude wave theory was developed by Stokes in 1847. It is applicable to steep waves in deep and transitional water depths. Following Stokes, Korteweg and de Vries(1895) developed a shallow water finite amplitude wave theory in 1895. They termed this Cnoidal theory, analogous to the sinusiodal Airy wave theory. Both of these theories relax the assumptions made in Airy theory which, as previously described, linearizes the kinematic and dynamic surface boundary conditions.

## Stokes' wave theory

The reader can refer to Dean and Dalrymple (1991) and Sorensen (1993) for a full mathematical description of all these theories. Hedges (1987,1995), provides useful guidance for the approximate regions of validity of analytical wave theories. It may be noted that the range of validity of linear theory is reassuringly wide, covering all of the transitional water depths for most wave steepnesses encountered in practice. For engineering design purposes, the main implication of using linear theory outside its range of validity is that wave celerity and wavelength are not strictly correct, leading to (some) inaccuracies in refraction and shoaling analysis. In addition, the presence of asymmetrical wave forms will produce harmonics in the Fourier analysis of recorded wave traces which could be incorrectly interpreted as free waves of higher frequency.

The linear or small-amplitude wave theory described in the preceding sections provides a useful first approximation to the wave motion. Ocean waves are generally not small in amplitude. In fact, it is usually the large waves are of interest since they result in the largest forces and greatest sediment movement. In order to approach the complete solution of ocean waves more closely, a perturbation solution using successive approximations may be developed to improve the linear theory solution of the hydrodynamic equations for gravity waves. Each order wave theory in the perturbation expansion serves as a correction and the net result is often a better agreement between theoretical and observed waves. The extended theories can also describe phenomena such as *mass transport* where there is a small net forward movement of the water during the passage of a wave. These higher-order or extended solutions for gravity waves are often called *nonlinear or finite amplitude wave theories*.

Development of the nonlinear wave theories has evolved for a better description of surface gravity waves. These include *cnoidal*, *solitary*, and *Stokes* theories.

A comparison of first- and second-order theories is useful to obtain insight about the choice of a theory for a particular problem. It should be kept in mind that linear (or first-order) theory applies to a wave that is symmetrical about the SWL and has water particles that move in closed orbits. On the other hand, Stokes' second-order theory predicts a waveform that is unsymmetrical about the SWL but still symmetrical about a vertical line through the crest and has water particle orbits that are open.

## 14.2. Stokes Finite amplitude wave theory

In Stokes' wave theory, (H/L) is assumed small and (d/L) is allowed to assume a wide range of values. The kinematic free surface boundary condition is then expressed as a power series in terms of H/L, and solutions up to and including the $n^{th}$ order of this power series is sought. Stokes derived the second order solution.

Stokes finite-amplitude wave perturbation theory assumed the wave slope '*ka*' is small where *k* is the wave number and *a* the amplitude of the wave. The perturbation solution, developed as a power series in terms of, $\varepsilon = ka$, is expected to converge as more and more terms are considered in the expansion. Convergence does not occur for steep waves unless a different perturbation parameter from that of Stokes is chosen.

In general, the perturbation expansion for velocity potential $\phi$ may be written as:

$$\phi = \varepsilon^1\phi_1 + \varepsilon^2\phi_2 \tag{14.1}$$

in which, $\varepsilon = ka$, is the *perturbation expansion parameter*. Each term in the series is smaller than the preceding term by a factor of order '$ka$'. In this expansion, $\phi_1$ is the first-order theory (linear theory), $\phi_2$ is the second-order theory, and so on.

Substituting the above Equation (14.1) and similar expressions for other wave variables (i.e., surface elevation '$\eta$', velocities '$u$'and '$w$', pressure '$p$', etc.) into the appropriate governing equations and boundary conditions describing the wave motion yields a series of higher-order solutions for ocean waves. Equating the coefficients of equal powers of '$ka$' gives recurrence relations for each order solution. A characteristic of the perturbation expansion is that each order theory is expressed in terms of the preceding lower order theories. The first-order Stokes theory is the linear (Airy) theory.

The Stokes expansion method is formally valid under the conditions that $H/d < (kd)^2$ for $kd < 1$ and $H/L < 1$. This condition restricts the wave heights in shallow water and the Stokes theory is not generally applicable to shallow water. For example, the maximum wave height in shallow water allowed by the second-order Stokes theory is about one-half of the water depth. The mathematics of higher-order Stokes theories is cumbersome. The detailed derivation of the Stokes second-order theory is given in the following pages as was given in Ippen (1966).

In the higher-order Stokes solutions, super harmonic components (i.e., higher frequency components at two, three, four, etc. times the fundamental frequency) arise. These are superposed on the fundamental component predicted by linear theory. Hence, wave crests are steeper and troughs are flatter than the sinusoidal profile (Fig. 14.1). The fifth-order Stokes expansion shows a secondary crest in the wave trough for high-amplitude waves (Peregrine 1972; Fenton 1985). In addition, particle paths for Stokes waves are no longer closed orbits and there is a *drift* or *mass transport* in the direction of wave propagation; called stokes drift.

The linear dispersion relation is still valid to second order, and both wavelength and celerity are independent of wave height to this order. At third and higher orders, wave celerity and wavelength depend on wave height, and therefore, for a given wave period, celerity and length are greater for higher waves. Some limitations are imposed on the finite-amplitude Stokes theory in shallow water both by the water depth and amplitude nonlinearities. For steeper waves in shallow water, higher-order terms in Stokes expansion may increase in magnitude to become comparable or larger than the fundamental frequency component (Fenton 1985; Chakrabarti 1987). When this occurs, the Stokes perturbation becomes invalid.

Higher-order Stokes theories include aperiodic (i.e., not periodic) terms in the expressions for water particle displacements. These terms arise from the product of time and a constant depending on the wave period and depth, and give rise to a continuously increasing net particle displacement in the direction of wave propagation. The distance a particle is displaced during one wave period when divided by the wave period gives a mean drift velocity U(z), called the *mass transport velocity*. To second-order, the mass transport velocity is

$$\bar{U}_z = \left(\frac{\pi H}{L}\right)^2 \frac{C}{2}\left[\frac{\text{Cosh}\, 2K(d+z)}{\text{Sinh}^2 Kd}\right] \tag{14.2}$$

This indicates that there is a net transport of fluid by waves in the direction of wave propagation. If the mass transport, indicated by equation (14.2) leads to an accumulation of mass in any region, the free surface must rise, thus generating a pressure gradient. A current, formed in response to this pressure gradient, will reestablish the distribution of mass. Earlier findings indicate that the vertical distribution of the mass transport velocity is modified so that the net transport of water across a vertical plane is zero.

Wave profiles obtained from different wave theories are sketched in Fig. 14.1 for comparison. The linear profile is symmetric about the SWL. The Stokes wave has higher more peaked crests and shorter, flatter troughs. The cnoidal wave crests are higher above the still water level (SWL) than the troughs are below the SWL. Cnoidal troughs are longer and flatter and crests are sharper and steeper than Stokes waves. The solitary wave, a form of the cnoidal wave described in the next section, has all of its profile above the SWL.

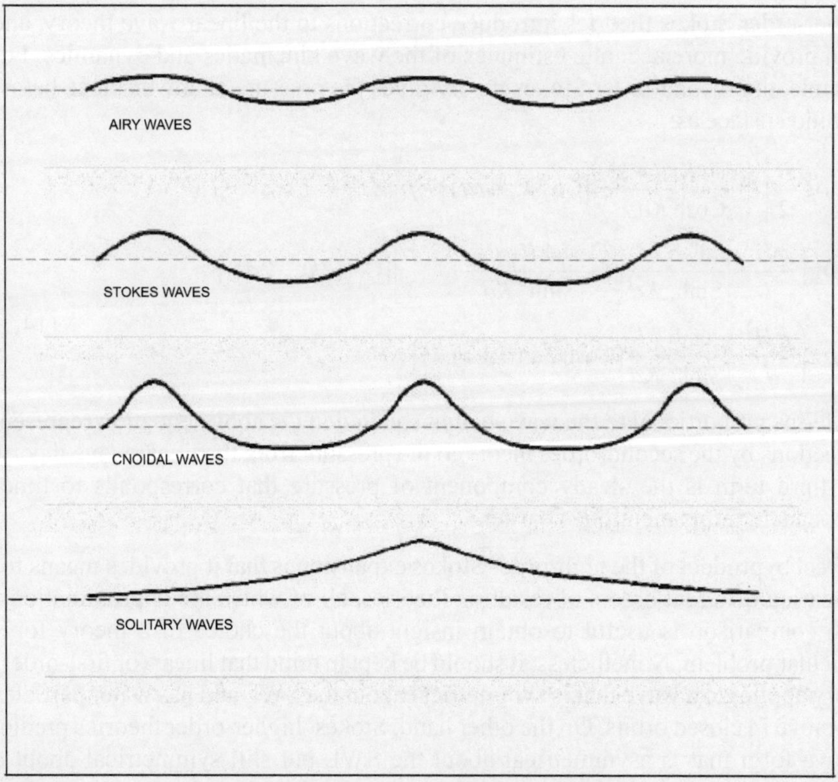


**Fig. 14.1.** Wave profile shape of different progressive gravity waves

Thus the Stokes wave has a broad trough and somewhat peaked crest (Fig. 14.1). Stokes presented a method for the solution of the above nonlinear equation of velocity potential. Hence, these are known as Stokes waves. The equation is solved as a series, using perturbation method. It is assumed that the solution can be represented in terms of a power series expansion of a small quantity and this quantity is related to wave steepness, which is the ratio of wave height to wavelength. The sum up to the n$^{th}$ order term is the n$^{th}$ order solution. Solution is, therefore, not unique and as the order increases, the resulting wave becomes more peaked. The solution satisfies the irrotationality condition. It may be noted that the method requires a number of numerical calculations, which increases as the order of the solution increases. Though Stokes presented the second order solution, subsequent researchers have worked out higher order solutions. The first order solution gives results for wave profile and wave celerity that are similar to those for small amplitude waves. In the case of second order solution, though the expression for wave celerity does not change, the wave profile is complicated and is given in following pages.

### 14.2.1. Subsurface pressure

Higher-order Stokes theories introduce corrections to the linear wave theory, and often provide more accurate estimates of the wave kinematics and dynamics. For example, the second-order Stokes theory gives the pressure at any distance below the fluid surface as:

$$\begin{aligned} P &= \rho g \frac{H}{2}\left[\frac{\operatorname{Cosh} K(d+z)}{\operatorname{Cosh} Kd}\right]\left[\operatorname{Cos}(kx-\sigma t)\right] - \rho g Z \\ &+ \frac{3}{8}\rho g\left(\frac{\pi H^2}{L}\right)\frac{\tanh kd}{\operatorname{Sinh}^2 Kd}\left[\frac{\operatorname{Cosh} 2K(d+z)}{\operatorname{Sinh}^2 Kd} - \frac{1}{3}\right]\left[\operatorname{Cos}(2kx-2\sigma t)\right] \\ &- \frac{1}{8}\rho g\left(\frac{\pi H^2}{L}\right)\frac{\tanh kd}{\operatorname{Sinh}^2 Kd}\left[\operatorname{Cosh} 2K(d+z)-1\right] \end{aligned} \tag{14.3}$$

The terms proportional to the wave height squared in the above equation represent corrections by the second-order theory to the pressure from the linear wave theory. The third term is the steady component of pressure that corresponds to time-independent terms mentioned earlier.

A direct byproduct of the high-order Stokes expansion is that it provides means for comparing different orders of resulting theories, all of which are approximations. Such comparison is useful to obtain insight about the choice of a theory for a particular problem. Nonetheless, it should be kept in mind that linear (or first-order) theory applies to a wave that is symmetrical about the SWL and has water particles that move in closed orbits. On the other hand, Stokes' higher-order theories predict a wave form that is asymmetrical about the SWL but still symmetrical about a vertical line through the crest and has water particle orbits that are open (Fig. 14.1).

### 14.2.2. Maximum wave steepness

(a) A progressive gravity wave is physically limited in height by depth and wavelength. The upper limit or breaking wave height in deep water is a function of the wavelength and, in shallow and transitional water, is a function of both depth and wavelength.

(b) Stokes (1880) predicted theoretically that a wave would remain stable only if the water particle velocity at the crest was less than the wave celerity or phase velocity. If the wave height were to become so large that the water particle velocity at the crest exceeded the wave celerity, the wave would become unstable and break.

Stokes determined that this breaking condition corresponds to a crest angle of $120^0$ (Fig.14.2). Mitchell (1893) found that for deep water this condition could also be expressed as limiting wave steepness

$$\left[\frac{H_0}{L_0}\right]_{max} = 0.142 \cong \frac{1}{7} \tag{14.4}$$

(c) Miche (1944) gives the limiting steepness for waves traveling in depths less than $L_0/2$ without a change in form as

$$\left[\frac{H}{L}\right]_{max} = \left[\frac{H_0}{L_0}\right]_{max} \tanh\left(\frac{2\pi d}{L}\right) = 0.142 \tanh\left(\frac{2\pi d}{L}\right) \tag{14.5}$$

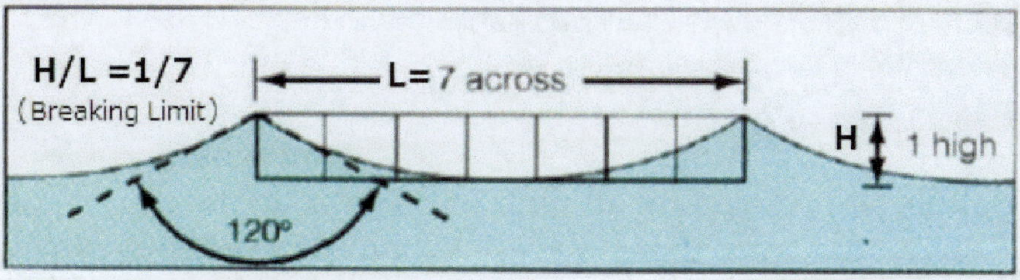


**Fig.14.2.** Wave limiting steepness in deep water (Solitary wave)

## 14.3. Derivation of Stokes first and second order wave theories

The required differential equation is:

The fluid is assumed to be incompressible and irrotational. So the Laplace equation applied is

$$\nabla^2\varphi = 0 \tag{14.6}$$

The required Boundary conditions:

The boundary condition applied on the bottom ( Z = -- h) is that no flow occurs across the bottom surface. So

$$w = -\frac{\partial \varphi}{\partial z} = 0 \text{ on } z = -h \tag{14.7}$$

At the free surface (z = η), two boundary conditions must be satisfied. The first one is dynamic free surface boundary condition (DFBC) which says that the total energy (Q) along the free surface must be constant. This is nothing but steady state Bernoullis equation:

$$\eta + \frac{1}{2g}\left[(u-c)^2 + w^2\right] = Q \tag{14.8}$$

Note that $u^2$ and $w^2$ in the above equation are nonlinear and were neglected in small amplitude wave theory.

The second boundary condition that must be satisfied on the free surface is Kinematic free surface boundary condition (KFBC). This says that no fluid be transported across the free surface. This can be formulated for a stationary free surface as the resultant velocity vector is tangent to the free surface. So

$$\frac{\partial \eta}{\partial x} = \frac{w}{u-c} \tag{14.9}$$

The derivation of finite amplitude waves uses the perturbation technique. This approach was first made by Stokes (1847). The procedure is to assume that the solution can be represented in terms of a power series expansion of a perturbation parameter '∈' for ϕ and η. The series expansion up to second order theory is:

$$\varphi(x,z) = \sum_{n=1}^{\infty} \varepsilon^n \varphi_n(x,z) = \varepsilon^1 \varphi_1 x + \varepsilon^1 \varphi_1 z + \varepsilon^2 \varphi_2 x + \varepsilon^2 \varphi_2 z$$

$$\eta = \sum_{n=1}^{\infty} \varepsilon^n \eta_n = \varepsilon^1 \eta_1 + \varepsilon^2 \eta_2$$

$$C = C_* + \sum_{n=1}^{\infty} \varepsilon^n C_n = C_* + \varepsilon^1 C_1 + \varepsilon^2 C_2 \tag{14.10}$$

Here n represents the n$^{th}$ order theory and represents the lowest order wave celerity. For first order $\varepsilon^1$ terms and for second order $\varepsilon^2$ terms will be taken and equated to zero to get the solution. The quantity ε is related to the ratio of wave height to wave length (H/L) or wave height to water depth (H/h).The above are velocity potential (ϕ), wave profile (η) and wave celerity (C).

## Derivation

In other words, if $\varphi_2(x,z) = \varepsilon^1 \varphi_1 x + \varepsilon^1 \varphi_1 z + \varepsilon^2 \varphi_2 x + \varepsilon^2 \varphi_2 z$ (14.11)

The second order Laplace equation can be written as:

$$\nabla^2 \varphi_2(x,z) = \varphi_1 xx + +\varphi_2 zz = 0 \tag{14.12}$$

## Boundary conditions

We know the bottom boundary condition as:

$$w = -\frac{\partial \varphi_2(x,z)}{\partial z} = 0 \text{ on z = - h or}$$

Where $\varphi_2(x,z) = \varepsilon^1\varphi_1 x + \varepsilon^1\varphi_1 z + \varepsilon^2\varphi_2 x + \varepsilon^2\varphi_2 z$

$$\varphi_2{}'(x,z) = \frac{\partial \varphi_2(x.z)}{\partial z} = \varepsilon^1\frac{\partial \varphi_{1z}}{\partial z} + \varepsilon^2\frac{\partial \varphi_{2z}}{\partial z}$$

Dynamic Boundary condition (DFBC) can be obtained from equation (14.8) as:

$$\eta + \frac{1}{2g}\left[(u-c)^2 + w^2\right] - Q = 0$$

We know u = $-\phi_x$; and w = $-\phi_z$, substituting these values

$$\eta - Q + \frac{1}{2g}\left[(\varphi_X)^2 + 2\varphi_X C + C^2 + (\varphi_Z)^2\right] = 0 \text{ on Z} = \eta \tag{14.13}$$

Where $Q = Q_O + \sum_{n=1}^{\infty}\varepsilon^n Q_n = Q_O + \varepsilon^1 Q_1 + \varepsilon^2 Q_2$

The Kinematic Free surface Boundary condition (KFSBC) can be written from equation (14.9) as:

$$\frac{\partial \eta}{\partial x} = \frac{w}{u-c}$$

Substituting the values of 'w' and u in terms of velocity potential we get

$$\eta_X = \frac{-\varphi_Z}{-\varphi_X - C} \text{ or } \eta_X\varphi_X - \varphi_Z + C\eta_X = 0 \tag{14.14}$$

Expansion of D.F.B.C (equation (14.13):

$$\left(\varepsilon^1\eta_1 + \varepsilon^2\eta_2\right) - \left(Q_O + \varepsilon^1 Q_1 + \varepsilon^2 Q_2\right) +$$

$$\frac{1}{2g}\{(\varepsilon^1\varphi_1 x + \varepsilon^2\varphi_2 x)^2 + (C_* + \varepsilon^1 C_1 + \varepsilon^2 C_2)^2 + 2(\varepsilon^1\varphi_1 x + \varepsilon^2\varphi_2 x)$$

$$(C_* + \varepsilon^1 C_1 + \varepsilon^2 C_2) + (\varepsilon^1\varphi_1 z + \varepsilon^2\varphi_{2Z})^2\} = 0$$

$$\left(\frac{C^2_*}{2g} - Q_0\right) + \varepsilon^1\left[\eta_1 - Q_1 + \frac{2C_*C_1}{2g} + \frac{2C_*\varphi_{1X}}{2g}\right] +$$

$$\varepsilon^2\left[\eta_2 - Q_2 + \frac{\varphi^2_{1X}}{2g} + \frac{C^2_1}{2g} + \frac{2C_*C_2}{2g} + \frac{\varphi^2_{1Z}}{2g} + \frac{C_*\varphi_{2X}}{g} + \frac{2C_1\varphi_{1X}}{2g} + \frac{2C_*\varphi_{1XZ}}{2g}\eta_1\right] + \varepsilon^3(...) = 0 \tag{14.15}$$

$\varepsilon^3$(...) implies other higher order terms which are not important upto second order waves.

The K.F.B.C can be determined from equation (14.14) as:

$$\{(\varepsilon^1\eta_{1X}+\varepsilon^2\eta_{2X})(\varepsilon^1\varphi_{1X}+\varepsilon^2\varphi_{2X})+(C_*+\varepsilon^1 C_1+\varepsilon^2 C_2)(\varepsilon^1\eta_{1X}+\varepsilon^2\eta_{2X})-(\varepsilon^1\varphi_{1Z}+\varepsilon^2\varphi_{2Z})\}=0$$

$$=> \varepsilon^1(C_*\eta_{1X}-\varphi_{1Z})+\varepsilon^2(\eta_{1X}\varphi_{1X}+C_*\eta_{2X}+C_1\eta_{1X}-\varphi_{2Z}-\varphi_{1ZZ}\eta_1)+\varepsilon^3(0)=0 \quad (14.16)$$

From equation (14.15), the constant or zero order term is: $\left(\frac{C_*^2}{2g}-Q_0\right)=o \ or \ Q_0=\frac{C_*^2}{2g}$

## 14.4. First order theory

Coefficient of $\in^1$ terms from equation (14.15) (DFBC) is:

$$\eta_1-Q_1+\frac{C_*C_1}{g}+\frac{C_*\varphi_{1X}}{g}=0 \text{ on } Z=0 \quad (14.17)$$

Coefficient of $\in^1$ terms from equation (14.16) (KFBC) is: $C_*\eta_{1X}-\varphi_{1Z}=0$ (14.18)

We know at bottom ( on Z = -- h), $\phi_1 x = 0$

Let the velocity potential as : $\varphi_1=-A_1\operatorname{Cosh}k(h+z)\operatorname{Sin}(kx-\sigma t)$ (14.19)

Then $\varphi_{1X}=-A_1K\operatorname{Cosh}K(h+z)\operatorname{Cos}(kx-\sigma t)$

### First order wave profile

Taking out constant terms in equation (14.17): $-Q_1+\frac{C_*C_1}{g}=0;\ Q_1=\frac{C_*C_1}{g}$ (14.20)

Hence equation (14.17) turns out to be: $\eta_1+\frac{C_*\varphi_{1X}}{g}=0$ (14.21)

Or $\eta_1=-\frac{C_*\varphi_{1X}}{g}=-\frac{C_*}{g}\left[-A_1K\operatorname{Cosh}K(h+z)\operatorname{Cos}(kx-\sigma t)\right]$

Or $\eta_1=\frac{C_*}{g}\left[A_1K\operatorname{Cosh}K(h+z)\operatorname{Cos}(kx-\sigma t)\right]$ (14.22)

### First order wave celerity

Solve equations (14.21 & 14.18), by taking the partial derivative of eqn (14.21) with respect to 'x' and substitute in equation (14.18):

Eqn (14.21): $\eta_1+\frac{C_*\varphi_{1X}}{g}=0$ ;Eqn (14.18): $C_*\eta_{1X}-\varphi_{1Z}=0$

Differentiating Eqn 14.21 w.r.t 'x' : $\eta_{1X}+\frac{C_*\varphi_{1XX}}{g}=0$ (14.23)

Solving equations (14.18) & (14.23) by eliminating $\eta_{ix}$, we get:

$$\frac{\varphi_{1Z}}{C_*}+\frac{C_*\varphi_{1XX}}{g}=0 \quad (14.24)$$

Double differentiate equation (14.19) with respect to 'X' and then differentiate equation (14.19) with respect to 'Z' and substitute in equation (14.24) to get the wave celerity as:

$$\frac{C_*^2}{g} A_1 K^2 \operatorname{Cosh} K(h+z) \operatorname{Sin}(kx-\sigma t) = A_1 K \operatorname{Sinh} K(h+z) \operatorname{Sin}(kx-\sigma t)$$

$$= C_*^2 = \frac{g}{K} \tanh K(h+z) \text{ or on } Z=0,\ C_*^2 = \frac{g}{K} \tanh Kh \qquad (14.25)$$

## 14.5. Second order theory

Take the coefficient of $\varepsilon^2$ terms in (DFBC) equation (14.15):

$$\eta_2 - Q_2 + \frac{C_* \varphi_{2X}}{g} + \frac{1}{2g}\left[\varphi^2_{1X} + \varphi^2_{1Z} + 2C_1\varphi_{1X} + 2C_*\varphi_{1XZ}\eta_1 + C^2_1 + 2C_*C_2\right] = 0 \text{ on } Z=0 \quad (14,26)$$

If the constant terms are taken out from eqn (14.26) : $Q_2 = \dfrac{C^2_1 + 2C_*C_2}{2g}$ (14.27)

Similarly the coefficient of $\varepsilon^2$ terms in equation (14.16) (KFBC):

$$C_*\eta_{2X} + \eta_{1X}\varphi_{1X} + C_1\eta_{1X} - \varphi_{2Z} - \varphi_{1ZZ}\eta_1 = 0 \text{ on } Z=0$$

$$\text{Or } \eta_{2X} = \frac{1}{C_*}\left[\varphi_{2Z} + \varphi_{1ZZ}\eta - \eta_{1X}\varphi_{1X} - C_1\eta_{1X}\right] \qquad (14.28)$$

Let the velocity potential for second order theory similar to first order theory can be taken as :

$$\varphi_2 = -A_2 \operatorname{Cosh} 2k(h+z) \operatorname{Sin} 2(kx-\sigma t) \qquad (14.29)$$

In order to get the second order celerity term $C_1$, solve equations (14.26) and (14.28) by taking a partial derivative of eqn (14.26) with respect to 'X' and substitute in equation (14.28) by using equations (14.19) and (14.29).

So the remaining of equation (14.26) after taking out the constant term (eqn 14.27) is

$$\eta_2 + \frac{C_*\varphi_{2X}}{g} + \frac{1}{2g}\left[\varphi^2_{1X} + \varphi^2_{1Z} + 2C_1\varphi_{1X} + 2C_*\varphi_{1XZ}\eta_1\right] = 0 \qquad (14.30)$$

Taking partial derivative of eqn (14.30) w.r.t 'X'

$$\eta_{2X} + \frac{C_*\varphi_{2XX}}{g} + \frac{1}{2g}\left[2\varphi_{1X}.\varphi_{1XX} + 2\varphi_{1Z}.\varphi_{1ZX} + 2C_1\varphi_{1XX} + 2C_*\{\varphi_{1XZ}\eta_{1X} + \eta_1\varphi_{1XXZ}\right] = 0 \qquad (14.31)$$

Eliminating $\eta_{2X}$ between eqns (14.28) & (14.31) we get:

$$-\frac{\varphi_{2Z}}{C_*} - \frac{C_*\varphi_{2XX}}{g}$$

$$= +\frac{1}{g}\left[\varphi_{1X}.\varphi_{1XX} + \varphi_{1Z}.\varphi_{1ZX} + C_1\varphi_{1XX} + C_*\{\varphi_{1XZ}\eta_{1X} + \eta_1\varphi_{1XXZ}\right] - \frac{\varphi_{1X}\eta_{1X}}{C_*} - \frac{C_1\eta_{1X}}{C_*} + \frac{\varphi_{1ZZ}\eta_1}{C_*}$$

on Z=0 (14.32)

From equation (14.19) we know: $\varphi_1 = -A_1 \operatorname{Cosh} k(h+z)\operatorname{Sin}(kx-\sigma t)$ ; Getting the different partial functions:

$$\varphi_{1X} = -A_1 K \operatorname{Cosh} K(h+z)\operatorname{Cos}(kx-\sigma t)$$
$$\varphi_{1XX} = A_1 K^2 \operatorname{Cosh} K(h+z)\operatorname{Sin}(kx-\sigma t)$$
$$\varphi_{1XZ} = -A_1 K^2 \operatorname{Sinh} K(h+z)\operatorname{Cos}(kx-\sigma t)$$
$$\varphi_{1Z} = -A_1 K \operatorname{Sinh} K(h+z)\operatorname{Sin}(kx-\sigma t)$$
$$\varphi_{1ZZ} = -A_1 K^2 \operatorname{Cosh} K(h+z)\operatorname{Sin}(kx-\sigma t)$$
$$\varphi_{1ZX} = -A_1 K^2 \operatorname{Sinh} K(h+z)\operatorname{Cos}(kx-\sigma t)$$
$$\varphi_{1XXZ} = A_1 K^3 \operatorname{Sinh} K(h+z)\operatorname{Sin}(kx-\sigma t)$$

Similarly from equation (14.29) We know: $\varphi_2 = -A_2 \operatorname{Cosh} 2k(h+z)\operatorname{Sin} 2(kx-\sigma t)$ and obtaining different partial functions:

$$So\,\varphi_{2Z} = -A_2 2K \operatorname{Sinh} 2k(h+z)\operatorname{Sin} 2(kx-\sigma t)$$
$$\varphi_{2X} = -A_2 2K \operatorname{Cosh} 2k(h+z)\operatorname{Cos} 2(kx-\sigma t)$$
$$\varphi_{2XX} = A_2 4K^2 \operatorname{Cosh} 2k(h+z)\operatorname{Sin} 2(kx-\sigma t)$$

Similarly we know from equation (14.22): $\eta_1 = \frac{C_*}{g}\left[A_1 K \operatorname{Cosh} K(h+z)\operatorname{Cos}(kx-\sigma t)\right]$

So $\eta_{1X} = -\frac{C_*}{g}\left[A_1 K^2 \operatorname{Cosh} K(h+z)\operatorname{Sin}(kx-\sigma t)\right]$

Substituting all the above $\varphi_1, \varphi_2, \eta_{1X}$ etc in equation (14.32) and with much simplification using equation (14.25) we can get:

$$A_2 \operatorname{Sinh}^2 kh \operatorname{Sin} 2(kx-\sigma t) = \frac{3}{8}\frac{A_1^2 K}{C_*}\operatorname{Sin} 2(Kx-\sigma t) - \frac{1}{2}\frac{C_1}{C_*}A_1 \operatorname{Cosh} Kh \operatorname{Sin}(Kx-\sigma t) \text{ on } Z=0 \quad (14.33)$$

Equating the terms of $\operatorname{Sin} 2(kx-\sigma t)$ : $A_2 \operatorname{Sinh}^2 kh = \frac{3}{8}\frac{A_1^2 K}{C_*}$

$$\text{Or } A_2 = \frac{3}{8}\frac{A_1^2 K}{C_*}\frac{1}{\operatorname{Sinh}^2 kh} \quad (14.34)$$

Similarly equating the terms of $\operatorname{Sin}(kx-\sigma t): -\frac{1}{2}\frac{C_1}{C_*}A_1 \operatorname{Cosh} Kh = 0$

This means $C_1 = 0$ (14.35)

This implies that the second order correction to the first order wave celerity is zero.

To get the second order velocity potential ($\phi_2$), substitute the value of $A_2$ (from eqn 14.34) in equation (14.29):

$$\varphi_2 = -\frac{3}{8}\frac{A_1^2 K}{C_*}\frac{1}{\operatorname{Sinh}^2 kh}\operatorname{Cosh} 2K(h+Z).\operatorname{Sin} 2(Kx-\sigma t) \quad (14.36)$$

The second order wave profile (η) is given by the equation (14.26):

$$\eta_2 = Q_2 - \frac{C_*\varphi_{2X}}{g} - \frac{1}{2g}\left[\varphi^2_{1X} + \varphi^2_{1Z} + 2C_1\varphi_{1X} + 2C_*\varphi_{1XZ}\eta_1 + C^2_1 + 2C_*C_2\right] = 0 \text{ on Z} = 0$$

From equation (14.35), $C_1 = 0$

$$\therefore \eta_2 = Q_2 - \frac{C_*\varphi_{2X}}{g} - \frac{1}{2g}\left[(\varphi_{1X})^2 + (\varphi_{1Z})^2 + 2C_*\varphi_{1XZ}\eta_1 + 2C_*C_2\right] = 0 \text{ on Z} = 0 \qquad (14.37)$$

Substituting the values of $\phi_{1x}$, $\phi_{1z}$, $\phi_{1xz}$, $\eta_1$ etc. (in eqn 14.37) and after much simplification we can get:

$$\eta_2 = Q_2 - \frac{1}{4}\frac{A_1^2K_1^2}{g} - \frac{C_*C_2}{g} + \frac{A_1^2K_1^2}{4g}\frac{\text{Cosh}^2 Kh(\text{Cosh}\,2Kh + 2)}{\text{Sinh}^2 kh}\text{Cos}\,2(Kx - \sigma t) \text{ on Z=0} \qquad (14.38)$$

Taking the non-hyperbolic constant terms out;

$$Q_2 = \frac{1}{4}\frac{A_1^2K_1^2}{g} + \frac{C_*C_2}{g} \qquad (14.39)$$

Hence the wave profile for second order waves is :

$$\eta_2 = \frac{A_1^2K_1^2}{4g}\frac{\text{Cosh}^2 Kh(\text{Cosh}\,2Kh + 2)}{\text{Sinh}^2 kh}\text{Cos}\,2(Kx - \sigma t) \qquad (14.40)$$

Now to estimate the perturbation parameter 'ε', take the equation (14.22) and as the wave profile is harmonic, it is reasonable to assume the first order as *aCos* (*Kx* - *σt*) where 'a' is amplitude which is half of the wave height (H).

$$\eta_1 = \varepsilon\frac{C_*}{g}\left[A_1K\,\text{Cosh}\,K(h+z)\text{Cos}(kx - \sigma t)\right]$$

At the sea surface, on z = 0,

$$\eta_1 = \varepsilon\frac{A_1KC_*}{g}\left[\text{Cosh}\,Kh\,\text{Cos}(kx - \sigma t)\right] = a\,\text{Cos}(Kx - \sigma t) = \frac{H}{2}\text{Cos}(Kx - \sigma t) \qquad (14.41)$$

$$\varepsilon\frac{A_1KC_*}{g}\left[\text{Cosh}\,Kh\,\text{Cos}(kx - \sigma t)\right] = a\,\text{Cos}(Kx - \sigma t)$$

$$\therefore \varepsilon = \frac{ag}{A_1KC_*\,\text{Cosh}\,Kh} \qquad (14.42)$$

Using all these above, the velocity potential for the second order wave system can be written as per the equation (14.1):

$$\phi = \varepsilon^1\phi_1 + \varepsilon^2\phi_2$$

Substituting the values from equations (14.42),(14.19) and (14.36) in equation (14.1) we get:

$$\therefore \varepsilon^1\varphi_1 = \frac{ag}{A_1KC_*\,\text{Cosh}\,Kh}\left[-A_1\,\text{Cosh}\,K(h+z)\text{Sin}(Kx - \sigma t)\right]$$

Using equation (14.25) and eliminate 'g/K' and put a = H/2 and simplify:

$$\therefore \varepsilon^1\varphi_1 = -\frac{H}{2}C_*\frac{\operatorname{Cosh} K(h+z)}{\operatorname{Sinh} Kh}\operatorname{Sin}(Kx-\sigma t)$$

Similarly $\varepsilon^2\phi_2$ can be estimated substituting equations (14.42) and (14.36)

$$\varepsilon^2\varphi_2 = -\left[\frac{ag}{A_1KC_*\operatorname{Cosh} Kh}\right]^2\left[\frac{3}{8}\frac{A_1^2K}{C_*}\frac{1}{\operatorname{Sinh}^2 kh}\operatorname{Cosh} 2K(h+Z).\operatorname{Sin} 2(Kx-\sigma t)\right]$$

$$\varepsilon^2\varphi_2 = -\frac{3}{8}\left[\frac{a^2g^2}{A_1^2K^2C_*^2\operatorname{Cosh}^2 Kh}\right]\left[\frac{A_1^2K}{C_*}\frac{1}{\operatorname{Sinh}^2 kh}\operatorname{Cosh} 2K(h+Z).\operatorname{Sin} 2(Kx-\sigma t)\right]$$

By eliminating g/K using equation (14.25) and taking K = 2π/L, where L is wave length, we get:

$$\varepsilon^2\varphi_2 = -\frac{3}{8}\frac{H^2}{4}\left[\frac{C_*^4\operatorname{Cosh}^2 Kh}{C_*^2\operatorname{Sinh}^2 Kh}\right]\left[\frac{1}{\operatorname{Cosh}^2 kh}\frac{K}{C_*}\frac{1}{\operatorname{Sinh}^2 Kh}\operatorname{Cosh} 2K(h+Z).\operatorname{Sin} 2(Kx-\sigma t)\right]$$

$$\varepsilon^2\varphi_2 = -\frac{3}{8}\frac{H^2}{4}\left[\frac{C_*^4}{C_*^3}\right]\left[\frac{2\pi}{L}\frac{1}{\operatorname{Sinh}^4 Kh}\operatorname{Cosh} 2K(h+Z).\operatorname{Sin} 2(Kx-\sigma t)\right] \because K = \frac{2\pi}{L}$$

$$\varepsilon^2\varphi_2 = -\frac{3}{8}\frac{H^2}{4}\left[\frac{2\pi C_*}{C_*T}\right]\left[\frac{1}{\operatorname{Sinh}^4 Kh}\operatorname{Cosh} 2K(h+Z).\operatorname{Sin} 2(Kx-\sigma t)\right] \because L = C_*T$$

$$\varepsilon^2\varphi_2 = -\frac{3}{16}\frac{\pi H^2}{T}\frac{\operatorname{Cosh} 2K(h+Z)}{\operatorname{Sinh}^4 Kh}.\operatorname{Sin} 2(Kx-\sigma t)$$

$$\varphi = \varepsilon^1\varphi_1 + \varepsilon^2\varphi_2 =$$

$$-\frac{H}{2}C_*\frac{\operatorname{Cosh} K(h+z)}{\operatorname{Sinh} Kh}\operatorname{Sin}(Kx-\sigma t) - \frac{3}{16}\frac{\pi H^2}{T}\frac{\operatorname{Cosh} 2K(h+Z)}{\operatorname{Sinh}^4 Kh}.\operatorname{Sin} 2(Kx-\sigma t) \tag{14.43}$$

The second order wave profile:

$$\eta = \varepsilon^1\eta_1 + \varepsilon^2\eta_2$$

From equation (14.42) and (14.22)

$$\varepsilon^1\eta_1 : \left[\frac{ag}{A_1KC_*\operatorname{Cosh} Kh}\right]\left[\frac{A_1KC_*}{g}\operatorname{Cosh} K(h+z)\operatorname{Cos}(kx-\sigma t)\right]$$

$$= \frac{H}{2}\left[\frac{\operatorname{Cosh} K(h+z)}{\operatorname{Cosh} Kh}\right]\left[\operatorname{Cos}(kx-\sigma t)\right] \because a = \frac{H}{2}$$

From equation (14.42) and (14.41)

$$\varepsilon^2\eta_2 = \left[\frac{ag}{A_1KC_*\operatorname{Cosh}Kh}\right]^2 \frac{A_1^2K_1^2}{4g}\frac{\operatorname{Cosh}^2 Kh(\operatorname{Cosh}2Kh+2)}{\operatorname{Sinh}^2 kh}\operatorname{Cos}2(Kx-\sigma t)$$

$$\varepsilon^2\eta_2 = \left[\frac{a^2g^2}{A_1^2K_1^2C_*^2\operatorname{Cosh}^2 Kh}\right]\frac{A_1^2K_1^2}{4g}\frac{\operatorname{Cosh}^2 Kh(\operatorname{Cosh}2Kh+2)}{\operatorname{Sinh}^2 kh}\operatorname{Cos}2(Kx-\sigma t)$$

$$\varepsilon^2\eta_2 = \frac{\pi}{8}\frac{H^2}{L}\frac{\operatorname{Cosh}Kh(\operatorname{Cosh}2Kh+2)}{\operatorname{Sinh}^3 kh}\operatorname{Cos}2(Kx-\sigma t);\because C_*^2 = \frac{g}{K}\tanh kh \;\&\; K = \frac{2\pi}{L}$$

$$\eta = \varepsilon^1\eta_1 + \varepsilon^2\eta_2 =$$

$$= \frac{H}{2}\left[\frac{\operatorname{Cosh}K(h+z)}{\operatorname{Cosh}Kh}\right][\operatorname{Cos}(kx-\sigma t)] + \frac{\pi}{8}\frac{H^2}{L}\frac{\operatorname{Cosh}Kh(\operatorname{Cosh}2Kh+2)}{\operatorname{Sinh}^3 kh}\operatorname{Cos}2(Kx-\sigma t)$$

Hence second order Stokes wave profile, on Z = 0 is

$$\eta = \frac{H}{2}\operatorname{Cos}(kx-\sigma t) + \frac{\pi}{8}\frac{H^2}{L}\frac{\operatorname{Cosh}Kh(\operatorname{Cosh}2Kh+2)}{\operatorname{Sinh}^3 kh}\operatorname{Cos}2(Kx-\sigma t) \tag{14.44}$$

For deep water (d/L > ½), the above wave profile equation reduces to

$$\eta = \frac{H_0}{2}\cos\left(\frac{2\pi x}{L_0} - \frac{2\pi t}{T}\right) + \frac{\pi H_0^2}{4L_0}\cos\left(\frac{4\pi x}{L_0} - \frac{4\pi t}{T}\right) \tag{14.45}$$

This equation (14.44) differs from the linear solution by the addition of the second order term. Its frequency is twice that of the first order term, which therefore increases the crest height, decreases the trough depth and thus increases the wave steepness.

## The second order wave celerity

From equation (14.10) upto second order wave is: $C = C_* + \sum_{n=1}^{\infty}\varepsilon^n C_n = C_* + \varepsilon^1 C_1$

From equation (14.35), $C_1 = 0$

$$\therefore C = C_* + \varepsilon^1 C_1 = C_* = \sqrt{\frac{g}{k}\tanh kh}$$

To second order, the wave celerity remains the same as linear theory. However, to third order the wave celerity increases with wave steepness and is approximately 20 higher than given by linear theory in deep water at the limiting steepness (1/7).

## Third order wave equations

The above equations evaluated up to third order are given by:

$$C = \frac{gT}{2\pi}\tanh\left(\frac{2\pi d}{L}\right)\left[1+\left(\frac{\pi H}{L}\right)^2\left\{\frac{5+2\cosh(4\pi d/L)+2\cosh^2(4\pi d/L)}{8\sinh^4(4\pi d/L)}\right\}\right] \tag{14.46}$$

$$L = \frac{gT^2}{2\pi}\tanh\left(\frac{2\pi d}{L}\right)\left[1+\left(\frac{\pi H}{L}\right)^2\left\{\frac{5+2\cosh(4\pi d/L)+2\cosh^2(4\pi d/L)}{8\sinh^4(4\pi d/L)}\right\}\right] \tag{14.47}$$

## 14.6. Finite amplitude wave theories

The generation of waves depends on the transference of energy from the wind to the sea. Some of the theories that explain how wind transfers energy for generation of waves is given below.

The simplest wave theory is the *first-order, small-amplitude*, or *Airy* wave theory which will hereafter be called *linear theory*. Many engineering problems can be handled with ease and reasonable accuracy by this theory. For convenience, prediction methods in coastal engineering generally have been based on simple waves. For some situations, simple theories provide acceptable estimates of wave conditions.

When waves become large or travel toward shore into shallow water, higher-order wave theories are often required to describe wave phenomena. These theories represent *nonlinear waves*.

There are five wave theories that are commonly applied to describe wave motion on the sea. The general wave profile, its application and related references are given below:

### 14.6.1. Airy waves (Sinusoidal waves)

This theory is applied to waves of small amplitude in deep water. References are Laplace (1776) and Airy (1845). Airy wave theory represents the simplest formulation but because of the assumptions involved in the derivation its application should be limited to waves of small amplitude. This theory is actually strictly true only for the limiting case of waves of zero height. For waves of appreciable height this theory does not hold good.

The linear theory that is valid when waves are infinitesimally small and their motion is small also provides some insight for finite-amplitude periodic waves (nonlinear). However, the linear theory cannot account for the fact that wave crests are higher above the mean water line than the troughs are below the mean water line. Results obtained from the various theories should be carefully interpreted for use in the design of coastal projects or for the description of coastal environment.

Laplace (1776) presented the first satisfactory treatment of waves of small amplitude in water of arbitrary depth. Airy (1845) later developed a theory for irrotational waves traveling over a horizontal bottom in any depth of water. In the derivation of this theory the equations are linearized and so this theory is often referred to as *linear wave theory.* Airy wave theory represents the simplest formulation but because of the assumptions involved in the derivation its application should be limited to waves of small amplitude. The theory is actually strictly true only for the limiting case of waves of least height. Where waves of appreciable height are involved more complex wave theories are required.

### 14.6.2. Stokes waves

As waves move into water of decreasing depth the form of the wave becomes distorted so that it is no longer symmetrical and the wave crest becomes narrower and more peaked while the wave trough becomes flatter and wider. This results in horizontal water velocity that is higher under the wave crest than the offshore velocity under the trough but of shorter duration. Wave steepness H/L increases and the relative depth h/L decreases, and as this occurs the wave amplitude can no longer be considered small relative to water depth instead we are dealing with finite amplitude waves. Historically, the first finite amplitude wave theory was that developed by Stokes (1847) which includes a perturbation solution that can be expanded to consider greater numbers of terms. Stokes' second order theory is the simplest of these.

Airy wave theory omits terms involving the wave height to the second ($H^2$) and higher orders. If the wave height is large, such an approximation is not adequate, and so it is necessary to retain the higher order terms. Stokes in 1847 considered small but finite height progressive wave over still water of finite depth which is a second order theory. The solutions he obtained are approximate and in the form of series with coefficients that require considerable detailed calculations. The method of Stokes has since been extended to third and fifth orders later by other workers.

To the second order approximation, the wave profile of the Stokes wave on Z =0 from equation (14.44) is given by

$$\eta = \frac{H}{2}\mathrm{Cos}(kx - ct) + \frac{\pi}{8}\frac{H^2}{L}\frac{\mathrm{Cosh}(kd\left[2+\mathrm{Cosh}(2kd)\right]}{\left[\mathrm{Sinh}(kd)\right]^3}\mathrm{Cos}\left[2(kx-\sigma t)\right] \tag{14.48}$$

The first part of the equation is the same as equation (14.41) the wave profile of small amplitude waves represents the linear contribution to the surface form, while the second part is the second-order contribution. Third-order theory would involve an additional set of expressions including $H^3$.

Which in deep water ($d/L_o > 1/2$) reduces to from (14.45)

$$\eta_0 = \frac{H_0}{2}\mathrm{Cos}\,2\pi\left(\frac{x}{L_o} - \frac{t}{T}\right) + \frac{\pi H_0^{\;2}}{4L_0}\mathrm{Cos}\,4\pi\left(\frac{x}{L_o} - \frac{t}{T}\right) \tag{14.49}$$

Another way:

$$\eta = \frac{H_0}{2}\cos\left(\frac{2\pi x}{L_0} - \frac{2\pi t}{T}\right) + \frac{\pi H_0^2}{4L_0}\cos\left(\frac{4\pi x}{L_0} - \frac{4\pi t}{T}\right)$$

It is seen that if H/L in equation (14.48) is small, the profile reduces to the sinusoidal form given by the Airy wave theory. For finite height waves there is an additional term added onto the basic sinusoidal shape. The effect of this term is shown in the Figure 14.3, where a finite wave form is compared to a sine wave with the same height. The added term enhances the crest amplitude and detracts from the trough amplitude, so that the Stokes wave profile has steeper crests separated by flatter troughs than does the sinusoidal Airy wave as shown in the Fig. 14.5. The waves are symmetrical about vertical planes through crests and troughs. The shallower the water, the more pronounced the peaks of the wave and flatter the toughs. The general shape of the Stokes wave more closely conforms to the profile of swell waves as they enter shallow water and begin to transform. To the second order of approximation, the equations of wave length and velocity are the same as those for the linear Airy wave theory.

To the third order of approximations, the wave velocity is dependent on the wave height as well as on the period and water depth which is given by:

$$C=\frac{gT}{2\pi}\tanh\left(\frac{2\pi d}{L}\right)\left[1+\left(\frac{\pi H}{L}\right)^2\frac{7+2\text{Cosh}^2\left(\frac{4\pi d}{L}\right)}{8\text{Sinh}^4\left(\frac{2\pi d}{L}\right)}\right] \tag{14.50}$$

This relationship corresponds to that developed by Hunt (1953). It is seen that if $(H/L)^2$ is small, the second term in the parenthesis vanishes and the relationship reduces to the equation 13.31of the last chapter.

In deep water the above equation (14.50) approximates to

$$C=\frac{gT}{2\pi}\left[1+\left(\frac{\pi H_0}{L_0}\right)^2\right] \tag{14.51}$$

A finite wave slope $H_0/L_0$ will cause a slight increase in $C_0$ over that calculated with Airy wave theory. The maximum value that the deep water steepness $H_0/L_0$ can achieve is approximately 1/7. At greater values the wave is unstable and breaks. For this limiting value of $H_0/L_0$ we have $\left[1+\left(\frac{\pi H_0}{L_0}\right)^2\right]=1.20$ which is 20% increase due to the finite wave height. At more steepness values the finite height produces only a slight increase in $C_0$.

### 14.6.2.1.Water particle velocities of the second order Stokes wave theory

The components of water particle velocities are given by

$$u=\frac{\pi H}{T}\frac{\text{Cosh}\left[k(z+d)\right]}{\text{Sinh}(kd)}\text{Cos}(kx-\sigma t)+\frac{3}{4}\left[\frac{\pi H}{L}\right]^2C\frac{\text{Cosh}\left[2k(z+d)\right]}{\left[\text{Sinh}(kd)\right]^4}\text{Cos}\left[2(kx-\sigma t)\right] \tag{14.52}$$

Another way:

$$u=\frac{HgT}{2L}\left[\frac{\text{Cosh}\,K(d+z)}{\text{Cosh}\,Kd}\right]\left[\text{Cos}(kx-\sigma t)\right]+\frac{3}{4}\left(\frac{\pi H}{L}\right)^2C\left[\frac{\text{Cosh}\,2K(d+z)}{\text{Sinh}^4\,Kd}\right]\left[\text{Cos}(2kx-2\sigma t)\right]$$

$$w = \frac{\pi H}{T}\frac{\text{Sinh}\left[k(z+d)\right]}{\text{Sinh}(kd)}\text{Sin}(kx-\sigma t)+\frac{3}{4}\left[\frac{\pi H}{L}\right]^2 C\frac{\text{Sinh}\left[2k(z+d)\right]}{\left[\text{Sinh}(kd)\right]^4}\text{Sin}\left[2(kx-\sigma t)\right] \tag{14.53}$$

Another way:

$$w = \frac{\pi H}{L}C\left[\frac{\text{Sinh}\,K(d+z)}{\text{Sinh}\,Kd}\right]\left[\text{Sin}(kx-\sigma t)\right]+\frac{3}{4}\left(\frac{\pi H}{L}\right)^2 C\left[\frac{\text{Sinh}\,2K(d+z)}{\text{Sinh}^4\,Kd}\right]\left[\text{Sin}(2kx-2\sigma t)\right]$$

The first terms in the relationships for 'u' and 'w' in equations (14.52 & 14.53) are the same as the equations obtained in Airy wave theory. The second term is positive under the wave crest and trough and negative ¼ and 3/4th wavelengths from the crest. The effect of this second term is to increase the magnitude but shorten the duration of the velocity under the crest, and decrease the magnitude but lengthen the duration of the velocity under the trough. This effect is observed in waves in shallow water.

#### 14.6.2.2. Water particle displacements of second order Stokes waves

Second order equations for water particle displacements from their mean position for a finite- amplitude wave are:

Horizontal displacement:

$$\xi = \frac{HgT^2}{4\pi L}\left[\frac{\text{Cosh}\,K(d+z)}{\text{Cosh}\,Kd}\right]\left[\text{Sin}(kx-\sigma t)\right]+\left(\frac{\pi H^2}{8L}\right)\frac{1}{\text{Sinh}^2\,kd}\left[1-\frac{3}{2}\frac{\text{Cosh}\,2K(d+z)}{\text{Sinh}^2\,Kd}\right]$$
$$\left[\text{Sin}(2kx-2\sigma t)\right]+\left(\frac{\pi H}{L}\right)^2\frac{ct}{2}\left[\frac{\text{Cosh}\,2K(d+z)}{\text{Sinh}^2\,Kd}\right] \tag{14.54}$$

and vertical displacement:

$$\zeta = \frac{HgT^2}{4\pi L}\left[\frac{\text{Sinh}\,K(d+z)}{\text{Cosh}\,Kd}\right]\left[\text{Cos}(kx-\sigma t)\right]+\frac{3}{16}\left(\frac{\pi H^2}{L}\right)\left[\frac{\text{Sinh}\,2K(d+z)}{\text{Sinh}^4\,Kd}\right]\left[\text{Cos}(2kx-2\sigma t)\right] \tag{14.55}$$

The last term in equation (14.54) is of particular interest; it is not periodic, but is the product of time (t) and a constant (c) depending on the given wave period and depth. The term predicts a continuously increasing net particle displacement in the direction of wave propagation. Equation (14.54) indicates that there is a net transport of fluid by waves in the direction of wave propagation.

#### 14.6.2.3. Surface Pressure

Pressure at any distance below the surface of second order waves:

$$p_2 = \frac{\rho gH}{2}\frac{\cosh k(d+z)}{\cosh kd}\cos\theta-\rho gz+\frac{3}{8}\rho g\frac{\pi H^2}{L}\frac{\tanh kd}{(\sinh kd)^2}\left\{\frac{\cosh 2k(d+Z)}{(\sinh kd)^2}-\frac{1}{3}\right\}$$
$$\cos 2(kx-\sigma t)-\frac{1}{8}\rho g\frac{\pi H^2}{L}\frac{\tanh kd}{(\sinh kd)^2}\left\{\frac{\cosh 2k(d+Z)}{L}-1\right\} \tag{14.56}$$

### 14.6.2.4. Stokes drift

An important difference in the orbital motion between linear wave theory and finite wave theories is that the orbital motion is no longer closed, so that as each wave passes a water particle moves a short distance in the direction of wave advance. The existence of this net motion, called Stokes drift, is well known and it can be predicted from Stokes second order theory

In the case of Stokes waves, there is a net transport of water as the wave passes. At this stage, it is necessary to consider the particle motion as presented earlier for small amplitude waves, i.e., the water particles return to their original position at the end of the passage of a sinusoidal wave. In other words, particle motion is in closed circles in deep water waves. However, on examining the horizontal and vertical components of particle motion for Stokes wave, it can be seen that the horizontal component has a non-periodic term. Hence, the circle is not a closed one and there is net transport of water in the direction of progress of wave (Fig.14.3).

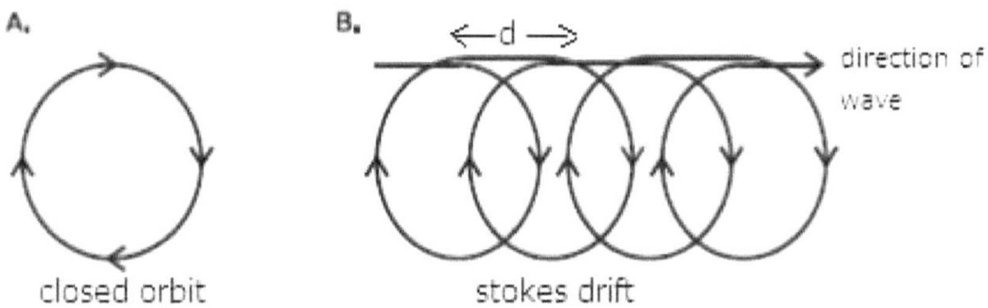


**Fig. 14.3.** Shows the net Stokes drift velocity in waves: the closed orbit produced and predicted by non-linear wave theory. At the end of each orbit the position of a water particle moves a short distance(d) in the direction of wave advance. Division of this distance by the wave period gives the net drift velocity.

One of the expressions for Stokes drift presented by Weigal (1964) is

$$u = \left(\frac{\pi H}{T}\right)\left(\frac{\pi H}{L}\right) e \frac{4\pi Z}{L}$$

where, $H$, $T$ and $L$ are height, period and length of the wave, respectively; and $z$ is the depth at which drift is estimated. The depth axis is positive upwards. Thus, the drift decreases rapidly with the depth. It can be said that the Stokes drift at surface is about 1% of the wind speed that generates the waves.

An interesting departure of the Stokes wave from the Airy wave is that the particle orbits are not closed (Fig.14.3). This leads to a non-periodic drift or mass transport in the direction of wave advance, the associated velocity comes from above:

$$\bar{U}_Z = \left(\frac{\pi H}{L}\right)^2 \frac{C}{2}\left[\frac{\text{Cosh}\, 2K(d+z)}{\text{Sinh}^2 Kd}\right] \tag{14.57}$$

which in deep water reduces to $\bar{U}_0 = \left[\frac{\pi H_0}{L_0}\right]^2 C_0 e^{2kz}$ (14.58)

Integration of the above equation (14.58) with depth yields to the discharge "q" which is the volume transported forward per unit wave crest length per unit time as:

$$q = \int_{-d}^{0} \bar{U}_o dz = \frac{\pi}{4}\frac{H_0^2}{T} \text{ (cm}^3\text{/cm.Sec)} \tag{14.59}$$

The above results for the Stokes mass transport were derived for a channel of infinite length and constant depth, with no consideration of viscosity. The discharge equation (14.59) is very low.

From the viewpoint of sediment transport and beach equilibrium, it is extremely important because it provides one mechanism by which net sediment transport can occur once sediment has been set in motion by the much higher instantaneous velocities associated with the purely oscillatory portion of wave motion. At the bed z = - d and Cosh 2K(d+z) = 1, so equation (14.57) reduces to

$$\bar{U}_{D(bed)} = \left(\frac{\pi H}{L}\right)^2 \frac{C}{2}\left[\frac{1}{\text{Sinh}^2 Kd}\right]$$

The net shoreward velocity near the bottom, given by Longuet –Higgins (1953) is

$$\bar{U}_0 = \frac{5}{4}\left[\frac{\pi H}{L}\right]^2 C \frac{1}{\left[\text{Sinh}(kd)\right]^2} \tag{14.60}$$

This is important for understanding transport of sediment toward the beach.

The net drift velocity at the bed is initially very small compared to the maximum orbital velocity, but it increases in magnitude as the wave shoals and near the breakpoint can be a substantial fraction of it.

For any given water depth and wave period there is an upper limit to the wave height of the Stokes wave at which the wave becomes unstable and breaks. The Stokes criterion for wave breaking is that the water particle velocity at the crest be just equal to the wave propagation velocity C. It is apparent that if the wave is so large that the crest water particle velocity exceeds C, then the wave would topple forward and break.

### 14.6.3. Gerstner or Trochiodal waves

This theory is applied to waves of finite amplitude in deep, intermediate and shallow water. More than half the wave height is above the mean water level for trochoidal wave as shown in Fig.14.2. Just before the time of breaking, the crest would form a cusp with an angle of 120° between the tangents to the two faces of the crest as shown in Fig.14.2. The references are Gerstner (1802), Stokes (1847), Froude (1862), Rankine (1863) and Rayleigh (1877).

The trochoidal theory (Gerstner, 1802) is the first theory to be developed for waves of finite height and is often used for engineering calculations. One reason for its use is the ease with which the equations may be used. It appears to represent the actual wave profiles as well as actually satisfying the pressure conditions at the surface and the continuity conditions. However, it requires rotation of the particles and does not predict any mass transport in the direction of wave propagation, while observations show that there is mass transport.

His solution is limited to waves in water of infinite depth. From the equations he developed, Gerstner concluded that the surface profile is trochoidal in shape as shown in Fig.14.2 and the development of trochoid is shown in Fig.14.4. Later Froude (1862) and Rankine (1863) developed further. This theory has wide applications to civil engineers and naval architects because the solutions are exact and equations are simple to use. The solutions also satisfy the pressure conditions at the water surface and continuity. The experimental studies also demonstrate that the surface profile of the trochoidal wave closely approximates the actual profile of waves on a horizontal bottom.

### 14.6.3.1. Formation of trochoid

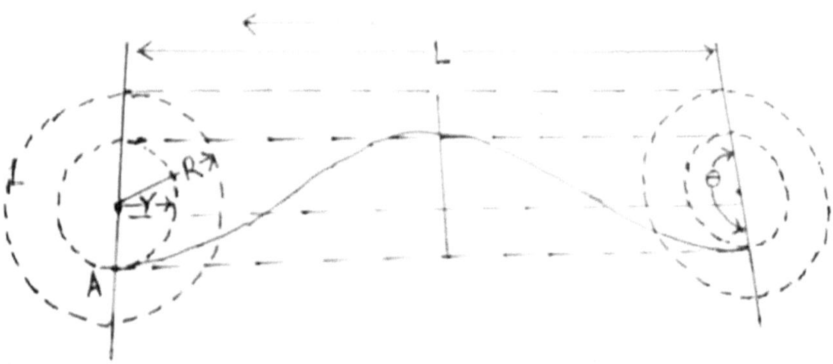


**Fig.14.4.**The trochoidal curve pattern generation due to the motion of point 'A' inside the circle, as the circle rolls along the underside of a straight line. (Adapted from Varma, 2014)

The trochoid curve is generated by the motion of a point 'A' interior to a circle, as the circle rolls along the underside of a line. Referring the Fig. 14.4, if R is the radius of the circle, then the wave length will be $L = L_0 = 2\pi R$ (the radius of the circle is 1/k). The wave height will be $H = H_0 = 2r$, where r is the radial distance from the circle center to the point'A'.

The equations of the surface profile (Fig. 14.4) are,

$$X = R\theta - r \sin \theta$$

$$Y = R - r \cos \theta$$

In order to plot the equation of wave shape in dimensionless form with the origin of the coordinates at the crest and the vertical dimension measured negatively, downward, these equations may be transformed to:

$$\frac{x'}{L_0} = 1 - \left[\frac{rad\theta}{2\pi} - \frac{H_0 \operatorname{Sin}\theta}{2L_0}\right]$$

$$\frac{y'}{H_0} = \frac{1}{2}(1-\cos\theta)$$

where, x' and y' are measured from the wave crest. It can be seen that as Ho/Lo approaches zero, the curve approaches a sine wave and the surface is nearly that as developed in the irrotational theory for waves of very small amplitude.

The positions of the crest and trough relative to the undisturbed water level (amplitudes) are:

Height (amplitude) of the crest $(a_c) = H_0 - \left[r_S - \frac{r_S^2}{2R}\right] = \frac{H_0}{2} + \frac{\pi H_0^2}{4L_0}$ (14.61)

Depth (amplitude) of the trough $(a_t) = \frac{2\pi R}{L_0}\left[r_S - \frac{r_S^2}{2R}\right] = \frac{H_0}{2} - \frac{\pi H_0^2}{4L_0}$ (14.62)

From the above, it is clear that the vertical distance from the mean level to the crest is more than half wave height, but the vertical distance from mean level to trough is less than half wave height. From the above equations of for the limiting wave steepness in deep water (1/7) we have 0:611H and 0:389H. Though this theory gives an exact solution, it does not satisfy the irrotationality condition.

The paths described by the water particles during one cycle are circles with the radii decreasing exponentially with depth (Fig.13.20 of Chapter 13 of small amplitude waves). This is expressed as,

$$a' = b' = \frac{r_S}{L_0} e^{2\pi z} = \frac{H_0}{2L_0} e^{2\pi z}$$

The energy of the wave is equally divided between kinetic and potential, with the total energy being,

$$\frac{WL_0 H_0^2}{8}\left[1 - \frac{1}{2}\left(\frac{\pi H_0}{L_0}\right)^2\right]$$

## Alternative theory

If the distance of the point from the center of the circle decreases, the shape tends to be sinusoidal, while if it increases to the radius of the circle, the crest becomes too pointed. It may be noted that these waves are also called Gerstner waves (Fig14.2). For an angle of rotation θ, the surface depression below crest level is

$$\eta = \frac{H}{2}(1 - \cos\theta)$$

While the horizontal distance of the surface from the origin at the crest is

$$x = L\left[\frac{\theta}{2\pi} + \frac{H}{2L}\text{Sin}\,\theta\right]$$

As H/L becomes small (as point 'A' approaches the center), the surface profile becomes nearly that of the Stokes wave, and with H nearly zero the shape tends toward a sine wave. Hence in the limit, the wave corresponds to a deep water Airy wave.

The positions of the crest and trough relative to the still-water level are:

$$\text{Height of the crest} = \frac{H}{2} + \frac{\pi H^2}{4L} \tag{14.61}$$

$$\text{Depth of the trough} = \frac{H}{2} - \frac{\pi H^2}{4L} \tag{14.62}$$

Thus it is clear that the crest is more than half the wave height above the still-water level and the trough is less than half below the still water level.

The particle orbits of the trochoidal wave are d = S = H $e^{kz}$ which is same as deep-water Airy waves whose orbits are circular and their diameters decrease exponentially with depth. The wave length and phase velocity expressions are also the same as those for a deep –water Airy wave. The energy is given by

$$E = \frac{1}{8}\rho g H^2\left[1 - \frac{1}{2}\left(\frac{\pi H}{L}\right)^2\right]$$

If the equations for the trochoid are expanded into a series, it is found that the first three terms are the same as those in the Stokes solution. It may be noted here that the trochoid theory is limited to water of infinite depth. Though this theory gives an exact solution, it does not satisfy the irrotationality condition.

### 14.6.4. Cnoidal wave theory

In Cnoidal theory, H/h is assumed small and $U_r$ of the order of unity. Korteweg and de Vries derived a first order solution. Much more recently (1960's to 1980's), these two theories have been extended to higher orders (third and fifth). The mathematics is complex and subsequently other researchers developed new methods whereby solutions could be obtained to any arbitrary order, by numerical solution.

If the depth decreases, the theories mentioned earlier are not the appropriate ones to be used. It is the Cnoidal wave theory that is more applicable. In this theory, Jacobian elliptical function (cn) is used. Hence, this is called as cnoidal, a word analogous to sinusoidal. Jacobian elliptical function is periodic and its modulus lies between 0 and 1, both inclusive. The shape of the cnoidal wave is characterized by very flat troughs and peaked crests (Fig. 14.5).

Some of the basic wave characteristics from cnoidal theory, such as the surface profle and the wave celerity, can be presented by diagrams that are based on two parameters, namely $k^2$ and $U_r$. The parameter $k^2$ is a function of the water depth, the wave length, and the vertical distance up from the bottom to the water surface at the wave crest and trough. It varies from 0 for the small-amplitude limit to 1.0 for the solitary wave limit as the ratio of the crest amplitude to wave height varies from 0 to 1.0 for the two wave theories. $U_r$, which is known as the Ursell number (Ursell, 1953), is a dimensionless parameter given as $L^2 H/d^3$ that is also useful for defining the range of application for various wave theories. From Hardy and Kraus (1987) the Stokes theory is generally applicable for $U_r < 10$ and the cnoidal theory for $U_r > 25$. The theories are equally applicable in the range $U_r = 10$ to 25. But the approximate range of validity for the cnoidal wave theory as determined by Laitone (1963) is $\frac{d}{L} < 1/8$. As wavelength becomes long and approaches infinity, cnoidal wave theory reduces to the solitary wave theory. Also, as the ratio of wave height to water depth becomes small (infinitesimal wave height), the wave profile approaches the sinusoidal profile predicted by the linear theory.

Wave characteristics are described in parametric form in terms of the modulus m of the elliptic integrals. While m itself has no physical significance, it is used to express the relationships between the various wave parameters. Tabular presentations of the elliptic integrals and other important functions can be obtained from the above references. The ordinate of the water surface $y_S$ measured above the bottom is given by

$$y_S = y_t + H(cn)^2\left[2K(m)\left(\frac{x}{L}-\frac{t}{T}\right),m\right] \tag{14.63}$$

Or the wave profile is given by:

$$\eta = y_t + H(cn)^2\left[2K(m)\left(\frac{2\pi}{L}x-\frac{2\pi}{T}t\right),m\right] \tag{14.64}$$

Here, *cn* is the Jacobian elliptical function, *K*(m) is the elliptical integral of first order with modulus 'm'. Both are dependent on elliptic parameter m. $y_t$ is the height of trough from sea bottom and *H* is the wave height. The argument of $(cn)^2$ is frequently denoted simply by ( ); thus, the equation (14.63) above can be written as:

$$y_S = y_t + H\,(c\,n)^2\,(\ \ )$$

The distance from the bottom to the wave trough $y_t$ as used in equation (14.63) and (14.64), is given by:

$$\frac{y_t}{d} = \frac{y_C}{d} - \frac{H}{d} = \frac{16d^2}{3L^2}K(m)[K(m)-E(m)]+1-\frac{H}{d} \tag{14.65}$$

Where $y_c$ is the distance from the bottom to the crest, and E(m) is the complete elliptic integral of the second kind. Wave length is given by:

$$L = mK(m)\sqrt{\frac{16d^3}{3H}} \tag{14.66}$$

And the wave period is given by:

$$T = \sqrt{\frac{g}{d}} = \sqrt{\frac{16y_t}{3H}\frac{d}{y_t}}\left[\frac{mK(m)}{1+\dfrac{H}{y_t m^2\left(\dfrac{1}{2}-\dfrac{E(m)}{K(m)}\right)}}\right] \tag{14.67}$$

Cnoidal waves are periodic and of permanent form. Thus L = CT is applicable.

Pressure under a cnoidal wave at any elevation y above the bottom depends on the local fluid velocity, and is therefore complex. However, it may be approximated in a hydrostatic form as (taking r = ρ):

$$p = \rho g\,(y_s - y) \tag{14.68}$$

i.e., the pressure distribution may be assumed to vary linearly from $\rho g y_s$ at the bed to zero at the surface.

As shown in Fig.14.5, when the modulus '*m*' is equal to zero, the wave profile becomes sinusoidal. At the other extreme, i.e., when m is 1, the period will be ∞ and the cnoidal wave gets peaked crests and (very) flat troughs. For values of *m* less than 0.95, the cnoidal function can be approximated with trigonometric functions. However, a small reduction in *m* from unity gives finite period (for example, if $m = 0.9999$, period becomes $7\pi$). This means there are two limits to the cnoidal wave theory. The first occurs when the period of the function cn is infinite when m=1. This corresponds to a solitary wave. As the wavelength becomes infinite, the cnoidal theory approaches the solitary wave theory. The second limit occurs for m = *0* where the cnoidal wave approaches the sinusoidal wave. This happens when the wave height is small compared to water depth and the cnoidal theory reduces to the linear theory.

Thus The elliptic cosine is a periodic function where $cn^2\,[2K(m)\,((x/L) - (t/T))]$ has a maximum amplitude equal to unity. The modulus *m* is defined over the range between 0 and 1. When $m = 0$, the wave profile becomes a sinusoid, as in the linear theory; when $m = 1$, the wave profile becomes that of a solitary wave.

The wave profile subtends a cone with two troughs as shown in Fig.14.7. This is also applied to waves of finite amplitude in deep, intermediate and shallow water. The references are Korteweg and Devries (1895) and Keller (1949)

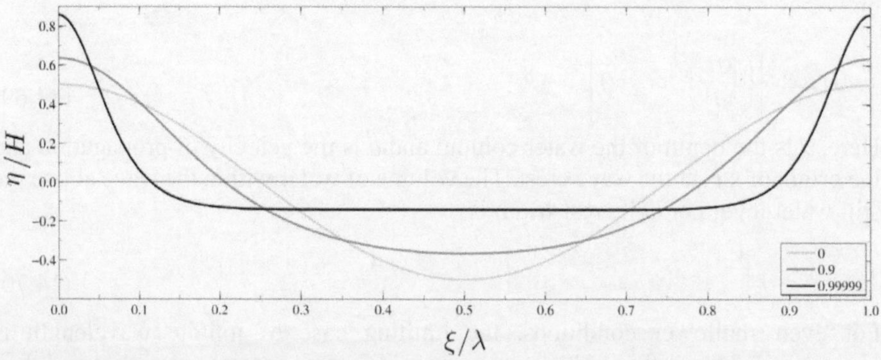


**Fig.14.5.** Cnoidal wave profiles for three values of the elliptic parameter *m*. blue : *m* = 0, red : *m* = 0.9 and black : *m* = 0.99999 (DNVGL, 2017)

### 14.6.5. Solitary wave theory

Waves considered in the previous sections are oscillatory in nature. That means the sea surface goes up and down as the wave moves. The water particles move backward and forward with the passage of each wave, and a distinct wave crest and wave trough are evident. A solitary wave is neither oscillatory nor does it exhibit a trough and so it is translatory. In the pure sense, the solitary waveform lies entirely above the still water level. Solitary waves consist essentially of a wave crest above the still water line (Figure 14.2) and they have infinite wave length and period. They are often described as waves of translation since the water particles are displaced quite a distance in the direction of the wave advance. This phenomenon can often be observed on sand bars at low tide when surf bores translate across the bar surface and into the trough landward of it.

Though in nature it is difficult to form a truly solitary wave, long waves such as tsunamis and waves resulting from landslides and earthquakes sometimes behave approximately like solitary waves. When an oscillatory wave moves into shallow water, it may often be approximated by a solitary wave (Munk, 1949). When an oscillatory wave moves into shoaling water, the wave amplitude becomes progressively higher, the crests become shorter and more pointed, and the trough becomes longer and flatter as shown in Fig.14.2.

The solitary wave is a limiting case of the cnoidal wave. When $m^2 = 1$, K($m$) = K(1) = ∞, and the elliptic cosine reduces to the hyperbolic secant function, $y_t = d$ and equation (14.63) reduces to

$$y_S = y_t + H(cn)^2\left[2K(m)\left(\frac{x}{L}-\frac{t}{T}\right),m\right]$$

$$y_s = d + H\,\text{sech}^2\left[\left(\frac{3H}{4d^3}\right)^{1/2}(x-ct)\right]$$

or

$$\eta = H\,\text{sech}^2\left[\left(\frac{3H}{4d^3}\right)^{1/2}(x-ct)\right] \tag{14.69}$$

Here, $d$ is the depth of the water column and $c$ is the velocity of propagation and the origin of x is at the wave crest. The volume of water within the wave above the still water level per unit crest width is

$$V = \left[\frac{16}{3}d^3 H\right]^{1/2} \tag{14.70}$$

For even shallower conditions, the limiting case of infinite wavelength is approached. This is the limiting case of cnoidal wave mentioned earlier and then cnoidal wave becomes a solitary wave. All the waves that have been considered earlier are oscillatory in nature. Solitary wave, on the other hand, is translatory. The profile of this consists of only a crest and the entire wave will be above mean level as shown in Figure (14.2). Thus, the wave translates forward.

The application is for isolated crests of finite amplitude moving in shallow water. The references are Scot-Russell (1844), Bousinesq (1871), Rayleigh (1876) and Mc Cowan (1891).The solitary wave, as its name suggests, is a progressive wave consisting of a single crest (Fig.14.1). So it is not oscillatory like that of the simple harmonic waves. Therefore there is no wave period and length associated with the solitary wave. This wave is not useful in describing the periodic wind waves and as it enters the shallow water sometimes their crests rise up and isolated with the separated wide flats (lines coinciding with mean water level as shown in Fig.14.1) and form solitary waves. As in the shallow water the wave depends on the water depth than on the period, so is the case in the case of solitary wave. Thus it has more applications in near shore region.

This kind of wave was first observed by John Scott Russell in 1834. Russell was an engineer and naval architect. On observing a boat being pulled rapidly by horses through a narrow channel, Scott saw that the water that accumulated just in front of the boat rose as a single wave crest and moved forward without change of shape or reduction of speed. He called it a 'wave of translation'. This chance of observation occurred in a channel in Edinburgh and was subsequently in 1844 verified.by him by producing in a laboratory tank by suddenly releasing a mass of water at one end of the tank.

For the orbital velocities, the solution of McCowan (1891) may be used. The solitary wave equation is

$$\eta = H\,\text{sech}^2\left[\left(\frac{3H}{4d^3}\right)^{1/2}(x-ct)\right] \Rightarrow \eta = H\,\text{sech}^2\left(\frac{x}{d}\sqrt{\frac{3}{4}\frac{H}{d}}\right)$$

Where η is the vertical coordinate above the still water line at a horizontal distance x from the crest; H is the wave height and d is the water depth below the still water level.

The wave velocity is given by Laiton (1959) to higher order as:

$$C = \sqrt{gd\left[1+\frac{1}{2}\frac{H}{d}-\frac{3}{20}\left(\frac{H}{d}\right)^2+.......\right]}$$

This is seen to be greater than the velocity of the shallow water Airy wave. The solitary wave takes the finite height as below by neglecting the higher order terms in the above equation

$$C = \sqrt{gd\left[1+\frac{H}{d}\right]} = \sqrt{g[d+H]} \tag{14.71}$$

But slight variations on Eq. (14.71) have also been developed for solitary wave celerity like

$$C = \sqrt{gd}\left[1+\frac{H}{2d}\right] \tag{14.72}$$

This equation was empirically determined by Russell (1844). The above two equations depart somewhat at higher values of H/d (r).

As the solitary wave advances into shoaling water, the wave height progressively increases until a condition is reached at which the wave becomes unstable and breaks. This instability is again reached when the particle velocity at the crest equals the wave velocity 'C. Also the angle at the crest is $120^0$.

As the solitary wave moves into shoaling water it eventually becomes unstable and breaks. McCown (1891) assumed that a solitary wave breaks when the water particle velocity at the wave crest becomes equal to the wave celerity. This occurs when

$$\left(\frac{H}{d}\right)_{max} = 0.78$$

However, laboratory investigations have shown that this equation agrees well with oscillatory waves than for the solitary waves. Camfield and Street (1969) found an empirical relationship between the slope and the breaker height to water depth as:

$$\left(\frac{H_b}{d_b}\right) = 0.75 + 25\varsigma - 112\varsigma^2 + 3870\varsigma^3$$

Where $\varsigma$ is the slope of the beach here.

Basing on the above solutions, the particle velocities obtained are

$$u = NC\frac{1+\cos\left(My/d\right)\text{Cosh}\left(Mx/d\right)}{\left[\cos\left(My/d\right)+\cosh\left(Mx/d\right)\right]^2} \tag{14.73}$$

$$w = NC\frac{\sin\left(\frac{My}{d}\right)\sinh\left(\frac{Mx}{d}\right)}{\left[\cos\left(\frac{My}{d}\right)+\cosh\left(\frac{Mx}{d}\right)\right]^2} \tag{14.74}$$

Where the coefficients N and M are defined by

$$\frac{H}{d} = \frac{N}{M}tan\frac{1}{2}\left[M\left(1+\frac{H}{d}\right)\right]$$

$$N = \frac{2}{3}sin^2\left[M\left(1+\frac{2}{3}\frac{H}{d}\right)\right] \tag{14.75}$$

Thus M and N are the functions of H/d shown in Figure 14.6 and y is measured from the bottom. The expression for horizontal velocity ‘u’ is often used to predict wave forces on marine structures found in the shallow water. The maximum velocity $U_{max}$ occurs when ‘x’ and ‘t’ are both equal to zero. Hence,

$$u_{max} = \frac{CN}{1+\cos(My/d)} \tag{14.76}$$

As the solitary wave is considered as a wave of translation, the water particles move only in the direction of wave advance and there is no return flow.

As a solitary wave passes a point the mass of water transported past that point is simply the integral of the water surface elevation above the still water level from x = minus to plus infinity (letting t = 0). For a unit width along the wave crest this yields the following volume of water:

$$V\ \left(\frac{16}{\ }d\ H\right)^{1/2} \tag{14.77}$$

Since the period of a solitary wave is infinite it is not possible to determine a mass transport in terms of a mass per unit time. Using the solitary wave theory, however, the mass transport can be estimated by dividing the water mass represented by Eq. (14.77) by the period of the wave in question.

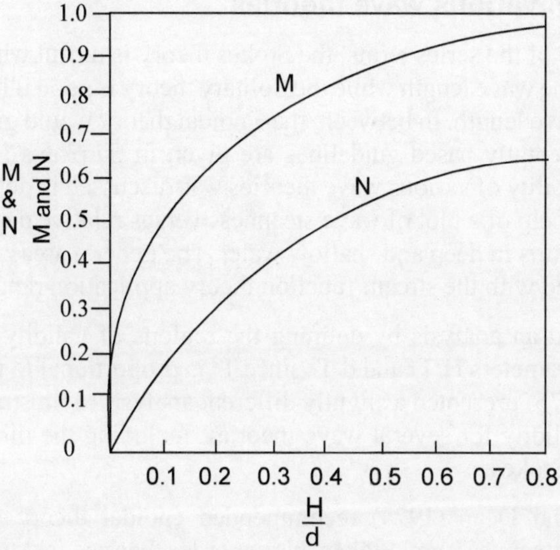


**Fig.14.6.** The values of M & N with H/d of solitary wave theory (Munk, 1949) (SPM, 1984)

Total energy in a solitary wave is about evenly divided between kinetic and potential energy. This means its energy is divided approximately half as potential energy and half as kinetic energy. Total wave energy per unit crest width is:

$$E = \frac{8}{3\sqrt{3}} \rho g H^{3/2} d^{3/2} \tag{14.78}$$

Since the length is infinite it is not possible to determine an energy density for a solitary wave. Since all of the energy in a solitary wave moves forward with the wave, the wave power is equal to the product of the wave energy and wave celerity.

And the pressure beneath a solitary wave depends on the local fluid velocity, as does the pressure under a cnoidal wave, however, it may be approximated by equation (14.68) which is identical to that used to approximate the pressure beneath a cnoidal wave.

$p = \rho g\ (y_1 - y)$ This Equation of 'p' is identical to that used to approximate the pressure beneath a cnoidal wave.

As a solitary wave moves into shoaling water it eventually becomes unstable and breaks. A solitary wave breaks when the water particle velocity at the wave crest becomes equal to the wave celerity. This occurs when (Miles 1980, 1981)

$$\left(\frac{H}{d}\right)_{max} = 0.78$$

## 14.7. The applicability of various wave theories

Considering the convergence of the series terms, the Stokes theory is useful when water is deeper than 10% of the wave length while the Solitary theory is good if it is shallower than 20% of the wave length. In between, the Cnoidal theory would give converging results. Experimentally based guidelines are given in Sarpkaya and Issacson (1981). The applicability of various wave theories was discussed in detail by DNVGL (2017) with the help of a plot of wave steepness versus relative depth with breaking wave cutoff limits in deep and shallow water. The general areas for use of each theory are denoted with the stream function theory application range.

Dean (1968, 1974) presented an analysis by defining the regions of validity of wave theories in terms of parameters $H/T^2$ and $d/T^2$ since $T^2$ is proportional to the wavelength. Le Méhauté (1976) presented a slightly different analysis to illustrate the approximate limits of validity for several wave theories, including the third- and fourth-order theories of Stokes.

Both Le Méhauté (1969) and Dean (1974) recommended cnoidal theory for shallow-water waves of low steepness, and Stokes' higher order theories for steep waves in deep water. Linear theory is recommended for small steepness $H/T^2$ and small $U_r$ values. For low steepness waves in transitional and deep water, linear theory is adequate but other wave theories may also be used in this region.

According to Laitone (1962) for an Ursell number ($U_r$) of approximately 20, Stokes' theory approaches the cnoidal theory. Stokes (1847) noted that this parameter ($U_r$, sometimes also called the Stokes number) should be small for long waves. An alternative, named the *Universal parameter* (Up), has recently been suggested (Goda 1983) for classification of wave theories.

## 14.8. Why are finite amplitude waves important?

The study of finite amplitude wave theories has different applications. One aspect is the waves in the generating area. The wind waves are generated by the transfer of energy of winds blowing over the sea. As the waves grow in size, the gravity tries to restore the sea surface back to normal condition. Gravity being the restoring force, these are known as gravity waves. The small amplitude and finite amplitude waves considered here are gravity waves. At the regions of wave generation, waves of different characteristics will be present simultaneously giving a chaotic appearance to the sea surface. Finite amplitude waves, such as the higher order Stokes waves, which break when they become unstable, will also be present here. This occurs when the breaking condition for deep water waves mentioned earlier is reached and then white caps appear on the sea surface. Sharp-crested waves are always present in the wave-generating area during strong winds. In order to understand the transfer mechanism of the energy of wind to waves and between different waves and also to delineate the wave generation processes, the finite amplitude wave theories are very important. Another aspect is the designing of marine structures. These structures are necessary for extraction of oil and natural gas, minerals, deep sea nodules, etc. Similarly, harnessing wave energy or for harvesting wind energy, by establishing wind mills in coastal waters also requires the construction of suitable structures. Use of the appropriate nonlinear wave theory depending on the area of application, gives better estimates of wave forcing, wave breaking, etc. Wave-induced forcing needs to be estimated according to the appropriate wave theory for the reliable design of marine structures. Using wave data and information on water depth, diagrams such as that of Le Méhauté (1976) can be used to decide which wave theory is appropriate.

## References


Airy, G.B., 1845, "Tides and waves", EncycLMetrop., London.

Bousinesq 1871Boussinesq, J. 1871. "Theorie de L'intumescence Liquide Appelee Onde Solitaire ou de Translation se Propageant dans un Canal Rectangulaire," Comptes Rendus Acad. Sci. Paris, Vol 72, pp 755-759.

Camfield & Street 1969CAMFIELD, F.E., and STREET, R.L., "Shoaling of Solitary Waves on Small Slopes," Journal of the Waterways and Harbors Division, Vol. 95, No. WW1, Feb. 1969.

Chakraborti 1987Chakrabarti, S. K. 1987. "Hydrodynamics of Offshore Structures," WIT Press, Southampton, UK.

DNVGL (2017): DNV-RP-C205 Environmental Conditions and Environmental Loads (www.sciencedirect.com/science/article/pii/s088997461930026#b6)

Dean, R.G. (1965), "Stream Function Representation of Nonlinear Ocean Waves,"Journal of Geophysical Research, Vol. 70, pp. 4561–4572.

Dean, R. G. 1968. "Relative Validity of Water Wave Theories," Proc. Civil Engr. in Ocean, ASCE, Vol 1, pp1-30.

Dean, R.G. (1974), "Evaluation and Development of Water Wave Theories for Engineering Application," Special Report No. 1, U.S. Army Coastal Engineering Research Center, Ft. Belvoir, VA.

Dean, R.F. and Dalrymple, R.A. (1984), Water Wave Mechanics for Engineers and Scientists, Prentice-Hall, Englewood Cliffs, NJ.

Dean, R.G. & Dalrymple, R.A., 1991. Water wave mechanics for engineers and scientists, Advanced Series on Ocean Engineering, Vol. 2, World Scientific, Singapore.

Fenton 1985Fenton, J. D. 1985. "A Fifth-Order Stokes Theory for Steady Waves," ASCE Jour. Waterw., Port, Coastal and Ocean Engr., Vol 111, pp 216-234.

Fenton J D, Nonlinear wave Theories, The Sea, Vol.9, Ocean Engineering Science, Eds. B Le Méhauté and D M Hanes, New York, Wiley, pp.1–17, 1990.

GERSTNER, F., "Theorie die Wellen," Abhandlungen der Konigichen Bohmischen Gesellschaft der Wissenschaften, Prague, Czechoslovakia, 1802.

Goda, Y. 1983. "A Unified Nonlinearity Parameter of Water Waves," Rept. Port and Harbor Res. Inst. of Japan, Vol 22, No. 3, pp 3-30.

Hardy, T.A. and Kraus, N.C. (1987), "A Numerical Model for Shoaling and Refraction of Second Order Cnoidal Waves Over an Irregular Bottom," Miscellaneous Paper CERC 87–9, U.S. Army Waterways Experiment Station, Vicksburg, MS.

Hedges, T.S., 1987. Combinations of waves and currents: an introduction. Proc. Inst. Civ. Engrs, Part 1,1987,June, 567-585.

Hedges, T.S., 1995. Regions of validity of analytical wave theories. Proc. Inst. Civ. Eng., Wat., Marit., & Energy, 112, June, 111-114.

Ippen,A.T (1966): Estuary and Coastline hydrodynamics, McGraw-Hill Book Company, Inc, New York.

Keller 1949Keller, J.B. (1949). The solitary wave and periodic waves in shallow water: Annals New York Acad. Sciences, vol. 51, pp. 345-350.

Komar, P. D. 1998. Beach Processes and Sedimentation, 2nd edn., Prentice-Hall, NJ, 544 pp.

Korteweg, D.J. and De Vries, G., 1895, "On the change of the form of long waves advancing in a rectangular canal, and on a new type of long stationary wave",Philosophical Magazine, 5th series, 422-443.

Laitone E.V. (1960) The second approximation to cnoidal and solitary waves. J. Fluid Mech. 9, 430–444

Laitone, E. V. 1962. "Limiting Conditions for Cnoidal and Stokes Waves," Journal of Geophysical Research, Vol 67, pp 1555-1564.

Laplace, P. S. (1775, 1776). Recherches sur quelques points de systeme du monde, Mem. Acad. Roy. Sci. (See also Mechanique celeste 4, no. 1, 1799.)

LeMehaute, B. (1969), "An Introduction to Hydrodynamics and Water Waves," Technical Report ERL 118-POL-3–2, U.S. Department of Commerce, Washington, DC.

LeM′ehaut′e, B., 1976, An Introduction to Hydrodynamics and Water Waves, New York, Springer-Verlag, 315pp. Laplace 1776

Miche 1944Miche, M. (1944), "Movements Ondulatoires des Mers en Profondeur Constante ou Decroissante," Annales des Ponts et Chaussees, pp. 25-78, 131-164, 270–292, 369–406.

Miles 1980, Miles, J. W. 1980. "Solitary Waves," Annual Rev. Fluid Mech., Vol 12, pp 11-43.

Miles, J. W. 1981. "The Korteweg de Vries Equation: A Historical Essay," Jour. Fluid Mech., Vol 106, pp 131-147.

Mitchell1893Michell, J.H. (1893). On the highest waves in water: Philosophical Magazine, (5), vol. XXXVI, pp. 430-437.

McGowan, J. (1891), "On the Solitary Wave," London, Edinburgh and Dublin Magazine and Journal of Science, Vol. 32, pp. 45–58.

Munk 1949Munk, H.W. (1949), "Solitary Wave and Its Application to Surf Problems," Annals, New York Academy of Science, Vol. 51, pp. 376–424.

Peregrine 1972Peregrine, D. H. 1972. "Equations for Water Waves and the Approximation Behind Them," Waves on Beaches and Resulting Sediment Transport, R. E. Meyer, ed., Academic Press, pp 95-121.

Rayleigh, L. 1876. "On Waves," Phil. Mag., Vol 1, pp 257-279.

Rayleigh, Lord (1877). On progressive waves: Proc. London Mathematical Society, vol. IX, pp. 21-26.

Russell, J. S. 1844. "Report on Waves," 14th Meeting Brit. Assoc. Adv. Sci., pp 311-390.

Sarpkaya, T. and Issacson, M. (1981). "Mechanics of Wave Induced Forces on Offshore Structures",Van Nostrand Reinhold.

Sorensen, R.M., 1993. Basic Wave Mechanics for Coastal and Ocean Engineers, John Wiley & Sons, New York.

SPM, 1973, 1984, Shore Protection Manual, U.S. Army Coastal Engineering Research Center, I

Stokes, G.G., 1847, "On the theory of oscillatory waves", Trans. Cambridge Philos. Soc., 8, 441-473.

Stokes,1880 G.C., "On the Theory of Oscillatory Waves," Mathematical and Physical Papers, Vol. 1, Cambridge University Press, London, England, 1880.

Ursell, F. (1953), "The Long Wave Paradox in the Theory of Gravity Waves," Proceedings, Cambridge Philosophical Society, Vol. 49, pp. 685–694.

USACE-SPM (1984): U.S. Army Coastal Engineering Research Center (1984), Shore Protection Manual, U.S. Government Printing OYce, Washington, DC.

USACE (2008): U S Army Corps of Engineers, Coastal Engineering Manual, Part II, Chapter 1. pp.II-1.1-1.121, 2008.

Varma, K,K (2014): Finite amplitude ocean waves-waves with peaked crests and broad troughs, Resonance, November, 2014, 1047-1057

The WAMDI Group (Hasselmann, S., Hasselmann, K., Bauer, E., Janssen, P.A.E.M., Komen, G.J., Bertotti, L., Lionello, P., Guillaume, A., Cardone, V.C., Greenwood, J.A., Reistad, M., Zambresky, L. and Ewing, J.A.), 1988,

"The WAM model - a third generation ocean wave prediction model", J. Phys. Oceanogr., 18, 1775-1810.

Wiegel, R. L., 1960, A presentation of cnoidal wave theory for practical application, J. Fluid Mech., 7, 2, 273–286

Wiegel R.L. 1964. Oceanographical Engineering. Englewood Cliffs, New Jersey: Prentice Hall

Williams J M, 1981: Limiting gravity waves in water of finite depth, Phil. Trans. Roy. Soc. London, Vol.A31, pp.139–188,.

## Exercises

### *Exercise 14.1*

A wave traveling in water depth d = 6 meters with a wavelength L = 60 meters and a height H = 1 meter. Find:

(a) Compare the wave profiles given by the first- and second-order theories.

(b) What is the difference between the first- and second-order horizontal velocities at the surface under both the crest and trough?

(c) How far in the direction of wave propagation will a water particle move from its initial position during one wave period when z = 0?

(d) What is the pressure at the bottom under the wave crest as predicted by both the first- and second-order theories?

(e) What is the wave energy per unit width of crest predicted by the first-order theory?

### *Solution*

a) First order wave profile equation is:

$$\eta_1 = \frac{H}{2}\text{Cos}(kx - \sigma t) = \frac{H}{2}\text{Cos}\left(\frac{2\pi x}{L} - \frac{2\pi t}{T}\right) = \frac{H}{2}\text{Cos}(\theta)$$

Where $\left(\frac{2\pi x}{L} - \frac{2\pi t}{T}\right) = (\theta)$

Second order wave profile equation is:

$$\eta_2 = \frac{H}{2}\text{Cos}(\theta) + \frac{\pi}{8}\frac{H^2}{L}\frac{\cosh\left(2\pi d/L\right)}{\left[\sinh\left(2\pi d/L\right)\right]^3}\left[2 + \cosh\left(\frac{4\pi d}{L}\right)\right]\text{Cos}[2(\theta)]$$

$$\cosh {2\pi d}/{L} = \cosh {2\pi 6}/{60} = 1.2040$$

Calculate: $\sinh {2\pi d}/{L} = \sinh {2\pi 6}/{60} = 0.6705$

$$\cosh {4\pi d}/{L} = \cosh {4\pi 6}/{60} = 1.8991$$

$$\eta_2 = \frac{H}{2}\text{Cos}(\theta) + \frac{\pi}{8}\frac{H^2}{L}\frac{\cosh\left({2\pi d}/{L}\right)}{\left[\sinh\left({2\pi d}/{L}\right)\right]^3}\left[2 + \cosh\left(\frac{4\pi d}{L}\right)\right]\text{Cos}[2(\theta)]$$

$$\eta_2 = \frac{1}{2}\text{Cos}(\theta) + \frac{\pi}{8}\frac{1m^2}{60m} \times \frac{1.2040}{(0.6705)^3}[2 + 1.8991]\cos 2\theta = 0.5\text{Cos}(\theta) + 0.102\text{Cos}(2\theta)$$

$$\eta_1 = 0.5\text{Cos}(\theta)$$
$$\eta_2 = 0.5\text{Cos}(\theta) + 0.102\text{Cos}(2\theta)$$

Please note the values of η:

at the crest : cosθ and cos 2θ = + 1 and

at the trough cos θ = –1 and cos 2θ = + 1

$$\eta_{1Crest} = 0.5\text{Cos}(\theta) = 0.5 \times 1 = 0.5; \eta_{1trough} = 0.5\text{Cos}(\theta) = 0.5 \times 1 = 0.5$$
$$\eta_{2crest} = 0.5\text{Cos}(\theta) + 0.102\text{Cos}(2\theta) = 0.5 \times 1 + 0.102 \times 1 = 0.602$$
$$\eta_{2trough} = 0.5\text{Cos}(\theta) + 0.102\text{Cos}(2\theta) = 0.5 \times (-1) + 0.102 \times 1 = 0.398$$

This means in the case of first order waves, crest and trough values are same = 0.5m

But in the case of second order waves, crest = 0.602m and trough = 0.398 m .

This means first order waves are simple sinusoidal curves and in the case of second order waves, crests are elongated and troughs are shortened with respect to mean water line.

b) Horizontal velocities: From equation (13.37a) of chapter 13:

i) First order:

$$u_1 = \frac{H}{2}\frac{gT}{L}\frac{\cosh\frac{2\pi}{L}(d+z)}{\cosh\frac{2\pi d}{L}}\cos\left(\frac{2\pi x}{L} - \frac{2\pi t}{T}\right)$$

At crest: Z= H/2; cos θ = 1

$$u_{1,C} = \frac{H}{2}\frac{gT}{L}\frac{\cosh\frac{2\pi}{L}(d+z)}{\cosh\frac{2\pi d}{L}}\cos\left(\frac{2\pi x}{L} - \frac{2\pi t}{T}\right)$$

At trough: Z = -- H/2; cos θ = – 1

$$u_{1,T}=-\frac{H}{2}\frac{gT}{L}\frac{\cosh\frac{2\pi}{L}(d+z)}{\cosh\frac{2\pi d}{L}}\cos\left(\frac{2\pi x}{L}-\frac{2\pi t}{T}\right)$$

ii) Second order:

$$u_2=\frac{HgT}{2L}\left[\frac{\text{Cosh}\,K(d+z)}{\text{Cosh}\,Kd}\right][\text{Cos}(kx-\sigma t)]+\frac{3}{4}\left(\frac{\pi H}{L}\right)^2 C\left[\frac{\text{Cosh}\,2K(d+z)}{\text{Sinh}^4\,Kd}\right][\text{Cos}(2kx-2\sigma t)]$$

According to second-order theory, the crest occurs at Z= $\eta_{2,C}$ = 0.602 and: cosθ and cos 2θ = + 1 and trough at cosθ = –1 and cos 2θ = +1

i) At crest , Z=H/2=1/2=0.5, Z+d = 0.5+6 = 6.5; (d+Z)/L = 6.5/60 = 0.108;

$$\cosh K(d+Z)=\cosh 2\pi\left(\frac{d+z}{L}\right)=\cosh 2\pi(0.108)=1{,}241$$

$$\cosh 2K(d+Z)=\cosh 4\pi\left(\frac{d+z}{L}\right)=\cosh 4\pi(0.108)=2.071$$

ii) At trough : Z = -- H/2 = -1/2= -0.5; Z+d = -0.5+6 = 5.5; (d+Z)/L = 5.5/60 = 0.092;

$$\cosh K(d+Z)=\cosh 2\pi\left(\frac{d+z}{L}\right)=\cosh 2\pi(0.092)=1.172$$

$$\cosh 2K(d+Z)=\cosh 4\pi\left(\frac{d+z}{L}\right)=\cosh 4\pi(0.092)=1.746$$

iii) $\cosh K(d)=\cosh 2\pi\left(\frac{d}{L}\right)=\cosh 2\pi(0.1)=1.204$

iv) $\cosh 2K(d)=\cosh 4\pi\left(\frac{d}{L}\right)=\cosh 4\pi(0.1)=1.8991$

v) $\sinh K(d)=\sinh 2\pi\left(\frac{d}{L}\right)=\sinh 2\pi(0.1)=0.671$

$$C^2=\frac{gL}{2\pi}\tanh\left(\frac{2\pi d}{L}\right)=\frac{9.8\times 60}{2\pi}\tanh(2\pi\times 0.1)=52.1$$

$$C=\sqrt{52.1}=7.22=\frac{L}{T}\Rightarrow\frac{T}{L}=\frac{1}{7.22}=0.14$$

$$u_{1,C}=\frac{H}{2}\frac{gT}{L}\frac{\cosh\frac{2\pi}{L}(d+z)}{\cosh\frac{2\pi d}{L}}\cos\theta=\frac{1\times 9.8\times 0.14}{2}\times\frac{1.241}{1.204}\times 1=0.707$$

$$u_{1,T}=\frac{H}{2}\frac{gT}{L}\frac{\cosh\frac{2\pi}{L}(d+z)}{\cosh\frac{2\pi d}{L}}\cos\theta=\frac{1\times 9.8\times 0.14}{2}\times\frac{1.172}{1.204}\times(-1)=0.668$$

For second order:

$$u_{2C} = \frac{HgT}{2L}\left[\frac{\text{Cosh}\,K(d+z)}{\text{Cosh}\,Kd}\right]\left[\text{Cos}(\theta)\right] + \frac{3}{4}\left(\frac{\pi H}{L}\right)^2 C\left[\frac{\text{Cosh}\,2K(d+z)}{\text{Sinh}^4\,Kd}\right]\left[\text{Cos}\,2(\theta)\right]$$

$$u_{2,C} = \frac{1\times 9.8\times 0.14}{2}\times\frac{1.241}{1.204}\times 1 + \frac{3}{4}\left(\frac{\pi\times 1}{60}\right)^2\times 7.22\times\frac{2.071}{(0.671)^4}(1) = 0.707 + 0.152 = 0.858ms^{-1}$$

$$u_{2,T} = \frac{1\times 9.8\times 0.14}{2}\times\frac{1.172}{1.204}\times(-1) + \frac{3}{4}\left(\frac{\pi\times 1}{60}\right)^2\times 7.22\times\frac{1.746}{(0.671)^4}(1) = -0.707 + 0.128 = -0.579ms^{-1}$$

c) How far in the direction of wave propagation will a water particle move from its initial position during one wave period when z = 0

Horizontal mass transport or Stokes drift equation(14.57)

$$\bar{U}_Z = \left(\frac{\pi H}{L}\right)^2 \frac{C}{2}\left[\frac{\text{Cosh}\,2K(d+z)}{\text{Sinh}^2\,Kd}\right]$$

The distance the water particle move when Z=0 is: $U_Z.T$

$$\Delta X = \bar{U}_Z T = T\left(\frac{\pi H}{L}\right)^2 \frac{C}{2}\left[\frac{\text{Cosh}\,2K(d+0)}{\text{Sinh}^2\,Kd}\right] = \frac{T}{L}\left(\frac{\pi^2 H^2}{L}\right)\frac{1}{2}\frac{L}{T}\left[\frac{\text{Cosh}\,2K(d+0)}{\text{Sinh}^2\,Kd}\right]$$

$$= \left(\frac{\pi^2 H^2}{L}\right)\frac{1}{2}\left[\frac{\text{Cosh}\,2K(d)}{\text{Sinh}^2\,Kd}\right] = \left(\frac{\pi^2 1^2}{60}\right)\frac{1}{2}\left[\frac{1.8991}{(0.671)^2}\right] = 0.347m$$

d) What is the pressure at the bottom under the wave crest as predicted by both the first- and second-order theories?

From chapter 13, equation: 13.39

First order pressure equation ($P_1$): $\frac{p}{\rho} = g\frac{\cosh k(d+z)}{\cosh kd}a\cos(kx-\sigma t) - gz$

$$p_1 = \frac{\rho g H}{2}\frac{\cosh k(d+z)}{\cosh kd}\cos\theta - \rho g z$$

The pressure at the sea bottom for first order waves: $Z = -\frac{d}{2} \Longrightarrow \cosh k(d+Z) = \cosh(0) = 1$

$$p_1 = \frac{\rho g H}{2}\frac{\cosh k(d - d/2)}{\cosh kd}\cos\theta - \rho g(-d) = \frac{1025\times 9.8\times 1}{2}\times\frac{1}{1.204}(1) - 1025\times 9.8\times(-6)$$

$$= 4171.5 + 60270 = 64441.5Nm^{-2}$$

$P_1$ = 64441.5 $Nm^{-2}$.

Pressure at the second order waves: From equation (14.56):

$$p_2 = \frac{\rho g H}{2}\frac{\cosh k(d+z)}{\cosh kd}\cos\theta - \rho g z + \frac{3}{8}\rho g\frac{\pi H^2}{L}\frac{\tanh kd}{(\sinh kd)^2}\left\{\frac{\cosh 2k(d+Z)}{(\sinh kd)^2} - \frac{1}{3}\right\}\cos 2(kx-\sigma t)$$

$$-\frac{1}{8}\rho g\frac{\pi H^2}{L}\frac{\tanh kd}{(\sinh kd)^2}\left\{\frac{\cosh 2k(d+Z)}{L} - 1\right\}$$

$P_2 = P_1$ + second order terms:

$$p_2 = p_1 + \frac{3}{8}(1025)\times 9.8\frac{\pi 1^2}{60}\frac{0.5569}{(0.671)^2}\left\{\frac{1}{(0.671)^2} - \frac{1}{3}\right\}(1)$$

$$-\frac{1}{8}(1025)\times 9.8\frac{\pi 1^2}{60}\frac{0.5569}{(0.671)^2}\{\cosh(0)-1\} = p_1 + 81.32\left\{\frac{3}{(0.671)^2} - \frac{3}{3}\right\} - 0 = p_1 + 460.5$$

$P_2 = 64441.5 + 460.5 = 64902\ Nm^{-2}$

a) What is the wave energy per unit width of crest predicted by the first-order theory?

The total wave energy in one wavelength per unit crest width is given by (equation 13.45a of previous chapter):

$$E = \frac{1}{8}\rho g H^2 L = \frac{1\times 1025 kgm^{-3}\times 9.8ms^{-2}\times 1m^2\times 60m}{8} = 75338 kgms^{-2} = 75{,}338 Nm^{-1}$$

### *Exercise 14.2*

FIND:(a) Using cnoidal wave theory, find the wavelength L and compare this length with the length determined using Airy theory.(b) Determine the celerity C. Compare this celerity with the celerity determined using Airy theory. (c) Determine the distance above the bottom of the wave crest yc and wave trough yt. (d) Determine the wave profile. Express both in C.G.S and F.P.S systems.

**Given**

A wave traveling in water depth $d = 3$ m (9.84 ft), with a period $T = 15$ sec, and a height $H = 1.0$ m (3.3 ft).

### *Solution*

Calculate

$$\frac{H}{d} = \frac{1}{3} = 0.33;\ T\sqrt{\frac{g}{d}} = 15\sqrt{\frac{9.8}{3}} = 27.11$$

From Figure 14.3.1, enter $H/d$ and $T$ to determine the square of the modulus of the complete elliptical integrals, $k^2$: $k^2 = 1 - 10^{-5}$

Entering both Figures 14.3.1 and 14.3.2 with the value of $k^2$ gives

$$\frac{L^2 H}{d^3} = 290 \rightarrow L = \sqrt{\frac{290 d^3}{H}} = \sqrt{\frac{290\times(3)^3}{1}} = 88.5m\quad (290.3\text{ft})$$

Thus L = 88.5 m (290.3 ft). The wavelength from the linear (Airy) theory is

$$L = \frac{gT^2}{2\pi} tanh\left(\frac{2\pi d}{L}\right) = 80.6\,m \quad (264.5\text{ft})$$

To check whether the wave conditions are in the range for which cnoidal wave theory is valid, calculate *d/L* and the *Ursell number* = $L^2 H/d^3$:

$$\frac{d}{L} = \frac{3}{88.5} = 0.0339 \rightarrow \text{As } 0.0339, < \frac{1}{8} \text{ it is O.K}$$

$$\frac{L^2 H}{d^3} = \frac{1}{\left(\frac{d}{L}\right)^2}\left(\frac{H}{d}\right) = 290 \rightarrow \quad \text{As } 290 > 26, \text{ it is O.K}$$

Therefore, cnoidal theory is applicable

(b) Wave celerity is given by

$$C = \frac{L}{T} = \frac{88.5}{15} = 5.90\,ms^{-1}(19.36\,ft/s)$$

while the linear theory predicts

$$C = \frac{L}{T} = \frac{80.6}{15} = 5.37\,ms^{-1}(17.63\,ft/s)$$

Thus, if it is assumed that the wave period is the same for cnoidal and Airy theories, then

$$\frac{C_{cnoiodal}}{C_{Airy}} = \frac{L_{cnoiodal}}{L_{Airy}} \approx 1$$

(c) The percentage of the wave height above the SWL may be determined from Figure 14.2.1 or 14.2.2.

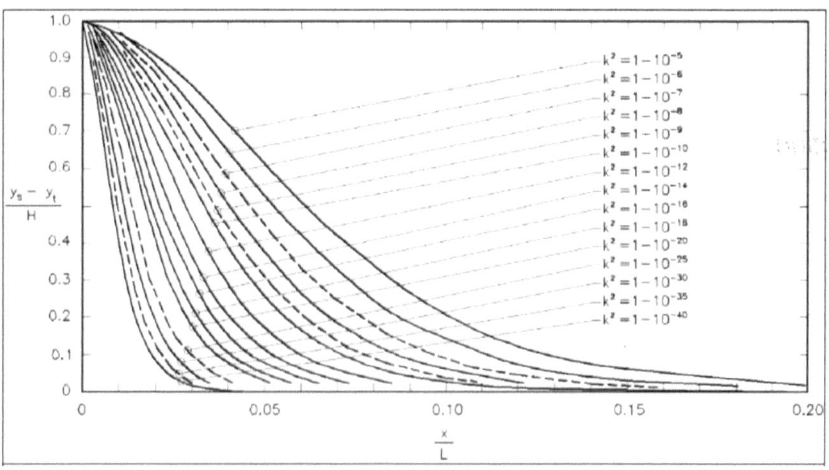


**Fig.14.2.1** Normalized surface profile of the cnoidal wave (Wiegel 1960).

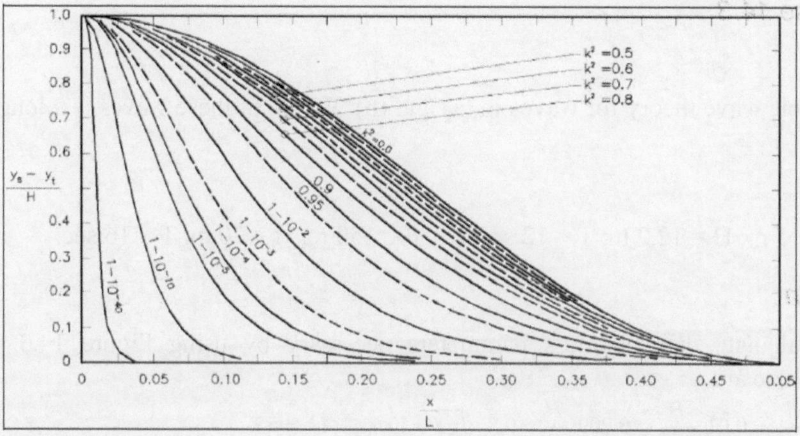


**Fig. 14.2.2.** Normalized surface profile of the cnoidal wave for higher values of and X/L (Wiegel 1960)

Entering these figures with $L^2 H/d^3 = 290$, the value of *(yc -d)/H* is found to be 0.865, or 86.5 percent. Therefore,

$y_c = 0.865\ H + d \rightarrow y_c = 0.865\ (1) + 3 = 3.865$ m (12.68 ft)

Also from Figure 14.2.1 or 14.2.2

$$\frac{(y_t - d)}{H} + 1 = 0.865 \rightarrow y_t = (0.865\text{-}1)\ (1) + 3 = 2.865\text{m}\ (9.40\text{ ft})$$

(d) The dimensionless wave profile is given in Figures 14.2.1 and 14.2.2 for $K^2 = 1 - 10^{-5}$. The results obtained in (c) above can also be checked by using Figures 14.2.1 and 14.2.2. For the wave profile obtained with $K^2 = 1 - 10^{-5}$, the SWL is approximately 0.14H above the wave trough or 0.86H below the wave crest.

The results for the wave celerity determined under (b) above can now be checked with the aid of Figure 14.3.3. Calculating

$$\frac{H}{y_t} = \frac{(1)}{2.865} = 0.349 \qquad (1)$$

Entering Figure 14.3.3 with

$$\frac{L^2 H}{d^3} = \frac{1}{2.865} = 0.349 \qquad (2)$$

and so from (1) and (2) it is found that: $\frac{C}{\sqrt{g y_t}} = 1.126$

Therefore

$$C = 1.126\sqrt{(9.8)(2.865)} = 5.97\,ms^{-1}\left(19.57\frac{ft}{s}\right)$$

The differences between this number and the 5.90 m /sec (18.38 ft/s) calculated under (b) above is the result of small errors in reading the curves.

### *Exercise 14.3*

FIND:

Applicable wave theory for waves in (a) and (b). Which of these waves is a long wave?

Given:

(a). d = 15 m, H = 12.2 m, T = 12 sec; (b). d = 150 m, H = 30 m, T = 16 sec.

### *Solution*

(a) Calculate dimensionless parameters necessary by using Figure 14.3.4 These are

$$\frac{d}{gT^2} = 0.01; \frac{H}{gT^2} = 0.009; \frac{H}{d} = 0.8; \sqrt{gd} = 12\,ms^{-1}; U_r = 55$$

From Figure 14.3.4, the applicable theory is cnoidal.

(b) In a similar fashion, computing the parameters:

$$\frac{d}{gT^2} = 0.06; \frac{H}{gT^2} = 0.01; \frac{H}{d} = 0.2; \sqrt{gd} = 40\,ms^{-1}; U_r = 1.5$$

With these values, Figure 14.3.4 indicates the applicable theory is Stokes third- or fifth-order. It is noted that the linear theory is also applicable.

Based on the values of Ursell parameter, neither wave (a) or (b) is a true long wave. Wave (a) may be considered a long wave in comparison to wave (b).

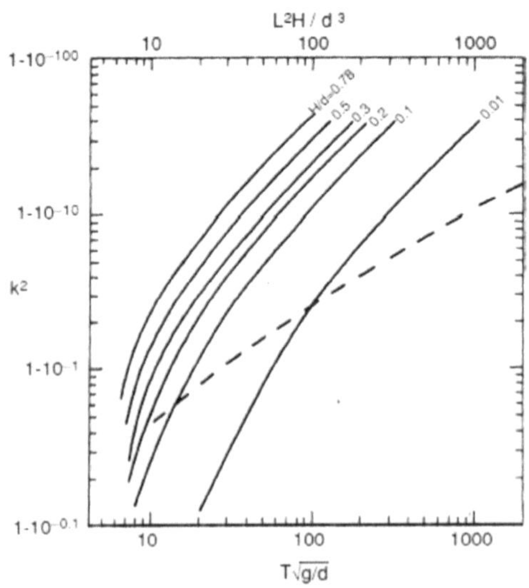


**Fig. 14.3.1.** $K^2$ versus $L^2$ H/d³ and $K^2$ versus $T\sqrt{g/d}$ and H/d (Modified from Wiegel, 1964)

Ref: CEM (2015): Coastal Engineering Manual - Part-II, U.S.Army corpse of Engineers Hydrodynamic analysis and Design Conditions 62 pp.

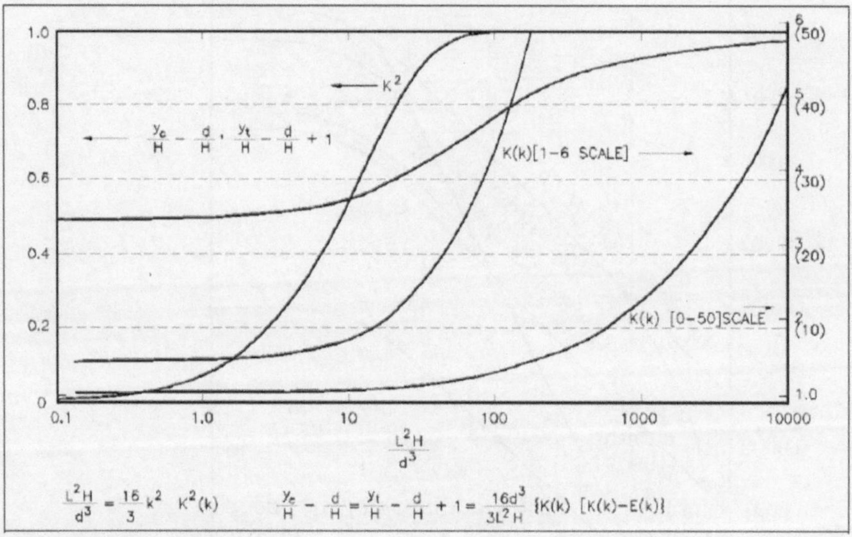


**Fig. 14.3.2.** Relationship among $L^2 H/d^3$ and the square of the elliptic modulus ($K^2$), $y_c/H$, $y_t/H$ and K(k) (Wiegel, 1960)

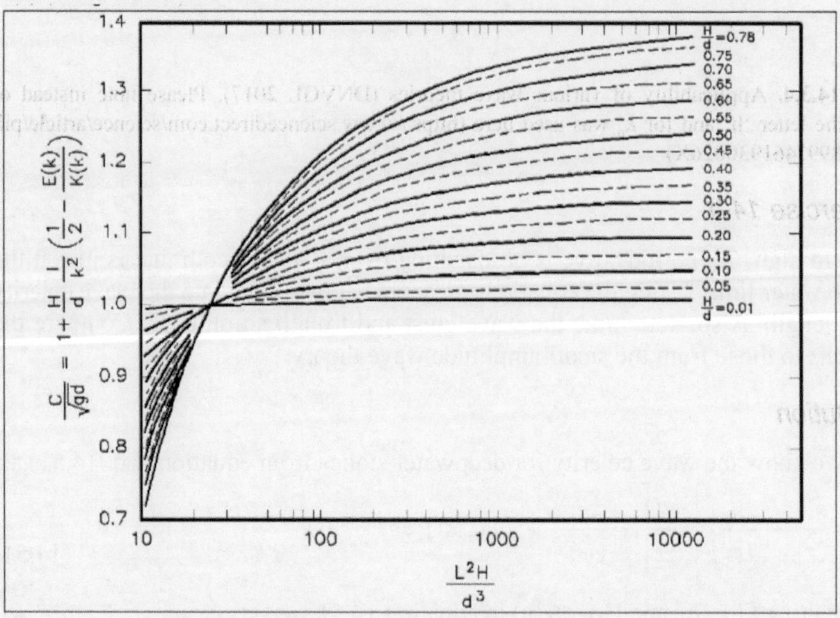


**Fig. 14.3.3.** Relationship between cnoidal wave velocity and $L^2 H/d^3$ (Wiegel, 1960)

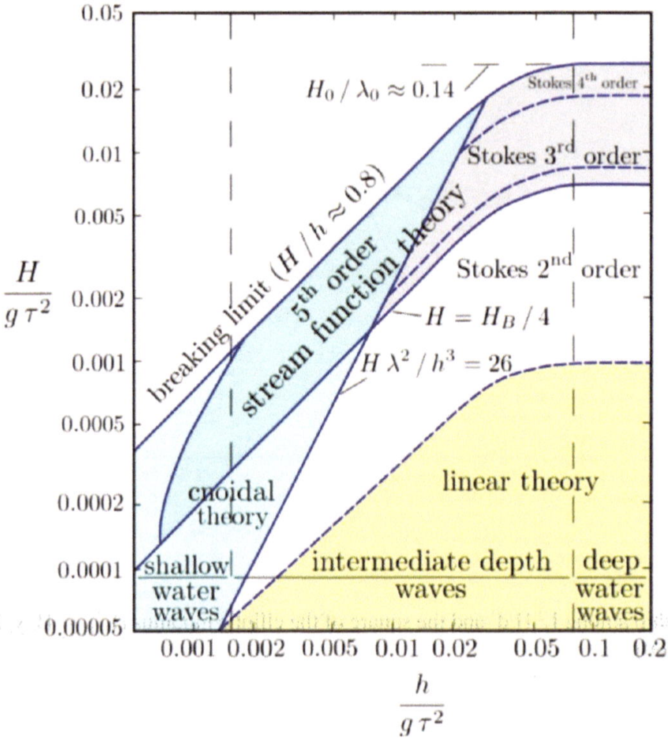


**Fig.14.3.4.** Applicability of various wave theories (DNVGL 2017). Please note instead of 'd' the letter 'h' and for $\lambda_0$ was used here (https://www.sciencedirect.com/science/article/pii/S088997461930026X)

### *Exercise 14.4*

A 6 m high, 7 s period wave is propagating over a water depth that is just at the deep water limit. Using the equations presented above, calculate the wave celerity and length. Also, determine the wave crest and trough amplitudes. Compare the results to those from the small-amplitude wave theory.

### *Solution*

We know the wave celerity for deepwater stokes from equation (Eq. 14.51) as:

$$Cos = \frac{gT}{2\pi}\left[1+\left(\frac{\pi H_0}{Los}\right)^2\right] = \frac{gT}{2\pi}\left[1+\left(\frac{\pi H_0}{CosT}\right)^2\right] (\because Lo = CosT) \tag{14.51}$$

By putting (T = 7 s and Ho = 6 m) in equation 14.51 we get: $C_{0s} = \frac{9.8\times 7}{2\pi}\left[1+\left(\frac{\pi\times 6}{C_{0s}\times 7}\right)^2\right]$

On solving we get $C_{0s} = 11.54\,\text{m/s}$

On substituting the $C_{os}$ value in the above equation we can get $L_{0s} = 80.78$ m

From small-amplitude wave theory we know from Eq (13.33) is: $C_0 = \frac{gT}{2\pi}$ $L_0 = \frac{gT^2}{2\pi} = C_0 T$

Similarly substituting the given value (T = 7 s) in equation (13.33) we can get the values for small amplitude waves as:

$C_0 = 10.93\, ms^{-1}$ and $L_0 = 76.50$.

Thus the Stokes Finite amplitude theory values ($C_{os}$ & $L_{os}$) are 5.5% higher than the small amplitude values $C_0$ &$L_0$). The steepness of this wave $H_0 / L_0$) is 6/80.78 = 0.074 or about half the limiting steepness of 1/7.

From Eq. (14.61 & 14.62) we know:

$$a_c = \frac{H_0}{2} + \frac{\pi H_0^2}{4L_0} \text{ and } a_t = \frac{H_0}{2} - \frac{\pi H_0^2}{4L_0}$$

$$a_c = \frac{6}{2} + \frac{\pi(6)^2}{4(80.78)} = 3.35\,m$$

$$a_t = \frac{6}{2} - \frac{\pi(6)^2}{4(80.78)} = 2.65\,m$$

### *Exercise 14.5*

Consider the wave discussed in Example 14.9.4. Calculate the mass transport velocity versus distance below the water surface level for z = 0, -0:1, - 0.2,-0.3, - 0.4, and -0.5 times the wave length. Compare this to the wave celerity and crest particle velocity.

### *Solution*

For the first or second order, the wave length is given by Eq. (13.33) or

$$L_0 = \frac{gT^2}{2\pi} \rightarrow L_0 = \frac{9.8(7)^2}{2\pi} = 76.5m$$

Thus, the water depth is — 76.5/2 = 38.25m and k =2π /76.5 = 0.0821. Then,

the mass transport velocity, given by Eq. (14.57), as

$$\bar{U}_Z = \left(\frac{\pi H}{L}\right)^2 \frac{C}{2}\left[\frac{\text{Cosh}\, 2K(d+z)}{\text{Sinh}^2 Kd}\right] = \frac{\pi^2 H^2}{2TL} \frac{\cosh 2k(d+z)}{\sinh^2(Kd)}$$

$$\bar{U}_z = \frac{\pi^2(6)^2}{2(7)(76.5)} \frac{\cosh 2(0.0821)(d+z)}{sinh^2(\pi)} = 0.0025 \cosh 0.164(d+z)$$

where (d + z) is the distance up from the bottom. Proceeding with the calculations yields:

(ii) Z= -0.1 times the wave length = 76.5 × (–0.1) = – 7.65; d + z = 38.25-7.65 = 30.60

(iii) Similarly Z= -0.2times the wave length = 76.5 × (–0.2) = – 15.30; d + z = 38.25 – 15.30 = 22.95. Like this if written and calculated $\bar{U}_z$ will be as given in the table below.

| Z(m) | (d+z)(m) | $\bar{U}$ z $ms^{-1}$ |
|---|---|---|
| 0 | 38.25 | 0.665 |
| -7.65 | 30.60 | 0.189 |
| -15.30 | 22.95 | 0.053 |
| -22.95 | 15.30 | 0.015 |
| -30.60 | 7.65 | 0.004 |
| -38.25 | 0 | 0.002 |

Note the rapid decay in the mass transport velocity with distance below the still water level from the third column of the above table. From Eq. (13.33) the wave celerity is

$$C_0 = \frac{gT}{2\pi} = \frac{9.81(7)}{2\pi} = 10.93\,ms^{-1}$$

for both the first and second order. Using the first-order crest particle velocity as sufficient for comparison purposes we have

$$u_c = \frac{\pi(6)}{7} = 2.69ms^{-1}$$

Thus, the celerity, crest particle velocity, and mass transport velocity at the water surface are 10.93 m/s, 2.69 m/s and 0.665 m/s respectively for this wave.

## *Exercise 14.6*

A wave having a period of 14 s and a height of 2 m is propagating in water 4 m deep. Using cnoidal wave theory determines the wave length and celerity and compares the results to the small-amplitude theory. Also plot the wave surface profile.

## *Solution*

To employ Figure (11.1-13) we need: $\frac{H}{d} = \frac{2}{4} = 0.5$

And $T\sqrt{\frac{g}{d}} = 14\sqrt{\frac{9.81}{4}} = 21.9$

This gives $k^2 = 1 - 10^{-5.3}$

And $U_r = 300$

So the cnoidal theory is quite appropriate for this wave condition. From the Ursell number the wave length is:

$$\frac{L^2H}{d^3} = 300 \rightarrow L = \sqrt{\frac{300(4)^3}{2}} = 98.0\,m$$

And $C = \frac{L}{T} = \frac{98}{14} = 7.0\,ms^{-1}$

Since d/L = 4/98 this is a shallow water wave. So for shallow water wave velocity

$$C = \sqrt{gd} = \sqrt{9.8 \times 4} = 6.26\,m/s$$

And L = CT = 6.26 × 14 = 87.6m

The difference between the results of the two theories is 11.7% with the small amplitude theory yielding smaller values of C and L for a given H, T, and d.

With the value of $k^2$ and the wave length and height, the surface profile can be determined from Figure (14.2.2). A plot of the surface profile (with a 10:1 vertical scale exaggeration) is given below:

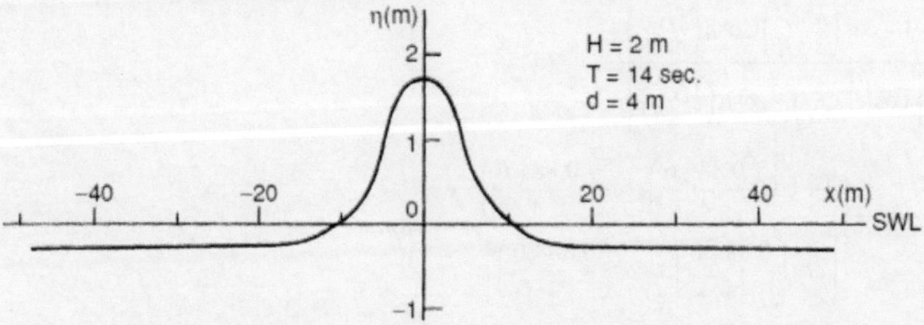


Note that the ratio of the crest amplitude to the wave height for this wave is 0.86.

### *Exercise 14.7*

Consider the same wave as in Exercise 14.9.6 (i.e., T = 14 s, H = 2 m, and d = 4 m). Using solitary wave theory, calculate the wave celerity and compare it to the results from that example. Also, calculate the crest particle velocity and compare it with the results from the small-amplitude wave theory.

### *Solution*

From Eq. (14.72) the wave celerity is

$$C = \sqrt{gd}\left[1 + \frac{H}{2d}\right] \rightarrow C = \sqrt{9.8 \times 4}\left[1 + \frac{2}{2 \times 4}\right] = 7.8\,m/s$$

which compares to 6.3 m/s and 7.0 m/s for the small-amplitude and cnoidal wave theories calculated in previous exercise, respectively.

Given H = 2 m and d = 4 m, Eqs. (14.75) can be solved simultaneously by trial and error to yield

$$\frac{H}{d}=\frac{N}{M}tan\frac{1}{2}\left[M\left(1+\frac{H}{d}\right)\right]$$

$$N=\frac{2}{3}sin^2\left[M\left(1+\frac{2}{3}\frac{H}{d}\right)\right]$$

After solving these two equations we can get M and N as:

$$\frac{2}{4}=\frac{N}{M}tan\frac{1}{2}\left[M\left(1+\frac{2}{4}\right)\right]$$

$$N=\frac{2}{3}sin^2\left[M\left(1+\left\{\frac{2}{3}\times\frac{2}{4}\right\}\right)\right]$$

M= 0.88 and N = 0.57

Then, with C = 7.8m/s, z = 2m, x = 0 we have for the particle velocity at the wave crest from equation (14.73)

$$u=NC\frac{1+\cos\left({}^{My}/_{d}\right)\mathrm{Cosh}\left({}^{Mx}/_{d}\right)}{\left[\cos\left({}^{My}/_{d}\right)+\cosh\left({}^{Mx}/_{d}\right)\right]^2}$$

$$u=(0.57\times7.8)\frac{1+\cos\left(\frac{0.88\times6}{4}\right)cosh\left(\frac{0.88\times0}{4}\right)}{\left[cos\left(\frac{0.88\times6}{4}\right)+cos\left(\frac{0.88\times0}{4}\right)\right]^2}=1.98\,m\,s^{-1}$$

From Eq. (13.37a), for the small-amplitude wave theory in shallow water

$$u=\frac{H}{2}\sqrt{\frac{g}{d}}\cos(kx-\sigma t)\rightarrow u=\frac{2}{2}\sqrt{\frac{9.81}{4}}(1)=1.6\,ms^{-1}$$

Thus, in shallow water there is a significant difference between the results from the two theories. For this wave the true value lies between the two results, but is probably closer to the result given by the solitary wave theory.

# 15

# Ocean Tides Long Waves and Sea Level Changes

## 15.1. Motions of the Earth-sun-moon

In order to understand the generation of tides first we should understand the motions of earth with respect to sun and moon. Tides are affected by the orbit of the moon around the earth and, less so, by the earth around the sun. The orbital motion of earth with respect to sun and the orbital motion of moon with respect to earth are as shown in Fig.15.1. Though earth and moon's motions are elliptical, they are not at the same distance on either side. Hence near position of the earth with respect to sun (perihelion), far position with respect to sun (aphelion) occurs. Similarly moon's position with respect to earth also occurs perigee (near position) and apogee (far position) as shown in Fig.15.1. Note that earth's axis always points out towards Pole Star irrespective of its position. The Earth's axis of rotation is tilted, at present at $24^0$ 27', to the plane of ecliptic.

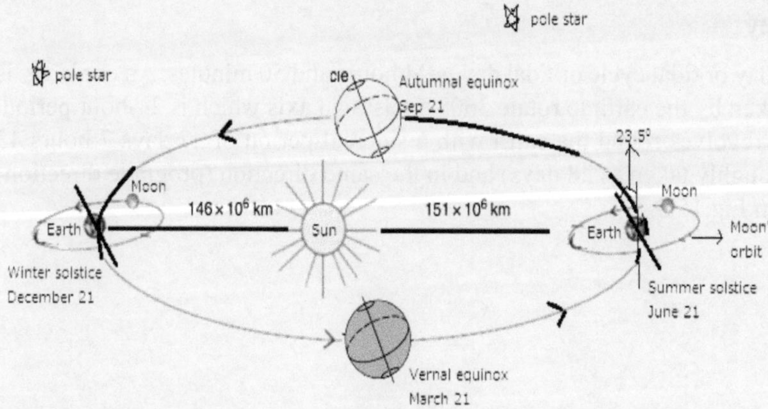


**Fig. 15.1.** Earth and moon rotations and revolutions around the sun

### Cycle of Sun-moon-earth apparent motions

Figure 15.2 shows the approximate orientation of the sun, moon, and earth at the quarter points of the moon's revolution around the earth relative to the position of the sun (29.5 days). At the first position (new moon) and third position (full moon) the solar and lunar tide generating forces reinforce and the highest or spring tides occur. At the second and fourth quarters the lowest or neap tides occur. Each quarter is 29.5/4 = 7.4 days long, so that spring and neap tides are 14.8 days or about 2 weeks apart.

At any place, the tide generally has two cycles in a lunar month (29.5 days). In the lunar month, the earth faces full or new moon and the moon is in quadrants. When the moon is full or new, we get spring tides (highest tides) and when the moon is in quadrants, we get neap tides (Lowest tides) as shown in Fig.15.2.

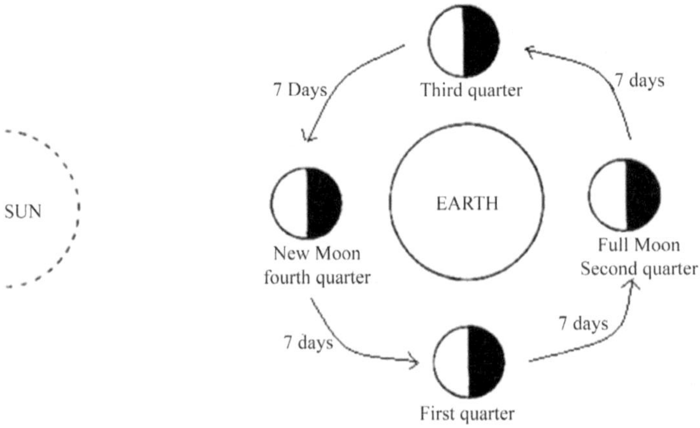


**Fig. 15.2.** Earth, sun, and moon positions during a lunar month.

## Lunar Day

The lunar day or tidal cycle or tidal day is 24 hours and 50 minutes. An earth day is the time taken by the earth to rotate 360° on its own axis which is 24-hour period. The moon revolves round the earth with a sidereal period of 27 days 7 hours 43 minutes (roughly taken as 28 days) and in the same direction (prograde direction) as shown in Fig.15.3.

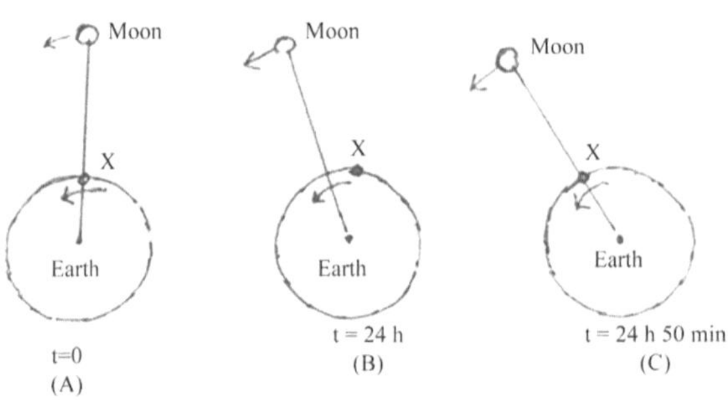


**Fig. 15.3.** The reason for the lag of the tide

By the time the point 'X' on earth completes one rotation (from A to B) (t=0 to t=24 h)(24 hours) the moon would complete the revolution round the earth and pass beyond the point X on earth (at t=24 h). The point 'X' will take 50 more minutes to reach the moon's axis (to align with the moon).This means the moon aligns with X at 'C' (at t=24 h 50 min). So the occurrence of tide on the next day will be 50 minutes later than the previous day. This is called as the 'lag of tide'.

Although the gravitational force exerted by the moon has the strongest influence on the tides and then followed by the gravitational force of the sun, there are other factors also that influence the tides. They are earth's axial tilt, moon's axial tilt with respect to plane of ecliptic and declination.

## Effects of Earth's axial tilt on tides

The Earth's axis is tilted 23.44° against the plane of the earth's orbit around the sun (the ecliptic). Besides giving rise to the seasons, this tilt of earth's axis influences the local tides as can be seen by the tidal bulges in the Fig.15.4 below. Note, that these two successive high tides at points 'A' and 'B' are unequal. A person anywhere on the given latitude (AB) would experience two different tidal heights due to this tilt of earth's axis. Similarly the inclination of the moon's orbit to the plane of ecliptic is 5.14°.

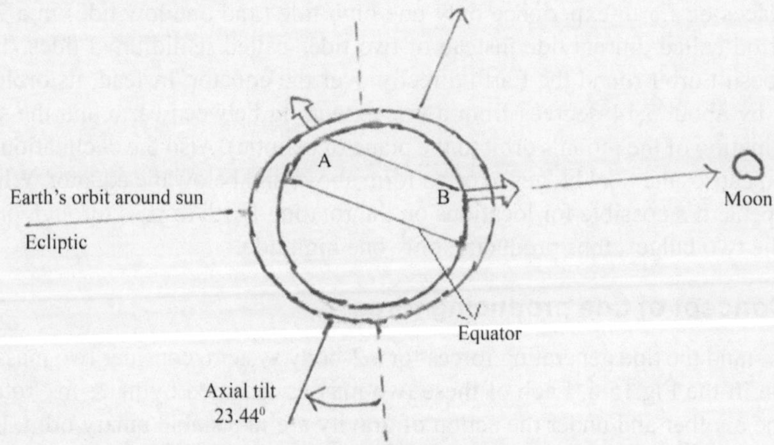


**Fig.15.4.** Effect of tides due to the earth's axial tilt

## Declination of moon

Other factor that influences the tides is the angle of the moon's orbit with respect to the earth-sun orbital plane called declination of moon. The moon's orbital path around the earth is not in line with the equator. Instead, the moon's orbit is tilted from the equator (Fig.15.5). As the moon completes its orbit around the earth, it is in north of the equator during half of the lunar month and south of the equator in the other half of the lunar month. The moon is only directly above the earth's equator for two days in the lunar month. The moon's angular distance north or south of the equator (the moon's declination) changes throughout the month as the moon moves in its orbit around the earth.

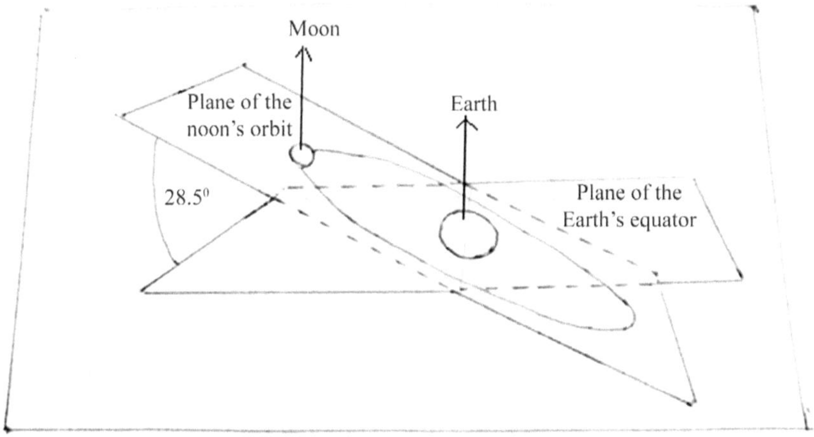


**Fig.15.5.** Influence of declination of moon on tide

Some places on Earth experience only one high tide (and one low tide) in a 24-hour period called diurnal tide instead of two tides called semidiurnal tides. The moon doesn't orbit round the Earth directly over the equator. Instead, its orbit is inclined by about 5.14 degrees from a line extending between earth and the sun (the inclination of the moon's orbit to the plane of ecliptic). Also the declination of the moon causes the tidal bulges to often form above and below the equator. When this happens, it's possible for locations on the rotating Earth to pass through only one of the two bulges, thus producing only one high tide.

## 15.2. Concept of tide producing force

To understand the tide generating forces for a 2-body system, consider two masses as shown in the Fig.15.6. Each of these two masses, denoted by $m_1$ & $m_2$, rotate about one another and under the action of gravity are in a stable binary orbit. Let the center of rotation is on a line between the two masses at a position denoted by "x".

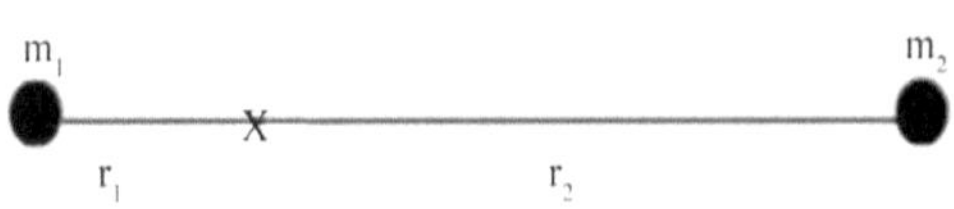


**Fig. 15.6.** Gravitational attraction between two bodies

Then the force balance on these two 'point' masses is given as:

$$F_g = \frac{Gm_1m_2}{r^2} \rightarrow F_g = \frac{m_1v_1^2}{r_1} = \frac{m_2v_2^2}{r_2}$$

The first is the gravitational force of attraction (G is a universal constant) and the second is the orbital balance reflecting a balance between centrifugal force and gravity, which must be in balance for a stable orbit. The total distance between the two masses having velocities v is denoted by r. If the rate of rotation (angular velocity) is given as ω, then the above equations can be rewritten as:

$$\frac{m_1v_1^2}{r_1} = m_1\omega^2 r_1 = F_g \; ; \frac{m_2v_2^2}{r_2} = m_2\omega^2 r_2 = F_g$$

$\therefore m_1 r_1 = m_2 r_2$

The second allows us to calculate the center of mass, where the center of rotation lies knowing the masses. For the case of the earth and moon, we know:

$m_e = 5.983 \times 10^{24}$ kg

$m_m = 1.228 \times 10^{-2}\, m_e$

$r_{em} = 3.8 \times 10^8$ m

Where $m_e$ and $m_m$ are the masses of earth and moon respectively and $r_{em}$ is the mean distance between earth and moon.

With these values, we can calculate the distance from the center of the earth to the center of rotation of the earth-moon binary system as 4500 km, which falls within the earth (since the earth's mean radius is 6371 km as detailed below in Fig.15.7. To know the real attraction and the associated tide producing force, computation should be made using three body system of earth, moon and sun, considering the rotation and revolution.

### 15.2.1. Common axis of revolution of earth-moon system

The gravitational attraction of the sun and moon on the oceans and the equal and opposite centrifugal forces owing to the rotation of the earth–sun–moon system are the primary tide generating forces. Although the sun's mass is approximately $2.7 \times 10^7$ times that of the moon, the closer proximity of the moon to the earth results in the moon having about twice the sun's gravitational force on the oceans. The resulting tide is a long period wave generated by these gravitational and centrifugal forces. Consequently, even in the deepest areas of the ocean the tide is a shallow water wave.

Consider an idealized earth–moon system with the earth covered by a uniform layer of water and the two bodies revolving around a common axis of rotation falls within the earth that is about 2900 miles from the earth's axis at the center (Figure

15.7). The gravitational forces $F_g$ and the centrifugal forces Fc are in balance at the center of mass of each body. But an element of water located on the side of the earth farthest from the moon would have $F_c > F_g$ and a resulting net outward force. On the half of the earth closest to the moon $F_g > F_c$ which also causes a net outward force. The result is two bulges as shown or two high and two low tides each day as the earth rotates. These bulges are the tide waves that propagate as shallow water waves.

While the earth rotates once (relative to the position of the sun), the moon makes 1/29.5 of a revolution (a lunar month is 29.5 solar days). Thus, the principal lunar tidal period $M_2 = 12.0\left(1 + \frac{1}{29.5}\right) = 12.42\,hours$

The principal solar tidal period $S_2$ =12:00 hours. Since the lunar force dominates, high and low tides progress 0.84 hours (about 50 min) each day.

The planes of the moon's and sun's rotation relative to the earth are not in line and change with regard to each other as time passes. Also, the orbits of the sun and moon are elliptical rather than circular. These plus other factors make the net tide generating force very complex.

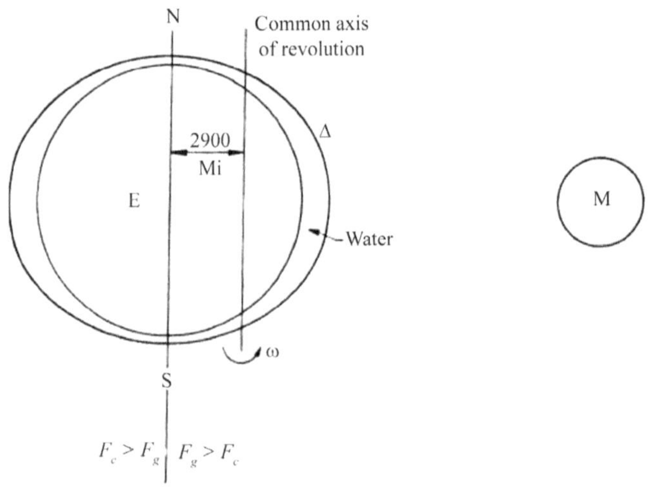


**Figure 15.7.** Tide generation—idealized earth/moon system (Ippen and Harleman, 1966; Parker, 2007)

### 15.2.2. Derivation of tide producing force

The earth and the moon form a single system, mutually revolving around a common center of mass due to the proximity when compared to the sun with a period of 27.3 days. At the centers of the earth and the moon there is a balance between the gravitational attraction and the centrifugal force.

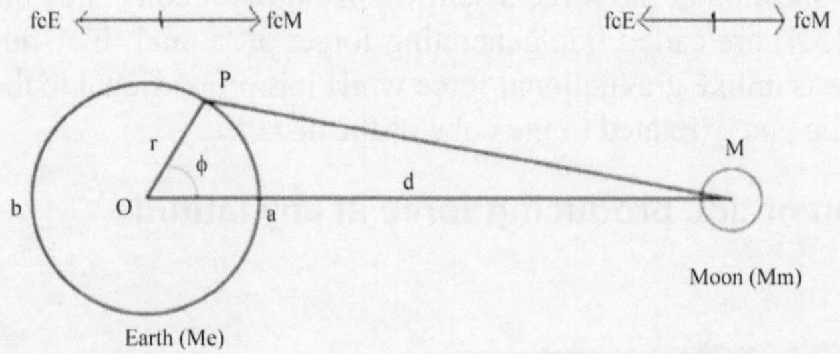


**Fig. 15.8.** Simple case of tide generating force

To begin with let us consider a simple case that the two points 'a' and 'b' are on the line joining the centers of earth and the moon (Fig.15.8). Let μ and 'm' are the masses of the particle on the earth and the moon and let $F_{ce}$ and $F_{cm}$ are the centrifugal forces respectively. Let 'd' is the distance between the centers of earth and moon.

As there is a balance between the gravitational attraction and the centrifugal acceleration of both with respect to a common axis of rotation

$$F_{ce} = F_{cm} = \frac{G\mu m}{d^2} \tag{15.1}$$

The tide generating force is defined as the difference between the gravitational force minus the centrifugal force.

Since according to Newton the force of gravitation varies directly with the product of masses and inversely with the square of the distance between them we can write the gravitational attraction at point 'a' on the surface of the earth on the particle 'μ' as

$$F_a = \frac{G\mu m}{(d-r)^2} \text{ towards the moon.} \tag{15.2}$$

Where 'r' is the radius of the earth.

The net (tide generating) force at point 'a' is

$$F_a - F_{ce} = G\mu m\left[\frac{1}{(d-r)^2} - \frac{1}{d^2}\right]$$

Since 'r' is very small compared with 'd' ( d = 60r), the squares of 'r' can be neglected. Then the equation turns out to be

$$F_a - F_{ce} = \frac{Gm\mu.2r}{d^3} \tag{15.3}$$

Which acts towards the moon as $F_a > F_{ce}$

Similarly we can write at point 'b' as

$$F_b - F_{ce} = -\frac{Gm\mu.2r}{d^3} \tag{15.4}$$

Which acts towards the (point b) as $F_c > F_b$. This is because in the denominator we get (d + r) in place of (d - r) in the equation (15.2) at point 'b'. The negative sign in equation (15.4) shows that the force acts away from the moon. Thus the equations (15.3) and (15.4) are called tide generating forces at 'a' and 'b'. Thus the tide generating force is unlike gravitational force while it is proportional to the product of masses it is inversely related to the cube of the distance.

### 15.2.3. Computation of tide producing force at any latitude

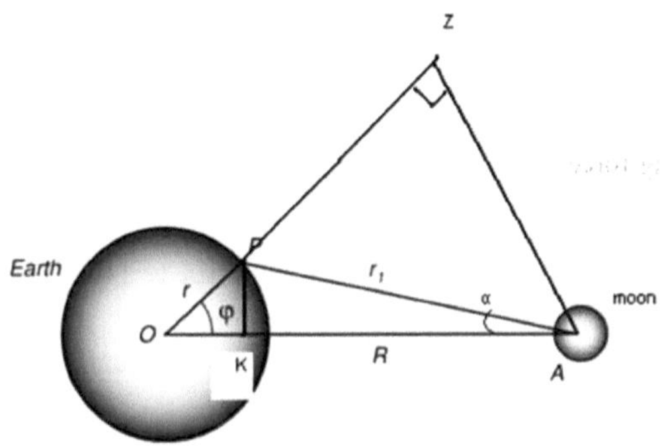


**Fig. 15.9.** Computation of Tide producing force at any latitude

From the Figure 15.9, let 'P' be any point on the earth's surface at which tide producing force is to be calculated. Let O and A denote the centers of the earth and the moon respectively. Other abbreviations are as mentioned in the figure.

Attraction of moon at point O of the particle 'μ' on the earth (OA) $= \dfrac{G\mu m}{R^2}$ (15.5)

Attraction of moon at point 'P' on mass 'μ' (PA) $= \dfrac{G\mu m}{r_1^2}$ (15.6)

Where 'R' and '$r_1$, are the distances of points O and P from A.

Let each of these forces are resolved into a vertical component along the radius OP and a horizontal component perpendicular to it in the plane OPA. Denoting the angles AOP and OAP as φ and α respectively, we can write equations (15.5) and (15.6) as

Attraction at O in the direction OP (radial or vertical component)

$$= \frac{G\mu m}{R^2}\cos\phi \qquad (15.7)$$

Attraction at O perpendicular to OP (tangential or horizontal component

$$= \frac{G\mu m}{R^2}\mathrm{Sin}\,\phi \qquad (15.8)$$

Attraction at P in the direction P to Z(where 'Z' is extension of OP) =

(Vertical component, PZ) = PM cos (α + φ) $= \dfrac{G\mu m}{r_1^2}\mathrm{Cos}(\alpha + \phi)$ (15.9)

Attraction at P perpendicular to the direction P to Z =

(Horizontal component PZ ) = PM Sin (α + φ) = $\frac{G\mu m}{r_1^2}\text{Sin}(\alpha+\phi)$ (15.10)

The attraction at point O, center of the earth of equations (15.7) and (15.8) shows the centrifugal forces of the earth. As the tide generating force is the difference between gravitational attraction and the centrifugal forces, we get tide generating forces by subtracting equation (15.7) from (15.9) for vertical component and equation (15.8) from (15.10) for horizontal component.

$$F_V = Gm\mu\left[\frac{\text{Cos}(\alpha+\phi)}{r_1^2}-\frac{\text{Cos}\,\phi}{R^2}\right] \quad (15.11)$$

$$F_H = Gm\mu\left[\frac{\text{Sin}(\alpha+\phi)}{r_1^2}-\frac{\text{Sin}\,\phi}{R^2}\right] \quad (15.12)$$

Since the angle α is very small for all practical purposes we can write PA = KA (Fig. 15.9) or if we substitute the projection of $r_1$ on R, we can write KA = R – r cosφ, where r is the radius of the earth taken as unity as it is very small compared to R.

Further from right angled triangle PZA,

$$\text{Cos}\,(\alpha+\varphi) = \frac{PZ}{PA} = \frac{OZ-OP}{KA} = \frac{R\cos\phi - r}{R - r\cos\phi} \quad (15.13)$$

Similarly, $\sin(\alpha+\phi) = \frac{ZO}{PO} = \frac{R\sin\phi}{R-r\cos\phi}$ (15.14)

Substituting equation (15.13) in (15.11) we get,

$$F_V\ Gm\mu\left[\frac{R\cos\phi - r}{(R-r\cos\phi)^3}-\frac{\text{Cos}\,\phi}{R^2}\right] =$$

$$\frac{Gm\mu}{R^3}\left[\frac{R\cos\phi - r}{\left(1-\frac{r^3}{R^3}\cos^3\phi - 3\frac{r}{R}\cos\phi + 3\frac{r^2}{R^2}\cos^2\phi\right)}-\frac{\text{Cos}\,\phi}{1}\right]$$

Neglecting the higher order terms of $\frac{r}{R}$ as '' is negligibly small compared to 'R'

$F_V = \frac{Gm\mu}{R^3}\left[\frac{r(3\cos^2\phi - 1)}{\left(1-3\frac{r}{R}\cos\phi\right)}\right]$ neglecting again $\frac{r}{R}$ term we get

$$= \frac{Gm\mu}{R^3}\left[\frac{r(3\cos^2\phi-1)}{(1)}\right]$$

$$F_V = \frac{Gm\mu}{R^3} 3\left(\cos^2\phi - \frac{1}{3}\right) \quad (\because \text{ r is taken as unity}) \tag{15.15}$$

Similarly substituting equation (15.14) in (15.12) and simplifying we get

$$F_H = \frac{Gm\mu}{R^3}\frac{3}{2}\sin 2\phi \tag{15.16}$$

These formulae (eqns 15.15 and 15.16) show that the tide producing force of the moon at any point on the earth depends on the zenith angle of the moon (φ) and the moon's distance from the earth (R). As the R value at perigee (357,000 km) and apogee (407,000) is different tidal forcing is not constant always.

The vertical component $F_V$ attains its maximum value at φ = 0 and 180° when the moon is in zenith and Nadir points ( from equation 15.15).

The horizontal component $F_H$ attains its maximum value when 2φ = 90° or 270° or φ = 45⁰ or 135° ( from equation 15.16).

The vertical component of tide producing force ($F_V$) is in the same direction of gravitation and therefore it doesn't have any effect on the water particle and so is neglected. Whereas the horizontal component of the tide producing force ($F_H$) will not be affected by gravity and so it will tend to drag the water particle in the ocean horizontally. This is called the tractive force.

The tide producing force (thick arrow in Fig.15.10) is the resultant force between gravitational and centrifugal forces between earth and moon.

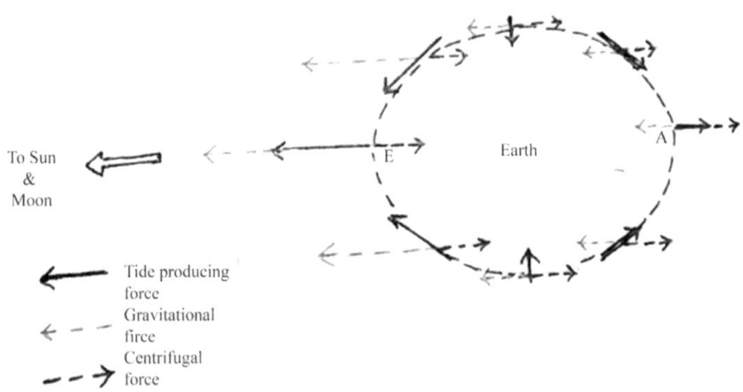


**Fig. 15.10.** Tide producing force is the resultant of gravitational and centrifugal forces between earth and moon

From Fig. 15.10 above it is clear that the magnitude of the gravitational force exerted by the moon on the earth is not the same at all points on earth, because all these points are not at the same distance from the moon. While the gravitational force of the moon is largest at point 'E', it is the weakest at point 'A' as Point E is closest and A is farthest from the moon. So while the gravitational force is dominant in one half of the earth facing the moon, the centrifugal force is dominant than gravitational force in the other half of the earth away from the moon. In addition, the direction of the moon's gravitational force at all points on the earth is directed towards the center of the moon. So only on the equatorial diameter the gravitational force (dashed arrow in Fig.15.10) is straight and at other places it is inclined. As the tide producing force (hatched thick arrow in Fig.15.10) is the difference between gravitational and centrifugal forces, it also is directed in the same way as that of gravitational force.

Similarly Sun also acts in the same way on the earth's surface. But as the Sun is far away its influence is relatively small when compared to that of the moon.

Fig.15.11a shows the distribution of horizontal component of the tide producing force over the earth's surface when the moon is on the equatorial plane. In Fig.15.11b the water particles facing the moon are dragged towards it and the other particles (shaded area) are dragged away from it. On any latitudinal circle ($A_1$ $A_5$) the variation of tractive force is significant (Fig.15.11b). At point $A_2$ facing the moon it is directed south. When the point has moved to point $A_4$ the force is zero and when it has moved to $A_5$ the force is again equal and opposite. This explains why two high tides ($A_2$ and $A_5$) occur and two low tides ($A_4$ and its opposite) occur in one full rotation of the earth.

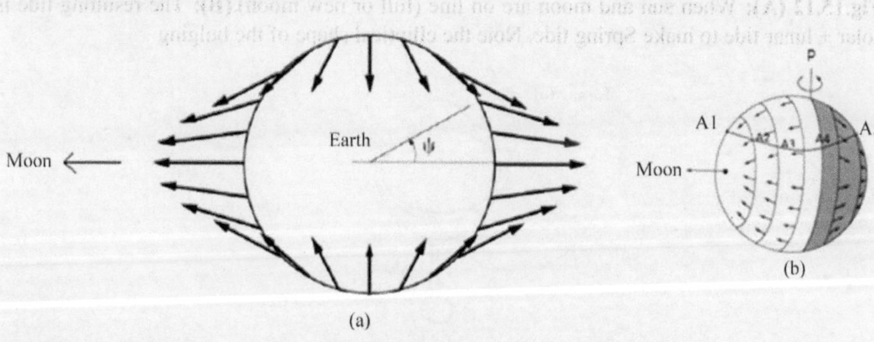


**Fig.15.11.a)** Distribution of the total tide generating forces for a cross section of the earth. **b)** the horizontal component of the tide generating force at the earth's surface

Though the tidal forces are weak compared to the local gravity (1part in 9 million), the horizontal component of the tide producing force can force the water particle to move laterally. This lateral movement of the water parcel will deform into an elliptical shape which under equilibrium theory will cause a bulge on either side of the earth of about 55 centimeters due to the lunar forcing as shown in Fig.15.12&15.13A.below

The sun produces a similar effect but relatively small tide producing force compared to the moon due to far off distance (Fig.15.12A). When the two forces of sun and moon coincide it produces a maximum tide generating force and it is called spring tide. This happens twice during a period of 29.5 days. When these two bodies act at right angles the tide generating force is least. This least (minimum) force produces neap tide. This also happens twice during 29.5 days.

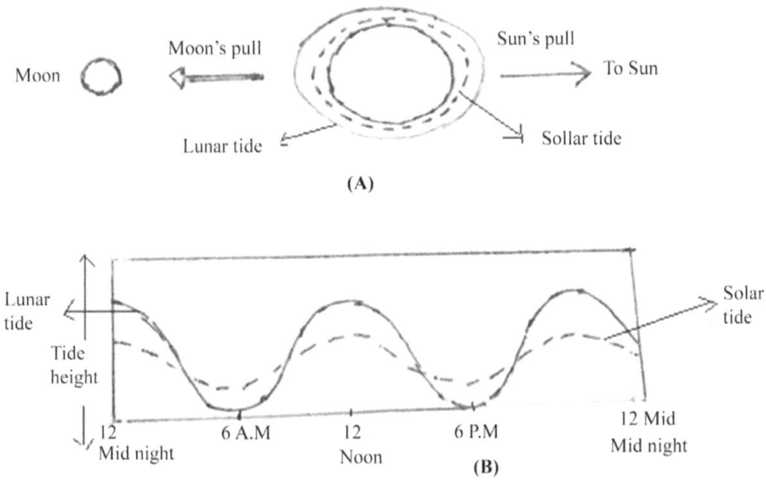


**Fig.15.12 (A):** When sun and moon are on line (full or new moon).**(B):** The resulting tide is solar + lunar tide to make Spring tide. Note the elliptical shape of the bulging

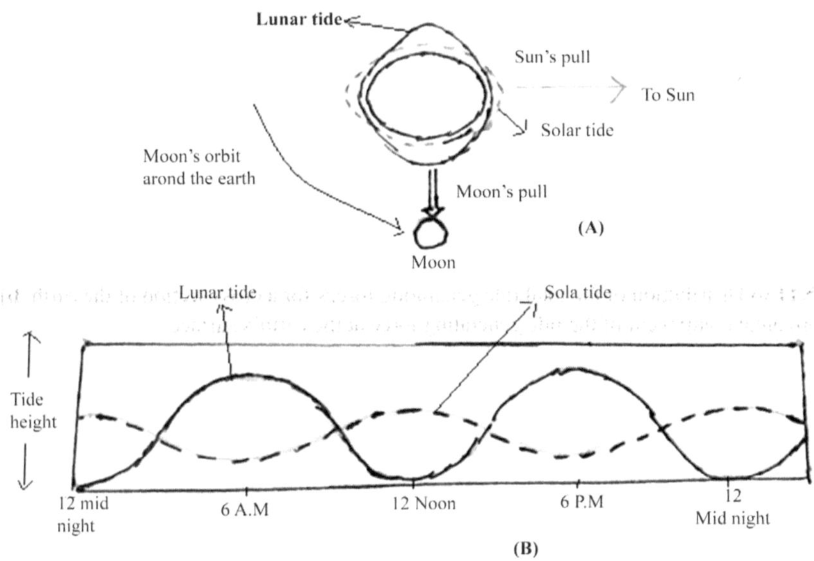


**Fig.15.13. (A):** when the sun and moon are at right angles, **(B):** The resulting tide is lunar tide minus solar tide (neap tide)

When the sun and moon are on line during new or full moon, the solar and lunar tides will reinforce each other which cause highest tides as shown in Fig.15.12B (Spring tide). Below denotes the two sinusoidal fluctuations of moon and sun with an amplitude 1(0.5) meter and periods of 12.42 (12) hours representing the semidiurnal tide of lunar (solar) origin respectively. The pattern in the Fig15.12B as a result has a maximum (crest) or spring tide when both are in phase.

Similarly when sun and moon are 90 or 270 degrees, minimum (trough) or neap tide occurs (Fig.15.13A &B), which are out of phase during 14 days period. The time difference between two successive spring or neap tides is approximately 14 days because each quarter is 7 days as mentioned earlier (Fig.15.2) which can be seen in the tidal curve of Visakhapatnam (Fig.15.14).

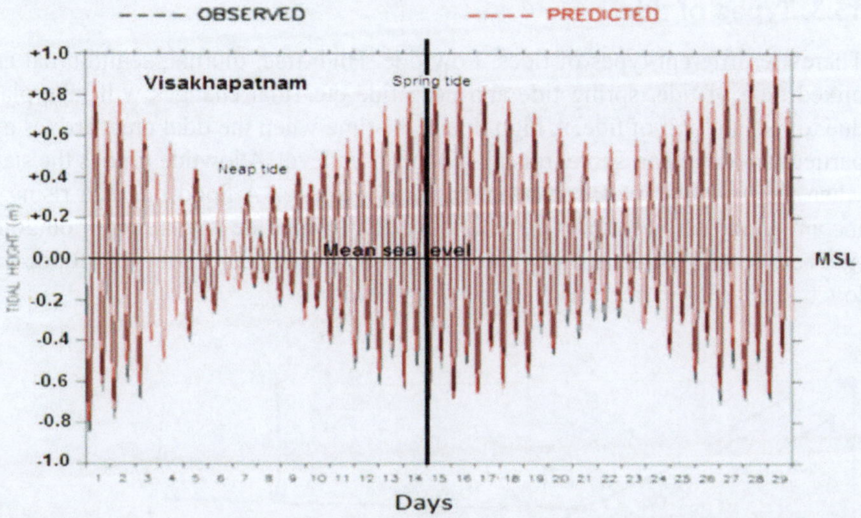


**Fig.15.14.** A typical 30-day semi diurnal (period 12.5 hours) tidal record showing oscillations in water level with a duration of about 7 day occurrence of spring to neap tide at Visakhapatnam https://incois.gov.in/documents/ITCOocean/OITC018/L13-Srinivas%20K_TIDES.pdf

At latitude $45^0$ N or S, the speed at which the equilibrium tide propagates as the earth revolves is quite large, over 300 m $s^{-1}$. This is faster than the phase speed of a surface gravity wave. However, friction will cause the earth to drag the tidal bulge in the direction of rotation causing a torque that acts to slow down the rotation of the earth. A further interesting fact is that the solid earth (lithosphere) also has tides and these lithospheric tides can reduce the ocean tides only by about 30%.

### 15.2.4. Age of the tide

There is usually a delay between the exact occurrence of new moon or full moon and the occurrence of spring tides. That delay is called the *age of the tide,* and is often on the order of a day or two, but it can be negative (that is, spring tide in some locations, due to hydrodynamics, can come before new moon or full moon).

The age of tide is the time required for the M2 and S2 tidal constituents to arrive at a phase agreement. It can therefore be obtained by dividing the difference in their epochs (epochs can be local or Greenwich) by the difference in their frequencies (i.e., their angular speeds). The resulting formula for the age of the tide is thus

Age of the tide (in hours) = $0.984(S_2^0 - M_2^0)$

## 15.3. Types of tide

There are different types of tides. Low tide, High tide, diurnal, semidiurnal and mixed type of tide, spring tide and neap tide etc.Tidal changes will take place due to rise and fall of tide. A high tide is the time when the tidal crest arrives at a particular location on shore, raising the local sea level. A lowtide time is the stage when the trough arrives, lowering the local sea level. A flood tide is a rising or incoming tide between low tide and high tide. An ebb tide is a falling or outgoing tide between high tide and low tide. The vertical distance between high tide and low tide is the tidal range as shown in Fig.15.15.

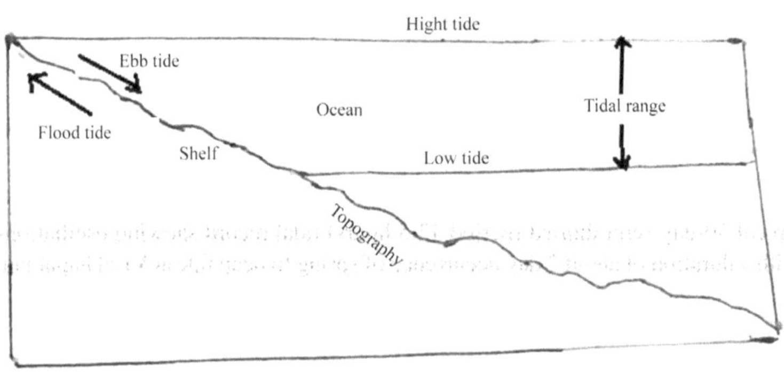


**Fig. 15.15.** Flood and ebb causes tidal range

With regard to the number of times of occurrences of tide they are classified as diurnal, semidiurnal and mixed type. If one High tide and one Low tide occur in a day (24 hours) it is called diurnal tide and if two high and two low tides occur in a day then they are called semidiurnal tides (Fig.15.16). If both are mixed they are called mixed tides.

### Harmonic method of classifying type of tide at a location

The important tidal constituents($M_2$, $S_2$, $K_1$ & $O_1$) can also be used to describe the type of tide (ie., semidiurnal, diurnal, or mixed) using tidal form number (TFN) which is given as:

$$TFN = \frac{K_1 + O_1}{M_2 + S_2}$$

The classification is as below:

if TFN < 0.25:semi diurnal

TFN: 0.25-1.5:mixed mainly semi diurnal

TFN: 1.5-3.0:mixed mainly diurnal

TFN:>3.0: diurnal

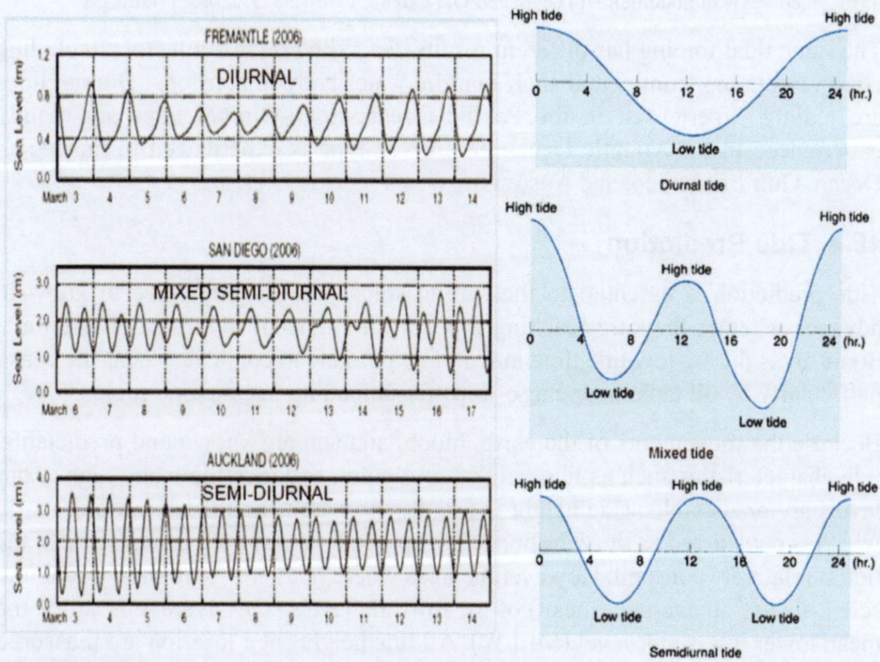


**Fig.15.16.** Shapes of semidiurnal and diurnal tidal curves in a day
https://incois.gov.in/documents/ITCOocean/OITC018/L13-Srinivas%20K_TIDES.pdf

The numbers along each coast indicate the type of tide occurs in that area. The different types of tides namely, Semidiurnal tides(1), diurnal tides (2) and mixed tides (3) occur as indicated in the Fig.15.17.

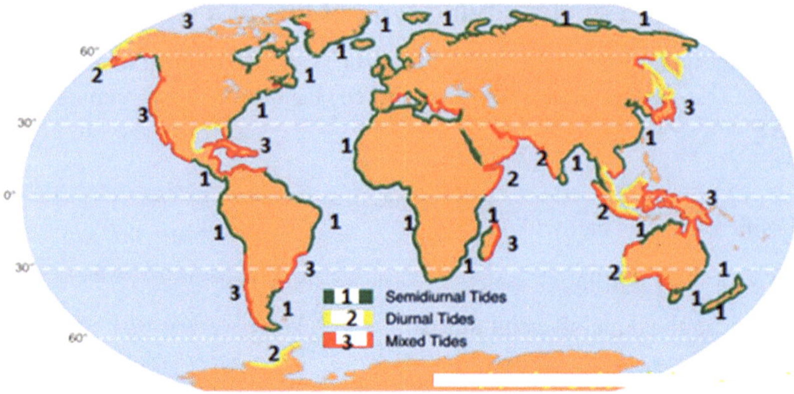


**Fig. 15.17.** Occurrence of different types of tides at each locality in the global oceans
https://incois.gov.in/documents/ITCOocean/OITC018/L13-Srinivas%20K_TIDES.pdf

The same tidal forcing has different results depending on many factors, including coast orientation, continental shelf margin, water body dimensions. Diurnal tides are mainly experienced in the Pacific Ocean. Semi-Diurnal tides are mainly experienced in the Atlantic Ocean and mixed type is experienced in the Indian Ocean, Gulf of Mexico, and Australia.

## 15.4. Tide Prediction

Tide prediction is essential to the sailing ships because they have to know in advance whether they are reaching a port at the time of low or high tide time. Some areas during low tide time may not be possible to come very near the coast, particularly by oil tankers or cargo ships for unloading in estuarine regions.

Because the movements of the earth, moon, and sun are known and predictable, tide characteristics such as tide height, zero point, and mean low and high-water levels are predictable. Tide height is usually measured from the low-water mark, which is considered as the zero point. This procedure is simple in areas where the tide is relatively constant. However, in areas where tides are continually changing, scientists use an average (mean) of the low tide levels. This average is called the mean lower low-water level (MLLW). All tide heights at a location are measured from the zero point. When the low-water level is lower than the zero point, tide height is measured with a negative number and is called a minus tide.

Tide predictions are based on harmonic analysis of tidal data. The harmonic analysis of tides is based upon an assumption that the rise and fall of the tide at any locality can be expressed mathematically by the sum of a series of harmonic terms having relation with movement of the astronomical bodies. The general harmonic equation of the height of (H) tide can be written as :

$$H = H_0 + A\cos(at+\alpha) + B\cos(bt+\beta) + C\cos(ct+\gamma) + etc. \tag{15.17}$$

where $H_0$ is the height of mean water level. Each cosine term in the equation is known as a tidal constituent. The coefficients A,B and C are the amplitudes of the constituents and are derived from observed or measured tidal data at each locality. The expression in parenthesis is called its phase angle which varies with time. The coefficient't' (of a,b,c) represents the rate of change in the phase and is called the speed of the constituent and is expressed in degrees per hour.

The purpose of tide analysis is to determine the amplitude and phase (the so-called tidal harmonic constants) of the individual cosine waves, each of which represents a tidal constituent identified by its period. Finding the tidal harmonic constants at a place allows one to predict tides at that place.

Doodson (1921) expanded the above harmonic equation (15.17) in a Fourier series using the cleverly chosen frequencies of the following table

$f = n_1 f_1 + n_2 f_2 + n_3 f_3 + n_4 f_4 + n_5 f_5 + n_6 f_6$

**Fundamental Tidal Frequencies**

| | Frequency (Degree/hour) | | Period | Source |
|---|---|---|---|---|
| $f_1$ | 14.49205211 | 1 | Lunar day | Local mean lunar time |
| $f_2$ | 0.54901653 | 1 | Month | Moon's mean longitude |
| $f_3$ | 0.04106864 | 1 | Year | Sun's mean longitude |
| $f_4$ | 0.00464184 | 8.847 | Years | Longitude of moon's perigee |
| $f_5$ | -0.00220641 | 18.613 | Years | Longitude of moon's ascending node |
| $f_6$ | 0.00000196 | 20.940 | Years | Longitude of sun's perigee |

The following constituents are independent of the place and are derived from astronomical data concerning the relative movements of the sun, moon and the earth:

$M_2$-Principal lunar semidiurnal constituent

$S_2$–Principal solar semidiurnal constituent

$N_2$-Larger Lunar elliptic semidiurnal constituent

$K_1$-Luni-solar declinational diurnal constituent

$O_1$-Lunar declinational diurnal constituent

$M_4$-First over tide of $M_2$ constituent (speed: $2 \times M_2$ speed)

$M_6$-Second over tide of $M_2$ constituent (speed: $3 \times M_2$ speed)

$S_4$-First over tide of $S_2$ constituent (speed: $2 \times S_2$ speed)

$MS_4$- A compound tide of $M_2$ and $S_2$ (speed: $M_2 + S_2$ speed)

The following table gives the period and speeds of each constituent

| Tidal Species | Name | $n_1$ | $n_2$ | $n_3$ | $n_4$ | $n_5$ | Equilibrium Amplitude † (m) | Period (hr) |
|---|---|---|---|---|---|---|---|---|
| **Semidiurnal** | | | | $n_1 = 2$ | | | | |
| Principal lunar | $M_2$ | 2 | 0 | 0 | 0 | 0 | 0.242334 | 12.4206 |
| Principal Solar | $S_2$ | 2 | 2 | -2 | 0 | 0 | 0.112841 | 12.0000 |
| Lunar elliptic | $N_2$ | 2 | -1 | 0 | 1 | 0 | 0.046398 | 12.6584 |
| Lunisolar | $K_2$ | 2 | 2 | 0 | 0 | 0 | 0.030704 | 11.9673 |
| **Diurnal** | | | | $n_1=1$ | | | | |
| Lunisolar | $K_1$ | 1 | 1 | 0 | 0 | 0 | 0.141565 | 23.9344 |
| Principal lunar | $O_1$ | 1 | -1 | 0 | 0 | 0 | 0.100514 | 25.8194 |
| Principal solar | $P_1$ | 1 | 1 | -2 | 0 | 0 | 0.046843 | 24.0659 |
| Elliptic lunar | $Q_1$ | 1 | -2 | 0 | 1 | 0 | 0.019256 | 26.8684 |
| **Long Period** | | | | $n_1=0$ | | | | |
| Fortnightly | Mf | 0 | 2 | 0 | 0 | 0 | 0.041742 | 327.85 |
| Monthly | Mm | 0 | 1 | 0 | -1 | 0 | 0.022026 | 661.31 |
| Semiannual | Ssa | 0 | 0 | 2 | 0 | 0 | 0.019446 | 4383.05 |
| †Amplitudes from Apel (1987) | | | | | | | | |

The prediction of tide at a given locality by means of the above formula involves a good deal of calculation. In order to make the computation easier Lord Kelvin devised a machine in 1872 called tide predicting machine (Fig.15.24A).

If tides in the ocean were in equilibrium with the tidal potential, tidal prediction would have been easy, but tides are far from equilibrium, because: i) the shallow-water wave (which is the tide) cannot move fast enough to keep up with sun and moon. On the equator, the tide would need to propagate around the world in one day. This requires a wave speed of around 460 m/s, which is only possible in an ocean of 22 km deep. ii) the continents interrupt the propagation of the wave

## 15.4.1. Prediction of tides in Ports where tides can be measured

i) The Harmonic Method typically uses 19 years of data from a coastal tide gauge from which the amplitude and phase of each tidal constituent (the tidal harmonics) in the tide-gage record are calculated.

ii) The Response Method, developed by Munk and Cartwright (1966), calculates the relationship between the observed tide at some point and the tidal potential. The relationship is the spectral admittance between the major tidal constituents and the tidal potential at each station: $Zf=G(f)/H(f)$, where $G(f)$and $H(f)$ are the Fourier transforms of the potential and the tide gage data, and f is frequency. The admittance is assumed to be a slowly varying function of frequency so that the admittance of the major constituents can be used for determining the response at nearby frequencies.

## 15.4.2. Prediction of tides in deep water where tides are measured by satellite altimeters

i) Prediction Using Hydrodynamic Theory

ii) Altimetry Plus Response Method

iii) Altimetry Plus Numerical Models

The declination angle (of sun or moon) and the proximity (of sun or moon) give rise to the various frequencies and magnitudes, known as the "tidal species" which can easily be done in TASK –2000 software as shown in Fig.15.18.

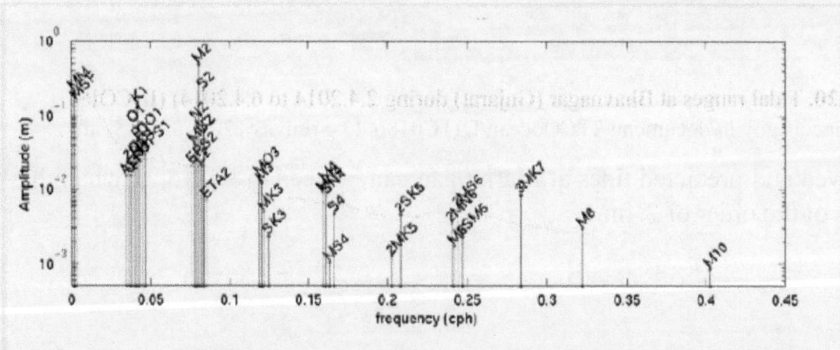


**Fig.15.18.** Tidal constituents (species) according to TASK-2000 software (INCOIS). https://incois.gov.in/documents/ITCOocean/OITC018/L13-Srinivas%20K_TIDES.pdf

Sum of individual tidal components gives the resultant predicted sea level using TASK 2000 software as given in the Fig.15.19 (up to 115 components are available in TASK 2000 software). Note the same tidal forcing has different results depending on many factors, including coast orientation, continental shelf margin, water body dimensions etc.

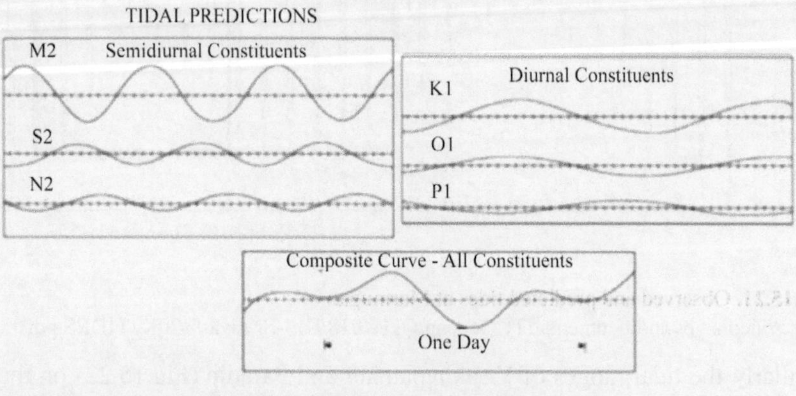


**Fig. 15.19.** Tidal curves using TASK 2000 software (INCOIS) https://incois.gov.in/documents/ITCOocean/OITC018/L13-Srinivas%20K_TIDES.pdf

### 15.4.3. Tidal prediction along the Indian coasts

While Bhavnagar experiences highest ranges of tides of 10 m as seen in table, other stations are 2.4m and less than that as in Fig.15.20.

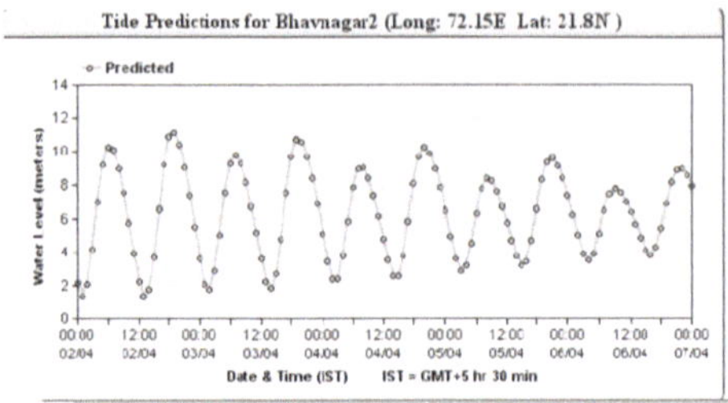


| High Tide | | Low Tide | |
|---|---|---|---|
| Time (IST) | Level (m) | Time (IST) | Level(m |
| 02-04-2014 06:18 AM | 10.34 | 02-04-2014 01:03 AM | 1.34 |
| 02-04-2014 06:43 PM | 11.17 | 02-04-2014 01:13 PM | 1.29 |
| 03-04-2014 06:56 AM | 9.80 | 03-04-2014 01:45 AM | 1.70 |
| 03-04-2014 07:22 PM | 10.79 | 03-04-2014 01:52 PM | 1.79 |
| 04-04-2014 07:34 AM | 9.15 | 04-04-2014 02:28 AM | 2.23 |
| 04-04-2014 08:01 PM | 10.25 | 04-04-2014 02:31 PM | 2.43 |
| 05-04-2014 08:14 AM | 8.45 | 05-04-2014 03:13 AM | 2.86 |
| 05-04-2014 08:45 PM | 9.62 | 05-04-2014 03:12 PM | 3.15 |
| 06-04-2014 09:00 AM | 7.77 | 06-04-2014 04:01 AM | 3.51 |
| 06-04-2014 09:36 PM | 8.98 | 06-04-2014 03:57 PM | 3.86 |

**Fig.15.20.** Tidal ranges at Bhavnagar (Gujarat) during 2.4.2014 to 6.4.2014) (INCOIS) https://incois.gov.in/documents/ITCOocean/OITC018/L13-Srinivas%20K_TIDES.pdf

Observed and predicted tides at Marmugao can be seen in Fig.15.21 which have ranges of the order of 2.4m.

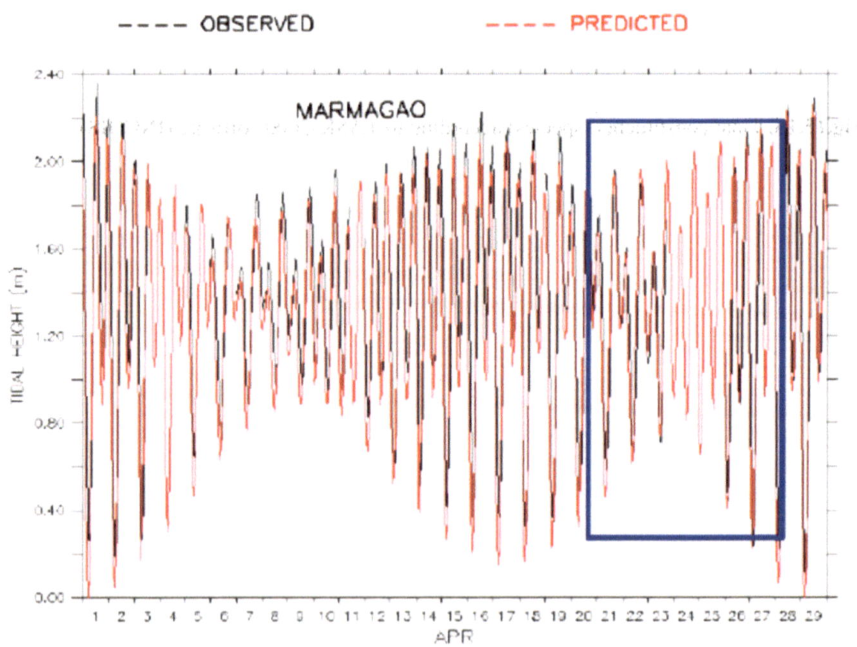


**Fig.15.21.** Observed and predicted tides at Marmugao https://incois.gov.in/documents/ITCOocean/OITC018/L13-Srinivas%20K_TIDES.pdf

Similarly the tidal ranges of Visakhapatnam and Paradip (Fig.15.22) on the east coast of India are very less (< 0.8 m) compared to Bhavnagar (10m) on the west coast of India and they are almost same.

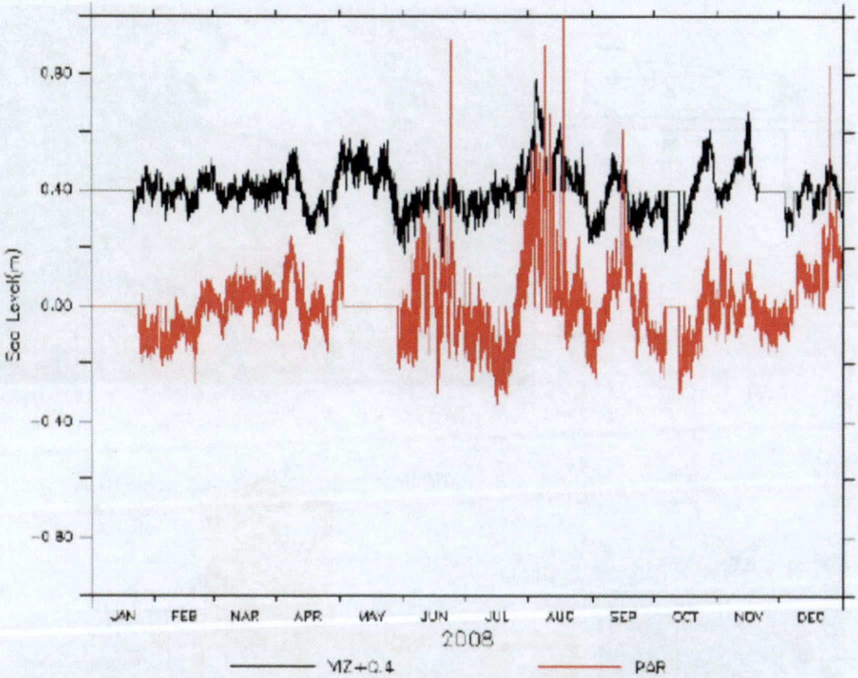


**Fig.15.22.** Tidal ranges of Visakhapatnam(VIZ-black)) and Paradip (PAR-Red)
https://incois.gov.in/documents/ITCOocean/OITC018/L13-Srinivas%20K_TIDES.pdf

## 15.5. Measurement of tide

Tides are measured using an instrument called tide gauge as shown in Fig..15.23

The tide gauge shown in Fig.15.23 consists of a float (F), which operates in the float well (R), to which the slow moving tide has free access. The more rapid waves caused by winds are filtered out by means of the relatively small size of the pipe (canal) (O) connecting the well to the ocean. The rising and falling of the float (F) turns a screw on the gauge which moves a pen (P) to and fro across a wide strip of paper on a drum (D) which is moved forward by clockwork. The combined motion of pen and paper gives a continuous graph showing the rise and fall of the tide. Generally this entire assembly will be housed in a shelter and built on coastal back waters as shown in Fig.15.23B. This shelter is built on a pier (a bridge type in the ocean bank). The graph drawn by the tide gauge is monthly and so every month the chart is to be changed. The above described methods are older as they used mechanical floats and recorders. The modern monitoring stations use advanced acoustics and electronics. To determine water level, modern recorders send an audio signal down a sounding tube and measure the time it takes for the reflected signal to travel back from the water's surface through satellite or Radar (Fig.15.23 C and D).

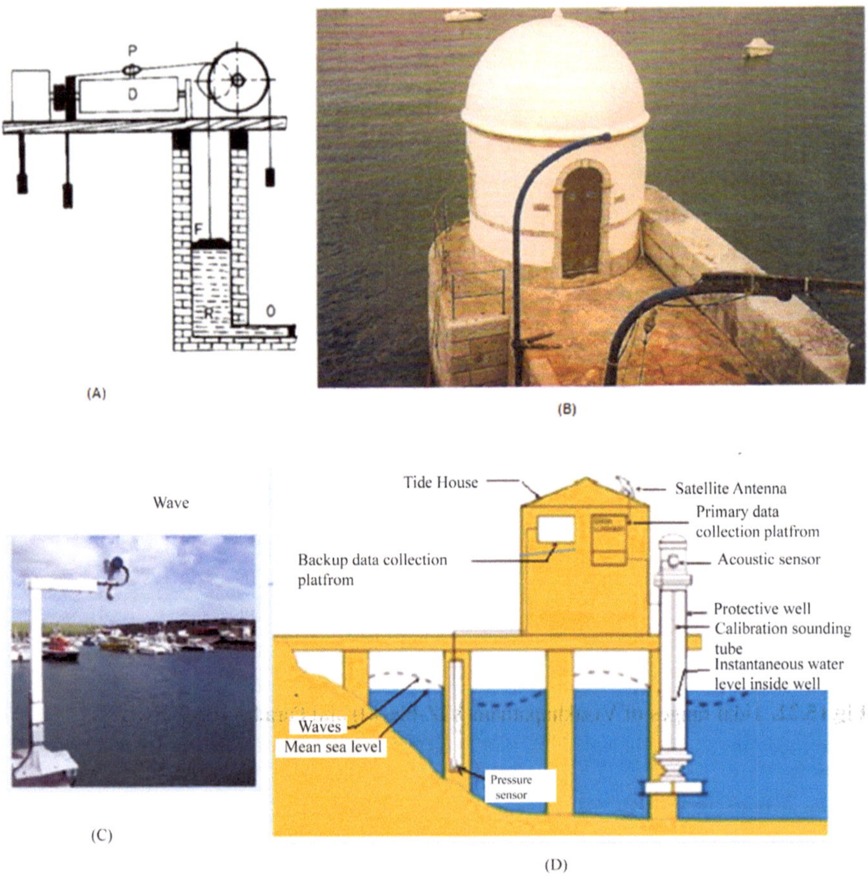


**Fig. 15.23.** Manuel Tide gauge (A) and its housing in the backwaters of the ocean (B),Non-Contact Water Level Monitoring: Radar tide gauge (C) and Satellite connectivity tide gauge (D).

## 15.5.1. Tide predicting machine

The prediction of tide at a given locality by means of the above formula involves a good deal of calculation. In order to make the computation easier in the pre-computer era Thomson came up with an ingenious way to automate tide predictions using harmonic tidal constituents. He invented a mechanical analog tide predicting machine, a machine made up of dozens of gears and pulleys, with a wire running over all the pulleys and then connected to a pen touching a moving roll of paper. Each tidal constituent was represented by one gear with a pin and yoke arrangement that turned the rotating motion of the gear into an up and down motion, moving a pulley up and down and thus providing that constituent's contribution to the tide curve being drawn by the pen on the moving roll of paper (see Figure 15.24A). This machine built to predict the tides was made out of brass and was finely crafted. Thomson's first tide predicting machine was built in London in 1872 by the Légé Engineering Company and summed the contributions of the ten most important tidal constituents.

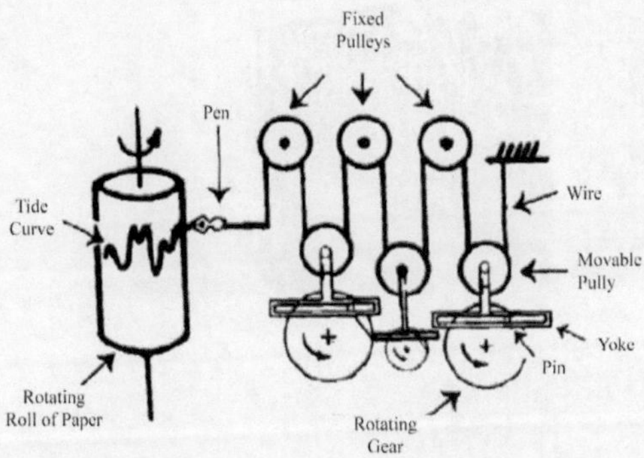


**Fig. 15.24A.** Gear and pulley system of an early analog tide predicting machine. Each gear and pulley combination represents one tidal harmonic constituent. The wire running over all the pulleys sums the motions and moves a pen on a moving roll of paper to draw the tide curve. (Parker 2007)

Later Lord Kelvin devised a refined machine in 1872 (Fig.15.24Bii). By means of this machine the time of occurrence of high and low waters at a place can be calculated and printed in a book in every year (tide tables) and can be distributed to sea ports, navigators, sailors, shipping authorities and researchers. However, this gives only astronomical tide and the actual tide at a locality is little different because of the influence of basin topography and bottom stress which varies time to time and is not included in astronomical tide. So the tide also should be measured with the help of a tide-gauge (Fig.15.23). The difference between the observed tide and the predicted tide is called 'the Residual Tide' (Fig.15.25). If the residual tide is positive, it reflects on the storm surge as that is caused due to some terrestrial force like wind, barometric pressure or density.

Using the tide predicting machine shown in Fig.15.24B, tides used to be predicted in advance during 1950s. Usually yearly tidal heights and times of occurrence at each important coastal place are predicted and published in a book. This information is highly useful for navigators in the ocean. It uses a harmonic equation for prediction. Each circle shown in the machine is a key for each partial tide $M_2$, $S_2$, $K_1$, $O_1$ etc. Depending upon the speed of each partial tide, each pulley is rotated that many times and finally a chart will be drawn by the pen and clock drum attached to the machine. The curve drawn on the chart by the machine is the tide predicting chart. Using this tide chart, time, height of the tide can be obtained and published in a book format. This tide predicting machine (Fig.15.24B$_i$) is located at Geodetic Branch of Survey of India, Dehradun, India. But now a days nobody is using this tide predicting machine. But the tidal prediction is done basing on a super computer.

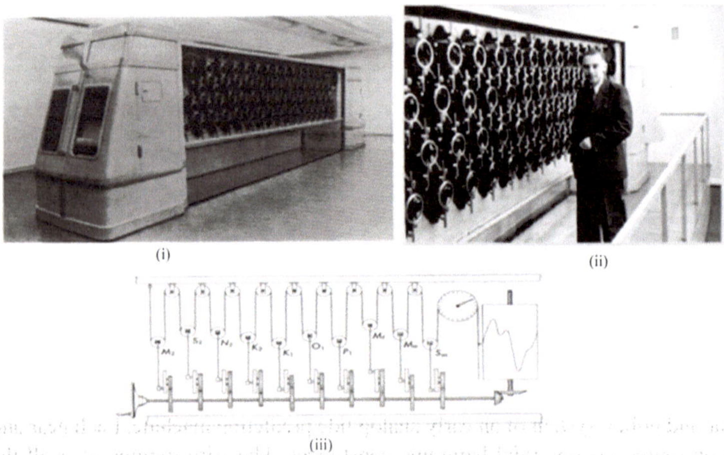

**Fig. 15.24B.** Large Kelvin's 61-component tide predicting machine at: (i): Survey of India, Dehradun (ii):Hydrographic Institute, Hamburg, Germany;(iii) Different pulleys for different constituents, handle and graphical drum of the machine.
https://incois.gov.in/documents/ITCOocean/OITC018/L13-Srinivas%20K_TIDES.pdf

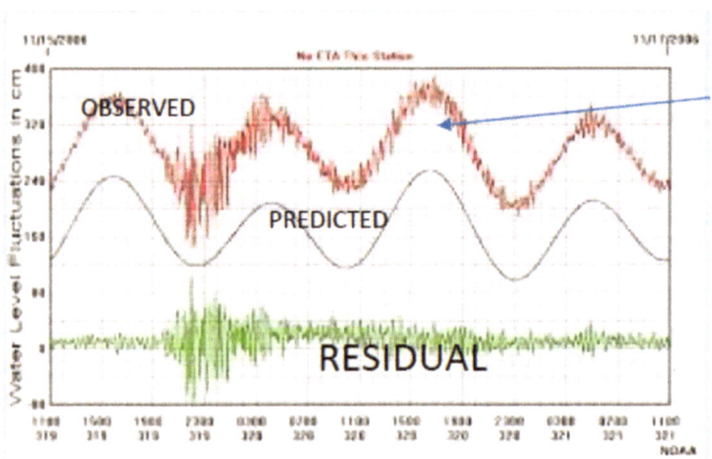


**Fig.15.25.** Observed minus predicted tide is residual tide. Note the abnormal fluctuation in observed tide is not found in predicted tide.
(https://incois.gov.in/documents/ITCOocean/OITC018/L13-Srinivas%20K_TIDES.pdf)

## 15.6. Amphidromic system

The combined constraint of ocean basin geometry and the influence of the coriolis force results in the development of amphidromic systems, in each of which the crest of the tidal wave at high water circulates around an amphidromic point once during each tidal cycle (Fig.15.26). The tidal range is zero at each amphidromic point, and increases outwards away from it (Fig.15.26 & Fig.15.27).

In each amphidromic system, lines of constant tidal range (*corange lines*), as well as lines of locations with the same time of high water (*cotidal lines*) are important. The successive co-tidal lines radiating outwards from the amphidromic point thus indicate the passage of the tidal wave crest around it.

Cutting across co-tidal lines, approximately at right angles to them are co-range lines, which join places having the same tidal range. Co-range lines form more or less concentric circles about the amphidromic point, representing larger and larger tidal ranges the further go away from it. Fig.15.25 shows the computed world-wide amphidromic systems for the dominant tidal component resulting from the diurnal influence of the Moon.

Amphidromic points are the characteristic features of tidal propagation. Regions where tidal ranges are nearly zero, but tidal currents can be present. Points where the tidal waves rotate due to the Corilois force and basin effects can be present. For example $M_2$ amphidromic points are found off Sri Lanka, southern part of the Arabian Sea (Fig.15.25).

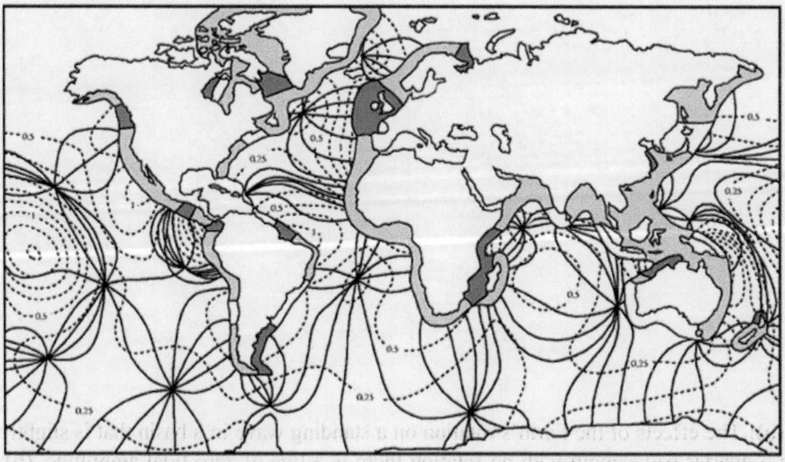

**Fig.15.25.** Co-tidal lines (solid lines) and lines of equal amplitude (dashed lines, amplitude in meters) of the semidiurnal lunar M2-tide in the oceans. The tidal range along the coast is indicated in dark for tidal ranges larger than 4 m, in grey for tidal ranges between 2 and 4 m and in white for tidal ranges below 2 m.

http://www.coastalwiki.org/wiki/Coriolis_and_tidal_motion_in_shelf_seas

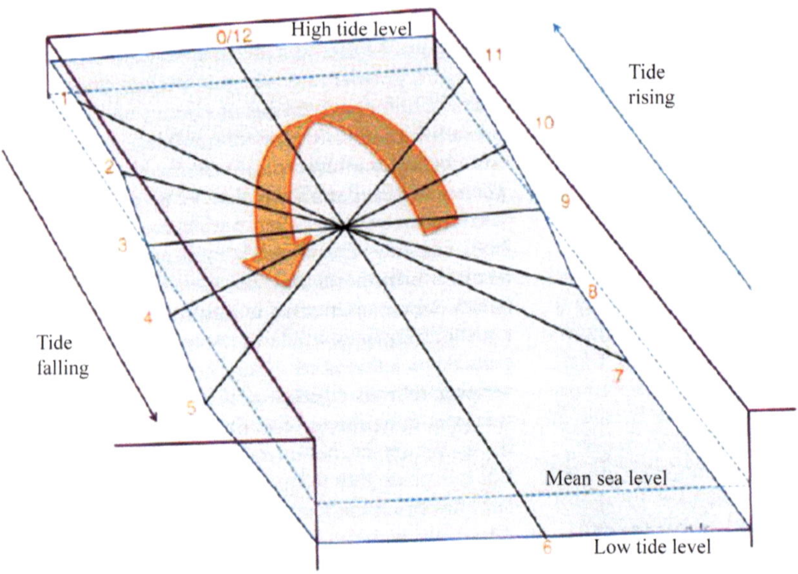


**Fig. 15.26.** A three dimensional drawing exaggerated to illustrate how a tidal wave progresses around an amphidrome in a basin in the northern hemisphere. The red numbers are hours in the semidiurnal cycle. (Sindhu and Unnikrishnan ,2013) https://incois.gov.in/documents/ITCOocean/ITC001/ppts/L15-Characteristics%20of%20tides..pdf

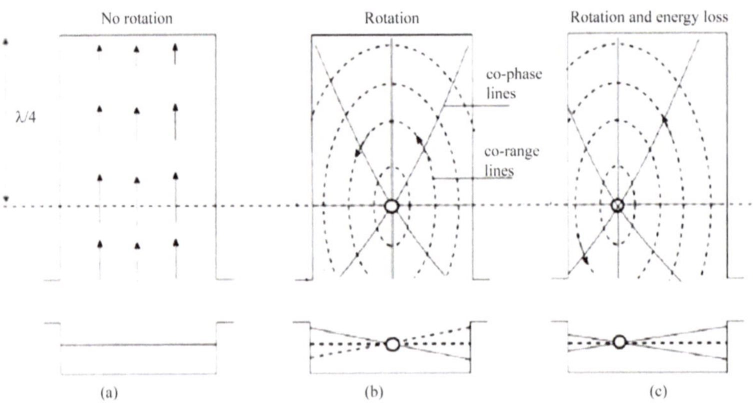


**Fig.15.27. (a):** The effects of the Earth's rotation on a standing wave in a basin that is slightly longer than a quarter wavelength with no rotation there is a line of zero tidal amplitude. (b): Because of the Earth's rotation the tidal wave rotates around a point of zero amplitude called an amphidromic point. (c ): In the third case because the reflected wave has lost energy through tidal friction, the amphidrome is displaced from the center line.( Sindhu and Unnikrishnan, 2013) https://incois.gov.in/documents/ITCOocean/ITC001/ppts/L15-Characteristics%20of%20tides..pdf

## 15.7. Tidal currents in estuaries

As the tide enters into the estuary during high tide, the water level advances and increase the sea level and flood on the banks of the estuary called flood tide which is deflected towards right hand side by the Coriolis force and so the direction of arrows bends towards right in Fig.15.28 a. On the other hand during low tide time after 12 hours the water start receding back towards the sea called ebb tide. Then the arrows bend towards west as shown in Fig.15.28b.

As the tide enters into the estuary during high tide, the water level advances and increase the sea level and flood on the banks of the estuary called flood tide which is deflected towards right hand side by the Coriolis force and so the direction of arrows bends towards right in Fig.15.28 a. On the other hand during low tide time after 12 hours the water start receding back towards the sea called ebb tide. Then the arrows bend towards west as shown in Fig.15.28b.

Fig.15.28 (a) shows the flood tide as bent arrows. This bending of tidal currents is due to deflection of Coriolis force to the right hand side in the northern hemisphere. As a result the water piles-up on the eastern side of the basin (Fig.15.28a at t=0). Conversely during ebb tide (Fig.15.28b) the water while returning in the basin bent opposite and so piles-up towards west of the bank or recedes in the east bank (Fig.15.28b at $\frac{6}{12}T$). Hence because the tidal wave is constrained (obstructed) by the basin land mass on either side, an anticlockwise amphidromic system is set-up (Fig.15.26, 15.27 b and c &15.28 c). Note in Fig.15.28c, the flooding in the east or right bank is at t=0, t= $\frac{1}{12}T$, $\frac{2}{12}T$, $\frac{10}{12}T$ and t= $\frac{11}{12}T$. Similarly the west bank or left hand bank flooding or east bank recession is at t= $\frac{4}{12}T$, $\frac{5}{12}T$, $\frac{7}{12}T$, $\frac{8}{12}T$ while the mean or null levels are at t= $\frac{3}{12}T$ and t= $\frac{9}{12}T$. Thus the motion of the tide is anticlockwise starting from t=0, t= $\frac{1}{12}T$, t= $\frac{2}{12}T$........ t= $\frac{11}{12}T$ and completes the cycle of rotation.

Thus tidal currents are the horizontal water movements that accompany the rise and fall of the tides as the tidal wave form rotates about the amphidromic point, and of course tidal currents change direction during the tidal cycle.

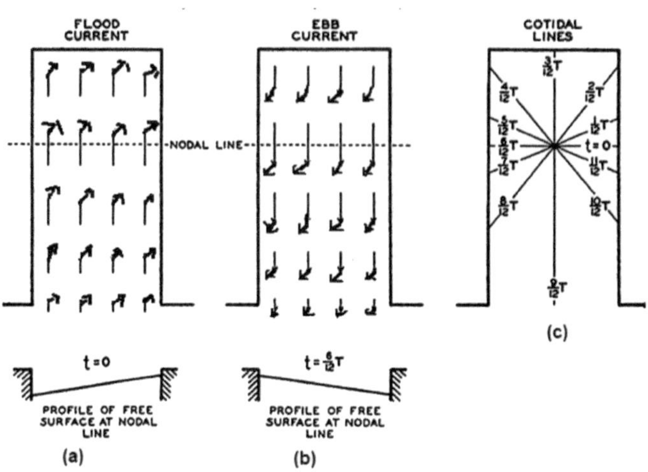


**Fig. 15.28.** Flood and ebb tides in an estuarine circulation due to high and low tides. a) Flood tide piles-up (sea level rise) on the right bank in N.H, b) Ebb tide piles-up (sea level rise) in return on the left bank. C) Anticlockwise development of the amphidromic system. Note the anticlockwise rotation of tidal period from $T_0$ such that total period T=12 hr 25min.

The tidal currents in shallow seas tend to follow an elliptical path and do not tend to and fro motion as shown in Fig15.29a below. The sense of rotation of the ellipse may be either clockwise or anticlock wise, but rotations cumsole tend to be favoured if there are non-constraining land masses. The elliptical path followed by water particles in a tidal current during a complete tidal cycle. The successive directions of the current are shown by arrows. The length of the arrows is proportional to current velocity (Fig.15.29 B & C) at the relevant time. Numbers refer to lunar hours (62 minutes) measured after an arbitrary starting time in the cycle

As mentioned earlier, the movements of the particles in long waves are for the most part horizontal. In narrow sounds and straits the tidal current will go approximately 6 hours each direction. Out in the open sea, the tidal current will normally not only have two directions but instead will gradually turn 360°. If one draws the velocity as an arrow from a fixed point, during the course of a tidal period the point of the arrow would in the ideal case describe an ellipse (the *tidal ellipse*). This is due to the earth's rotation. The presence of more than one tidal component will, however complicate the picture. At hour 3 water at the surface is moving in one direction while near the bed is moving in the other direction (Fig. 15.29 A &D).

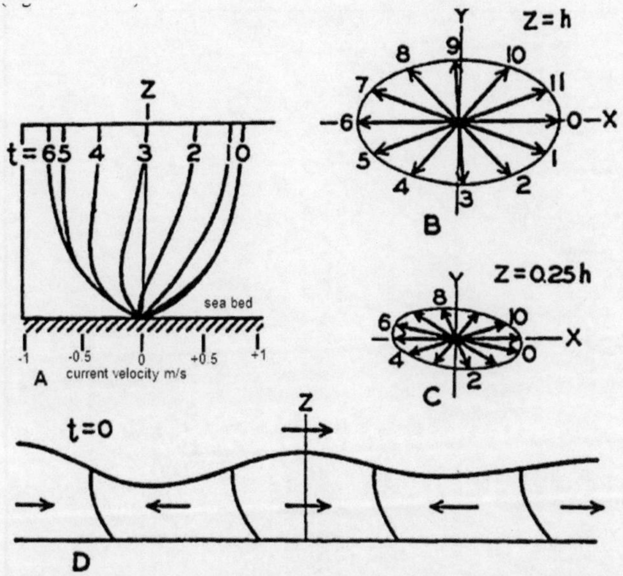


**Fig.15.29 . (A) & (D):** Note that at hour 3 water at the surface is moving in one direction while after near the bed is moving in the other direction. **(B):** shows a series of vertical tidal current profiles, showing retardation of the currents close to the sea bed (Z=h). The numbers refer to time in lunar hours after an arbitrary starting time. Only half tidal cycle is shown here **(C):** Tidal current pattern at one fourth depth (Z=0.25h) (UC press E-book collection, 1982-2004, University of California Press)

## 15.8. Tides in North Indian Ocean

India has a large coastline of 7516 km. This vast coastline along nine coastal states and union territories of India experiences tides of varying ranges starting from 0.5 m to 11 m as shown in Figure 15.30a. In India the highest tidal range is observed along the coastline of Gujarat, which is in the range of 9 m to 11 m.

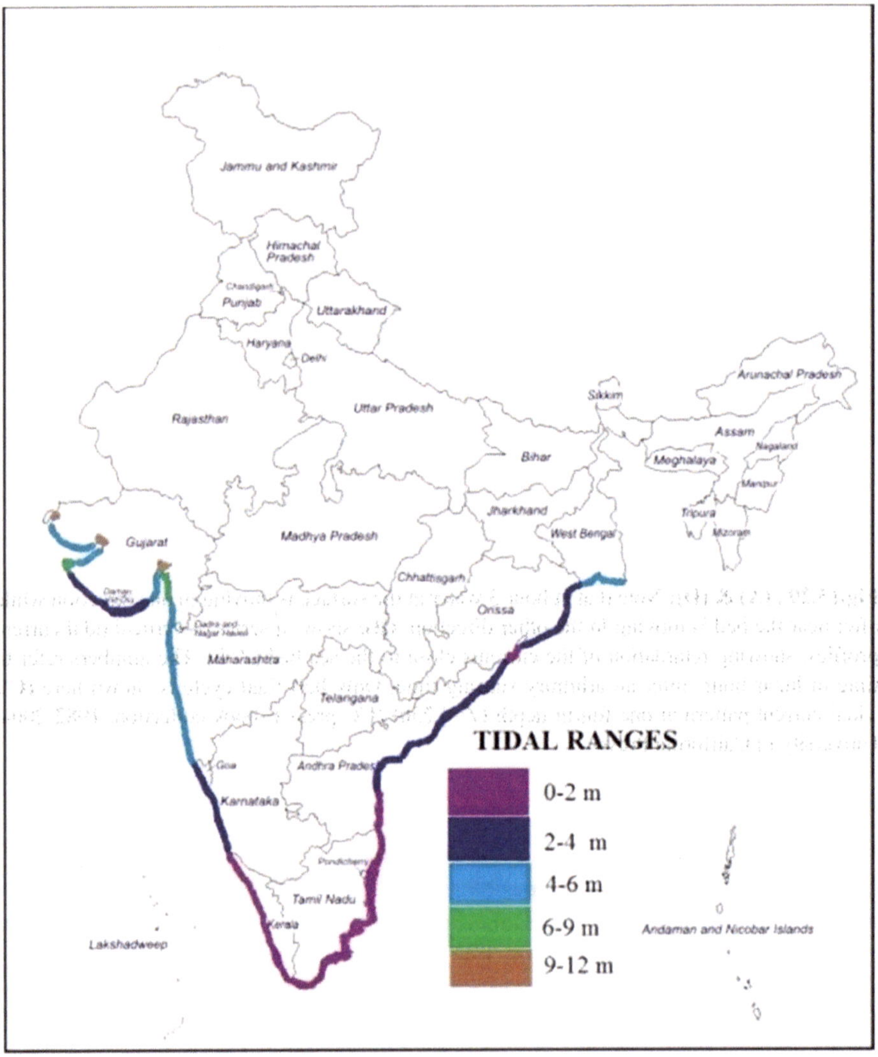


**Fig. 15.30a:** Tidal Ranges in India (Kashyap et al, 2017).

## 15.8.1. Tides in Bay of Bengal

A two-dimensional depth averaged model was used for the simulation of tides in the Bay of Bengal by Sindhu and Unnikrishnan (2013). The co-tidal and co-range charts were prepared for the constituents $M_2$, $S_2$, $N_2$, $K_1$ and $O_1$ for the Bay of Bengal, which were analyzed in detail. The distribution of semi-diurnal constituents ($M_2$ and $S_2$) shows (Fig.15.30b) that they amplify nearly twice in the head bay, along the southern coast of Myanmar and in the Gulf of Martaban. The diurnal constituent $K_1$ amplifies significantly only in the north eastern end of the head bay and in the Gulf of Martaban. The continental shelf in the head bay and along the southern coast of Myanmar is wide enough to cause significant amplification of semidiurnal tides but do not affect the diurnal tides. In the north eastern end of the head bay, the geometrical configuration of the coastline could also contribute to the amplification of semi-diurnal and diurnal constituents. The funnel shaped configuration of the Gulf of Martaban and the presence of wide continental shelf in this region could be the factors responsible for the highest amplification of semi-diurnal tides and relatively significant amplification of diurnal tides inside the Gulf. In the Malacca Strait, the amplitudes of both semi-diurnal and diurnal tides are found to increase from the northern end to 2.5$^0$N, which could be due to shallow bathymetry and its funnel shaped configuration. The highest amplitudes of diurnal tides, $K_1$ and $O_1$ are observed in the Malacca Strait. The characteristic feature observed in the Malacca Strait is the decrease in the amplitude of semi-diurnal and diurnal tides towards the southern boundary and presence of a virtual amphidromic point. The co-tidal and co-range charts of $M_2$ and $S_2$ tidal constituents show the presence of two degenerate amphidromic points; one in the central part of head bay and the other in the north eastern end of the head bay. A virtual amphidromic point for $M_2$ is identified in the Malacca Strait.

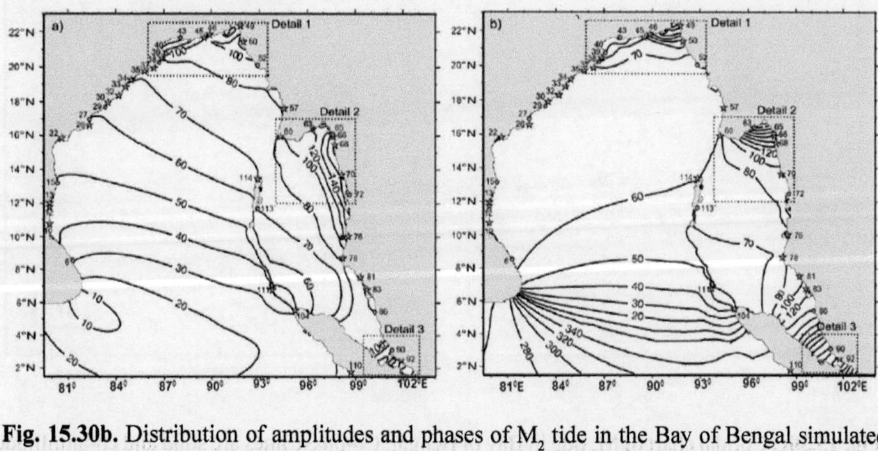


**Fig. 15.30b.** Distribution of amplitudes and phases of $M_2$ tide in the Bay of Bengal simulated using the tidal model showing **(a)** co-range lines and **(b)** co-tidal lines. Amplitudes are given in centimetres and phase is given in degrees (Sindhu and Unnikrishnan, 2013).

Jithin et al (2016) studied the tidal currents in the continental shelf region of east coast of India using Acoustic Doppler Current Profiler (ADCP) measurements at four locations for about 9 months. According to them the tidal currents in this region are mainly semidiurnal and the dominant constituent is $M_2$ and the semi-major axis of barotropic tidal ellipse of $M_2$ is about 1.9 $cms^{-1}$ at southern location (off Cuddalore ) and it is about 4.0 $cms^{-1}$ in the northern location (off Gopalpur), which indicates a northward increase of tidal currents.

According to Wan et al (2019), as shown in Figure 15.30c, the amplitudes for $M_2$ tide has a NE increase from the amphidromic point toward the head of the bay. However, tidal propagation is affected upon reaching Andaman Island, where it splits into two parts; one of which, continues to propagate into the head of the bay with its direction gradually changing to North, and the other enters the Gulf of Martaban, propagating North East. As the $M_2$ tide propagates into the head of the bay, it amplifies nearly two fold. The cophase lines are densely distributed near the head of the bay and the Gulf of Martaban because, in such areas, the local water depth is shallower, which reduces the wave speed.

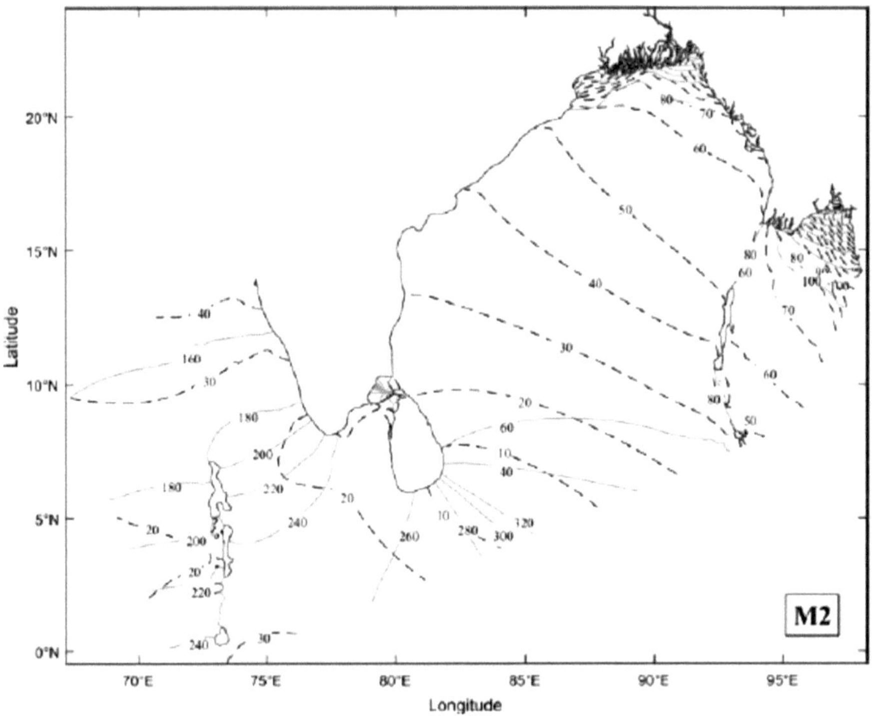


**Fig. 15.30c.** Cotidal chart of $M_2$ tide in Bay of Bengal. Co-phase lines are solid and co-amplitude lines are dashed. Amplitudes are in centimeters, and phase is in degrees (Wan et al 2019).

## 15.8.2. Tides in Arabian Sea

Tidal characteristics of the Gulf of Khambhat (Gujarat) were described by Mitra et al (2020) based on measured and modelled sea-level data. A northward increase in tidal amplitude is noticed from Daman (eastern side) and Diu (western side) and attains maxima at Bhavnagar. A similar trend is followed by the amplitude of the major tidal constituents, although there are discrepancies for that of the minor constituents. Harmonic analysis of sea-level data shows that M2 is the major tidal constituent which propagates in a non-linear fashion inside the Gulf. The MIKE21- Global Tide Model-simulated amplitudes of the major constituents ($K_1$, $M_2$, $O_1$ and $S_2$) are presented in Fig. 15.30d. The contour lines of $M_2$ suggest that the propagation of $M_2$ is along the channel. The trend of propagation is similar in the case of $S_2$, even though the amplitude of $S_2$ is much lesser while cross-channel propagation has been noticed in the case of the diurnal constituents ($O_1$ and $K_1$). Amplification of tides was noticed from offshore towards the Gulf but it is quite slower for the diurnal tide than the semi-diurnal. Gradual decreases in the phase lag are obtained from outer Gulf of Khambat to inner Gulf of Khambat for all other tidal constituents, except $O_1$, where higher phase lag values are obtained inside the Gulf.

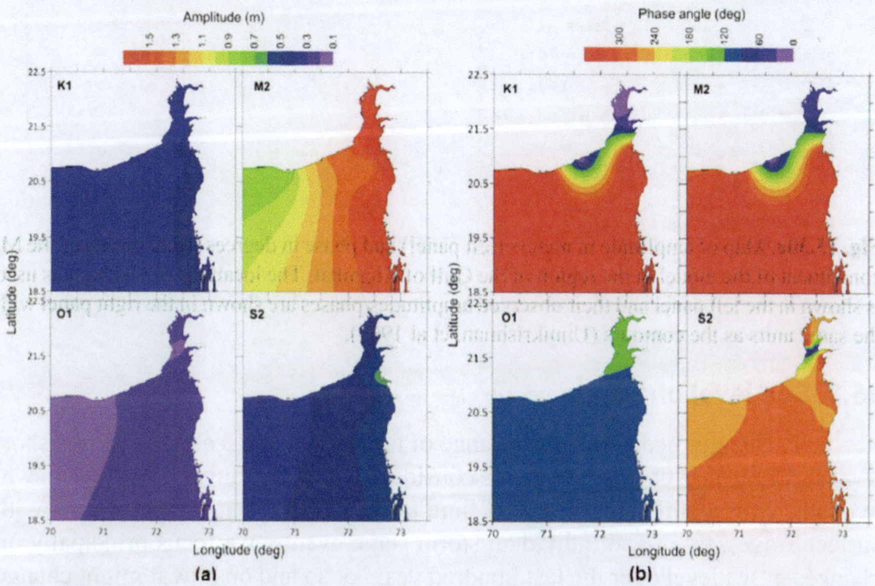


**Fig.15.30d:** Contours of major tidal constituents ($K_1$,$M_2$, and $S_2$) of Gulf of Khambhat and surroundings amplitudes **(a)** and phase lags **(b)** (Mitra et al 2020).

While global tidal models provide information on tides accurately in deep-sea, near the coast (< 20 km), they generally are not accurate. In order to overcome this, there are two ways (i) assimilate tide gauge data (Matsumoto et al, 2000) or (ii) develop regional tidal models. Fig. 15.30e gives the detailed picture of $M_2$ constituent in the Gulf of Khambat of west coast of India basing on the analysis of a regional model (Unnikrishnan et al,1999 and Sinha et al 2000).

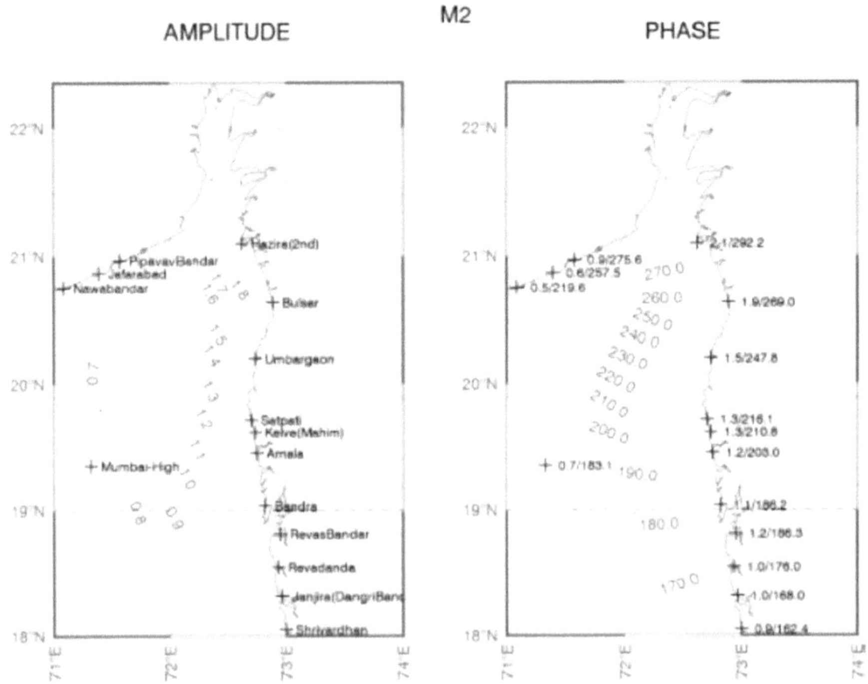


**Fig. 15.30e.** Map of amplitude in meters (left panel) and phase in degrees (right panel) of the $M_2$ constituent of the model in the region of the Gulf of Khambat. The locations of tide gauges used is shown in the left panel and their observed amplitudes/phases are shown in the right panel with the same units as the contours (Unnikrishnan, et al 1999).

## 15.9. Sea level changes

Sea-level changes occur on a wide range of time and space scales. There are short term and long term changes. Here we consider long term changes. That is sea level averaged over a sufficient period of time to remove fluctuations associated with surface waves, tides, and individual storm surge events. We focus principally on changes in sea level over the last hundred years or so and on how it might change over the next one hundred years. However, to understand these changes we need to consider what has happened since the last glacial maximum 20 000 years ago. The implications for these long-term changes are due to changes in the earth's climate arising from changes in atmospheric greenhouse gas concentrations.

Changes in mean sea level can be measured by a coastal tide gauge with respect to the nearby land (relative sea level) or a fixed reference frame. A global data set is maintained by the Permanent Service for Mean Sea Level (PSMSL) located at Bidston, U.K. During the 1990s, sea level has been measured globally with satellites.

The world ocean, which has an average depth of about 3800 m, contains over 97% of the earth's water. The Antarctic ice sheet, the Greenland ice sheet, and the hundred thousand nonpolar glaciers/ ice caps, presently contain water sufficient to raise sea level by 61 m, 7 m, and 0.5m respectively if they were entirely melted.

On the time-scales of millions of years, continental drift and sedimentation change the volume of the ocean basins, and hence affect sea level. A major influence is the volume of mid-ocean ridges, which is related to the arrangement of the continental plates and the rate of sea floor spreading.

Sea level also changes when mass is exchanged between any of the terrestrial, ice, or atmospheric reservoirs and the ocean called Eustatic changes. During glacial occurrence times (ice ages), water is removed from the ocean and stored in large ice sheets in high-latitude regions. Variations in the surface loading of the earth's crust by water and ice change the shape of the earth as a result of the elastic response of the lithosphere and viscous flow of material in the earth's mantle and thus change the level of the land and relative sea level.

Relative sea level can also be affected by local tectonic activities as well as by the land sinking when ground water is extracted or sedimentation increases. Sea water density is a function of temperature. As a result, sea level will change if the ocean's temperature varies due to thermal expansion without any change in mass.

### 15.9.1. Sea level changes since the Last Glacial Maximum

On timescales of thousands to hundreds of thousands of years, the most important processes affecting sea-level are those associated with the growth and decay of the ice sheets through glacial–interglacial cycles. Sea-level variations during a glacial cycle exceed 100m in amplitude, with rates of up to tens of millimeters per year during periods of rapid decay of the ice sheets (Figure 15.31). At the last glacial maximum (about 21 000 years ago), sea level was more than 120m below current levels. The largest lowering of this sea-level was due to the ice that formed in the North American (Laurentide) and European (Fennoscandian) ice sheets. In addition, the Antarctic ice sheet was larger than at present and there were smaller ice sheets in presently ice-free areas.

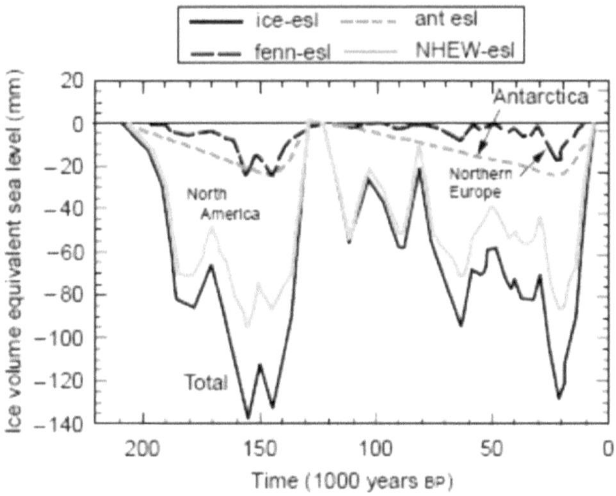


**Fig. 15.31.** Change in ice sheet volume over the last 200 000 years. Fenn= Fennoscandian; ant= Antarctic; NHEW= North America; esl= equivalent sea level. (Reproduced from Lambeck, 1998.)

## 15.9.2. Tide-gauge observations & present trend of sea level

Unfortunately, determination of global-averaged sea-level rise is severely limited due to the small number of gauges with long records). Most of the long term tide gauge records (up to several hundred years, Figure 15.32) are present in Europe and North America. A widely accepted estimate of the current rate of global-average sea-level rise is about 1.8mm $y^{-1}$. This estimate is based on a set of 24 long tide-gauge records, corrected for land movements resulting from deglaciation.

The various assessments of the global-average rate of sea-level change over the past century are not all consistent within stated uncertainties, indicating further sources of error. The treatment of vertical land movements remains a source of potential inconsistency, perhaps amounting to 0.5 mm $y^{-1}$.

Comparison of the rates of sea-level rise over the last 100 years (1.0–2.0mm $y^{-1}$) and over the last two millennia (0.1–0.0mm $y^{-1}$) suggests that the rate has accelerated fairly well recently. From the few tide-gauges of very long records (Figure 15.32), it appears that an acceleration of about 0.3–0.9mm $y^{-1}$ per century occurred over the nineteenth and twentieth century. However, there is little indication that sea-level rise accelerated during the twentieth century.

The most recent estimates of global-average sea level rise based on the (short period-since 1992) TOPEX/POSEIDON time series range from 2.1mm $y^{-1}$ to 3.1 mm $y^{-1}$ .

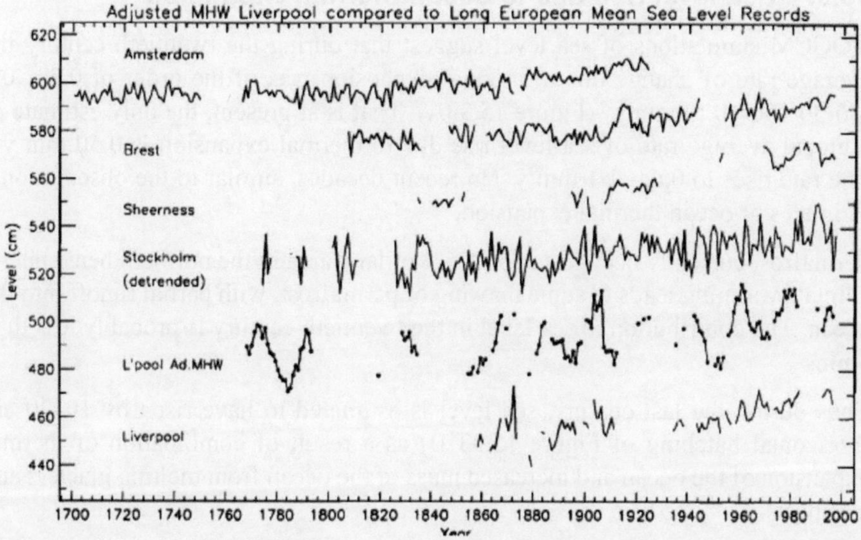


**Fig. 15.32.** Time series of relative sea level over the last 300 years from several European coastal locations. For the Stockholm record, the trend over the period 1774 to 1873 has been removed from the entire data set. For Liverpool two series are given. These are mean sea level and, for a longer period, the mean high water (MHW) level. (Woodworth, 1999.)

## 15.9.3. Processes determining present rates of sea level change

The major factors determining sea-level change during the twentieth and twenty-first century are ocean thermal expansion, the melting of nonpolar glaciers and ice caps, variation in the mass of the Antarctic and Greenland ice sheets, and changes in terrestrial storage.

Projections of climate change caused by human activity rely principally on detailed computer models referred to as atmosphere–ocean general circulation models (AOGCMs). These models simulate the global three dimensional behavior of the ocean and atmosphere by numerical solution of equations representing the underlying physics. For simulations of the next hundred years, future atmospheric concentrations of gases that may affect the climate (especially carbon dioxide from combustion of fossil fuels) are estimated on the basis of assumptions about future population growth, economic growth, and technological change. AOGCM experiments indicate that the global-average temperature may rise by $1.4 - 5.81^0$ C between 1990 and 2100, but there is a great deal of regional and seasonal variation in the predicted changes in temperature, sea level, precipitation, winds, and other parameters.

### 15.9.4. Sea level rise due to ocean thermal expansion

AOGCM simulations of sea level suggest that during the twentieth century the average rate of change due to thermal expansion was of the order of 0.3 – 0.8 (80-30/100= 0.5) mm $y^{-1}$ (Figure 15.33A). That is at present, the only estimate of a global average rate of sea-level rise due to thermal expansion is 0.50 mm $y^{-1}$. The rate rises to 0.6 – 1.1 mm $y^{-1}$ in recent decades, similar to the observational estimates of ocean thermal expansion.

Permafrost currently occupies about 25% of land area in the northern hemisphere. Climate warming leads to some thawing of permafrost, with partial runoff into the ocean. The contribution to sea level in the twentieth century is probably less than 5mm.

Thus during the last century, sea level is estimated to have risen by 10–20 cm (horizontal hatching of Figure 15.33 B) as a result of combination of thermal expansion of the ocean and increased mass of the ocean from melting glaciers and ice sheets.

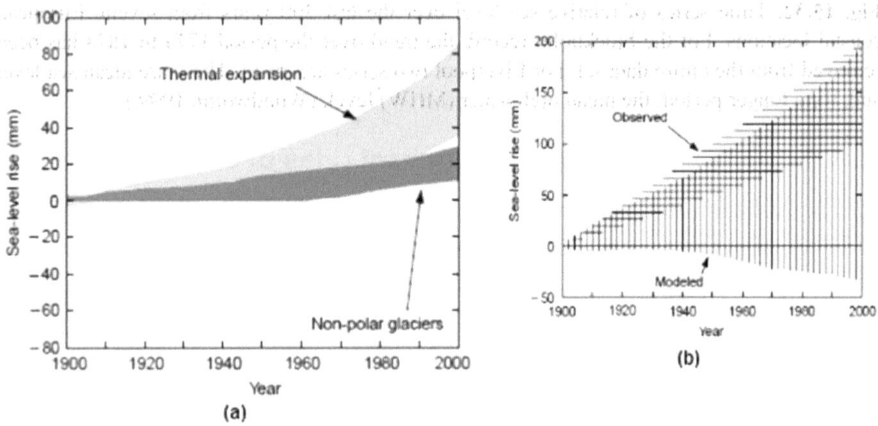


**Fig. 15.33.** Computed sea-level rise from 1900 to 2000 AD. (A) The estimated sea level rise due to thermal expansion is shown by the light stippling; the estimated sea level rise due to nonpolar glacial contribution is shown by the thick stippling. (B) The computed (modeled) total sea level change during the twentieth century is shown by the vertical hatching and the observed sea level change is shown by the horizontal hatching.

### 15.9.5. Projected sea-level changes for the 21st century

The major components of the detailed projections of changes in sea level derived from AOCGM results are thermal expansion of the ocean (a few tens of centimeters), melting of nonpolar glaciers (about 10–20 cm), melting of Greenland ice sheet (several centimeters), and increased storage in the Antarctic (several centimeters).

Over the twenty-first century, sea level is expected to rise as a result of anthropogenic climate change. The main contributors to this rise are expected to be thermal expansion of the ocean and the partial melting of nonpolar glaciers and the Greenland ice sheet. After allowance for the continuing changes in the ice sheets since the last glacial maximum, the total projected sea-level rise over the twenty-first century is currently estimated to be between about 9 and 88 cm (Figure 15.34).

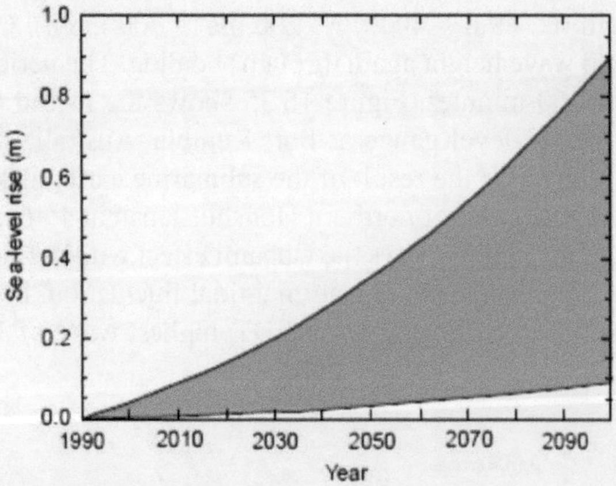


**Fig. 15.34.** The estimated range of future global-average sea level rise from 1990 to 2100 AD for a range of plausible projections in atmospheric greenhouse gas concentrations.

## 15.10. Long waves

Short waves (deep water waves) show normal dispersion, i.e. wave speed depends on the period, with the longer period waves moving faster than the shorter period waves (and the longer period waves have the longer wave lengths).

In contrast, long waves (shallow water waves) are non-dispersive: Their wave speed is independent of their period. It depends only on the water depth, in the form $c = \sqrt{gh}$ (where $c$ is wave speed, $h$ water depth and $g$ gravity).

The velocity structure in a long wave is described by $u = \dfrac{g\zeta}{\sqrt{gh}}$ :

where $\zeta$ is the time dependent surface elevation (wave amplitude) and $u$ the horizontal particle velocity. It follows that $u$ is independent of depth and the vertical particle velocity varies linearly with depth. Particles move on very flat elliptic paths in nearly horizontal motion.

### 15.10.1. Tsunamis

Tsunamis are long waves generated by submarine earthquakes. Tsunami is a Japanese word and it means for "harbour wave". They are often called tidal waves; but this is a misnomer, since tsunamis have nothing to do with tides.

The strongest tsunami in the history was produced due to the eruption of the Krakatau in the Sunda Island group in 1883. The height of tsunami wave was 35 m and claimed 36,830 lives. Four tsunamis with heights in excess of 30 m have also been documented in the Pacific Ocean since 684 A.D. A strong tsunami in the Atlantic Ocean was observed in 1755 after an earthquake near Lisbon (Portugal).

In the vicinity of the epicenter of an earthquake, tsunamis can result an extreme wave heights. Once they reach the open ocean and travel through deep water, tsunamis have extremely small amplitudes but travel fast with a speed of about 700 km/h in 4000 m water depth ($C = 9.8 \times 4000)^{1/2} = 200\ \mathrm{ms^{-1}} = 700$ km/h). On approaching a coast they build up wave height again through shoaling. The period of tsunamis is in the range of 10-60 minutes. Figure 15.35 shows the record of a tsunami from the output of the sea-level gauge at Port Kembla, Australia on Saturday 12 March 2011. This tsunami is the result of the submarine earth quake of magnitude 9.0, centred off the east coast of northern Honshu, Japan at 4:46pm EDT on Friday 11 March 2011. The red line shows the tsunami effect with normal tidal fluctuations removed. The blue line shows the normal tidal fluctuation. Port Kembla, NSW - tsunami effects observed from 7:11am EDT, highest wave of 35 cm observed at 11:20am EDT.

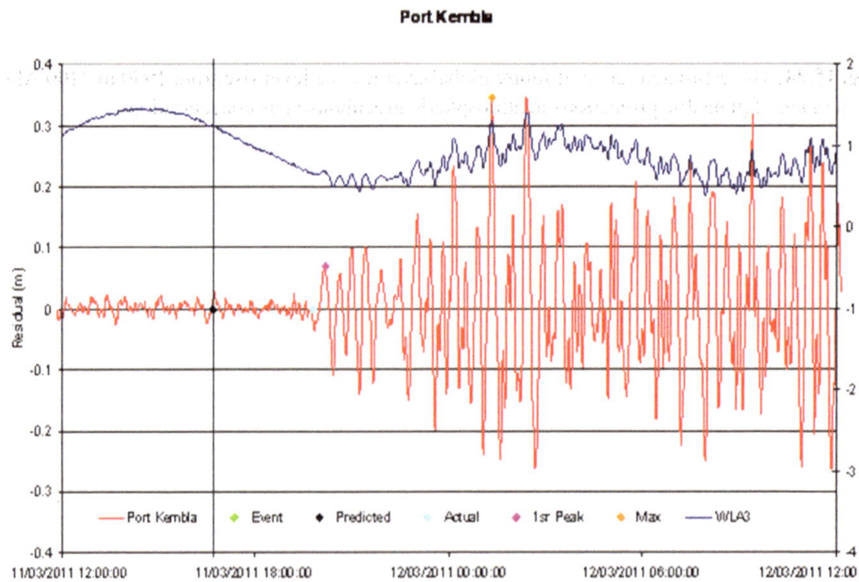


**Fig.15.35:** Passage of a tsunami as seen in a sea level record at Port Kembla, Australia on Saturday 12 March 2011.The slowly varying kinked line indicates the rise and fall of the tide as it would have been observed without the tsunami. The observed sea level shows high frequency variations with a period of approximately 20 minutes and initial amplitude of nearly two meters (total tsunami wave height 3.7 m). Bureau of Meteorology, Australia

Tsunamis were used to estimate the depth of the ocean by observing their phase speed, when direct depth measurements were not available. The depth calculation for the North Pacific in 1856 using the speed of tsunami was 4200 - 4500 m, which was a considerable improvement when compared to the older estimate of 18,000 m.

The most destructive tsunami known to have been occurred in Indian Ocean is on 26 December 2004. On the night of December 26, 2004, a massive quake of magnitude 9.1-9.3 with its epicentre in the vicinity of the Andaman Islands and northern Sumatra and caused death and destruction in 15 countries around the Indian Ocean. The death toll is estimated to have been between 265,000 and 320,000. In India nearly 10,136 people died with majority of the casualties reported from Tamil Nadu due to this tsunami. Figure 15.36 shows the passage of the tsunami of 26 December 2004 through Seychelles.

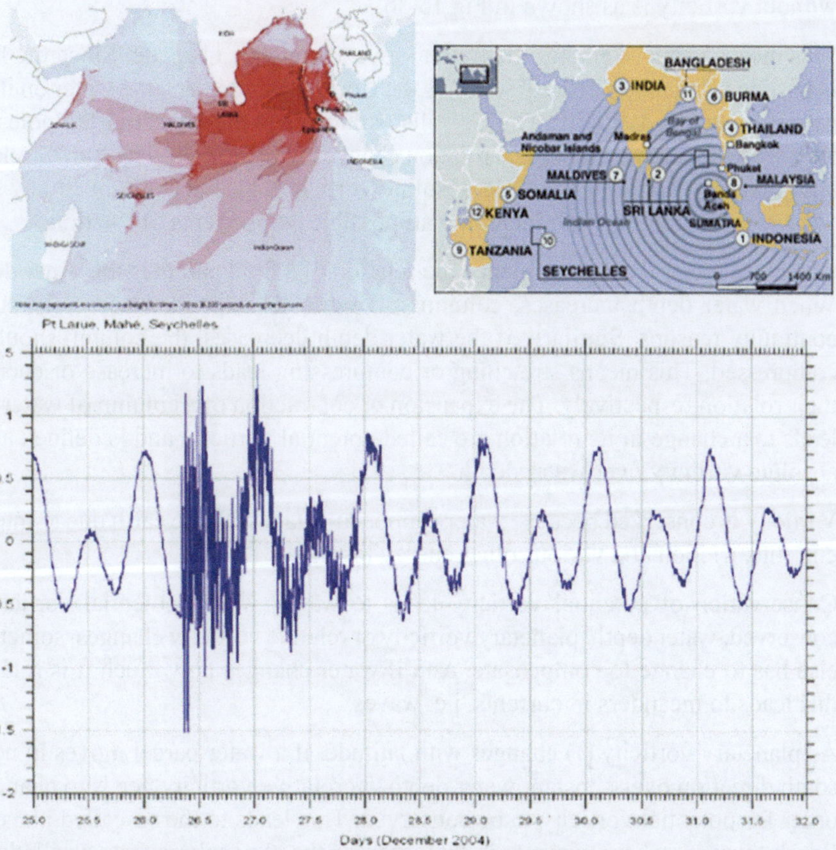


**Fig. 15.36.** The 4-minute water-level record from the tide gauge at Mahé (Seychelles) on 26 December 2004, showing the arrival of the tsunami wave train and ongoing oscillations over several days (data courtesy of SOEST, University of Hawaii).

## 15.10.2. Rossby waves

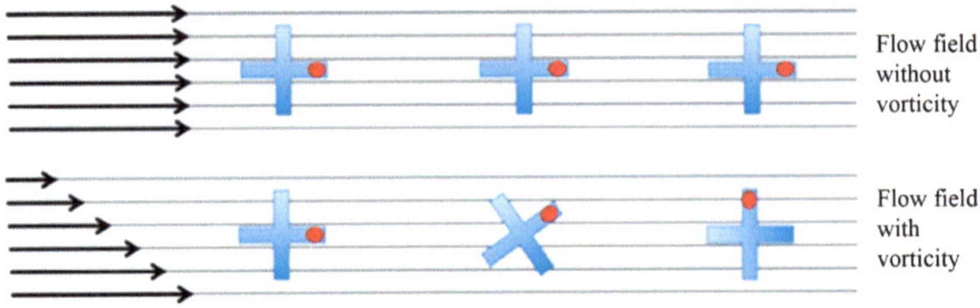


**Fig. 15.36.** Flow field with and without vorticity. (Adapted from Dr.Mirjam S. Glessmer, Adventures in Oceanography Teaching blog).

Vorticity is a measure of how much a fluid is rotating. The flow field with and without vorticity is as shown in Fig.15.36.

This means vorticity makes the fluid to give a tendency to rotate. Since the Earth is turning (planetary vorticity which is the Coriolis parameter), the water on it also carries planetary vorticity, The Coriolis parameter is the largest at the poles and zero at the equator, meaning the rotation changes with latitude. The sum of relative and planetary vorticity is called absolute vorticity: This is how much any fluid column is rotating in total, including all possible components of rotation.

One more factor that can influence the rotation of a fluid parcel is the water depth. When water depth increases, columns of water get expanded vertically due to continuity reasons. Similarly if the water depth decreases, the column should be compressed. This means stretching or compressing leads to increase or decrease their rotation respectively. The expansion or contraction of a column of water thus leads to a change in its rotation are called potential vorticity and is defined as the absolute vorticity over water depth.

Vorticity is conserved because water cannot co-rotate with the earth due to internal currents, friction and various other energy losses.

Conservation of potential vorticity leads to waves. When potential vorticity is conserved, water depth, planetary vorticity or relative vorticity changes, something else has to change to compensate. And if water changes how much it is rotating, this leads to meanders in currents, i.e. waves.

As planetary vorticity (f) changes with latitude, if a water parcel moves in north-south direction over constant water depth, its relative vorticity needs to change in order for potential vorticity to be conserved. This leads to the so-called planetary Rossby waves, where currents in the ocean or the atmosphere start oscillating in north-south direction (Fig. 15.37).

For example, from position 1 in Fig.15.37, if a fluid parcel is pushed northward, its planetary vorticity increases and its relative vorticity therefore has to decrease to compensate for the absolute vorticity at position 2. This leads to southward movement, but at position 3 (the initial latitude) a little overshot happens and so swings upto position 4. This again leads to a change in relative vorticity at position 4, which brings the water parcel back to its initial latitude position 5, but it overshoots again. So this mechanism leads to a wave-like motion in the horizontal plane, with the phase of the wave propagating westward. This is how Rossby wave generates.

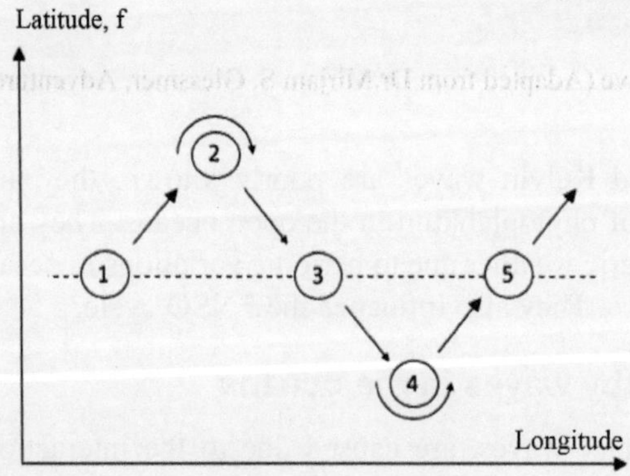


**Fig. 15.37.** Generation of Rossby waves

## 15.10.2.1. Equatorial waves

This can happen at any latitude. If it happens at the equator "equatorial planetary Rossby waves" occur. At the equator f = 0, but as soon as the water column has moved slightly north or south from the equator due to overshooting effect, the particle swings north and south of the equator.

## 15.10.2.2. Topographic waves

If latitude is constant but depth changes, the waves generated with the same mechanism are called Topographic waves. If a current encounters a ridge, the water depth changes and so the current gets deflected. This motion is called topographic wave. When a water column gets compressed (goes from higher topography to lower topography), it loses relative vorticity, making it rotate anticyclonically (position 1) in the Northern Hemisphere (Fig.15.38). On the other hand when it runs into deeper water, it gains relative vorticity and so rotates cyclonically (position 2) as shown in Fig. 15.38 below.

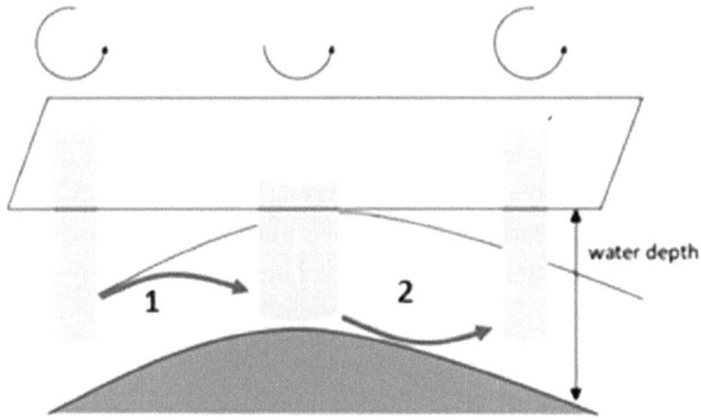


**Fig.15.38.** Generation of topographic wave (Adapted from Dr.Mirjam S. Glessmer, Adventures in Oceanography Teaching blog).

Though the effects of Rossby and Kelvin waves are poorly known, they are believed to affect the distribution of phytoplankton in the open oceans. They are also believed to affect the atmospheric weather due to periodic variations in ocean currents and sea surface temperature. They also influence the ENSO cycle.

### 15.10.2.3. Derivation of Rossby waves in the oceans

Rossby waves, often called planetary waves, are caused due to the interaction of the latitudinal gradient in the Coriolis Effect with geostrophic flow of water around the sea surface or with the flow of air around a depression or elevation of atmospheric pressure. These waves occur both in the ocean and in the atmosphere. Their generation is somewhat complex. In the ocean their wavelengths range from hundreds to thousands of kilometers and the wave heights are just a few centimeters. As they are very long and low, they are not directly observed by naked eye. But after the advent of satellite and remote sensing technology, the sea surface height was able to be measured very precisely and so they are detected.

Rossby waves move only from east to west. However, they appear to move from west to east if they are carried by a much faster moving air or water mass. While Rossby wave speeds depend on the latitude, oceanic Rossby waves generally move on the order of magnitude of a few kilometers per day. So a single wave can take months or even years to the whole Pacific Ocean at mid latitudes.

For large wave formations (such as cyclonic and anticyclonic vortices and Gulf Stream meanders), it is necessary to consider the meridional change in $f$.

We know $f = 2\Omega sin\phi$

If the y coordinate is oriented in the meridional direction and is measured from a reference latitude $\phi_0$ (somewhere in the middle of the wave) then

$$\varphi = \varphi_0 + \frac{y}{a} \tag{15.18}$$

Where'a' is the radius of the earth (6371 km). Then equation (15.18) can be expanded using Taylor's series as

$f = f_0 + \frac{\partial f_0}{\partial \varphi}\Delta\varphi + \frac{\partial^2 f_0}{\partial \varphi^2}\frac{(\Delta\varphi)^2}{\angle 2} + .....$ Limiting upto terms in the right hand side

$$f = f_0 + \frac{\partial f_0}{\partial \varphi}\Delta\varphi = 2\Omega \sin\varphi_0 + 2\Omega\cos\varphi_0 \frac{y}{a}$$

$$f = f_0 + \beta_0 y \tag{15.19}$$

Where $f_o = 2\Omega sin\phi_o$ and $\beta_0 = \frac{2\Omega\cos\varphi_0}{a}$ is called β parameter or Rossby parameter.

Typical mid latitude values on earth are $f_0 = 8 \times 10^{-5}\ S^{-1}$, $\beta_0 = 2 \times 10^{-11}\ m^{-1}\ S^{-1}$

The Cartesian frame work where β plane is not retained and considered along latitude circle is called f-plane.

Note that the β plane representation is validated at mid latitudes only if the $\beta_o y$ term is small compared to the leading $f_0$ term. In terms of meridional length scale L, this implies

$\beta = \frac{\beta_0 L}{f_0} << 1 (\because L << a)$ This dimensionless ratio is called planetary number. In such case $\frac{\partial f_0}{\partial y} = \beta$ can be taken as constant. This is called β plane approximation.

## Derivation starting

$$\frac{du}{dt} = -\alpha\frac{\partial p}{\partial x} + fv \text{ or } \frac{\partial u}{\partial t} + u\frac{\partial u}{\partial x} + v\frac{\partial u}{\partial y} + w\frac{\partial u}{\partial z} = fv - g\frac{\partial z}{\partial x}$$

Neglecting the non linear terms and putting (z = η) we can write

$$\frac{\partial u}{\partial t} - fv = -g\frac{\partial \eta}{\partial x} \tag{15.20}$$

Substitute eqn (15.19) in eqn (15.20)

$$\frac{\partial u}{\partial t} - (f_0 + \beta_0 y)v = -g\frac{\partial \eta}{\partial x} \tag{15.21}$$

$$f_0 v = g\frac{\partial \eta}{\partial x} + (\frac{\partial u}{\partial t} - \beta_0 yv) \tag{15.22}$$

Similarly 'y' equation can be written as

$$f_0 u = -g\frac{\partial \eta}{\partial y} - (\frac{\partial v}{\partial t} + \beta_0 yu) \tag{15.23}$$

Writing equations (15.23) and (15.22) as

$$u=-\frac{g}{f_0}\frac{\partial\eta}{\partial y}-\frac{1}{f_0}(\frac{\partial v}{\partial t}+\beta_0 yu) \tag{15.24}$$

$$v=\frac{g}{f_0}\frac{\partial\eta}{\partial x}+\frac{1}{f_0}(\frac{\partial u}{\partial t}-\beta_0 yv) \tag{15.25}$$

Substitute equations (15.24) and (15.25) in equation (15.21)

$$\frac{\partial}{\partial t}\left[-\frac{g}{f_0}\frac{\partial\eta}{\partial y}-\frac{1}{f_0}\frac{\partial v}{\partial t}-\frac{1}{f_0}\beta_0 yu\right]-f_0v-\beta_0 y\left[\frac{g}{f_0}\frac{\partial\eta}{\partial x}+\frac{1}{f_0}\frac{\partial u}{\partial t}-\frac{1}{f_0}\beta_0 yv\right]=-g\frac{\partial\eta}{\partial x}$$

On further simplification

$$\frac{1}{f_0}\frac{\partial}{\partial t}\left[-g\frac{\partial\eta}{\partial y}-(\frac{\partial v}{\partial t}+\beta_0 yu)\right]-f_0v-\frac{\beta_0 y}{f_0}\left[g\frac{\partial\eta}{\partial x}+(\frac{\partial u}{\partial t}-\beta_0 yv)\right]=-g\frac{\partial\eta}{\partial x} \tag{15.26}$$

Neglecting the terms in the small brackets, we can write the eqn (15.26) as

$$\frac{g}{f_0}\frac{\partial^2\eta}{\partial t\partial y}-f_0v-\frac{\beta_0 g}{f_0}y\frac{\partial\eta}{\partial x}=-g\frac{\partial\eta}{\partial x} \tag{15.27}$$

Similarly writing the ‘y’ component we get

$$\frac{g}{f_0}\frac{\partial^2\eta}{\partial t\partial x}+f_0u-\frac{\beta_0 g}{f_0}y\frac{\partial\eta}{\partial y}=-g\frac{\partial\eta}{\partial y} \tag{15.28}$$

Substituting (15.24) and (15.25) into eqn (15.28) we get

$$\frac{\partial}{\partial t}\left[\frac{g}{f_0}\frac{\partial\eta}{\partial x}+\underbrace{\frac{1}{f_0}\left\{\frac{\partial u}{\partial t}-\beta_0 yv\right\}}_{\text{I}}\right]+f_0u+\beta_0 y\left[-\frac{g}{f_0}\frac{\partial\eta}{\partial y}-\underbrace{\frac{1}{f_0}\left\{\frac{\partial v}{\partial t}-\beta_0 yu\right\}}_{\text{II}}\right]=-g\frac{\partial\eta}{\partial y} \tag{15.29}$$

(Neglecting the small terms I & II) ,Equations 15.27 and 15.28 can be written as

$$u=-\frac{g}{f_0}\frac{\partial\eta}{\partial y}-\frac{g}{f_0^2}\frac{\partial^2\eta}{\partial t\partial x}+\frac{\beta_0 g}{f_0^2}y\frac{\partial\eta}{\partial y} \tag{15.30}$$

$$v=\frac{g}{f_0}\frac{\partial\eta}{\partial x}-\frac{g}{f_0^2}\frac{\partial^2\eta}{\partial t\partial y}-\frac{\beta_0 g}{f_0^2}y\frac{\partial\eta}{\partial x} \tag{15.31}$$

The first terms on the right hand side are geostrophic velocities and the other two small terms are called ageostrophic terms.

Substituting equations (15.30) and (15.31) in equation of continuity (eqn 15.32)

$$\frac{\partial\eta}{\partial t}+h\left[\frac{\partial u}{\partial x}+\frac{\partial u}{\partial y}\right]=0 \tag{15.32}$$

For substitution of eqn (15.30) and (15.31) in eqn (15.32):

Differentiate eqn (15.30) with respect to 'x'

$$\frac{\partial}{\partial x}(u)=\frac{\partial}{\partial x}\left[-\frac{g}{f_0}\frac{\partial \eta}{\partial y}-\frac{g}{f_0^2}\frac{\partial}{\partial t}\left(\frac{\partial \eta}{\partial x}\right)+\frac{\beta_0 g}{f_0^2}y\frac{\partial \eta}{\partial y}\right]$$

Similarly differentiating eqn (15.31) with respect to 'y':

$$\frac{\partial}{\partial y}(v)=\frac{\partial}{\partial y}\left[\frac{g}{f_0}\frac{\partial \eta}{\partial x}-\frac{g}{f_0^2}\frac{\partial}{\partial t}\left(\frac{\partial \eta}{\partial y}\right)-\frac{\beta_0 g}{f_0^2}y\frac{\partial \eta}{\partial x}\right]$$

On simplifying we get $\frac{\partial u}{\partial x}+\frac{\partial v}{\partial y}=-\frac{g}{f_0^2}\frac{\partial}{\partial t}\left(\frac{\partial^2 \eta}{\partial x^2}+\frac{\partial^2 \eta}{\partial y^2}\right)-\frac{\beta_0 g}{f_0^2}\frac{\partial \eta}{\partial x}$ (15.33)

Substituting eqn (15.33) in eqn (15.32) we get

$$\frac{\partial \eta}{\partial t}+h\left[-\frac{g}{f_0^2}\frac{\partial}{\partial t}\left(\frac{\partial^2 \eta}{\partial x^2}+\frac{\partial^2 \eta}{\partial y^2}\right)-\frac{\beta_0 g}{f_0^2}\frac{\partial \eta}{\partial x}\right]=0$$

Put $\frac{gh}{f_0^{\,2}}=R^2$ (Baroclinic Rossby radius of deformation)

$$\frac{\partial \eta}{\partial t}-R^2\frac{\partial}{\partial t}\left(\frac{\partial^2 \eta}{\partial x^2}+\frac{\partial^2 \eta}{\partial y^2}\right)-\beta_0 R^2\frac{\partial \eta}{\partial x}=0$$

or $\frac{\partial \eta}{\partial t}-R^2\frac{\partial}{\partial t}\left(\nabla_H{}^2\eta\right)-\beta_0 R^2\frac{\partial \eta}{\partial x}=0$ (15.34)

To solve eqn (15.34), assume the wave profile (η) solution as

$\eta=\eta_0 e^{i(kx+ly-\sigma t)}$ (15.35)

Where k = east-west wave number, and = north-south wave number.

Differentiate equation (15.35) with respect to 't' then double differentiate with respect to 'x' and 'y' and then substitute in eqn (15.34) and simplify.

Differentiation of (15.35) w.r.t 't':

$\frac{\partial \eta}{\partial t}=\eta_0(-i\sigma)e^{i(kx+ly-\sigma t)}$ (15.36)

Double differentiation of (15.35) w.r.t 'x':

$\frac{\partial \eta}{\partial x}=\eta_0(ik)e^{i(kx+ly-\sigma t)},\ \frac{\partial^2 \eta}{\partial x^2}=\eta_0\left(i^2k^2\right)e^{i(kx+ly-\sigma t)}$ (15.37)

Double differentiation of (15.35) w.r.t 'y':

$\frac{\partial^2 \eta}{\partial y^2}=\eta_0\left(i^2l^2\right)e^{i(kx+ly-\sigma t)}$ (15.38)

Substituting 15.36,15.37 and 15.38 in 15.34:

$$\underset{\text{I}}{\eta_0(-i\sigma)e^{i(kx+ly-\sigma t)}}$$

$$\underset{\text{II}}{-R^2\left[\frac{\partial}{\partial t}\left\{\eta_0\left(i^2k^2\right)e^{i(kx+ly-\sigma t)}\right\}+\eta_0\left(i^2l^2\right)e^{i(kx+ly-\sigma t)}\right]}-\underset{\text{III}}{\beta_0R^2\left[\eta_0\left(ik\right)e^{i(kx+ly-\sigma t)}\right]}=0 \quad \text{.....(15.39)}$$

Further simplification of (15.39) : I + II = III

Simplification of II:

$$-R^2\left[\frac{\partial}{\partial t}\left\{\eta_0\left(-k^2\right)e^{i(kx+ly-\sigma t)}\right\}+\eta_0\left(-l^2\right)e^{i(kx+ly-\sigma t)}\right](\because i^2=-1)$$

On differentiation w.r.t 't' and simplifying

$$-R^2\eta_0e^{i(kx+ly-\sigma t)}\left[i\sigma\left(k^2+l^2\right)\right] \tag{15.40}$$

Substituting 15.40 in 15.39 and simplifying:

$$\eta_0e^{i(kx+ly-\sigma t)}\left[(-i\sigma)-R^2\left(i\sigma k^2+i\sigma l^2\right)-\beta_0R^2(ik)\right]$$

since $\eta=\eta_0e^{i(kx+ly-\sigma t)}$ according to (15.35)

$$\text{or}\quad \sigma=-\frac{\beta_0k}{\left(1/R^2\right)+(k^2+l^2)} \tag{15.41}$$

Note that if β corrections were not retained ($\beta_0 = 0$) the frequency σ would have been zero which corresponds to steady state geostrophic flow on the *f*-plane.

By tracing the time derivatives as small terms ($\frac{\partial u}{\partial t}-\beta_0yv$) and II in eqn (15.29) ( i.e., assumed a small temporal Rossby number R <<1), we have retained only the low frequency (< $f_0$) for these waves to develop.

The zonal phase speed of Rossby wave can be obtained from the eqn (15.41) as

$$C_x=\frac{\sigma}{k}=-\frac{\beta_0R^2}{1+R^2(k^2+l^2)} \tag{15.42}$$

Which can also be expressed as

$$\omega=-\beta_oR^2\frac{k_x}{1+R^2(k_x^2+k_y^2)}$$

Where $k_x$ and $k_y$ are the zonal and meridional wave numbers (Fig.15.39). This $C_x$ is always negative implying the phase propagation is in the negative x direction. This means its propagation is towards west.

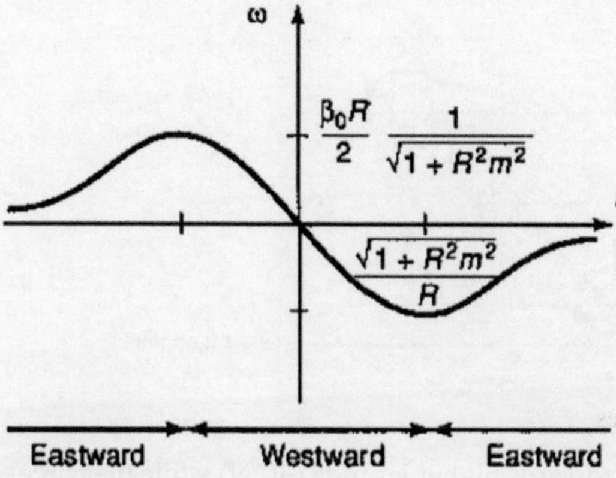


**Fig.15.39.** Dispersion relation of Rossby waves. The frequency ω is plotted against the zonal wave number 'l' ($k_x$) at constant meridional wave number m ($k_y$). As the slope of the curve reverses, the direction of zonal propagation and its associated energy also changes.

For both cases of long waves and short waves $\omega \ll f_0$ for sub inertial wave. $C_x = C_x(k)$ for dispersive wave and for $C_x = \frac{\omega}{k_x} < 0$ westward propagation. The group velocity of Rossby wave is $C_g = \left(\frac{\partial\omega}{\partial k_x}, \frac{\partial\omega}{\partial k_y}\right)$ is westward for long waves and eastward for short waves.

The meridional phase speed $C_y = \frac{\sigma}{l}$ is undetermined since the wave number ($l$) has either sign under the root. Thus for very long waves for which $\frac{1}{k}$ and $\frac{1}{l}$ are larger then R, equation (15.42) will become:

$$C_{x(\max)} = -\beta_0 R^2 = -\beta_0 \frac{gh}{f_0^2} \tag{15.43}$$

This is the maximum speed allowed for Rossby waves

## Mechanism of propagation of Rossby waves

Let us consider five parcels of water on the same latitude moving on a β-plane as shown in Fig.15.40.

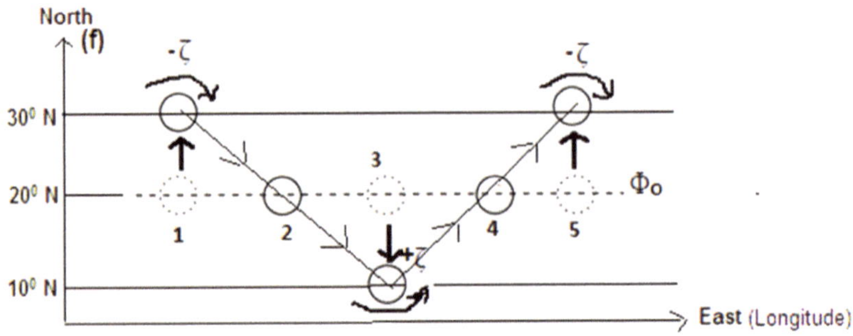


**Fig.15.40.** Propagation of Rossby waves

Let two parcels 1 and 5 have gone towards higher latitude (30°N) while the parcel 3 has moved toward lower latitude to 10°N. Let the water depth (h) is constant. Thus the water column length is constant. Then the potential vorticity $\left[\frac{\zeta + f}{h} = const\right]$ is conserved.

The two parcels moved to higher latitudes (30°N) develops negative relative vorticity (-ζ) as 'f' increased. This means parcel (1 & 5) acquires anticyclonic (clockwise) rotation. Whereas parcel 3 acquires cyclonic rotation as it acquires positive relative vorticity (+ζ) due to decrease of '$f$'.

Thus 1,3,5 parcels develop an eastward tending wave field. This flow field disturbs the parcels 2 and 4. When the parcels 1, 3, 5 are restored under stability considerations, the parcels 2 and 4 will tend to move on the β-plane. Let suppose the parcel '2' has a tendency to go to lower latitude ( toward 10°N) and the parcel 4 has a tendency to go toward higher latitude (30°N) side in the next instant. Then the wave profile shifts toward west progressing westward as shown by solid profile in the Fig. 15.41.

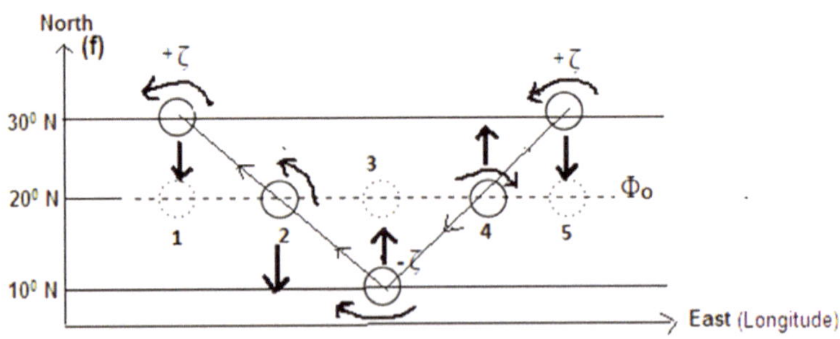


**Fig. 15.41.** Reverse propagation

Note that these waves are called Rossby waves and they are transverse waves. This means the particles move perpendicular to the direction of propagation. The restoring force of this wave is the north-south gradient of potential vorticity

$\left[\dfrac{\zeta + f}{h}\right]$. The gradient can be either due to variation of '*f*' or 'h'.

The Rossby waves are two kinds. They are tropospheric Rossby waves and the other are continental shelf waves.

When the variation of 'h' dominate the variation of 'f', topography comes into picture and so they are called *topography waves*. However, variations in topography may lead to wave trapping (concentration of wave energy in between shallow depths). Then these are called shelf waves. The gravity waves also can be trapped near the shelf. Then they are called *edge waves*.

The variation of '*f*' also under shallow depths of long waves may lead to trapping near the equator. Then they are called *equatorial waves*.

## Salient points of Rossby waves

1. The phase speed $C_x$ is always negative (eqn 15.42). This means the wave propagates westward.
2. The group velocity at which the energy of a wave packet propagates is defined by $(\dfrac{\partial \sigma}{\partial k}) = \dfrac{\partial \omega}{\partial k_y}$ and $(\dfrac{\partial \sigma}{\partial l}) = \dfrac{\partial \omega}{\partial k_x}$ is the gradient of function σ (= ω) in k (= $k_y$) and '*l*'(= $k_x$) wave number plane. It is thus perpendicular to the circles of constant σ (= ω). The group velocity vector is directed inward toward the center. Therefore longwaves (small 'k' and '*l*' point near the origin) have westward propagating group velocities, whereas shorter waves (larger 'k' and '*l*' point on the opposite side of the circle) have eastward energy propagation as shown in Fig.15.42.

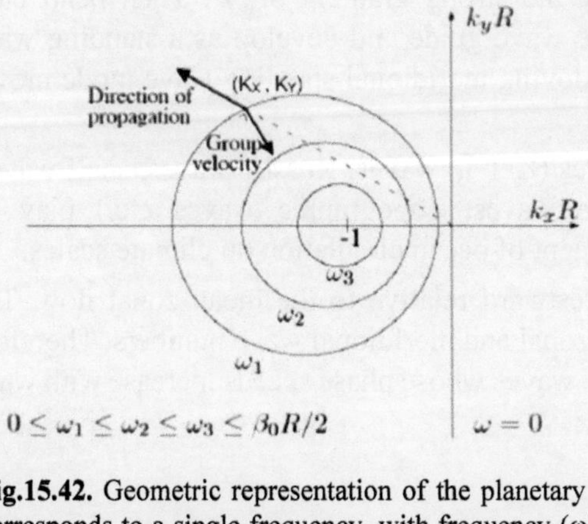


**Fig.15.42.** Geometric representation of the planetary – wave dispersion relation. Each circle corresponds to a single frequency, with frequency (ω) increasing with decreasing radius. The group velocity of ($k_x$, $k_y$) wave is a vector perpendicular to the circle at point ($k_x$, $k_y$) and directed toward its center.

3. Thus as far as energy propagation is concerned it can be either direction: i) i.e., when $\frac{\partial \sigma}{\partial k} < 0$ for low 'k' (long waves) = westward ; ii) when $\frac{\partial \sigma}{\partial k} > 0$ for short waves = eastward. In other words with reference to the Fig.15.42, the group velocity of Rossby wave is $C_g = \left( \frac{\partial \omega}{\partial k_x}, \frac{\partial \omega}{\partial k_y} \right)$ is westward for long waves and eastward for short waves.
4. Lines of constant frequency (σ) in the (K, $l$) wave number space are circles defined by $\left( K + \frac{\beta_0}{2\sigma} \right)^2 + l^2 = \left( \frac{\beta_0^{\,2}}{4\sigma^2} - \frac{1}{R^2} \right)$ These circles are illustrated in Fig.15.42. Such circles exist only if their radius is a real number. i.e., $\beta_0^{\,2} > 4\sigma^2$ this implies the existence of maximum frequency $|\sigma|_{\max} = \frac{\beta_0 R}{2}$ beyond which planetary waves do not exist.
5. The wave length limits for long waves is k, l ≪ 1
6. They strictly propagate westward with a maximum speed of $C_{x(\max)} = -\beta_0 R^2$ (when,l≪1) as mentioned in equation 15.43.
7. These are non dispersive (frequency linear function of wave number) waves.
8. Rossby wave phase speed decrease as it moves away from the equator (as β decreases). Wave crest also changes to crescent shaped.
9. These are considered as sub-inertial waves because their maximum frequency is always less than '$f$'. So the periods of these waves may be weeks to months.
10. Right close to the equator within (± $5^0$), they have greater acceleration. The equator acts as a wave guide like Kelvin waves. These Rossby waves sometimes are refracted by the strong gradient of '$f$'. Then bend back towards the equator into the wave guide and develop as a standing wave moving westward. Fastest moving meridional standing wave mode moves westward at C/3.
11. All these types of long waves (Kelvin waves, Rossby waves, shelf waves, equatorial waves, Poincare waves, Topographic waves etc.) play an important role in the adjustment of ocean circulation on climate scales.
12. Rossby wave propagates westward relative to the mean zonal flow. The wave speed depends on the zonal and meridional wave numbers. Therefore Rossby waves are dispersive waves whose phase speeds increase with wave length.

The frequency of Rossby waves is compared with other waves as given below:

| S.No | Type of wave | Frequency | |
|---|---|---|---|
| | | Waves per second | Waves per day |
| 1. | Sound waves | $10^3$ | $10^8$ |
| 2. | Long gravitational waves | $10^{-1}$ | $10^4$ |
| 3. | Inertial waves | $10^{-2}$ | $10^3$ |
| 4. | Cyclone | $10^{-5}$ | $10^0$ |
| 5. | Rossby | $10^{-6}$ | $10^{-1}$ |

The above table shows that in the case of a cyclone the frequency $10^0(=1)$ per day which means one wave pass in a day (i.e., 10 waves in 10 days). Whereas the frequency in the case of Rossby waves is $10^{-1}$ per day (1/10) which means to pass one wave 10 days it will take.

## Further analysis on Rossby wave characteristics

We know Rossby wave frequency as: $\sigma = -\dfrac{\beta_0 k}{\left(\frac{1}{R^2}\right) + (k^2 + l^2)}$

Let us ignore the north south wave number '*l*' for simplicity as it has not much effect on wave propagation.

$$\sigma = -\frac{\beta_0 k}{(k^2 + \frac{1}{R^2})} = \frac{-\beta_0 R^2 k}{(R^2 k^2 + 1)} = -\beta R^2 \left[\frac{k}{R^2 k^2 + 1}\right] \tag{15.44}$$

### 1. Rossby frequencies for long and short waves

From equation (15.44) , we can distinguish Rossby frequencies for long and short waves as:

i) If KR << 1 , $(1+ R^2K^2) \sim 1$; So $\sigma_{Long} = -\beta R^2 k$ (15.45)

ii) If KR >> 1, $(1+ R^2K^2) \sim R^2K^2$; then $\sigma_{Short} = -\dfrac{\beta}{k}$ (15.46)

From equation (15.44), we can write phase velocity, $C_p = \dfrac{\sigma}{k} = \dfrac{-\beta_0 R^2}{(R^2 k^2 + 1)}$ (15.47)

### 2. Rossby phase velocities for long and short waves

From equation (15.47) we can write phase velocities for long and short waves as:

i) If KR << 1 , $(1+ R^2K^2) \sim 1$; So $C_{PLong} = -\beta R^2$ (15.48)

ii) If KR >> 1, $(1+ R^2K^2) \sim R^2K^2$; then $C_{PShort} = -\dfrac{\beta}{k^2}$ (15.49)

### 3. Rossby group velocity for long and short waves

Similarly we know the Group velocity as:

$$C_g = \frac{\partial \sigma}{\partial k} = -\frac{\partial}{\partial k}\left\{\beta_0 R^2\left[\frac{k}{R^2k^2+1}\right]\right\} = -\beta_0 R^2\left[\frac{(R^2k^2+1)\frac{\partial}{\partial k}(k) - k\frac{\partial}{\partial k}(R^2k^2+1}{(R^2k^2+1)^2}\right]$$

$$C_g = -\beta_0 R^2\left[\frac{1-k^2R^2}{(1+R^2k^2)^2}\right] \tag{15.50}$$

### Conditions for long and short waves in the case of group velocity of Rossby waves

i) If KR << 1 , $R^2K^2$ = negligible; So $C_{GLong}$ = - $CR^2$ (15.51)

ii) If KR >> 1, RK is finite; then $C_{gshort} = -\beta_0 R^2\left[\frac{-k^2R^2}{(R^2k^2)^2}\right] = +\frac{\beta}{k^2}$ (15.52)

### The condition of Rossby waves relative to the deformation radius 'R':

i) It may be noted that KR << 1, means R << $\frac{1}{k} \Rightarrow R << \frac{\lambda}{2\pi} \rightarrow longwaves$

ii) If KR >> 1, means R >> $\frac{1}{k} \Rightarrow R >> \frac{\lambda}{2\pi} \rightarrow shortwaves$

### Comparison of phase and group velocities of Rossby waves

i) From equations (15.48) and (15.51) $C_{PLong} = C_{GLong} = -\beta R^2$. This implies phase speed and group speed are same in the case of Rossby long waves which are non-dispersive.

ii) From equations (15.49) and (15.52) $C_{PShort} = -\frac{\beta}{k^2}$ and $C_{GShort} = +\frac{\beta}{k^2}$. This implies phase speed and group speed are different in the case of Rossby short waves which means they are dispersive.

So Rossby long waves propagate faster as they are non- dispersive. The negative sign indicates Long waves propagate westward. However, the group velocity of short waves indicates they are not only slow due to dispersive nature but also move eastward as the sign is opposite. Short waves means their wavelengths are short relative to the Rossby radius 'R'.

### Estimations giving suitable values

i) Estimation of Rossby deformation radius 'R':

$$R = \frac{\sqrt{g'H}}{f} = \frac{\sqrt{10^3 \times 10^{-2}}}{10^{-4}} = 3 \times 10^4 m = 30km$$

ii) Estimation of Rossby long wave frequency and time period:

Let the longest wave length is 300 km, From equation (15.45): $\sigma_{Long} = -\beta R^2 k$

$$\sigma_{Long} = -\beta_0 R^2 k \Rightarrow \frac{2\pi}{T} = -\beta_0 R^2 \frac{2\pi}{\lambda} \Rightarrow T = \frac{-\lambda}{\beta_0 R^2}$$

Let $\lambda = 300$km$=3 \times 10^5$ m; R =30km = $3 \times 10^4$ m; $\beta_0 = 2 \times 10^{-11}$ m$^{-1}$ s$^{-1}$

a) $$\sigma_{Long} = -\beta_0 R^2 \frac{2\pi}{\lambda} = 2\times10^{-11} m^{-1}s^{-1} \times (3\times10^4)^2 m^2 \times \frac{2\pi}{3\times10^5 m} = 36\times10^{-8} \approx 10^{-7} s^{-1}$$

b) $$T = \frac{-\lambda}{\beta_0 R^2} = \frac{3\times10^5 m}{2\times10^{-11} m^{-1}s^{-1} \times (3\times10^4)^2 m^2} = \frac{1}{6}\times10^8 s = 192 days \approx 200 days$$

c) Long wave phase velocity: From equation (15.48): $C_{GLong} = -\beta R^2 = 2 \times 10^{-11}$ m$^{-1}$ s$^{-1}$ $\times (3 \times 10^4)^2$ m$^2$ $= 1.8 \times 10^{-2}$ ms$^{-1}$ = 1.8cms$^{-1}$

iii) Estimation of Rossby short wave frequency and time period:

Let $\lambda = 30$km $= 3 \times 10^4$m (max value taken)as $\lambda \leq$ R for short waves; $\beta_0 = 2 \times 10^{-11}$ m$^{-1}$ s$^{-1}$

a) From equation (15.46) short wave frequency :

$$\sigma_{Short} = -\frac{\beta_0}{k} \Rightarrow \frac{2\pi}{T} = -\frac{\beta_0 \lambda}{2\pi} = \frac{2\times10^{-11} m^{-1}s^{-1} \times 3\times10^4 m}{2\pi} = 10^{-7} s^{-1}$$

b) $$\Rightarrow T = -\frac{4\pi^2}{\beta_0 \lambda} = -\frac{4\pi^2}{2\times10^{-11} m^{-1}s^{-1} \times 3\times10^4 m} = -6.6\times10^7 s = 1.9 yrs \approx 2 yrs$$

c) Short wave Phase velocity: From equation (15.49):

$$C_{PShort} = -\frac{\beta}{k^2} = -\frac{\beta\lambda^2}{4\pi^2} = \frac{2\times10^{-11} m^{-1}s^{-1} \times (3\times10^4 m)^2}{4\pi^2} = 0.9\times10^{-3} ms^{-1} = 9\times10^{-2} cms^{-1}$$

This means short waves are very slow. Almost stationary. Thus long waves propagate westward very fast (1.8 cms$^{-1}$)and short waves propagate slowly (0.09 cms$^{-1}$) eastward.

### 15.10.3. Kelvin waves

The Kelvin wave is a large scale wave motion of great practical importance in the Earth's atmosphere and oceans. The Kelvin wave is a special type of gravity wave that is affected by the earth's rotation and trapped at the equator or along lateral vertical boundaries such as coastlines. Coastlines and the equator can support this special type of hybrid wave which includes both gravity wave and Coriolis effects. Kelvin waves are "trapped" to the coastlines and trapped at the equator, which means that their amplitude is highest at the coast (or equator) and decays exponentially with offshore (or poleward) distance. Hence Kelvin waves are trapped gravity waves. A trapped wave is one that decays exponentially in some direction. Kelvin waves need a boundary to exist. Thus there are two types of Kelvin waves, coastal and equatorial, and they are both gravity driven and non-dispersive. Kelvin waves are of particular importance on eastern boundaries since they transfer information poleward from the equator. They are often excited by an abrupt change in the overlying wind field, such as the shift in the trade winds at the start of El Niño. They are also central to how the equatorial ocean adjusts to changes in wind forcing, such as during an El Nin˜o.

The existence of Kelvin wave relies on a) gravity and stable stratification for sustaining a gravitational oscillation, b) significant Coriolis acceleration and c) the presence of vertical boundary or equator.

An important property of Kelvin wave is its unidirectional propagation. The Kelvin wave moves equatorward along a western boundary, poleward along an eastern boundary and cyclonically (N.H) and anticyclonically (in S.H) in a closed boundary.

The two types of Kelvin waves, boundary trapped and equatorially trapped, each type is further subdivided into surface and internal Kelvin waves. Surface Kelvin waves are barotropic and internal Kelvin waves are baroclinic. Internal Kelvin waves are often found in the layer of large density gradients. For example in the oceanic thermocline zone these internal waves exist. Whereas in the layer of inversion zone, these waves can exist in the atmosphere

### 15.10.3.1. Coastal Kelvin wave

In the case of coastal Kelvin waves amplitudes decay away from the boundary (coastline) as shown in Fig.15.43. The offshore decay scale is the Rossby deformation radius. Lastly, Kelvin wave water velocities in the direction perpendicular to the coast are exactly zero. The water velocities are, therefore, exactly parallel to the coast. Moreover, the alongshore velocities are geostrophic, so they are associated with pressure differences (pressure gradient force) in the across-shore direction. Kelvin waves propagate with the coast to the right in the Northern Hemisphere and to the left in the Southern Hemisphere.

Coastal Kelvin waves propagate around the northern hemisphere oceans in a counterclockwise direction using the coastline as a wave guide. These waves, especially the surface waves, are very fast moving, typically with speeds of ~2.8 m/s, or about 250 kilometers in a day. A Kelvin wave would take about 2 months to cross the Pacific from Papua New Guinea to South America (Fig. 15.46A).

The Coastal Kelvin waves have sufficiently long periods and slow periods that they are deflected and constrained to move along a boundary due to Coriolis effect. Coastal Kelvin waves flow along the coastal margin in a counter clockwise direction in the Northern Hemisphere and a clockwise direction in the Southern Hemisphere. However, at the equator where Coriolis Effect is zero, Kelvin waves flow directly along the equator from west to east as a special case (Fig.15.46). As equatorial Kelvin waves are three times faster than the Rossby waves, they can cross the Pacific Ocean in about 60 days.

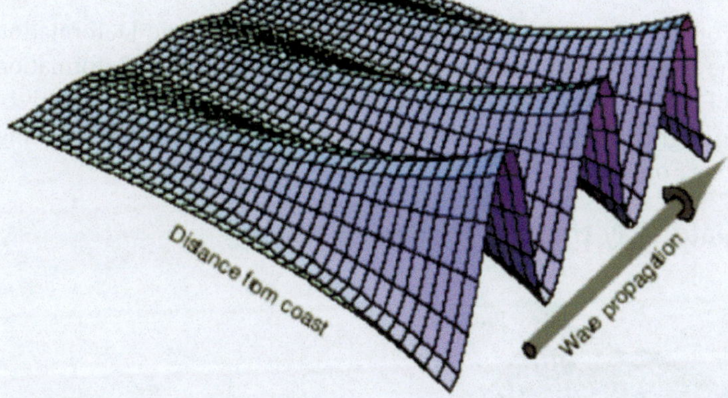


**Fig.15.43.** Coastal Kelvin wave

Coastal Kelvin Waves balance the Coriolis Force against a Topographic Boundary (i.e., Coastline). They always propagate with the shoreline on the right (Fig.15.44a) in the northern and the left (Fig.15.44c) in the southern hemisphere. A Coastal Kelvin Wave moving northward along the coast is deflected to the right, but the coast prevents the wave from turning right and instead causes water to pile up on the coast. The piling-up of water creates a pressure gradient directed offshore (Fig.15.44b) and a geostrophic current directed northward. Kelvin Wave Amplitude is negligible at a distance offshore given by the Rossby Radius of Deformation (R). For mid-latitude Kelvin Waves traveling on the ocean surface, R is about 200 km. For mid-latitude Kelvin Waves traveling in the thermocline, R is about 25 km. Because of this rapid decay Coastal Kelvin waves appear to be trapped Close to the Coast as shown in Fig.15.44.

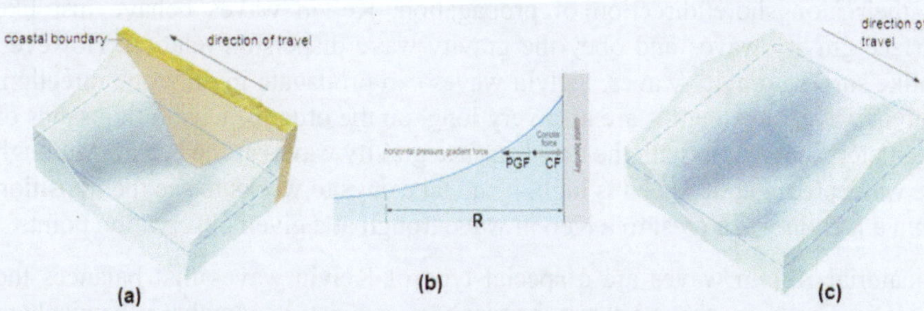


**Fig. 15.44.** Trapped coastal Kelvin waves. Wave of depression greatest onshore (a) https://slideplayer.com/ slide/5241975/16/ images/9/ Kelvin+Waves+ Travel+ eastward+along+ the+ equator+as+a+double+wave+ %E2%80% 98equatorial+wave+guide% E2%80%99.jpg

**Rossby Radius of Deformation** (R) is the distance that a particle or wave travels before being significantly affected by the earth's rotation.

$$R = \frac{U}{f} = \frac{C}{f} = \frac{\sqrt{gh}}{f}$$

If h is the depth of the upper ocean layer, then L is the Baroclinic Deformation Radius. If h is the depth of the ocean, we call L as the Barotropic Deformation Radius. Note that L decreases with latitude (increasing *f*) so that a wave (or current for that matter) at high latitude need only to travel a short distance before being affected by Coriolis Force.

### 15.10.3. 2. Equatorially trapped Kelvin waves

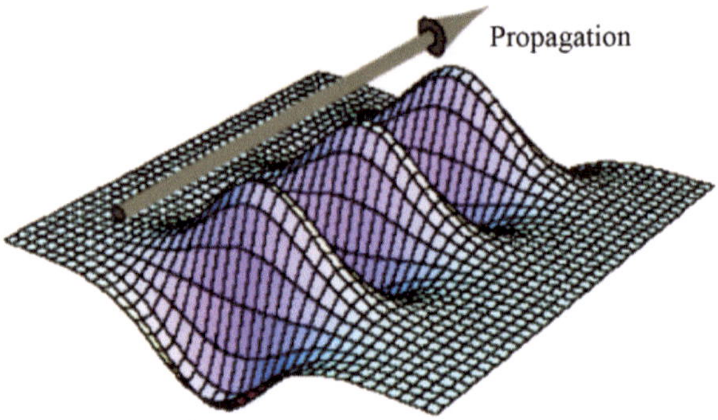


**Fig.15.45.** Equatorial Kelvin wave

In the case of equatorial Kelvin waves, amplitudes decay away from the "boundary" (equator) as shown in Fig. 15.45. Equatorial waves propagate to the east in the northern hemisphere, using the equator as a wave guide.

At the equator, which acts like a boundary, Kelvin waves propagate only eastward. In their alongshore direction of propagation, Kelvin waves behave just like surface gravity waves and obey the gravity wave dispersion relation. However, unlike surface gravity waves, Kelvin waves can propagate in only one direction. Kelvin wave wavelengths are also very long, on the order of tens to thousands of kilometers, compared with the usual surface gravity waves at the beach. Although the wave propagation speed is high, it can take days to weeks to see the transition from a Kelvin wave crest to a Kelvin wave trough at a given observation point.

Equatorial Kelvin waves are a special type of Kelvin waves that balances the Coriolis Force in the northern hemisphere against its southern hemisphere counterpart. This wave always propagates eastward and generates due to sharp change in wind field and only exists on either side of the equator (Fig15.46).

Equatorial Kelvin Waves propagating in the thermocline have wave speeds slow enough to give a Rossby Radius of Deformation that is on the order of 250 km and thus they appear to be trapped close to the equator as shown in Fig.15.46. A sharp change in wind field causes a wind disturbance that generates the propagation of Kelvin waves

(A)

(B)

(C)

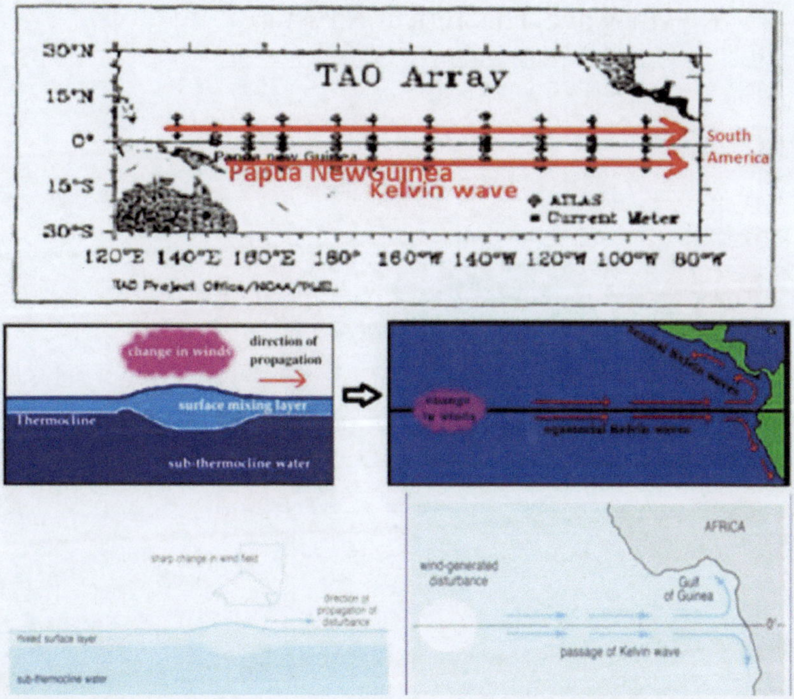


**Fig. 15.46. (A)** Kelvin Wave propagation from Papua New Guinea to south America (Boulanger and Menkes, 1995) **(B)**: Equatorial Kelvin wave generation due to wind shear: Kelvin Waves off South America and **(C)**: off Africa

https://images.slideplayer.com/16/5216284/slides/slide_14.jpg

Matsuno in 1966 showed that the eastward propagating Kelvin wave is a possible free solution to the perturbation equations of the shallow water on an equatorial β plane, provided that meridional velocity (v = 0) vanishes. This is called equatorially trapped Kelvin wave because it is similar to the coastally trapped Kelvin wave but the equator serve as a boundary.

The equatorial Kelvin wave moves eastward only. The speed of the Kelvin wave is like the shallow water gravity wave $C = \sqrt{gh}$ . Because the Coriolis parameter changes sign on either side of the equator, eastward flow occurs on both sides of the equator. This would induce equatorward Ekman Net transport causing convergence and meridional pressure gradient at the equator.

Because of the conservation of momentum in the north-south direction requires a geostrophic balance between the eastward velocity and the meridional pressure gradient force. This geostrophic balance results in the perturbation zonal velocity reaching a maximum on the equator and decay with increasing distance from the equator as shown in Fig.15.47.

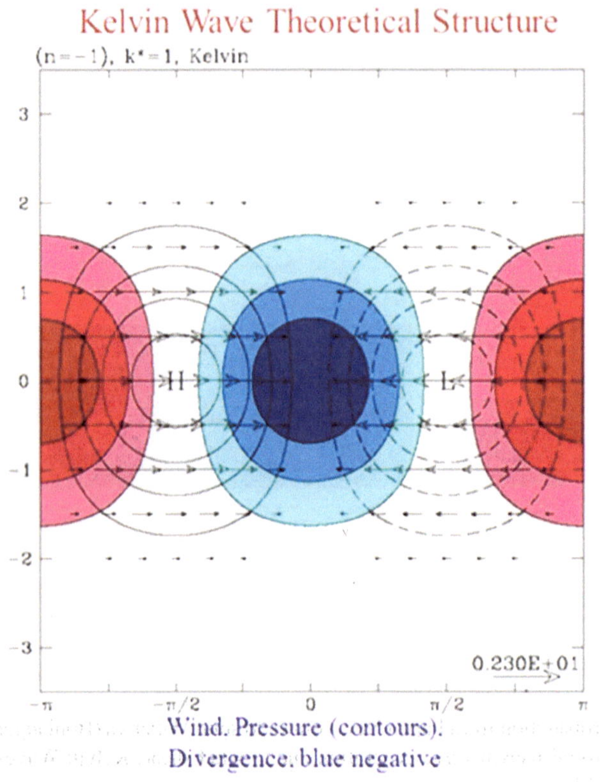


**Fig. 15.47.** Equatorial kelvin wave. Hatched semi circles near the equator denotes convergence. Centre circle denotes divergence. Unshaded contours are for geopotential with a contour interval of 0.5 units. Negative contours are dashed.

The tide waves traveling along continents is transmitted in the form of barotropic Kelvin waves such that the cotidal lines rotate anticlockwise (cyclonic) in the higher latitudes side in the N.H and clockwise in the S.H. For shallow seas and coastal waters the Rossby radius is about 200 km.

When a Kelvin wave moves through a region in which the fluid depth or the coriolis parameter varies and the wave energy flux remains constant then the amplitude of the wave varies in proportion to $\left[\frac{f}{h}\right]^{1/2}$. Thus the wave amplitude increases when it moves into shallow water.

Variable longshore winds, internal density gradients play a vital role in generating Kelvin waves. Equatorial Kelvin waves play a vital role in thermocline adjustment.

The equatorial Kelvin waves play an essential role in the ENSO cycle. A westerly wind anomaly (lateral wind shear) excites downwelling Kelvin waves, which propagate into the eastern Pacific suppressing thermocline and ceases upwelling during El Nino as shown in Fig.15.48.

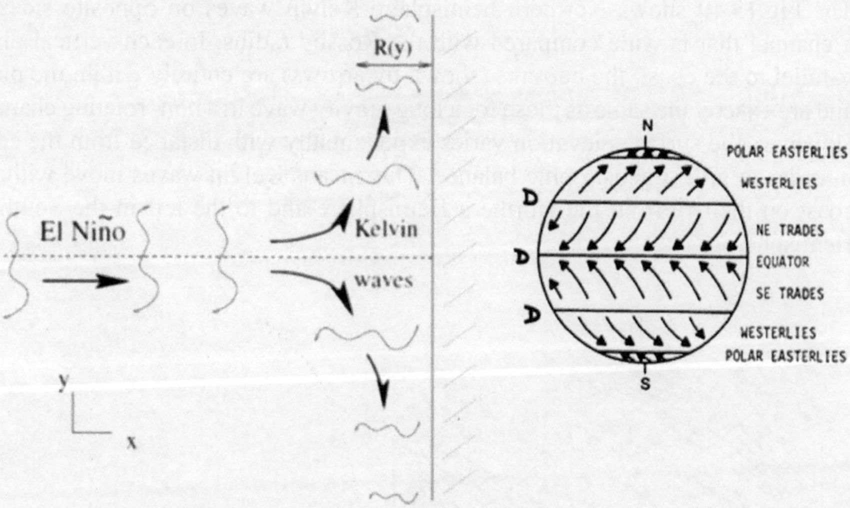


**Fig.15.48.** The generation of poleward propagating Kelvin wave along eastern oceanic boundary by equatorial variability (during ElNino). The thick solid arrows indicate the Kelvin wave propagation direction. The curved arrows indicate the width of the Kelvin wave in the perpendicular direction. The deformation radius $R = \sqrt{gh}/f$ decreases with increase of latitude since f increases. Note the offshore decay scale (shown by dashed curve) of the wave decreases exponentially. The balance of horizontal pressure gradient and coriolis gives the flow in the meridional. Convergence occurs near the boundary due to Ekman Net transport.

## Decay scales of kelvin waves

Taking Pressure gradient from reduced gravity, phase speed (C ) = $\sqrt{gh}$

Scale of amplitude decay (Rossby Deformation Radius, R) = $\dfrac{C}{f} = \dfrac{\sqrt{gh}}{f}$

## Example of Decay Scales - Barotropic Kelvin Wave

Taking in deep Water: h = 5km, C = $\sqrt{gh} = \sqrt{10m/s^2 \times 5000m} = 2.23 \times 10^2 m/s \approx 220m/s$

Taking f = $10^{-4}$, C/|f| = $220ms^{-1} / 10^{-4}/s$ = 2200km ~ 2000 km

In Shallow Water: h = 100m, C = $\sqrt{gh}$ = $\sqrt{10\times 100m} = 3.16\times 10m \approx 30\,m/s$, f = $10^{-4}$,

$\frac{C}{f} = \frac{30}{10^{-4}} = 3\times 10^{5} \sim$ 300km

## Example of Decay Scales - Internal Kelvin Wave

h = 150m, $c = \sqrt{gh}$ = $\sqrt{.02\times 150m} = 0.3\times 10m \approx$ g*=.02, C = 3m/s, f = $10^{-4}$, C/|f| ~ 30 km

The Fig.15.49 shows Northern hemisphere Kelvin waves on opposite sides of a channel that is wide compared with the Rossby radius. In each vertical plane parallel to the coast, the currents (shown by arrows) are entirely within the plane and are exactly the same as those for a long gravity wave in a non- rotating channel. However, the surface elevation varies exponentially with distance from the coast in order to give a geostrophic balance. This means Kelvin waves move with the coast on their right in the Northern Hemisphere and to the left in the Southern Hemisphere.

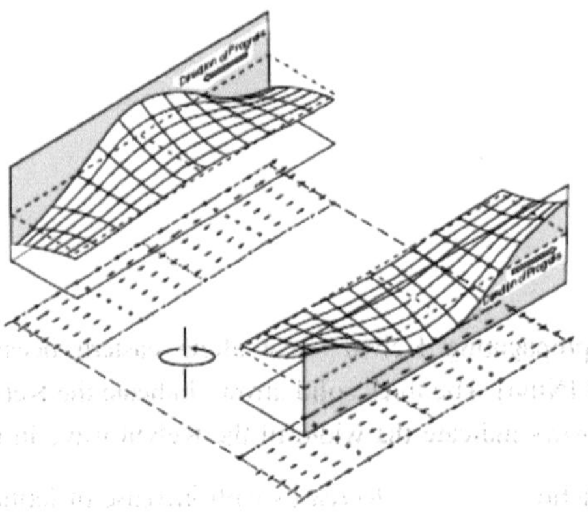

**Fig. 15.49.** Northern hemisphere Kelvin waves (Mar555 lect 23)

### 15.10.3.3. Internal Kelvin wave

The horizontally propagating internal or baroclinic Kelvin wave behaves in the same manner as the surface wave except that the motion varies with depth as shown in Fig.15.50.

For internal Kelvin waves, the pressure gradient force normal to the lateral boundary arises from the tilt of the interface and is balanced by the coriolis force such that the vertical differential flow is parallel to the boundary. In the ocean, the typical speed of internal coastal boundary Kelvin wave is of the order of 1 m/s and the Rossby radius of deformation ($C_0/f$) is of the order of 10 km in the mid latitude.

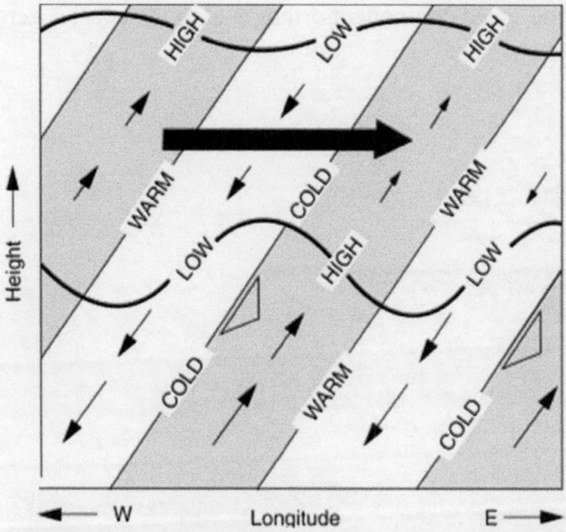


**Fig. 15.50.** Longitude-height section along the equator showing pressure, temperature and wind pertrubation for a thermally damped Kelvin wave. Heavy wavy lines indiacate material lines, short blunt arrows show phase propagation. Areas of high pressure are shaded. Length of small thin arrows is proportional to the wave amplitude which decreases with height due to damping. The large shaded arrow indicates the net mean flow acceleration due to the wave stress divergence.(Fedorov and Brown,2009)

## Derivation of kelvin wave

Take the equation of motion and continuity:

$$\frac{du}{dt} = -\alpha\frac{\partial p}{\partial x} + fv$$

$$\frac{dv}{dt} = -\alpha\frac{\partial p}{\partial y} - fu$$

$$\frac{\partial u}{\partial x} + \frac{\partial v}{\partial y} + \frac{\partial w}{\partial z} = 0$$

Assumptions

1. Let us assume the ocean is barotropic ( $\rho$ = constant)
2. The horizontal pressure gradients are expressed in terms of surface elevation $\eta$ gradients to incorporate the rise and fall of the free surface.
3. An eastern boundary exists such that the meridional velocity, v = 0
4. In the horizontal equations, w = 0
5. non linear terms are neglected.

It is a special solution, where the speed component orthogonal to the coast is zero (Fig.15.51)

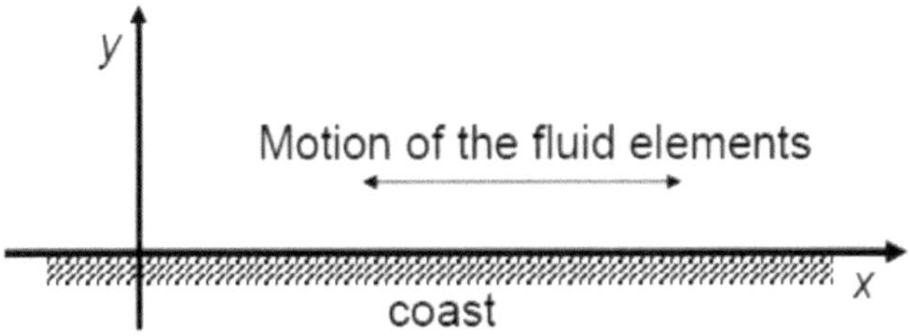


**Fig. 15.51.**

Zonal component:

$$\frac{\partial u}{\partial t}+u\frac{\partial u}{\partial x}+v\frac{\partial u}{\partial y}=fv-\alpha\frac{\partial p}{\partial x}$$

Using the assumptions 2, 3,4 and 5 and using hydrostatic equation s $\alpha\partial p=g\partial z$, this zonal component reduces to

$$\frac{\partial u}{\partial t}=-g\frac{\partial z}{\partial x}=-g\frac{\partial \eta}{\partial x} \tag{15.53}$$

Meridional component:

$$\frac{\partial v}{\partial t}+u\frac{\partial v}{\partial x}+v\frac{\partial v}{\partial y}+w\frac{\partial v}{\partial z}=-fu-\alpha\frac{\partial p}{\partial y}$$

Using the assumption of 3 and hydrostatic assumption it becomes

$$fu=-g\frac{\partial z}{\partial y}=-g\frac{\partial \eta}{\partial y} \tag{15.54}$$

Equation of continuity:

$$\frac{\partial u}{\partial x}+\frac{\partial v}{\partial y}+\frac{\partial w}{\partial z}=0$$

on integrating we get

$$\int_{-h}^{\eta}\left(\frac{\partial u}{\partial x}+\frac{\partial v}{\partial y}\right)dz=-\int_{-w_h}^{w_\eta}dw$$

$\left[\frac{\partial u}{\partial x}+\frac{\partial v}{\partial y}\right][z]_{-h}^{\eta}=-\left[w_\eta-w_{-h}\right]=-w_\eta$ since at z = -h, $w_{-h}$ =0 which means vertical velocity vanishes at the sea bottom.

$\therefore w_\eta=\frac{dz}{dt}=\frac{d\eta}{dt}=\frac{\partial \eta}{\partial t}$ ($\because$ non- linear components neglected)

$$\therefore \frac{\partial \eta}{\partial t} = -(h+\eta)\left[\frac{\partial u}{\partial x} + \frac{\partial v}{\partial y}\right] \quad (15.55a)$$

Putting (h + η) = D, equation 15.55a can be written as

$$\therefore \frac{\partial \eta}{\partial t} = -D\left[\frac{\partial u}{\partial x}\right] \ (\because \ v = 0 \text{ is assumed}) \quad (15.55b)$$

So the three required equations for the derivation of Kelvin waves are: 15.53,1554 and 15.55b).

$$\frac{\partial u}{\partial t} = -g\frac{\partial \eta}{\partial x} \quad (15.56)$$

$$fu = -g\frac{\partial \eta}{\partial y} \quad (15.57)$$

$$\frac{\partial \eta}{\partial t} = -D\left[\frac{\partial u}{\partial x}\right] \quad (15.58)$$

## Solution

Find one equation for 'η' using (15.58) and (15.56).

Differentiate eqn. (15.58) with respect to (w.r.t) 't':

$$\frac{\partial}{\partial t}\left[\frac{\partial \eta}{\partial t}\right] = -D\frac{\partial}{\partial t}\left[\frac{\partial u}{\partial x}\right] \text{ which can be written as } \frac{\partial^2 \eta}{\partial t^2} = -D\left[\frac{\partial^2 u}{\partial x \partial t}\right] \quad (15.59)$$

Differentiate eqn. (15.56) with respect to (w.r.t) 'x':

$$\frac{\partial}{\partial x}\left(\frac{\partial u}{\partial t}\right) = \frac{\partial}{\partial x}\left[-g\frac{\partial \eta}{\partial x}\right] \text{ which can be written as } \frac{\partial^2 u}{\partial x \partial t} = -g\frac{\partial^2 \eta}{\partial x^2} \quad (15.60)$$

From eqn (15.59) and (15.60) we can write:

$$\frac{\partial^2 \eta}{\partial t^2} = -gD\left[\frac{\partial^2 \eta}{\partial x^2}\right] = 0, \text{ put } gD = C^2 \text{ then we can write}$$

$$\frac{\partial^2 \eta}{\partial t^2} = -C^2\left[\frac{\partial^2 \eta}{\partial x^2}\right] = 0 \quad (15.61)$$

The solution of eqn (15.61) can be written as $\eta = a_y e^{i(kx-\sigma t)}$ (15.62)

Where a = amplitude of the wave, k = wave number ($\frac{2\pi}{L}$) and σ = wave radian frequency ($\frac{2\pi}{T}$) such that L = wave length and T = time period of the wave.

Equation (15.62) is a wave equation. For Kelvin waves it is further assumed that the wave travels along the wall (coast in the case of oceans) $e^{ikx}$ with the asymptote $e^{i(kx-\sigma t)}$.

## Solving equation (15.62)

Combine equation (15.56) and (15.57): To do this

Differentiate eqn (15.57) w.r.t "t"

$$f\frac{\partial}{\partial t}(u) = \frac{\partial}{\partial t}(-g\frac{\partial \eta}{\partial y}) \text{ which is equal to } f\frac{\partial u}{\partial t} = -g\frac{\partial^2 \eta}{\partial y \partial t} \tag{15.63}$$

Substitute eqn (15.56) in (15.63) $f\left[-g\frac{\partial \eta}{\partial x}\right] = -g\frac{\partial^2 \eta}{\partial y \partial t}$ which is equal to

$$f\left[\frac{\partial \eta}{\partial x}\right] - \frac{\partial^2 \eta}{\partial y \partial t} = 0 \tag{15.64}$$

Using equation (15.62) and (15.64) solve for $\eta$:

Differentiate eqn (15.62) w.r.t "x"

$$\frac{\partial \eta}{\partial x} = a(ik)e^{i(kx-\sigma t)} \tag{15.65}$$

Differentiate eqn (15.62) w.r.t "t":

$$\frac{\partial \eta}{\partial t} = a(-i\sigma)e^{i(kx-\sigma t)} \tag{15.66}$$

Further differentiate eqn (15.66) w.r.t 'y'

$$\frac{\partial^2 \eta}{\partial y \partial t} = (-i\sigma)e^{i(kx-\sigma t)}\frac{\partial a}{\partial y} \tag{15.67}$$

Substituting (15.65) and (15.67) in (15.64) and simplifying:

$$fkdy = -\sigma\frac{da_{(y)}}{a} \text{ on integrating we get } \int\frac{da_{(y)}}{a} = \left[-\frac{fk}{\sigma}\right]\int dy \tag{15.68}$$

$$\therefore \ln(a_y) = \left[-\frac{fk}{\sigma}\right]y + \ln C \tag{15.69}$$

The additive integration constant 'C' can be eliminated by a proper reexamination of 'y' coordinate. Eqn (15.69) can be written as

$$\therefore \ln\left(\frac{(a_y)}{C}\right) = \left[-\frac{fk}{\sigma}\right]y \text{ which is equal to } a_{(y)} = Ce^{-\frac{fk}{\sigma}y}$$

to find 'C', when y $y \to 0$, $C = a_{(0)} = \eta_0$ (say)

$$\therefore a_{(y)} = \eta_0 e^{-\frac{fk}{\sigma}y} \tag{15.70}$$

Substituting eqn (15.70) in eqn (15.62)

$$\eta = \eta_0 e^{-\frac{fk}{\sigma}y} e^{i(kx-\sigma t)} \tag{15.71}$$

Expanding (15.71) further

$$\eta=\eta_0 e^{-\frac{fk}{\sigma}y}\left[\cos(kx-\sigma t)+i\sin(kx-\sigma t)\right] \tag{15.72}$$

Eqn (15.72) has real and imaginary parts. The imaginary part is zero (as in left hand side there is no imaginary part). So taking only the real part we can write $\eta=\eta_0 e^{-\frac{fk}{\sigma}y}\left[\cos(kx-\sigma t)\right]$ .We know $C=\frac{\sigma}{k}$ is the wave celerity. Then we can write

$$\eta=\eta_0 e^{-\frac{y}{R}}\left[\cos(kx-\sigma t)\right] \tag{15.73}$$

Equation (15.73) is called the Kelvin wave profile equation.

Where $R=\frac{C}{f}$ (15.74)

is called the Rossby radius of deformation (Fig.15.52).

We already put v = 0 (15.75)

And $C^2$ = gD or $C=\sqrt{gD}$ (15.76)

<u>To find 'u':</u>

<u>Differentiate eqn (15.71) w.r.t to 'y'</u>

$$\frac{\partial\eta}{\partial y}=\eta_0\left[-\frac{fk}{\sigma}\right]e^{-\frac{fk}{\sigma}y}e^{i(kx-\sigma t)}+e^{-\frac{fk}{\sigma}y}e^{i(kx-\sigma t)}\frac{\partial\eta_0}{\partial y}$$

The second term is zero, since $\eta_0$ is constant ($\eta_0 = a_0$)

$$\therefore\frac{\partial\eta}{\partial y}=\eta_0\left[-\frac{fk}{\sigma}\right]e^{-\frac{fk}{\sigma}y}e^{i(kx-\sigma t)} \tag{15.77}$$

Substitute eqn (15.77) in eqn (15.57) and simplfy

$u=\eta_0\left[\frac{gk}{\sigma}\right]e^{-\frac{fk}{\sigma}y}e^{i(kx-\sigma t)}$, substituting eqn (15.71) in this we get $u=\eta\left[\frac{gk}{\sigma}\right]=\eta\left[\frac{g}{c}\right]$

But w know from eqn (15.76) $C=\sqrt{gD}$

$$\therefore u=\eta\left[\frac{g}{\sqrt{gD}}\right]=\eta\sqrt{\frac{g}{D}} \tag{15.78}$$

So from eqns 15.73, 15.74, 15.75,15.76 and 15.78 we can write Kelvin wave parameters as

$$\eta=\eta_0 e^{-\frac{y}{R}}\left[\cos(kx-\sigma t)\right] \tag{15.73a}$$

$R=\frac{C}{f}$ Rossby radius of deformation (15.73a)

v = 0

$C = C_g = \sqrt{gD}$ where $C_g$ is group velocity

$$u = \eta\sqrt{\frac{g}{D}} \quad a_{(y)} = \eta = \eta_0 e^{\pm\frac{y}{R}}\left[\cos\left(kx \pm \sigma t\right)\right]$$

$$a_{(y)} = \eta = ae^{\pm\frac{y}{R}}\left[\cos\left(kx \pm \sigma t\right)\right] \tag{15.73b}$$

$$a_{(x)} = \eta = ae^{\pm\frac{y}{R}}\left[\cos\left(ky \pm \sigma t\right)\right] \tag{15.73b}$$

The whole derivation briefly is in the box:

(1) $\frac{\partial u}{\partial t} - fv = -g\frac{\partial \eta}{\partial x}$ ⟶ (1k) $\frac{\partial u}{\partial t} = -g\frac{\partial \eta}{\partial x}$

(2) $\frac{\partial v}{\partial t} + fu = -g\frac{\partial \eta}{\partial y}$ ⟶ (2k) $fu = -g\frac{\partial \eta}{\partial y}$

(3) $\frac{\partial \eta}{\partial t} + H\left(\frac{\partial u}{\partial x} + \frac{\partial v}{\partial y}\right) = 0$ ⟶ (3k) $\frac{\partial \eta}{\partial t} + H\frac{\partial u}{\partial x} = 0$

substituting $\frac{\partial}{\partial x}$ (1k) in $\frac{\partial}{\partial t}$ (2k) ⇒ $\frac{\partial^2 \eta}{\partial t^2} = gH\frac{\partial^2 \eta}{\partial x^2}$

The solution is: $\eta(x, y, t) = a(y)\cos(\pm kx - \sigma t)$ ; $k = \frac{2\pi}{\lambda}$ ; $\sigma = \frac{2\pi}{T}$ (we)

And the dispersion relation: $\sigma = \sqrt{gH}k$

So that using (1k) $u(x, y, t) = \pm\sqrt{\frac{g}{H}}a(y)\cos(\pm kx - \sigma t)$ (wu)

sign+/- correspond to the propagation of the wave crests in the positive/negative direction
$a(y)$ is a function of $y$, and is obtained replacing (we) and (wu) in (2k)

$a(y) = \mp\frac{a(y)}{R_R}$, $R_R = \frac{\sqrt{gH}}{f}$ = Rossby radius

In the table above $\eta_0$ = a and R = $R_R$ is used as defined

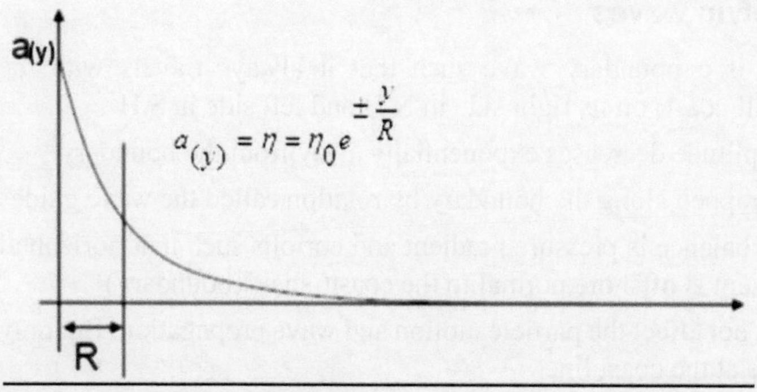


**Fig. 15.52.** Kelvin wave profile

## Explanation

Actually four possible solutions can be realized basing on equation (15.73b) as given below:

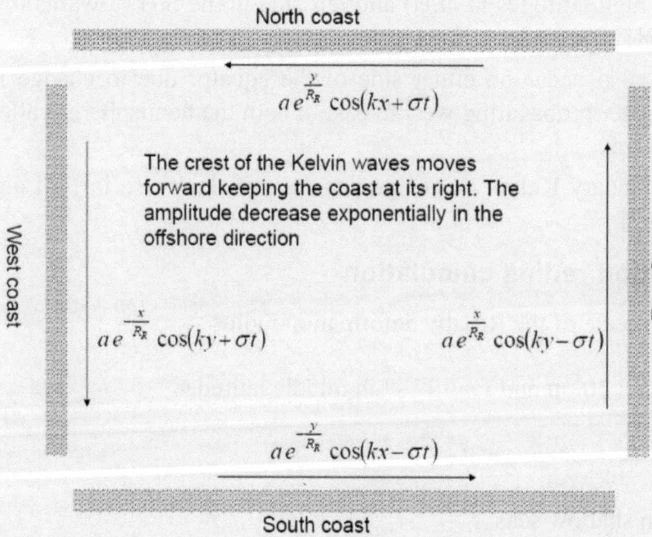


Case i): If the wave is on the y > 0 side as per equation (15.73b) it becomes

Limit y $\rightarrow \infty, \eta = 0$ (finite) $\therefore C = +\sqrt{gD} \Rightarrow \frac{\sigma}{k} > 0$ or k > 0 (+ve)

This means the wave travels in the positive (+ve) x direction.

Case ii): If the wave is on the y < 0 side then

Limit y $\rightarrow \infty, \eta = 0$ (finite) which means the wave travels in the same mode of direction in either hemispheres.

Case iii) In either case in N.H, f > 0

## Properties of Kelvin waves

1. Kelvin wave is a boundary wave such that it always travels with the boundary (wall/coast) on its right side in N.H and left side in S.H
2. The wave amplitude decreases exponentially away from the boundary
3. The wave is trapped along the boundary by rotation called the wave guide
4. The forces of balance is pressure gradient and coriolis such that horizontal pressure gradient is offshore normal to the coast( shelf)(boundary)
5. Rotation does not affect the particle motion and wave propagation. But only traps the wave at the coast line.
6. The decay scale is $\frac{C}{f}$ and is called as the Rossby deformation radius which is a measure of its decrease in amplitude as it goes northward.
7. The amplitude is greatest at the boundary and decays exponentially away from it.
8. It propagates meridionally such that the coast is on its right side in the N.H (towards high latitudes in N.H) and left side in the S.H (towards low latitudes in S.H)
9. Kelvin waves also occur on either side of the equator due to change in Coriolis parameter propagating west to east in both the hemispheres called double Kelvin waves.
10. In a closed boundary Kelvin waves propagates anticlockwise in N.H and clockwise in S.H.

## Rossby deformation radius calculation

Calculation of length scale of the Rossby deformation radius $\frac{C}{f} = \frac{\sqrt{gh}}{f}$ :

i) If h = 5 km = $5 \times 10^3$ m and f = $10^{-4}$ $s^{-1}$ in middle latitudes

$$\therefore R = \frac{\sqrt{gh}}{f} = \frac{\sqrt{9.8 \times 5 \times 10^3}}{10^{-4}} = 2200 km$$

ii) If h = 100 m in shallow seas

$$R = \frac{\sqrt{gh}}{f} = \frac{\sqrt{9.8 \times 10^2}}{10^{-4}} = 300 km$$

The graphic design of dispersion relation of kelvin wave is shown in Fig. 15.53.

**Dispersion relations**

Gravity wave in a rotating reference frame in hydrostatic balance (Poincarè-Sverdrup waves) ⟶ $\sigma^2 = f^2 + gH(k^2 + l^2)$

Non hydrostatic gravity wave ⟶ $\sigma^2 = gk\tanh(kH)$

Kelvin waves (tides) ⟶ $\sigma = \sqrt{gHk}$

Rossby waves (planetary waves) ⟶ $\sigma = -\beta k \dfrac{R_R^2}{1 + R_R^2\left(k^2 + l^2\right)}$ $\quad \beta = \dfrac{df}{dy}$

Internal waves (reduced gravity and coriolis in the interior of the ocean) ⟶ $\sigma^2 = f^2\cos\alpha + N^2\sin\alpha$

$$N^2 = \sqrt{-\left(\frac{g}{\rho}\frac{\partial\rho}{\partial z} + \frac{g^2}{c_s^2}\right)}$$

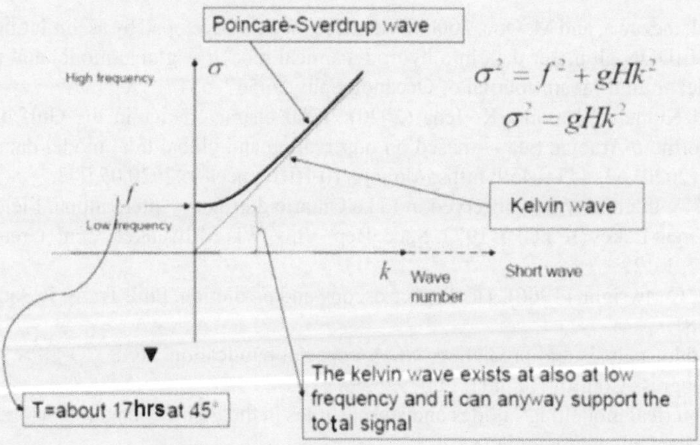


**Fig. 15.53.** Graphic design of dispersion relation of kelvin wave

## References


Boulanger J.P and Menkes C (1995): Propagations and reflections of long equatorial waves in the tropical Pacific Ocean during the 1986-1989 ENSO events as observed from GEOSAT, TOPEX/POSEIDON and TOGA-TAO data. https://www.researchgate.net/publication/32973234

Doodson, A.T., 1922. The harmonic development of the tide-generating potential. Proc. Royal Society A, 100:305-329.

Eckart, C.,1952 "The Propagation of Gravity Waves from Deep to Shallow Water," Gravity Wavs, Circular No. 521, National Bureau of Standards, Washington, D.C., 1952.

Fedorov A.V. , J.N. Brown, 2009: Encyclopedia of Ocean Sciences (Second Edition), https://www.sciencedirect.com/topics/earth-and-planetary-sciences/equatorial-wave

Ippen, A.T. and D.R.F Harleman, 1966. Tidal dynamics in estuaries. In: Estuary and Coastline Hydrodynamics, (Ed.) Arthur T. Ippen, McGraw-Hill Book Company, New York, pp. 493-545.

Jithin, A.K.; Unnikrishnan, A.S.; Fernando, V.; Subeesh, M.P.; Fernandes, R.; Khalap, S.; Narayan, S.; Agarvadekar, Y.; Gaonkar, M.; Taria, P.; Kankonkar, A., and Vernekar, S., 2017. Observed tidal currents on the continental shelf off the east coast of India. Continental Shelf Research, 141, 51–67.http://www.elsevier.com/open-access/userlicense/1.0/

Jithin, A. K., Francis P. A., Chatterjee A., Suprit, K., & Fernando, V. (2017). Validation of the simulations by the high-resolution operational ocean forecast and reanalysis system (HOOFS) for the Bay of Bengal, Tech. rep., INCOIS.

Kashyape Parag A, H. B. Jagadeesh, T. Nagendra (2017): Design of multipurpose tidal basin model for estuarine port development studies, Hydro-2017 International, L. D. College of Engineering Ahmedabad, India.. https://www.researchgate.net/publication/322635960

Lambeck, K., C. Smither, and M. Ekman, 1998: Tests of glacial rebound models for Fennoscandinavia based on instrumented sea- and lake level records. Geophysical Journal International, 135, 375-387.

Mar 555 lect 23: http://fvcom.smast.umassd.edu/Courses/MAR555/Lectures_pdf/MAR555_Lec_23.pdf

Matsumoto, K., T. Takanezawa, and M. Ooe, 2000. Ocean tide models developed by assimilating TOPEX/POSEIDON altimeter data into hydrodynamical model: A global model and a regional model around Japan. Journal of Oceanography 56:567–581

Mitra Aditi, V. Sanil Kumar , Basanta K. Jena (2020): Tidal characteristics in the Gulf of Khambhat, northern Arabian Sea — based on observation and global tidal model data, Oceanologia (2020) 62, 443—459; https://doi.org/10.1016/j.oceano.2020.05.002.

Mortimer, C. H. (1977). Internal waves observed in Lake Ontario during the lnternational Field Year for the Great Lakes (IFYGL), 1972. Spec. Rep.-Vniv. Wis-Milwaukee, Cent. Great Lakes Stud.32, 1-122.

Munk, W.H. and D.E. Cartwright, (1966). Tidal spectroscopy and prediction. Phil. Trans. R.Soc. A, 259:533-581.

Parker, B.B (2007): tide analysis and prediction, NOAA special publication, NOS CO-OPS 3, Library of Congress Control Number: 2007925298

Sindhu, B. 2012. Numerical modelling of tides and storm surges in the Bay of Bengal. Doctoral dissertation, Goa University, India.

Sindhu, B.; Suresh, I.; Unnikrishnan, A.S.; Bhatkar, N.V.; Neetu, S., and Michael G.S., (2007). Improved bathymetric datasets for the shallow water regions in the Indian Ocean. Journal of Earth System Science, 116(3), 261–274.

Sindhu, B. and Unnikrishnan, A.S., (2016). Characteristics of tides in the Bay of Bengal. Marine Geodesy, 36(4), 377–407. DOI: 10.1080/01490419.2013.781088

Sinha, P. C., S. K. Dube, A. K. Mitra, and T. S. Murty, 2000. A tidal flow model for the Gulf of Kachchh, India. Marine Geodesy 23:117–132.

TASK-2000: Production Oceangraphic Laboratory/Permanent Services to Mean Sea Level Tidal analysis software kit 2000 by (C.Bell, J.M. Vassive and P.I. Woodworth)

UC press E-book collection, 1982-2004, University of California Press (https//publishing.edlib/org/ucpressebooks/vise?=kt167nb66r&chunk.id=ch14&toc. depth=1&toc.id=ch14&brand=eschol)

Unikrishnan, a.S., S.R. shetye, and G.S. Michael, 1999. Tidal propagationin the Gulf of Science (Earth and Plantary Sciences) 108(3)" 157-177.

Wan,W.; Feng, X.; Liu, Q; YIn, B.; Yang, D., and Gao, G. 2019. Study of the tide and tidal currents in the Bay of Bengal based on refined simulations. *Journal of Coastal Research* 35(1); 33-40. *Coconut Greek* (Florida), ISSN 0749-0208.

Woodworth, P.L., 1999. High waters at Liverpool since 1768: the UK's longest sea level record Geophys. Res Lett. 26, 1589-1592. https//doi.org/10.1029/1999GL900323

## Exercises

### *Exercise 15.1*

If the moon orbits the Earth with a period of 28 days at an inclination of $28.5^0$(meaning the plane of the satellite's orbit is at $28.5^0$to the equatorial plane), what are the balance of forces act on the moon in the inertial frame and in the frame rotating with the solid Earth. (Assume that the orbit is circular and take the required standard values of Earth and Moon).

### *Solution*

We know: Mass of the Earth as: 5.983 $10^{24}$ kg; Moon's mass ($M_m$) as = 7.347 × $10^{22}$ kg (1.2% of earth's mass); The distance from Earth to Moon ($r_m$ ) as: 3.8 × $10^8$ m.

Given time period = 28 days; i.e. $T = \frac{2\pi r_m}{V_m} \Rightarrow V_m = \frac{2\pi r_m}{T} = \frac{2\pi \times 3.8 \times 10^8\, m}{28 \times 24 \times 60 \times 60s} \cong 987 ms^{-1}$

i) We know the forces of attraction at inertial frame as:

$$F_g = \frac{GM_E M_m}{r^2} = \frac{M_m V_m^{\ 2}}{r_m^{\ 2}} = \frac{7.34 \times 10^{22}\, kg \times (987)^2 m^2 s^{-2}}{3.8 \times 10^{22}} = 1.8 \times 10^{20}\, kgms^{-2} = 1.8 \times 10^{20}\, newtons$$

ii) For the inclination of $28.5^0$, $r_m = r_m \cos 28.5$, so force of attraction=

$$F_g = \frac{GM_E M_m}{r^2} = \frac{M_m V_m^{\ 2}}{r_m^{\ 2}} = \frac{7.34 \times 10^{22}\, kg \times (987)^2\, m^2 s^{-2}}{3.8 \times 10^{22} \cos 28.5} = 2.05 \times 10^{20}\, kgms^{-2} = 2.05 \times 10^{20}\, newtons$$

### *Exercise 15.2*

Calculate the length scale of the Rossby deformation radius taking $\frac{C}{f} = \frac{\sqrt{gD}}{f}$ :

i) If D = 5 km = 5 × $10^3$ m and f = $10^{-4}$ $S^{-1}$ in middle latitudes

### *Solution*

$$\therefore R = \frac{\sqrt{gD}}{f} = \frac{\sqrt{9.8 \times 5 \times 10^3}}{10^{-4}} = 2200 km$$

ii) If D = 100 m in shallow seas

$$R = \frac{\sqrt{gD}}{f} = \frac{\sqrt{9.8 \times 10^2}}{10^{-4}} = 300km$$

### Exercise 15.3

The Red Sea between Arabian Peninsula and North East African continent has an average depth of 500 m and a coastal perimeter of 2100 km. What is the trapping distance (Rossby deformation radius) and frequency of this wave at $30^0$ N latitude?

### Solution

$$C = \sqrt{gD} = \sqrt{9.8 \times 500} = 70m/s$$

L/T = 70 m/s which means T = L/70 = $2100 \times 10^3$ m ÷ 70 = $30 \times 10^3$ seconds = 8hours 20 minutes. This means this Kelvin wave will take 8 hours 20 minutes to travel around the shores of red Sea at $30^0$ N latitude. The frequency is given by σ = wave radian frequency ($\frac{2\pi}{T}$).

$$\therefore \sigma = \frac{2\pi}{T} = \frac{2\pi}{30 \times 10^3} = \frac{12}{10^3} \text{ rad/sec}$$

Trapping distance: $R = \frac{\sqrt{gD}}{f} = \frac{\sqrt{9.8 \times 500}}{2 \times 7.3 \times 10^{-5} \times \sin 30} = 10^6 m = 10^3 km$

### Exercise 15.4

If the Ocean is 4 km deep, what is the Rossby wave speed at latitude $45^0$ for a wave of 10,000 km zonal wave length? (Take $\beta_0 = 2 \times 10^{-11}$ m$^{-1}$ s$^{-1}$ and $f_0 = 8 \times 10^{-5}$ s$^{-1}$)

### Solution

We know the Rossby wave velocity equation as,

$C = -\beta_0 \frac{gh}{f_0^2}$ substitute all the values in the equation and simplify.

$$C = \frac{2 \times 10^{-11} 9.8 \times 4 \times 10^3}{\left(8 \times 10^{-5}\right)^2} = 122.5 m/s$$

### Exercise 15.5

i) Show that the phase velocity and group velocity are same in the case of Rossby Long waves. ii) Show that the phase and group velocities of Rossby short waves are opposite.

### Solution

We know Rossby wave frequency as: $\sigma = -\frac{\beta k}{\left(\frac{1}{R^2}\right) + (k^2 + l^2)}$

Let us ignore the north south wave number 'l' for simplicity as it has not much effect on wave propagation.

$$\sigma = -\frac{\beta k}{(k^2 + \frac{1}{R^2})} = \frac{-\beta R^2 k}{(R^2k^2+1)} = -\beta R^2 \left[\frac{k}{R^2k^2+1}\right] \quad (1)$$

## A) Rosbby frequencies for long and short waves

From equation (1) , we can distinguish Rosbby frequencies for long and short waves as:

iii) If KR << 1, $(1+R^2K^2) \sim 1$; So $\sigma_{Long} = -\beta R^2 k$ (2)

iv) If KR >> 1, $(1+R^2K^2) \sim R^2K^2$; then $\sigma_{Short} = -\frac{\beta}{k}$ (3)

From equation (1), we can write phase velocity, $C_p = \frac{\sigma}{k} = \frac{-\beta R^2}{(R^2k^2+1)}$ (4)

## B) Rosbby phase velocities for long and short waves

From equation (4) we can write phase velocities for long and short waves as:

iii) If KR << 1 , $(1+R^2K^2) \sim 1$; So $C_{PLong} = -\beta R^2$ (5)

iv) If KR >> 1, $(1+R^2K^2) \sim R^2K^2$; then $C_{PShort} = -\frac{\beta}{k^2}$ (6)

## C) Rosbby Group velocity for long and short waves

Similarly we know the Group velocity as:

$$C_g = \frac{\partial \sigma}{\partial k} = -\frac{\partial}{\partial k}\left\{\beta R^2\left[\frac{k}{R^2k^2+1}\right]\right\} = -\beta R^2 \left[\frac{(R^2k^2+1)\frac{\partial}{\partial k}(k) - k\frac{\partial}{\partial k}(R^2k^2+1)}{(R^2k^2+1)^2}\right]$$

$$C_g = -\beta R^2 \left[\frac{1-k^2R^2}{(1+R^2k^2)^2}\right] \quad (7)$$

Conditions for long and short waves in the case of group velocity of Rossby waves:

iii) If KR << 1 , $R^2K^2$ = negligible; So $C_{GLong} = -\beta R^2$ (8)

iv) If KR >> 1, RK is finite; then $C_{gshort} = -\beta R^2 \left[\frac{-k^2R^2}{(R^2k^2)^2}\right] = +\frac{\beta}{k^2}$ (9)

## D) The condition of Rossby waves relative to the deformation radius, R

It may be noted that KR << 1, means R << $\frac{1}{k} \Rightarrow R << \frac{\lambda}{2\pi} \rightarrow longwaves$

If KR >> 1, means R >> $\frac{1}{k} \Rightarrow R >> \frac{\lambda}{2\pi} \rightarrow shortwaves$

## E) Comparison of phase and group velocities

i) From equations (5) and (8) $C_{PLong} = C_{GLong} = -\beta R^2$. This implies phase speed and group speed are same in the case of Rossby long waves which are non-dispersive.

ii) From equations (6) and (9) $C_{PShort} = -\frac{\beta}{k^2}$ and $C_{Gshort} = +\frac{\beta}{k^2}$. This implies phase speed and group speed are different in the case of Rossby short waves which means they are dispersive.

So Rossby Long waves propagate faster as they are non- dispersive. The negative sign indicates Long waves propagate westward. However, the group velocity of short waves indicates they are not only slow due to dispersive nature but also move eastward as the sign is opposite. Short waves means their wavelengths are short relative to the Rossby radius 'R'.

### *Exercise 15.6*

i) Estimate Rossby long wave frequency and time period taking standard values

ii) Estimate Rossby short wave frequency and time period taking standard values

iii) Estimate the speeds of Rossby Long and short waves and Compare taking standard values

(take $\beta_0 = 2 \times 10^{-11}$ m$^{-1}$ s$^{-1}$; R = 30 km; $\lambda_{Longwave}$ = 300km; $\lambda_{Shortwave}$ = 30 km

### *Solution*

i). Estimation of Rossby long wave frequency and time period:

Let the longest wave length taken is 300 km, From equation (2) of Exercise (1): $\sigma_{Long} = -\beta R^2 k$

$$\sigma_{Long} = -\beta R^2 k \Rightarrow \frac{2\pi}{T} = -\beta R^2 \frac{2\pi}{\lambda} \Rightarrow T = \frac{-\lambda}{\beta R^2}$$

Given $\lambda$ = 300km = $3 \times 10^5$ m; R =30km = $3 \times 10^4$ m; $\beta = 2 \times 10^{-11}$ m$^{-1}$ s$^{-1}$

d) $\sigma_{Long} = -\beta R^2 \frac{2\pi}{\lambda} = 2\times10^{-11} m^{-1} s^{-1} \times (3\times10^4)^2 m^2 \times \frac{2\pi}{3\times10^5 m} = 36\times10^{-8} \approx 10^{-7} s^{-1}$

e) $T = \frac{-\lambda}{\beta R^2} = \frac{3\times10^5 m}{2\times10^{-11} m^{-1} s^{-1} \times (3\times10^4)^2 m^2} = \frac{1}{6}\times10^8 s = 192 days \approx 200 days$

f) Long wave phase velocity: From equation (5)of Exercise (1):

$C_{PLong} = -\beta R^2 = 2 \times 10^{-11}$ m$^{-1}$ s$^{-1}$ $\times(3 \times 10^4)^2$ m$^2$ = $1.8 \times 10^{-2}$ ms$^{-1}$ = 1.8cms$^{-1}$

D) Estimation of Rossbyshort wave frequency and time period

Let λ = 30km = 3 × $10^4$m (max value taken)as λ ≤ R for short waves; $\beta_0 = 2 \times 10^{-11}$ $m^{-1}$ $s^{-1}$

d) From equation (3) short wave frequency :

$$\sigma_{Short} = -\frac{\beta_0}{k} \Rightarrow \frac{2\pi}{T} = -\frac{\beta_0 \lambda}{2\pi} = \frac{2\times 10^{-11} m^{-1} s^{-1} \times 3\times 10^4 m}{2\pi} = 10^{-7} s^{-1}$$

e)

$$\Rightarrow T = -\frac{4\pi^2}{\beta_0 \lambda} = -\frac{4\pi^2}{2\times 10^{-11} m^{-1} s^{-1} \times 3\times 10^4 m} = -6.6\times 10^7 s = 1.9 yrs \approx 2 yrs$$

f) Short wave Phase velocity: From equation (6):

$$C_{PShort} = -\frac{\beta}{k^2} = -\frac{\beta \lambda^2}{4\pi^2} = \frac{2\times 10^{-11} m^{-1} s^{-1} \times (3\times 10^4 m)^2}{4\pi^2} = 0.9\times 10^{-3} ms^{-1} = 9\times 10^{-2} cms^{-1}$$

This means short waves are very slow. Almost stationary

Thus long waves propagate westward very fast (1.8 $cms^{-1}$ )and short waves propagate slowly (0.09 $cms^{-1}$) eastward.

### Exercise 15.7

If the Ocean is 4 km deep, what is the Rossby wave speed at latitude $45^0$ for a wave of 10,000 km zonal wave length? (Take $\beta_0 = 2 \times 10^{-11}$ $m^{-1}$ $s^{-1}$ and $f_0 = 8 \times 10^{-5} s^{-1}$)

### Solution

We know the Rossby wave velocity equation as,

$C = -\beta_0 \frac{gh}{f_0^2}$ substitute all the values in the equation and simplify.

$$C = \frac{2\times 10^{-11} 9.8\times 4\times 10^3}{\left(8\times 10^{-5}\right)^2} = 122.5 m/s$$

### Exercise 15.8

Calculate the speed of tsunami in deep water at a depth of 5 km.

### Solution

Tsunami is a shallow water wave. So shallow water wave equation is to be used in the deep ocean as: $C = \sqrt{gd} = \sqrt{9.8\times 5000}$ = 70 m/s = 252 km/hr.

### Exercise 15.9

Calculate the speed and time period of semidiurnal tides in the deep ocean at a depth of 10,000 m if their wave length is 31,590 km.

## *Solution*

As tides are shallow water waves: the equation used is $L = \frac{gT^2}{2\pi}$.

So the time period is $T = \sqrt{\frac{2\pi L}{g}} = \sqrt{\frac{2\pi \times 31590 \times 10^3 m}{9.8m}} = 45 \times 10^2 \sec onds$

$$C = \frac{L}{T} = \frac{31590 \times 10^3 m}{45 \times 10^2} = 2527.2km / hr = 702ms^{-1}$$

# 16

# Estuaries, Seiches Internal Waves and Storm Surges

## 16.1. Estuaries

The word estuary is derived from the latin word 'aestus' meaning tide and the adjective 'aestuarium' means tidal. This means low ground covered by the sea at high water.

If the coastal ocean is the region where the continents and the ocean overlap, the estuaries are the place where they really meet. In estuaries, freshwater collected over vast regions of the land pours into an ocean, which sends salt water upstream far beyond the river mouth. Various definitions of estuaries given by different scientists are as below:

Pritchard (1967) - a semi-enclosed coastal body of water which has a free connection with the open sea and within which seawater is measurably diluted with freshwater derived from land drainage.

Fairbridge (1980) – an inlet of the sea reaching into a river valley as far as the upper limit of tidal rise, usually being divisible into three sectors: a) a marine or lower estuary, in free connections with the open sea; b) a middle estuary subject to strong salt and freshwater mixing; and c) an upper or fluvial estuary, characterized by freshwater but subject to strong tidal action. The limits between these sectors are variable and subject to constant changes in the river discharges.

Day (1981) - a partially-enclosed body of water which is either permanently or periodically open to the sea and within which there is a measurable variation of salinity due to the mixture of seawater with freshwater derived from land drainage.

Kjerfve (1989) - a coastal indentation that has a restricted connection to the ocean, remains open at least intermittently, and has a tidal river zone, a mixing zone, and a nearshore turbid zone.

A classical widely accepted definition of an estuary according to Cameron and Pritchard (1963) is 'an estuary is a semi enclosed coastal body of water which has free connection with the open sea and within which sea water is measurably diluted with fresh water derived from land drainage'.

A typical estuary has most of the fresh water being discharged at its head (river joining place) and has a transitional section (mouth) between the body of the estuary and the coastal ocean and the circulation is clockwise from head to mouth along the rim of body of the estuary, while in the vertical cross section as the river water flows at the surface ( as it is relatively light), sea water flows below along the bottom(as it is heavy) as illustrated in Fig.16.1.1.

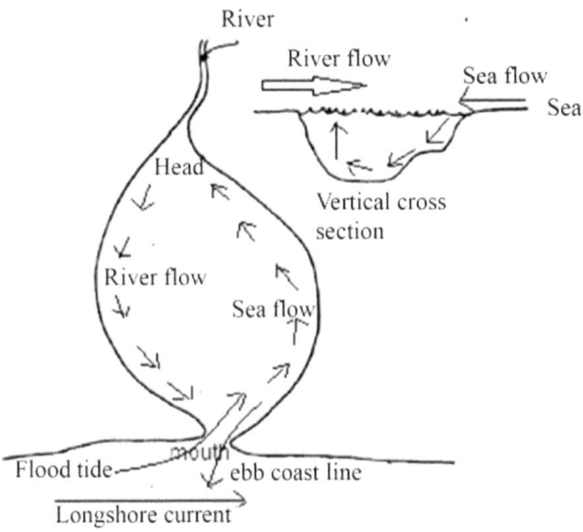


**Fig. 16.1.1.** A typical sketch of an estuary

However, actually the estuaries are located at the mouths of the rivers joining the sea. Most estuaries are geologically very young. They have developed since the last post glacial rise in the places of sea level inundated coast lines and drowned river valleys. They are now being progressively filled with sediment. The life of an estuary is expected to be more if the river transports small amount of sediment and more amount of fresh water where wave and tide action are strong. However, if the sediment discharge is high and there is limited wave and tide action then in course of time estuary vanishes and a delta is formed.

As estuaries are considered long and narrow with poor vertical dimensions, the water movement in the vertical is considered zero normal to estuary axis. Hence Coriolis force need not be considered in estuarine circulation but the pressure gradient here balances the frictional forces along the estuary that is caused due to strong tidal currents. In a way it is similar to the circulation in a laboratory tank.

The estuarine salinity field is determined by the balance between advection by the mean flow and turbulent tidal diffusion.

Mathematically this balance is expressed in a nonlinear differential equation. Nonlinear dynamic systems are much more difficult to describe than linear systems. This is because the solution to a linear problem can be found by solving simpler problems first and then adding the results. Nonlinear dynamic systems do not have this additive property, so their solution has to be found in a single attempt including all relevant forces. Thus analyses of dynamics of estuaries are more complicated than that of the open ocean.

So to solve this problem estuaries are classified into different types basing on their circulation characteristics or topography.

### 16.1.1. Motion in Estuaries

The typical circulation in estuaries is shown in Fig.16.1.2. $P_4$ is the line of Level of no motion or it is the demarcation line between the less saline riverine water above and more saline sea water below.

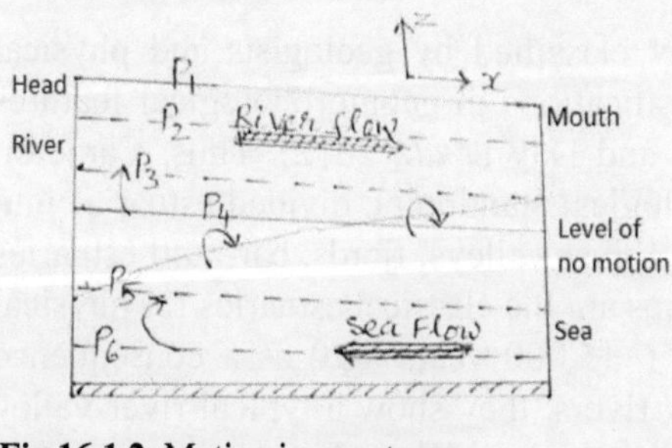


**Fig.16.1.2.** Motion in an estuary

#### 16.1.1.1 Flood and ebb tides in the Estuaries

As the tide enters into the estuary during high tide, the water level advances and increase the sea level and flood on the banks of the estuary called flood tide which is deflected towards right hand side by the Coriolis force and so the direction of arrows bends towards right as in Fig.15.28A in the chapter 15 of Tides. On the other hand during low tide time after 12 hours the water start receding back towards the sea called ebb tide. Then the arrows bend towards west as shown in Fig.15.28B in the chapter 15 of Tides in the previous tides chapter.

### 16.1.2 Some important physical processes in the Estuaries

**Entrainment:** It is a process of vertical mixing between two layers of water which flow at different velocities in the water column. Entrainment is a one-way process, so no fresh water is mixed downward. This results in a salinity increase for the upper layer, while the salinity in the lower layer remains unchanged,

**Riverine plume** : It is a surface layer of water which flows out of the river channel into the estuary, usually distinct in color (brown), and bounded by the riverine front.

**Front:** It is an interface or zone between two different water bodies which move toward each other

**Riverine Front:** It is a front where the river ouflow meets the saline water, usually distinct in color, debris gathering line.

**Salt wedge:** It is a body of saline water invading the river channel underneath the fresh water flow.

**Estuarine Plume:** It is a surface layer of water which flows out of the estuary into the open region of the coastal ocean.

**Estuarine Front:** It is a front where the estuarine plume meets the oceanic water. This front is usually much wider than the riverine front.

### 16.1.3. Classification of Estuaries

At the beginning, estuaries were mostly classified by geologists and physical oceanographers who centered their classifications in geomorphological features and salinity stratification (Perillo, 1995 and Day *et al.*, 2012). Thus, Cameron and Pritchard (1963) from a geomorphological standpoint, divided estuaries into four categories: coastal plains or drowned river valleys, fjords, bar-built estuaries and tectonic origin estuaries. Coastal plains are the classical estuaries for physical geographers, formed in the Pleistocene (~15,000 years ago) as a consequence of sea-level rise; as they were originally rivers, they show a typical river valley shape with several kilometers wide and relatively shallow. Fjords are associated with high latitudes where erosive glacial activity is intense; they are characterized by a deep elongated U-shape deep channel (several hundreds of meters), with a sill (a shallow lip) at their mouths as a consequence of moraine transport-deposit activity. Bar-built estuaries were embayments that became semi-enclosed as a consequence of littoral drift, causing the formation of a sand bar or spit (broken by one or more inlets) between the coast and the ocean; originally, they were river valleys drowned by sea-level rise; thus, this type of estuary is considered a composite system. The inlets are relatively small compared to the dimensions of the sound within the barrier; tidal action is weak, and these estuaries are usually shallow systems where the wind is the main mixing mechanism. Finally, tectonic estuaries are a consequence of geological events such as earthquakes, fractures, subsidence, creases or faults that generated deformations on the Earth's crust in adjacent regions to the ocean, forming a hollow basin; an estuary is formed when this hollow basin is drowned by the ocean.

Pritchard (1955) and Cameron and Pritchard (1963) presented another approach to the classification of estuaries, considering the vertical structure of salinity or estuarine circulation. According to water column stratification estuaries can be classified as salt wedge, strongly stratified, weakly stratified or vertically mixed (Fig. 16.1.3). There are three basic processes that produce motion and mixing in an estuary: the wind, the tide and the inflow of fresh water (Pritchard, 1967). In a wind dominated estuary (such as bar-built estuaries), the wind provides all the energy for motion and mixing of the water. On the other hand, in a tide-dominated estuary the main driver for sea and freshwater mixing is the turbulence associated to tidal currents. When mixing is low, the fresh river water flows on top of the salt water forming a well stratified estuary. Strongly stratified estuaries are the result of moderate to large river discharge and weak to moderate tidal force; these estuaries have a similar stratification profile to salt wedge estuaries, but the stratification remains stable during the whole tidal cycle; fjords and other deep estuaries are included in this category (Fig. 16.1.3A).

When moderate mixing energy is present, the estuary is slightly stratified Weakly stratified estuaries result from moderate to strong tidal action and weak to moderate river discharge; their mean salinity and density profile show a weak halocline or continuous stratification from surface to bottom except for a near bottom mixed layer (Fig.16.1.3B).

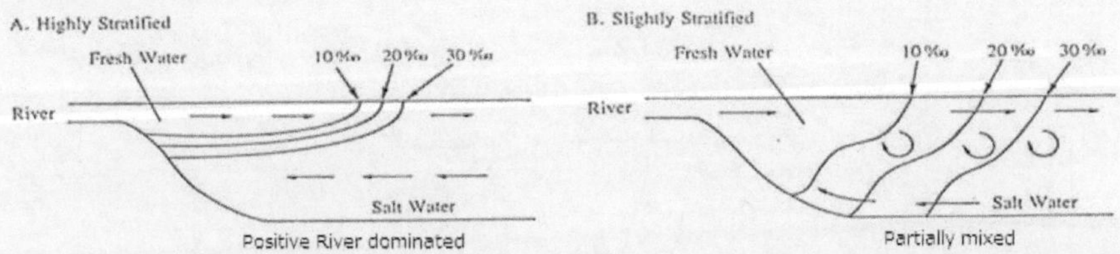


**Fig. 16.1.3A,B.** Classification of estuaries basing on mixing of river and sea water. (A): When mixing is low, the fresh river water flows on top of the salt water forming a well stratified estuary. (B): When moderate mixing energy is present, the estuary is slightly stratified.

When mixing energy is high, the estuary is vertically mixed. Strong tidal forcing and weak river discharge result in vertically mixed estuaries (minimal vertical stratification); salinity profiles in mixed estuaries are practically uniform and mean flows are unidirectional with depth; in wide (and shallow) estuaries, inflows may develop on one side across the estuary and outflow on the other side, especially during the dry season (Fig.16.1.3C).

When river water flow is low to moderate, the salt water can intrude as a salt wedge Salt wedge estuary is the simplest circulation pattern where a large, fast flowing river enters the ocean in an area where the tidal range is low to moderate. The rapidly entering freshwater holds back a wedge of salt water. The density difference allows the fresh water to flow over the salt water. The wedge of seawater will rise and fall with the tide. Thus salt wedge estuaries are the result of weak tidal force and large river discharge. In such systems, the water column is stratified due to the density difference between fresh and sea water layers (Fig. 16.1.3D).

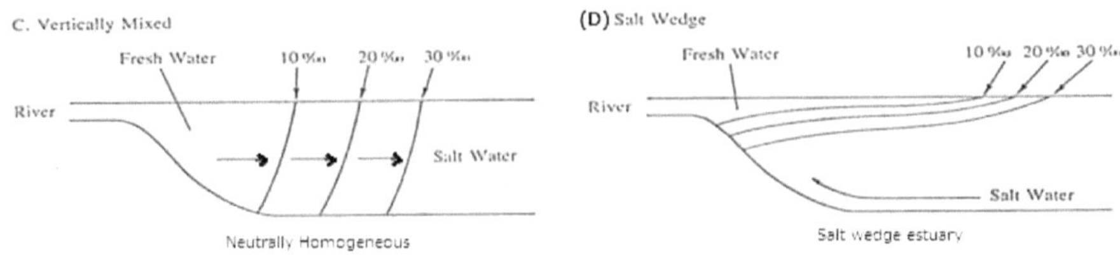


**Fig.16.1.3 C, D.** Classification of estuaries basing on mixing of river and sea water. (C) : When mixing energy is high, the estuary is vertically mixed. (D): When river water flow is low to moderate, the salt water can intrude as a salt wedge.

When the river water is poor and sea water dominates the whole estuary, there will be reverse circulation, called negative estuary (Fig.16.1.3E). In a river-dominated estuary (those from microtidal seas) the mixing is mostly produced by the breaking of unstable interfacial waves at the upper boundary of the salt wedge (halocline). This classification mainly considers competition between river discharge and tidal action.

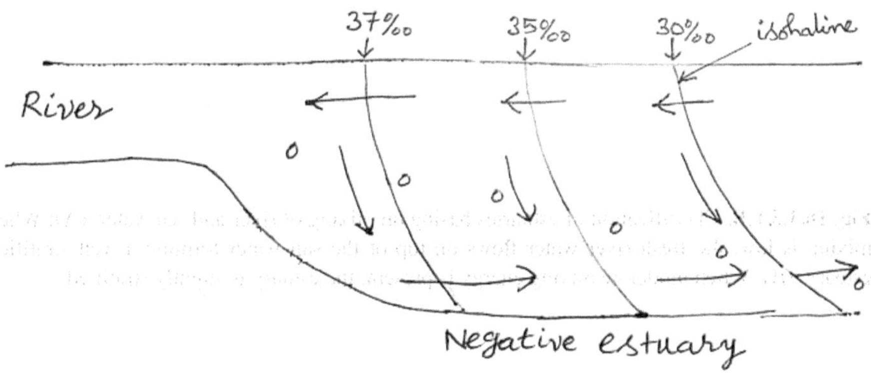


**Fig. 16.1.3E.** Classification of estuaries basing on mixing of river and sea water. (E): When the river water is poor and sea water dominates the whole estuary, there will be reverse circulation, called negative estuary.

The range of tidal amplitude varies in a constant pattern in the seas of the world. For this reason, Hayes (1975) proposed the classification of estuaries based on the tidal range that led to three different types of estuaries: microtidal, mesotidal and macrotidal. More recently, McLusky and Elliott (2004) added one more category to this classification called hyper tidal. Hence, the final classification is microtidal, tidal amplitude less than 2 meters; mesotidal, tidal amplitude between 2 and 4 meters; macrotidal, tidal amplitude between 4 and 6 meters and hypertidal, tidal amplitude higher than 6 meters.

The importance of tidal amplitude relies on the power of wave action, and therefore, on the mixing of waters and vertical profile of salinity. In addition, the tidal amplitude may considerably influence the mudflats and vegetation of an estuary. While microtidal estuaries have only limited intertidal areas, mesotidal estuaries often develop extensive intertidal areas, which are covered with vegetation, such as mangroves; whereas macrotidal and hypertidal estuaries typically have bare mudflats without large plants (McLusky and Elliot, 2004, Ibáñez *et al.*, 2012). Regarding estuarine hydrodynamics, one of the most accepted classifications was the diagram proposed by Hansen and Rattray (1966). This classification is based on two hydrodynamic parameters (stratification and circulation); in particular, by plotting the stratification versus the estuarine circulation (Fig. 16.1.4). These two parameters refer to tidal averaged (salinities are first averaged over one or more complete tidal cycle) and flow cross-sectional averaged (to smooth lateral circulation). In this way, the stratification parameter is the ratio of the salinity difference between surface and bottom ($\delta$S) and the cross-section averaged salinity ($S_0$). The circulation is simply the ratio between net surface flow-speed ($u_s$) and freshwater sectional averaged flow ($U_f$). Hansen and Rattray (1966) found that most estuaries could be grouped into four main regions on their diagram (Fig. 16.1.4). Type 1 estuaries are lagoons or bar-built estuaries; subtype 1a estuaries are vertically mixed or slightly-stratified, whereas subtype 1b estuaries show 'appreciable' vertical stratification. Both subtypes (1a and 1b) have scarce gravitational circulation and they are mainly dominated by diffusive processes. In general, type 1 estuaries describe well-mixed estuaries with net seaward flows (outflows) and no vertical structure. Type 2 includes most temperate estuaries; these systems are characterized by a reasonable well-developed gravitational and longitudinal circulation, with contribution of adventive and diffusive processes to landwards salt transfer. Again this type is subdivided in analogous subtypes like Type 1, well-mixed or weakly-stratified (Subtype 2a) and stratified estuaries (Subtype 2b). Subtype 2a are well mixed or weakly-stratified estuaries; whereas, Subtype 2b estuaries are strongly stratified. Type 3 is distinguished from Type 2, primarily by the dominance of advection (well-developed gravitational circulation) accounting for over 99% of the upstream salt transfer. This type of estuaries corresponds to fjords (deep basins with strong surface outflow and very small depth-averaged flows). Subtype 3a estuaries are moderately stratified and Subtype 3b estuaries are highly stratified. In Type 4 (salt wedge or highly stratified) the stratification is still greater than that for Type 3 estuaries; as the freshwater flow grades from a thick layer (upstream) to a narrow surface layer (close to the river mouth), the salt water flows under the freshwater layer and change from a thin upstream layer to a deep lower layer (at river mouth). Vertical mixing is limited and the gravitational circulation is weak or nonexistent. The influence between layers is scarce and restricted to a thin contact layer called halocline or pycnocline, where salt transfer by advection and diffusion processes is limited.

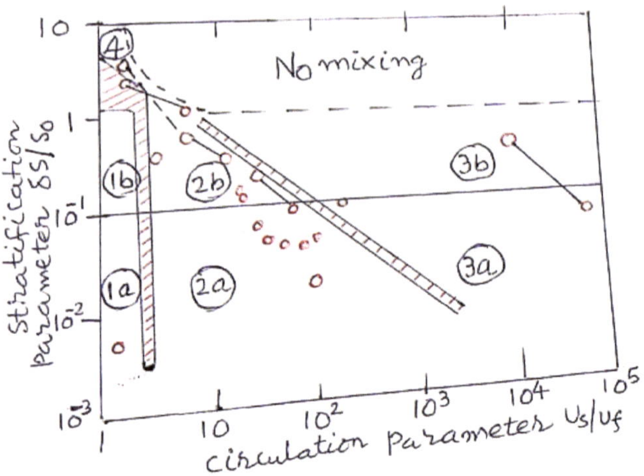


**Fig. 16.1.4.** Estuarine classification diagram using non-dimensional stratification and circulation parameters. Red dots represent examples of estuaries. The numbers denote different sub types of estuaries.

Finally, there is a classification based on water balance. Therefore, estuaries can be classified as positive, inverse or negative and low-inflow (Valle-Levinson, 2010) (Fig. 16.1.5). Positive estuaries are those in which freshwater inputs from land drainage and rain exceed freshwater losses from evaporation; these estuaries show a well-established longitudinal density gradient, with net surface outflow seawards due to freshwater contribution. Negative estuaries are typically found in arid regions where there is a scarce river discharge into the system because freshwater losses from evaporation exceed freshwater inputs, for this reason, they are called negative (Fig.16.1.3E); because the longitudinal density and salinity gradient increases landwards. Water renewal rate in negative estuaries is very low; thus, they are prone to water quality problems. Low-inflow estuaries also occur in arid regions, but show a small influence from river discharge (in the order of a few $m^3\ s^{-1}$). During dry periods, evaporation processes may cause a salinity maximum zone or 'salt plug' within the estuary. Seawards of the salt plug, the salinity decreases, as in negative estuaries; whereas, upstream of the plug, the salinity decreases as in positive estuaries. The salt plug (maximum density zone) acts as a barrier avoiding marine intrusion landwards and river discharge seawards; because of this, low-inflow estuaries are also prone to water quality problems.

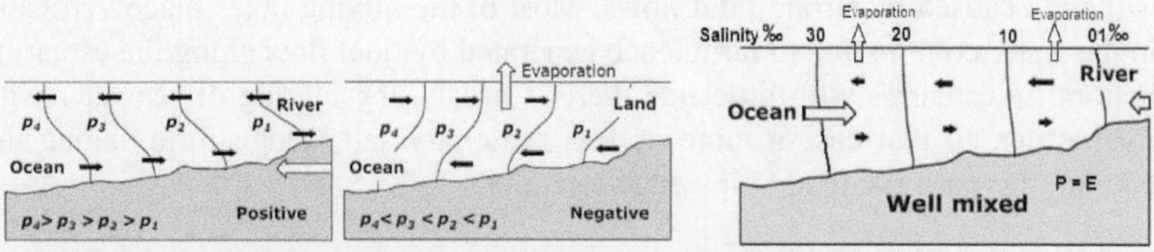


**Figure 16.1.5.** Classification of estuaries based on water balance (redrawn from Valle-Levinson, 2010). Positive (P>E), negative (inverse)(P<E), & well mixed type(P=E).

The Fig 16.1.6 below is a salt plug estuary. This type of estuary receives some fresh water from river inflow but evaporation is so strong that at some intermediate position all river water evaporates and salinity becomes higher than in the open sea. Upstream from this position the circulation is "normal"; downstream it is "reverse".

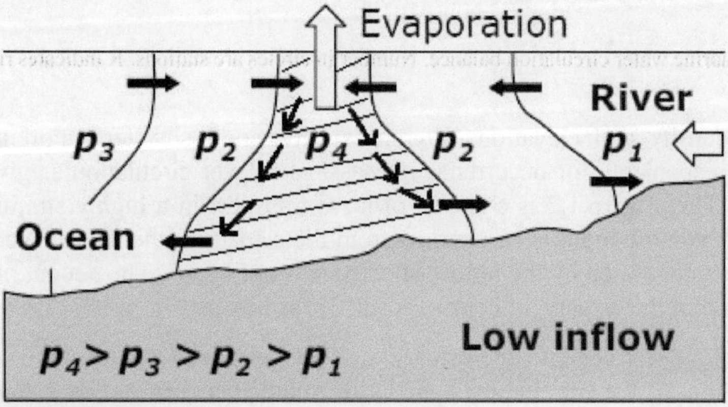


**Fig. 16.1.6** Salt plug estuary

## 16.1.4 Vertical circulation in the Estuaries

Seawater has a higher density than river water, due to its high salinity. The density difference is of the order of 2.5%. In estuaries where seawater and river water meet, seawater will tend to intrude beneath the river outflow. Part of the intruding seawater will mix with river water, whereby seawater is returned to the sea by estuarine outflow in the upper water layer. This allows 'new' seawater to enter the estuary via the lower water layers. In this way, a circulation of seawater is created, which is directed inland along the bottom and seaward along the surface, (Fig. 16.1.7). This circulation is called estuarine circulation or gravitational circulation. It is driven by the longitudinal density gradient along the estuary: higher-density seawater in the downstream part of the estuary 'diving' under lower-density mixed water in the inner estuary. The circulation is kept going by mixing of seawater with outflowing river water. Mixing is the result of turbulence, which in estuaries is mainly caused by strong tidal flows. Most of the mixing takes place vertically in the water column due to turbulence generated by tidal flow along the estuarine bottom. In estuaries with little tide there is hardly any mixing of seawater with river water. In that case a more or less stationary salt wedge forms along the bottom; this case is called Salt wedge estuaries.

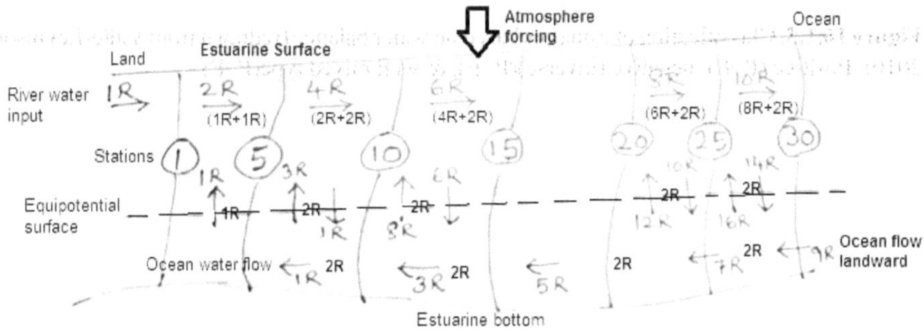


**Fig.16.1.7.** Estuarine water circulation balance. Number in circles are stations. R indicates river water input.

Induced by density and elevation differences between freshwater runoff and salt water is responsible for occurrence of classical 2-layer circulation as given in Fig.16.1.7 This Fig.16.1.7 is a sketch of mass transport in a highly stratified estuary. River volume input is R. Outflow from the estuary in the upper layer is 10R. This is compensated by the inflow of oceanic water by 9R. The net out flow at the outer end of the estuary, of course, is still 1R as below:

Let initial inflow of river into the estuary is 1R. Ocean water inflow is in the estuarine bottom. At the entrainment zone in the middle between station 1 and 5, 1R is going up and mixing with the 1R coming from the river water input. So the 1R in put before station 1 becomes 2R between stations 1 and 5. Similarly between stations 5 and 10, 2R is entrained. So the 2R before station 5 becomes 4R between stations 5 and 10. Thus this continues to add up 2R at the other remaining stations till stations between 25 and 30 and at station 30, the outflow is 10R. However, ocean water inflow at station 30 from below starting at 9R, gradually losing 2R at respective stations through entrainment and finally between station 1 and 5 only 1R entrains. Thus there is a net balance in the circulation with 1R losing to ocean. The direction of halocline tilt or curvature relative to the vertical indicates the direction of flow of water in the Fig.16.1.7, surface along which net flow is zero is called Equipotential Surface (red dashed line).

In the Fig.16.1.8 the fresh water flow is in the near surface and salt water flow is below. As the salt water entrainment continues, the upper layer gets thicker. Let the pool of light water with a density of, $\rho_1$, lies on top of the higher density of sea water, $\rho_2$. If AB is the equipotential surface, the pressure at A and B are same. But the layer above A is higher than that at B as shown in Fig.16.1.8 (Mann & Lazier 2006).

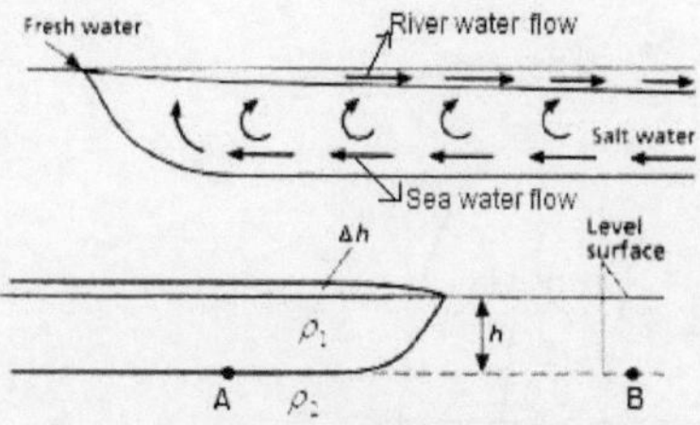


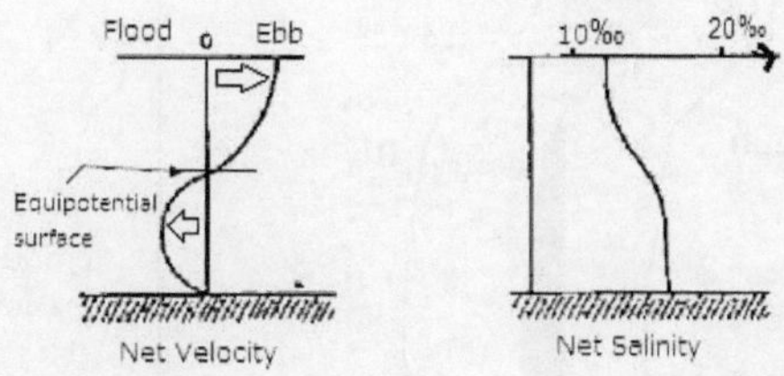


**Fig:16.1.8.** The thickness of the layer

$$\Delta h = \frac{h(\rho_2 - \rho_1)}{\rho_1}$$

Where Δh is the difference in the height of the surface water in the two regions of A and B, $\rho_1$ is the density in the fresh water layer and $\rho_2$ is the density of sea water layer and 'h' is the thickness of the equipotential surface at B. For example if $\rho_1$ = 1000 kg m$^{-3}$, $\rho_2$ = 1025 kg m$^{-3}$, h= 2m. Then

$$\Delta h = \frac{h(\rho_2 - \rho_1)}{\rho_1} = \frac{2(1025 - 1000)}{1000} = 0.05\,m = 5\,cm$$

Referring to the classification basing on salinity stratification in estuaries (Fig.16.1.3 A to D), the salt content of the estuary depends on the difference in volume of river water (R) that enters at one end of the estuary and leaves at the other end (mouth). In other words the salinity field is controlled by the balance between advection of fresh water and diffusion of salt. This diffusion is achieved through turbulence caused by tides.

The type of turbulence that occurs in an estuary is determined by the strength of the tide. This can be expressed by tidal volume (V). The tidal volume is defined as the difference in volume of water brought into the estuary by the flood tide to the ebb tide during a tidal cycle.

By comparing V and R over the same tidal period we derive the first four types of estuaries as mentioned above (Fig.16.1.3 A to D).

The vertical integral of the average velocity profile multiplied by the cross sectional area of the estuary gives net volume transport (M) which is equal to R divided by the tidal period. So the unit of M is m$^3$s$^{-1}$. The cross sectional average of the vertically averaged mean velocity is given by R/AT, where A is the cross sectional area of the estuary and T is the tidal period. For classification of estuaries the absolute values of R and V is not important but their ratio R/V is important (Fig.16.1.9).

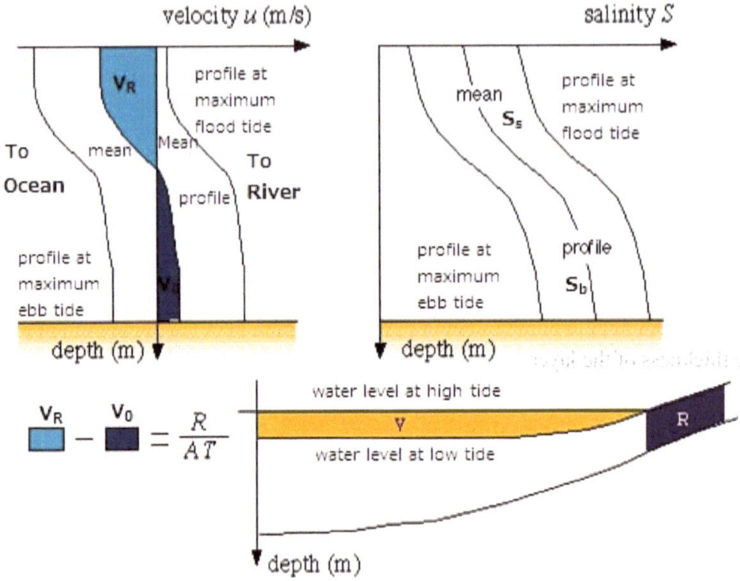


**Fig.16.1.9.** Sketch of variations of current and salinity over a tidal period *T*. Top diagrams: velocity and salinity profiles at maximum flood and ebb tides and after averaging over *T*. Bottom diagram: the relationship between tidal volume *V* and fresh water volume *R*.

As Fresh water flows in the upper layer from River to Ocean in Fig.16.1.9, the profile is concave. Similarly sea water flows from ocean to river side in the sub surface so profile is convex below. Same situation appears in salinity profile also. The difference between the shaded areas of mean velocity profile corresponds to $V_R - V_0 = \frac{R}{AT}$ where $V_R$ is Ocean volume in the above (top hatching area) and $V_0$ is ocean volume below (bottom hatching area), R is river volume, A cross sectional area of estuary and T tidal period.

The estuary type is determined by the ratio *R*: *V*. Varying this ratio produces a range of salinity distributions, which can be classified by the ratio of surface salinity $S_s$ against bottom salinity $S_b$. The ratio $S_s$: $S_b$ can therefore be used as a substitute for the ratio *R*: *V*. As salinities are easier to measure than tidal and river volume, a ratio based on salinities is therefore of more practical value (Fig. 16.1.9 salinity profile).

$$\frac{R}{V} \approx \frac{S_s}{S_b}$$

## Classification zones

While most classification schemes focused on vertical stratification. Day (1964) proposed to classify sections of the estuary based on average salinity concentrations.

Table below illustrates the most significant estuary zones.

**Table:** Zonation of estuary based on salinity

| Zone | Physiographic area | Classification | Salinity |
|---|---|---|---|
| I | Head of estuary | Oligihaline | 0.5-5.0 |
| II. | Upper reaches | Mesohaline | 5.0-18.0 |
| III | Middle-lower reaches | Polyhaline | 18.0-30.0 |
| IV | Mouth | Euhaline | >30.0 |

## 16.1.5. The classical analytical model for estuarine circulation

The model applies to a straight prismatic estuary (uniform depth h and width b); the width is small enough for neglecting the influence of earth's rotation (Rossby radius $\frac{U}{fb}<1$).

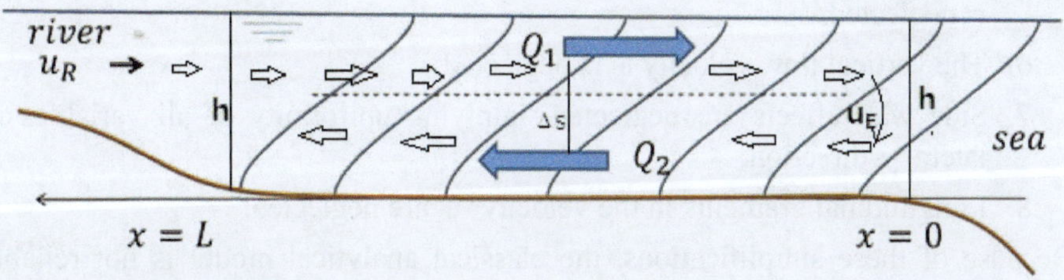


**Fig. 16.1.10:** Schematic representation of estuarine circulation in a prismatic channel with isohalines

The following notations are used (see figure 16.1.10):

'x', is the coordinate along the estuarine axis (x=0 at the sea boundary, x = L at the landward boundary of salt intrusion);

'u'is the velocity in the x-direction, $u(z) = u_c(z) + U\cos\omega t$ where $u_R$ is the velocity related to river runoff (river discharge divided by estuarine cross-section), $u_c(z)$ is the estuarine circulation, U is the depth-averaged tidal velocity amplitude and $\omega = \frac{2\pi}{T}$ is the tidal radial frequency; $S = s/s_0$ is the ratio of the local salinity s and the seawater salinity $s_0$; S is related to the density ρ by the relation $\rho = \rho_0 (1 + \beta s_0 S)$ where $\rho_0$ is the fresh water density and the expansion coefficient $\beta = 7.7 \times 10^{-4}$

$p(z) = g\int_z^{\eta} \rho\, dz$ is the hydrostatic pressure; is the surface elevation relative to a horizontal plane and (-z) is the depth relative to the surface;

$K_m$ is the eddy viscosity (vertical momentum diffusion coefficient ($m^2/s$) and $K_s$ the eddy diffusivity (vertical salinity diffusion coefficient).

Depth-averaging is denoted by an overbar.

## The model

The model is based on the following simplifications:

1. The diffusion coefficients, $K_m$, $K_s$ do not depend on the spatial and temporal coordinates x, z, t; this means that the classical model is realistic only for continuously stratified estuaries (no layered salinity structure); such estuaries are often called "well-mixed", see Fig. 16.1.10);
2. The estuarine geometry is represented by a prismatic channel;
3. The river flow velocity $u_R$ and the tidal velocity amplitude do not depend on the spatial coordinates x, z ;
4. The river flow velocity is at least an order of magnitude smaller than the tidal velocity, $u_R << U$;
5. The water surface variation is much smaller than depth, $\eta \ll h$; Stokes drift is neglected;
6. The vertical flow velocity ω is neglected;
7. Side wall effects are neglected, implying uniformity of all variables in lateral y-direction;
8. Longitudinal gradients in the velocity 'u' are neglected;

Because of these simplifications, the classical analytical model is not reliable for predicting seawater intrusion into real estuaries. In particular, the first and second simplifications violate the conditions that are representative of real estuaries and imply that important mechanisms for seawater intrusion are ignored, as discussed above. The importance of the classical analytical model lies in the simple illustration it offers for one of the major processes responsible for seawater intrusion into estuaries.

We consider the influence of a constant river flow '$u_R$' on the salinity distribution in a frame moving with the velocity $u_R + U\cos\omega t$. In the moving frame the velocity u is time-independent and depends only on the vertical coordinate z; we thus have $u = u_c(z)$. The only driving force acting on the fluid in the moving frame is the pressure gradient $\frac{\partial p}{\partial x}$, which is a linearly increasing function of depth (-z) below mean water level

$$\frac{1}{\rho g}\frac{\partial p}{\partial x} = \frac{\partial \eta}{\partial x}(\eta - z)\beta S_0 \frac{\partial s}{\partial x} \tag{16.1.1}$$

The fluid acceleration caused by this pressure gradient is balanced by momentum dissipation through turbulent diffusion:

$$\frac{1}{\rho}\frac{\partial p}{\partial x} = \frac{\partial}{\partial z}\left(K_M \frac{\partial u_c}{\partial z}\right) \tag{16.1.2}$$

This equation can be solved considering the following boundary conditions:

No slip at the bottom: $u_c$ (-h)=0, no momentum diffusion through the surface: $\frac{du_c}{dz}=0$ at and no net discharge: $\int_{-h}^{0} u_c dz = 0$

Because the river runoff is constant and the tide cyclic, the salinity distribution should not change with time in the moving frame. However, the estuarine circulation modifies the salinity distribution, because salt advection $u_c z$ S(x,z) has a gradient in x-direction. According to the classical theory, the salt balance is restored by vertical diffusion of salinity:

$$u_c \frac{\partial S}{\partial x} = \frac{\partial}{\partial z}\left(K_S \frac{\partial S}{\partial z}\right) \tag{16.1.3}$$

This equation can be solved with the boundary conditions: no salt flux through bottom and surface, $\frac{\partial S}{\partial z}=0$ at z = -h, z = 0

The solution of equations (16.1.1 to 16.1.3) is given by:

$$u_c(z) = u_E(-1 + 9\zeta^2 + 8\zeta^3);\ S = \overline{S} + s_c(z);\ s_c(z) = s_E(-5 + 30\zeta^2 - 45\zeta^4 - 24\zeta^5)$$

$$\zeta = \frac{z}{h} \tag{16.1.4}$$

$$u_E = -\frac{c^2h^2}{48K_M}\frac{\partial \overline{S}}{\partial x},\ S_E = -\frac{h^2 u_E}{60K_S}\frac{\partial \overline{S}}{\partial x} \tag{16.1.5}$$

Where $c = \sqrt{\beta s_0 g h}$ is twice the maximum internal wave speed in a layered channel with a density difference $\beta s_0$ and total depth 'h'.

## Salt balance

The motion in the moving frame is not exactly cyclic; after one tide the moving frame has moved seaward over a distance $Tu_R$. In order to prevent complete flushing of the salinity a net landward salt transport should exist. Many seawater intrusion mechanisms can contribute to landward salt transport. In this case, we assume that estuarine circulation is the primary mechanism for seawater intrusion. This means that the loss of salt content from the estuary due to the net shift of the moving frame (moving from any start position in the estuary), $hu_r S$, should be compensated by salt transport related to estuarine circulation:

$$hu_R\overline{S} = \int_{-h}^{0} u_c(z) S(z) dz \tag{16.1.6}$$

This balance equation holds for all cross-sections landward of the estuarine mouth. Using Eqs. (16.1.4 and 16.1. 5) this equation can be rewritten as a nonlinear differential equation for the depth-averaged salinity

$$\overline{S}(x) = -\left(L\frac{d\overline{S}}{dx}\right)^3, \text{ with solution } \overline{S}(x) = \left(1 - \frac{2x}{3L}\right)^{\frac{3}{2}},\ L = 0.024h^2c\left(\frac{c}{u_R K_S K_M{}^2}\right)^{\frac{1}{3}} \tag{16.1.7}$$

This equation shows that -1/L is a fair estimate for the salinity gradient $\frac{d\bar{S}}{dx}$ due to estuarine circulation in the range $0 \leq x \leq L/2$. Substitution in Eq. (16.1.5) gives the maximum estuarine circulation velocity $u_E$:

$$u_E \approx 0.87\left(u_R c^2 \frac{K_S}{K_M}\right)^{\frac{1}{3}}, S_E \approx 0.6\left(\frac{u_R}{c}\right)^{\frac{2}{3}}\left(\frac{K_M}{K_S}\right)^{\frac{1}{3}} \tag{16.1.8}$$

## Estuary number

The estuary number was introduced by Harleman and Abraham (1966) to provide a rapid estimate for the degree of stratification of an estuary. It is defined as

$$E = \frac{E_{tide}}{E_{river}}$$

$E_{tide}$ = Turbulent energy dissipated by tidal currents during a tidal cycle

$E_{river}$ = Energy needed for vertical mixing of river water spread during one tidal cycle over the estuarine surface

In formulas:

$$E_{tide} = \rho b_0 l \int \left|u_*^3(t)\,dt\right| = \text{—}\, \rho b_0 l T u_{*max}^3$$

Where T is the tidal period, ρ the estuarine water density, $b_0$ is the estuarine width at the mouth and is a representative estuarine length scale, for example, the basin length for short estuaries or the convergence length for exponentially converging estuaries ($b = b_0 \exp(-x/l)$). The turbulent energy dissipation is represented by $\rho\left|u_*^3(t)\right|$, where $\left|u_*(t)\right| \approx u_{*max}\left|sin\frac{2\pi t}{T}\right|$ is a measure of the turbulent velocity fluctuations. The turbulent velocity fluctuations are a fraction of the tidal velocity. The proportionality factor $\sqrt{cD}$ is of the order of 0.05-0.07 for well-mixed conditions, but smaller for partially mixed or stratified conditions.

$$E_{river} = g\frac{\Delta\rho}{2}\frac{h}{2}TQ_{river}$$

Where Δρ is the density difference between sea and river water (which is assumed as the twice the difference between estuary and river water), $Q_{river}$ is the river discharge.

The estuary number is then given by

$$E = \frac{16}{3\pi}\frac{\rho}{\Delta\rho}\frac{b_0 l u_{*max}^3}{ghQ_{river}}$$

If '*E*' is at least an order of magnitude larger than 1, they are called well mixed estuaries. If $E < 1$ are stratified estuaries and they are partially mixed when $E$ is less than an order of magnitude larger than unity.

## Stratification

Turbulent mixing over the water column is suppressed by the effect of density stratification, also in the case of continuously stratified (non-layered) systems. This implies that the diffusion coefficients $K_M$, $K_S$ depend not only on the turbulence generated by the tidal flow, but also on the damping by stratification. This dependence can be related to the so-called Richardson number, $R_i = -\frac{g}{\rho}\frac{\partial\rho/\partial z}{(\partial u/\partial z)^2}$.

In accordance with the strong assumption underlying the classical model (well mixed, continuously stratified estuaries, no layered salinity structure), the Richardson number can be approximated by:

$$R_i \approx \left(\frac{c}{U}\right)^2 \Delta S,\ c^2 = \beta s_0 gh,\ \Delta S = 9S_E, \tag{16.1.9}$$

For small Richardson numbers (typically much smaller than 0.25) the diffusion coefficients are not strongly influenced by vertical density gradients. In this case an estimate of the turbulent diffusion coefficients is given by $K_M \approx K_S \approx K = c_D hu$.The friction coefficient $C_D$ is of the order of 0.001 – 0.004 (Bowden 1967). Substitution in the expressions for (Eq. 16.1.7), (Eq. 16.1.8) and (Eq. 16.1.9) yields

$$u_E \approx 0.87\left(u_R c^2\right)^{\frac{1}{3}},\ \Delta S \approx 5.4\left(\frac{u_R}{c}\right)^{\frac{2}{3}},\ L \approx \frac{0.024h}{c_D}\left(\frac{c}{u_R}\right)^{\frac{1}{3}}\frac{c}{U} \tag{16.1.10}$$

The estuarine circulation and the stratification increase with increasing river run off $u_R$; they do not depend on the tidal velocity amplitude 'U'. The sea water intrusion length 'L' decreases with increasing river runoff and with increasing tidal velocity.

For situations with strong (but continuous non-layered) stratification the Richardson number is larger than 1. In this case one may use the expressions for $K_M$, $K_S$ established experimentally by Munk-Anderson.

$$K_M = K\left(1+10R_i\right)^{-\frac{1}{2}} \approx 0.316K\left(R_i\right)^{-\frac{1}{2}}, K_S = K\left(1+3.3R_i\right)^{-\frac{3}{2}} \approx 0.167K\left(R_i\right)^{-\frac{3}{2}}, K = c_D hu \tag{16.1.11}$$

With these approximations we find

$$u_E \approx 0.27U, \Delta S \approx 17\frac{u_R}{U},\ L \approx \frac{h}{c_D}\left(\frac{c}{U}\right)^3\left(\frac{u_R}{U}\right)^{\frac{1}{2}} \tag{16.1.12}$$

The estuarine circulation $u_E$ now depends only on the tidal velocity amplitude U. The situation U=0 corresponds to an arrested salt wedge. However, in this case the present model loses its validity (the bottom-surface salinity difference ΔS becomes larger than 1). The present model for estuarine circulation requires $u_R \ll U$. Assuming that this condition is satisfied, the seawater intrusion length L increases with an increase of runoff $u_R$. This counter intuitive result can be explained by the fact that the river runoff loses its 'grip' on the near-bottom seawater intrusion with increasing stratification.

### 16.1.6. Salt balance technique in estuaries

This topic discusses the methods to derive volume transports from the observations of salinity distribution. Such methods are based on the principle of salt conservation and therefore known as salt balance technique.

The basis for salt balance techniques is the diffusion equation for salt. For a two dimensional estuary it can be written as:

$$\frac{\partial}{\partial x}(us)+\frac{\partial}{\partial z}(ws)-\frac{\partial}{\partial x}\left(K_h\frac{\partial s}{\partial x}\right)-\frac{\partial}{\partial z}\left(K_v\frac{\partial s}{\partial z}\right)=0 \tag{16.1.13}$$

Where u and w are horizontal and vertical component of velocity, S is salinity, $K_h$ and $K_v$ are horizontal and vertical turbulent diffusion (Austausch) coefficients for salt, x is the horizontal coordinate measured positive from the inner end of estuary towards the sea and z is the vertical coordinate downward.

The first term in the equation (16.1.13) expresses transport of salt by horizontal current (horizontal advection). The second term expresses transport of salt by vertical water movement (vertical advection). The third term represents the net transport of salt from horizontal turbulent mixing and the last term gives the contribution to the salt transport from vertical turbulent mixing. Note that all turbulent mixing is the result of tidal action. Therefore all this depends on strength of the tide.

The equation can be simplified for different classes of positive estuaries. In the case of salt wedge and highly stratified estuary all salt transport is achieved by advection only. So the diffusion equation changes as:

$$\frac{\partial}{\partial x}(us)+\frac{\partial}{\partial z}(ws)=0 \tag{16.1.14}$$

The vertical velocity w represents the effect of entrainment from the lower layer into the upper layer.

In the case of the slightly stratified estuary, the movement of salt between upper and lower layer is achieved by turbulent mixing. Horizontal mixing is negligibly small.

$$\therefore \frac{\partial}{\partial x}(us)+\frac{\partial}{\partial z}(ws)-\frac{\partial}{\partial z}\left(K_v\frac{\partial s}{\partial z}\right)=0 \tag{16.1.15}$$

In the case of vertically mixed estuary, it is characterized by vertically uniform property. This means vertical changes vanish. So in this case horizontal advection and tidal turbulent mixing are important.

$$\frac{\partial}{\partial x}(us)-\frac{\partial}{\partial x}\left(K_h\frac{\partial s}{\partial x}\right)=0 \tag{16.1.16}$$

In the case of vertically mixed wide estuaries where lateral salinity variations are significant, then the diffusion equation is:

$$\frac{\partial}{\partial x}(us)+\frac{\partial}{\partial y}(vs)-\frac{\partial}{\partial y}\left(K_h\frac{\partial s}{\partial y}\right)=0 \qquad (16.1.17)$$

## 16.1.7. Knudsen's hydrographical theorem

The salt transport in estuaries is a balance between two processes horizontal and vertical advection of salt. This makes it possible to derive some simple calculation scheme for the determination of mean circulation. They are based on the principle of conservation of volume and salt known as Knudsen's Hydrographical Theorem (Fig.16.1.11).

Let $A_1$, $u_1$ and $S_1$ are the cross sectional area, velocity and salinity at the entry point and $A_2$, $u_2$ and $S_2$ are the corresponding points at the exit point of the area of the estuary body 'B' as shown in Fig.16.1.11. The cross sectional area is the product of width and depth. If the flow is from $A_1$ to $A_2$ it is positive and if it is reverse negative. Then

i) the volume transport per unit time at the entry point (Voloume flux)($Q_1$): $A_1u_1$

ii) the volume transport per unit time at the exit point (Volume flux)($Q_2$) : $A_2u_2$

iii) the amount of salt entering the estuary per unit time (Salt flux) : $Q_1S_1$

iv) the amount of salt leaving the estuary per unit time (Salt flux at exit point) : $Q_2\,S_2$

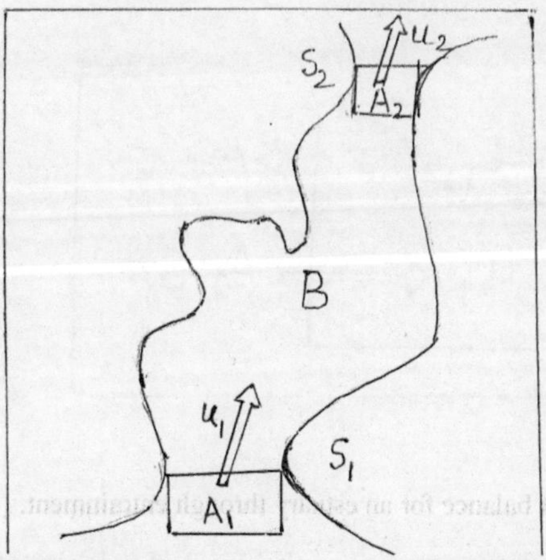


**Fig.16.1.11.** Sketch to illustrate the relationship between through flow velocity and salinity.

If there is a rainfall in the estuary, then let the rate of rainfall is r and let the area of the estuary is B. Then the volume of rain water (R) added is Br. In the steady state condition the amount of water entering the estuary is equal to the amount of water leaving the estuary.

$$\therefore Q_1 + R = Q_2 \qquad (16.1.18)$$

Similarly for conservation of salt: $Q_1S_1 = Q_2 S_2$ (16.1.19)

As rain does not add salt rainfall term is not included here. The units used here are length in meters, time in seconds. $A_1$, $A_2$ and B are in m$^2$; $u_1$, $u_2$ are in ms$^{-1}$ and convert r from mm/day to meters /second. Then solving for $Q_1$, and $Q_2$ from equations (16.1.18) and (16.1.19), we get as:

$$Q_1 = \frac{R}{\frac{S_1}{S_2} - 1}, \; Q_2 = \frac{R}{1 - \frac{S_2}{S_1}} \qquad (16.1.20)$$

A quick inspection of this result shows that if R > 0 (rain added to the system), the water moves from the region of high salinity to low salinity because in order to be R positive, $Q_1$ must be positive. $Q_1$ will be positive when $S_1 > S_2$. So water moves from $A_1$ to $A_2$. Similarly if $S_2 > S_1$, $Q_2$ is negative and hence R < 0. Which means evaporation is more than the rainfall. Then the flow is from $A_2$ to $A_1$.Which means the flow is from the river (low salinity) to ocean (high salinity side). Under this condition the concentration of salt (salt plug) will be somewhere on the way.

### 16.1.8. Entrainment

In a highly stratified estuary higher salinity sea water flows below and lower salinity river water mixture goes above as shown in Fig.16.1.12. However mixing at the interface takes place through entrainment.

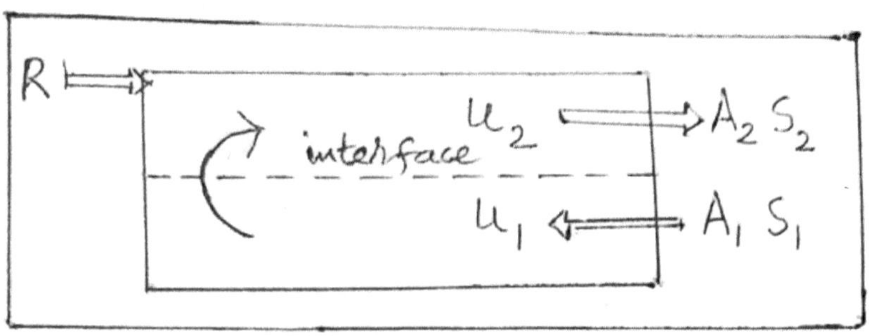


**Fig.16.1.12.** A sketch to derive the mass and salt balance for an estuary through entrainment.

If $B_e$ is the area of the estuary, the entrainment velocity or the upward velocity at the interface (w, the upward velocity at the interface) is given by:

$$wB_e = Q_2 = -\frac{S_1}{(S_1 - S_2)} R = \frac{S_1}{(S_2 - S_1)} R \qquad (16.1.21)$$

The negative sign is assigned as it is rising from high denser water to low denser water. For example, Fjords are good examples of highly stratified estuaries. The ratio $\frac{S_1}{(S_2 - S_1)}$ in Hardanger fjord of Norway varies between 2 to 6 during summer.

### 16.1.9. Iterative Knudsen Formula

Using the iteration method we can find the volume transport at second station if we know initial or first station. Let there are two stations upper and lower as shown in Fig.16.1.13. Let the through flow occurs in upper (top) and lower layer in opposite directions.

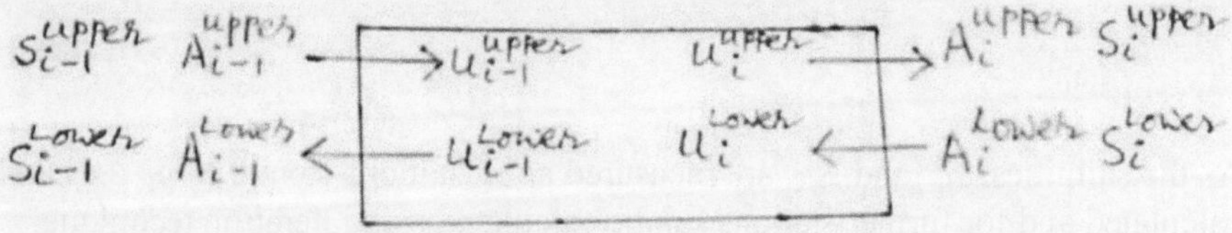

**Fig. 16.1.13.** Sketch diagram for the iterative Knudsen formula

Then according to steady state principle, what goes in must be equal to what goes out. So we can write:

$$Q_i^T - Q_i^L = Q_{i-1}^T - Q_{i-1}^L \tag{16.1.22}$$

Similarly for conservation of salt we can write as:

$$Q_i^T S_i^T - Q_i^L S_i^L = Q_{i-1}^T S_{i-1}^T - Q_{i-1}^L S_{i-1}^L \tag{16.1.23}$$

In equation (16.1.23) there are four unknowns. If we know two values at one station, we can find Q values at the other station. Solving the equations 16.1.22 and 16.1.23 we get:

$$Q_i^T = Q_{i-1}^T \frac{1 - \frac{S_{i-1}^T}{S_{i-1}^L}}{1 - \frac{S_i^T}{S_i^L}} \; similarly \; Q_i^L = Q_{i-1}^L \frac{\frac{S_{i-1}^L}{S_{i-1}^T} - 1}{\frac{S_i^L}{S_i^T} - 1} \tag{16.1.24}$$

These are known as iteration equations. Once you know the volume transport at any station (i-1), you can obtain at another station (i).

Similarly this technique can be applied if we have salinity data for {0,1,2…(n-1)n} stations starting from estuary head (river entry point) to estuary mouth. We can find the volume transport at station 'n' from the volume transport at (n-1), which we find from (n-2). The volume transport at station '0' represents the starting point or initial condition of the iteration scheme which is to be found from the independent argument as below:

Let the volume flux from the river is R and it enters on the upper layer of the estuary. So

$$Q_0^T = R, S_0^T = 0 \quad (16.1.25)$$

Because there is now lower layer at station '0' we cannot apply equation (16.1.24) to determine $Q_1^L$. But we know that $Q_1^L$ has to supply the volume flux necessary to increase the volume flux of the upper layer from R at station '0' to $Q_1^T$ at station 1.

$$\therefore Q_1^L = Q_0^T - R \quad (16.1.26)$$

Then $Q_1^T$ can be calculated from equation (16.1.24) taking i = 1 and using equation (16.1.25). This gives

$$Q_1^T = R\frac{1}{1-\frac{S_1^T}{S_1^L}} \quad (16.1.27)$$

As the salinities, $S_1^T$ and $S_1^L$, are measured at all stations from 0 to n, $Q_1^T$ can be calculated and for further stations can be calculated using iteration technique.

### 16.1.10.1. The flushing time

Increased use of estuaries by human population causes pollution in estuaries. So one of the tools used for estuarine management is the concept of flushing time. It is often used to determine how much of a potentially harmful substances an estuary can tolerate before its ecosystem is adversely affected to a significant degree.

### Definition of Flushing Time

In its simplest form, it is defined as the time needed to drain a volume of water (V) through an outlet area A with a current velocity 'u'. Let 'B' is the surface area of the estuary and 'H' is the depth and 'A' is the cross section area of the mouth of the estuary and 'T' is the tidal cycle. Then the volume of water drained out in a tidal cycle is

$$V = BH = AuT$$

Quantitatively, the flushing time ($t_F$) of an estuary can be defined as the time needed to replace its fresh water volume $V_F$ at the rate of the net flow through the estuary by the river water discharge rate R.

$$\therefore t_F = \frac{V_F}{R}$$

As the measurement of fresh water volume is not easy, it can be estimated as

$$\therefore t_F = \frac{f^*V}{R} \quad (16.1.28)$$

Where $f^* = \int f dv = \frac{S_0 - S}{S_0}$ (16.1.29)

Here 'V' is the volume of the estuary, $S_0$ is the oceanic salinity found outside the estuary mouth and S is mixed water in the estuary. Also here while 'f' represents fresh water fraction in a water sample, 'f*'represents the average fresh water fraction or freshness distribution.

Calculation of flushing time by equation (16.1.28) requires the knowledge of the volume of the estuary (V), rate of river discharge (R) and salinity distribution through the entire estuary. The volume of the estuary can be estimated by measuring the bathymetry. The river discharge can be measured at a single point at the inner end of the estuary by a flow meter or current meter.

Note while f* =1 for pure fresh water, f* =0 for pure sea water

Total amount of fresh water in an estuary between x=L and x = 0 (mouth) is $V = \int_0^L \int f \, dA \, dX$ :

Flushing time: $T_f = \dfrac{V}{Q_F}$ ..........(16.1.29A)

Where, $T_f$ = time required for the freshwater discharge (river), and $Q_F$ is to completely replace the fresh water in the estuarine volume.

If a tracer (pollutant) is injected into the estuary by an industry, the replacement time (time required for replacement of a tracer throughout a distance *L* landward from the mouth in an estuary with dispersion coefficient (*K) or* time required for complete mixing between $x = L$ and $x = 0$ is

$$T = 0.4\frac{L^2}{K} \qquad (16.1.29B)$$

Where T is time required for a slug of material initially concentrated at one end of basin to reach an approximately uniform concentration throughout the basin; L is length of the basin and K is dispersion coefficient.

### 16.1.10.2. Tidal prism method

The flushing time derived from the tidal prism method represents the shortest possible time during which the entire fresh water fraction of an estuary can be removed.

The tidal prism is the volume of water flowing up the estuary through a given cross section during flood time. In other words, it is the net combined sea and fresh water volume entered into estuary during a tidal cycle as shown in Fig.16.1.14.

This method starts from the concept that a sea water volume $V_T$ enters the estuary with the rising tide, while a fresh water volume $V_R$ enters the estuary during a tidal cycle (rising to falling tide). It assumes that sea water $V_T$ is completely mixed with fresh water $V_R$ at high tide and that the mixed water ($V_T + V_R$) leaves the estuary during low tide.

If $S_o$ is sea water salinity and $S^*$ is mixed water salinity we can write

$$(V_T + V_R)S^* = V_T S_0$$

$$\therefore S^* = \frac{S_0 V_T}{(V_T + V_R)} \tag{16.1.30}$$

Substituting the value of $S^*$ from the equation (16.1.29) in (16.1.30) we get

$$f^* = 1 - \frac{V_T}{(V_T + V_R)} = \frac{V_R}{(V_T + V_R)} \tag{16.1.31}$$

The flushing time according to equation (16.1.28) is $t_F = \frac{f^* V}{R}$ where R is rate of river discharge or fresh water volume per unit time. In the tidal prism concept the unit of time is tidal cycle (T).

$$\therefore R = \frac{V_R}{T} \tag{16.1.32}$$

Substituting equation (16.1.31) and (16.1.32) in equation (16.1.28)

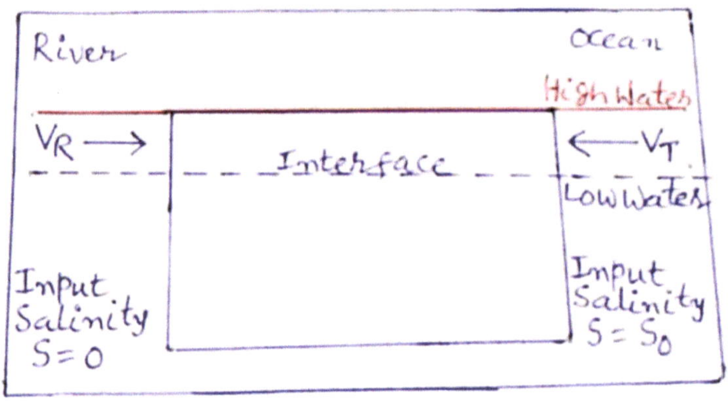


**Fig.16.1.14.** Sketch diagram for the definition of the fresh water and salt water volumes in the tidal prism method. The combined volume $V_T + V_R$ represents the difference between low water and high water.

$$t_F = \frac{VT}{(V_T + V_R)} \tag{16.1.33}$$

This method gained importance as the only two quantities required to be estimated are $(V_T+V_R)$ and V. Whereas in the case of general method extensive observations are needed for measurement of salinity, bathymetry and current or flow and river discharge season wise.

### Limitations

The assumptions of the tidal prism method are never completely met in real estuaries. Mixing of the two waters $V_T$ and $V_R$ is never complete. Some of the mixed water that leaves the estuary during ebb tide may enter again during flood tide.

Tidal prism method is more suitable for slightly stratified and vertically mixed estuaries.

### 16.1.10.3. Knudsen's method

Another method of estimation of flushing time is due to Knudsen's method. This can be achieved from the salt balance equation of Knudsen's formula.

If R denotes the river discharge rate, $Q_B$ and $Q_T$ are the volume transports of water entering and leaving the estuary at its mouth with respective salinities $S_B$ and $S_T$, then we can write the continuity of volume and mass (salt) as

$$Q_T - Q_B = R \qquad (16.1.34)$$
$$Q_T S_T = Q_B S_B$$

Eliminating $Q_B$ from these two equations we can write $Q_B = Q_T - R$. Substituting this in second equation we get $Q_T\, S_T = (Q_T - R)S_B$

$$\therefore Q_T = \frac{RS_B}{S_B - S_T} \qquad (16.1.35)$$

But we know the flushing time as $t_F = \frac{V}{Q_T}$ (16.1.36)

substituting the value of $Q_T$ from equation (16.1.35) in equation (16.1.36) we get

$$t_F = \frac{V}{R}\left[1 - \frac{S_T}{S_B}\right]$$

Where V is volume of the estuary, R River discharge $S_T$ salinity at the surface and $S_B$ at bottom near any station at the mouth of the estuary.

The main assumption of Knudsen's formula is that all the sea water that enters the estuary through the bottom leaves it through upper layer after complete mixing with fresh water.

### Limitation

If the interface is weak and more diffuse the upper layer will not maintain zero salinity at the head then the above formula underestimates the true flushing time.

In practice Knudsen's formula appears more appropriate for highly stratified and salt wedge estuaries.

The following table gives the summary of the three methods:

| Method | Formula | Input data | Remarks |
|---|---|---|---|
| General method | $t_F = \frac{f^* V}{R}$ | • Estuary volume, bathymetry<br>• Salinity<br>• Rate of river discharge | |
| Tidal prism method | $t_F = \frac{VT}{(V_T + V_R)}$ | • Estuary topography(volume)<br>• Sea level (tidal record) | • Under estimates<br>• Suitable for slightly stratified estuaries |
| Knudesen's method | $t_F = \frac{V}{R}\left[1 - \frac{S_T}{S_B}\right]$ | • Estuary topography<br>• Salinity profile at the mouth<br>• Rate of river discharge | • Under estimates<br>• Suitable for highly stratified estuaries |

### 16.1.11. Estuarine modelling

A conceptual model basically consists of one or more notions of how the system responds to external factors. For instance: the shoal level will tend to follow a rise in high water level. The essence of empirical models is that they are based on observations / data. They can range from a simple line fitted through data in a scatter diagram to relationships based on artificial intelligence, such as neural networks. A simple example is the empirically established relationship between the cross-sectional area below mean sea level of a tidal channel and the tidal discharge that moves through it.

Semi-empirical are based on a combination of data or observed behaviour and basic conservation laws, such as a sediment mass balance. An example from coastal morphology is a diffusion-type profile model, which is based on observed behaviour and satisfies the sediment balance equation, while the diffusion coefficient is derived from observations or more sophisticated model results. Genetic algorithms partly based on physical knowledge, such as a known property of a sediment transport relationship, also come under this category.

Physical models can be physical scale models or analogs. An example of the latter is the Deltar, an electrical analog formerly used by "Rijkswaterstaat" to produce forecasts of tides and surges (see https://en.wikipedia.org/wiki/Deltar).

There is multitude of mathematical models, in one, two or three spatial dimensions. Most of the hydrodynamic models in this category are process-based, i.e. based on the conservation laws of mass and momentum from classical physics. The constituting equations, such as the Navier-Stokes equations for turbulent flow, are solved in various degrees of simplification.

Analytical models seem to be regaining ground now that techniques become available to tackle systems of nonlinear equations (e.g. Schuttelaars and de Swart,2000).

Lumped or aggregated models basically involve a reduction of the number of spatial dimensions. Box models have no spatial dimension at all, i.e. they consider processes within a bounded volume or at an arbitrary point in space. They are often used in ecology (e.g. food chain models). Pots & pipes models lump processes into a number of interconnected containers or spatial boxes, and describe how the fluxes between those boxes influence the state of the dependent variables within them, and how the state of the dependent variables influences the fluxes. An example of such a model is ASMITA, which has been used extensively to describe the morphological behaviour of short tidal embayments such as those in the Wadden Sea (Stive and Wang, 2003 ).

Spatially integrated models reduce the number of dimensions by formally integrating the constituting equations over one or more spatial dimension. An example is ESTMORF, which describes the morphological evolution of a tidal channel and shoal system as a 1-D network with the shoals as storage areas for water and sediment (Wang et al.1998).

Grid-based models form a very large category of mathematical models. They exist in a variety of forms, to be distinguished by the type of grid and by the way the constituting equations are discretized and solved. In the case of morphological models, an important aspect is how the hydrodynamic, sediment transport and sediment balance components of the model are coupled, at time-step level (like Delft3D; see Lesser et al.2004) or at intertidal level. A special category of grid models are cellular automata (e.g. Dearing et al.2006), in which objects can move from one box to another according to prescribed rules of behaviour and driven by the conditions in the box where they are and its neighbours. Thus it is a mixed form of grid-based and particle-based models.

Particle-based models are increasingly used now that computer power has sufficiently increased. Spectacular results have been obtained with Smooth Particle Hydrodynamics, especially with highly non-linear water motion at small scales (e.g. breaking waves). Random walk models were formerly used only at the analytical level, in order to come up with parameterizations of e.g. turbulence or turbulent sediment transport in turbulent flow. Random-walk numerical simulation models seem to be coming within sight now. In sediment transport, grain-by-grain transport models are gaining ground, as they circumvent part of the empiricism of traditional sediment transport models. Especially in complex areas such as estuaries, this may become an important asset. Swarm models describe the behaviour of large number of mutually interacting particles, such as emergent pattern formation. They may become of interest for cohesive sediment (e.g. flocculation), but this kind of application still seems to be beyond the horizon.

#### 16.1.11.1. Numerical modelling of Estuaries

Man's most intense utilization of the marine environment occurs in the estuary. The estuary has long been considered a reservoir for man's wastes, but its capacity to absorb the waste materials is quickly being reduced. A significant portion of the world's food supply is harvested from the estuary. The distance of inland penetration of seawater varies with the relative strength of the fresh water discharge and the co- oscillating tidal flow resulting from the dynamical interaction with an adjacent coastal sea. So the knowledge of the dynamics that govern the estuarine circulation pattern becomes essential.

To simulate tidal circulation, sedimentation, salinity distribution etc. in an estuary, various modelling techniques such as physical modelling, analytical modelling etc. are used. Though physical modelling techniques are useful for the predictions of the estuarine processes, this approach is expensive and has limitations in scaling some physical parameters. Analytical modelling is another tool to study the estuarine systems and many studies were carried out in estuaries using analytical models (Ketchum, 1950, 1951a,b, 1954, 1955; Ippen and Harleman, 1961; Hansen and Rattray, 1965).

Simulation of estuarine processes requires proper representation of bottom topography, irregular coastline etc. The limitation of analytical modelling is that it assumes a regular geometry of systems. The geometry of an estuarine system has an important role on circulation and phase speed of waves. An improper representation of geometry does not show the important characteristics of estuarine dynamics. Unnikrishnan et al. (1997) used a network numerical model in the Mandovi and Zuari estuaries to simulate the tidal decay during wet season.

The development of new generation computers since 1970s paved the way for development of two dimensional simulations of estuarine systems fast and less costly. Two dimensional models can simulate circulation in wide bays and estuaries much better than one dimensional model. Leendertse, (1967) developed a two dimensional model and he used a finite difference scheme to solve partial differential equations. Unnikrishnan et al. (1999) developed a two dimensional barotropic numerical model based on shallow water wave equations to simulate the sea level and tidal circulation in the shelf regions of the central west coast of India including Gulf of Khambhat, west coast of India. In the study of Mandovi and Zuari estuaries by Manoj (2008), for the simulation of tides and salinity, a hybrid network numerical model consisting of 2D and 1D models was used. As the Mandovi and Zuari estuaries are very shallow and narrow a 1D formulation is used at the upstream regions and a 2D formulation is used for the wider regions near the mouth. Later the 2D and 1D models were coupled at a distance of about 15 and 12 km for the Mandovi and Zuari estuaries respectively from the mouth.

### 16.1.11.2. Equations in 2D Model

The vertically averaged equations of momentum, continuity and advection diffusion equation of salinity were used in the 2D model. A Cartesian coordinate system was used for the present model domain where x and y-axis are positive toward east and north respectively. The momentum equation in the transport form can be written as follows:

$$\frac{\partial U}{\partial t}+\frac{\partial(uU)}{\partial x}+\frac{\partial(vU)}{\partial y}=-g(h+\eta)\frac{\partial \eta}{\partial x}+fv+A_H\nabla^2U-C_DU\frac{\sqrt{U^2+V^2}}{(h+\eta)^2} \quad (16.1.37)$$

$$\frac{\partial V}{\partial t}+\frac{\partial(uV)}{\partial x}+\frac{\partial(vV)}{\partial y}=-g(h+\eta)\frac{\partial \eta}{\partial y}-fu+A_H\nabla^2V-C_DV\frac{\sqrt{U^2+V^2}}{(h+\eta)^2} \quad (16.1.38)$$

The continuity equation can be written as

$$\frac{\partial \eta}{\partial t}+\frac{\partial U}{\partial x}+\frac{\partial V}{\partial y}=0 \quad (16.1.39)$$

Advection diffusion equation of salinity is written as follows

$$\frac{\partial S}{\partial t}+\frac{\partial(uS)}{\partial x}+\frac{\partial(vS)}{\partial y}=K_x\frac{\partial^2 S}{\partial x^2}+K_y\frac{\partial^2 S}{\partial y^2} \quad (16.1.40)$$

where U = u (h + η) and V = v (h + η)

### 16.1.11.3. Equations in 1D model

The equations used in the one dimensional model are area averaged momentum and continuity equations and advection diffusion equation of salinity (Harleman and Lee, 1969; Thatcher and Harleman, 1972).

The one dimensional momentum equation can be written as

$$\frac{\partial Q}{\partial t}+\frac{2Q}{A}q-\frac{2bQ}{A}\frac{\partial \eta}{\partial t}=-gA\frac{\partial}{\partial x}(z_0+h+\eta)-g\frac{Q|Q|}{AC^2R} \quad (16.1.41)$$

The continuity equation is written as follows

$$b\frac{\partial \eta}{\partial t}+\frac{\partial Q}{\partial x}-q=0 \quad (16.1.42)$$

Chezy coefficient is calculated using the formula given below.

$$C=\frac{1.49}{\eta}R^{\frac{1}{6}} \quad (16.1.43)$$

Where η is the Manning's frictional coefficient

Area averaged advection diffusion equation of salinity can be written as

$$A\frac{\partial S}{\partial t}+\frac{\partial(QS)}{\partial x}=AK_x\frac{\partial^2 S}{\partial x^2} \quad (16.1.44)$$

### 16.1.11.4. Finite difference scheme

There are many numerical methods such as finite difference method, finite element method etc. for solving the partial differential equations. Finite difference schemes are widely used numerical schemes because they give less complexity compared to other numerical schemes.

With regard to finite element method, Galerkin finite element method is famous. In this method the water body to be modelled is divided into smaller regions called elements. An element can be either triangular or quadrangular in shape that can be easily arranged to fit complex boundaries. The elements are defined by a series of mode points located at the element vertices, mid-side points and in the case of nine-node quadrangular elements at their centers. Values of dependent variables are approximated within each element using the values and a set of interpolation functions.

Hydrodynamic modelling of estuaries normally requires the physical dimensions to be approximated by a computational grid (Martin and McCutcheon, 1999). There are various types of model grids and a grid is selected for approximating the solutions of differential equations based on what kind of numerical scheme is used. In the present days many people use Arakawa C grid (Arakawa and Lamb, 1977; Thubum, 2007). Arakawa C grid is a staggered grid and u and v velocities are computed at the grid boundaries and free surface elevation and salinity are computed at the centre of the grid. The advantage of using this grid is that it enables the user to include barriers, across which no flow is allowed. Also, a zero velocity boundary condition can be applied at the closed boundaries (Grzechnik, 2000).

The leapfrog scheme was used to march in time. The leapfrog scheme is a three time level scheme having second order accuracy. Despite its three time level structure, the leapfrog scheme is efficient because it requires derivative computations only once at each time step. It is neutrally stable, with no dissipation error (Zhou, 2002).

For example the continuity equation in finite difference form can be written as:

$$\frac{\eta_{i,j}^{n+2} - \eta_{i,j}^{n}}{2\Delta t} = -\left(\frac{U_{i,j}^{n+1} - U_{i-1,j}^{n+1}}{\Delta x}\right) - \left(\frac{V_{i,j}^{n+1} - V_{i,j-1}^{n+1}}{\Delta y}\right) \tag{16.1.45}$$

Similarly the advection diffusion equation of salinity and the momentum equation along the x and y axes in finite difference form can be written.

### 16.1.12. Governing equations for a partially mixed estuary

The set of equations that describe the dynamics of a partially mixed estuary is the vertically integrated continuity equation, momentum balance equations, the thermodynamic equation, the salt balance equation, turbulent energy density equation and an equation of state. Here the parameterization of the turbulent processes is considered by the application of a transport equation for the turbulent energy density. The set of equations has been simplified to eliminate the lateral (y) momentum balance equation and all the other lateral derivatives. The simplification process assumes that all variables except the lateral velocity (v) are independent of y and the lateral average of v is zero. These assumptions not only reduce the dimensionality of the problem but permit a physical interpolation for both the turbulent stresses and salt fluxes. The governing equations contain the Bousinesq approximation and the hydrostatic assumption.

Therefore the equation of continuity in the channel is written as:

$$b\frac{\partial H}{\partial t}+\frac{\partial}{\partial x}\left(bH\bar{u}\right)=0 \tag{16.1.46}$$

Where (H = ζ + h) is total depth of the channel, b is the breadth of the channel, ζ the elevation of water level over mean level, h depth of the channel (Fig.16.1.15) and $\bar{u}$ is depth averaged velocity given by:

$$\bar{u}=\frac{1}{H}\int_{-h}^{\zeta}udz$$

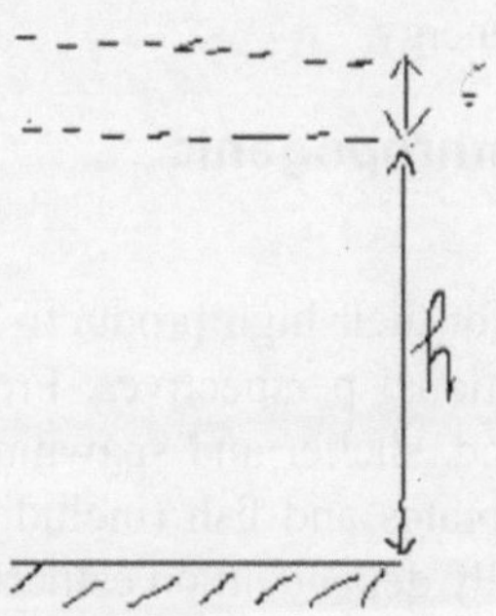

Fig. 16.1.15. Channel dimensions

Here z = $\zeta_{(x,t)}$ and $-h_{(x)}$ refer to the location of the disturbed free surface to the impermeable floor of the channel and its breadth is denoted by $b_x$.

The momentum equation has the form:

$$\frac{\partial}{\partial t}(bu)+\frac{\partial}{\partial x}\left(bu^2\right)+\frac{\partial}{\partial z}(buw)=-gb\frac{\partial\zeta}{\partial x}-\frac{b}{\rho_0}\int_z^{\zeta}\frac{\partial}{\partial x}(\rho g)dz+\frac{\partial}{\partial z}[K_M\frac{\partial}{\partial z}(bu)]+\ell(u) \tag{16.1.47}$$

The letter $\ell$ is an operator such that $\ell(u)$ simulates the effect of horizontal momentum exchange by turbulent scale motions. The inclusion of such a term is required if there is any anticipated formation of an internal hydraulic jump in the estuary.

The salinity S is determined from the transport equation which has the form:

$$\frac{\partial}{\partial t}(bs)+\frac{\partial}{\partial x}(bus)+\frac{\partial}{\partial z}(bws)=\frac{\partial}{\partial z}[K_S\frac{\partial}{\partial z}(bs)]+\ell(s) \tag{16.1.48}$$

Where $K_S$ is exchange coefficient. The term $\ell(S)$ represents the effect of turbulent horizontal salinity flux which is expected to be of secondary importance when compared to the advective salinity flux.

Similarly we can write the transport equation for turbulence energy density, E as:

$$\frac{\partial}{\partial t}(bE)+\frac{\partial}{\partial x}(buE)+\frac{\partial}{\partial z}(buE)=bK_M\left(\frac{\partial u}{\partial z}\right)^2+P+\frac{g}{\rho_0}bK_S\frac{\partial\rho}{\partial z}+\frac{\partial}{\partial z}[K_E\frac{\partial}{\partial z}(bE)+\ell(E)-b\varepsilon \tag{16.1.49}$$

The terms on the right hand side are: the first term represents a production of turbulent energy consequent upon vertical shear in the Reynolds averaged flow. The second term represents a production of turbulent energy due to the action of normal eddy stresses. These two terms describe the ways in which the energy is transported from the Reynolds averaged flow to the turbulent energy budget.

For $\frac{\partial\rho}{\partial z}<0$, the third term represents a sink of turbulent energy resulting from the presence of stable density stratification. The fourth term represents the vertical diffusion of turbulent energy with an eddy coefficient $K_E$. The fifth term represents the horizontal redistribution of turbulent energy by the turbulence itself. The final term involving ε simulates the dissipation of turbulent energy.

### 16.1.13. Estuaries: relevance, benefits and anthropogenic pressures

Estuaries are complex ecosystems, largely recognized for their high productivity and their interest from social, economic and conservational perspectives. From an ecological point of view, estuaries are source of food, shelter and spawning-nursery areas for many organisms, *e.g.* macro invertebrates and fish (including commercial species), or waders and waterfowls especially dependent on estuaries for breeding, feeding or sheltering. Furthermore, estuarine related habitats such as mudflats, coastal lagoons, barrier beaches, deltaic and flood plains, salt marshes or sea-grass beds, mangroves and meadows, are also valuable ecosystems that host unique biological communities.

#### 16.1.13.1. Issues of Indian Estuarine Ecosystem

The Country has 14 major, 44 medium and 162 minor rivers drains into the sea through various estuaries. Major estuaries occur in the Peninsular region of India. Many estuaries are the locations of some of the major seaports. Most of India's major estuaries occur on the east and west coasts. In contrast, the estuaries on the west coast are smaller. Two typical examples of estuaries on the west coast are the Mandovi and Zuari estuaries.

Modifications of the estuarine catchments result in changes in water flow in various estuaries, either far in excess or much lower than required (E.g. Hooghly, Godavari,

Pulicat etc.). The Indian estuaries are subjected to the following pressures today: Pollution through industries and combined city sewage discharge; Recreational boating and fishing; Navigation, dredging and shipping (e.g. Hooghly); Expansion of urban and rural settlements; mining & industries; agriculture and dumping of solid wastes; Overexploitation of target fish stock due to increased demand; Reclaiming the fringed areas for intensive aquaculture in pens; Obstructing the migratory routes of fish and prawn recruitment (e.g., Chilka, Pulicat); Polluting the environment through feeding of stocked fish and prawn in pens (Chilka); Destruction of biodiversity through prawn seed collection and operation of small-meshed nets (e.g., Hooghly, Chilka, Pulicat); and submergence of catchment areas due to rising in water level.

### 16.1.14. Mangroves

Mangroves represent a characteristic littoral (near the seashore) forest ecosystem. These are mostly evergreen forests that grow in sheltered low lying coasts, estuaries, mudflats, tidal creeks backwaters (coastal waters held back on land), marshes and lagoons of tropical and subtropical regions. They are shrubs or small trees that grow in coastal saline or brackish water. They grow below the high water level of spring tides. The best locations are where abundant silt is brought down by rivers or on the backshore of accreting sandy beaches. Mangroves are highly productive ecosystems, and the trees may vary in height from 8 to 20 m. They protect the shoreline from the effect of cyclones and tsunamis. They are the breeding and spawning grounds for many commercially important fishes. Since mangroves are located between the land and sea, they represent the best example of ecotone. As they are salt tolerant trees, they are also called halophytes, and are adapted to harsh coastal conditions. This kind of vegetation facilitates more water loss. Leaves are thick and contain salt-secreting glands. Some block absorption of salt at their roots itself. They contain a complex salt filtration system and complex root system to cope with salt water immersion and wave action. They are adapted to the low oxygen (anoxic) conditions of waterlogged mud. They produce pneumatophores (blind roots) to overcome the respiration problem in the anaerobic soil conditions. Mangroves occur worldwide in the tropics and subtropics, mainly between latitudes 25° N and 25° S. They require high solar radiation to filter saline water through their roots. This explains why mangroves are confined to only tropical and sub-tropical coastal waters. Mangroves occur in a variety of configurations. Some species (e.g.Rhizophora) send arching prop roots down into the water. While other (e.g. Avicennnia) send vertical "Pneumatophores" or air roots up from the mud.

The largest single block of tidal halophytic mangroves of the world are at Sundarbans (India). The other famous mangrove areas in India are at Bhitarkanika (Orissa), Godavari-Krishna deltaic regions of Andhra Pradesh, while Pichavaram and Vedaranyam in Tamilnadu are degraded mainly due to the construction of aquaculture ponds and salt pans on the east coast and on the west coast of India, mangroves, mostly scrubby and degraded occurring along the intertidal region of estuaries and creeks of Maharashtra, Goa and Karnataka, in Gulf of Kutch and the Kori creek. On the Andaman & Nicobar Islands, in the small tidal estuaries and the lagoons dense and diverse undisturbed mangrove flora are available.

## 16.2. Seiches

A Seiche is a standing wave, which can be considered as the sum of two progressive waves travelling in opposite directions as shown in Fig. 16.2.1. Seiches can occur in Lakes, bays, estuaries and harbours which are open to the sea at one end.

In most of the coastal bodies like bays and estuaries, the water is relatively shallow compared to the wavelength of the seiche (L) and the period of the Seiche is determined by the length of the basin and the depth of the water. Figure 16.2.1 &16.2.2 shows the idealized water motions in a seiche of a closed basin.

At either end of the container, water level is alternately high and low, whereas in the middle the water level remains constant. The length of the container ( 1 ), corresponds to half the wavelength (L) of the Seiche in Fig.16.2.1.

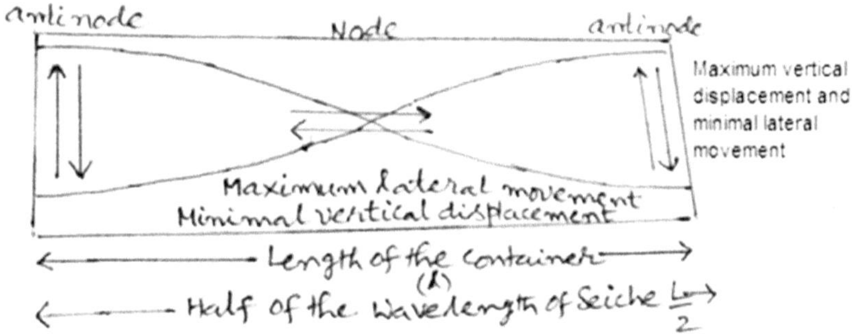


**Fig. 16.2.1.** Idealized profile of a seiche in a closed basin

The above Fig. 16.2.1 is an idealized vertical profile of a seiche. At either end of the basin alternately the water level rises and falls, while at the middle the water level remains constant. The length of the basin ($l$) corresponds to half the wave length (L) of the seiche. While at node the horizontal flow is maximum, at antinode the vertical flow is maximum.

Seiches are standing waves in closed or semi-enclosed basins. Consider a basin of length $l$ and depth 'd'. Then the long wave speed is : $C = \sqrt{gd}$ .

Then the time it takes for the wave to travel distance of $l$ is: $T = \frac{l}{\sqrt{gd}}$

As reflection occurs at the closed end of the wall and it takes the same time to return to the starting point, the basic period of a standing wave in the basin is

$$T_1 = \frac{2l}{\sqrt{gd}}$$

Where l = length of the basin, d = depth of the basin, g = 9.8m/sec²

This is the period for the free oscillation of lowest (first) order. Higher order waves are possible with periods $T_1/n$ for order *n*. The order is given by the number of nodes in the surface oscillation. Fig. 16.2.2 shows the first order seiche (of period $T_1$).

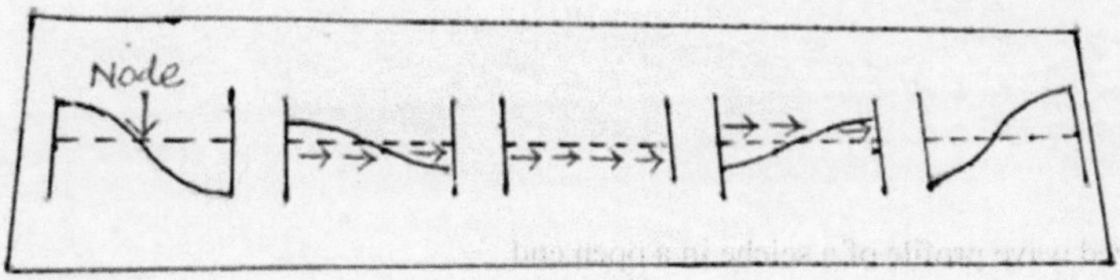


**Fig.16.2.2:** Sketch of a first order seiche at time intervals of 1/8 $T_1$ (i.e over 1/2 of its period). Arrows indicate the direction of water movement. Water movement is in the opposite direction during the other half of the wave's period.

Fig. 16.2.3A shows the second order seiche (of period $T_2 = 1/2\ T_1$). If the basin is open (Fig.16.2.3B), the connecting line (dashed line of Fig. 16.2.3B) to the open sea has to be a node. The corresponding period for the lowest order wave is, therefore, twice that of the lowest order Seiche that would exist if the basin were closed (The effective wave length is twice the length of the basin).

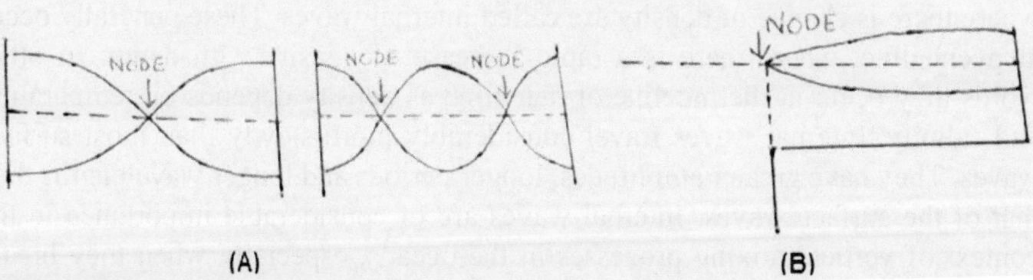


**Fig.16.2.3.** Sketch of seiches of different order, showing the position of the sea surface at the two opposite phases of the wave (1/2 period apart).

In other words, the node can occur at the entrance to the basin and an antinode at the landward end in some basins open to sea at one end as shown in Fig.16.2.4. In this case, the length of the basin (l), corresponds to a quarter of the wave length of the seiche (L). Then the period of oscillation of seiche is:

$$T = \frac{4l}{\sqrt{gd}}$$

Here 'T' is also known as the resonant period. For standing waves to develop, the resonant period of the basin must be equal to the period of the wave or to a small whole number of multiples of that period as:

$$T = \frac{4l}{n\sqrt{gd}}$$

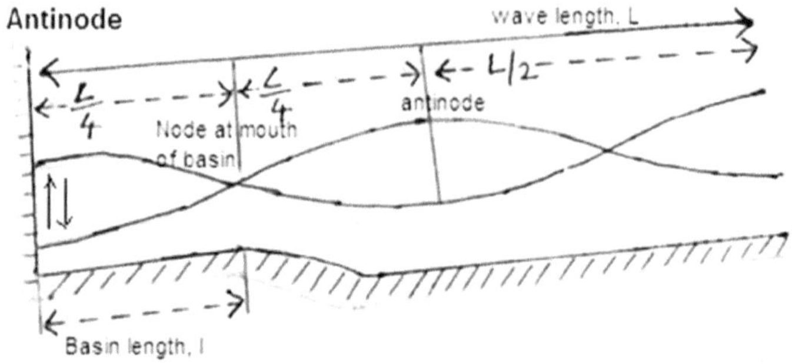


**Fig.16.2.4.** Idealized wave profile of a seiche in a open end

## 16.3. Internal waves

Internal waves occurring in the stratified waters of the sea have a variety of applications. On mesoscale (few hours to few days) fish concentrations, offshore drilling operations, underwater acoustic intensity fluctuations, density currents, ocean thermohaline fine structure and pollution dispersal are some of the areas that are directly affected by internal waves.

Sometimes the gravity waves can be generated at an interface between two layers of ocean water of differing densities. The waves occurring at subsurface where there is change of density are called internal waves. These generally occur at pycnocline, where there is a rapid increase of density with depth. In other words they occur at thermocline or halocline as density depends on temperature and salinity. Internal waves travel considerably more slowly than most surface waves. They have greater amplitudes, longer periods and longer wavelengths than that of the surface waves. Internal waves are of considerable importance in the context of vertical mixing processes in the oceans, especially when they break. The occurrence of internal waves can be detected from surface by the presence of slicks that are formed due to the convergence of water above the internal wave troughs in the mixed layer.

If a pycnocline is displaced up or down, the displacement will set the pycnocline in motion and create internal waves. Though relatively little is known about internal waves, their principal causes are apparently tidal motions and shear stress between layers as the layers flow across one another in wind driven currents or in water mass movements. Sometimes if the pycnocline is shallow, internal waves can be generated by ships in motion.

The wave height of internal waves can be great (upto 100 m) when compared to that of surface waves. The reason for this is at internal interface the density difference is much smaller than the density difference between air and water (by several orders of magnitude) so internal waves can attain much larger amplitudes than surface waves. It also takes longer for the restoring force to return particles to their average position, and internal waves have periods much longer than surface gravity waves (from 10 - 20 minutes to several hours, compared to several seconds or minutes for surface gravity waves). The horizontal water movement (particle velocities) in internal waves is largest near the surface and bottom and minimal at mid-depth. The restoring force for waves is proportional to the product of gravity and the density difference between the two layers (the relative buoyancy).

Internal waves travel approximately one-eighth as fast as surface waves with similar period, and they typically have periods of 5 to 8 minutes and wave length of 0.6 to 1 km. They have much less energy than surface waves, but they can be very important to sea life, submarines and offshore platforms.

Internal waves behave just like surface waves as they enter shallow water and interact with sea floor. Their speeds and wavelengths will reduce and heights will increase until they break. Due to their long wave lengths, they break on the outer part of the continental shelf. In some places, breaking internal waves mix nutrient rich cold water from below the pycnocline into warmer, nutrient poor surface waters. This will enrich the nutrients to phytoplankton which in turn support larger populations of fishes and other sea animals. Internal waves may also move the sediments up and down the canyons. Internal waves can substantially change the depth of the submarine. Drilling and oil production platforms also can be affected by internal waves.

Above the internal wave troughs at the surface particles and floating matter is collected due to convergence of particles. This collection of matter is visible as slick marks. These slicks may act as a hindrance for the moving ship and so internal waves are also sometimes called as 'dead water'. Particularly when the ship moves against the direction of the internal wave, the effect of dead water is much stronger. This phenomenon of "dead water" is most common in fjords, where the interface is produced by a shallow layer of freshwater from glacier runoff overlying oceanic water underneath. The different propagation speeds of the internal waves can be seen by semi-circular shapes in RADARSAT image bands. Some bands are nearly straight indicating uniform propagation speed in these regions. If the internal waves are seen bending, it indicates reduction in speed.

### 16.3.1 Internal wave equations

In a fluid consisting of two layers of infinite thickness, one lower layer of density $\rho_2$ and one upper layer of density $\rho_1$, waves at the boundary surface between the two layers will have a velocity of progress as given by

$$C^2 = \frac{gL}{2\pi}\frac{\rho_2 - \rho_1}{\rho_2 + \rho_1} \tag{16.3.1}$$

These waves are short waves because it is assumed that both layers of fluid are of infinite thickness, on which assumption the wave length L is always negligible compared to the thickness of the layers. If $\rho_1$ means the density of the air and $\rho_2$ the density of the water, the equation gives the velocity of progress of ordinary surface waves, as $\rho_1$ is very small relative to $\rho_2$.

When dealing with internal waves in water which has a free surface but consists of two homogeneous layers of different density, the kinematic and dynamic boundary conditions must be fulfilled both at the free surface and at the internal boundary surface, and the equation of continuity must be satisfied. This leads to a quadratic equation for $C^2$ which, for short waves, has the approximate root

$$C_1^2 = \frac{gL}{2\pi};\ C_2^2 = \frac{gL}{2\pi}\left[\frac{\rho_2 - \rho_1}{\rho_2 \coth kh_1 + \rho_1}\right] \tag{16.3.2}$$

assuming that the thickness of the lower layer *h is* great compared to the wave length. Here $\rho_2$ represents the density of the lower layer and $\rho_1$ and $h_1$ represent density and thickness of the upper layer. If the wave length is great compared to $h_1$, $kh_1$ is a small quantity, coth $kh_1$ can be replaced by $\frac{1}{kh_1}$, and equations (16.3.2) are reduced to

$$C_1^2 = \frac{gL}{2\pi};C_2^2 = \left(gh_1\frac{\rho_2 - \rho_1}{\rho_2}\right) \tag{16.3.3}$$

Where $h_1$ is the thickness of the upper layer and $\rho_1$ and $\rho_2$ are the mean densities of the upper and lower layers. The wave length is given as $\lambda = \frac{C}{f}$ where $f$ is the frequency.

This means the equation (16.3.3) implies that wherever there exists a thin top layer of water of small density, two types of waves are possible: the ordinary surface waves that progress with velocity $C_1$; and the internal waves at the boundary between the light top layer and the heavier water underneath, that progress with velocity $C_2$.

The short internal waves may be present anywhere in the ocean, but escape observation on the high seas where the variation of density with depth is less' conspicuous. In the open ocean long internal waves exist, however, and these have in recent years received much attention. When dealing with two layers and neglecting the effect of the earth's rotation the velocities of progress of the ordinary long wave and of the internal wave are obtained from the equations:

$$C_1^2 = g(h_2 + h_1);\ C_2^2 = \left(\frac{gh_2h_1}{h_2 + h_1}\left\{\frac{\rho_2 - \rho_1}{\rho_2}\right\}\right) \tag{16.3.4}$$

It is assumed that ($\rho_2 - \rho_1$) *is a* small quantity and that the wave length is long compared to the total depth,. *($h_2 + h_l$)* Evidently $C_1$ represents the velocity of progress of an ordinary long wave and need not be considered here. The velocity $C_2$, on the other hand, represents the velocity of progress of the internal wave. If $h_2$ is great relative to $h_1$, the formula is reduced to (eqn 16.3.4).

The internal wave is characterized by its maximum amplitude at the boundary surface. At the free surface the amplitude of the internal wave does not entirely disappear but is reduced to

$$\eta_0 = -\frac{Z(\rho_2 - \rho_1)}{\rho_2} \qquad (16.3.5)$$

where the minus sign indicates that at the surface the phase is opposite to the phase at the boundary Z. At an internal boundary surface in the open sea the difference in density $\rho_2 - \rho_1$ hardly ever exceeds $2 \times 10^{-3}$.

With this difference and with Z = 10m, one obtains $\eta_0$ = 2 cm, meaning that at the free surface the amplitude of the wave is so small that for all practical purposes it can be disregarded. At the bottom, no motion normal to the bottom can exist and there the vertical displacement must also disappear.

In the simple case, as shown in Fig. 16.3.1B, the amplitude of the internal wave increases linearly from the free surface to the boundary surface and decreases linearly from the boundary surface to the bottom. The change in amplitude with depth is therefore equal to $\frac{Z}{h_1}$ in the upper layer, and to $-\frac{Z}{h_2}$ in the lower layer.

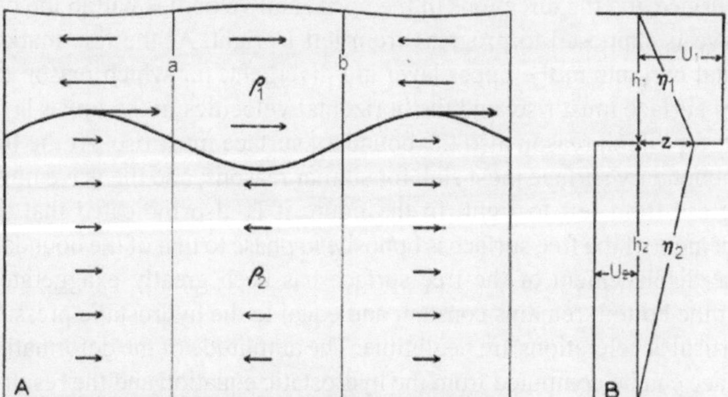


**Fig. 16.3.1.** (A) Schematic representation of an internal wave at the boundary between two liquids of densities $\rho_2$ and $\rho_1$. (B) Schematic representation of the variation with depth of the amplitudes of the vertical displacements $\eta_2$ and $\eta_1$, and of the maximum horizontal velocities $U_1$ and $U_2$.

The amplitude of the horizontal particle velocities can be derived from the vertical amplitudes because the relation exists

$$v = C_2 \frac{\partial \eta}{\partial z} \tag{16.3.6}$$

giving $V_1 = \frac{C_2 Z}{h_1}$ and $V_2 = -\frac{C_2 Z}{h_2}$, respectively. Here $V_1$ and $V_2$ represent the amplitudes of the horizontal velocities and in the present case the amplitude is evidently constant within each layer but it changes abruptly at the boundary surface. The opposite sign indicates that the velocities are in opposite directions in the two layers and, as $V_1 h_1 = V_2 h_2$, the velocities are inversely proportional to the thickness of the two layers (Fig. 16.3.1B). Introducing the velocity $C_2$ of the internal wave, one obtains

$$V_1 = Z\left[\left\{\frac{gh_2}{h_1(h_2+h_1)}\right\}\left\{\frac{\rho_2-\rho_1}{\rho_2}\right\}\right]^{1/2};\ V_2 = Z\left[\left\{\frac{gh_1}{h_2(h_2+h_1)}\right\}\left\{\frac{\rho_2-\rho_1}{\rho_2}\right\}\right]^{1/2} \tag{16.3.7}$$

With $g = 981\ cms^{-2}$; $\rho = 1.025g\ cm^{-3}$; $(\rho_2 - \rho_1) = 2 \times 10^{-3}\ gcm^{-3}$; $h_1 = 40m$; $h_2 = 160m$ and $Z = 10m$, we can get

$V_1 = 19.5\ cms^{-1}$, $V_2 = 4.8cms^{-1}$.

This numerical example shows that internal waves are characterized by large horizontal particle velocities. The corresponding velocity of progress of the internal wave is 78 $cms^{-1}$, whereas the ordinary long wave velocity is 4430 $cms^{-1}$.

The character of the internal wave at the boundary between two liquids of different density is illustrated in Fig. 16.3.1A, which shows the deformation of the boundary surface and the directions of the horizontal velocities within the two layers. The wave is supposed to progress from left to right. At the line marked 'a' the horizontal currents in the upper layer are divergent, for which reason the lower boundary surface must rise, and the horizontal velocities in the lower layer are convergent, for which reason also the boundary surface must rise. At the line marked 'b' the boundary surface must sink for similar reasons, and the wave must, therefore, progress from left to right. In the figure it is also indicated that the vertical displacement of the free surface is opposite in phase to that of the boundary surface, but the displacement of the free surface has been greatly exaggerated. The pressure at the bottom remains constant and equal to the hydrostatic pressure because the vertical accelerations are negligible. The amplitude of the deformation of the free surface can be computed from the hydrostatic equation and the result is as shown in equation (16.3.5)

### 16.3.2 Internal waves in Arabian Sea and Bay of Bengal

Murthy et al (1992) studied the internal wave characteristics in the Eastern Arabian Sea during Southwest monsoon using the time series of Bathy Thermograph profiles and surface winds in the deep waters off Ratnagiri and Karwar along south west coast of India. They found low frequency (< 2 cycles per day, cpd) internal waves with a speed of 83 cms$^{-1}$, wavelength of 45 km and wave heights of 40m off Ratnagiri and off Karwar these correspond for a high frequency of (< 2 cpd) to 99 cms$^{-1}$, 3km and 23m respectively. Propagation of internal waves off Karwar is shown in Fig. 16.3.2

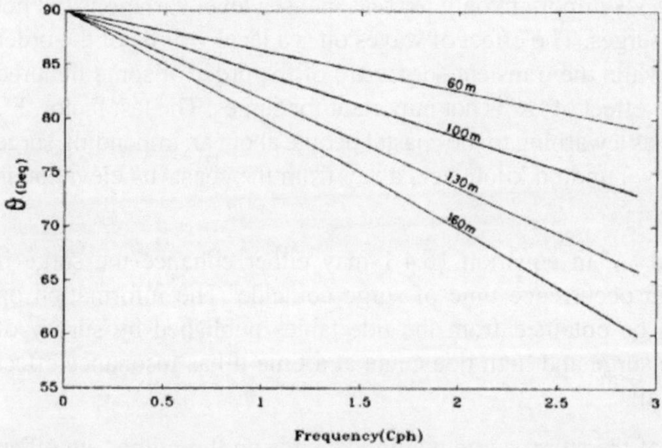


**Fig. 16.3.2.** Propagation of internal waves at different depth off Karwar

Internal wave features in the northwest Bay of Bengal were observed by Suryanarayana et al (2020) between 18.5°–19.5° N and 84.5°–85.5° E using ENVISAT ASAR imageries. The direction of propagation of Internal waves is generally toward the coast with wavelengths ranging from 0.4 to 1.1 km. The phase speeds of the Internal waves vary from 0.12 to 1.3 m/s. The origin of generation of internal waves according to them is tidal forcing over local topographic features. Internal waves of different types like wave packets or wave trains and single solitons of weak manifestations were observed by them depending on the phase and the tidal amplitude as well as the stratification of the ocean, which in turn vary due to the dynamic and complex nature of Bay of Bengal

## 16.4 Storm surges

A storm surge is defined by Heaps (1967) as a 'raising or lowering of sea level produced by wind and by changes in atmospheric pressure over the sea associated with a storm'.

Actually the height to which the water level near the coast rises consequent up on the passage of a storm depends on various factors as given by the equation (Murty, 1984):

$$H = A + B + P + X + F \qquad (16.4.1)$$

Where 'A' is contribution due to astronomical tides, 'B' is static rise due to barometric pressure drop commonly known as inverted barometric effect which is generally one centimeter of sea level for every one millibar pressure drop, 'P' is the rise due to piling up of water against the coast by winds, 'X' is the mean height of the crests of the individual waves superimposed on the general rise of the sea level and 'F' is the effect of the fore runners in equation 16.4.1.

The effect of waves (X) is important only for seasonal sea level variations but not to the transient storm surges. The effect of waves on sea level will be of the order of a few centimeters while the transient surges are of the order of some hundred centimeters. Hence the effect of 'X' is not important for surges. The last factor 'F', the fore runner, serves as a warning to the coastal people about an impending surge when the storm is a few hundred kilometers away from the coast. Its elevation is also very small.

The astronomical tide 'A' in equation 16.4.1 may either enhance the surge if the high tide time and occurrence time of surge coincide. The information on astronomical tide can be obtained from the tide tables published by survey of India which when the surge and high tide come at a time it has resonance effect otherwise its effect is nil.

Thus the final height of the storm surge mainly depends on the piling –up effect (P) due to winds of the storm which need to be computed using the gradient wind equation and the isobar pattern of the cyclone.

## 16.4.1. The computation of Piling-up effect (P)

Welander (1957) gave the slope of the equation caused due to wind as:

$$\frac{dZ}{dx}=\frac{\lambda\tau}{\rho_w gh} \tag{16.4.2}$$

Where Z is the elevation of the water surface, x is the horizontal coordinate in the direction of wind, $\tau$ is wind stress, g is acceleration due to gravity, $\rho_w$ is the density of water (1025 kg $m^{-3}$), h is the average depth of the basin and $\lambda$ is the Mannings frictional factor which varies between 1 and 1.5 depending upon the frictional conditions at the sea bottom.

The wind stress $\tau$ can be computed using the equation: $\tau = \rho_a C_d w^2$ where $\tau$ is N $m^{-2}$, $\rho_\alpha$ density of air (1.3 kg $m^{-3}$) and $C_d$ is the drag coefficient which depends on the height and speed of wind at which it is measured, the stability of the lowest few meters of the atmosphere, the roughness of the sea surface. For Indian Ocean region we can take $C_d = 2.6\times10^{-3}$ when wind speeds are greater than 7 $ms^{-1}$ and it is equal to $1.5\times10^{-3}$ for wind speeds less than 7 $ms^{-1}$.

Substituting all the above values taking $C_d = 2.6\times10^{-3}$ we can write the Welander's equation (16.4.2) as: $\frac{dZ}{dx}=4.2\times10^{-7}\left(\frac{W^2}{h}\right)$. The value of P is obtained by the summation of the slopes $\left(\frac{dZ}{dx}\right)$ generated by the wind stress starting from the origin of the cyclone to the landfall point along the track of the cyclone. Therefore, P is calculated by step wise integration from the initial position to landfall from the equation 16.4.3 (Murty, 1984), Sharma and Murty 1988):

$$\therefore P = \sum dZ = 4.2\times10^{-7} W^2 \sum \frac{dx}{h} \tag{16.4.3}$$

### 16.4.2. Methods of forecasting storm surges

#### 16.4.2.1. Empirical methods

Prior to the development of numerical models, empirical methods were devised for forecasting the surge level at a particular coastal location by correlating it with meteorological conditions over an area of sea. For example, a formula was obtained by Heaps (1967) for the surge height at Southend, at the mouth of the Thames estuary, in terms of northerly and easterly pressure gradients over the North Sea.

#### 16.4.2.2. Mathematical methods

The methods for forecasting storm surges now use a suitable model obtained from the basic dynamic equations. The model may be two or three dimensional. If the variation of velocity with depth is retained in the depth integrated equation, then it is called three dimensional model. Two dimensional models may be classified as linearized when several approximations are made and nonlinear if the full equations are used. Though in many cases linearized models give an adequate representation of conditions, but this is not always true. Because in shallow areas where the tidal amplitude or surge height is not small compared with the depth of water and bottom friction effects are considerable, there is significant interaction between tide and surge and misleading results will come unless the nonlinear terms are retained.

##### 16.4.2.2.1. Linearized equations for depth integrated flow

Assumptions required for linearizing the basic storm surge equations.

1. the elevation $\zeta$ remains small compared with the undisturbed depth of water 'h'
2. the convective acceleration terms $\frac{\partial}{\partial x}\left(\frac{U^2}{h+\zeta}\right), \frac{\partial}{\partial y}\left(\frac{U\,V}{h+\zeta}\right)$ are negligible compared with $\frac{\partial U}{\partial t}, \frac{\partial V}{\partial t}$ terms.
3. The components of bottom stress can be linearly related to the components of integrated transport U and V such that setting $\tau_{bx} = \lambda\rho U$ and $\tau_{by} = \lambda\rho V$ where $\lambda = \frac{k}{h} \approx \frac{kC}{h}$ where k = ARC
4. Tide generating potential $\zeta$ is neglected

Where λ is a coefficient having the dimensions of velocity. The admittance coefficient, A, is a factor of order of unity, R the drag coefficient, C is the depth mean current. 'A' is due to the ratio of bottom velocity to depth mean velocity and includes the factor $\frac{4}{\pi}$.

With these assumptions momentum equations and equation of continuity reduces to

$$\frac{\partial U}{\partial t} + \lambda U - fV = -\frac{h}{\rho}\frac{\partial P_a}{\partial x} - gh\frac{\partial \zeta}{\partial x} + \frac{\tau_{sx}}{\rho} \quad \text{.................(16.4.4)}$$

$$\frac{\partial V}{\partial t} + \lambda V + fU = -\frac{h}{\rho}\frac{\partial P_a}{\partial y} - gh\frac{\partial \zeta}{\partial y} + \frac{\tau_{sy}}{\rho} \quad \text{.................(16.4.5)}$$

$$\frac{\partial U}{\partial t} + \frac{\partial V}{\partial y} + \frac{\partial \zeta}{\partial t} = 0 \quad \text{.........................................(16.4.6)}$$

Note that the coefficient λ which depends on the tidal current amplitude and the depth h is a function of x and y.

## Analytical and numerical solutions

### a) Analytical solutions:

A number of analytical solutions have been obtained for areas of simple geometrical shape like shelf adjacent to a straight coast, a closed basin or a channel closed at one end. The wind stress is assumed to be applied suddenly and then maintained at a constant value.

An example of an investigation using analytical solutions is that by Heaps (1965) dealing with storm surges on a continental shelf. A shelf of uniform width and depth was assumed bounded by a straight coast on one side and connected with an infinitely deep ocean on the other (Fig.16.4.1). A moving wind field was applied and the following cases considered.

i) the response on the shelf to an input surge from the ocean, specified by $\zeta = f(t)$ at x= 1
ii) the generation of a surge on the shelf by a steady wind stress suddenly created
iii) the elevations at the coast produced by a wind field with a moving boundary.

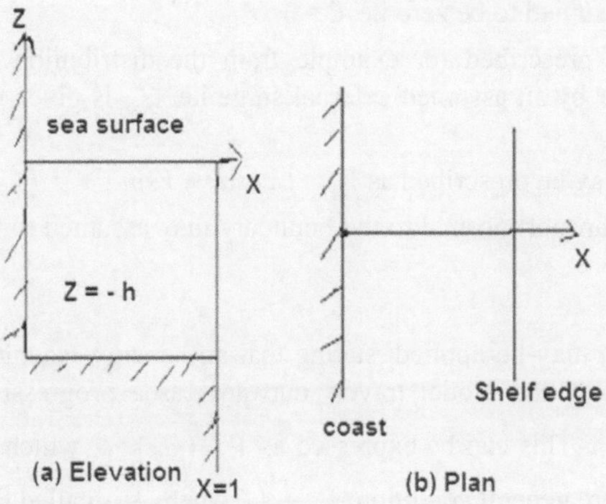


**Fig. 16.4.1.**

## b) Numerical solutions

For detailed studies of surges in a particular region numerical models are more suitable. The region can be represented by a net-work of elementary areas which approximate to the actual coastline at a boundary and which use the actual depths of water in the offshore areas.

An important point in formulating a numerical model is the specification of appropriate boundary conditions. The coast is usually represented as a vertical wall with the condition that transports normal to it, denoted by $V_n$, is zero. Thus if the normal to the coast is inclined at an angle α to the x-axis as in the figure 16.4.2, the condition is:

$$V_n = U\cos a + V\sin a = 0$$

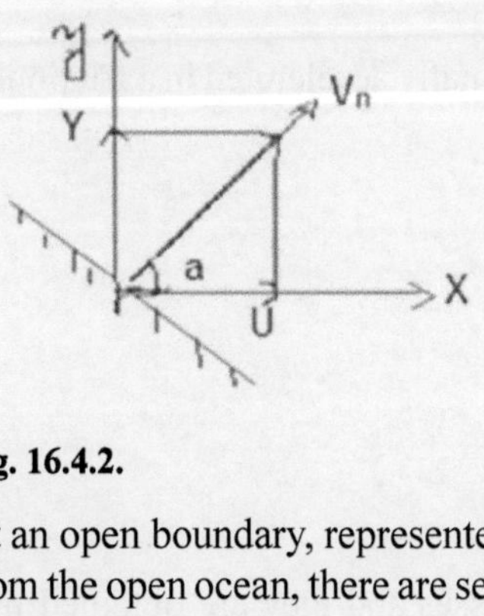


**Fig. 16.4.2.**

At an open boundary, represented by a straight line separating the modeled region from the open ocean, there are several alternative conditions which may be applied.

a) the elevation may be assumed to be zero i.e. $\zeta = 0$

b) The elevation may be prescribed for example from the distribution of atmospheric pressure or by an assumed external surge i.e. $\zeta_{x,y,t}$ is given on the boundary.

c) The normal transport may be prescribed as $V_n = Ucosa + Vsina + = F_{x,y,t}$

d) The derivative of the transport normal to the boundary may assumed to be zero i.e. $\frac{\partial V_n}{\partial y_n} = 0$

e) The radiation condition may be applied, stating that any energy reaching the boundary from inside the model travels outwards as a progressive wave, without reflection. This can be expressed as $V_n = (gh)^{1/2}\zeta$ which is a special case of the more general condition: $V_n = A\,\zeta$ where A is called the Admittance Coefficient.

### 16.4.2.2.2. Nonlinear effects on depth integrated flow

When the elevation of the water surface ($\zeta$) is no longer small compared with the depth of water (h) the effects of the non linear terms in the equation of motion become significant. Hence to obtain the results sufficiently accurate, the full equations must be used.

The basic hydrodynamic equations of motion are:

$$\frac{du}{dt} = -\alpha\frac{\partial p}{\partial x} + fv + F_x$$
$$\frac{dv}{dt} = -\alpha\frac{\partial p}{\partial y} - fu + F_y \qquad (16.4.7)$$

The vertical component equation of motion is

$$\frac{dw}{dt} = -\alpha\frac{\partial p}{\partial z} - g + F_z \qquad (16.4.8)$$

In equation (16.4.8) $\frac{dw}{dt} = 0$ as the surge is horizontally accelerated in a frictionless medium

$$\therefore \frac{\partial p}{\partial z} = -\rho g$$

$$\int_{p_a}^{p} dp = -\rho g\int_{\zeta}^{z} dz$$

$$\therefore (p - p_a) = -\rho g(z - \zeta)$$

$$p = p_a + \rho g(\zeta - z) \ldots\ldots\ldots\ldots\ldots\ldots\ldots\ldots\ldots\ldots (16.4.9)$$

Where $p_a$ is atmospheric pressure

The friction terms here describe the bottom stresses and they are obtained from

Reynolds shearing stresses as

$$F_x = \alpha \frac{\partial \tau_x}{\partial z} and F_y = \alpha \frac{\partial \tau_y}{\partial z} \tag{16.4.10}$$

Substituting (16.4.9) and (16.4.10) in (16.4.7) we can write as:

$$\begin{aligned} \frac{du}{dt} - fv &= -\alpha \frac{\partial}{\partial x}\left[ p_a + \rho g(\zeta - z). \right] + \alpha \frac{\partial \tau_x}{\partial z} \\ \frac{dv}{dt} + fu &= -\alpha \frac{\partial}{\partial y}\left[ p_a + \rho g(\zeta - z). \right] + \alpha \frac{\partial \tau_y}{\partial z} \end{aligned} \tag{16.4.11}$$

Taking z = $\bar{\zeta}$ and αρ =1 we can write as

$$\begin{aligned} \frac{du}{dt} - fv &= -\alpha \frac{\partial p_a}{\partial x} - g\frac{\partial}{\partial x}\left( \zeta - \bar{\zeta} \right) + \alpha \frac{\partial \tau_x}{\partial z} \\ \frac{dv}{dt} + fu &= -\alpha \frac{\partial p_a}{\partial y} - g\frac{\partial}{\partial y}\left( \zeta - \bar{\zeta} \right) + \alpha \frac{\partial \tau_y}{\partial z} \end{aligned} \tag{16.4.12}$$

Expanding these equations:

$$\frac{\partial u}{\partial t} + u\frac{\partial u}{\partial x} + v\frac{\partial u}{\partial y} - fv = -\alpha \frac{\partial p_a}{\partial x} - g\frac{\partial}{\partial x}\left( \zeta - \bar{\zeta} \right) + \alpha \frac{\partial \tau_x}{\partial z} \text{............(16.4.13a)}$$

$$\frac{\partial v}{\partial t} + u\frac{\partial v}{\partial x} + v\frac{\partial v}{\partial y} + fu = -\alpha \frac{\partial p_a}{\partial y} - g\frac{\partial}{\partial y}\left( \zeta - \bar{\zeta} \right) + \alpha \frac{\partial \tau_y}{\partial z} \text{............(16.4.13b)}$$

Consider the volume transports by integrating through a vertical column from the bottom z = -h to the surge elevation z = ζ as

$$\begin{aligned} U &= \int_{-h}^{\zeta} u dz = u(\zeta + h), u = \frac{U}{\zeta + h} \\ V &= \int_{-h}^{\zeta} v dz = v(\zeta + h), v = \frac{V}{\zeta + h} \end{aligned} \quad \text{................................(16.4.14)}$$

On substituting (16.4.14) in (16.4.13) we get :

$$\frac{\partial}{\partial t}\left( \frac{U}{\zeta + h} \right) + \frac{\partial}{\partial x}\left[ \frac{U^2}{(\zeta + h)^2} \right] + \frac{\partial}{\partial y}\left[ \frac{UV}{(\zeta + h)^2} \right] - f\left[ \frac{V}{\zeta + h} \right] = -\alpha \frac{\partial p_a}{\partial x} - g\frac{\partial}{\partial x}\left( \zeta - \bar{\zeta} \right) + \alpha \frac{\partial \tau_x}{\partial z} \text{.......(16.4.15a)}$$

$$\frac{\partial}{\partial t}\left( \frac{V}{\zeta + h} \right) + \frac{\partial}{\partial x}\left[ \frac{UV}{(\zeta + h)^2} \right] + \frac{\partial}{\partial y}\left[ \frac{V^2}{(\zeta + h)^2} \right] + f\left[ \frac{U}{\zeta + h} \right] = -\alpha \frac{\partial p_a}{\partial y} - g\frac{\partial}{\partial y}\left( \zeta - \bar{\zeta} \right) + \alpha \frac{\partial \tau_y}{\partial z} \text{.......(16.4.15b)}$$

Multiply both sides by (ζ + h)

$$\frac{\partial U}{\partial t} + \frac{\partial}{\partial x}\left[ \frac{U^2}{(\zeta + h)} \right] + \frac{\partial}{\partial y}\left[ \frac{UV}{(\zeta + h)} \right] - fV = -\alpha(\zeta + h)\frac{\partial p_a}{\partial x} - g(\zeta + h)\frac{\partial}{\partial x}\left( \zeta - \bar{\zeta} \right) + \alpha(\zeta + h)\frac{\partial \tau_x}{\partial z} \text{..(16.4.16a)}$$

$$\frac{\partial V}{\partial t} + \frac{\partial}{\partial x}\left[ \frac{UV}{(\zeta + h)} \right] + \frac{\partial}{\partial y}\left[ \frac{V^2}{(\zeta + h)} \right] + fU = -\alpha(\zeta + h)\frac{\partial p_a}{\partial y} - g(\zeta + h)\frac{\partial}{\partial y}\left( \zeta - \bar{\zeta} \right) + \alpha(\zeta + h)\frac{\partial \tau_y}{\partial z} \text{..(16.4.16b)}$$

The stress terms $\frac{\partial \tau_x}{\partial z} and \frac{\partial \tau_y}{\partial z}$ can be integrated with the limits –h to ζ taking $\tau_{sx}$, $\tau_{bx}$, $\tau_{sy}$, $\tau_{by}$ respectively and writing

Let $\frac{\partial \tau_x}{\partial z} = K_x$ then $\int_{\tau_{bx}}^{\tau_{sx}} \partial \tau_x = \int_{-h}^{\zeta} K_x dz$ where s = surface and b= bottm

$$\therefore K_x(\zeta + h) = (\tau_{sx} - \tau_{bx}) \therefore K_x = \frac{(\tau_{sx} - \tau_{bx})}{\zeta + h}$$

Similarly $\frac{\partial \tau_y}{\partial z} = K_y$ then $\int_{\tau_{by}}^{\tau_{sy}} \partial \tau_y = \int_{-h}^{\zeta} K_y dz$

Substituting these in the above equations we get

$$\frac{\partial U}{\partial t} + \frac{\partial}{\partial x}\left[\frac{U^2}{(\zeta + h)}\right] + \frac{\partial}{\partial y}\left[\frac{UV}{(\zeta + h)}\right] - fV = -\frac{1}{\rho}(\zeta + h)\frac{\partial p_a}{\partial x} - g(\zeta + h)\frac{\partial}{\partial x}\left(\zeta - \bar{\zeta}\right) + \frac{(\tau_{sx} - \tau_{bx})}{\rho} \quad ..(16.4.17a)$$

$$\frac{\partial V}{\partial t} + \frac{\partial}{\partial x}\left[\frac{UV}{(\zeta + h)}\right] + \frac{\partial}{\partial y}\left[\frac{V^2}{(\zeta + h)}\right] + fU = -\frac{1}{\rho}(\zeta + h)\frac{\partial p_a}{\partial y} - g(\zeta + h)\frac{\partial}{\partial y}\left(\zeta - \bar{\zeta}\right) + \frac{(\tau_{sy} - \tau_{by})}{\rho} \quad ..(16.4.17b)$$

And the continuity equation remains as it is : $\frac{\partial U}{\partial x} + \frac{\partial V}{\partial y} + \frac{\partial \zeta}{\partial t} = 0$

These equations represent storm surge equations. In these equations the elevation $\zeta$, the transport components U, V refer to the resultant values due to the tidal constituents and the meteorological storm effects of wind stresses at surface and bottom. In coastal waters of limited extent the direct effects of the tide generating potential, represented by $\bar{\zeta}$ can often be neglected.

In these non- linear equations $\zeta$ U,V refer to the elevation and transport components due to the tide surge combined.

### 16.4.3. Effect of Atmospheric Pressure of the cyclone

From equation 16.4.13a neglecting the terms acceleration, coriolis, stress and mean $\zeta$ terms we can write:

$$-\alpha \frac{\partial p_a}{\partial x} - g\frac{\partial}{\partial x}\left(\zeta - \bar{\zeta}\right) = 0$$

Or

$$\frac{\partial}{\partial x}\left(\zeta - \bar{\zeta}\right) = -\frac{\alpha}{g}\frac{\partial p_a}{\partial x}$$

$$\Delta\zeta = \frac{1}{\rho g}\Delta P_a \qquad (16.4.18)$$

Thus the increase in elevation above mean level $(\zeta - \bar{\zeta})$ is accomplished by a decrease in pressure of the cyclone. The amount is proportional to $\left(\frac{1}{\rho g}\right)$. This roughly comes to for every one millibar drop of pressure one centimeter rise of sea level.

### 16.4.4. Wind stress effect

Considering stress and elevation terms of equation 16.4.17a ignoring $\bar{\zeta}$ terms

$$-g(\zeta+h)\frac{\partial}{\partial x}\left(\zeta-\bar{\zeta}\right)+\frac{(\tau_{sx}-\tau_{bx})}{\rho}.=0.$$

$$\frac{\partial}{\partial x}\left(\zeta-\bar{\zeta}\right)=\frac{(\tau_{sx}-\tau_{bx})}{\rho g(h+\zeta)}.$$

Generally for a channel closed at one end such as a gulf, the longitudinal transport U = 0 in the steady state but there will be surface flow in the direction of wind with a compensatory flow in the lower layer, then $\tau_{bx}$ can be neglected as the return flow is weak.

$$\therefore \frac{\partial}{\partial x}\left(\zeta-\bar{\zeta}\right)=\frac{(\tau_{sx})}{\rho g(h+\zeta)}$$

However, Manning suggested, in shallow basins $\frac{\partial}{\partial x}\left(\zeta-\bar{\zeta}\right)=\frac{\lambda(\tau_{sx})}{\rho g(h+\zeta)}$ where λ lies between 1 and 1.5. If the elevation ζ remains small compared to the mean depth of water, the above equation becomes

$$\frac{\partial}{\partial x}(\delta\zeta)=\frac{\lambda(\tau_{sx})}{\rho gh} \qquad (16.4.19)$$

where $(\zeta-\bar{\zeta})=\delta\zeta$ represents the elevation above mean sea level as in the Fig.16.4.3 below

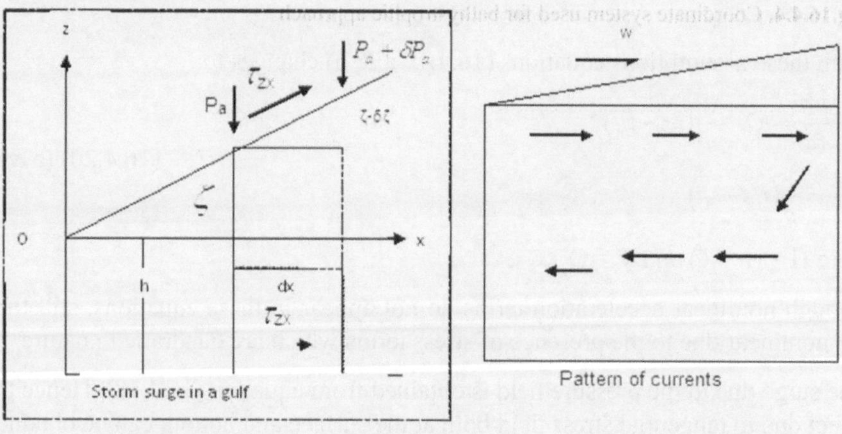


**Fig.16.4.3.** Representation of elevation above sea level

## 16.4.5. Bathystrophic approach for calculation of storm surge

The technique for the determination of wind induced surge along a straight open coast using the bathystrophic approximation was given by Freeman, et al (1957).

The Fig. (16.4.4) shows the coordinate system used in this discussion. We shall consider a steady and uniform wind 'W = ($w_x$, $w_y$) blowing at an angle θ to the x-axis which is normal to a straight open coast.

The basic assumptions involved in the bathystrophic approximations are

a) there is no sustained onshore transport i.e. $Q_x = 0$ and hence $t_{bx} = 0$
b) space derivatives of surge height and volume transport among the y-axis are negligible i.e. $\zeta_y = 0$ and $Q_{y,y} = 0$
c) the advective acceleration terms are negligible
d) the pressure effect is small as only wind effect is considered here. Hence pressure gradient is not considered.

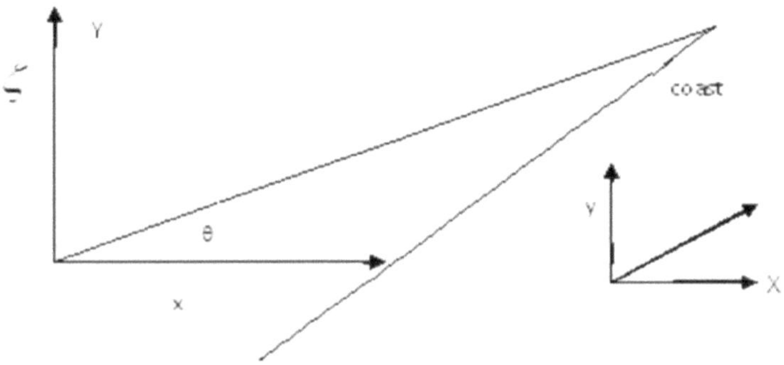


**Fig.16.4.4.** Coordinate system used for bathystrophic approach

With these assumptions equations (16.4.17 a & b) changes to:

$$gH\frac{\partial \zeta}{\partial x} = fQ_y + \frac{1}{\rho}(\tau_{sx} - \tau_{bx})$$
$$\frac{\partial Q_y}{\partial t} = \frac{1}{\rho}(\tau_{sy} - \tau_{by}) \qquad (16.4.20)\ (a\ \&b)$$

Here H = (h + ζ) and V = $Q_y$

Though nonlinear acceleration terms do not appear in these equations, still they are nonlinear due to the presence of stress terms which are quadratic in nature.

The surge due to the pressure field is obtained from equation (16.4.18). Hence the effect due to tangential stress field both at the surface and bottom can be obtained by solving equations (16.4.20).

The tangential stresses are assumed as follows using quadratic law:

$$\tau_S = K_a\rho|W|W$$
$$\therefore \tau_{Sy} = K_a\rho W^2 \operatorname{Sin}\theta \qquad (16.4.21)$$

From Fig.16.4.4, the component where $K_a$ is drag coefficient of air and along y-axis it is W Sinθ.

The bottom stress $\tau_b$ can be related to the resultant bottom current $U_b$ by the quadratic law:

$\tau_b = K_\rho \rho |U_b| U_b$

Where K is drag coefficient and $U_b$ is measured at a height of 1 m above the bottom. If $U_b$ is resolved into x and y components as $u_b$ and $v_b$ we can write:

$U_b^2 = u_b^2 + v_b^2$ ; $U_b = \sqrt{u_b^2 + v_b^2}$

$\therefore \tau_{bx} = K_\rho \rho |U_b| u_b \; and \; \tau_{by} = K_\rho \rho |U_b| v_b$

So the bottom stress is assumed as : $\tau_b = K^* \rho H^{-2} |Q_y| Q_y|$ (16.4.22)

Where H = (D + ζ) and D includes the undisturbed mean depth, the astronomical tide height and the height of water corresponding to the pressure effect of the storm. However, often D is considered as h by neglecting astronomical tide and the pressure effect. While K* is bottom drag coefficient (Freeman et al 1957), Q is bottom transport. Freeman et al (1957) gave bottom drag coefficient as:

$K^* = \gamma(D+\zeta)^{\frac{-1}{3}}$ …….(16.4.23)

Where γ is Mannings frictional coefficient. Equation 16.4.22 can be written using 16.4.23 as:

$$\tau_b = \gamma(D+\zeta)^{-1/3} \rho (D+\zeta)^{-2} Q_y^2 = \rho\gamma(D+\zeta)^{-7/3} Q_y^2 \quad (16.4.24)$$

Using equations (16.4.21) and (16.4.24), equation (16.4.20b) can be written as :

$$\frac{\partial Q_y}{\partial t} = \frac{1}{\rho}(\tau_{Sy} - \tau_{by}) = \frac{1}{\rho}\left[K_a \rho W^2 \operatorname{Sin}\theta - \rho\gamma(D+\zeta)^{-7/3} Q_y^2\right]$$

$$\frac{dQ_y}{dt} = K_a W^2 \operatorname{Sin}\theta - \gamma(D+\zeta)^{-7/3} Q_y^2$$

$$\therefore dQ_y = \left\{K_a W^2 \operatorname{Sin}\theta - \gamma(D+\zeta)^{-7/3} Q_y^2\right\} dt \quad \text{on integrating}$$

$$Q_y = [K_a W^2 \operatorname{Sin}\theta - \gamma(D+\zeta)^{-7/3} Q_y^2]\, t$$

Taking the initial condition $Q_y = 0$ at t = 0 we can write

$$K_a W^2 \operatorname{Sin}\theta - \gamma(D+\zeta)^{-7/3} Q_y^2 = 0$$

$$\therefore Q_y^2 = \frac{K_a W^2 \operatorname{Sin}\theta}{\gamma(D+\zeta)^{-7/3}} \; ; or \therefore Q_y = \sqrt{\frac{K_a W^2 \operatorname{Sin}\theta}{\gamma(D+\zeta)^{-7/3}}} \quad (16.4.25)$$

This is for steady state condition. Substituting eqn (16.4.25) in (16.4.20a), we get the Bretschneider's (1967) storm surge Bathystrophic equation as :

$$gH\frac{\partial\zeta}{\partial x} = fQ_y + \frac{1}{\rho}(\tau_{sx} - \tau_{bx})$$

$$\therefore \frac{\partial\zeta}{\partial x} = \frac{f}{gH}Q_y + \frac{1}{gH\rho}(\tau_{sx} - \tau_{bx}) \quad (16.4.26)$$

Substituting eqn (16.4.25) in (16.4.26) we can get

$$\frac{\partial \zeta}{\partial x} = \frac{f}{gH}\left[\frac{K_a W^2 \operatorname{Sin}\theta}{\gamma (D+\zeta)^{-7/3}}\right]^{1/2} + \frac{1}{gH\rho}(\tau_{sx})$$ since $\tau_{bx}$ is assumed as zero in the beginning.

$$\frac{\partial \zeta}{\partial x} = \frac{f}{g(D+\zeta)}\left[\frac{K_a W^2 \operatorname{Sin}\theta}{\gamma (D+\zeta)^{-7/3}}\right]^{1/2} + \frac{1}{gH\rho}\left(K_a \rho W^2 \operatorname{Cos}\theta\right)$$

Since $\tau_{SX} = K_a \rho \left|\bar{W}\right| \bar{W}$ along x direction from Fig (16.4.4) we know $\bar{W} = W\cos\theta$

We can write $$\frac{\partial \zeta}{\partial x} = \frac{f}{g}\left[\frac{K_a W^2 \operatorname{Sin}\theta}{\gamma}\right]^{1/2} (D+\zeta)^{1/6} + \frac{K_a}{g(D+\zeta)}\left(W^2 \operatorname{Cos}\theta\right) \tag{16.4.27}$$

Now the surge height due to the tangential wind stresses can be determined using numerical integration this equation assuming that the surge is zero at the seaward boundary that is at the shelf break.

The coefficients $K_a$ and $\gamma$ used for east coast of India are

$K_a = 1.1 \times 10^{-6}$ and $\gamma = 0.0093554$ (meters)$^{1/3}$

## 16.4.6. Numerical modelling of storm surges in India

With regard to the modelling of storm surges in the Indian seas is concerned, many studies are available, for example for the Bay of Bengal (Das et al 1974, Johns and Ali 1980, Johns et al 1981 and Ghosh 1981) and Arabian Sea (Dube et al 1984 and Indu Jain 2007).

First the coordinate system used depends on the coast i.e east coast or west coast. For example, for west coast of India (Arabian Sea coast), the coordinates chosen are as below.

For the model formulation a system of rectangular Cartesian coordinates is used in which the origin, 0, is within the equilibrium level of the sea-surface; OX points westward, OY points southward and OZ is directed vertically upwards.

The Navier-Stokes equations for an incompressible, homogeneous fluid in a rotating plane may then be expressed in the following form:

$$\frac{\partial u}{\partial x} + \frac{\partial v}{\partial y} + \frac{\partial w}{\partial z} = 0 \tag{16.4.28}$$

$$\frac{\partial u}{\partial t} + u\frac{\partial u}{\partial x} + v\frac{\partial u}{\partial y} + w\frac{\partial u}{\partial z} - fv = -\frac{1}{\rho}\frac{\partial p}{\partial x} + \frac{1}{\rho}\frac{\partial \tau_x}{\partial z} \tag{16.4.29}$$

$$\frac{\partial v}{\partial t} + u\frac{\partial v}{\partial x} + v\frac{\partial v}{\partial y} + w\frac{\partial v}{\partial z} + fu = -\frac{1}{\rho}\frac{\partial p}{\partial y} + \frac{1}{\rho}\frac{\partial \tau_y}{\partial z} \tag{16.4.30}$$

$$\frac{\partial w}{\partial t}+u\frac{\partial w}{\partial x}+v\frac{\partial w}{\partial y}+w\frac{\partial w}{\partial z}=-\frac{1}{\rho}\frac{\partial \rho}{\partial z}-g \tag{16.4.31}$$

where:

u, *v, w* components of fluid velocity; *f* Coriolis parameter; *P* pressure; ρ density of sea water $\tau_x$, $\tau_y$ components of frictional stress.

In the above equations the molecular viscosity has been neglected. The terms in $\tau_x$ and $\tau_y$ are included to model vertical turbulent diffusion.

If z = 0 is the undisturbed sea-surface, z = ζ (x, y, t) is the free surface, z = - h(x, y) the seabed, ($(\tau_x^\zeta, \tau_y^\zeta)$) denote the wind-stress components and *pa* is the surface pressure, the boundary conditions become:

$$\left.\begin{array}{l} u=v=w=0 \\ \tau_x=\tau_x^\zeta \\ \tau_y=\tau_y^\zeta \end{array}\right] \text{at } Z=-h \tag{16.4.32}$$

$$\left.\begin{array}{l} p=p_a \\ \dfrac{\partial \zeta}{\partial t}+u\dfrac{\partial \zeta}{\partial x}+v\dfrac{\partial \zeta}{\partial y}=w \end{array}\right] at\, z=\zeta \tag{16.4.33}$$

The last condition expresses the fact that the free surface is materially following the fluid.

Since the main interest here lies in predicting long gravity waves (- 100 km) in shallow coastal waters, one can make the reasonable assumption of shallow water theory, i.e. the pressure varies hydrostatically with depth:

$$p=\rho g(\zeta-z)+p_a \qquad -h\le z\le\zeta \tag{16.4.34}$$

For the purpose of numerical prediction, equations (16.4.28) to (16.4.30) are further simplified by integrating them vertically. The unknown dependent variables are then the transport components:

$$\left(\bar{u},\bar{v}\right)=\frac{1}{\zeta+h}\int_{-h}^{\zeta}(u,v)\,dz \tag{16.4.35}$$

and the sea-surface elevation ζ. Straightforward manipulation of (16.4.28) to (16.4.30) then leads to:

$$\frac{\partial \zeta}{\partial t}+\frac{\partial}{\partial x}\left[(\zeta+h)\bar{u}\right]+\frac{\partial}{\partial y}\left[(\zeta+h)\bar{v}\right]=0 \tag{16.4.36}$$

$$\begin{aligned}&\frac{\partial \bar{u}}{\partial t}+\bar{u}\frac{\partial \bar{u}}{\partial x}+\bar{v}\frac{\partial \bar{u}}{\partial y}-f\bar{v}+\frac{1}{\zeta+h}\left[\frac{\partial}{\partial x}\left\{(\zeta+h)\left(\overline{u^2}-\bar{u}^2\right)\right\}+\frac{\partial}{\partial y}\left\{(\zeta+h)\left(\overline{uv}-\bar{u}\bar{v}\right)\right\}\right]\\&=-g\frac{\partial \zeta}{\partial x}-\frac{1}{\rho}\frac{\partial p_a}{\partial x}+\frac{1}{\rho(\zeta+h)}\left(\tau_x^\zeta-\tau_x^{-h}\right)\end{aligned} \tag{16.4.37}$$

$$\frac{\partial \bar{v}}{\partial t}+\bar{u}\frac{\partial \bar{v}}{\partial x}+\bar{v}\frac{\partial \bar{v}}{\partial y}+f\bar{u}+\frac{1}{\zeta+h}\left[\frac{\partial}{\partial x}\left\{(\zeta+h)\left(\overline{uv}-\bar{u}\bar{v}\right)\right\}+\frac{\partial}{\partial y}\left\{(\zeta+h)\left(\overline{v^2}-\bar{v}^2\right)\right\}\right]$$
$$=-g\frac{\partial \zeta}{\partial y}-\frac{1}{\rho}\frac{\partial p_a}{\partial y}+\frac{1}{\rho(\zeta+h)}\left(\tau_y^{\zeta}-\tau_y^{-h}\right) \tag{16.4.38}$$

where ( $\tau_x^{-h},\tau_y^{-h}$ ) denote the components of the bed stress. It may be seen that retention of terms such as ( $\overline{u^2}-\bar{u}^2$ ) in the above equations leads to a fundamental difficulty as they cannot be evaluated within the framework of a vertically integrated model. To overcome this difficulty, the following assumption is usually made:

$$(\overline{u^2}-\bar{u}^2)=0 \tag{16.4.39}$$

In view of the strong associated wind in a cyclone, forcing due to barometric changes, i.e. $\frac{\partial P_a}{\partial x}$ and $\frac{\partial P_a}{\partial y}$ is usually neglected in surge prediction models.

For subsequent numerical treatment, the predictive equations (16.4.36) to (16.4.38) are expressed in flux form as (dropping the overbars for convenience):

$$\frac{\partial \zeta}{\partial t}+\frac{\partial \bar{u}}{\partial x}+\frac{\partial \bar{v}}{\partial y}=0 \tag{16.4.40}$$

$$\frac{\partial \bar{u}}{\partial t}+\frac{\partial}{\partial x}(u\bar{u})+\frac{\partial}{\partial y}(v\bar{u})-f\bar{v}=-g(\zeta+h)\frac{\partial \zeta}{\partial x}+\frac{1}{\rho}\left(\tau_x^{\zeta}-\tau_x^{-h}\right) \tag{16.4.41}$$

$$\frac{\partial \bar{v}}{\partial t}+\frac{\partial}{\partial x}(u\bar{v})+\frac{\partial}{\partial y}(v\bar{v})+f\bar{u}=-g(\zeta+h)\frac{\partial \zeta}{\partial y}+\frac{1}{\rho}\left(\tau_y^{\zeta}-\tau_y^{-h}\right) \tag{16.4.42}$$

where, $\bar{u}=u(\zeta+h)$ and $\bar{v}=v(\zeta+h)$. The surface and bed stresses are parameterized by a conventional quadratic law as follows:

$$\left(\tau_x^{\zeta},\tau_y^{\zeta}\right)=C_d\rho_a\left(u_a^2+v_a^2\right)^{\frac{1}{2}}(u_a,v_a)$$
$$\left(\tau_x^{-h},\tau_y^{-h}\right)=C_f\rho\left(u^2+v^2\right)^{\frac{1}{2}}(u,v) \tag{16.4.43}$$

where $C_d = 2.8\times10^{-3}$ and $C_f = 2.6\times10^{-3}$ are empirical surface and seabed friction coefficients respectively, $\rho_a$ is the density of air and $(u_a, v_a)$ are the components of the surface wind.

Since the eastern lateral boundary of the analysis area (west coast of India) is represented by vertical side-walls in the x or y direction, the condition of zero normal velocity at these yields:

$\bar{u} = 0$ at meridional boundaries

$\bar{v} = 0$ at latitudinal boundaries (16.4.44)

At the open-sea boundaries, the radiation-type conditions (Johns et al 1981) are applied for Indian west coast by Dube et al (1984) as:

$$v+\left(\frac{g}{h}\right)^{\frac{1}{2}}\zeta = 0 \, at \, y = 0 (\, northern\, open\, sea\, boundary) \tag{16.4.45}$$

$$v-\left(\frac{g}{h}\right)^{\frac{1}{2}}\zeta = 0 \, at \, y = L (\, southern\, open\, sea\, boundary) \tag{16.4.46}$$

$$u-\left(\frac{g}{h}\right)^{\frac{1}{2}}\zeta = 0 \ \ (along\, western\, open\, sea\, boundary \tag{16.4.47}$$

The physical meaning of the radiation boundary conditions is that disturbances generated within the analysis area, and propagating towards the open-sea boundaries, are transmitted across those boundaries, in the form of outgoing progressive waves. They also help to eliminate the transient response more quickly as a result of the frictional dissipation in the system. Concerning their effectiveness Flather (1976) noted that application of a radiation condition in the numerical model may remove the unrealistically large currents and grid scale oscillations in the vicinity of the open boundary, which may be produced by application of the conventional open-sea boundary condition (i.e. $\zeta = 0$).

### 16.4.6.1. Finite difference formulation

The predictive equations (16.4.40) to ( 16.4.42) are now solved numerically by considering a discrete set of grid points defined by:

$x = x_i = (i - 1)\Delta x \qquad i = 1,2,\ldots\ldots,m$

$y = y_j = (j - 1)\Delta y \qquad j = 1,2,\ldots\ldots,n \qquad (16.4.48)$

where $\Delta x$ and $\Delta y$ are the grid increments. A sequence of time instants is defined by:

$t = t_p = p\Delta t, \; p = 0,1,\ldots\ldots, \qquad (16.4.49)$

The computations are performed on a staggered grid in the (x, y) plane (Johns and Ali,1980). The grid lines are parallel to the coordinate axes and form a uniform network with a rectangular mesh having sides of length Ax in the x-direction and Ay in the y-direction.

With: ***i*** even, j odd, the point is a $\zeta$-point at which $\zeta$ is computed; i odd j odd the point is a u-point at which u is computed; i even j even, the point is a v-point at which v is computed.

The rectangular grid system described above is chosen to cover the whole analysis area with ***m*** chosen to be even and ***n*** to be odd. The lateral boundaries of the finite difference grid are so constructed that the northern and southern open-sea boundaries (y = 0 and y = L) consist of $\zeta$–points and u-points. The western open-sea boundary consists of $\zeta$-points and v-points. Meridional side-walls consist of u points and the latitudinal side-walls consist of v-points.

The coastal boundary of the model region is formed by line segments connecting u, v-points so that the natural coast is approximated as closely as possible by a step-wise boundary configuration. The manner of the boundary construction used for the west coast of India by Dube et al (1984), for example, is illustrated in Fig. 16.4.5. The grid points lying within the analysis area and located at the coastal boundary or outside the analysis area are separated by defining an integer array. This array is formed by the numbers, 0, 1 representing land points and sea points respectively. During each step of computation reference is made to these numbers by an appropriate conditional statement in the program.

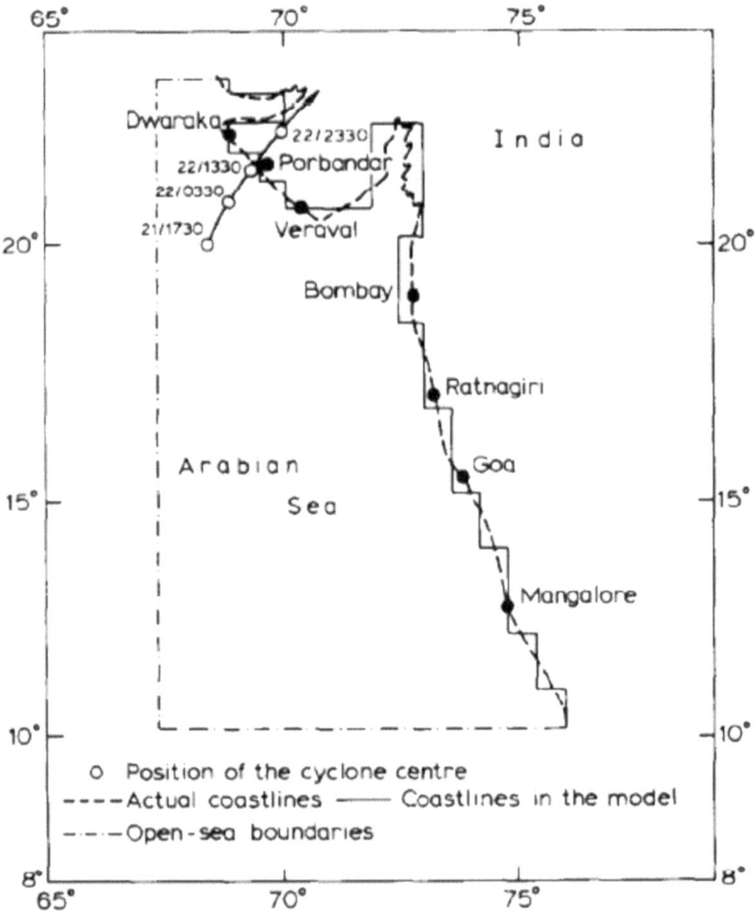


**Fig.16.4.5.** Boundary construction used for west coast of India (Dube et al 1984)

Any variable X, at a grid point ***(i, j)*** may be represented by:

$$X(x_i, y_j, t_p) = X^p_{i,j} \tag{16.4.50}$$

In order to describe the finite difference equations, difference operators are defined by:

$$\Delta_t X = \frac{\left(X_{i,j}^{p+1} - X_{i,j}^{p}\right)}{\Delta t}$$

$$\delta_x X = \frac{\left(X_{i+1,j}^{p} - X_{i-1,j}^{p}\right)}{2\Delta x}$$

$$\delta_y X = \frac{\left(X_{i,j+1}^{p} - X_{i,j-1}^{p}\right)}{2\Delta y} \tag{16.4.51}$$

Averaging operations are defined by:

$$\bar{X}^x = \frac{1}{2}\left(X_{i+1,j}^{p} + X_{i-1,j}^{p}\right)$$

$$\bar{X}^y = \frac{1}{2}\left(X_{i,j+1}^{p} + X_{i,j-1}^{p}\right)$$

$$\bar{X}^{xy} = \bar{X}^{xy} \tag{16.4.52}$$

And a shift operator is defined by

$$E_t X = X_{i,j}^{p+1} \tag{16.4.53}$$

The discretization of (16.4.40) is based on

$$\Delta_t(\zeta) + \delta_x(\bar{u}) + \delta_y(\bar{v}) = 0 \tag{16.4.54}$$

Equation (16.4.54) yields an updating procedure to compute the elevation at all the interior ζ-points and is consistent with the mass conservation in the system. The elevations at j = 1 and $i$ = 2,4, . . . , (m - 2) are determined by equation (16.4.45); in practice, this is applied at j = 2 and replaced by:

$$\frac{1}{2}\left(\frac{g}{h_{i,2}}\right)^{1/2}\left\{\zeta_{i,1}^{p+1} + \zeta_{i,3}^{p+1}\right\} + v_{i,2}^{p} = 0 \tag{16.4.55}$$

thus, leading to an updating procedure for the elevation on the northern open-sea boundary y = 0) in the form of:

$$\zeta_{i,1}^{p+1} = -\zeta_{i,3}^{p+1} - 2\left(\frac{h_{i,2}}{g}\right)^{\frac{1}{2}} v_{i,2}^{p} \tag{16.4.29}$$

Similarly, along the open-sea boundaries at j = n and $i$ = $m$ the elevations are updated by application of (16.4.46) and (16.4.47), which leads to:

$$\zeta_{i,1}^{p+1} = -\zeta_{i,n-2}^{p+1} + 2\left(\frac{h_{i,n-1}}{g}\right)^{\frac{1}{2}} v_{i,n-1}^{p} \tag{16.4.57}$$

$$\zeta_{m,j}^{p+1} = -\zeta_{m-2,j}^{p+1} + 2\left(\frac{h_{m-1,j}}{g}\right)^{\frac{1}{2}} u_{m-1,j}^{p} \tag{16.4.58}$$

The discretization of (16.4.41) is based on:

$$\Delta_t(\bar{u}) + \delta_x(\bar{u}^x\bar{\bar{u}}^x) + \delta_y(\bar{v}^x\bar{\bar{u}}^y) - f\bar{\bar{v}}^{xy} = -gE_\tau[(\bar{\zeta}^x + h)\delta_x\zeta] + \frac{1}{\rho}E_\tau\left(\tau_x^\zeta\right) - \frac{C_f[(u^2+(v^{xy})^2)^{\frac{1}{2}}E_\tau(u)]}{E_\tau(\zeta^x+h)} \quad (16.4.59)$$

Equation (16.4.59) is used to update ū at the u-points. Updated values of u may then be deduced from ū and ζ by applying:

$$u = \frac{\bar{u}}{\left(\bar{\zeta}^x + h\right)} \quad (16.4.60)$$

It can be seen that when equation (16.4.59) is applied at i = *m* - 1, the averaging operator references the value of u outside the analysis area, and to overcome this difficulty, an appropriate one-sided definition of $\delta_x$, is then used. Similarly, when j = 3 and j = n - 2 the averaging operator references values of u at j = 1 and j = n and at these positions, a one-sided definition of $\delta_y$, is used. With this procedure, values of fi (and hence u) may be updated at all the interior u-points.

A similar discretization of (16.4.42) is based on:

$$\Delta_t(\bar{v}) + \delta_x(\bar{u}^y\bar{\bar{v}}^x) + \delta_y(\bar{v}^y\bar{\bar{v}}^y) + E_\tau f\bar{\bar{u}}^{xy} = -gE_\tau[(\bar{\zeta}^y + h)\delta_y\zeta] + \frac{1}{\rho}E_\tau\left(\tau_y^\zeta\right) - \frac{C_f[\{(u^{xy})^2+v^2\}^{\frac{1}{2}}E_\tau(v)]}{E_\tau(\zeta^y\ +h)} \quad (16.4.61)$$

Equation (16.4.61) is used for updating $\overline{v}$ at the v-points. Updated values of v are then deduced from those of v̄ and ζ by using:

$$v = \frac{\overline{v}}{\left(\bar{\zeta}^y + h\right)} \quad (16.4.62)$$

It may be seen that when (16.4.61) is applied at j = 2 and at j = n - 1, the averaging operator references a value of v outside the analysis area. In this case, appropriate one-sided definition of $\delta_y$, is used.

The following general points are made about the discretizations. In equations (16.4.59) and (16.4.61) the pressure gradient terms are evaluated at the advanced time-level. This is possible explicitly using values of j- previously updated by application of (16.4.54) and, following Sielecki (1968), ensures computational stability subject only to the time-step being limited by the space increment and gravity wave speed. In (16.4.59) the Coriolis term is evaluated explicitly at the old time-level whereas in (16.4.61) it is evaluated at the advanced time-level using the previously updated value of u. In both (16.4.59) and (16.4.61) the friction term is evaluated partly implicitly, the resulting difference equations being solved algebraically before incorporating them into the updating scheme. This ensures unconditional computational stability with reference to the treatment of the dissipative terms.

While updating u and **v** at points adjacent to the stair-step boundaries, the central-difference scheme is used for the horizontal advection terms { $\frac{\partial(u\bar{v})}{\partial y}$ and $\frac{\partial(u\bar{v})}{\partial x}$ } which automatically reduce to their one-sided approximation. This is because central-differences of these terms always contain a component of the horizontal velocity, either u or v depending on the position of the boundary, normal to the boundary which should, by definition, vanish. Later numerical experimentation will be made for a specific cyclone and for a specific coast using the above model.

ADCIRC is a (parallel) ADvanced CIRCulation model for oceanic, coastal and estuarine waters developed by Dr. R.A. Luettich, Jr, University of North Carolina at Chapel hill, Institute of Marine Sciences (email: rick_luettich@unc.edu) and Dr. J.J. Westerink, Department of Civil Engineering and Geological Sciences, University OF Notre Dame (email: jjw@photius.ce.nd.edu). The complete information of ADCIRC model can be seen from the link: http://www.unc.edu/ims/adcirc/document/Introduction.html

## References

### Estuaries

Cameron, W.M. and Pritchard, D.W., 1963. Estuaries, In The Sea, Vo1ume II, Edited by M.N. Hi11, John Wi1ey and Sons, New York, PP. 306-324.

Day J.H. (1964). "The Origin and Distribution of Estuarine Animals in South Africa," Monogr. Biol. 14:159-173.

Day, J.H. (1981). Estuarine ecology with particular reference to southern Africa.

Day Jr, J.W., Kemp, W.M., Alejandro, Y., Crump, B.C., 2012. Estuarine ecology. John Wiley and Sons, New York.

Dearing JA, Richmond N, Plater AJ, Wolf J, Prandle D & Coulthard TJ (2006). Modelling approaches for coastal simulation based on cellular automata: the need and potential. Phil. Trans. R. Soc. A 364: 1051–1071. doi: 10.1098/rsta.2006.1753.

Fairbridge, R.W. (1980): The estuary: its definition and geodynamic cycle. In: F. Olausson & I. Cato (eds.): Chemistry and biogeochemistry of estuaries. Chichester: John Wiley, p. 1-35.

Grzechnik ,M. P. 2000. Three-Dimensional Tide and Surge Modelling and Layered Particle Tracking Techniques Applied to Southern Australian Coastal Seas. Ph.D. thesis, Department of Applied Mathematics, The University of Adelaide, Australia.

Hansen, D. V. and M. Rattray. 1965. Gravitational circulation in straits and estuaries. Journal of Marine Research, 23:104–122.

Hansen, D. V. and M. Rattray. 1966. New dimensions in estuary classification. Limnol. Oceanography, 11:319–326.

Harleman, D. R. F, and C. H. Lee. 1969. The computation of tides and currents in estuaries and canals. Massachusetts Institute of technology, Cambridge.

Harleman, D. R. F. and Abraham, G. 1966. One-dimensional analysis of salinity intrusion in theRotterdam Waterway, Delft HydraulicsLaboratory Publication 44, October 1966.

Hayes, M.O., 1975. Morphology and sand accumulations in estuaries, ed. LE Cronin, Estuarine Research. Academic Press, New York 2, 3-22.

Ibáñez, C., Alcaraz, C., Caiola, N., Rovira, A., Trobajo, R., Alonso, M., Duran, C., Jiménez, P.J., Munné, A., Prat, N., 2012. Regime shift from phytoplankton to macrophyte dominance in a large river: top-down versus bottom-up effects.Science of the Total Environment 416, 314-322.

Ippen, A.T. and Harleman, D.R.F., 1961, One dimensional analysis of salinity intrusion in estuaries, US Army Corps of Engineers, Waterways Experiment Station, Vicksburg, Mississippi Technical Bulletin Number 5.

Ketchum, B. H. 1950. Hydro-graphic factors involved in the dispersion of pollutants introduced into tidal waters. J. Boston Soc. Civil Eng, 37:296–314.

Ketchum B. H. 1951a. The exchanges of fresh and salt water in a tidal estuary. J. Mar. Res, 10: 18–38.

Ketchum B. H. 1951b. The flushing of tidal estuaries. Sewage and Industr. Wastes, 23:198–208.

Ketchum B. H. 1954. Relation Between Circulation and Planktonic Populations in Estuaries. Ecology, 35:191–200.

Ketchum, B. H. 1955. Distribution of coliform bacteria and other pollutant in tidal estuaries. Sewage and Industrial Wastes, 27:1288–1296.

Kjerfve, B., 1989. Estuarine geomorphology and physical oceanography. In Day, J. W., Jr., Hall, C. A. S., Kemp, W. M., and Yáñez-Arancibia, A. (eds.), Estuarine Ecology. New York: John Wiley & Sons, pp. 47–78.

Leendertse, J. J. 1967. Aspects of a computational model for long-period water-wave propagation. United States Air Force Project Rand, The Rand Corporation, Santa Monica, CA.

Lesser GR, Roelvink JA, van Kester JATM & Stelling GS (2004). Development and validation of a three-dimensional morphological model. Coastal Engineering 51: 883-915. doi: 10.106/jcoastaleng.2004.07.014

Martin, J. L. and S. C. McCutcheon,1999. Hydrodynamics and transport for water quality modelling. Lewis Publishers, Boca Raton, USA.

McLusky, D.S., Elliott, M., 2004. The estuarine ecosystem: ecology, threats and management. Third edition. Oxford University Press, New York.

Munk, W.H. and Anderson, E.R. (1948). Notes on the theory of the thermocline. J. Marine Res. 7(3): 276-295.

Perillo, G.M.E., 1995. Definitions and geomorphologic classifications of estuaries. Geomorphology and sedimentology of estuaries. Developments in Sedimentology 53, 17-47.

Pritchard, D.W. (1955): "Estuarine circulation patterns". Proc. Amer. Soc. Civil Eng. 81: 1-11.

Pritchard, D.W. (1967): "What is an estauary: physical viewpoint". In: Estuaries (Ed. G.H. Lauff). Publ. No. 83 , American Association for the Advancement of Science, Washington, D.C. pp 3-5.

Schuttelaars HM & de Swart HE (2000). Multiple morphodynamic equilibria in tidal embayments. JGR-Oceans 105(C10): 24105-24118.

Stive MJF & Wang ZB (2003). Morphodynamic modeling of tidal basins and coastal inlets. In: Lakhan V.C. (ed.), Advances in Coastal Modeling, Elsevier Oceanography Series 67: 367-392. doi: 10.1016/s0422-9894(03)80130-7

Thatcher, M. L. and D. R. F. Harleman. 1972.Prediction of unsteady salinity intrusion in estuaries: Mathematical model and user's Manual. Massachusetts Institute of technology, Cambridge.

Thubum, J. 2007. Rossby wave dispersion on the C-grid. Atmospheric Science Letters, 8: 37–42.

Unnikrishnan A. S., Shetye S. R. and Gouveia A. D. (1997) Tidal propagation in the Mandovi–Zuari estuarine network, west coast of India: Impact of freshwater influx; Estuarine, Coastal and Shelf Science 45 737–744.

Unnikrishanan, A. S., Shetye, S. R. and Michael, G. S. (1999). Tidal propagation in the Gulf of Khambhat, Bombay High, and surrounding areas. Proceedings of the Indian Academy of Science (Earth & Planetary Sciences), 108, 155-177.

Valle-Levinson Arnoldo. 2010. Contemporary issues in estuarine physics. Cambridge University Press

Wang ZB, Karssen B, Fokkink RJ & Langerak A (1998). A dynamic-empirical model for estuarine morphology. In: Dronkers J & Scheffers MBAM. (Eds.), Physics of Estuaries and Coastal Seas, Balkema, Rotterdam, pp. 279-286.

Zhou, W. 2002. An Alternative Leapfrog Scheme for Surface Gravity Wave Equations. Journal of Atmospheric and Oceanic Technology, 19:1415–1423.

Storm surges:

Bretschneider C. L., 1967. Storm surges, Advances in Hydroscience 4, 341–418, ed. Chow V. T., Academic Press, New York.

Das, P. K., Sinha, M. C. and Balasubrahmanyam, V. 'Storm surges in the Bay of Bengal', Quart. J. Roy. Met. Sot. 1974, 100,431-449

Flather. R. A. 'Results from a storm surge prediction model of the north-west European continental shelf for April, November and December 1973'. ZOS Reu. No. 24, Bidston Observatory, England, 33 pp, 1976.

Freeman John C Baer Ledolph, and Jung, G. H. (1957) "The Bathystropic Storm Tide", published in the Journal of Marine Research, Vol. 16, no.1

Ghosh, S. K. 'Objective prediction of storm surges on Indian coasts', Proc. Symp. on Meteorological and Oceanic Fluid Dynamics, 7-8 February, Jadavpur University, India, 1981

Heaps, N. S., 1965. Storm surges on a continental shelf, Phil. Trans. R. Soc. A, 257, 351–383.

Heaps, N. S., 1967. Storm surges, in Oceanography and Marine Biology Annual Review, 5, 11–47, ed. Barnes H., Allen & Urwin, London.

Indu Jain, P. Chittibabu, Neetu Agnihotri, S. K. Dube, P. C. Sinha, A. D. Rao. 2007. Numerical storm surge model for India and Pakistan. Nat Hazards. 42:67–73. DOI 10.1007/s11069-006-9060-7

Johns, B. and Ah, A. 'The numerical modelling of storm surges in the Bay of Bengal', Quart. J. Roy. Met. Sot. 1980, 106, 1-18

Johns, B., Dube, S. K., Mohanty, IJ. C. and Sinha, P. C. 'Numerical simulation of the surge generated by the 1977 Andhra cyclone', Quart. J. Roy. Met. Sot. 1981, 107, 919-934

Murty, A.S.N. (1984): Studies on Storm surges and Sea level variations in the Bay of Bengal. Ph.d thesis awarded by Cochin University of Science and Technology, 141pp.

Sharma G.S. and A.S.N. Murty (1988): Storm Surges along the cast coast of India: An article appeared in the book, "Natural and Man made Hazards", Published by D. Reidel Publishing Company, Dordhech, The Netherlands, 1998, pp 257-277.

Sielecki, A. 'An energy-conserving difference scheme for the storm surge equations', Mon. Wea. Rev. 1968, 96, 150-156

## Internal waves

Murthy, P.G.K, G.S.Sharma, V.V.James and K.V.Susheela (1992): Internal wave characteristics in th eastern Arabian Sea during summer monsoon. Proc.Indian Acad.Sci (Earth Planet. Sci),Vol.1, No.4, pp 317-327.

Suryanarayana, A.P, C.Venkateswarlu, B.Gireeshsh, K.V.S.R.Prasad (2020): Study of internal wave characteristics off northwest Bay of Bengal using synthetic aperture radar.Natural Hazards, Vol.104, pp 2451-2460

## Exercises

### *Exercise 16.1*

If the depth averaged tidal velocity amplitude (U) is 1 m/s), the river run off ($u_R$) is 0.05m/s and the depth of the estuary is (h) is 10m, and $C_D$=0.002, find estuarine circulation velocity ($u_E$), bottom to surface salinity difference ($\Delta S$), Richardson number ($R_i$) and the salt water intrusion length (L).

### *Solution*

Given : (U = 1 m/s, $u_R$ = 0.05m/s, h = 10m; $C_D$=0.002

From equation 16.1.12: $u_E \approx 0.27U$, $\Delta S \approx 17\frac{u_R}{U}, L \approx \frac{h}{c_D}\left(\frac{c}{U}\right)^3\left(\frac{u_R}{U}\right)^{\frac{1}{2}}$

$$u_E \approx 0.27U = 0.27 \times 1ms^{-1} = 0.27ms^{-1}$$

$$\Delta S \approx 17\frac{u_R}{U} = 17\times\frac{0.05ms^{-1}}{1ms^{-1}} = 0.85$$

From equation 16.1.10: $\Delta S \approx 5.4\left(\frac{u_R}{c}\right)^{\frac{2}{3}} \rightarrow 0.85 = 5.4\left(\frac{0.05ms^{-1}}{c}\right)^{\frac{2}{3}} \rightarrow c = 7.91$

Using this value of c, put in equation 16.1.12 and find L.

$$L \approx \frac{h}{c_D}\left(\frac{c}{U}\right)^3\left(\frac{u_R}{U}\right)^{\frac{1}{2}} = \frac{10m}{0.002}\left(\frac{7.91}{1ms^{-1}}\right)^3\left(\frac{0.05ms^{-1}}{1ms^{-1}}\right)^{\frac{1}{2}} = 5049.74m \approx 5\,km$$

From eqn 16.1.9 : $R_i \approx \left(\frac{c}{U}\right)^2 \Delta S \rightarrow R_i \approx \left(\frac{7.91}{1}\right)^2 0.85 = 53.2$

### *Exercise 16.2*

Calculate volume transports for the following stations using Knudsen's hydrographical theorem for the given salinity data in a highly stratified estuary. Take the river flow rate as 12 $m^3S^{-1}$ when lower layer salinity is constant.

| Stn No/ salinity | 0 | 1 | 2 | 3 | 4 | 5 | 6 |
|---|---|---|---|---|---|---|---|
| $S^T$ | 0 | 5 | 10 | 15 | 20 | 25 | 30 |
| $S^L$ | - | 35 | 35 | 35 | 35 | 35 | 35 |

### Solution

| Stn No/ Vol transport | 0 | 1 | 2 | 3 | 4 | 5 | 6 |
|---|---|---|---|---|---|---|---|
| $Q^T$ $m^3$ $s^{-1}$ | 12 | 14 | 16.8 | 21 | 28 | 42 | 84 |
| $Q^L$ $m^3$ $s^{-1}$ | - | 2 | 4.8 | 9 | 16 | 30 | 72 |

### Exercise 16.3

In continuation of Exercise 2, calculate the volume transport for the given salinity data in a slightly stratified estuary when lower salinity varies and river flow is same.

| Stn No/ salinity | 0 | 1 | 2 | 3 | 4 | 5 | 6 |
|---|---|---|---|---|---|---|---|
| $S^T$ | 0 | 5 | 10 | 15 | 20 | 25 | 30 |
| $S^L$ | - | 20 | 25 | 28 | 30 | 33 | 35 |

### Solution

| Stn No/ Vol transport | 0 | 1 | 2 | 3 | 4 | 5 | 6 |
|---|---|---|---|---|---|---|---|
| $Q^T$ $m^3$ $s^{-1}$ | 12 | 16 | 20 | 25.85 | 36 | 49.5 | 84 |
| $Q^L$ $m^3$ $s^{-1}$ | - | 4 | 8 | 13.85 | 24 | 37.5 | 72 |

### Exercise 16.4

Lake Erie is prone to remarkable seiches — standing waves that slosh water back and forth in the lake basin from the west end at Toledo to the east end at Buffalo. Figure below shows smoothed data for the displacement from normal water levels along the lake at the high point of one particular seiche. After 3 hours the water was at normal levels throughout the basin; 6 hours later the water was high in Toledo and low in Buffalo. If so calculate length of the basin and seiche wave length, the wavelength, frequency, speed, amplitude and wave height of the standing wave.

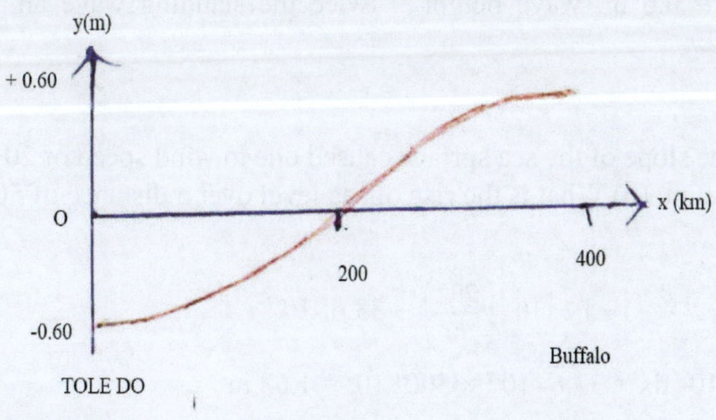

### Solution

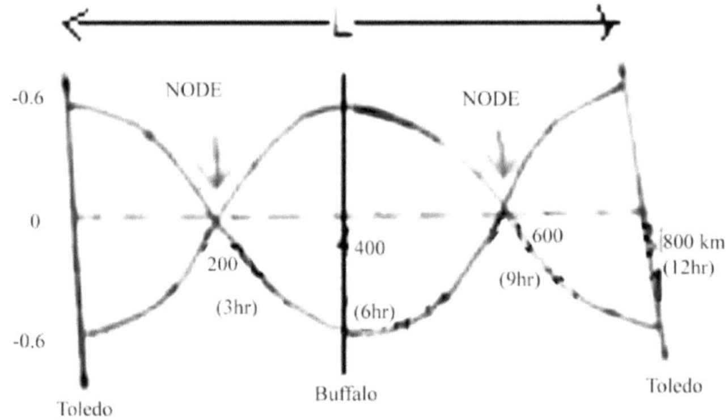


i) Length of the basin ($l$) from Toledo to Buffalo is 400 km.

ii) Referring to Fig.16.2.1 the Seiche wave length = Standing wave length $(L) \div 2 \rightarrow L = 800\,km \div 2 = 400\,km = 4\times10^5\,m$ as shown in Figure above.

iii) Standing wave length = Seiche wave length $\times 2 = 400\times2 = 800\ km$

iv) Time period of the standing wave = the time taken to pass the full length of the standing wave (800 km) =12 hours = 12×60×60S= 432 seconds

Frequency $\frac{1}{T} = \frac{1}{432\,S} = 0.0023\,S^{-1}$ or $\frac{1}{12\,hr} = 0.08\,hr^{-1}$

v) Speed of the standing wave $C = \frac{L}{T} = \frac{8\times10^5\,m}{432\,s} = 18.51\,ms^{-1} \approx 19ms^{-1}$

vi) Amplitude = 0.6 m, wave height = twice the standing wave amplitude $= 0.6\times2 = 1.2m$

### Exercise 16.5

(a) Calculate the slope of the sea surface caused due to wind speed of 20 m/s at a depth of 50 m. (b) What is the rise of sea level over a distance of 500 km.

### Solution

a) $\frac{dZ}{dx} = 4.2\times10^{-7}\left(\frac{W^2}{h}\right) = 4.2\times10^{-7}\left(\frac{20^2}{50}\right) = 33.6\times10^{-7}$

b) $dZ = 33.6\times10^{-7}\,dx = 33.6\times10^{-7} \times 500\times10^3 = 1.68$ m

### Exercise 16.6

What is estuarine Richardson number? Explain its uses.

### Solution

In a partially mixed estuary while the river water flows above in one direction, sea water moves from below in the opposite direction. If there is some tide, the wedge moves back and forth. The more the wedge moves, the more kinetic energy is available to break down the interface and turbulently mix the fresh and saline layers. Thus river is a source of buoyancy ($\Delta\rho\ g\ Q_f$) and the tide is a source of kinetic energy to overcome this buoyancy ($W\ U_t^3$). From this Estuarine Richardson Number, *R can be written as*

$$R = \frac{\left(\frac{\Delta\rho}{\rho}\right) g Q_f}{W U_t^3}$$

Where $\Delta\rho$ = the difference in density between the river and ocean water; $Q_f$ = discharge of fresh water; $U_t$ = the rms tidal velocity; W = the channel width

### Its uses

i) If *R is* large → the estuary is strongly stratified and flow is dominated by density currents

ii) If *R* is small → the estuary is well mixed, density effects are negligible.

iii) If $0.08 < R < 0.8$ → transition from a well mixed to a strongly stratified estuary occurs

### Exercise 16.7

The following figure shows the circulation pattern in two different types of estuaries.

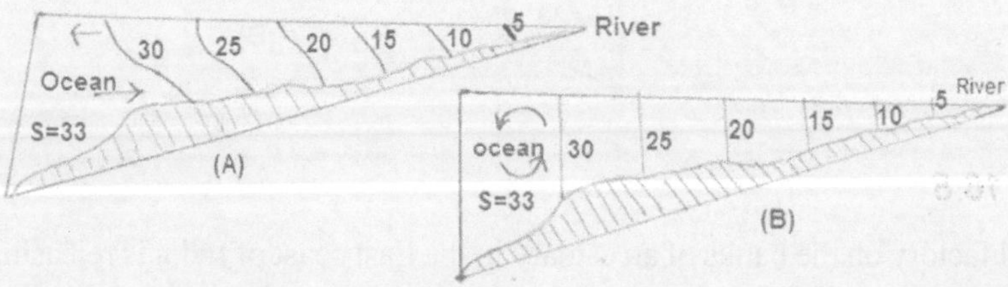


i)Where is the head and mouth of the estuary? ii)What is the type of estuary (a) and (b); iii) What kind of tidal pattern you expect in each category.iv) Demarcate the entrainment zone in each category. v) what will be the direction of pressure gradient?

### Solution

i) The river entering point is the head and the sea joining estuarine point is the mouth as shown in Fig. 3 below.

ii) As Isohalines are curved in the case of (A), it is a partially stratified estuary and as isohalines are vertically straight in the case of (B), it is a well-mixed estuary

iii) As (A) is partially stratified tidal pattern is not strong, so it may look like the Fig.1, and as (B) is well mixed, tidal pattern is very strong or tide dominated, so tidal pattern may be like the Fig.2

iv) Red demarcation dashed line indicates the zone of entrainment (Fig. 3). The line joining of maximum curvature of isohalines is the entrainment in the case of (A) and mean circulation area in the case of (B), while the rest of (B) no entrainment as isohalines are vertically straight indicates well mixed between 30ppm to the Head (5 ppm) of the estuary.

v) Pressure gradient will be from mouth to the head as higher salinity is found at the mouth than the head as shown by yellow line in Figs 3.

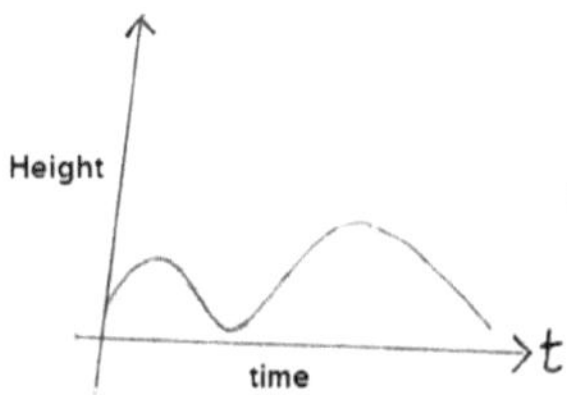


Fig.1

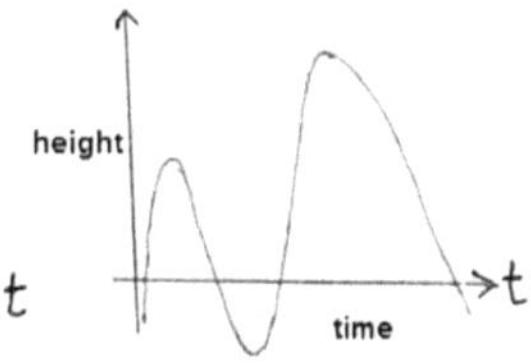


Fig.2

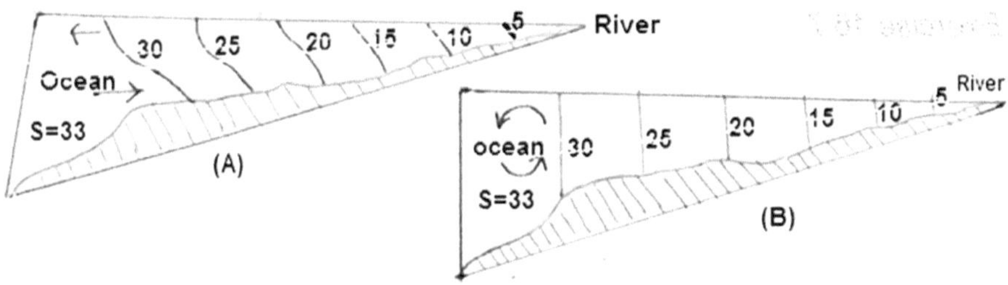


## *Exercise 16.8*

A chemical factory on the banks of an estuary in the East coast of India is releasing effluents. The estuary has a constant cross-sectional area. With the following values, find i) the freshness distribution (f), ii) flushing time ($T_f$) and iii) the mean detention or replacement time

A= 10,000 m$^2$; K (longitudinal dispersion coefficient.) = 100 m$^2$/s; $Q_f$ (fresh water inflow)=30m$^3$ s$^{-1}$ ; L = 30 km..

## *Solution*

Analytical solution for 1D equation of conservative pollutant for steady state for salinity distribution is:

$$S = S_0 exp\left[-\left(\frac{Q_f}{A}\right)\left(\frac{x}{K}\right)\right] = S_0 exp\left[-\left(\frac{30m^3s^{-1}}{10^4m^2}\right)\left(\frac{(x)m}{10^2m^2s^{-1}}\right)\right] = S_0\exp\left(-3\times10^{-5}x\right)$$

$$\frac{S}{S_0} = \exp\left(-3\times10^{-5}x\right) \rightarrow 1-\frac{S}{S_0} = 1-\exp\left(-3\times10^{-5}x\right)$$

i) ∴ Fresh ness distribution (f)from eqn 16.1.29= $\frac{S_0 - S}{S_0}$ =1 - exp(-3 × $10^{-5}$ x)

Volume of fresh water ($V_f$) in the estuary between x= 0 to x=L is given as:

$$V_f = \int_0^L fAdx = A\int_0^{30,000}\left[1-exp\left(-3\times10^{-5}x\right)\right]dx = 1.02\times10^8 m^3$$

ii) Flushing time (from equation 16.1.29A) is =

$$T_f = \frac{V_f}{Q_f} = \frac{1.02\times10^8 m^3}{30m^3s^{-1}} = 3.41\times10^6\ seconds = 39.4\ days$$

iii) Replacement time by equation (16.1.29B)as:

$$T = 0.4\frac{L^2}{K} = 0.4\frac{(30,000\,m)^2}{100\,m^2s^{-1}} = 3.6\times10^6\ Seconds = 41.7\ days$$

### *Exercise 16.9*

Obtain an expression for tidal exchange ratio (R) in an estuary

### *Solution*

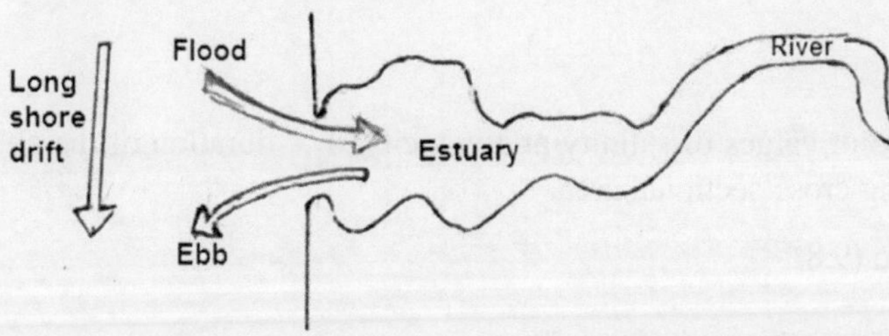


Volume of water entering an estuary during flood tide = water that left the estuary on the previous ebbs + new ocean water. The longshore current deflects the ebb flow downcoast and delivers the supply of new ocean water for the flood.

$V_f = V_{f0} + V_0$ (9.1)

Tidal exchange ratio R is given as: $R = \frac{V_0}{V_f}$ (9.2)

Where $V_0$ = volume of new ocean water entering the estuary during the flood tide; $V_f$ = total water volume entering the estuary on the flood tide; $V_{f0}$ = part of $V_f$ which flowed out of the estuary on the previous ebb.

As total salt and water content of the estuary are to remain constant, we can write the following relations:

$$S_f V_f = S_e V_e \tag{9.3}$$

$$V_f + V_Q = V_e \tag{9.4}$$

$$S_f V_f = S_e V_{f0} + S_0 V_0 \tag{9.5}$$

Combining all the above equations in the following way we can write:

Multiplying equation 9.1 with $\{S_e\}$:

$$\{(9.1) \times S_e\} = S_e V_f = S_e V_{f0} + S_e V_0 \tag{9.6}$$

Subtracting (9.6) from (9.5) we can write:

$$S_f V_f - S_e V_f = S_0 V_0 - S_e V_0 \tag{9.7}$$

We can write equation (9.7) as $V_f (S_f - S_e) = V_0 (S_0 - S_e) \rightarrow \frac{V_0}{V_f} = \frac{(S_f - S_e)}{(S_0 - S_e)} = R$

$$\therefore \quad = \frac{S_f \quad S_e}{S \quad S} \tag{9.8}$$

Where R is tidal exchange ratio.

Accurate determination of average flood ($S_f$) and ebb ($S_e$) salinities require complete cross-sectional measurements of both salinity (S) and velocity (u) throughout the tidal cycle with the following equation.

$$S_e = \frac{\int_0^{T_e} \int Su\, dA\, dt}{\int_0^{T_e} \int u\, dA\, dt}$$

Where S and u are point values of salinity and velocity; $T_e$ = duration of the ebb flow; A = time-varying cross-sectional area

Substituting (9.3) into (9.8):

$$R = \frac{S_e \frac{V_e}{V_f} - S_e}{S_0 - S_e} = \frac{S_e}{S_0 - S_e}\left[\frac{V_e}{V_f} - 1\right] \tag{9.9}$$

Substituting equation 9.4 in equation (9.9)

$$R = \frac{S_e}{S_0 - S_e}\left[\frac{V_e - V_f}{V_f}\right] = \frac{S_e}{S_0 - S_e}\left[\frac{V_Q}{V_f}\right] \rightarrow$$

$$\therefore R = \frac{S_e}{S_0 - S_e}\frac{V_Q}{V_f} \tag{9.10}$$

### *Exercise 16.10*

Obtain an expression for the average concentration of the toxic material in the estuary in the vicinity of the discharge point in relation to the total flow and salt balance.

### *Solution*

The Salt balance in the estuary can be written as:

$$Q_0 S_0 = S(Q_0 + Q_e + Q_f) \tag{10.1}$$

Where $Q_0$ = circulating flow of ocean water; $Q_f$ = tributary discharge from all tributaries upstream of the effluent discharge point; $Q_e$ = effluent flow; $S_0$ =ocean salinity; S= salinity in the estuary

Re arranging equation (10.1) we can write:

$$Q_0 = \frac{S\left(Q_e + Q_f\right)}{(S_0 - S)} \tag{10.2}$$

The total flow available for diluting the effluent in the estuary is:

$$Q_d = (Q_0 + Q_e + Q_f) = \frac{S_0\left(Q_e + Q_f\right)}{(S_0 - S)} \tag{10.3}$$

Where $Q_d$ is return discharge

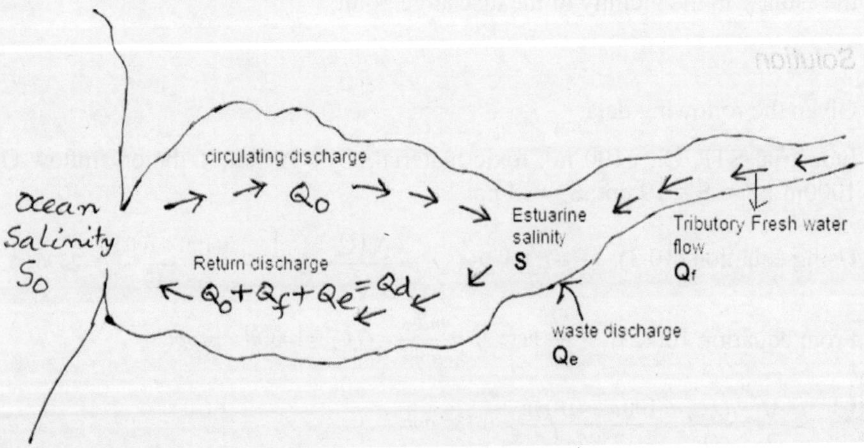


Assuming complete mixing near the discharge point in a short time, the mean concentration of effluent near the point of discharge ($C_d$) is

$$C_d = \frac{\bar{M}}{Q_d} \tag{10.4}$$

where $\bar{M}$ = discharge rate of material (mass/time)

Tidal exchange analysis from equation 9.2 of exercise 9 can be written as:

$$Q_0 T = RV_f \tag{10.5}$$

Where T is duration of tidal period

Substituting equation 9.10 of exercise 9 in 10.5, we can write:

$$Q_0 = \frac{S_e}{S_0 - S_e}\frac{V_Q}{T} = \frac{S_e}{S_0 - S_e}\left(Q_e + Q_f\right) \tag{10.6}$$

Note here discharge of river water = $\frac{V_Q}{T} = Q_Q = \left(Q_e + Q_f\right)$

The equation (10.6) is the same as Eq. (10.2), if S is taken to be equal to the average ebb flow salinity, $S_e$.

In bays with low tributary inflow, the salinity may be near oceanic, then we can write: $Q_0 = \frac{RP}{T}$ Where $P$ = tidal prism, $V_f$

### Exercise 16.11

An industrial plant is discharging 100 m$^3$ of effluent containing 10ppt of toxic material into an estuary. The minimum tributary inflow upstream of the discharge point is 1000m$^3$. The salinity observed at the discharge point is 19ppt and the ocean salinity is 33ppt Estimate the average concentration of the toxic material in the estuary in the vicinity of the discharge point.

### Solution

Given the following data:

Industrial STP, $Q_e$ = 100 m$^3$; toxic material, $C_e$ = 10 ppt; Tributory inflow $Q_f$ = 1000m$^3$; $S_e$ = S = 19 ppt; $S_0$ = 33 ppt

Using equation (10.3): $Q_d = (Q_0 + Q_e + Q_f) = \frac{S_0\left(Q_e + Q_f\right)}{(S_0 - S)} = \frac{33(100+1000)}{33-19} = 2596m^3$

From equation 10.4, $C_d = \frac{\bar{M}}{Q_d}$ here $\bar{M} = \frac{mass}{time} = Q_e C_e = 100\,m^3 \times 10\,ppt$

$$\therefore C_d = \frac{\bar{M}}{Q_d} = \frac{Q_e C_e}{Q_d} = \frac{100\,m^3 \times 10\,ppt}{2596m^3} = 0.385\,ppt$$

### Exercise 16.12

Assuming the same values and conditions as for the Exercise 11, find the concentration downstream of the outfall at a point 'A' where the salinity is 24 ppt and upstream of the outfall at a point 'B' where the salinity is 5 ppt.

### Solution

Given are: Industrial STP, $Q_e$ = 100 m$^3$; toxic material, $C_e$ = 10 ppt; Tributory inflow $Q_f$ = 1000m$^3$; $S_e$ = S = 19 ppt; $S_0$ = 33 ppt

For a conservative material,

i) Downstream of the outfall → the effluent continues to be diluted as it approaches the ocean, just as fresh water is diluted.

$$C_{Xd} = C_d \frac{S_0 - S_{Xd}}{S_0 - S_e} \tag{12.1}$$

ii) Upstream of the discharge point → concentration of the effluent is reduced in the same way that the salinity is reduced by mixing with the tributary fresh water.

$$C_{Xu} = C_d \frac{S_{Xu}}{S_e} \tag{12.2}$$

Here are some assumptions:

(i) The salinity observed at some instant is representative of steady-state conditions.

(ii) The results are valid only for conservative substances

(iii) The results give cross-sectional averages, not peak concentrations.

Given

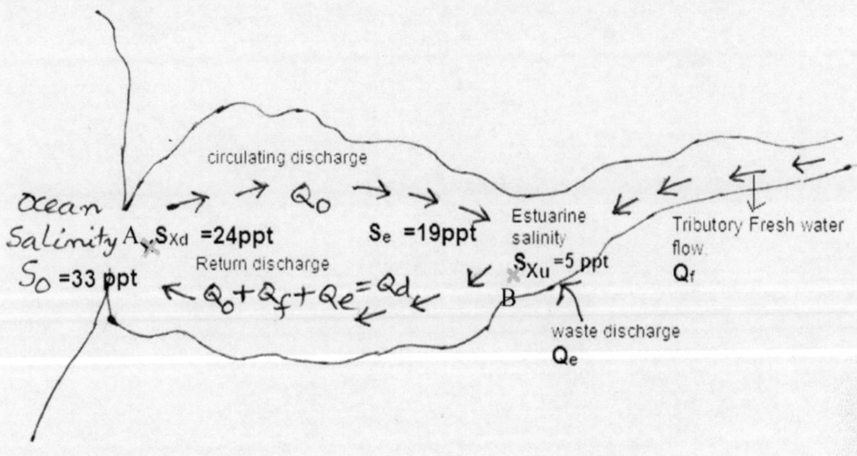


The concentration for down stream ($C_{Xd}$) at point "A' in the figure is given as:

$$C_{Xd} = C_d \frac{S_0 - S_{Xd}}{S_0 - S_e} \tag{12.1}$$

Here 'd' stands for downstream and 'X' stands for point required to be calculated. Where $S_0$ is salinity in the sea beyond estuarine mouth $S_{Xd}$, is the salinity at the down stream out fall point (A) and $S_e$ is the salinity in the estuary proper. Similarly

$C_{Xd}$ is the concentration at the down stream point (A), $C_d$ is the concentration at the discharge point.

Hence given values are : $S_0$ = 33 ppt; $S_{Xd}$ = 24 ppt; $S_e$ = 19 ppt; $C_d$ = 0.385 (calculated in exercise 11). Substituting all these values in the equation(12.1)

$$C_{Xd} = C_d \frac{S_0 - S_{Xd}}{S_0 - S_e} = 0.385 \frac{33 - 24}{33 - 19} = 0.248\, ppt$$

Similarly at the upstream point (B), the concentration ($C_{Xu}$) =

$$C_{Xu} = C_d \frac{S_{Xu}}{S_e} \tag{12.2}$$

Note here $S_0$ is not included in equation (12.2) because sea is far away at point 'B'. and 'u' stands for upstream.

Known values are: $S_{Xu}$ = 5 ppt; $S_e$ = 19 ppt ; $C_d$ = 0.385 (calculated in exercise 11). Substituting all these values in the equation (12.2)

$$C_{Xu} = C_d \frac{S_{Xu}}{S_e} = 0.385 \frac{5}{19} = 0.101\, ppt$$

# Appendex

Important constants required to do the Exercises in all the chapters of this book.

Earth Properties

| **Characteristic** | **Value** |
|---|---|
| Average radius of the earth | $6.37 \times 10^6$ m; 6357-6378 km ≈ 6400 km |
| Angular velocity of earth ($\Omega = 2\pi/24$hrs) | $7.27 \times 10^{-5}$ rads$^{-1}$ |
| Volume of earth | $1.37 \times 10^9$ km$^3$ |
| Mass of earth ($m_e$) | $5.97 \times 10^{24}$ kg |
| Surface area of earth ($m_e$) | $5.10 \times 10^{14}$ m$^2$ |
| Ocean surface area | $3.61 \times 10^{14}$ m$^2$ |
| Mean Ocean depth | $3.8 \times 10^3$ m |
| Mean Land elevation | $8.4 \times 10^2$ m |
| Mass of moon ($m_m$) | $1.228 \times 10^{-2}$ $m_e$ = 1.2% of earth |
| Distance between earth and moon ($r_{em}$) | $3.8 \times 10^8$ m |
| One degree of latitude in SI units | 110 km |
| Earth's gravitational acceleration (g) | 9.81 m s$^{-2}$ |

## Physical dimensions, conversions

### Time (T)

| | |
|---|---|
| 1 minute | 60 seconds |
| 1 hour | 3600 seconds |
| 1day | 86400 seconds |
| 1 year | 365.25 days |

### Length (L)

| | | |
|---|---|---|
| 1 meter (m) | 100 centimeters ( cm) | |
| 1 Kilo meter (km) | $10^3$ m | |
| 1 inch (in) | 2.54 centi meter ( cm) | |
| 1meter | 39.37 inches | 3.281 feet (ft) |
| 1 statute mile | 5280 feet | 1.609 km |
| 1 nautical mile(n.m) | 6000 ft=1.136 mi | 1.829 km |
| 1 fathom (fm) | 6 feet | |

## Velocity ($LT^{-1}$)

| | | |
|---|---|---|
| 1 $ms^{-1}$ | 2.2369 mi $hr^{-1}$ | |
| 1 mi $hr^{-1}$ | 1.4667 $fts^{-1}$ =0.8690 knots | 1.6093 km $hr^{-1}$ = 0.447041 $ms^{-1}$ |
| I knot | 1 nautical mile $hr^{-1}$ =1.1508 mi $hr^{-1}$ | 0.51477 $ms^{-1} \approx$ 1.9 km $hr^{-1}$ |

## Angle

| | | |
|---|---|---|
| 1 radian (rad) | $360^0/2\pi$ | 57.29578 degrees |
| 1 degree | 60 min | 0.01745 rad |
| 1 min | 60seconds | $2.90888 \times 10^{-4}$ rad |
| 1 steradian (sr) | 0.079577 total solid angle | |

## Angular velocity and frequency ($T^{-1}$)

| | | |
|---|---|---|
| 1 rad $s^{-1}$ | 0.1592 rev $s^{-1}$ | 9.549 rpm = 57.296 deg $s^{-1}$ |
| 1 cycle per second (cps) | 1 hertz | $2\pi$ rad $s^{-1}$ |

## Mass (M)

| | | |
|---|---|---|
| 1 kilo gram (Kg) | 1000 grams (gm) | |
| 1 kg | 2.20462 pound mass (lbm) | 0.068522 slugs |
| 1 slug | 32.17398 lbm | 14.5939 kg |

## Force *($MLT^{-2}$)*

| | | |
|---|---|---|
| 1 newton (N) | 1 $kgms^{-2}$ | |
| 1 dyne (dyn) | 1 gm cm $s^{-2}$ | |
| 1 newton | $1 \times 10^5$ dynes | 0.22481 pound force (lbf) |
| 1 dyn | $2.2481 \times 10^{-6}$ lbf | |
| 1 lbf | 4.4482 N | |

## Pressure (*$FL^{-2}$*)

| | | |
|---|---|---|
| 1 dyn cm $^{-2}$ | $10^{-1}$ $Nm^{-2}$ | $10^{-1}$ Pascal(Pa) |
| 1 Pascal (Pa) | 1 $Nm^{-2}$ | 10 dyn $cm^{-2}$ |
| 1 bar | $10^5$ $Nm^{-2}$ | |
| 1 mb | $10^{31}$ dyn $cm^{-2}$= $10^{-3}$ bar | $10^2$ $Nm^{-2}$ |

## Work & Energy (FL)

| | | | | |
|---|---|---|---|---|
| 1 Joule (J) | 1 watt.sec | 1 newton meter | $1 \times 10^7$ ergs | 1volt. coulomb |
| | =0.738 ft.lbf | =$9.48 \times 10^{-4}$ Btu | =$2.778 \times 10^{-7}$ kW.hr | |
| 1 calorie (cal) | 4.186 J | 0.003968 British thermal unit (Btu) | | |
| 1 kw.hr | 3413 Btu | $2.655 \times 10^6$ ft.lbf | 1.341 hp.hr | |
| 1 electron volt (ev) | $1.60210 \times 10^{-19}$ J | | | |

## Power ($FL\,T^{1}$)

| 1 watt (W) | 1 J s$^{-1}$ | $1.341 \times 10^{-3}$ hp | 0.0569 Btu min$^{-1}$ | 0.738 ft.lbf s$^{-1}$ |
|---|---|---|---|---|
| 1 kilowatt (kW) | 1000 W | 1.340 horse power(hp) | | |
| 1 horse power (hp) | 550 ft.lbf s$^{-1}$ | 746 W | 0.707 Btu s$^{-1}$ | |
| 1 langley (ly) min$^{-1}$ | 0.697 kW m$^{-2}$ | | | |

## Properties of Air

| | |
|---|---|
| Air density ($\rho_a$) at 20$^0$ C, Zero humidity and 1 atmospheric pressure | 1.21 or 1.3 kg m$^{-3}$ |
| Specific heat | 1006 J kg$^{-1}$ K$^{-1}$ |
| Molecular kinematic viscosity | $14.8 \times 10^{-6}$ m$^2$ s$^{-1}$ |
| Molecular heat diffusivity | $20.8 \times 10^{-6}$ m$^2$ s$^{-1}$ |
| Drag coefficient ($C_D$) | $1.2 \times 10^{-3}$ m$^2$ s$^{-1}$ |
| Average mass of air molecule | $4.2 \times 10^{-26}$ kg |

## Properties of Sea Water

| | |
|---|---|
| Surface (or reference) Sea water density ($\rho_0$) at 35 psu, 20$^0$C and at atmospheric pressure | 1026.95 kg m$^{-3}$ ≈ 1027kgm$^{-3}$ |
| Density range of sea water | 1020-1028 kgm$^{-3}$ |
| Specific heat | 3986 J kg$^{-1}$ K$^{-1}$ |
| Molecular kinematic viscosity | $1.8 \times 10^{-6}$ m$^2$ s$^{-1}$ |
| Molecular heat diffusivity | $1.3 \times 10^{-7}$ m$^2$ s$^{-1}$ |
| Molecular salt diffusivity | $0.74 \times 10^{-9}$ m$^2$ s$^{-1}$ |
| Heat capacity of sea water $C_p$ | $= 4.2 \times 10^3$Jkg$^{-1}$K$^{-1}$ |
| Thermal expansion coefficient, $\alpha_T$ | $1.0 \times 10^{-4}$ K$^{-1}$ |
| Haline expansion coefficient, $\alpha_S$ | $7.6 \times 10^{-4}$ K$^{-1}$ |
| Mean temperature of sea water | 3.5$^0$ C |
| Salinity range at the ocean surface | 33 to 37‰ |
| Mean salinity of sea water | 34.7‰ |
| Haline contraction coefficient $\left(\beta = \dfrac{1}{\rho_0}\dfrac{\partial \rho}{\partial s}\right)$ | $8 \times 10^{-4}$ |
| Freezing point of sea water | -1.8$^0$ C |
| Average pH of sea water | 7.8 |
| Temperature range of ocean surface | -2 to 30$^0$ C |
| Range of sea level atmospheric pressure | 930-1030 hPa |

## Potential temperature

**The following table contains values of pressure (p, decibars), salinity (S, psu), temperature (T $^0$C), potential temperature ($\theta^0$C) and density ($\rho_w$.1000kg m$^{-3}$) of sea water (after Mellor 1991).**

| P | S | T | θ | ρ.1000 | P | S | T | θ | ρ.1000 |
|---|---|---|---|---|---|---|---|---|---|
| 0 | 35 | 0 | 0.00 | 28.106 | 0 | 37 | 0 | 0.00 | 29.722 |
| 0 | 35 | 4 | 4.00 | 27.786 | 0 | 37 | 4 | 4.00 | 29.378 |
| 0 | 35 | 8 | 8 | 27.274 | 0 | 37 | 8 | 8 | 28.846 |
| 0 | 35 | 12 | 12 | 26.590 | 0 | 37 | 12 | 12 | 28.144 |
| 0 | 35 | 16 | 16 | 25.748 | 0 | 37 | 16 | 16 | 27.288 |
| 0 | 35 | 20 | 20 | 24.763 | 0 | 37 | 20 | 20 | 26.290 |
| 0 | 35 | 24 | 24 | 23.643 | 0 | 37 | 24 | 24 | 25.158 |
| 0 | 35 | 28 | 28 | 22.397 | 0 | 37 | 28 | 28 | 23.902 |
| 1000 | 35 | 0 | -0.05 | 32.618 | 1000 | 37 | 0 | -0.05 | 34.416 |
| 1000 | 35 | 4 | 3.92 | 32.393 | 1000 | 37 | 4 | 3.92 | 33.969 |
| 1000 | 35 | 8 | 7.89 | 31.791 | 1000 | 37 | 8 | 7.89 | 33.348 |
| 1000 | 35 | 12 | 11.87 | 31.029 | 1000 | 37 | 12 | 11.06 | 32.569 |
| 2000 | 35 | 0 | -0.12 | 37.427 | 2000 | 37 | 0 | -0.11 | 39.010 |
| 2000 | 35 | 4 | 3.83 | 36.903 | 2000 | 37 | 4 | 3.83 | 38.464 |
| 2000 | 35 | 8 | 7.78 | 36.215 | 2000 | 37 | 8 | 7.77 | 37.757 |
| 4000 | 35 | 0 | -0.28 | 46.351 | 4000 | 37 | 0 | -0.29 | 47.904 |
| 4000 | 35 | 4 | 3.61 | 45.642 | 4000 | 37 | 4 | 3.60 | 47.175 |
| 6000 | 35 | 0 | -0.51 | 54.902 | 6000 | 37 | 0 | -0.52 | 56.430 |
| 6000 | 35 | 4 | 3.34 | 54.024 | 6000 | 37 | 4 | 3.32 | 55.533 |
| 10000 | 35 | 0 | -1.10 | 70.965 | 10000 | 37 | 0 | -1.12 | 72.446 |
| 10000 | 35 | 4 | 2.60 | 69.790 | 10000 | 37 | 4 | 2.46 | 71.252 |

## Some useful constants

| Symbol | Meaning, cross reference | Units, Typical value |
|---|---|---|
| | Boltzman constant | $1.38 \times 10^{-23}$ $Jk^{-1}$ |
| | Stefan-Boltzman constant | $5.67\times10^{-8}$ $Wm^{-2}$ $K^{-4}$ |
| $f=2\Omega \sin\phi$ | Coriolis parameter at latitude $\phi = 45^0$ | $10^{-4}$ $s^{-1}$ |
| X | Stress | $Nm^{-2}$ |
| $\beta = \frac{df}{dy} = \frac{2\Omega \cos\phi}{R_E}$ | Planetary beta | $2\times10^{-11}$ $m^{-1}$ $s^{-1}$ |
| u,v,w | Typical velocity vectors | $(10^{-3}\text{-}1)$m $s^{-1}$ |
| $u_g$, $v_g$ | Geostrophic velocity vector | $(10^{-3}\text{-}1)$m $s^{-1}$ |
| $C_s = \left[\frac{\partial p}{\partial \rho}\right]^{-1/2}$ | Speed of sound in the oceans | 1500 $ms^{-1}$ |
| q | Potential vorticity | $m^{-1}$ $s^{-1}$ |
| (U,V) | Depth integrated current | $m^{-2}$ $s^{-1}$ |
| $U_E$ | Ekman transport | $m^{-2}$ $s^{-1}$ |
| $U_P$ | Geostrophic mass transport(Sverdrup-Munk) | $m^{-2}$ $s^{-1}$ |
| $N = \left[-\frac{g}{\rho_0}\frac{\partial \rho}{\partial z}\right]^{1/2}$ | Brunt-Vaisala frequency | $10^{-2}$ $s^{-1}$ |
| $\nu$, $\kappa_\theta$, $\kappa_s$ | Kinematic viscosoity, thermal and saline diffusivity | $m^2$ $s^{-1}$ |
| $\alpha = -\frac{1}{\rho_0}\frac{\partial \rho}{\partial T}$ | Thermal expansion coefficient (aspect ratio $\alpha = \frac{H}{L}$ ) | $2 \times 10^{-4}$ $^0C^{-1}$ |
| $h_E$ | Ekman layer thickness | ~100 m |
| $\frac{\lvert u\rvert}{fL}$ | Rossby number | |
| $\frac{\lvert u\rvert}{C_s}$ | Mach number | |
| $\frac{\lvert u\rvert L}{\nu}$ | Reynolds number | |

Properties of Fresh water and ice:

| | |
|---|---|
| Boiling point of fresh water | $100^0$C |
| Freezing point of fresh water | $0^0$C |
| Typical density of fresh water | 1000 $kgm^{-3}$ |
| Typical density of ice | 917 $kgm^{-3}$ |
| Speed of sound in air | 330 $ms^{-1}$ |
| Specific heat capacity of water ice | $2.11\times10^3$ $Jkg^{-1}\,{}^0C^{-1}$ |
| Latent heat of fusion of water | $3.34\times10^5$ $Jkg^{-1}$ |

## Temperature conversion

| | |
|---|---|
| $^0$C | = 273.15 K. |
| $^0$F | =1.8 $^0$C +32.0 |

## Ocean properties

| | |
|---|---|
| Temperature range of ocean surface | -2 to$30^0$ C |
| Range of sea level atmospheric pressure | 930-1030 hPa |
| Ocean pressure of 1 meter depth | = 1 decibar |
| Depth of seawater corresponding to 1 atmosphere pressure | 10 m |
| Average depth of the oceans | 3.8 km |
| Deepest portion of the oceans (Mindanao trench) | >11 km |
| Depth of Euphotic zone | 100m |
| Period of the semi-diurnal tide | 12h 25 m |
| One Sverdrup(Sv) | $10^6$ $m^3$ $s^{-1}$ |
| Specific heat capacity of water and sea water | $4.18\times10^3$ $Jkg^{-1}\,{}^0C^{-1}$ |
| Boltzman constant | $1.38\times10^{-23}$ $Jk^{-1}$ |
| Stefan Boltzman constant | $5.67\times10^{-8}$ $Wm^{-2}$ $K^{-4}$ |